PRINCIPLES AND APPLICATIONS
IN ENGINEERING SERIES

# OCCUPATIONAL ERGONOMICS

## Engineering and Administrative Controls

PRINCIPLES AND APPLICATIONS IN ENGINEERING SERIES

# OCCUPATIONAL ERGONOMICS

## Engineering and Administrative Controls

EDITED BY

## Waldemar Karwowski

*University of Louisville*
*Louisville, Kentucky*

## William S. Marras

*The Ohio State University*
*Columbus, Ohio*

CRC Press
Taylor & Francis Group
Boca Raton  London  New York

CRC Press is an imprint of the
Taylor & Francis Group, an **informa** business

This material was previously published in *The Occupational Ergonomics Handbook.* © CRC Press LLC 1999.

CRC Press
Taylor & Francis Group
6000 Broken Sound Parkway NW, Suite 300
Boca Raton, FL 33487-2742

First issued in paperback 2019

© 2003 by Taylor & Francis Group, LLC
CRC Press is an imprint of Taylor & Francis Group, an Informa business

No claim to original U.S. Government works

ISBN-13: 978-0-8493-1800-9 (hbk)
ISBN-13: 978-0-367-39529-2 (pbk)

### Library of Congress Cataloging-in-Publication Data

Occupational ergonomics : engineering and administrative controls / edited by Waldemar Karwowski, William S. Marras.
    p. cm. — (Principles and applications in engineering ; 14)
    Includes bibliographical references and index.
    ISBN 0-8493-1800-9
    1. Human engineering. 2. Musculoskeletal system—Wounds and injuries—Prevention. 3. Industrial hygiene.  I. Karwowski, Waldemar. II. Marras, William S. (William Steven), 1952- III. Series.

    TA166.O257 2003
    620.8'2—dc21

                    2002041399

Library of Congress Card Number 2002041399

**Visit the Taylor & Francis Web site at
http://www.taylorandfrancis.com**

**and the CRC Press Web site at
http://www.crcpress.com**

# The Editors

Waldemar Karwowski, Ph.D., P.E., C.P.E., is Professor of Industrial Engineering and Director of the Center for Industrial Ergonomics at the University of Louisville, Kentucky. He holds an M.S. (1978) in production management from Technical University of Wroclaw, Poland, and a Ph.D. (1982) in Industrial Engineering from Texas Tech University. His research, teaching, and consulting activities focus on prevention of low back injury and cumulative trauma disorders, human and safety aspects of advanced manufacturing, fuzzy sets and systems, and theoretical aspects of ergonomics.

Dr. Karwowski served as President (2000–2003) of the International Ergonomics Association. He is editor of many international journals, including *Human Factors and Ergonomics in Manufacturing*, and *Theoretical Issues in Ergonomics Science*, and consulting editor of *Ergonomics*. He is the author or co-author of more than 200 scientific publications, including 25 books.

Dr. Karwowski is founder and chairman of the International Conference on Human Aspects of Advanced Manufacturing and Hybrid Automation. He was the recipient of the Outstanding Young Engineer of the Year Award, given by the Institute of Industrial Engineering. He was also a Fulbright Scholar at Tampere University of Technology in Finland. He received the President's Award for Outstanding Scholarship, Research, and Creative Activity in the category of Basic and Applied Science at the University of Louisville.

William S. Marras, Ph.D., C.P.E., holds the Honda Endowed Chair in Transportation in the Department of Industrial, Welding, and Systems Engineering at The Ohio State University, Columbus. He is also the director of the biodynamics laboratory and holds appointments in the departments of physical medicine and biomedical engineering. Professor Marras is also the co-director of the Ohio State University Institute for Ergonomics.

Dr. Marras received his Ph.D. in bioengineering and ergonomics from Wayne State University in Detroit, Michigan. His research centers around biomechanical epidemiologic studies, laboratory biomechanic studies, mathematical modeling, and clinical studies of the back and wrist.

His findings have been published in more than 100 refereed journal articles and 12 book chapters, and he holds two patents, including one for the lumbar motion monitor (LMM). His work has also attracted national and international recognition. He has won the prestigious Swedish Volvo Award for Low Back Pain Research and Austria's Vienna Award for Physical Medicine.

# Contributors

**Elsayed Abdel-Moty**
Department of Industrial
  Engineering
University of Miami
Coral Gables, Florida

**W. Gary Allread**
The Ohio State University
Columbus, Ohio

**Charles K. Anderson**
Advanced Ergonomics, Inc.
Dallas, Texas

**Gunnar B.J. Andersson**
Department of Orthopedic Surgery
St. Luke's Medical Center
Chicago, Illinois

**Michele C. Battié**
Department of Orthopaedics
University of Washington
Seattle, Washington

**Patricia Bertsche**
The Ohio State University
Columbus, Ohio

**Ram Bishu**
IMSE Department
University of Nebraska
Lincoln, Nebraska

**A. Kim Burton**
Department of Clinical
  Biomechanics
Huddersfield Polytechnic
Huddersfield, England

**Don B. Chaffin**
Center for Ergonomics
University of Michigan
Ann Arbor, Michigan

**Patrick G. Dempsey**
Liberty Mutual Research Center for
  Safety and Health
Hopkinton, Massachusetts

**Bradley Evanoff**
School of Medicine
Washington University
St. Louis, Missouri

**Paul Gaddie**
University of Louisville
Louisville, Kentucky

**Katharyn A. Grant**
Robert A. Taft Laboratories
National Institute for Occupational
  Safety and Health
Cincinnati, Ohio

**Thomas Hales**
National Institute for Occupational
  Safety and Health
Cincinnati, Ohio

**Simon M. Hsiang**
Liberty Mutual Research Center
  for Safety and Health
Hopkinton, Massachusetts

**Renliu Jang**
University of Louisville
Louisville, Kentucky

**Waldemar Karwowski**
Department of Industrial
  Engineering
University of Louisville
Louisville, Kentucky

**Glenda L. Key**
Key Functional Assessments
Minneapolis, Minnesota

**Tarek M. Khalil**
Department of Industrial
  Engineering
University of Miami
Coral Gables, Florida

**Jung-Yong Kim**
Department of Industrial
  Engineering
Hanyang University
Ansan, Korea

**Stephan Konz**
Department of IMSE
Kansas State University
Manhattan, Kansas

**Steven A. Lavender**
Department of Orthopedic
  Surgery
St. Luke's Medical Center
Chicago, Illinois

**Wook Gee Lee**
University of Louisville
Louisville, Kentucky

**Chris J. Main**
Department of Behavioural
  Medicine
Hope Hospital
Salford, England

**Richard W. Marklin**
Department of Mechanical and
  Industrial Engineering
Marquette University
Milwaukee, Wisconsin

**William S. Marras**
ISE Department
The Ohio State University
Columbus, Ohio

**Stuart M. McGill**
University of Waterloo
Waterloo, Ontario, Canada

**Raymond M. McGorry**
Liberty Mutual Research Center
    for Safety and Health
Hopkinton, Massachusetts

**Donald R. McIntyre**
Interlogics
Hillsborough, North Carolina

**Stephen J. Morrissey**
State of Oregon
OSHA
Portland, Oregon

**A. Muralidhar**
University of Nebraska
Lincoln, Nebraska

**Robert W. Norman**
Department of Kinesiology
University of Waterloo
Waterloo, Ontario, Canada

**Ewa Nowak**
Department of Ergonomics
    Research
Institute of Industrial Design
Warsaw, Poland

**Mohamad Parnianpour**
Department of ISE
The Ohio State University
Columbus, Ohio

**Malcolm H. Pope**
Department of Orthopaedic Surgery
Iowa Spine Research Center
University of Iowa
Iowa City, Iowa

**Vern Putz-Anderson**
Applied Psychology and Ergonomics
National Institute for Occupational
    Safety and Health
Cincinnati, Ohio

**Robert G. Radwin**
Department of Industrial
    Engineering
University of Wisconsin
Madison, Wisconsin

**David Rempel**
University of California
San Francisco, California

**Valerie J. Rice**
U.S. Army Research Institute of
    Environmental Medicine
Occupational Physiology Division
Natick, Massachusetts

**Richard G. Ried**
The Ohio State University
Columbus, Ohio

**Stephen N. Robinovitch**
San Francisco General Hospital
San Francisco, California

**Hubert L. Rosomoff**
Comprehensive Pain and
    Rehabilitation Center
University of Miami
Coral Gables, Florida

**Renee Steele-Rosomoff**
Comprehensive Pain and
    Rehabilitation Center
University of Miami
Coral Gables, Florida

**Aboulfazl Shirazi-Adl**
École Polytechnique
Montreal, Quebec, Canada

**Carolyn M. Sommerich**
ISE Department
The Ohio State University
Columbus, Ohio

**Leon M. Straker**
Curtin University of Technology
Shenton Park, Australia

**Carol Stuart-Buttle**
Stuart-Buttle Ergonomics
Philadelphia, Pennsylvania

**Eira Viikari-Juntura**
Department of Physiology
Finnish Institute of Occupational
    Health
Topeliuksenkatu, Finland

**Thomas R. Waters**
National Institute for Occupational
    Safety and Health
Cincinnati, Ohio

**Richard Wells**
Department of Kinesiology
University of Waterloo
Waterloo, Ontario, Canada

# Contents

## PART II ADMINISTRATIVE CONTROLS

### SECTION I   Ergonomics Surveillance

## SECTION II  Medical Management Prevention

# Preface

Ergonomics (or human factors) is defined by the International Ergonomics Association (www.iea.cc) as *the scientific discipline concerned with the understanding of interactions among humans and other elements of a system, and the profession that applies theory, principles, data and methods to design in order to optimize human well-being and overall system performance. Ergonomists contribute to the design and evaluation of tasks, jobs, products, environments, and systems in order to make them compatible with the needs, abilities, and limitations of people.*

Currently, there is substantial and convincing evidence that the proficient application of ergonomics knowledge, in a system context, will help to improve system effectiveness and reliability, increase productivity, reduce employee healthcare costs, and improve the quality of work processes, products and working life for all employees. As ergonomics promotes a holistic approach in which considerations of physical, cognitive, social, organizational, environmental and other relevant factors are taken into account, the professional ergonomists should have a broad understanding of the full scope of the discipline. Development of this book was motivated by the quest to facilitate a wider acceptance of ergonomics as an effective methodology for work-system design aimed at improving the overall quality of life for millions of workers with a variety of needs and expectations.

This book focuses on prevention of work-related musculoskeletal disorders with emphasis on engineering and administrative controls. This volume contains a total of 36 chapters divided into two parts, each of which is divided into two sections.

Part I focuses on engineering factors relevant to management of work-related musculoskeletal disorders. Section I provides knowledge about risk factors for upper and lower extremities at work, while Section II concentrates on risk factors for work-related low back disorders. The knowledge presented in Section I includes epidemiology, biomechanics, and analysis of upper extremity disorders. This section also includes discussion of occupational risk factors, shoulder, design and evaluation of handtools, gloves, and industrial mats. In addition, information about injuries to the foot and leg is provided. The section on low back disorders includes knowledge on epidemiology of back pain in industry, static and dynamic low back biomechanical modeling, quantitative assessment of trunk performance, revised NIOSH equation, and population-based limits for manual lifting. In addition, this section discusses psychophysical basis and psychosocial factors in preventing musculoskeletal disorders. It also includes a method for assessment of risk of occupational low back disorders, occupational injuries due to falls, and provides a useful glossary of low back pain terminology.

Part II focuses on administrative controls in prevention and management of musculoskeletal disorders. Section I discusses fundamentals of surveillance of such disorders, requirements for surveillance database systems, OSHA record keeping system, and surveillance methods based on assessment of body discomfort. Section II focuses on medical management of work-related musculoskeletal disorders, including programs for post-injury management, testing of physical ability for employment decisions, assessment of worker strength and other functional capacities, and applications of ergonomics knowledge in rehabilitation.

The use of back belts and supporting devices for upper extremities is also considered. Finally, the influence of psychosocial factors and implications of back pain in the workplace is provided.

We hope that this volume will be useful to a large number of professionals, students, and practitioners who strive to improve product and process quality, worker health and safety, and productivity in a variety of industries and businesses. We trust the knowledge presented in this volume will help the reader learn and apply the principles of ergonomics in prevention of work-related musculoskeletal disorders.

**Waldemar Karwowski**
*University of Louisville*

**WIlliam S. Marras**
*The Ohio State University*

# Part I

# Musculoskeletal Disorders

# Section I

## Disorders of the Extremities

# 1

# Epidemiology of Upper Extremity Disorders

Bradley Evanoff
*Washington University School of Medicine*

David Rempel
*University of California San Francisco*

This chapter summarizes findings from epidemiologic studies that address workplace and individual factors associated with upper extremity musculoskeletal disorders. These disorders are not new: epidemics and clinical case series of work-related upper extremity problems were reported throughout the 1800s and early 1900s (Conn, 1931; Thompson et al., 1951). Although there are almost no prospective studies in this area, within the last 20 years a number of well-designed, cross-sectional studies have focused on disorders of the hand, wrist, and elbow as related to work. These studies point to the multifactorial nature of work-related upper extremity disorders. The severity of these disorders is influenced not only by biomechanical factors, but also by other work organizational factors, the worker's perception of the work environment, and medical management.

From an epidemiologic point of view, this topic is problematic because there are many specific disorders that can occur in the hand, arm, and shoulder, ranging from arthritis to nerve entrapments. To complicate the matter further, there are few accepted criteria for case definitions for these many disorders. In their early stages, these disorders usually present with nonspecific symptoms without physical examination or laboratory findings. In fact, the only laboratory tests consistently of value in diagnosing these disorders are nerve conduction studies for nerve entrapment disorders and radiographs for osteoarthritis. Finally, symptoms at the hand or wrist may be due to nerve compression or vascular pathology in the neck or shoulder.

## 1.1 Frequency, Rates, and Costs

Rates of hand and wrist symptoms and associated disability among working adults were assessed by a 1988 national interview survey of 44,000 randomly selected U.S. adults (National Health Interview Survey) (Park et al., 1993). Of those who had worked anytime in the past 12 months, 22% reported some finger, hand, or wrist discomfort that fit the category "pain, burning, stiffness, numbness, or tingling" for one or more days in the past 12 months. Only one-quarter were due to an acute injury such as a cut, sprain, or broken bone. Nine percent reported having prolonged hand discomfort that was not due to an acute injury; that is, discomfort of 20 or more days or 7 or more consecutive days during the last 12 months. Of those with prolonged hand discomfort, 6% changed work activities and 5% changed jobs due to the hand discomfort.

**TABLE 1.1**   Examples of Disorders of the Hand, Wrist, and Elbow
Observed in Workplace Studies

| | |
|---|---|
| Non-specific hand and wrist pain | Hand Arm Vibration Syndrome |
| Tendinitis | Osteoarthritis |
|    Tenosynovitis | Hypothenar hammer syndrome |
|    Finger tendinitis | Gamekeeper's thumb |
|    Wrist tendinitis | Digital neuritis |
|    Stenosing tenosynovitis | Nerve entrapments |
| Lateral epicondylitis |    Carpal Tunnel Syndrome |
| Medial epicondylitis |    Ulnar neuropathy at the wrist |
| Ganglion cysts |    Ulnar neuropathy at the elbow |

Elbow pain and epicondylitis are common in working populations. Symptoms of elbow pain are reported by 7 to 21% of workers in industrial populations (Chiang et al., 1993; Ohlsson, 1989; Buckle, 1987). Epicondylitis is seen in 0.7 to 2.0% of workers in jobs with low levels of physical demands to the arms and hands, and in 2 to 33% of worker groups with high levels of demands.

In the U.S., hand and wrist disorders account for 55% of all work-related repeated motion disorders reported by U.S. private employers (Bureau of Labor Statistics, 1993). This category excludes low back pain. A similar percentage is also reported in industrial (McCormack et al., 1990) and other national studies (Kivi, 1984). A similar rise in work-related hand/forearm problems has been observed in other countries such as Finland (Kivi, 1984), Australia (Bammer, 1987), and Japan (Ohara et al., 1976).

Costs for work-related musculoskeletal disorders are difficult to estimate reliably. Webster and Snook (1994) analyzed 1989 insurance claims data from 45 states, restricting their analysis to upper extremity claims classified as cumulative trauma disorders. They estimated that the total compensable cost for upper extremity cumulative trauma disorders in the U.S. was $563 million in 1989. The National Institute for Occupational Safety and Health has estimated that the annual workers' compensation costs for neck and upper extremity disorders is $2.1 billion, plus $90 million in indirect costs (NIOSH, 1996).

## 1.2   Disorder Types and their Natural History

Table 1.1 lists the most common workplace hand, wrist, and elbow problems. Nonspecific hand/wrist pain is the most common problem, followed by tendinitis, ganglion cysts, and carpal tunnel syndrome (Silverstein et al., 1987; McCormack et al., 1990; Hales et al., 1994). In many workplace studies, rates of nonspecific symptoms, tendinitis, and CTS appear to track each other, that is, a number of specific disorders typically occur together. For example, in a pork processing plant, the rank order of hand and wrist problems, as a percentage of all morbidity, was: nonspecific hand/wrist pain (39%), CTS (26%), trigger finger (23%), trigger thumb (17%), and DeQuervain's tenosynovitis (17%) (Moore and Garg, 1994). Similar ratios of disorders have been observed in manufacturing (Armstrong et al., 1982; Silverstein et al., 1986; McCormack et al., 1990), food processors (Kurppa et al., 1991; Luopajärvi et al., 1979), and among computer operators (Hales et al., 1994; Bernard et al., 1993).

Tendinitis is the most common specific, work-related hand disorder (McCormack, 1990; Luopajarvi, 1979). For the purposes of this chapter tendinitis will include hand, wrist, and distal forearm tendinitis or tenosynovitis, and trigger finger. Tendinitis occurs at discrete locations; the most common site is the first extensor compartment (De Quervain's Disease), followed by the five other pulley sites on the extensor side of the hand and three on the flexor side. The diagnosis is based on history, symptom location, and palpation and provocative maneuvers on physical exam. There has been no association of tendinitis with age or gender, but work-related tendinitis is higher among workers with less than 3 years of employment (McCormack et al., 1990).

Lateral epicondylitis is the most common specific elbow disorder; medial epicondylitis is less common. The diagnosis is based on pain and tenderness over the lateral or medial elbow and pain on movement of the wrist or fingers against resistance. Other disorders of the elbow which may be related to occupational activities include olecranon bursitis, triceps tendinitis, and osteoarthritis.

Studies of carpal tunnel syndrome have generated considerable controversy. While there is agreement that this disorder results from compression of the median nerve at the wrist, there are no universally accepted diagnostic criteria for carpal tunnel syndrome. Some consider an abnormal nerve conduction study a gold standard (Katz et al., 1991; Nathan et al., 1992; Heller et al., 1986). However, relying exclusively on nerve conduction studies can lead to reporting very high prevalence rates — 28% (Nathan et al., 1992) and 19% (Barnhart et al., 1991) in low-risk working populations. A case definition incorporating typical symptoms and signs has been proposed by NIOSH for surveillance purposes (CDC, 1989); however, the usual signs have relatively poor sensitivities and specificities (Katz et al., 1991; Heller et al., 1986; Franzblau et al., 1993). Therefore, this definition may have limited value in distinguishing CTS from other hand disorders. Hand diagrams completed by patients are reproducible and sensitive, but may lack specificity (Katz et al., 1990; Franzblau et al., 1994). Only in the later stages are weakness and thenar atrophy a noticeable feature. In approximately 25% of cases, CTS is accompanied by other disorders of the hand or wrist (Phalen, 1966).

Few studies have evaluated the work-relatedness of osteoarthritis of the hand and wrist (Hadler et al., 1978; Williams et al., 1987). Hadler et al. (1978) assessed the hands of 67 workers at a textile plant in Virginia. Significant differences in finger and wrist joint range of motion, joint swelling, and X-ray patterns of degenerative joint disease were observed between three different hand intensive jobs; the observed differences matched the pattern of hand usage.

Hand arm vibration syndrome or Vibration White Finger disease occurs in occupations involving many years of exposure to vibrating hand tools (NIOSH, 1989). This is a disorder of the small vessels and nerves in the fingers and hands presenting as localized blanching at the fingertips with numbness on exposure to cold or vibration. The symptoms are largely self-limited if vibration exposure is eliminated at an early stage (Ekenvall and Carlsson, 1987; Futatsukal and Ueno, 1986).

Hypothenar hammer syndrome or occlusion of the superficial palmar branch of the ulnar artery has been associated in clinical series and case-control studies with habitually using the hand for hammering (Little and Ferguson, 1972; Nilsson et al., 1989) and with exposure to vibrating hand tools (Kaji et al., 1993). The mean years of exposure before presentation were 20 to 30 years.

Small case-control studies or clinical series have described factors associated with less common disorders such as Gamekeeper's thumb (Campbell, 1955; Newland, 1992), digital neuritis, and ulnar neuropathy at the wrist (Silverstein et al., 1986).

## 1.3  Individual Factors

Some data on individual risk factors, such as age and gender, are available for carpal tunnel syndrome but not for other disorders of the hand and wrist. The risk of CTS increases with age (Stevens et al., 1988), but in a cross-sectional study of an industrial cohort, age explained only 3% of the variability in median nerve latency (Nathan et al., 1992). Although CTS is more common among women in the general population, in workplace studies, when employees perform similar hand activities, the ratio of female to male rates is close to 1.2:1 (Franklin et al., 1991; Nathan et al., 1992; Silverstein et al., 1986). Certain female-specific factors, such as pregnancy (Eckman-Ordeberg et al., 1987) are clearly associated with CTS; however, the role of other female factors such oophorectomy, hysterectomy (Cannon et al., 1981; Bjorkquist et al., 1977; de Krom et al., 1990), or use of oral contraceptives (Sabour, 1970), is less certain. Other individual factors have strong associations with carpal tunnel syndrome based on multiple studies: diabetes mellitus (Phalen, 1966; Yamaguchi et al., 1965; Stevens et al., 1987), rheumatoid arthritis (Phalen, 1966; Yamaguchi et al., 1965; Stevens et al., 1987), and obesity (Nathan et al., 1992; de Krom et al., 1990; Falck and Aarnio, 1983; Vessey et al., 1990; Werner et al., 1994). For some putative risk factors, the associations are based on single studies on studies presenting conflicting results: thyroid disorders (Phalen, 1966; Hales et al., 1994), vitamin B6 deficiency (Amadio, 1985; Ellis et al., 1982; McCann, 1978), wrist size and shape (Johnson et al., 1983; Armstrong and Chaffin, 1979; Bleeker et al., 1985), and general de-conditioning (Nathan et al., 1988, 1992).

**TABLE 1.2**  Work-Related Factors Associated with Disorders of the Hands and Wrists

| | |
|---|---|
| Repetition | Mechanical contact |
| Force | Duration |
| Posture extremes | Work organization |
| Vibration | |

**TABLE 1.3**  Controlled Epidemiologic Workplace Studies Evaluating the Association between Work and Wrist, Hand or Distal Forearm Tendinitis*

| Authors | Exposed Population | Control Population | Rate in Exposed | Rate in Control |
|---|---|---|---|---|
| Luopajärvi et al., 1979[5] | 152 bread packaging | 133 shop attendants | 53%[1] | 14% |
| Silverstein et al., 1986[2,3] | industrial | industrial | | |
| | 143 low force/high repetition | 136 low force/low repetition | 3% | 1.5% |
| | 153 high force/low repetition | 136 low force/low repetition | 4%[1] | 1.5% |
| | 142 high force/high repetition | 136 low force/low repetition | 20%[1] | 1.5% |
| McCormack et al., 1990 | manufacturing | manufacturing | | |
| | 369 packers/folders | 352 knitting workers | 3.3%[1] | 0.9% |
| | 562 sewers | 352 knitting workers | 4.4%[1] | 0.9% |
| | 296 boarding workers | 352 knitting workers | 6.4%[1] | 0.9% |
| Kurppa et al., 1991[4,5] | 102 meat cutters | 141 office workers | 12.5% | 0.9% |
| | 107 sausage makers | 197 office workers | 16.3%[1] | 0.7% |
| | 118 packers | 197 office workers | 25.3%[1] | 0.7% |

* Case criteria are based on history and physical examination.
[1] significant difference from control
[2] adjusted for age, sex, and plant
[3] analysis includes other disorders, although tendinitis was most common
[4] cohort study with 31-month follow-up
[5] all exposed and control subjects are female

From Rempel, D. and Punnet, L., Epidemiology of wrist and hand disorders, in *Musculoskeletal Disorders in the Workplace: Principles and Practice,* eds. M. Nordin et al., Mosby-Year Book, Inc., St. Louis, Missouri, 1997. With permission.

# 1.4  Work-Related Factors

Table 1.2 summarizes the characteristics of work that have been associated with elevated rates of upper extremity symptoms and specific disorders, including carpal tunnel syndrome and tendinitis. These associations have been observed in multiple studies and in different population groups, while dose–response trends have been seen in several studies. Most studies have been cross-sectional in design, limiting our ability to draw conclusions about causation. The preponderance of evidence, however, suggests strongly that there is a causal relationship between work exposures and upper extremity disorders. Carpal tunnel syndrome and hand–wrist tendinitis have been the best studied; several recent reviews have evaluated the work-relatedness of these disorders and concluded that there is a causal relationship (Stock, 1991; Hagberg et al., 1992; Kuorinka and Forcier, 1995). Tables 1.3, 1.4, and 1.5 summarize selected studies of wrist and hand tendinitis, carpal tunnel syndrome, and epicondylitis.

Studies using crude measures of exposure have reported associations between repetition and hand/wrist pain and disorders. In a study relying exclusively on nerve conduction measurements, median nerve slowing occurred at a higher rate among assembly line workers than among administrative controls (Nathan et al., 1992; Hagberg et al., 1992). Although no systematic assessment of exposure was carried out, the assembly line work was considered more repetitive than the control group. Rate of persistent wrist and hand pain was higher in garment workers performing repetitive hand tasks than in the control group, hospital employees (Punnett et al., 1985). Persistent wrist pain, or that lasting most of the day for

**TABLE 1.4**  Selected Controlled Epidemiologic Workplace Studies Evaluating the Association between Work and Carpal Tunnel Syndrome*

| Authors | Exposed Population | Control Population | Criteria | Rate in Exposed | Rate in Control |
|---|---|---|---|---|---|
| Silverstein et al., 1987[2] | industrial high force, high repetition | industrial low force, low repetition | history & physical exam | 5.1%[1] | 0.6% |
| Nathan, 1988[3,4] | 22 keyboard operators | 147 administrative/clerical | electrodiagnostic | 27% | 28% |
|  | 164 industrial assembly line | 147 administrative/clerical | electrodiagnostic | 47%[1] | 28% |
|  | 115 general plant | 147 administrative/clerical | electrodiagnostic | 38% | 28% |
|  | 23 grinders | 147 administrative/clerical | electrodiagnostic | 61%[1] | 28% |
| Chiang, 1990[2] | frozen food factory | frozen food factory | history and signs |  |  |
|  | 37 high repetition | 49 low repetition & cold |  | 46% | 4% |
|  | 121 high repetition & cold | 49 low repetition & cold |  | 47%[1] | 4% |
| Barnhart, 1991[3] | 106 ski manufacturing repetitive jobs | 67 ski manufacturing nonrepetitive jobs | electrodiagnostic and signs | 15.4%[1] | 3.1% |
| Osorio, 1994[2,3] | supermarket workers high exposure | supermarket workers low exposure | history & signs electrodiagnostic | 63%[1] 33%[1] | 0.0% 0.0% |

* Diagnosis based on history and physical exam or nerve conduction study.
[1] significantly different from control group
[2] control for age, gender, years on job
[3] control for age and gender
[4] low participation rate and limited exposure assessment
From Rempel, D. and Punnet, L., Epidemiology of wrist and hand disorders, in *Musculoskeletal Disorders in the Workplace: Principles and Practice*, eds. M. Nordin et al., Mosby-Year Book, Inc., St. Louis, Missouri, 1997. With permission.

**TABLE 1.5**  Selected Epidemiologic Workplace Studies Evaluating the Association between Work and Epicondylitis*

| Authors | Exposed Population | Control Population Criteria | Rate in Exposed | Rate in Controls |
|---|---|---|---|---|
| Kurppa, 1991[1] | 107 female sausage makers | 197 female office workers and supervisors | 11.1 | 1.1 |
|  | 118 female meatpackers | 197 female office workers and supervisors | 7.0 | 1.1 |
|  | 102 male meat cutters | 141 male office workers, maintenance men an d supervisors | 6.4 | 0.9 |
| Chiang, 1993[2] | 28 fish processors with high repetition and high force movements of the arms | 61 fish processors without high repetition or high force | 21.4% | 9.8% |
|  | 118 fish processors with high repetition or high force movements of the arms | 61 fish processors without high repetition or high force | 15.3% | 9.8% |
| Roto and Kivi, 1984[2] | 90 male meat cutters | 77 male construction foremen | 8.9% | 1.4% |
| McCormack, 1990[2] | 369 manufacturing workers | 352 knitting workers | 2.2% | 1.4% |
|  | 562 manufacturing workers | 352 knitting workers | 2.1% | 1.4% |
|  | 468 manufacturing workers | 352 knitting workers | 1.9% | 1.4% |
|  | 296 manufacturing workers | 352 knitting workers | 1.0% | 1.4% |
| Viikari-Juntura, 1991[2] | 332 meat plant workers | 288 office workers, maintenance workers and supervisors | 0.6% | 0.5% |
| Luopajärvi, 1979[2] | 152 female packers | 133 female shop assistants | 2.6% | 2.3% |

* Diagnosis based on history and physical exam.
[1] prospective cohort study: rates are incidence of epicondylitis per 100 workers/yr
[2] cross-sectional study: rates are prevalence of epicondylitis observed in active workers

at least one month in the last year, occurred in 17% of garment workers and 4% of hospital controls, while persistent hand pain occurred in 27% of garment workers and 10% of controls. Others have observed a similar link between high hand/wrist repetition and carpal tunnel syndrome (Chiang et al., 1990; Barnhart et al., 1991) and tendinitis (Kurppa et al., 1991). The link to repetition may be that these are jobs that require high velocity or accelerations of the wrist (Marras and Schoenmarklin, 1993).

Rates of wrist tendinitis among scissors makers was compared to shop attendants in department stores in Finland. Examinations and histories were systematized and performed by one person. The rates between the groups were not significantly different; however, among the scissors makers the rate of tendinitis increased with increasing number of scissors handled (Kuorinka and Koskinen, 1979). Luopajärvi et al. (1979) compared packers in a bread factory to the same control group. The packers' work involved repetitive gripping, up to 25,000 cycles per day, with maximum extension of thumb and fingers to handle wide bread packages. Approximately half of the packers had wrist/hand tenosynovitis compared to 14% among the controls. The most common disorder of the hand or wrist was thumb tenosynovitis followed by finger/wrist extensor tenosynovitis. CTS was diagnosed in four packers and no controls.

The force applied to a tool or materials during repeated or sustained gripping are also predictors of risk for tendinitis and carpal tunnel syndrome. For example, in a study of the textile industry the risk of hand and wrist tendinitis was 3.9 times higher among packaging and folding workers than among knitters (McCormack et al., 1990). The packing and folding workers were considered to be performing physically demanding work compared to the knitting workers. Armstrong et al. (1979) observed that women with carpal tunnel syndrome applied more pinch force during production sewing than did their job- and sex-matched controls. It is possible that those with carpal tunnel syndrome altered their working style as the carpal tunnel syndrome progressed; however, it is unlikely that they would increase the pinch force because this would also trigger symptoms. In a study by Moore et al. (1994) at a pork processing plant, the jobs that involved high grip force or long grip durations, such as Wizard knife operator, snipper, feeder, scaler, bagger, packer, hanger, and stuffer, affected almost every employee. Others have observed a similar relationship with work involving sustained or high-force grip in grinders (Nathan et al., 1992), meatpackers and butchers (Kurppa et al., 1991; Falck and Aarnio, 1983), and other industrial workers (Thompson et al., 1951; Welch, 1972).

The most comprehensive study of the combined factors of repetition and force was a cross-sectional study of 574 industrial workers by Silverstein et al. (1986, 1987; Armstrong, 1982). Disorders were assessed by physical exam and history and were primarily tendinitis followed by carpal tunnel syndrome, Guyon tunnel syndrome, and digital neuritis. Subjects were classified into four exposure groups based on force and repetition. The "high-force" work was that requiring a grip force on average of more than 4 kg-force, while "low-force" work required less than 1 kg of grip force. The "high-repetition" work involved a repetitive task in which either the cycle time was less than 30 seconds (greater than 900 times in a work day) or more than 50% of cycle time was spent performing the same kind of fundamental movements. The high-risk groups were compared to the low-risk group after adjusting for plant, age, gender, and years on the job. The odds ratio of all hand/wrist disorders for just high force was 4.9, and it increased to 30 for jobs which required both high-force and high-repetition. The identical analysis of just carpal tunnel syndrome revealed an odds ratio of 1.8 for force and 14 for the combined high-force and high-repetition group. A meta-analysis of Silverstein's data and Luopajärvi study concluded that for high-force and high-repetition work the common odds ratio for carpal tunnel syndrome was 15.5 (95% C.I. 1.7–141) and for hand/wrist tendinitis it was 9.1 (95% C.I. 5–16) (Stock, 1991). Estimates of the percentage of CTS cases among workers who perform repetitive or forceful hand activity that can be attributed to work range from 50 to 90% (Hagberg et al., 1992; Cummins, 1992; Tanaka et al., 1994).

With regard to epicondylitis, the individual roles played by force and repetition are less clear. One cohort study and six cross-sectional studies have evaluated the incidence or prevalence of epicondylitis in relation to specific jobs, which were characterized by high force, high repetition, or both. Kurppa et al. (1991) found a relative risk of 6.4 for epicondylitis in jobs with high repetition, some of which also involved high force. One cross-sectional study found a significantly elevated risk of epicondylitis only among recently employed workers in high-repetition or high-repetition/high-force jobs (Chiang et al.,

1993). Another cross-sectional study found an odds ratio of 6.9 epicondylitis in a high-repetition, high-force job (Roto and Kivi, 1984). This odds ratio was not statistically significant. Four other cross-sectional studies found little or no increase in risk for epicondylitis in workers involved in jobs characterized by high force and/or high repetition (McCormack et al., 1990; Luopajärvi et al., 1979).

Work involving increased wrist deviation from a neutral posture in either the extension, flexion or ulnar, radial direction has been associated with carpal tunnel syndrome and other hand and wrist problems (Thompson et al., 1951; Hoffman et al., 1981; Tichauer, 1966). De Krom et al. (1990) conducted a case-control study of 156 subjects with carpal tunnel syndrome compared to 473 controls randomly sampled from the hospital and population registers in a region of the Netherlands. After adjusting for age and sex, a dose–response relationship was observed for increasing hours of work with the wrist in extension or flexion. No risk was observed for increasing hours performing a pinch grasp or typing. Some studies of computer operators have linked awkward wrist postures to severity of hand symptoms (Faucett and Rempel, 1994), risk of tendinitis or carpal tunnel syndrome (Seligman et al., 1986), arm and hand discomfort (Sauter et al., 1991; Duncan and Ferguson, 1974; Hünting et al., 1981).

Prolonged exposure to vibrating hand tools, such as chain saws, has been linked in prospective studies to Hand Arm Vibration Syndrome (Ekenvall and Carlsson, 1987; Futatsuka and Ueno, 1986). The risks are primarily vibration acceleration amplitude, frequency, hand coupling to tool, hours per day of exposure, and years of exposure. However, based on existing studies, there is no clear vibration acceleration/frequency/duration threshold that would protect most workers. Therefore, medical surveillance is recommended to identify cases early while the disease can still be reversed (NIOSH, 1989). Use of vibrating hand tools may also increase the risk of CTS (Seppalainen, 1970; Cannon et al., 1981) indirectly by increasing applied grip force through a reflex pathway (Radwin et al., 1987).

Prolonged or high-load localized mechanical stress over tendons or nerves from tools or resting the hand on hard objects have been associated with tendinitis (Tichauer, 1966) and nerve entrapments (Phalen, 1966; Hoffman and Hoffman, 1985) in case studies.

The average total hours per day that a task is repeated or sustained has been a factor in predicting hand problems (Margolis, 1987; Macdonald, 1988). Among computer operators increasing self-reported hours of computer use has been a predictor of symptom intensity or disorder rate in all (Faucett and Rempel, 1994; Burt et al., 1990; Bernard et al., 1993; Oxenburgh et al., 1985; Maeda et al., 1982; Hünting et al., 1981). De Krom et al. (1990) did not observe a relationship between CTS and hours of computer use.

Work organizational (work structure, decision control, work load, deadline work, supervision) and psychosocial factors (job satisfaction, social support, relationship with supervisor) appear to have some influence on hand and wrist symptoms among computer users. Among newspaper reporters and editors, work organizational factors modified the expected relationship between workstation design and hand and wrist symptoms. Symptom intensity increased as keyboard height increased among those with low decision latitude but among those with high decision latitude (Faucett and Rempel, 1994). In another study of newspaper employees, the risk of hand and wrist symptoms was increased among those with increasing hours on deadline work and less support from the immediate supervisor (Bernard et al., 1993). Among directory assistance operators at a telephone company, high information processing demands were associated with an elevated rate of hand and wrist disorders (Hales et al., 1994). On the other hand, in the industrial setting, Silverstein et al. (1986) observed no effect on job satisfaction.

## 1.5  Summary

The lack of prospective studies and an uncertainty about the precise pathophysiologic mechanisms involved limits our ability to definitively identify causative factors. Nonetheless, current studies point to a multifactorial relationship between work exposures and disorders of the hand, wrist, and elbow. Symptom severity and disorder rate appear to be influenced by work organizational factors, such as decision latitude and cognitive demands. Some disorders, such as tendinitis and carpal tunnel syndrome, are clearly associated with work involving repetitive and forceful use of the hands. It seems likely that there is a causal relationship between some work exposures and these disorders. For other disorders, such as epicondylitis and osteoarthritis, the relationship to work exposures is less clear, although current data are suggestive. Carpal tunnel syndrome has been linked to individual factors in population-based studies and in clinical case series. However, in workplace studies where workplace exposures are adequately quantified, individual factors play a limited role relative to workplace factors (Cannon et al., 1981; Silverstein et al., 1987; Armstrong and Chaffin, 1979; Franklin et al., 1991; Faucett and Rempel, 1994; Hales et al., 1994).

# References

Adams ML, Franklin GM, Barnhart S. Outcome of carpal tunnel surgery in Washington State workers' compensation. *Am J Ind Med* 1994; 25:527-536.

Al-Qattan MM, Bowen V, Manktelow RT. Factors associated with poor outcome following primary carpal tunnel release in non-diabetic patients. *J Hand Surg* (Br Volume) 1994; 19B:622-625.

Amadio PC. Pyridoxine as an adjunct in the treatment of carpal tunnel syndrome. *J Hand Surg* 1985; 10A:237-241.

Armstrong TJ, Langolf GD. Ergonomics and occupational safety and health, in *Environmental and Occupational Medicine,* ed WN Rom, Little, Brown Co, Boston, 1982, pp. 765-784.

Armstrong TJ, Chaffin DB. Carpal tunnel syndrome and selected personal attributes. *J Occup Med* 1979; 21:481-486.

Armstrong TJ, Foulke JA, Joseph BS, Goldstein SA. Investigation of cumulative trauma disorders in a poultry processing plant. *Am Ind Hygiene Assoc J* 1982; (43)2:103-116.

Armstrong TJ, Buckle P, Fine LJ et al. A conceptual model for work-related neck and upper-limb musculoskeletal disorders. *Scand J Work Environ Health* 1993; 19:73-84.

Bammer G. VDUs and musculoskeletal problems at the Australian National University, in *Work with Display Units 86,* Eds Knave B and Wideback PG, Elsevier Science Publishers B.V. North-Holland, 1987.

Barnhart S, Demers PA, Miller M, Longstreth WE, Rosenstock L. Carpal tunnel syndrome among ski manufacturing workers. *Scand J Work Environ Health* 1991; 17:46-52.

Bernard B, Sauter S, Peterson M, Fine L, Hales T. Health Hazard Evaluation Report: Los Angeles Times. U.S. Department of Health and Human Services, Public Health Service, Centers for Disease Control, National Institute for Occupational Safety and Health, NIOSH Report No. 90-013-2277. 1993.

Birkbeck MQ, Beer TC. Occupation in relation to the carpal tunnel syndrome. *Rheumatol Rehabil* 1975; 14:218-221.

Bjorkqvist SE, Lang AH, Punnonen R, Rauramo L. Carpal tunnel syndrome in ovariectomized women. *Acta Obstet Gynecol Scand* 1977; 56:127-130.

Bleeker MQ, Bohlman M, Moreland R, Tipton A. Carpal tunnel syndrome: role of carpal canal size. *Neurology* 1985; 35:1599-1604.

Burt S, Hornung R, Fine L, Silverstein B, Armstrong T. Health hazard evaluation report: Newsday. U.S. Department of Health and Human Services, Public Health Service, Centers for Disease Control, National Institute for Occupational Safety and Health, NIOSH Report No. 89-250-2046. 1990.

Campbell, CS. Gamekeeper's Thumb. *Journal of Bone and Joint Surgery* 1955; 37(B) 1:148-149.

Cannon LJ, Bernacki EJ, Walter SD. Personal and occupational factors associated with carpal tunnel syndrome. *J Occup Med* 1981; 23:255-258.

Centers for Disease Control: Occupational disease surveillance — carpal tunnel syndrome. *MMWR* 1989; 38:485-489.

Cheadle A, Franklin G, Wolfhagen C, Savarino J, Liu PY, Salley C, Weaver M. Factors influencing the duration of work-related disability: a population-based study of Washington State Workers' Compensation. *Am J Pub Health* 1994; 84:190-196.

Chiang HC, Chen SS, Yu HS, Ko YC. The occurrence of carpal tunnel syndrome in frozen food factory employees. *Kaohsiung J Med Sci* 1990; 6:73-80.

Chiang HC, Ko YC, Chen SS, Yu HS, Wu TN, Chang PY. Prevalence of shoulder and upper-limb disorders among workers in the fish-processing industry. *Scand J Work Environ Health* 1993; 19:126-131.

Conn HR. Tenosynovitis. *Ohio State Med J* 1931; 27:713-716.

de Krom M, Kester A, Knipschild P, Spaans F. Risk factors for carpal tunnel syndrome. *Am J Epi* 1990, 132:1102-1110.

Duncan J, Ferguson D. Keyboard operating posture and symptoms in operating. *Ergonomics* 1974; 17:651-662.

Ekenvall L, Carlsson A. Vibration white finger: a follow up study. *Br J Ind Med* 1987; 44:476-478.

Ekman-Ordeberg G, Salgeback S, Ordeberg G. Carpal tunnel syndrome in pregnancy: A prospective study. *Acta Obstet Gynec Scand* 1987; 66:233-235.

Ellis J, Folkers K, Watanabe T et al. Clinical results of a cross-over treatment with pyridoxine and placebo of the carpal tunnel syndrome. *J Clin Nutr* 1979; 2046-2070.

Falck B and Aarnio P. Left-sided carpal tunnel syndrome in butchers. *Scand J Work Environ Health* 1983; 9:291-297.

Faucett J and Rempel D. VDT-related musculoskeletal symptoms: Interactions between work posture and psychosocial work factors. *Am J Ind Med* 1994; 26:597-612.

Fine LJ, Silverstein BA, Armstrong TJ et al. Detection of cumulative trauma disorders of upper extremities in the workplace. *J Occup Med* 1986; 28:674-678.

Franklin GM, Haug J, Heyer N, Checkoway H, Peck N. Occupational carpal tunnel syndrome in Washington State, 1984-1988. *Am J Pub Health* 1991, 81:741-746.

Franzblau A, Werner R, Valle J, Johnston E. Workplace surveillance for carpal tunnel syndrome: a comparison of methods. *J Occup Rehab* 1993; 3:1-14.

Franzblau A, Werner RA, Albers JW, Grant CL, Olinski D, Johnston E. Workplace surveillance for carpal tunnel syndrome using hand diagrams. *J Occup Rehab* 1994; 4:185-198.

Futatsuka M, Ueno T. A follow-up study of vibration-induced white finger due to chain-saw operation. *Scand J Work Environ Health* 1986; 12:304-306.

Hadler N, Gillings D, Imbus H et al. Hand structure and function in an industrial setting. *Arthritis and Rheum* 1978, 21:210-220.

Hagberg M, Morgenstern H, Kelsh M. Impact of occupations and job tasks on the prevalence of carpal tunnel syndrome. *Scand J Work Environ Health* 1992; 18: 337-345.

Hales TR, Sauter SL, Peterson MR, Fine LJ, Putz-Anderson V, Schleifer LR, Ochs TT, Bernard BP. Musculoskeletal disorders among visual display terminal users in a telecommunications company. *Ergonomics* 1994; 10:1603-1621.

Heller L, Ring H, Costeff H, Solzi. Evaluation of Tinel's and Phalen's signs in diagnosis of carpal tunnel syndrome. *Eur Neurol* 1986; 25:40-42.

Hoffman J, Hoffman PL. Staple gun carpal tunnel syndrome. *J Occup Med* 1985; 27:848-849.

Hünting W, Läubli T, Grandjean E. Postural and visual loads at VDT workplaces. *Ergonomics* 1981; 24:917-931.

Johnson EW, Gatens T, Poindexter D, Bowers D. Wrist dimensions: correlation with median sensory latencies. *Arch Phys Med Rehab* 1983; 64:556-557.

Katz JN, Stirrat CR, Larson MG, Fossel AN, Eaton HM, Liang MH. A self-administered hand symptom diagram for the diagnosis and epidemiologic study of carpal tunnel syndrome. *J Rheumatol* 1990; 3:1-14.

Katz JN, Larson MG, Fossel AH, Liang MH. Validation of a surveillance case definition of carpal tunnel syndrome. *Am J Public Health* 1991; 81:189-193.

Kaji H, Honma H, Usui M, Yasuno Y, Saito K. Hypothenar Hammer Syndrome in workers occupationally exposed to vibrating tools. *J Hand Surg* (Br Volume) 1993; 18B:761-766.

Kivi P. Rheumatic disorders of the upper limbs associated with repetitive occupational tasks in Finald in 1975-1979. *Scand J Rheum* 1984; 13:101-107.

Kuorinka I, Koskinen P. Occupational rheumatic diseases and upper limb strain in manual jobs in a light mechanical industry. *Scand J Work Environ Health* 1979, 5:39-47.

Kuorinka I, Forcier L. (eds.) *Work Related Musculoskeletal Disorders: A Reference Book for Prevention.* London: Taylor & Francis, 1995.

Kurppa K, Viikari-Juntura E, Kuosma E, Huuskonen M, Kivi P. Incidence of tenosynovitis or peritendinitis and epicondylitis in a meat processing factory. *Scand J Work Environ Health* 1991; 17:32-37.

Little JM, Ferguson DA. The incidence of Hypothenar Hammer Syndrome. *Arch Surg* 1972; 105:684-685.

Luopajärvi T, Kuorinka I, Virolainen M, Holmberg M. Prevalence of tenosynovitis and other injuries of the upper extremities in repetitive work. *Scand J Work Environ Health* 1979. 5:48-55.

Marras WS, Schoenmarklin RW. Wrist motions in industry. *Ergonomics* 1993; 36:341-351.

Maeda K, Hunting W, Grandjean E. Factor analysis of localized fatigue complaints of accounting-machine operators. *J Human Ergol* 1982; 11:37-43.

Magnusson M, Ortengren R. Investigation of optimal table height and surface angle in meatcutting. *Applied Ergonomics* 1987; 18.2:146-152.

Masear VR, Hayes JM, Hyde AG. An industrial cause of carpal tunnel syndrome. *J Hand Surg* 1986; 11A:222-227.

McCormack RR Jr., Inman RD, Wells A, Berntsen C, Imbus HR. Prevalence of tendinitis and related disorders of the upper extremity in a manufacturing workforce. *J Rheumatol* 1990; 17:958-964.

Moore JS, Garg A. Upper extremity disorders in a pork processing plant: relationships between job risk factors and morbidity. *Am Ind Hyg Assoc J* 1994, 55:703-715.

Muffly-Elsey D, Flinn-Wagner S. Proposed screening tool for the detection of cumulative trauma disorders of the upper extremity. *J Hand Surg* 1987, 12A: 2(2), 931-935.

Nathan PA, Keniston RC, Myers LD, Meadows KD. Obesity as a risk factor for slowing of sensory conduction of the median nerve in industry. *J Occup Med* 1992; 34:379-383.

Nathan PA, Meadows KD, Doyle LS. Occupation as a risk factor for impaired sensory conduction of the median nerve at the carpal tunnel. *J Hand Surg* 1988; 13B: 167-170.

NIOSH Criteria for a recommended standard. Occupational exposure to hand-arm vibration. DDHS Publication No. 89-106. 1989. National Institute for Occupational Safety and Health. Cincinnati, Ohio.

NIOSH National Occupational Research Agenda. DDHS Publication No. 96-1115. 1996. National Institute for Occupational Safety and Health. Cincinnati, Ohio.

Newland, CC. *Gamekeeper's Thumb.* Orthopedic Clinics of North America 1992; 23(1):41-48.

Nilsson T, Burström L, Hagberg M. Risk assessment of vibration exposure and white fingers among platers. *Int Arch Occup Environ Health* 1989; 61: 473-481.

Ohara H, Aoyama H, Itani T. Health hazards among cash register operators and the effects of improved working conditions. *J Human Ergology* 1976; 5:31-40.

Osorio AM, Ames RG, Jones JR, Rempel D, Castorina J, Estrin W, Thompson D. Carpal tunnel syndrome among grocery store workers. *Am J Ind Med* 1994, 25:229-245.

Oxenburgh M, Rowe S, Douglas D. Repetitive strain injury in keyboard operators. *J Occup Health and Safety — Australia and New Zealand* 1985; 1:106-112.

Park CH, Wagener DK, Winn DM, Pierce JP. Health conditions among the currently employed: United States, 1988. National Center for Health Statistics. *Vital Health Stat* 1993; 10(186).

Phalen GS. The carpal-tunnel syndrome. *J Bone Joint Surg* 1966; 48A:211-228.

Punnett L, Robins JM, Wegmen DH, Keyserling WM. Soft tissue disorders in the upper limbs of female garment workers. *Scand J Work Environ Health* 1985; 11:417-425.

Radwin RG, Armstrong TJ, Chaffin DB. Power hand tool vibration effects on grip exertions. *Ergonomics* 1987; 30:833-855.

Rempel D, Punnet L, Epidemiology of wrist and hand disorders, in *Musculoskeletal Disorders in the Workplace: Principles and Practice,* eds M Nordin et al., Mosby-Year Book, Inc., St. Louis, Missouri, 1997.

Roto P, Kivi P. Prevalence of epicondylitis and tenosynovitis among meatcutters. *Scand J Work Environ Health* 1984; 10:203-205.

Sauter SL, Schleifer LM, Knutson SJ. Work posture, workstation design, and musculoskeletal discomfort in a VDT data entry task. *Human Factors* 1991; 33:151-167.

Seligman P, Boiano J, Anderson C. Health Hazard Evaluation of the Minneapolis Police Department. NIOSH HETA 84-417-1745, 1986. U.S. Department of Commerce, NTIS, Springfield, Virginia.

Seppalainen AM. Nerve conduction in the vibration syndrome. *Scand J Work Environ Health* 1970; 7:82-84.

Silverstein BA, Armstrong T, Longmate A, Woody D. Can in-plant exercise control musculoskeletal symptoms? *J Occup Med* 1988; 30: 922-927.

Silverstein BA, Fine LJ, Armstrong TJ. Hand wrist cumulative trauma disorders in industry. *Br J Ind Med* 1986. 43:779-784.

Silverstein BA, Fine LJ, Armstrong TJ. Occupational factors and carpal tunnel syndrome. *Am J Ind Med* 1987. 11:343-358.

Silverstein BA, Fine LJ, Stetson D. Hand-wrist disorders among investment casting plant workers. *J Hand Surg* 1987; 12A (5 part 2): 838-844.

Stevens JC, Sun S, Beard CM, O'Fallon WM, Kurland LT. Carpal tunnel syndrome in Rochester, Minnesota. 1961 to 1980. *Neurology* 1988; 38:134-138.

Stock SR. Workplace ergonomic factors and the development of musculoskeletal disorders of the neck and upper limbs: A meta-analysis. *Am J Ind Med* 1991; 19:87-107.

Tanaka S, Wild DK, Seligman PJ, Behrens V, Cameron L, Putz-Anderson V. The U.S. prevalence of self-reported carpal tunnel syndrome. *Am J Public Health* 1994; 84:1846-1848.

Thompson A, Plewes L, Shaw E. Peritendinitis crepitans and simple tenosynovitis: a clinical study of 544 cases in industry. *Br J Ind Med* 1951; 8:150-160.

Tichauer E. Some aspects of stress on forearm and hand in industry. *J Occup Med* 1966; 8:63-71.

Viikari-Juntura E. Neck and upperlimb disorders among slaughterhouse workers. *Scand J Work Environ Health,* 1983; 9:283-290.

Vessy MP, Villard-MacIntosh L, Yeates D. Epidemiology of carpal tunnel syndrome in women of child-bearing age. Finding in a large cohort study. *Am J Epi* 1990; 19:655-659.

Webster BS, Snook SH. The cost of compensable upper extremity cumulative trauma disorders. *J Occup Med* 1994; 36:713-7.

Welch R. The causes of tenosynovitis in industry. *Indust Med* 1972; 41:16-19.

Werner RA, Albers JW, Franzblau A, Armstrong TJ. The relationship between body mass index and the diagnosis of carpal tunnel syndrome. *Muscle & Nerve* 1994; 17:632-636.

Williams, Cope et al. Metacarpo-phalangeal arthropathy associated with manual labor (Missouri metacarpal syndrome). *Arthritis and Rheum* 1987; 30:1362-1371.

Yamaguchi D, Liscomb P, Soule E. Carpal tunnel syndrome. *Minn Med J* 1965; January 22-23.

# 2

# Integrated Analysis of Upper Extremity Disorders

Richard Wells
*University of Waterloo*

## 2.1   Introduction

Work and activity-related musculoskeletal disorders (WMSD) have a complex multifactorial etiology including not only the physical aspects of the activities that people perform but also the psychosocial aspects. These disorders may involve muscular, tendinous, ligamentous, nervous tissues and include both acute (overexertion) as well as chronic (overuse) onset. A number of sources of information ranging from biomechanics, epidemiology, and clinical case series have identified a number of major risk factors associated with the development of upper limb musculoskeletal disorders. (For reviews, see Stock, 1992 or Hagberg et al., 1995.) These include forcefulness, adverse posture, repetition or continuous activity, angular velocity and acceleration, or joints and duration of exposure. Plausible biological mechanisms by which these risk factors may result in disorders of the musculoskeletal system have been proposed. Despite this, our best evidence points to a complex interaction of physical, psychosocial, and individual factors in the development of musculoskeletal disorders at work.

    An integrated approach to the causation of WMSD helps us understand the many simultaneous and interacting physical stressors which act on the upper limb during activity. These approaches help form a bridge between the performance of work and the cellular and other descriptions of the degenerative/inflammatory processes involved in work and activity-related musculoskeletal disorders. An integrated approach also guides us in the construction and evaluation of workplace assessment tools. In the sections which follow, concepts important to the assessment of the risk factors of force and posture are reviewed prior to analyzing the features of a number of assessment tools.

## 2.2   Site and Types of Upper Limb Work-Related Musculoskeletal Disorders

Terminology describing work-related musculoskeletal disorders (WMSD) has become extremely convoluted; for example, in the U.S., where the term of preference is cumulative trauma disorders (CTD), disorders in visual display terminal (VDT) operators are called repetitive strain injuries (RSI). In this chapter WMSD refers to all disorders of the musculoskeletal system (both upper extremity and low back and limbs) both to specific tissue as well as nonspecific symptoms and syndromes where associations with work have been found (cf. Hagberg et al., 1995).

An examination of Figure 2.1 reveals that a large number of types of tissues have been identified as being affected by work: tendon, muscle, nerve, and joint. The disorders identified are found in a wide variety of locations in the hand, forearm, arm, shoulder, and neck. How can we possibly devise methods which will allow us to predict injury risk in such a wide variety of sites and tissues? Fortunately, quantification of the external loads applied to the upper limb and its posture have been successful in describing the differences between jobs and tasks with high versus low risks of developing WMSD. Technically this is known as a low specificity of effect; a specific work factor can cause a number of different musculoskeletal disorders in a number of anatomic sites (Hagberg et al., 1995). This is likely so because increasing the external demand, in terms of increased force or frequency of exertion, increases the demands on most of the tissues (internal) of the arm and shoulder. While this makes the development of workplace assessment tools simpler, it makes it more difficult to examine causation and the mechanisms of disorders.

## 2.3   Risk Factors for Upper Limb Work-Related Musculoskeletal Disorders

Sources of information ranging from biomechanics, epidemiology and clinical case series have identified a number of major extrinsic (external) risk factors associated with the development of upper limb musculoskeletal disorders. These include forcefulness, adverse posture, repetition or continuous activity, joint angular velocity and acceleration, and duration of exposure. In addition, there are a number of potentiating factors which are commonly mentioned including, cold, vibration, and use of gloves (Hagberg et al., 1995). The following sections explore some concepts useful in the quantification of time and posture.

### Time as an Integral Part of Risk Factor Description

The time or frequency characteristics of tasks have typically been described by the term "repetitiveness." Unfortunately, this word is so often used and overused as to make such terms as "repetitive job" and "highly repetitive" almost meaningless. No clear definition of the term is usually offered, which compounds the lack of clarity.

In general the word is used in three main ways. First, it is used as a qualitative term to describe both the high frequency of actions as well as the sameness or monotony of the job. Second, it has been used to describe fast manual work with little apparent rest between movements. Third, repetitive work can be quantified by the number of parts, efforts, keystrokes or wrist movements/per unit time. Perhaps the most widely used operational definition of repetitive is that of Silverstein and colleagues: work with a cycle time of less than 30 seconds or having a repeated sub-cycle lasting more than 50% of the main cycle was categorized as being highly repetitive (Silverstein et al., 1986). Marras and colleagues have

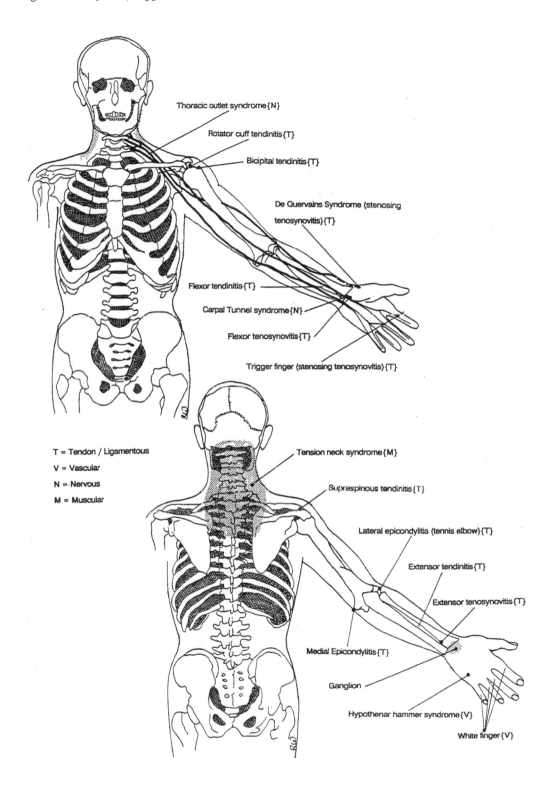

**FIGURE 2.1** Schematic of the upper limb showing examples of the sites and tissues potentially involved in work- and activity-related musculoskeletal disorders: {T} = tendon-related disorders; {N} = nerve-related disorders; {M} = muscle-related disorders; {V} = vascular disorders.

developed approaches to quantifying the time-varying nature of body motions using angular velocity and acceleration of both the wrist and trunk which have also been shown to be related to risk of injury (Marras and Schoenmarklin, 1993; Marras et al., 1993; Schoenmarklin et al., 1994).

It can be noted that many of these definitions rate the frequency of both motions and force generation; in general, each risk factor has an associated time variation. In addition, the phrase, "repetitive job," ignores the various functions of the different parts of the body. For example, a keyboard data entry task is thought of as repetitive; true, the fingers have a high frequency of movement; however, the forearms, shoulders, and back have almost constant and unchanging (static) posture and muscle activity. Westgaard and Winkel have argued that each risk factor should be described by its intensity, time variation, and its duration (Westgaard and Winkel, 1994; Winkel and Mathiassen, 1994). Ideally, the time dimension should allow the effects of different work organizations to be predicted: the effects of micropauses, of different work/rest ratios, of different break schedules, of rotation and work enlargement. At this time our knowledge does not permit us to deal with this important dimension at more than a rudimentary level. This argues for research into better ways of characterizing the time-varying nature of the major risk factors.

In the assessment of injury risk, tools must account for one more aspect of time. The estimation of risk factors which relate either to the highest demand or to some measure of cumulative or average loading. Despite the common notion that WMSD's are related to the accumulated exposure over months or years, i.e., cumulative trauma disorders, there are surprisingly few examples where cumulative exposure-response relationships have been demonstrated; most associations found are between exposure intensity and WMSD. For example, Stenlund et al. (1992) found relationships between the cumulative load lifted and arthrosis of the acromioclavicular joint in the shoulder, and Kumar (1990) found that workers with back pain had higher cumulative loads on the spine. In a similar manner, long periods in non-neutral postures have been associated with back pain (Punnett et al., 1991). Relationships have also been found for maximum loads; for example, one of the stronger predictors found by Marras et al. (1993) for low back pain was from maximum hand load.

## Posture as a Risk Factor

Postures of the limbs and trunk have a long history in characterizing tasks because, unlike many other risk factors, they are often observable and quantifiable without instrumentation. Posture is an important element of task analysis because it can be related to a number of injury mechanisms. In general, posture can give information about four kinds of stressors on the musculoskeletal system. First, if a limb segment is inclined with respect to the line of gravity a *joint moment of force* is required about the proximal end with the necessity for muscular or ligamentous forces to support it. Second, a joint angle close to the end range of motion ("extreme" posture) will load ligaments and may compress blood vessels and nerves. Third, joint angles away from the joint's optimal working range will change the geometry of the muscles crossing the joint, possibly impairing the optimal functioning of joints or tendons around the joint. Fourth and last, the change (or lack of change) in posture may be used to characterize the frequency (repetitiveness) or the static nature of the task.

### Posture as a Predictor of Joint Moment of Force

As the previous section illustrated, the joint moment of force gives important insights into tissue loads. As body segments deviate from the vertical, the ever-present force of gravity acts on the mass of the body segment: the "hidden load" of the arm mass about the shoulder and particularly the trunk mass about the low back are important, especially in sedentary tasks where the posture may be maintained for substantial periods of time. This is frequently termed postural load. If weights are held in the hand, a moment is usually created by the load in non-neutral postures.

### "Extreme" Posture as a Predictor of Soft Tissue Loads

Usually the extremes of a joint's motion are constrained by ligaments: use of extreme posture during work may not be desirable. For example, in the low back during "stoop" lifts, flexion of the lumbar spine creates tension in the posterior ligaments of the spine, and in many people a "flexion relaxation" phenomenon is seen whereby the extensor muscles of the spine become inactive and the ligaments support the moment (McGill and Norman, 1993). The drawback and potential risk in this for low back injury is that if there are unexpected loads or slipping, the only structures which can support the extra loads are the ligaments. If the posture is held for long periods of time, for example in steel reinforcement workers or gardeners, creep of the spinal ligaments and a change in the stability of the spine may result.

### Joint Posture and Optimal Musculoskeletal Geometry

For each joint there is a range of posture which minimizes possible adverse features of work and which allows effective force application with minimum fatigue and injury potential. Even before an "extreme" posture is reached, there are changes in the function of the musculoskeletal system which usually make the postures less than optimal and which may elevate tissue loads.

For example, at the wrist, extension of greater than about 30 degrees increases intracarpal pressure, even in normal people, above 30 mmHg (Rempel et al., 1995). This pressure, if maintained for substantial periods of time, likely decreases microcirculation of the structures in the tunnel, including the median nerve. This may be one of the mechanisms by which work activities cause carpal tunnel syndrome. Another example at the wrist involves grasping a small object with the wrist in flexion. This can require large effort, and forcing the wrist into maximal flexion will usually cause the object to be dropped. This is the basis of a number of actions in self-defense. This example shows that nonoptimal postures require higher efforts to perform a given task. Large deviation from approximately neutral postures can also affect blood supply: looking upward, as during the picking of fruit, can compromise cerebral blood flow especially if coupled with neck twist (Sakakibara et al., 1987).

Each joint has an optimal position for different work activities; it is often near the midpoint of the range of motion, but this rule has sufficient exceptions to make it unreliable. For example, the knee functions very well close to the extreme straight position during most locomotor tasks.

### Change of Posture

Work involves changes in posture, and the changes can be used to quantify the frequency of movements. Frequency of activity is described further in a later section. If postures do not change for long periods of time, such as shoulder and trunk posture during computer (VDT) work, the task may be called static.

### Force as a Risk Factor

Despite the existence of a large number of external risk factors (Figure 2.2), it can be argued that the final common pathway by which work causes or contributes to the development of WMSD is force. External loads and postures give rise to "internal exposures" in the tissues of the upper limb. Thus a fingertip force may give rise to tensile force in the finger flexor tendons; simultaneous wrist flexion stretches the wrist and finger extensors, increasing their passive tensile force (Keir et al., 1996); the wrist flexion also increases the hydrostatic pressure in the carpal canal. These forces have a different effect on each tissue which will be discussed in more detail shortly.

Externally we may wish to measure the force required or the force exerted by the hand (which may be considerably more, depending on the friction and size of the object). We may wish to measure the absolute force or the force relative to an individual's capacity.

It can be seen that posture is most frequently a modifier of or a predictor of the loads experienced by the tissue. The foregoing shows why posture is such a valuable measure of workplace risks: only when external loads are applied is it of limited use in workplace evaluation. The internal exposure to force, however, remains central; other risk factors affect this directly or indirectly.

## 2.4  Integrated Approach to Evaluate Potential for Upper Limb WMSD

Figure 2.2 illustrates an integrated approach to evaluate the potential for upper limb WMSD: the "external" factors of force, posture, time variation (repetitiveness), and duration of exposure act on the

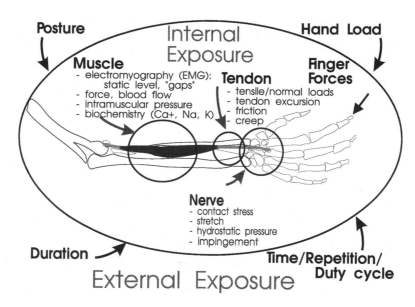

**FIGURE 2.2**   Illustration of the distinction between "external exposures" to the body and "internal exposures" to individual tissues. Under each tissue some of the measures for evaluating the potential for developing work-related musculoskeletal disorders are listed.

musculoskeletal system. These create "internal exposures" to the tissues of the body. It is these internal exposures which stress the individual tissues and which must be resisted: it is at this level which injury mechanisms can be tested using histological, physiological, or electrophysiological techniques. For example, through the use of external forces (An et al., 1987) and limb accelerations (Marras and Schoenmarklin, 1993) moments of force at the elbow during work can be predicted. Because the musculoskeletal system is mechanically indeterminate, i.e., there are more force-producing structures than equations of equilibrium to describe the system, analysis to predict the load in the individual tissues demands either that assumptions be made concerning how muscles are recruited be made or some criterion measure is minimized through optimization approaches (An et al., 1987; Wells et al., 1995). In addition, through the incorporation of the biological materials properties the response of the tissue to load can be grossly predicted. The models available to study activity-related musculoskeletal disorders are in preliminary stages of development. This is complicated by the lack of a good animal model of these disorders and the delicate balance of physical strain and restorative responses.

In general these "internal exposures" are the subject of laboratory-based research rather than workplace assessment, but they are important because they help us conceptualize the best external variables to measure and the best way in which to evaluate them. For example, in the depiction of manual material handling tasks to elucidate the link between work and low back pain, one could describe the load lifted by a person, and separately, the distance away from the body of this mass. It has been found more useful to compute the joint moment; the product of the force and the distance (Marras et al., 1993). This is done based on a biomechanical model which demonstrates that tissue loads (internal exposures) are better reflected by moment than either load or posture separately. Similar arguments can be made at the shoulder.

Figure 2.3 depicts a conceptual model of the factors influencing the development of work-related carpal tunnel syndrome. The pathways indicated a way in which the multiple demands of work may combine to reduce or elevate potential for trauma to the tissue. Take, for example, work in the cold with a vibrating tool; gloves will also likely be worn. The diagram illustrates how the presence of gloves may increase grip force; the presence of cold and vibration may decrease tactile sensation and further increase grip force.

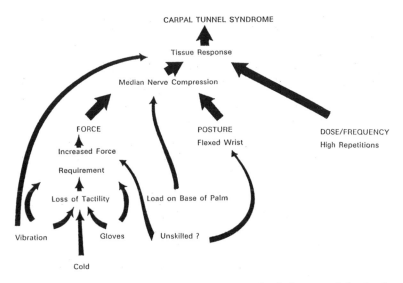

**FIGURE 2.3** Conceptual model of the relationship between external risk factors and the development of carpal tunnel syndrome due to both compression of the median nerve by the flexor tendons and changes in carpal tunnel geometry due to posture alone. Across the center are listed the major (external) risk factors of force, posture and time. Below are listed risk factors which may elevate the effect of these primary factors, and above are noted possible pathways by which the external exposures (risk factors) create internal exposures that could plausibly lead to the development or aggravation of CTS. Many additional factors can alter the forces in the flexor tendons such as the grip type, number of digits used, and the angular acceleration of the wrist. Lack of skill may lead to poorer postures, higher grip forces, or increased coactivation of muscles, more movements to perform a given task, or jerkier (higher acceleration) motions.

Building on this concept, Moore et al. (1991) described an integrated approach using biomechanical modeling to predict internal exposure variables likely related to injury. This is achieved by the synthesis of posture, force, movement, and muscle loading data. The model produced a profile of measures for use in industrial settings, that reflect the loading on the different tissues affected by WMSD (nerves, tendons, muscles) and the different loading mechanisms (e.g., highly static postures, repeated extreme postures, dynamic movements). The measures involve continuous monitoring of hand/wrist postures, forces, and muscle activations (electromyographic signals) over the duration of the task both in the arms and shoulders. Using the previously described measures as input to a biomechanical model of the forearm and hand, a profile of 12 risk factors was created which characterized the demand of the task on the distal upper limb. The 12 variables were: peak tendon force, cumulative tendon force, cumulative tendon excursion, peak tendon excursion velocity, average pressure on the flexor retinaculum (and thus the median nerve), peak pressure on the flexor retinaculum, cumulative pressure on the flexor retinaculum, cumulative frictional work on the flexor tendons, peak frictional power, and three measures of the flexor myographic signal, the 10th or static, 50th, and 90th percentiles of the amplitude probability distribution function (APDF).

The models above integrated information from anatomical, biomechanical, and epidemiological studies to produce a profile of measures to characterize tasks and which reflected injury mechanisms for different tissue types. These will now be briefly reviewed.

## Approaches to Investigating Tendon Disorders

Etiologically, reduced lubrication between tendons and tendon sheaths due to excess relative movement has been suggested in tenosynovitis (Rowe, 1987) while high peak loads and cumulative strain have been suggested for tendinitis (Goldstein et al., 1987). Norman and Wells (1990) have proposed a model for assessment of tenosynovitis which was operationalized by Moore et al. (1991). This is seen in Figure 2.4.

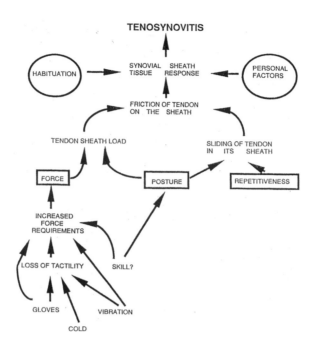

**FIGURE 2.4**   Conceptual model of the relationship between external risk factors and the development of tenosynovitis. Across the center are listed the major risk factors of force and posture. Below are listed risk factors which may elevate the effect of these primary factors, and above are noted possible pathways by which the external exposures (risk factors) create internal exposures that could plausibly lead to the development or aggravation of tenosynovitis. Many additional factors can alter the forces in the flexor tendons, such as the grip type, number of digits used, and the angular acceleration of the wrist. (Adapted from Norman, R.K. and Wells, R.P. 1990. Biomechanical aspects of occupational injury, *Proceedings of the 23rd Annual Conference of the Human Factors Association of Canada*. With permission.)

In this approach the frictional work done by the tendon sliding through its sheath is calculated. One type of frictional work is present due to a "belt-pulley" interaction when the wrist deviates from a straight position (Armstrong and Chaffin, 1979). In addition to this type of frictional work, it is suggested that a non-negligible resistance to movement is present to move the tendons through the carpal tunnel even in the straight position. Estimates from the work of Goldstein et al. (1987) and Smutz et al. (1995) put this resistance at around 5N in the neutral position. Excursion of the tendons at the wrist (caused by finger and wrist movement) in both deviated, and straight postures will therefore create an energy input, possibly beyond the recovery capability of the tissue.

## Approaches to Investigating Nerve Disorders

Insult to the median nerve, whether due to increased hydrostatic pressures in the carpal canal (Rempel et al., 1995) or due to mechanical insult upon the nerve by overlying tendon (Keir and Wells, 1995), has often been suggested as a likely mechanism of work-induced carpal tunnel syndrome. More controversially, it may be caused by hypertrophy or edema of the synovial sheaths and thus be secondary to tenosynovitis.

Mechanical stress to the median nerve can be predicted by modified belt-pulley models of the wrist and is an output of a biomechanical model (Moore et al., 1991). Hydrostatic pressures are measurable using *in vivo* and *in vitro* techniques (Keir and Wells, 1995).

## Methodological Approaches to Predicting Muscle Disorders in the Upper Limb

Recent clinical findings have suggested that forearm muscle pain may be an overlooked problem in studying work-related chronic musculoskeletal injuries (Ranney et al., 1995). While work-related muscle pain is well accepted in the shoulder area (e.g., Veiersted et al., 1993), pain in the forearm is usually attributed to tendinitis or epicondylitis. The major approach to investigating muscle-related occupational disorders is electromyography. Jonsson (1982) described a technique in which the frequency of any particular level of EMG occurring is calculated. From this, an amplitude probability curve is developed. The static level describes the ability of the muscle to rest at least 10% of the time and appears important in the development of chronic work-related muscle problems. If the value is greater than zero, the muscle is not given a chance to completely rest at least 10% of the time during a task. While this is a useful technique for quantifying muscle usage throughout the duration of a task, it gives no indication of the duration of each rest pause, i.e., whether the rests came as numerous pauses or one big pause. Veiersted et al. (1990, 1993) addressed this by using a "gaps" analysis. This analysis looks at the number of times the muscle is turned off (an EMG "gap" is defined as a muscle activation of less than 0.5% MVC lasting for more than 0.2 seconds), and it appeared that people with pain had fewer gaps. More recently it has been shown that workers likely to require neck/shoulder sick leave can be predicted from measures of "gap" frequency (Veiersted et al., 1993).

## 2.5   Workplace Assessment Tools

### Scope of Workplace Assessment Tools Reviewed

The previous sections have described some of the main risk factors which increase the risk of developing upper limb musculoskeletal disorders and how these risk factors may produce internal exposures to tissues of the body potentially leading to premature fatigue or WMSD. It falls to this section to describe and review some of the tools in the literature which have been developed to assess workplace risk of WMSD. Tools included met two conditions: (1) the tool or method not only recorded risk factors but rated them and (2) rated the injury risk to the upper limbs or the intervention priority.

The workplace assessment tools reviewed fell into two main categories: those using mainly observational methods to identify, rate, and combine a number of risk factors to produce an estimate of risk for upper limb musculoskeletal disorders (e.g., Strain Index) or intervention priority (e.g., RULA) and which are often suggested as screening tools, and those which measured (sometimes called technical methods) the time course of a more complex risk factor (e.g., the trapezius electromyogram) and produced an estimate of risk. The tools are described in Table 2.1 and evaluated on a number of criteria which are important either from a conceptual, measurement, or usability viewpoint. The following sections describe these criteria.

### Stated Purpose

The tools can only be reviewed on the basis of the stated purpose; this purpose differed between the tools. For example, the RULA output is in terms of an intervention priority, whereas trapezius EMG has been associated with neck/shoulder musculoskeletal sick leave. Some tools were designed for screening purposes whereas others were intended to produce more definitive analyses. The precise area of the body for which the tool predicts risk also differs between tools and methods; for example, RULA has upper limb and trunk scores and defines an intervention priority based upon the combined whole body score, whereas the Strain Index is specific to the distal upper limb.

### Input Variables

As noted in the previous sections, there is a wide range of potential stressors for the upper limb; how many there are and how they are measured is important information concerning the potential generalizability of the method. The handling of the time-varying nature of the risk factors is, to this author, of prime importance both for its usefulness and its ability to answer the frequently asked questions concerning line balancing, job rotation, and break schedules. The inputs are described by the forces and postures adopted, the characterization of their time course, as well as the treatment of other risk factors.

**TABLE 2.1**  Comparison of Workplace Assessment Tools in the Literature which Have Been Developed to Assess Workplace Risk of WMSD and Priority for Intervention

| Workplace Assessment Tool | Reference(s) | Stated Purpose | Input Variables | | | |
|---|---|---|---|---|---|---|
| | | | Force | Posture | Time | Other |
| Cumulative Trauma Disorder Checklist *Upper Extremity CTD* | Lifschitz and Armstrong (1986) | "…to indicate easily perceived problems in the workplace from the design point of view with respect to upper extremity cumulative trauma disorders; … to rapidly give readable feedback to the user." | "Weight of the tool below 10 lbs" | "…bending the wrist" | "is the cycle time above 30 seconds" | contact stress, vibration, temperature, seating, adjustability, tool characteristics |
| Electromyography *Neck/shoulder* | Aaras and Westgaard (1987); Veiersted et al. (1990, 1993); and Veiersted (1996) | Use of electromyographic evaluation as a predictor of neck/shoulder sick leave and myalgia. | Muscle activation in %Maximum Voluntary Contraction (MVC) | — | Continuous record | — |
| Wrist Goniometry *Upper Limb CTD* | Marras and Schoenmarklin (1993); Schoenmarklin, Marass, and Leurgans (1994) | "… to determine *quantitatively* the association between specific wrist motion parameters and the incidence of CTD as a group." | — | Wrist flexion/extension, ulnar/radial deviation and pronation/ supination | Continuous record | — |
| Psychophysical *Hand/wrist* | Snook et al. (1995) | "…to investigate the feasibility of using psychophysical methods to determine maximum acceptable forces for various types and frequencies of repetitive wrist motion." | Wrist moment | Testing performed over ROM, ± 45 deg. | Frequency (#/min) and duration (hrs) | — |

| Workplace Assessment Tool | Evaluation of Risk Factors | | Equipment Required | Measurement Characteristics | Validity | Study base/ Generalizability |
|---|---|---|---|---|---|---|
| | Rated | Combined | | | | |
| Cumulative Trauma Disorder Checklist | Factor present or absent | Sum of number of risk factors present (NO's). | — | — | Number of NO scores increased with the incidence of reported CTD | Auto plant |
| Electromyography | Raw surface EMG processed to form APDF or "gaps." | — | Portable EMG system | Reliability; "static" APDF (0.59) and gaps (0.85) in a field setting | Static level of trapezius EMG predictive of neck/shoulder sick leave in an inception cohort of packers | Packaging workers and workers from telephone assembly |
| Goniometry | Wrist posture differentiated to form angular velocity and acceleration | — | 3-D goniometer system and software | — | In a comparison of high and low risk jobs flexion accelerations were significantly different between groups. | 40 workers in 8 automotive sector manufacturers |
| Psychophysical | Acceptability of a combination of wrist moment and frequency judged by psychophysical adjustment experimentation | — | Applied force and its moment arm required | — | Such a method has been validated for manual materials handling, Snook (1978). | Laboratory study, 29 female participants drawn from a working population |

**TABLE 2.1 (continued)**  Comparison of Workplace Assessment Tools in the Literature which Have Been Developed to Assess Workplace Risk of WMSD and Priority for Intervention

| Workplace Assessment Tool | Limit/Guideline Level Proposed | Information for Intervention | Limitations | Other |
|---|---|---|---|---|
| Cumulative Trauma Disorder Checklist | No | Address NO items in checklist | — | |
| Electromyography | Yes, some suggestion for static load of 2–5% and less than 1% | No clear link to help work station redesign except reducing static loading. May be valuable in evaluation of prototypes. Could be used as EMG biofeedback to improve work methods | May not be sensitive to some postural effects, impingement and postures likely to promote nerve entrapment. Not tested for sensitivity to short duration, high load situations | Could be generalized to other areas of the upper limb, (e.g., Moore et al., 1991) |
| Goniometry | Yes, acceleration characteristics of high and low risk jobs described | Reduce acceleration | Not tested for activities involving constant gripping | One of the few approaches to allow evaluation of rapid dynamic manual tasks |
| Psychophysical | Yes, acceptable for x % of the population | Can determine a combination of wrist moment and frequency to be acceptable to any percentage of the population | Unknown relationship between acceptability and risk of musculoskeletal disorders. A "library" of tasks is needed to allow analysis of many jobs. | This approach uses psychophysical experimentation to determine acceptable (guideline) values for wrist moment. Other tasks are planned to be added to the "library" of tasks. |

| Workplace Assessment Tool | Reference(s) | Stated Purpose | Input Variables | | | Other |
|---|---|---|---|---|---|---|
| | | | Force | Posture | Time | |
| OWAS (Ovako Working Posture Analysis System) *Whole Body* | Karhu et al., (1977), Mattilla et al., (1992), Leskinen and Tönnes (1994) | "...OWAS has been planned to meet the criteria. (a) simple enough to be used by ergonomically untrained personnel, (b) it must provide unambiguous answers even if it results in over-simplification, (c) it must also offer possibilities for correcting the over-simplified ergonomic approach...so that a systematic guide to corrective action can be constructed." | Effort of <10 kg, <20 kg, and >20 kg. | Observed shoulder posture, limb above or below shoulder. (Other categories for trunk and lower limbs) | Posture sampling. OWASCO/OWASAN allow calculation of sampling statistics | — |
| RULA (Rapid Upper Limb Assessment) *Upper Limbs + Trunk and Lower Limbs* | McAttamney and Corlett (1993) | "...investigate the exposure of individual workers to risk factors associated with work-related musculoskeletal disorders." | Categories for 0–2 kg, 2–10 kg, and more than 10 kg. | Observed posture at wrist, lower arm and shoulder from predefined postures. | Action repeated more than 4/min or mainly static and held for more than 1 minute. Analysis performed at an instant "...held for the greatest amount of time or where highest loads occur." | — |
| Strain Index *Distal Upper Extremity* | Moore and Garg (1995) | "...semi-quantitative job analysis that ... appears to identify accurately jobs associated with distal upper extremity disorders...." | Observed intensity of exertion on a five point scale from light to near maximal | Hand/wrist posture on a scale from very good to very bad | Three temporal descriptors: Duration of cycle(%), efforts/min and duration/day all on a five point scale. | — |
| Upper Extremity Checklist *Upper Limb* | Keyserling et al., (1993) | "...designed to function as a rapid screening tool which could be used by persons with relatively little training and experience in ergonomics to identify jobs with potentially harmful exposures to upper extremity risk factors." | Presence of forceful exertions; manual materials handling >4.5 kg, poor grip friction, pressing with fingertip, hold object >2.7 kg. | Postures of the arms and shoulder; pinch, wrist deviation, twisting 2 adverse shoulder postures. | Duration of risk factor as none, up 1/3 and greater than a 1/3 of a cycle. Presence of a cycle time of 30 sec or less. | Contact stresses, gloves, jerk. |

**TABLE 2.1 (continued)**    Comparison of Workplace Assessment Tools in the Literature which Have Been Developed to Assess Workplace Risk of WMSD and Priority for Intervention.

| Workplace Assessment Tool | Evaluation of Risk Factors | | Equipment Required | Measurement Characteristics | Validity | Study base/ Generalizability |
|---|---|---|---|---|---|---|
| | Rated | Combined | | | | |
| OWAS (Ovako Working Posture Analysis System) | Postures rated by 32 experienced steel workers from "....no discomfort and no effect on health ..." to "...short exposure leads to discomfort; ill effect on posture possible." Ranking also by small group of international ergonomists | — | Paper form or video/computer system for OWASCO/ OWASAN | Inter-observer (2 observers); 74–99% agreement. Test-retest (morning/afternoon) (2 observers) 70–100% agreement | — | Steel Works |
| RULA (Rapid Upper Limb Assessment) | Individual risk factor weights by expert judgment | Additive model. "Look-up" tables to obtain intervention priority by expert judgment. | Clipboard and pen | "...high consistency of scoring among ..." 120 persons | Statistically significant association between upper and lower body scores (A and B) and discomfort in that area in 16 VDT operators | Ratings developed from production line, VDT data entry and sewing tasks. Validity data determined on VDT operators |
| Strain Index | Intensity of effort raised to power of 1.6 (psycho-physical power law). Other rating using professional judgment | Multiplicative model | None | Sensitivity analysis | Increase in mean incidence rate with Strain Index in 25 jobs | One pork processing plant |
| Upper Limb Checklist | By time duration and expert judgment as minor(√) or significant(*) | Potential(_) and significant risk (*) factors summed separately. | None | Inter-observer; moderate, intra-observer; moderate | Agreement between trained Health and Safety Personnel and experts: | Auto-industry |

| Workplace Assessment Tool | Limit or Guideline Level Proposed? | Information for Intervention? | Limitations | Other |
|---|---|---|---|---|
| OWAS (Ovako Working Posture Analysis System) | Posture ranking used to create 4 action categories; Class 1 = Normal postures which do not need any special attention Class 4 = postures need immediate attention. | "...postures can be evaluated in terms of their desirability and need of attention..." | Only shoulder posture rated and at only two postures. No information on distal upper limbs | |
| RULA (Rapid Upper Limb Assessment) | Intervention priority indicated (for whole body) | Address highest upper limb risk factor scores | Little effect of time course of risk factors on ratings | Scores for the upper limbs as well as the trunk and lower limbs. Both are combined to form a grand score which is used to indicate an intervention priority. |
| Strain Index | Preliminary suggestion of 5 as limit value | Address highest upper limb risk scores. Equation also allows impact of changing risk factor levels to be assessed. | | Contact stresses, cold, and vibration not accounted for |
| Upper Limb Checklist | "...any job receiving one or more 'stars' was considered to have a high priority for additional investigation and analysis." | "...any job receiving one or more 'stars' was considered to have a high priority for additional investigation and analysis." | | The period between the number of √ and * is a separator with no meaning. A job with ten * is not necessarily ten times as risky as a job with one *; it simply has more separate risk factors. These risk factors may or may not be synergistic or act at the same body joint. |

The tools differed considerably in their treatment of the time course of the input variables; some used only a single instant while others used mathematical or electronic processing to extract information about the time variation of the risk factor.

## Rating of Individual Risk Factors

As previously noted, only tools which evaluated the size of the risk factor(s) have been reviewed here; how this is done and the quality of this assessment is key to the usefulness of the tool. It is difficult to imagine any work without some risk factors present and one can quickly fall into the mindset that work is inherently dangerous and the observation of a bent wrist during work implies hazard. A distinction is made in industrial hygiene between toxicity and hazard. Benzene is highly toxic, yet if used infrequently where the concentration is small (a person's exposure is low), the hazard is small. Similarly, even for wrist flexion close to an individual's range of motion (ROM), the risk is also low if the motion is infrequent. In fact the adoption of "extreme postures" for short periods of time is probably beneficial; they are called stretch breaks.

A number of approaches are seen ranging from statistical treatments based on epidemiological studies to expert and consensus judgments. For single risk factors epidemiological approaches are possible; however, some element of expert judgment becomes necessary to "fill-in" the holes in the epidemiological literature.

## Combination of Risk Factors

Very few epidemiological studies allow the interactions of a number of risk factors to be examined. For example, Silverstein and colleagues did study a simple $2 \times 2$ interaction of force and repetitiveness (Silverstein et al., 1986), while the psychophysical approach allows combinations of multiple dimensions to be rated. These studies are not common and so the majority of studies combine rating of the individual risk factors with additive or multiplicative models to arrive at a risk estimate.

The combination of risk factors to produce an estimate of risk is perhaps the most difficult issue in workplace evaluation and tool construction. Should the individual risk factor ratings be added or multiplied or even considered completely separately? For example, in evaluating the risk of low back pain on a job one could measure the risk factor of posture and load separately. Does one add the posture and load score or multiply them? Biomechanical models indicate that multiplying the load and its moment arm about the low back (in effect, posture) gives the low back moment of force (or torque). This has been found to give the best single prediction of low back pain risk (Marras et al., 1993). Clearly the integrated approach advocated here uses biological and mechanical arguments to help in this decision.

In industrial hygiene exposure levels are considered separately except when the agent of interest has the same target organ or pathway. For example, in a given job there is exposure to work overhead and hand/arm vibration from a hand tool. Does this mean the job has two risk factors which need improvement or is there more risk than if either of these exposures occurred separately? The first approach is supported by different target "organs," the shoulder and forearm, while the low specificity of effect (Hagberg et al., 1995) could argue for the second interpretation.

Based on the discussion of the importance of time as a descriptor of risk factors, it would appear that time must be considered along with the primary risk factor, force. In some cases both force and posture are considered and in these cases, both force and posture may be considered with their time variation. In some cases posture may be used as a surrogate of force and in such cases its time variation must be considered. In all these cases, this may be done additively or, more commonly, multiplicatively (cf. NIOSH, 1981; Moore and Garg, 1995).

Definitive answers to the above conceptual issues are not available, and so the magnitude of the total score calculated and its relation to risk must be cautiously interpreted. In the case of tools whose purpose is to calculate intervention priority, these questions are not as critical because the scores are used as a summary of the size of the individual risk factors and their number combined.

## Equipment Required

The equipment required is important in the choice of a tool; some methods utilize observations of work, while others use various "technical" methods, such as electromyography or goniometry, in the measurement process. This obviously affects the time and cost of the assessment. While it is sometimes heard that ergonomics assessments must be simple and cheap, the value of more costly yet precise technical measures with epidemiologically determined relationships to risk must not be undervalued. The time, training required, and cost of using the tools are rarely reported.

## Measurement Characteristics

The tools developer can arrive at an estimate of risk in a large number of ways; what is important, however, is the quality of the tool's predictions. This can be assessed in terms of the measurement characteristics and validity. The measurement characteristics refer to such qualities as intra-observer reliability (or test–retest reliability) as well as inter-observer agreement or reliability. A tool with poor reliability will be of limited usefulness. It may still, however, be able to distinguish jobs with many risk factors and high risks from those with few risk factors and low risks, but it may not reliably distinguish between jobs with less extreme contrasts. Good measurement characteristics become of even greater concern if the tool is used for guidelines or legislative purposes.

## Validity

The term validity can be used in a number of senses. Content validity refers to the completeness of the assessment. Questions such as "are all important risk factors rated" are asked here. Most tools reviewed, however, used some variant of criterion-related validity. The output of the tool was compared to some health-related output on jobs or individuals. The stated purpose is important here in evaluating the appropriate comparisons.

## Study Base/Generalizability

It is not possible to test a tool under all possible conditions; the range of workplaces used to develop and test the tool are useful in judging the applicability of the tool to a given target workplace. For example, it is likely problematic to use a tool developed in an office environment to apply to a construction site.

## Proposed Limit of Guideline Level?

Although not universal, many of the tools produced some recommendations or guidelines in terms of the score or output of the tool. A number of tools have screening as their stated purpose. In this case a two- (or more) step process is assumed, and those jobs exceeding some criterion score are further analyzed. In this case a high sensitivity is desirable; a moderate number of "false positive" findings are accepted so as not to miss potentially risky jobs in the first step.

For those tools whose purpose is to define risk, it is unreasonable to imagine that a single threshold divides risky jobs from non-risky ones. Where the risk of developing various WMSDs has been produced against a continuous exposure measure, it has been found that the risk increases steadily from the nonexposed state, i.e., there is no obvious step or threshold evident (e.g., Punnett et al., 1991). The threshold chosen is then dependent on the increased risk which is to be accepted (a societal judgment). While a guideline value can be useful in the interpretation of an instrument's score, the measurement characteristics (the validity and the generalizability) of the tool must be of high quality before reliance can be placed on these recommendations.

## Information for Intervention

The assessment of risk is but one step in process of workplace improvement; a good tool provides direction on which risk factors need addressing and also provides material and suggestions for solutions. Ideally, it might also allow "what if" scenarios to be explored and predict what the level of risk will be for the new combination of risk factors.

## Limitations

No tool is perfect; the limitations of a given tool need to be understood, however, so that undue reliance on the output is not made where the tool's predictions are likely not to be of high quality. Each tool could have a large number of limitations; the focus of this section is on major areas where the predictive power of the tool is suspect or untested.

## 2.6  Summary

This chapter has reviewed some concepts important in the measurement and evaluation of the major risk factors for the development of WMSDs in the upper extremity. An "integrative" approach is followed whereby anatomical, physiological, and biomechanical information is used to conceptualize upper limb function and as a possible means to learn how work might cause WMSDs. These concepts are then used to inform a review of some of the major upper limb workplace evaluation tools; each has different purposes, strengths, and weaknesses. As these tools are further developed, the framework used in this chapter should provide a springboard from which the reader can assess existing and new tools in the light of their needs and available resources.

## References

Åaras, A. 1987. Postural load and the development of musculo-skeletal illness: *Scand. J. Rehab. Med;* Suppl. 18, 1-35.

Åaras, A. and Westgaard, R.H. 1987. Further studies of postural load and musculo-skeletal injuries of workers at an electro-mechanical assembly plant. *Applied Ergonomics;* 18(3):211-219.

An, K-N, Chao, E.Y., Cooney, W.P., and Lischeid, R.L. 1985. Forces in the normal and abnormal hand. *J. Orthop. Res.;* 3:202-211.

Armstrong, T.J., Castelli, W.A., Evans, F.G., and Diaz-Perez, R. 1984. Some histological changes in the carpal tunnel contents and their biomechanical implications. *J. Occup. Med;* 26(3):197-201.

Armstrong, T.J. and Chaffin, D.B. 1979. Some biomechanical aspects of the carpal tunnel. *J. Biomechanics;* 12:567-570.

Goldstein, S.A. 1981. *Biomechanical Aspects of Cumulative Trauma to Tendons and Tendon Sheaths.* PhD Thesis, University of Michigan.

Goldstein, S.A., Armstrong, T.J., Chaffin, D.B., and Matthews, L.S. 1987. Analysis of cumulative strain in tendons and tendon sheaths. *J. Biomechanics;* 20(1):1-6.

Hagberg, M., Silverstein, B., Wells, R., Smith, R., Carayon, Hendrick, H.P., Perusse, M., Kourinka, I., and Forcier, L. (eds.). 1995. *Work-Related Musculoskeletal Disorders (WMSD): A Handbook for Prevention,* Taylor & Francis, London.

Jonsson B. 1982. Measurement and evaluation of local muscular strain in the shoulder during constrained work. *J. Human Ergon.;* 11:73-88.

Karhu, O., Kansi, P., and Kuorinka, I. 1977. Correcting working postures in industry: a practical method for analysis. *Applied Ergonomics;* 8(4):199-201.

Keir, P., Wells, R., and Ranney, D. 1996. Passive stiffness of the forearm musculature and functional implications; A pilot study, in press. *Clin. Biomech.,* 11(7):401-409.

Keir, P.J. and Wells R. 1995. The effect of tendon loading and wrist posture on carpal tunnel pressure in cadavers. *Proceedings of the 19th Annual Meeting of the American Society of Biomechanics,* Stanford University, August 1, 1995, pp.129-130.

Keyserling, W.M., Stetson, D.S., Silverstein, B.A., and Brouwer, M.L. 1993. A checklist for evaluating ergonomic risk factors associated with upper extremity cumulative trauma disorders. *Ergonomics;* 36(7):807-831.

Kumar, S., 1990. Cumulative load as a risk factor for low-back pain. *Spine* 15:1311-1316.

Lifshitz, Y. and Armstrong, J. 1986. A design checklist for control and prediction of cumulative trauma disorder in intensive manual jobs. *Proceedings of the Human Factors Society*, 30th Annual Meeting; 837-841.

Larsson S.E., Bengtsson, A Bodegård, L., Henriksson, K.G., and Larsson, J. 1988. Muscle changes in work related chronic myalgia, *Acta Orthop. Scand.*; 59(5):552-6.

Leskinen, T. and Tönnes, M. 1994. Utilization of a video-computer system for analyzing postural load-evaluation of observation. *Proceedings of the 12th Triennial Congress of the International Ergonomics Association*, Toronto, Canada, August 15, 1994; Vol. 2 pp. 383-385.

Marras, W.S. and Schoenmarklin, R.W. 1993. Wrist motions in industry. *Ergonomics*; 36(4):341-351.

Marras, W.S., Lavender, S.A., Leurgans, S.E., Rajulu, S.L., Allread, W.G., Fathallah, F.A., and Ferguson, S.A. 1993. The role of dynamic three-dimensional trunk motion in occupational-related low back disorders: the effects of workplace factors trunk position and trunk motion characteristics on risk of injury. *Spine*; 18(5):617-28.

Mattilla, M., Vilkki, M., and Tiilikainen, I. 1992. A computerized OWAS analysis of work postures in the papermill industry, in: Mattila, M. and Karwowski, W. (eds.), *Computer Applications in Ergonomics, Occupational Safety and Health* Elsevier, Amsterdam, 1-11.

Mathiassen, R. and Winkel, J. 1991. Quantifying variation in physical load using exposure-vs-time data. *Ergonomics*; 34(12):1455-68.

McGill, S.M. and Norman, R.W. 1992. Low back biomechanics in industry: the prevention of injury through safer lifting, in Grabiner, M. (ed.), *Current Issues in Biomechanics*, Human Kinetics Publishers, Champaign, Ill, 69-120.

McAtamney, L. and Corlett, E.N. 1993. RULA: a survey method for the investigation of work-related upper limb disorders. *Applied Ergonomics*; 24(2):91-99.

Moore, J.S. and Garg, A. 1995. The strain index: A proposed method to analyze jobs for risk of distal upper extremity disorders. *Am. Ind. Hyg. Assoc. J.*; 56:443-458.

Moore, A., Wells, R., and Ranney, D. 1991. Quantifying exposure in occupational manual tasks with cumulative trauma disorder potential. *Ergonomics*; 34(12):1433-1453.

NIOSH 1981. *Work Practices Guide for Manual Lifting*, Cincinnati, OH: U.S. Department of Health and Human Services. Technical Report No. 81-122.

Norman, R.K. and Wells, R.P. 1990. Biomechanical aspects of occupational injury, *Proceedings of the 23rd Annual Conference of the Human Factors Association of Canada*.

Punnett, L., Robins, J.M., Keyserling, W.M., Herrin, G., and Chaffin, D.B. 1991. Back disorders and non-neutral trunk postures of automobile assembly workers. *Scand. J. Work Environ. Health*, 17(5):337-346.

Ranney, D., Wells, R., and Moore, A. 1995. Upper limb musculoskeletal disorders in highly repetitive industries: precise anatomical physical findings. *Ergonomics*, 38(7):1408-1423.

Rempel, D. 1995. Musculoskeletal loading and carpal tunnel pressure, in Gordon, S., Blair, S., and Fine L. (eds.), *Repetitive Motion Disorders of the Upper Extremity*, American Academy of Orthopedic Surgeons, Rosemont Il, 123-133.

Rowe M. 1987. The diagnosis of tendon and tendon sheath injuries: *Sem. Occup. Med.*; 2(1):1-6.

Sakakibara, H., Miyao, M., Kondo, T., Yamada, S., Nakagawa, T., and Kobayashi, F. 1987. Relationship between overhead work and complaints of pear and apple orchard workers. *Ergonomics*; 30(5):805-815.

Schoenmarklin, R.W, Marass, W.S., and Leurgans, S. 1994. Industrial wrist motions and incidence of hand/wrist cumulative trauma disorders, *Ergonomics*; 37(9):1449-1460.

Silverstein, B.A, Fine, L.J., and Armstrong, T.J. 1986. Hand wrist cumulative trauma disorders in industry. *Br. J. Ind. Med.*; 43:779-784.

Skie, M., Zeiss, J., Ebraheim, N.A., and Jackson, W.T. 1990. Carpal tunnel changes and median nerve compression during wrist flexion and extension seen by magnetic resonance imaging. *J. Hand Surg.*; 15-A(6):934-939.

Smutz, W.P, Bishop, A., Niblock, H., and Drexler, M. 1995. Measurement of creep strain in flexor tendons during low-force, high-frequency activities such as computer keyboard use. *Clin. Biomech.*; 10(2):67-72.

Snook, S., 1978. The design of manual materials handling tasks. *Ergonomics*; 21:963-985.

Snook, S., Vaillancourt, D.R., Ciriello, V.M., and Webster, B.S. 1995. Psychophysical studies of repetitive wrist flexion and extension. *Ergonomics;* 38(7):1488-1507.

Stenlund, B., Goldie, I., Hagberg, M., Hogstedt, C., and Marions, O., 1992, Radiographic osteoarthrosis in the acromioclavicular joint resulting from manual work or exposure to vibration. *Br. J. Ind. Med.;* 49:588-593.

Stock, S.R., 1991, Workplace ergonomic factors and the development of musculoskeletal disorders of the neck and upper limbs: A meta-analysis, *Am. J. Ind. Med.*, 19:87-107.

Veiersted, K., Westgaard, R., and Andersen, P. 1990. Pattern of muscle activity during stereotyped work and its relation to muscle pain. *Int. Arch. Occup. Environ. Health*; 62:31-41.

Veiersted, K.B., Westgaard, R.H., and Andersen, P. 1993. Electromyographic evaluation of muscular work pattern as a predictor of trapezius myalgia. *Scand. J. Work Environ. Health*; 19:284-290.

Veiersted, K.B. 1996. Reliability of myoelectric trapezius muscle activity in repetitive light work. *Ergonomics*; 39(5):797-807.

Wells, R., Ranney, D., and Keir, P. 1994. Passive force length properties of cadaveric human forearm musculature, in *Advances in the Biomechanics of the Hand and Wrist*. Schuind, F., An, K-N., Cooney, W.P., and Garcia-Elias, M. (eds.), Kluwer, Dordrecht, pp. 31-40.

Wells, R., Keir, P.J., and Moore, A.E, 1995. Applications of biomechanical hand and wrist models to work-related musculoskeletal disorders of the upper extremity, in Gordon, S.L., Blair, S.J., and Fine, L.J. (eds.) *Repetitive Motion Disorders of the Upper Extremity*, American Academy of Orthopaedic Surgeons, Rosemount, IL.

Winkel, J. and Westgaard, R., 1992. Occupational and individual risk factors for shoulder-neck complaints, part II: the scientific basis (literature review) for the guide. *Int. J. Ind. Erg.*; 10:85-104.

Winkel, J and Mathiasen, S-E. 1994. Assessment of physical work load in epidemiologic studies, concepts, issues and operational considerations. *Ergonomics*; 37:979-988.

# 3

# Biomechanical Aspects of CTDs

Richard W. Marklin
*Marquette University*

## 3.1 Introduction

The purpose of this chapter is to explain the biomechanical *etiology* of *cumulative trauma disorders* (*CTDs*) that affect the hand, wrist, elbow, and shoulder. The assumption that these CTDs are caused, in part, by work-related activity is based on biomechanical mechanisms that are consistent with epidemiological findings. CTDs affect the soft tissues in the body, namely tendons, ligaments, muscles, and nerves, and in general not bone tissue. Although some authors include bone tissue within the umbrella of CTDs (Kuorinka and Forcier, 1995), this chapter will focus only on those CTDs affecting soft tissue. A brief description of the anatomy of the upper extremity will be provided to familiarize the reader with anatomical terms. Then the three major classes of CTDs, namely those involving muscle, the muscle–tendon unit, and nerve compression, will be discussed.

## 3.2 Anatomy of the Upper Extremity

### Skeletal System

The bones of the upper extremity, which are illustrated in Figure 3.1, are of two types, long and short bones. The long bones connecting the shoulder to the elbow (humerus), the elbow to the wrist (radius

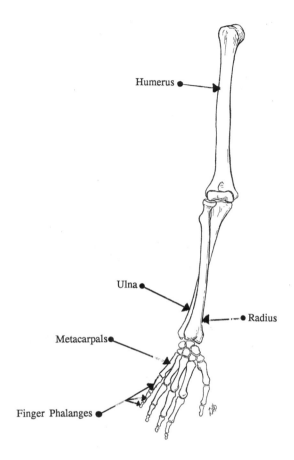

**FIGURE 3.1**    The long and short bones of the right upper extremity. (From Basmajian, J.V. 1982. *Primary Anatomy,* 8th edition, p. 57, Williams and Wilkins, Inc. With permission.)

and ulna), and the wrist to the fingers (metacarpals and phalanges) are adapted for weight bearing and for sweeping, speedy movements that allow the hand to move in space and grasp and touch objects (Rasch, 1989). The movement of the radius (thumb side) around the ulna (little finger side) in the forearm permits the hand to be turned up (supination) or down (pronation), as illustrated in Figure 3.2. The *proximal* and *distal* parts of the long bones display flared ends that act as attachment points for other bones and for connective tissue, such as *tendons* and *ligaments.*

The cluster of small cubical bones comprising the wrist are the eight carpal bones, which are categorized as short bones (Rasch, 1989). The carpal bones move with respect to each other to *flex* (palm side) and *extend* (back side of hand) the wrist joint, while also allowing the wrist to move side to side, from a neutral position to *radial deviation* (thumb side) and to *ulnar deviation* (little finger side), as shown in Figure 3.3.

## Muscular System

The *muscles* of the body are the generators of internal force that convert energy chemically stored in the body into mechanical work (Rasch, 1989). Skeletal muscle, also called *striated muscle,* is composed of longitudinal fibers that follow the direction in which a muscle exerts a force, as seen in Figure 3.4. A muscle is like a rope in that it can only pull or exert a force in tension, and it cannot push or exert a weight-bearing force (compression force). As shown in Figure 3.5, a muscle exerts a tensile force by contracting its thread-like fibers, which shortens the length of the muscle and in fusiform muscles creates a bulge at its center.

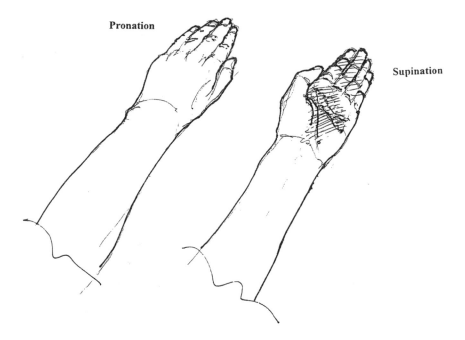

**FIGURE 3.2** The forearm in a pronated and supinated posture. (From Marklin, R.W. original artwork. With permission.)

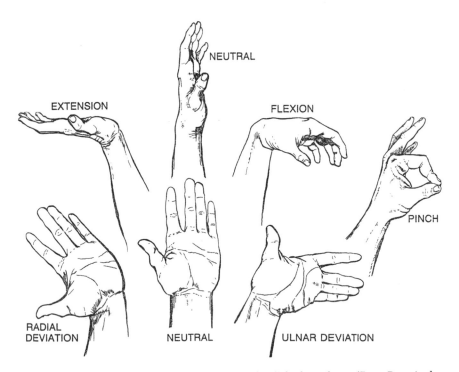

**FIGURE 3.3** Postures of the wrist in the flexion–extension and radial–ulnar planes. (From Putz-Anderson, V. 1988. *Cumulative Trauma Disorders: A Manual for Musculoskeletal Diseases of the Upper Limbs*, p. 54, Taylor & Francis. With permission.)

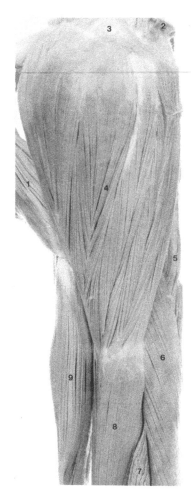

**FIGURE 3.4** A photograph of the left shoulder muscles from a cadaver, as seen from the side. (From McMinn, R.M.H. and Hutchings, R.T. 1977. *Color Atlas of Human Anatomy*, p. 113, Year Book Medical Publishers, Inc., Chicago, IL. With permission.)

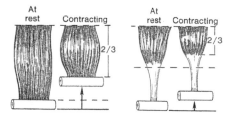

**FIGURE 3.5** The shortening of a muscle as it contracts, generating a pulling force. (From Basmajian, J.V. 1982. *Primary Anatomy*, 8th edition, p. 113, Williams and Wilkins, Inc. With permission.)

The muscles that flex and extend the elbow, which are shown in Figure 3.6, are the biceps and triceps. The group of muscles that flex and extend the wrist are the forearm flexors and extensors, as shown in Figure 3.7. The flexors and extensors located on the thumb side of the forearm also radially deviate the wrist; likewise, the forearm flexors and extensors on the little finger side of the forearm ulnarly deviate the wrist. The muscles in the forearm, which are the primary generators of hand pinch and grasp forces, are called *extrinsic muscles*, while the much smaller muscles located within the hand are called *intrinsic muscles*. One of the main functions of the intrinsic muscles in the hand is to cooperate with the extrinsic muscles to generate hand movements that require dexterity and fine motor control.

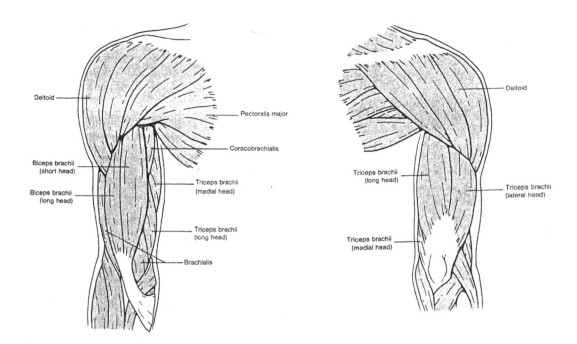

A) Front of Right Arm        B) Back of Right Arm

**FIGURE 3.6** The muscles that flex (a) and extend (b) the elbow. Figures (a) and (b) are the front and back views of the right arm, respectively. (From Van de Graaff, K.M. and Rhees, R.W. 1987. *Human Anatomy and Physiology*, p. 107, Schaum's Outline Series, McGraw-Hill Book Co. With permission.)

## Connective Tissue and Carpal Tunnel

As shown in Figure 3.7, the extrinsic muscles of the forearm are attached to the fingers with strong cord-like collagen structures called tendons. The tendons attached to the flexor and extensor forearm muscles are constrained within the wrist area by thick bands called the *flexor retinaculum* and *extensor retinaculum*, as illustrated in Figure 3.8. The flexor and extensor retinacula are ligaments that attach carpal bones on one side of the wrist to bones on the other side. The flexor retinaculum and carpal bones form a canal called the *carpal tunnel*, through which nine tendons from the forearm flexor muscles and the *median nerve* pass, as shown in Figure 3.8. As the flexor tendons course through the carpal tunnel on their way to the fingers, they travel through a network of *synovial sheaths*, as shown in Figures 3.9 and 3.10. These sheaths reduce the friction between the tendons and their adjacent structures as they wrap around tendons in articulating joints of the wrist and fingers. The structure of a synovial sheath is an elongated and double-walled *bursa* that contains *synovial fluid*, as illustrated in Figures 3.11 and 3.12. The inner wall of the sheath is attached to the tendon, and the outer wall is attached to a fibrous sheath moored to a bone or ligament. The inside surfaces of the sheath's inner and outer walls are lined with synovial fluid, which acts as a lubricant as the tendon traverses inside the tunnel formed by the fibrous sheath.

## Nervous System

The primary purposes of the *peripheral nervous system* (PNS), which serves voluntary skeletal muscles of the extremities, head, neck, and torso, are first, to receive sensory information from outlying parts of the body and relay this information to the *central nervous system* (CNS), which consists of the brain and spinal cord. The second major purpose of the peripheral nervous system is to send motor signals that activate muscles in the outlying area(s) in response to the sensory input. Of the several nerves traveling through the arm, the nerve most often associated with CTDs is the median nerve. As shown in

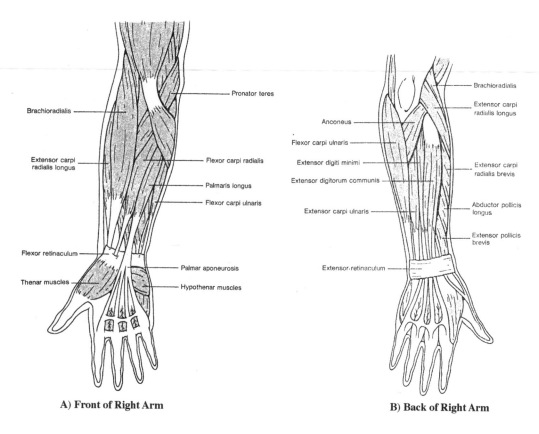

A) Front of Right Arm                    B) Back of Right Arm

**FIGURE 3.7**   The muscles that flex (a) and extend (b) the wrist. Figures (a) and (b) are the front and rear views of the right arm, respectively. (From Van de Graaff, K.M. and Rhees, R.W. 1987. *Human Anatomy and Physiology,* p. 109, Schaum's Outline Series, McGraw-Hill Book Co. With permission.)

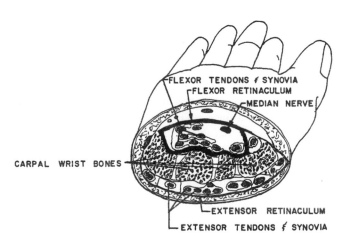

**FIGURE 3.8**   Cross-sectional anatomy of the wrist. The area highlighted is the carpal tunnel. (From Chaffin, D.B. and Andersson, G.B.J. 1991. *Occupational Biomechanics,* 2nd edition, p. 240, John Wiley & Sons Publishers. With permission.)

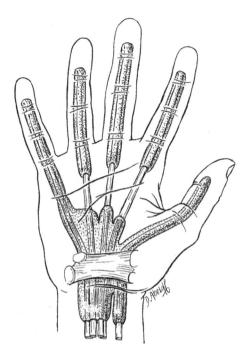

**FIGURE 3.9**  The system of synovial sheaths that lubricate the flexor tendons as they bend around the wrist and finger joints (palmar view of the right hand). (From Basmajian, J.V. 1982. *Primary Anatomy,* 8th edition, p. 158, Williams and Wilkins, Inc. With permission.)

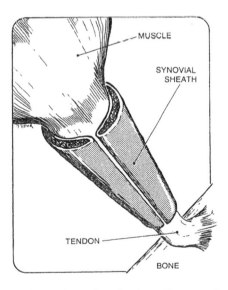

**FIGURE 3.10**  A magnified end view of a muscle, tendon, sheath, and bony attachment point. (From Putz-Anderson, V. 1988. *Cumulative Trauma Disorders: A Manual for Musculoskeletal Diseases of the Upper Limbs,* p. 12, Taylor & Francis. With permission.)

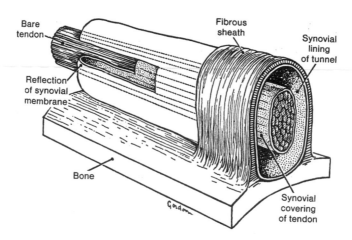

**FIGURE 3.11**   Structure of a synovial sheath. An area of the sheath has been cutaway to expose its double-walled structure. Synovial fluid lines the inside of the sheath's inner and outer walls and reduces friction as the tendon moves within its tunnel. **Note:** normally the tendon fits snugly in its tunnel, but is shown having a loose fit in this figure for illustration purposes. (From Basmajian, J.V. 1982. *Primary Anatomy,* 8th edition, p. 119, Williams and Wilkins, Inc. With permission.)

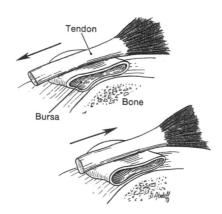

**FIGURE 3.12**   A bursa, which is a collapsed bag of connective tissue filled with synovial fluid, forms whenever a tendon rubs against a hard structure, such as a bone. (From Basmajian, J.V. 1982. *Primary Anatomy,* 8th edition, p. 118, Williams and Wilkins, Inc. With permission.)

Figure 3.13a, the median nerve starts at the shoulder, provides motor inputs to muscles in the forearm and thumb region, and provides sensory feedback from the palm region and from the thumb to the center of the ring finger. The median nerve is the "nerve of precision" because it supplies motor function to the extrinsic muscles in the forearm that flex the fingers and the intrinsic muscles in the thumb that exert a precision grip (Feldman et al., 1983). Figure 3.14 indicates the sensory regions of the hand served by the median nerve and the radial and ulnar nerves, which travel down the radial and ulnar sides of the forearm, respectively, as illustrated in Figures 3.13b and 3.13c. The radial nerve is the "nerve of stability" because it *innervates* the forearm extensor muscles that oppose and stabilize the precision and power muscles on the forearm's flexor side. The ulnar nerve is the "nerve of power" because it innervates the muscles that provide wrist flexor power, but little precision (Feldman et al., 1983).

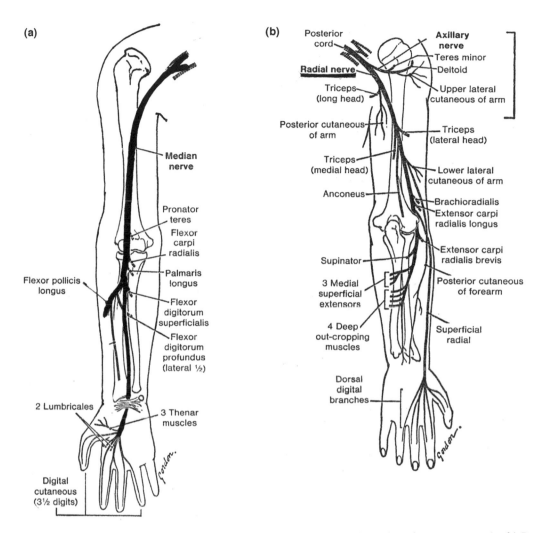

**FIGURE 3.13**  a) Front view of the paths of the *median nerve* as it travels down the right upper extremity. b) Rear view of the paths of the *radial nerve*. c) Front view of the paths of the *ulnar nerve*. (From Basmajian, J.V. 1982. *Primary Anatomy,* 8th edition, p. 340, Williams and Wilkins, Inc. With permission.)

## 3.3  Work-Related Muscle Disorders

After frequent or prolonged contractions, a muscle can feel painful for a relatively short period of time and recover to full function, or it could develop a more serious chronic condition. If the pain disappears after a relatively short period of time, the cause was probably temporary fatigue of the muscular tissues. However, if the pain persists, the worker could have developed a muscle CTD.

The medical term describing muscle pain is *myalgia*, which includes a few specific muscle pain syndromes. Myalgia can occur after vigorous or unaccustomed exercise, and also from work-related activity. A worker can develop a *myopathy* called *myofascial syndrome*, which is characterized by "the presence of one or more discrete areas (or trigger points) that are tender and hypersensitive and from which pain may radiate when pressure is applied" (Kuorinka and Forcier, 1995, p. 81). Myofascial syndrome could be associated with work-related activity, and a common work-related myofascial syndrome is *tension neck syndrome* (also called *shoulder–neck myofascial syndrome*). Tension neck syndrome

**(c)**

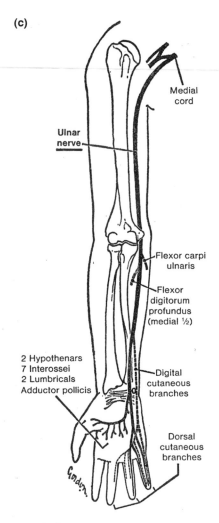

**FIGURE 3.13**    (continued)

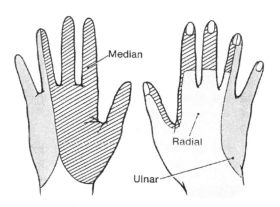

**FIGURE 3.14**    Sensory regions of the right hand served by the median, radial, and ulnar nerves. (From Basmajian, J.V. 1982. *Primary Anatomy*, 8th edition, p. 341, Williams and Wilkins, Inc. With permission.)

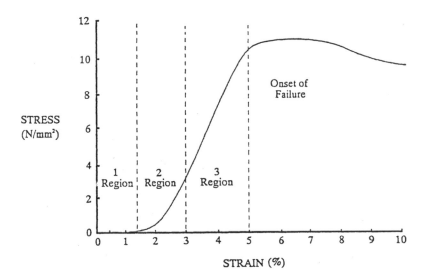

**FIGURE 3.15**  A typical stress-strain curve of a tendon. (From Abrahams, M. 1967. Mechanical analysis of tendon *in vitro*: A preliminary report, *Med Biol Eng*, Vol. 5, p. 435. With permission.)

is a myofascial syndrome localized in the shoulder and neck region with tenderness descending into the trapezius muscle. Occupational groups cited in the literature that have been associated with high rates of tension neck syndrome are those requiring repetitive arm movements and constrained postures (Kuorinka and Forcier, 1995).

The *pathogenesis* of myofascial syndromes is unknown; however, several hypotheses have been offered in the literature, which include a lower capillary-to-fiber ratio for the slow twitch fibers (Type I), severe depletion of ATP in the muscle, and dysfunctional energy metabolism (Kuorinka and Forcier, 1995). For a more thorough discussion of muscle CTDs, the reader is referred to Kuorinka's and Forcier's (1995) book, which has a comprehensive description and discussion of biomechanical mechanisms of muscle CTDs.

## 3.4  Biomechanical Aspects of Muscle–Tendon Disorders

As a muscle shortens during contraction and lengthens during stretching, its tendon acts like a rope and transmits the muscle force to the bony attachment site. As a tendon moves with a muscle, the length of the tendon does not necessarily stay constant. A tendon has elastic properties and is analogous to a rubber band. The muscle force applied to the tendon is a tensile force, which is commonly converted to the units of stress (force divided by cross-sectional area of tendon). As the tensile force increases, the tendon elongates, which is measured by strain (the percentage of change in length). Figure 3.15 shows a typical stress–strain curve of a tendon with its three characteristic regions (Abrahams, 1967). In Region 1, the crumpled collagen fibers of a relaxed tendon merely straighten under negligible loads. Then, as the tensile force increases, the tendon passes through the "toe" region (region 2), and then has a linear relationship between stress and strain in region 3. Although the tendon can elongate up to 5% strain before onset of failure, normal tendon strain is below 3% (Abrahams, 1967; Elliott, 1965; Rigby et al., 1959).

The loading and unloading of a tendon can change its elastic properties depending on whether the tensile force is increasing or decreasing. As shown in Figure 3.16, the amount of stress required to elongate a tendon to a specific strain level is greater when a tendon is loaded (increase in tensile load) then when it is unloaded (decrease in tensile load). This change in stress–strain curve is called *hysteresis*, which results from a loss of energy, probably as heat, during the unloading phase (Moore, 1992).

Several CTDs reported in the literature occur at sites where a tendon wraps around a deviated joint. As a muscle contracts and moves its tendon accordingly, the tendon can rub against its adjacent surface,

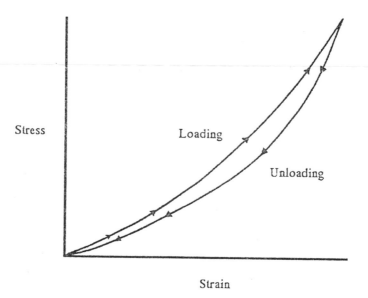

**FIGURE 3.16**   The stress–stress curve of a tendon depends on whether the tensile force is increasing or decreasing. The difference in the stress–strain curves is called hysteresis. The units of stress and strain are N/mm² and % elongation, respectively. (From Moore, J.S. 1992. Function, structure, and responses of components of the muscle–tendon unit, in *State of the Art Reviews in Occupational Medicine,* Vol. 7, No. 4, p. 721. With permission.)

usually a bone or ligament, as a rope rubs against a nonrotating pulley. Likewise, when the muscle lengthens, the tendon moves in the opposite direction against its adjacent structures. In the wrist area, the repetitive rubbing of the tendons against the carpal bones and flexor retinaculum can cause CTDs known as *tendinitis* and *tenosynovitis*, which are inflammation of the tendon and its sheath, respectively.

Based on Landsmeer's (1962) model, Armstrong and Chaffin (1979) developed a static model of a tendon wrapping around a joint. Figure 3.17 depicts Landsmeer's model of a tendon, which is analogous to a rope bent around a nonrotating pulley, and Figure 3.18 illustrates the Armstrong and Chaffin (1979) model as a reasonable representation of Landsmeer's model. When the wrist is flexed, the flexor tendons bend around the flexor retinaculum that is assumed to have a constant radius. When the wrist is extended, the flexor tendons are supported on the *dorsal* side by the carpal bones that are assumed to have a constant radius. Armstrong and Chaffin (1978) found that the radius in a flexed posture is larger than in an extended posture.

The arc length of the tendon wrapping around the pulley is defined in Equation (3.1).

$$X = R \times \theta \tag{3.1}$$

where
$X$ = tendon arc length around pulley (mm)
$R$ = radius of curvature of supporting tissues (mm)
$\theta$ = angle of deviation of wrist from neutral (in radians)

The reaction forces acting normal to the tendon are shown in Figure 3.18 and defined in Equation (3.2).

$$F_n = \left(F_t \times e^{[\mu \times \theta]}\right) / R \tag{3.2}$$

where
$F_n$ = normal supporting force per unit of arc length (N/mm)
$F_t$ = average tendon force in tension (N)

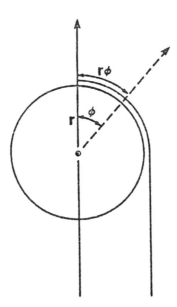

**FIGURE 3.17** Landsmeer's (1962) model of a tendon wrapping around a joint. (From Chao, Y.S., An, K.N., Cooney, W.P., and Linscheid, R.L. 1982. *Biomechanics of the Hand: A Basic Research Study,* p. 15, Biomechanics Laboratory, Department of Orthopaedic Surgery, Mayo Clinic/Mayo Foundation, Rochester, MN. With permission.)

$\mu$ = coefficient of friction between tendon and supporting synovia
$\theta$ = wrist deviation angle (radians)
R = radius of curvature of supporting tissues (mm)

Since $\mu$ is considered small (approximately 0.0032 [Fung, 1981]), it can be approximated by zero. This changes Equation (3.2) to Equation (3.3).

$$F_n = F_t / R \tag{3.3}$$

Equation (3) reveals $F_n$ is a function of the tendon force and radius of curvature. As the radius of curvature decreases, the normal supporting force per unit of arc length increases. The normal supporting force for women would be greater than for men because women have smaller wrists. Also, as the tendon force increases, the normal supporting force increases.

The total supporting force $F_r$ in Figure 3.18 is the force of the ligaments, bones, and median nerve in the carpal tunnel acting on the flexor tendons. $F_r$ is defined in Equation (3.4).

$$F_r = 2 \times F_t \times \sin(\theta/2) \tag{3.4}$$

where
$F_r$ = resultant force exerted by adjacent wrist structures on the flexor tendons (N)
$F_t$ = tendon force (N)
$\theta$ = wrist deviation angle (in radians)

Equation (4) indicates that $F_r$ is a function of the tendon force and wrist deviation angle, but is independent of radius of curvature. Figure 3.19 illustrates this relationship in that as the tendon force and wrist angle increase, the resultant force $F_r$ increases linearly.

The significance of $F_n$ and $F_r$ is based on the theory that increased normal forces place greater stress on the tendon and its surrounding structures. The increase in normal force could cause the tendon and

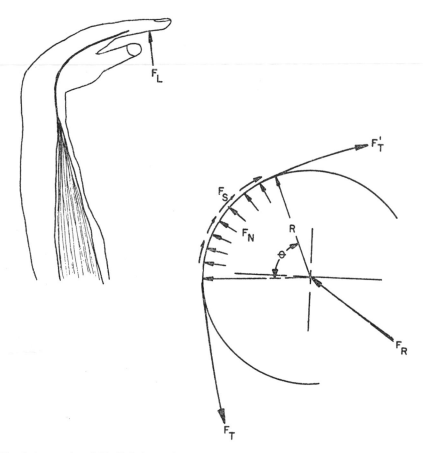

**FIGURE 3.18** Armstrong's and Chaffin's (1979) biomechanical model of a flexor tendon wrapping around the flexor retinaculum. $F_t$ is the tendon force, and $F_r$ is the resultant reaction force exerted against the tendon. (From Chaffin, D.B. and Andersson, G.B.J. 1991. *Occupational Biomechanics,* 2nd edition, p. 243, John Wiley & Sons Publishers. With permission.)

its sheath or the fibrous sheath moored to bone or ligament to *hypertrophy* or *inflame*. If these structures were to hypertrophy or inflame, then the coefficient of friction ($\mu$ in Equation (3.2)) would increase, thereby placing even greater $F_n$ on the tendons.

Dynamic movements that accelerate and decelerate the tendons around a nonrotating pulley could exacerbate the trauma imposed on the tendons. Schoenmarklin and Marras (1990) developed a dynamic model of a flexor tendon bent around the carpal bones or flexor retinaculum, taking into account the acceleration and deceleration of a tendon's movements. This model analyzes the effects of peak angular acceleration on the resultant reaction force that the wrist bones and ligaments exert on tendons and their sheaths in the flexion/extension plane. Like the Landsmeer (1962) and Armstrong and Chaffin (1979) models, Schoenmarklin and Marras (1990) model the tendon as a rope bent around a fixed pulley.

The quantitative effects of the wrist's peak angular acceleration on resultant reaction forces were based on the free body diagram (FBD) and mass × acceleration diagram (MAD) approach in engineering dynamics (Meriam and Kraige, 1986). Figure 3.20 illustrates the FBD and MAD approach applied to a wrist and hand in midposition (neither pronated or supinated). There is no externally applied load in the hand. The hand is rotated in the horizontal plane around a vertical **z** axis, so the effects of gravity do not play a role in this example. All the flexor tendons are grouped together as one tendon force vector in order to maintain static determinacy. The hand is assumed to accelerate from a stationary posture, so the angular velocity is theoretically zero, resulting in zero centripetal force.

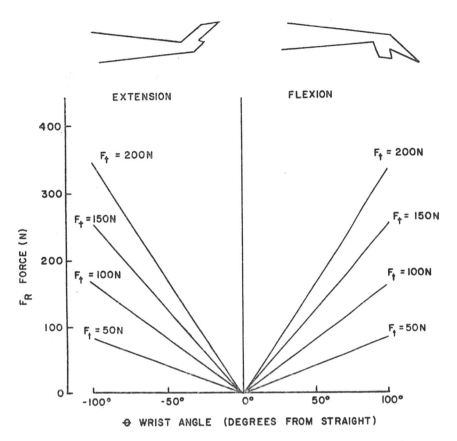

**FIGURE 3.19** The resultant reaction force ($F_r$), as modeled by Armstrong and Chaffin (1979), that is exerted against the flexor tendons as a function of wrist angle and tendon force. (From Chaffin, D.B. and Andersson, G.B.J. 1991. *Occupational Biomechanics*, 2nd edition, p. 247, John Wiley & Sons Publishers. With permission.)

The maximum tendon force ($F_{t\text{-max}}$) was computed as a function of five peak angular accelerations ($\theta$ = 3000, 6000, 9000, 12000, and 15000°/sec²). Based on empirical data from normal subjects, 15000°/sec² was found to be about 50% of peak wrist acceleration in the flexion/extension plane (Schoenmarklin and Marras, 1993). The $F_{t\text{-max}}$ is depicted in Figure 3.21 and is derived from Equation (3.5) (LeVeau, 1977).

$$F_{t\text{-max}} = F_{t\text{-min}} \times e^{[\mu \times \theta]} \tag{3.5}$$

where
$F_{t\text{-max}}$=maximum force in flexor tendons, which is the force that the extrinsic flexor muscles in the forearm exert on their tendons (N)
$F_{t\text{-min}}$=minimum force in flexor tendons, which is the force that the flexor tendons transmit to the hand and fingers (N)
$\mu$  = coefficient of friction between tendons and their sheaths
$\theta$  = wrist deviation angle (radians)

Since the coefficient of friction for human synovial joints bones is estimated to be very low (0.0032 according to Fung, 1981), then the calculation of the $F_{t\text{-max}}$ force is very close to $F_{t\text{-min}}$. The $F_r$ depicted in Figure 3.21 and expressed in Equation (3.4) was calculated as the resultant force necessary to resist $F_{t\text{-max}}$ and $F_{t\text{-min}}$.

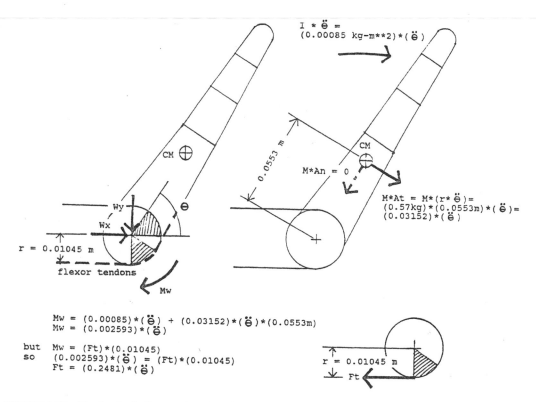

FREE BODY DIAGRAM                    MASS*ACCELERATION DIAGRAM

**FIGURE 3.20** The free body diagram (FBD) and mass acceleration diagram (MAD) approach used by Schoenmarklin and Marras (1990) to calculate the peak reaction force ($F_r$ in Figures 3.21 and 3.22) on the wrist when the wrist is accelerated (in the flexion direction) at an extension angle of θ. (From Schoenmarklin, R.W. and Marras, W.S. 1990. In *Proceedings of the 34th Meeting of the Human Factors Society,* p. 807. With permission.)

As shown in Figure 3.22, $F_r$ increases approximately linearly as wrist angle or angular acceleration increases, resulting in a curved plane that signifies an interactive effect between wrist angle and angular acceleration. The greatest $F_r$ occurs when the wrist is accelerated at a deviated wrist posture. The large peak reaction forces exerted on the flexor tendons and their sheaths are due solely to wrist motion without any externally applied load in the hand. If loads were applied in the hand (e.g., power grip or pinch grip) while the hand was accelerated in deviated postures, then $F_r$ would increase even more, resulting in even more stress on flexor tendon tissue. The large peak $F_r$ in Figure 3.22 could possibly cause the tendon and its sheath or the fibrous sheath moored to bone or ligament to hypertrophy or inflame, which could result in tendinitis or tenosynovitis. The occurrence of either tendinitis or tenosynovitis would most likely increase μ in Equation (3.5), thereby increasing $F_{t\text{-max}}$ and $F_r$ even more (refer to Figure 3.21 and Equation (3.4)).

The large resultant reaction forces on the tendons from wrist deviation and accelerations could possibly explain the findings of Armstrong et al. (1984), who investigated the histological changes in the flexor tendons as they pass through the carpal tunnel. These investigators found hypertrophy and increased density in the synovial tissue in the carpal tunnel area. These authors suggested that biomechanical factors, such as repeated exertions with a flexed or extended wrist posture, could have partially caused degenerative changes in tendon tissue. In addition to reaction forces from supporting structures, the hypertrophy of the tendon tissue could have been caused by differences in strain within a tendon. In an investigation of the viscoelastic properties of tendons and their sheaths, Goldstein et al. (1987) found

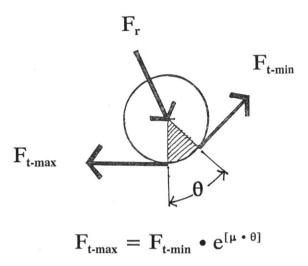

$$F_{t\text{-max}} = F_{t\text{-min}} \bullet e^{[\mu \bullet \theta]}$$

**FIGURE 3.21** Relationship between the maximum ($F_{t\text{-max}}$) and minimum ($F_{t\text{-min}}$) forces of a tendon and the resultant reaction force ($F_r$). $F_{t\text{-max}}$ is the force emanating from the forearm flexor muscles, and $F_{t\text{-min}}$ is the force transmitted to the hand. The flexor tendon is wrapped around the wrist's carpal bones. $F_{t\text{-max}}$ and $F_{t\text{-min}}$ are the maximum and minimum tendon forces in Schoenmarklin's and Marras's (1990) dynamic model of a flexor tendon passing through the wrist joint. The equation for $F_{t\text{-max}}$ and $F_{t\text{-min}}$ is from Leveau (1977). (From Schoenmarklin, R.W. and Marras, W.S. 1990. In *Proceedings of the 34th Meeting of the Human Factors Society*, p. 807. With permission.)

that flexion/extension wrist angle increased the shear traction forces between tendons, their sheaths, and bones and ligaments that form the anatomical pulley. As depicted in Figure 3.23, when the wrist is extended approximately 10°, the strain in the flexor digitorum profundus (FDP) tendons, which pass through the carpal tunnel and move the fingers, is approximately 10% to 15% lower on the side distal (hand side) to the flexor retinaculum than the proximal side (forearm side). This difference in strain within a tendon creates shear traction forces, which are magnified when the wrist angle is deviated to 65° flexion or extension.

## 3.5 Work-Related CTDs Involving the Muscle–Tendon Unit

The pathogenesis of the most frequently studied CTDs involving the muscle–tendon unit will be discussed below.

### Tendinitis

Although tendinitis is defined as inflammation of the tendon, Moore (1992) contends there is scant scientific evidence that the collagenous fibers that comprise the tendon actually inflame. According to Moore (1992), tendinitis is often used as a term that implies soreness localized to a muscle–tendon unit that increases with tensile load from either muscle contraction or passive stretch. Moore (1992) further states that these clinical findings of soreness "may represent no more than a normal pattern to varying degrees of use, rather than inflammation." Because clinicians do not have sensitive diagnostic tools to differentiate between tendinitis and tenosynovitis (inflammation of the tendon's sheath), soreness in joints where the tendons do not have sheaths, such as in the elbow and shoulder, is usually diagnosed as tendinitis, whereas soreness in joints with sheathed tendons is commonly diagnosed as tenosynovitis.

One theory of the pathogenesis of tendinitis is the physical disruption of a small number of collagen fibers within a tendon and the ensuing repair process. According to Moore (1992), the body responds to this disruption in a manner similar to that of a partial tendon laceration, which is the partial cutting or severing of a tendon. The healing process of the tendon occurs in three stages: inflammatory stage,

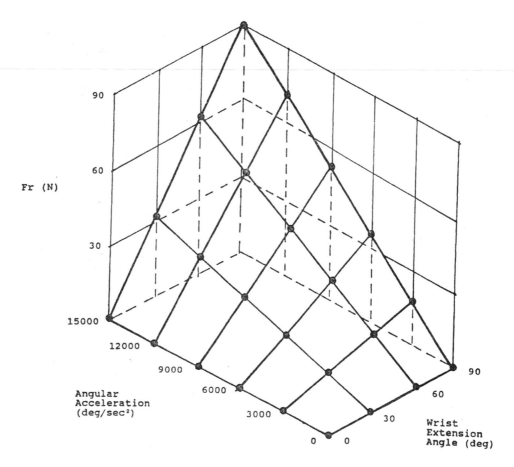

**FIGURE 3.22**  The resultant reaction force ($F_r$) exerted by the carpal bones or flexor retinaculum against a flexor tendon and its sheath as a function of wrist angle and acceleration. (From Schoenmarklin, R.W. and Marras, W.S. 1990. In *Proceedings of the 34th Meeting of the Human Factors Society*, p. 809. With permission.)

reparative or collagen-production stage, and a remodeling stage (Gelberman et al., 1988). As the body rebuilds its tendon tissue, the collagen content increases and the tendon could hypertrophy, or increase in size. In addition, the body may not repair all of the disrupted fibers, resulting in permanent fraying of the tendon. Due to hypertrophy and fraying, the tendon may be biomechanically different, possibly deficient, after the body's attempt to completely restore the disrupted tissues.

Moore (1992) relates the effects of partial tendon laceration to CTDs in that when a joint is deviated from a neutral position, the tendons could react to resulting reaction forces from the joint structures ($F_r$ in Equation (3.4) and Figures 3.21 and 3.22) in a manner similar to a partial tendon laceration. As shown by Armstrong and Chaffin's (1979) static model, as the wrist deviates, the resulting reaction forces from the flexor retinaculum or carpal bones could cause physical disruption to the tendons, much like the effects of partial tendon laceration. Theoretically, physical disruption of the tendon tissue would be exacerbated by even greater reaction forces if the wrist were accelerated or decelerated, particularly at extreme wrist deviation angles (Schoenmarklin and Marras, 1990). If the wrist and fingers were deviated excessively in repetitive motions, hypertrophy of the tendons from the healing process or permanent fraying of the tendons could cause soreness at the wrist. Depending on the specific tendon, this soreness may be diagnosed as tendinitis or tenosynovitis. Soreness in the wrist flexor muscles' tendons (flexor carpi radialis and ulnaris) would probably be diagnosed as tendinitis because these tendons do not have sheaths, whereas soreness in tendons passing through the carpal tunnel (flexor digitorum superficialis and profundus) would commonly be diagnosed as tenosynovitis because these tendons are sheathed.

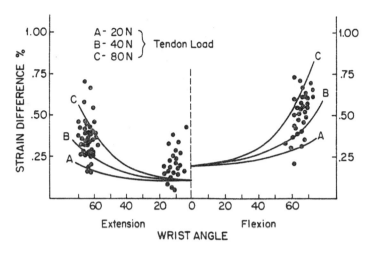

**FIGURE 3.23**   The difference in strain within the flexor digitorum profundus tendon between measurements taken proximal and distal to the flexor retinaculum. Even at an extended wrist angle of 10°, there is a 10% to 15% difference in strain between the FDP tendon proximal and distal to the flexor retinaculum. The difference in strain is magnified as wrist deviation angle increases. (From Goldstein, S.A., Armstrong, T.J., Chaffin, D.B., and Matthews, L.S. 1987. Analysis of cumulative strain in tendons and tendon sheaths. *J Biomed,* 20, p. 4. With permission.)

## Lateral Epicondylitis (Tennis Elbow)

*Lateral epicondylitis,* which is also called tennis elbow in lay parlance, is tendinitis of the forearm extensor and supinator muscles at the *lateral epicondyle* of the elbow. The lateral epicondyle is the small bony attachment point on the outside of the elbow where the group of forearm extensor and supinator muscles originate. The extrinsic extensor and supinator muscles fuse into an *aponeurosis,* or a broad, flat tendon, which is attached to the lateral epicondyle in the elbow. Soreness and pain occur at the point where the aponeurosis of the extensor and supinator muscles pull on the lateral epicondyle. The small size of the lateral epicondyle and the relatively large mass of extensor and supinator muscles create high stresses on the lateral epicondyle and its attached aponeurosis. Patients who have lateral epicondylitis report their pain is particularly acute when they extend their wrist or supinate the forearm against resistance, which occurs when one is hitting a tennis ball with a backhand stroke.

Lateral epicondylitis is a CTD in that it is directly related to the motions that tense the wrist's extensor and supinator muscles (Nirschl, 1983). In a study of 113 patients, Goldie (1964) found that repeated wrist extensions or alternating pronating and supinating movements of the forearm were causal factors in 83 of the cases.

Review of the medical literature reveals several hypotheses regarding the pathogenesis of lateral epicondylitis, although all of them do agree that the basic mechanism is deterioration of the aponeurotic tendinous tissue at the lateral epicondyle. Cyriax (1936), who treated 20 patients with lateral epicondylitis, concluded it is caused by a tear between the tendinous origin of the extensor muscles and the *periosteum* of the lateral epicondyle. Goldie (1964) suggested that lateral epicondylitis is due to a buildup of lesions in a space under the tendon and distal to the epicondyle. Microscopically, Nirschl (1985) found that the affected tendon in lateral epicondylitis had a characteristic appearance of hypertrophy that was grayish, edematous, and friable. Nirschl (1985) interpreted this medical description as a "thick unhappy gray tendon, weeping with *edema.*" A normal tendon has collagen fibers that run parallel, but the tendons afflicted with lateral epicondylitis look coarse and granular.

Often, a patient with lateral epicondylitis will wear a brace around the forearm near the elbow (Froimson, 1971; Moore, 1992), as is often seen on tennis players. Although it has not been validated experimentally, one of the theories of why the forearm brace is beneficial is based on biomechanics. Figure 3.24 shows a free body diagram of the elbow as viewed from the head position. The tendons of the forearm extensor and supinator muscles are modeled as a single vector. When tightened, the brace may keep the aponeurosis from vibrating against underlying bony tissue during repeated extending or supinating exertions. In addition, when the forearm brace is tightened it compresses the aponeurosis against the underlying structures, thereby creating a *frictional force* that resists, albeit partially, the pull of the forearm extensor and supinator muscles. Theoretically, this frictional force would relieve the lateral epicondyle of carrying the full tensile load of the aponeurosis. However, because the coefficient of friction among the musculoskeletal tissues underlying the brace is probably very low, the reduction in tensile load on the lateral epicondyle may be small and negligible or it may be large enough to retard lateral epicondylitis. Experimental research is needed to determine whether this biomechanical theory can explain the efficacy of forearm braces.

## Supraspinatus Tendinitis (Rotator Cuff Syndrome)

*Supraspinatus tendinitis*, which is often called rotator cuff syndrome, is tendinitis of the muscle that elevates the shoulder. Elevation of the shoulder in the *frontal plane* is called shoulder *abduction*. Pain is felt on the *acromion process*, or bony top of the shoulder, when one abducts the shoulder, particularly when the arm is holding a load or exerting a pushing force.

The pathogenesis of supraspinatus tendinitis is impingement of the bursa and supraspinatus tendon as the shoulder is abducted. The superficial and deep muscles of the shoulder are shown in Figure 3.25. The deltoid muscle, which covers the outside of the shoulder, and the supraspinatus muscle, which is a smaller muscle under the deltoid and trapezius muscles and acromion, are the major abductors of the shoulder. As shown in Figures 3.26a and 3.26b, abduction of the shoulder compresses the acromion downward, thereby pinching the underlying bursa and supraspinatus tendon (Chaffin and Andersson, 1991). As illustrated in Figure 3.12, the bursa is a tubular synovial sheath whose purpose is to lubricate the contact between the deltoid muscle and acromion and the supraspinatus tendon. However, if the shoulder is abducted repeatedly, and particularly under heavy loads, the resulting impingement could damage the bursa and supraspinatus tendon fibrils and produce fraying of the tendon. The relative *avascular* nature of the supraspinatus tendon diminishes its capability to repair itself, thereby leading to degeneration, as shown in Figure 3.26c (Moore, 1992). In addition, intramuscular pressure from muscle fibers attached to the tendon can also diminish the reparative process of the tendon.

## Tenosynovitis

Although tenosynovitis is defined as inflammation of the tendon sheath (Stedman, 1982), any tendon sheath disorder is called tenosynovitis, regardless of the presence or absence of inflammation (Moore, 1992). As shown in Figures 3.9 and 3.10, the tendon sheath is a tubular structure that wraps around a tendon and contains synovial fluid to "provide lubrication, protection, and repair assistance for the surrounded tendon" (Moore, 1992). Tenosynovitis is diagnosed only where tendons are sheathed, whereas tendinitis could occur in a tendon regardless of whether it is sheathed. Usually, soreness in a sheathed tendon area is diagnosed as tenosynovitis, whereas soreness in a tendon without sheathing is diagnosed as tendinitis.

## DeQuervain's Tenosynovitis

*DeQuervain's disease* is the *stenosing tenosynovitis* of the tendons that abduct and extend the thumb (abductor pollicis longus [APL] and extensor pollicis brevis [EPB]) (Williams and Ward, 1983). This disease was named after a Swiss surgeon who observed the condition in 1895. The practical importance of the muscles that flex, extend, and abduct the thumb cannot be overestimated. According to Bunnel (1956), "a hand without a thumb is no more than a hook." The APL and EPB are two of the thumb's muscles that are necessary for dexterity and fine manipulations.

As shown in Figure 3.27, the tendons of the APL and EPB pass underneath the extensor retinaculum of the wrist, and then they share the same synovial sheath on their way to the dorsal and lateral side of

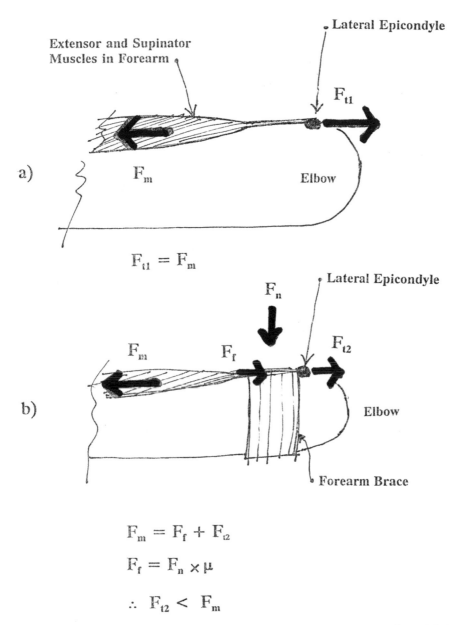

**FIGURE 3.24** Free body diagram (FBD) analysis of the forearm extensor and supinator muscles and their tendinous attachment (aponeurosis) to the lateral epicondyle. The view of the elbow is from the head looking down, with the forearm in midposition (neither supinated or pronated). a) FBD of the elbow without a forearm brace. The force the aponeurosis has to exert, $F_{t1}$, is equal to the tensile pull of the extensor and supinator muscles, $F_m$. b) The tightening of the forearm brace around the forearm creates a frictional force, $F_f$, that opposes $F_m$, thereby lessening the force on the aponeurosis, $F_{t2}$. (From Marklin, R.W. original artwork. With permission.)

the thumb (Lamphier, 1965). The APL's and EPB's common sheath, which is about 5 cm long, passes over a bony depression called the radial styloid.

The APL and EPB tendons and their common sheath are subject to cumulative trauma because of their position in the bony groove in the radial styloid. DeQuervain's disease is caused by the friction of the two tendons rubbing against each and against the long bony groove (Lamphier et al., 1965). DeQuervain's disease is a stenosing tenosynovitis in that the common synovial sheath thickens (refer to

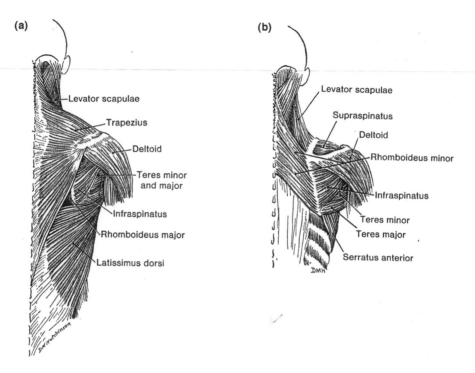

**FIGURE 3.25**    a) Superficial muscles of the shoulder complex. The deltoid muscle is a major shoulder abductor. b) Deep muscles of the shoulder complex. The supraspinatus muscle is responsible for initiating abduction of the shoulder, after which the deltoid provides most of the abduction force. (From Basmajian, J.V. 1982. *Primary Anatomy*, 8th edition, p. 141, Williams and Wilkins, Inc. With permission.)

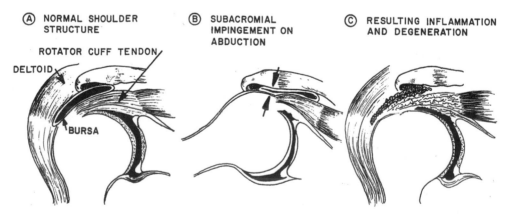

**FIGURE 3.26**    a) Normal shoulder structure with the arm hanging at the side. Bursa separates the deltoid muscle and acromion from the supraspinatus tendon (rotator cuff tendon). b) When the shoulder is abducted, bursa and supraspinatus tendon are pinched between the acromion and humerus bone (arm bone). c) With repeated abductions, both the bursa and tendon could swell and degenerate and the tendon could fray. (From Chaffin, D.B. and Andersson, G.B.J. 1991. *Occupational Biomechanics*, 2nd edition, p. 381, John Wiley & Sons Publishers. With permission.)

Figure 3.27), thereby increasing the friction between the APL and EPB tendons within their common sheath. In his review of the medical literature, Moore (1992) described the pathogenesis of DeQuervain's disease. In mild cases, the synovial layer within the synovial sheath thickens up to twice the normal

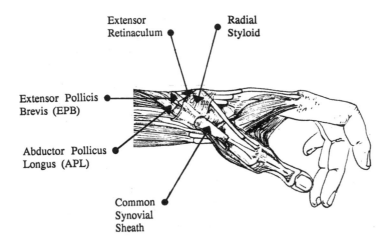

**FIGURE 3.27** Abductor pollicis longus (APL) and extensor pollicis brevis (EPB) tendons as they proceed under the extensor retinaculum and through their common synovial sheath. DeQuervain's disease is stenosing tenosynovitis of the APL and EPB in their common sheath. (From Lamphier, T.A., Crooker, C., and Crooker, J.L. 1965. DeQuervain's disease. *Industrial Medicine and Surgery,* p. 848. With permission.)

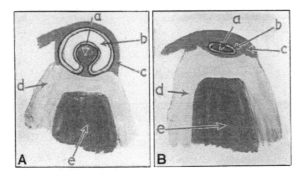

**FIGURE 3.28** A) Cross-section of a normal fibroosseus canal as it passes over the radial styloid. The tendon (a) and its synovial sheaths (b) are moored to the radius bone (d) with a fibrous ligamentous sheath (c). B) Cross-section of a fibroosseus canal with stenosing tenosynovitis. The tendons (a) are flattened, the synovial sheath (b) is thinned, and the fibrous ligamentous sheath (c) is thickened. (From Finkelstein, H. 1930. Stenosing tendovaginitis at the radial styloid process. *J Bone Jt Surg,* Vol. 12, p. 515. With permission.)

thickness. However, at the point of constriction where the tendons rub against the radial styloid, the synovial sheath of the APL and EPB tendons thin and the tendons flatten (Finkelstein, 1930), as illustrated in Figure 3.28. The thinning of the tendon sheath and flattening of the tendons is caused by hypertrophy of the fibrous ligamentous sheath, namely the extensor retinaculum, that holds the APB and EPB tendons and their common sheath to the radial styloid bone. In severe cases, the fibrous ligamentous sheath thickens three to four times (Lamphier et al., 1965), and the tendon could swell, forming a bulbous shape adjacent to the site of constriction, as illustrated in Figure 3.29. This bulbous swelling could cause popping of the APL and EPB tendons as the wrist is ulnarly deviated with the thumb flexed inside the palm.

## Trigger Finger

*Trigger finger* is stenosing tenosynovitis of the tendons that flex the fingers and is manifested by painful locking of the finger during finger flexion. As illustrated in Figure 3.30, the finger flexor tendons and

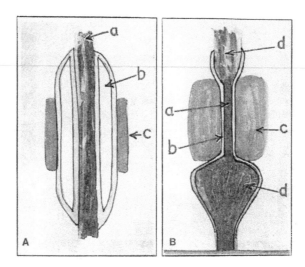

**FIGURE 3.29** A) Longitudinal section of a normal fibroosseus canal. The tendon (a) is covered by it sheath (b). The fibrous ligamentous sheath (c) forms a canal through which the tendon can travel unimpeded. B) Longitudinal section of a fibroosseus canal with stenosing tenosynovitis. The tendon (a) and its synovial sheath (b) are thinned by a thickened fibrous ligamentous sheath (c), forming a nodule in the tendon (d). (From Finkelstein, H. 1930. Stenosing tendovaginitis at the radial styloid process. *J Bone Jt Surg*, Vol. 12, pp. 514 and 515. With permission.)

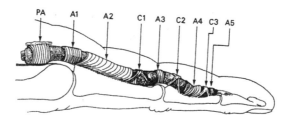

**FIGURE 3.30** A finger flexor tendon as it traverses from the knuckle (left) to the tip (right). Fibrous ligamentous sheaths comprise the five annular pulleys (A1–A5) and three cruciate pulleys (C1–C3). These pulleys hold the tendons close to the finger joints to avoid bowstringing during finger flexion. (From Doyle, J.R. 1989. Anatomy of the finger flexor tendon sheath and pulley system: A current review. *J Hand Surg*, 14A, p. 350. With permission.)

their synovial sheaths are moored to the bones of the finger with fibrous ligamentous sheaths to avoid bowstringing during finger flexion. These fibrous ligamentous sheaths are called pulleys, of which there are two types: annular (A1–A5) and cruciate (C1–C3). Repetitive and forceful flexing of the fingers could cause trigger finger, whose pathogenesis is related to that of DeQuervain's disease. As depicted in Figure 3.29A, a normal fibroosseus canal allows the tendon to glide with no obstruction. However, in the case of trigger finger, bulbous swellings, such as those shown in Figure 3.29B, will restrict the finger flexor tendons from traversing through their fibrous ligamentous sheaths. If large enough, the bulbous swellings could immobilize the tendon, thereby locking the finger in a fixed flexed position. In order to "unlock" the finger, the bulbous swelling has to snap to move beyond the constriction in the fibroosseus canal. External aid from the other hand may be required to extend the finger back to a straight, neutral position (Rowe, 1985).

Although trigger finger tends to occur at the creases of the finger (Caillet, 1984), it occurs most frequently near the knuckle (metacarpophalangeal joint) (Quinnel, 1980). The middle and ring fingers are the predominant sites of trigger finger in the dominant hand, accounting for over 40% of the cases in Quinnel's (1980) study.

# 3.6 Biomechanical Aspects of Nerve Compression Disorders

Nerve compression disorders are a group of disorders in which a peripheral nerve is compressed or pinched, causing trauma to the immediate area served by the nerve and sometimes distal to the site of impingement. The effects from the trauma could be temporary or long term. An example of a widely known nerve disorder is sciatica, which is compression of the sciatic nerve in the lower back but whose pain is felt throughout the areas of the lower leg served by the sciatic nerve.

Nerve compression disorders of the upper extremity can affect all three major nerves of the upper extremity, namely the radial, ulnar, and median nerves (refer to Figure 3.13a). The most widely publicized nerve compression disorder of the upper extremity is *carpal tunnel syndrome*, which is compression of the median nerve at the site where it passes through the carpal tunnel in the wrist. Examples of lesser known nerve compression disorders of the upper extremity are cubital tunnel syndrome and posterior interosseous nerve syndrome, which are caused by compression of the ulnar and radial nerves, respectively.

The anatomy of a *neuron* is illustrated in Figure 3.31. Although neurons vary in size and shape, they generally consist of a cell body (soma), *dendrites*, and an *axon*. The axon is the shaft of the nerve through which electrical impulses travel. When the impulse reaches the axon terminal, it then crosses over to the dendrites of an adjacent neuron. As shown in Figure 3.31, there are small gaps, called the *nodes of Ranvier*, between segments of the axon. These segments are covered with a *myelin sheath* and *neurilemma* (or Schwann cells), which insulate the nerve fibers from adjacent cellular compartments and allow the electrical impulse to travel from node to node (Van de Graaf and Rhees, 1987). The speed of an impulse traveling through a neuron is called *conduction velocity*. A cross-sectional view of the axon of a neuron reveals three layers of connective tissue that hold the nerve fibers together and protect them. As illustrated in Figure 3.32, the epineurium is the most external layer, holding together several *fasciculi*. Several fasciculi surrounded by epineurium is called a single *nerve* (Spence, 1986). The perineurium, which consists of fibrous collagen, surrounds each fasciculus or bundle of nerve fibers. The endoneurium is the connective tissue within each fasciculus and forms a tube-like membrane around each *nerve fiber* (Szabo and Gelberman, 1987).

Although the pathogenesis of nerve entrapment syndromes is controversial, two prominent theories have been hypothesized (Moore, 1992). These two theories are first, mechanical compression of the nerve, and second, inadequate blood supply serving the nerve. The first theory, mechanical compression, is described in detail in Feldman et al. (1983) and Szabo and Gelberman (1987) and is summarized below. Any physical disturbance to the nerve can cause motor or sensory dysfunction. An impingement upon the *efferent* portion of the nerve — carrying impulses to the peripheral nervous system from the brain and spinal cord — can result in loss of muscular strength. Likewise a disturbance to the *afferent portion* — carrying impulses to the central nervous system from receptors located throughout the body — can decrease sensory feedback. Myelinated nerves are more susceptible to the effects of pressure than unmyelinated nerves. Since motor nerves consist predominantly of thick myelinated fibers, the motor nerves are theoretically more susceptible to the effects of compression than sensory cutaneous nerves (Feldman et al., 1983). However, based on clinical observations of nerve entrapment syndromes, the sensory function appears to show decrements before motor function (Szabo and Gelberman, 1987). The pathogenesis of why sensory decrements are manifested before motor decrements is still unclear.

Compression can cause neural dysfunction in the following manner. First, compression causes bulbous swellings on the fibers (Aguayo, 1975), which can block conduction of electrical impulses. If compression continues, the myelin between nodes on the nerve becomes thinner, and the fibers start to segmentally demyelinate, which can further decrease the conduction velocity of nerve impulses (Feldman et al., 1983).

The second prominent theory posed to explain the pathogenesis of nerve entrapment syndromes is insufficient blood supply to the nerves (Moore, 1992). *Ischemia*, or inadequate circulation due to mechanical obstruction in the nerve's path, can cause symptoms typical of nerve entrapment syndromes, such as *paresthesia* or acute pain and possibly a reduction in conduction velocity. The mechanical obstruction could take the form of pinching or entrapment of the nerve as the nerve travels around tendons, ligaments,

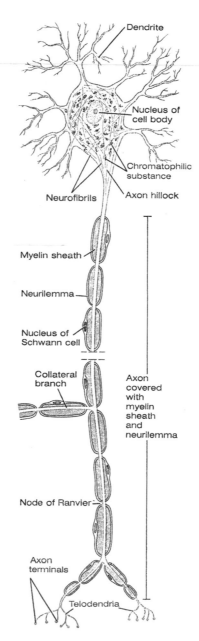

**FIGURE 3.31**   Structure of a typical motor neuron. (From Spence, A.P. 1986. *Basic Human Anatomy,* The Benjamin Cummings Publishing, Inc., p. 348. With permission.)

or bones in a joint. Vascular deficiencies and changes in blood pressure could account for the variation in symptoms noted in patients with approximately equal levels of nerve conduction delay (Shivde et al., 1981; Moore, 1992). In one study, vascular *sclerosis* was observed in 98% of the carpal tunnel cases (Fuchs et al., 1991).

   The nerve fiber distal to the site of compression or physical trauma, whether by compression or inadequate blood supply, can also be detrimentally affected. *Wallerian degeneration* is the deterioration of the myelin sheath distal to the site of trauma and can lead to *atrophy* and destruction of a neuron. Wallerian degeneration is caused by impaired flow of electrical impulses down the nerve's axon and ischemia (Feldman et al., 1983). Ogata and Naito (1986) investigated the effects of compression and

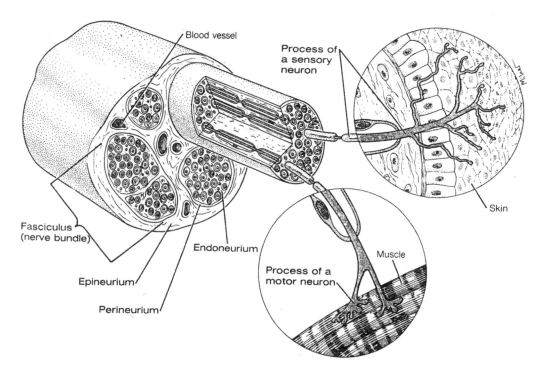

**FIGURE 3.32** Cross section of a single peripheral nerve, which is composed of several fasciculi wrapped in epineurium connective tissue. The perineurium wraps each fasciculus, and the endoneurium encases each single nerve fiber. (From Spence, A.P. 1986. *Basic Human Anatomy,* The Benjamin Cummings Publishing, Inc., p. 353. With permission.)

stretching of nerves, and they found that compression and stretching do restrict blood flow to the nerves, even to the point of arresting blood flow. In addition to Wallerian degeneration, compression of a nerve can impede nerve conduction and lower conduction velocity distal to the site of trauma, resulting in decreased motor and sensory function such as muscle atrophy and paresthesia, respectively (Feldman et al., 1983).

In the upper extremity, peripheral nerves can be compressed at any point along its path, ranging from the wrist (carpal tunnel syndrome) to the shoulder (thoracic outlet syndrome). These two syndromes are described and discussed below.

## 3.7 Work-Related CTDs Involving the Nerve

### Carpal Tunnel Syndrome

Carpal tunnel syndrome (CTS) is a nerve entrapment disorder arising from compression of the median nerve at the site where it travels through the carpal tunnel in the wrist. Although CTS was first recognized by Sir James Paget more than a century ago (1860), the name CTS was not uniformly applied to compression of the median nerve until the late 1950s (Phalen, 1981). Prior to and through the early 1950s, CTS was referred to with long, clumsy names, such as "spontaneous compression of the median nerve in the carpal tunnel" and "compression secondary to trauma, tumor, or systemic disease." The much shorter, more facile name, "carpal tunnel syndrome," started to gain acceptance in the late 1950s to include all conditions that might cause compression of the median nerve, regardless of the source of compression. In 1957, Phalen and Kendrick stated, "The term carpal tunnel syndrome is now used to describe all cases of compression neuropathy of the median nerve at the wrist" (Moore, 1992).

The symptoms of CTS involve the motor, sensory, and *autonomic* functions of the median nerve (Armstrong, 1983):

1. Motor nerve impairment: reduced motor control and atrophy of the abductor pollicis brevis, a major thumb abductor located at the base of the thumb, and weakness in precision grip (Feldman et al., 1983). A patient with severe and long-standing CTS may show a severely diminished musculature at the *thenar eminence*, which is the protruding area at the base of the thumb.
2. Sensory nerve impairment: paresthesia (burning, prickling, and tingling) and *hypesthesia* (diminished sensitivity to stimulation) in the thumb and fore, middle, and one half of the ring fingers and in the thenar eminence on the palmar side; paresthesia and hypesthesia on the distal *phalanges* of the thumb and same fingers on the dorsal side of the hand (refer to Figure 3.14).
3. Autonomic nerve impairment: diminished sweat function, resulting in dry and shiny skin in areas noted in sensory nerve impairment.

Of all the tissues that pass through the carpal tunnel at the wrist, the median nerve is the softest and most vulnerable to pressure. As illustrated in Figure 3.8, the median nerve along with nine forearm flexor tendons pass through the carpal tunnel, which is formed by the carpal bones on the dorsal side of the hand and the flexor retinaculum on the palmar side. In 1963, Robbins conducted a systematic study, considered by some to be the most thorough, of the anatomy of the carpal tunnel (Moore, 1992). He analyzed cross sections of cadaveric wrists dissected at distances 2 to 4 cm proximal and distal to the wrist crease. Among Robbins's (1963) major findings were:

1. The cross-sectional area of the carpal tunnel decreases from its proximal entrance to a point 2 cm distal from its origin. He said the effect is "to have a canal with a slightly narrowed waist."
2. All the structures are crowded in the canal, and in 6 of the 7 specimens, the median nerve was flattened and directly beneath the flexor retinaculum.
3. When the wrist was in a neutral position in the cadaveric specimens, Robbins (1963) inserted and withdrew a piece of rubber easily underneath the flexor retinaculum. However, when the wrist was flexed or extended, the resistance increased concomitantly with the angle of deviation. Considerable force was required to withdraw the rubber proximally in either of the extreme flexed or extended positions. Robbins (1963) concluded the volume of the carpal tunnel decreases as the wrist is deviated in either flexion or extension.

Patients with CTS show a significant elevation of pressure in the carpal tunnel. At a neutral posture, Gelberman et al. (1981) found that the mean pressure in the carpal tunnel of patients with CTS was 32 mmHg, compared to 2.5 mmHg for control subjects. These researchers also showed that 90° of wrist flexion and extension increased the pressure precipitously to approximately 100 mmHg and 32 mmHg for the CTS patients and healthy subjects, respectively.

Although the pathophysiology of CTS is unknown, researchers have postulated thickening of the flexor tendon sheaths in CTS patients as contributing to the increase in pressure in the carpal tunnel. In a study of 212 wrists surgically treated for CTS, Phalen (1966) observed thickening or fibrosis of the flexor synovium in 203 of the cases. Biopsy specimens of the flexor synovium from 181 of the 212 wrists revealed chronic fibrosis or thickening in 91 specimens, chronic inflammation compatible with symptoms of rheumatoid arthritis in 64, and no pathologic change in 26. Yamaguchi et al. (1965) observed microscopically an aging effect on the flexor tendon sheaths, manifested by fibrous thickening of the sheaths. They compared the sheath anatomy of CTS patients vs. healthy controls, and they found that almost 90% of the patients exhibited a greater increase in thickening and fibrosis of the sheaths than the healthy control subjects. Kerr et al. (1992) observed hypertrophy of the synovium of the flexor tendons, with little or no evidence of inflammation, in patients with CTS. Schuind et al. (1990) observed fibrous hypertrophy and necrotic lesions in flexor synovium "typical of a connective tissue undergoing degeneration under repeated mechanical stresses." The experimental findings of Armstrong et al. (1984) support Schuind's (1990) association between degeneration of tendon tissue and mechanical stress. Synovial hypertrophy and the mean densities of subsynovium and adjacent connective tissue were significantly greater at the wrist crease as compared to locations proximal and distal to the crease. Armstrong et al. (1984) concluded that repeated exertions with a flexed or extended wrist are an important factor in the etiology of the degeneration and hypertrophy of tissue surrounding the tendon.

## Occupational Sources of Median Nerve Compression in CTS

The three main occupational risk factors of CTS (wrist repetition, deviated wrist angle, and tendon force), along with the basic anatomical structure of the wrist, can increase the pressure in the carpal tunnel, compress the median nerve, and result in the following:

1. *Increase in carpal tunnel pressure at deviated wrist angles due to reduction in tunnel volume.* According to Robbins (1963), extreme flexion and extension of the wrist reduced the volume of the carpal tunnel, thereby augmenting the pressure on the median nerve. This increase in tunnel pressure would, theoretically, affect the median nerve first because it is the softest and most vulnerable tissue in the carpal tunnel.

2. *General increase in tunnel pressure from wrist deviation.* Deviation of the wrist in the flexion/extension plane has been shown repeatedly in the anatomical and physiological literature to increase the pressure in the carpal canal (Phalen, 1966; Smith et al., 1977). In a flexed or extended wrist posture, the median nerve is squeezed between the flexor retinaculum and the overlying flexor tendons, thereby exposing a worker to CTS. Recently, Rempel et al. (1994) measured the carpal tunnel pressure of subjects typing on a computer keyboard elevated at various slopes to extend the wrist at five different angles. As shown in Figure 3.33, these researchers found that the pressure in the carpal tunnel was lowest at a neutral position compared to postures up to 50° extension and 20° flexion. The approximately 100 mmHg maximum pressures Rempel et al. (1994) measured were in the same range as in Gelberman's et al. (1981) study. In addition to the flexion/extension plane, carpal tunnel pressure has been shown to increase as the wrist radially and ulnarly deviates from a neutral posture. Sommerich (1994) measured the carpal tunnel pressures of four subjects typing on a standard QWERTY computer keyboard and an alternative keyboard split and angled to reduce ulnar deviation. She found that all subjects showed a decrease in carpal tunnel pressure with a concomitant decrease in ulnar deviation. The maximum carpal tunnel pressures of 80 mmHg measured in Sommerich's (1994) study were similar to those measured in the studies of Gelberman et al. (1981) and Rempel et al. (1994).

3. *Thickening and fibrosis of synovium and hypertrophy of synovial sheaths.* The well-documented reporting of thickening of the flexor sheaths and synovium in the carpal tunnel (Phalen, 1966; Yamaguchi et al., 1965; Armstrong et al., 1984; Schuind et al., 1990; and Kerr et al., 1992) could possibly be explained by the biomechanical models of Armstrong and Chaffin (1979) and Schoenmarklin and Marras (1990). The resultant reaction force on the flexor tendons and the median nerve passing through the carpal tunnel increases concomitantly, not only with deviation angle ($F_r$ in Figures 3.18 and 3.19 from Armstrong and Chaffin [1979]), but also with acceleration of the wrist ($F_r$ in Figures 3.21 and 3.22 from Schoenmarklin and Marras [1990]). Wrist deviation and acceleration are the static and dynamic components of repetitive movements of the wrist, which have been associated with CTS.

In order to accelerate the wrist, the extrinsic muscles in the forearm have to exert force which is transmitted to the tendons. As modeled by Schoenmarklin and Marras (1990), some of the force transmitted through the tendon is lost to friction against the ligaments and bones that form the carpal tunnel (refer to Figure 3.21). This frictional force could irritate the tendons and their sheaths and possibly cause the synovitis and hypertrophy found experimentally in the carpal tunnel. Armstrong et al. (1984) found sizeable increases in synovium and synovial density in the carpal tunnel area, which they attributed to repeated flexion/extension exertions. From a modeling point of view, Tanaka and McGlothlin (1989)

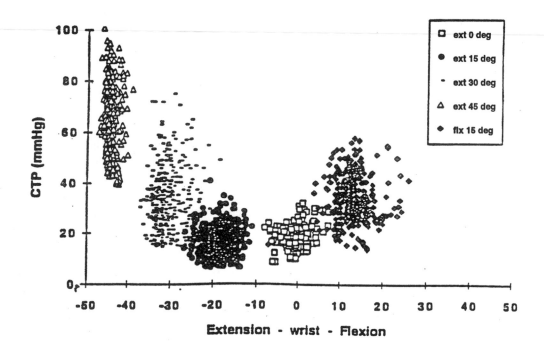

**FIGURE 3.33** Carpal tunnel pressure in one subject's wrist while typing on a computer keyboard set at different wrist extension angles. (From Rempel, D. and Horie, S. 1994. Effect of wrist posture during typing on carpal tunnel pressure. In *Proceedings of Working With Display Units: Fourth International Scientific Conference,* Milan, Italy, p. C27. With permission.)

hypothesized that the friction between tendons and adjacent structures is a major cause of CTDs and CTS, and Moore and Wells (1989) and Moore et al. (1991) showed that the frictional work generated in the carpal tunnel supported Silverstein's et al. (1986, 1987) dose–response relationship between repetition and CTD risk.

In theory, the deleterious effects of frictional work generated between the tendons and their sheaths or supporting structures is exacerbated by coactivation of the forearm extensor muscles during movements of the wrist and hand. In order for the hand to maintain the same flexor torque or power/pinch force, the flexor muscles have to exert more force to overcome the extensor force. Greater forces in the flexor muscles will generate, in theory, increased frictional work between the flexor tendons and their adjacent structures, thereby exposing workers to an increased risk of CTS.

Coactivation of *antagonist* muscles during static and dynamic contractions of the *agonist* muscles have been found experimentally and modeled (Schoenmarklin and Marras, 1992; Marras and Sommerich, 1991a,b; Marras, 1992). With regard to hand grip exertions, Grant et al. (1992) measured the electromyographic (EMG) signal of the forearm flexor and extensor musculature while subjects gripped various diameters of handle, and these researchers found contractions of the extensor muscles (which act as antagonists in this case) up to 30% maximum voluntary contraction (MVC). Coactivation of the extensors stabilizes the wrist during flexion movements (and vice versa, coactivation of the flexors stabilizes the wrist during extension movements), and coactivation of the extensors also helps guide and stabilize the hand while it exerts a power or pinch force.

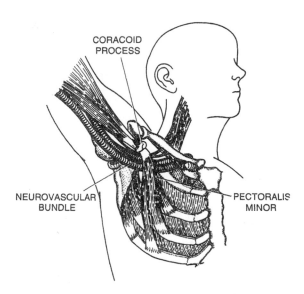

CORACOID
PROCESS

NEUROVASCULAR
BUNDLE

PECTORALIS
MINOR

**FIGURE 3.34** Abduction of the shoulder causes the neurovascular bundle, containing the brachial plexus nerve and subclavian artery, to be stretched under the pectoral muscles. Hypertrophy of the pectoral muscles would impinge the neurovascular bundle, thereby causing deficiencies in the neural and circulatory systems of the upper extremity. (From Putz-Anderson, V. 1988. *Cumulative Trauma Disorders: A Manual for Musculoskeletal Diseases of the Upper Limbs*, p. 20, Taylor & Francis. With permission.)

## Thoracic Outlet Syndrome

*Thoracic outlet syndrome* (TOS) is a *neurovascular* disorder that affects the bundle carrying nerves, arteries, and veins from the neck through the shoulder and into the arm, as illustrated in Figure 3.34. This neurovascular bundle, which contains the brachial plexus nerve and the subclavian artery and vein, could be compressed by adjacent muscles in the shoulder region, such as the scalene, subclavius, or pectoralis minor, or the bones forming the thoracic outlet in the shoulder. The motions and tasks that are associated with TOS are repetitive shoulder abduction and *adduction*, carrying heavy loads on the shoulder, and working overhead (Feldman et al., 1983).

The pathogenesis of TOS is hypertrophy of the subclavius or pectoral muscles, which can pinch the neurovascular bundle and produce symptoms in the neural and circulatory system throughout the arm. Compression of the brachial plexus, which branches out into the median and ulnar nerves, can cause numbness or paresthesia in the lower parts of arm and hand served by the median and ulnar nerves. An impinged subclavian artery, which is the major artery supplying the upper extremity (refer to Figure 3.35), will result in ischemia, a reduction of blood flow to shoulder and arm (Basmajian, 1982). This diminished blood flow will reduce the amount of oxygen and nutrients available for dynamic arm movements and exacerbate the fatiguing effect from static (anaerobic) contractions by reducing the amount of blood needed to carry away lactate and metabolites.

## 3.8   Summary

In this chapter, major CTDs that affect the soft tissues of the body and have been associated with work-related activity are categorized into three groups: CTDs affecting muscle, the muscle–tendon unit, and nerve. The anatomy of the upper extremity's musculoskeletal system was described, followed by a discussion of biomechanical mechanism(s) that theoretically explain, in part, the epidemiological associations between respective categories of CTD and work-related activity. In addition, each CTDs is described in detail along with its pathogenesis.

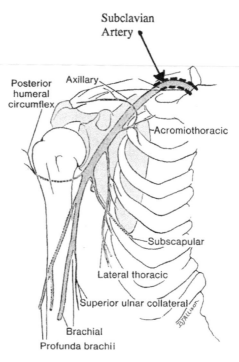

**FIGURE 3.35**   The subclavian artery, which branches off into several arteries, is the major artery that serves the upper extremity. Compression of the subclavian artery can cause ischemia and paresthesia in the arm and hand. (From Basmajian, J.V. 1982. *Primary Anatomy,* 8th edition, p. 274, Williams and Wilkins, Inc. With permission.)

## Defining Terms

**Abduct (abduction):**   Literally means to "to lead away from." In anatomical parlance, abduction is moving a joint away from the center of the body. Shoulder abduction is elevating the arm in the frontal plane.

**Acromion process:**   The bony top of the shoulder. The acromion is actually the highest part of the scapula bone, which spans the back of the shoulder.

**Adduct (adduction):**   Literally means "to lead towards." Adduction of a joint is moving the joint toward the center of the body. Shoulder adduction is moving the shoulder towards the side of the torso.

**Afferent nerves:**   Nerves that travel from the peripheral nervous system to the central nervous system, namely the brain and spinal cord.

**Agonist muscle:**   Muscle that initiates and carries out motion (Chaffin and Andersson, 1991).

**Antagonist muscle:**   Muscle that opposes the action of the agonist muscle (Chaffin and Andersson, 1991).

**Aponeurosis:**   A broad, flat tendon. An aponeurosis usually occurs at the point where the tendons from several muscles fuse and connect to a bone.

**Atrophy:**   A wasting of tissues or organs. A severe case of carpal tunnel syndrome will show atrophy of the muscles in the thenar eminence.

**Autonomic:**   Relating to the autonomic nervous system, which is part of the efferent division of the peripheral nervous system. The autonomic nervous system is the involuntary system functioning below the conscious level, that controls the heart, organs, and glands of the body (Spence, 1982). Patients with carpal tunnel syndrome may display dry and shiny palms, resulting from insufficient activity of the sweat glands in the hand, which are controlled by the autonomic nervous system.

**Avascular:**   Nonvascular; without blood vessels (Stedman, 1982).

**Axon:**   The shaft of a neuron that transmits electrical impulses (Spencer, 1986).

**Bursa:** A structure resembling a collapsed bag with cellophane-thin walls whose inner surfaces are extremely smooth, moist, and slippery. A bursa usually forms where a tendon rubs back and forth on a hard structure. For example, rotator cuff syndrome is the deterioration of the bursa that lubricates the shoulder bone from a tendon whose muscle raises the shoulder (Basmajian, 1982).

**Carpal tunnel:** The tunnel formed by eight small bones of the wrist and the transverse carpal ligament (flexor retinaculum). The tendons that connect the forearm flexor muscles to the fingers pass through the carpal tunnel.

**Carpal tunnel syndrome:** Compression of the median nerve as it passes through the carpal tunnel.

**Central nervous system (CNS):** The nervous system consisting of the brain and spinal cord. The CNS is the integrative and control center of the body. The CNS receives sensory input from the peripheral nervous system (PNS) and develops response strategies to the input (Spence, 1986).

**Conduction velocity:** The velocity at which an electrical impulse travels through a nerve. Conduction velocity, which for a motor nerve is normally 50 to 60 m/s (Johnson, 1989). When the impulse reaches muscle fiber, its conduction velocity slows down to about 3 to 6 m/s (Basmajian and DeLuca, 1985).

**Cumulative trauma disorders (CTDs):** Any of a class of pathologies affecting soft tissues (muscles, tendons, and nerves) created from excessively frequent use of a particular joint or tissue, especially in combination with awkward positioning, inadequate or no rest periods, or excessive loads. Also called repetitive strain injury (RSI), repetitive stress injury, repetitive motion injury, and overuse syndrome (Stramler, 1993).

**Dendrites:** Thin extensions emanating from the cell body of a neuron that receive electrical impulses from adjoining neurons (Spence, 1986).

**DeQuervain's tenosynovitis:** A narrowing of the passage (stenosing tenosynovitis) of the tendons and their sheaths that extend the thumb in the palmar plane.

**Disease:** Illness, sickness, or cessation of bodily functions, systems, or organs. A disease is characterized usually by at least two of the following criteria: a recognized causal agent(s), an identifiable group of signs and symptoms, or consistent anatomical alterations (Stedman, 1982). Compare to *Syndrome* and *Disorder*.

**Distal:** Located away from the center of the body or point of origin. For example, the wrist is distal to the elbow, and the elbow is distal to the shoulder (Stedman, 1982).

**Disorder:** A disturbance of the functions, structure, or both resulting from a failure in development or from external factors such as physical contact, injury, or disease (Stedman, 1982). Compare to *Disease* and *Syndrome*.

**Dorsal:** Pertaining to the back of an anatomical structure, as in the dorsal (back) side of the hand.

**Edema:** An accumulation of an excessive amount of watery fluid in cells, tissues, or cavities (Stedman, 1982).

**Efferent nerves:** Nerves that travel from the brain and spinal cord in the central nervous system to the outlying areas in the peripheral nervous system.

**Endoneurium:** Thin connective-tissue sheath that wraps each individual nerve fiber (Spence, 1986).

**Epicondyle:** A projection from a long bone above an articulating joint.

**Epineurium:** Connective-tissue sheath that surrounds several fasciculi in a nerve. Several fasciculi surrounded by epineurium constitute a single nerve (Spence, 1986).

**Etiology:** The science and study of the causes of disease and their mode of operation (Stedman, 1982). Compare to *Pathogenesis*.

**Extend (extension):** Movement of a body part that increases the angle of its adjacent joint. For example, moving your hand away from your shoulder requires elbow extension.

**Extensor retinaculum:** A strong fibrous ligament that stretches across the back of the hand at the wrist. The extensor retinaculum holds the forearm extensor tendons close to the carpal bones of the wrist.

**Extrinsic muscles:** Flexor and extensor muscles in the forearm that generate much of the movement and force production of the hand and fingers. The extrinsic muscles can also be recruited for hand movements and forces requiring fine motor control.

**Fasciculus (fasciculi):**   A bundle of nerve fibers surrounded by connective tissue (perineurium) (Spence, 1986).

**Fibroosseus canal:**   A canal formed by a fibrous ligamentous sheath and an underlying bone. DeQuervain's disease and trigger finger are thickening of the fibrous ligamentous sheath, resulting in a thinning of the tendon and its synovial sheath that pass through the canal.

**Flex (flexion):**   Movement of a body part that decreases the angle of its adjacent joint. For example, moving your hand toward your shoulder requires elbow flexion.

**Flexor retinaculum:**   The transverse carpal ligament that stretches across the palmar side of the wrist. The flexor retinaculum, which forms the top of the carpal tunnel, holds the flexor tendons and their sheaths inside the wrist.

**Frictional force:**   A force that impedes impending motion.

**Frontal plane:**   The plane of the body that travels through the chest and arms. The frontal plane shows the front of the body. Also called coronal plane.

**Humerus:**   The arm bone connecting the shoulder to the elbow.

**Hypertrophy:**   General increase in the bulk of a part or organ that is not due to tumor formation (Stedman, 1982). Compare to *Inflame.*

**Hypesthesia (hypoesthesia):**   An abnormal sensation characterized by diminished sensitivity to stimulation (Stedman, 1982).

**Hysteresis:**   The difference in the stress–strain response of a material to an increasing load or decreasing load. With tendons, the stress required to maintain a certain amount of strain is less when the force is decreasing than when increasing.

**Innervate:**   To supply nerve function to a specific muscle group.

**Insert (insertion):**   The distal attachment point of a muscle to bone, via a tendon.

**Inflame:**   Pathologic process consisting of a histologic reaction to affected blood vessels and adjacent tissues in response to an injury or abnormal stimulation caused by a physical, chemical, or biologic agent (Stedman, 1982). Compare to *Hypertrophy.*

**Intrinsic muscles:**   The small muscles located within the hand that move the thumb and fingers. The intrinsic muscles, which are much smaller than the extrinsic muscles in the forearm, are used primarily for motions and forces requiring dexterity and fine motor control.

**Ischemia:**   Inadequate circulation of the blood due to mechanical obstruction, mainly arterial narrowing (Stedman, 1982).

**Lateral epicondyle:**   A bony protrusion located on the lateral side of the elbow. Viewed from the side, the lateral epicondyle is located close to the pivot point of the elbow as it flexes and extends.

**Lateral epicondylitis (tennis elbow):**   Tendinitis of the forearm extensor and supinator muscles at the lateral epicondyle of the elbow. Lateral epicondylitis is colloquially dubbed "tennis elbow" because patients with lateral epicondylitis report pain when the wrist is extended and supinated — a movement similar to that of a back stroke in tennis.

**Ligament:**   Connective tissue resembling a tendon except a ligament attaches bone to bone (Basmajian, 1982).

**Median nerve:**   The great "flexor" nerve of the upper extremity. It supplies motor function to the forearm flexor muscles and the thumb's (thenar) muscles and sensory function to the palm and the digits from the thumb to the center of the ring finger (Basmajian, 1982).

**Morbidity:**   A diseased state (Stedman, 1982).

**Muscle:**   A contractile organ of the body that moves various body parts and internal organs. A muscle can only pull (tensile force) and not push (compression force). The origin of a muscle is the end that is more fixed and the insertion of a muscle is the end that is more movable. For example, the flexor muscles of the forearm originate at the elbow and insert at the wrist and fingers (Stedman, 1982).

**Myalgia:**   Muscular pain.

**Myelin sheath:**   A sheath that surrounds the axon of a neuron. The purpose of the myelin sheath is to insulate the neuron from adjacent cellular fluids so an electrical impulse can travel down the axon jumping from one node of Ranvier to the next.

**Myofascial syndrome:**   Term referring to regional muscle pain syndromes (Kuorinka and Forcier, 1995).

**Myopathy:**   Term for measurable pathological changes in a muscle with or without symptoms (Kuorinka and Forcier, 1995).

**Nerve:**   A collection of nerve fibers in the peripheral nervous system (Spence, 1986).

**Nerve fiber:**   Any long process of a neuron. The term usually refers to axons, but also includes the peripheral processes of sensory neurons (Spence, 1986).

**Neurilemma:**   The thin membrane between the myelin and connective tissue (endoneurium) in a neuron. Neurilemma is also called the sheath of Schwann (Spence, 1986).

**Neuron:**   A nerve cell (Spence, 1986).

**Neurovascular:**   Pertaining to the nervous and circulatory systems of the body.

**Nociceptor:**   A peripheral nerve organ or mechanism that senses pain or injurious stimuli and transmits them (Stedman, 1982).

**Nodes of Ranvier:**   Gaps in the myelin sheath and neurilemma of a neuron's axon. Electrical impulses known as potentials travel down the axon from one node to the next, which is called "saltatory conduction." (The etymology of "saltatory" is from the Latin word "saltare," meaning "to dance.")

**Origin (originate):**   The proximal attachment point of a muscle to bone, via a tendon.

**Paresthesia:**   An abnormal sensation, such as of burning, prickling, tickling, or tingling (Stedman, 1982).

**Pathogenesis:**   The mode of origin or development of any disease or morbid process (Stedman, 1982). Compare to *Etiology*.

**Periosteum:**   Thick fibrous membrane covering the surface of a bone (Stedman, 1982).

**Peripheral nervous system:**   All nervous structures located outside the central nervous system (CNS), which consists of the brain and spinal cord. The PNS consists of nerves that connect the outlying parts of the body and their receptors with the brain and spinal cord (Spence, 1986).

**Phalanx (phalanges):**   One of the long bones of the fingers (Stedman, 1982). A finger has three phalanges:   distal (tip), middle, and proximal (connected to the knuckle).

**Proximal:**   Located toward the center of the body or point of origin. For example, elbow is proximal to the wrist, and the shoulder is proximal to the elbow (Stedman, 1982).

**Radial deviation:**   Rotation of the wrist joint toward the radius bone or thumb side. Sometimes radial deviation is referred to as abduction of the wrist joint.

**Sclerosis:**   The process of becoming hard or firm. In the case of arteriosclerosis, the walls of the arteries harden and become less elastic. The blood vessels are unable to expand and recoil in response to pressure changes, thereby elevating one's maximum blood pressure (Spence, 1986).

**Shoulder-neck myofascial syndrome:**   See *Tension neck syndrome*.

**Stenosing tenosynovitis:**   A narrowing of the canal through which a tendon and its sheath pass.

**Striated muscle:**   Muscle that appears striped with dark and light bands under a microscope. Also called skeletal muscle, which are the muscles that move the limbs, head, neck, and torso.

**Supraspinatus tendinitis (rotator cuff syndrome):**   Tendinitis of the supraspinatus muscle, which abducts the shoulder from the side of the trunk. Supraspinatus tendinitis is often called rotator cuff syndrome.

**Syndrome:**   The collection of signs and symptoms associated with any disease process and constituting the picture of the disease (Stedman, 1982). Compare to *Disease* and *Disorder*.

**Synovial fluid:**   A clear fluid whose purpose is to lubricate a tendon within a sheath or a joint (Stedman, 1982).

**Synovial sheath:**   An elongated and double-walled tubular structure (bursa) that surrounds a tendon and allows the tendon to travel with little friction. Synovial sheaths are located where tendons move around joints, such as in the wrist (Basmajian, 1982).

**Tendon:**   Connective tissue resembling a tough cord or band and always part of a muscle, usually forming an attachment of muscle to bone (Basmajian, 1982).

**Tendinitis** (also spelled **tendonitis**):   Inflammation of a tendon (Stedman, 1982).

**Tenosynovitis:**   Inflammation of a tendon's sheath (Stedman, 1982). Clinically, any tendon sheath disorder is called tenosynovitis, regardless of the presence or absence of inflammation (Moore, 1992).

**Tension neck syndrome (TNS):** Myalgia in the shoulder–neck region of the body. TNS is synonymous with shoulder–neck myofascial syndrome, and TNS is defined by symptoms of pain in the shoulder–neck region with simultaneous findings of tenderness over the shoulder–neck muscles (Kuorinka and Forcier, 1995).

**Thenar eminence:** The area at the base of the thumb that is raised above the general level of the palm. The thenar eminence contains the intrinsic musculature that controls the thumb.

**Thoracic outlet syndrome:** A neurovascular disorder that affects the brachial plexus nerve and the subclavian artery and vein as they traverse through the thoracic outlet in the shoulder. The early symptoms of thoracic outlet syndrome are found in the areas in the forearm and hand served by the median and ulnar nerves.

**Trigger finger:** Painful locking of a finger caused by narrowing of the canal through which a finger flexor tendon and its sheath pass.

**Ulnar deviation:** Rotation of the wrist joint toward the ulna bone or little finger side. Sometimes ulnar deviation is referred to as adduction of the wrist joint.

**Wallerian degeneration:** Degeneration of the myelin sheath of a neuron's axon caused by compression of the nerve and ischemia. Wallerian degeneration leads to the atrophy and destruction of the neuron (Stedman, 1982).

# References

Abrahams, M. Mechanical analysis of tendon *in vitro*: A preliminary report. *Med Biol Eng*, 5:433-443, 1967.

Aguayo, A.J. Neuropathy due to compression and entrapment, in *Peripheral Neuropathy*, Dyck, P.J. and Lambert, E.H., editors, 688-713, W.B. Saunders Co., 1975.

Armstrong, T.J. An ergonomics guide to carpal tunnel syndrome. *Ergonomics Guides*, American Industrial Hygiene Association, 1983.

Armstrong, T.J., Castelli, W.A., Evans, F.G., and Diaz-Perez, R.D. Some histological changes in carpal tunnel contents and their biomechanical implications. *Journal of Occupational Medicine*, 26, No. 3, 197-201, 1984.

Armstrong, T.J. and Chaffin, D.B. Some biomechanical aspects of the carpal tunnel. *Journal of Biomechanics*, 12, 567-570, 1979.

Basmajian, J.V. *Primary Anatomy, Eighth edition*. Williams and Wilkins, 1982.

Basmajian, J.V. and DeLuca, C.J. *Muscles Alive: Their Functions Revealed by Electromyography*, Fifth edition. Williams and Wilkins, 1985.

Brand, P.W. *Clinical Mechanics of the Hand*. The C.V. Mosby Co., 1985.

Bunnel, S. *Surgery of the Hand*, J.B. Lippincott Co., 1956.

Caillet, R. *Hand Pain and Impairment*. Philadelphia, F.A. Davis Co., 1984.

Chaffin, D.B. and Andersson, G.B.J. *Occupational Biomechanics*, Second Edition. John Wiley & Sons, 1991.

Cyriax, J.H. The pathology and treatment of tennis elbow, *Journal of Bone and Joint Surgery*, 18, 921-940, 1936.

Elliot, D.H. Structure and function of mammalian tendon. *Biol Rev*, 40:392-421, 1965.

Feldman, R.G., Goldman, R., and Keyserling, W.M. Peripheral nerve entrapment syndromes and ergonomic factors. *American Journal of Industrial Medicine*, 4, 661-681, 1983.

Finkelstein, H. Stenosing tendovaginitis at the radial styloid process. *Journal of Bone and Joint Surgery*, 12, 509-540, 1930.

Froimson, A.I. Treatment of tennis elbow with forearm support band. *Journal of Bone and Joint Surgery*, 53A:183-184, 1971.

Fuchs, P.C., Nathan, P.A., and Myers, L.D. Synovial histology in carpal tunnel syndrome. *Journal of Hand Surgery*, 16A, 753-758, 1991.

Fung, Y.C. *Biomechanics: Mechanical Properties of Living Tissues*. Springer-Verlag, 1981.

Gelberman, R.H., Hergenroeder, P.T., Hargens, A.R., Lundbor, G.N., and Akeson, W.H. The carpal tunnel syndrome: A study of carpal canal pressures. *Journal of Bone and Joint Surgery*, 63A(3), 380-383, 1981.

Gelberman, R., Goldberg, V., and An, K.N. Tendon, in *Injury and Repair of the Musculoskeletal Soft Tissues*, Woo, S.L.Y. and Buckwalter, J.A., editors, pp. 1-40, American Academy of Orthopaedic Surgeons, 1988.

Goldie, I. Epicondylitis lateralis humeri (Epicondylitis of tennis elbow): A pathogenetic study. *Acta Chir Scand Suppl*, 339, 1-119, 1964.

Goldstein, S.A., Armstrong, T.J., Chaffin, D.B., and Matthews, L.S. Analysis of cumulative strain in tendons and tendon sheaths. *Journal of Biomechanics*, 20, No. 1, 1-6, 1987.

Grant, K.A., Habes, D.J., and Steward, L.L. An analysis of handle designs for reducing manual effort: The influence of grip diameter. *International Journal of Industrial Ergonomics*, 10, 199-206, 1992.

Johnson, E.W. *Practical Electromyography*, 2nd edition, Williams and Wilkins, 1989.

Kerr, C., Sybert, D.R., and Albarracin, N.S. An analysis of the flexor synovium in idiopathic carpal tunnel syndrome: Report of 625 cases. *Journal of Hand Surgery*, 17A(6), 1028-1030, 1992.

Kuorinka, I. and Forcier, L., editors. *Work-Related Musculoskeletal Disorders (WMSDs): A Reference Book for Prevention*. Taylor & Francis, 1995.

Lamphier, T.A., Crooker, C., and Crooker, J.L. *Industrial Medicine and Surgery*, 847-856, 1965.

Landsmeer, J.M.F. Power grip and precision handling. *Annals Rheumatoid Diseases*, 21, 164-170, 1962.

LeVeau, B. *Biomechanics of Human Motion*. Baltimore: Williams and Lissner, 1977.

Marras, W.S. Towards an understanding of dynamic variables in ergonomics. *Occupational Medicine: State of the Art Reviews*, Vol. 7, No. 4, 655-677, 1992.

Marras, W.S. and Sommerich, C.M. A three-dimensional motion model of loads on the lumbar spine: I. Model structure. *Human Factors*, 33(2), 123-137, 1991a.

Marras, W.S. and Sommerich, C.M. A three-dimensional motion model of loads on the lumbar spine: II. Model validation. *Human Factors*, 33(2), 139-149, 1991b.

Merriam, J.L. and Kraige, L.G. *Engineering Mechanics: Dynamics*, Second edition, John Wiley & Sons, 1984.

Moore, J.S. and Garg, A. State of the art reviews: Ergonomics: low-back pain, carpal tunnel syndrome, and upper extremity disorders in the workplace. *Occupational Medicine*, Vol. 7, No. 4, 1992.

Moore, A.E. A system to predict internal load factors related to the development of cumulative trauma disorders of the carpal tunnel and extrinsic flexor musculature during grasping. Master's thesis, University of Waterloo, Waterloo, Canada, 1988.

Moore, A., Wells, R., and Ranney, D. Quantifying exposure in occupational manual tasks with cumulative trauma disorder potential. *Ergonomics*, 34, 1433-1453, 1991.

Nirschl, R. Muscle and tendon trauma: tennis elbow. Chapter 28 in *The Elbow and Its Disorders*, W.B. Saunders Co., 1985.

Ogata, K. and Naito, M. Blood flow of peripheral nerve effects of dissection, stretching, and compression. *The Journal of Hand Surgery*, 11B(1), 10-14, 1986.

Paget, J. *Lectures in Surgical Pathology*, Philadelphia: Lindsay & Blakiston, p. 42, 1860.

Phalen, G.S. The carpal tunnel syndrome: seventeen years' experience in diagnosis and treatment of six hundred fifty-four hands. *Journal of Bone and Joint Surgery*, 48-A(2), 211-228, 1966.

Phalen, G.S. The birth of a syndrome, or carpal tunnel revisited, Guest Editorial. *Journal of Hand Surgery*, 109-110, 1981.

Phalen, G.S. and Kendrick, J.I. Compression neuropathy of the median nerve in the carpal tunnel, *JAMA*, 164:524-530, 1957.

Putz-Anderson, V., Editor. *Cumulative Trauma Disorders: A Manual for Musculoskeletal Diseases of the Upper Limbs*. Taylor & Francis, 1988.

Quinnel, R.C. Conservative management of trigger finger. *The Practitioner*, 224, 187-190, 1980.

Rasch, P.J. *Kinesiology and Applied Anatomy*, 7th edition, Lea and Febiger, 1989.

Rempel, D. and Horie, S. Effect of wrist posture during typing on carpal tunnel pressure, in *Proceedings of Working With Display Units: 4th International Scientific Conference*, Milan, Italy, p. C27, 1994.

Rigby, B.J., Hirai, N., and Spikes, J.D. The mechanical properties of rat tail tendon. *J Gen Physiol*, 43:265-283, 1959.

Robbins, H. Anatomical study of the median nerve in the carpal tunnel and the etiologies of carpal tunnel syndrome. *Journal of Bone and Joint Surgery*, 45A, 953-966, 1963.

Rowe, M.L. *Orthopaedic Problems at Work,* Perinton Press, 1985.

Schoenmarklin, R.W. and Marras, W.S. A dynamic biomechanical model of the wrist joint, in *Proceedings of the 34th meeting of the Human Factors Society,* Orlando, FL., 805-809, 1990.

Schoenmarklin, R.W. and Marras, W.S. Dynamic capabilities of the wrist joint in industrial workers, *International Journal of Industrial Ergonomics,* 11, 207-224, 1993.

Schoenmarklin, R.W. and Marras, W.S. An EMG-assisted biomechanical model of the wrist joint. *Advances in Industrial Ergonomics and Safety IV,* edited by Kumar, S., 777-781, Taylor & Francis Publishers, 1992.

Schivde, A.J., Dreizen I., and Fisher, M.A. The carpal tunnel syndrome: A clinical-electrodiagnostic analysis. *Electromyogr Clin Neurophysiol,* 21, 143-153, 1981.

Schuind, P.F., Garcia-Elias, M., Cooney, W.P., and An, K.N. Flexor tendon forces: *In vivo* measurements. *Journal of Hand Surgery,* 17A(2), 291-298, 1992.

Silverstein, B.A., Fine, L.J., and Armstrong, T.J. Hand wrist cumulative trauma disorders in industry. *British Journal of Industrial Medicine,* 43, 779-784, 1986.

Silverstein, B.A., Fine, L.J., and Armstrong, T.J. Occupational factors and carpal tunnel syndrome. *American Journal of Industrial Medicine,* 11, 343-358, 1987.

Smith, E.M., Sonstegard, D.A., and Anderson, W.H. Contribution of flexor tendons to the carpal tunnel syndrome. *Archives of Physical Medicine and Medical Rehabilitation,* 58, 379-385, 1977.

Sommerich, C.M. Carpal tunnel pressure during typing: Effects of wrist posture and typing speed, in *Proceedings of the Human Factors and Ergonomics Society 38th Annual Meeting,* 611-615, 1994.

Spence, A.P. *Basic Human Anatomy,* Second Edition, The Benjamin/Cummings Publishing Co., 1986.

Stedman, T.L. *Stedman's Medical Dictionary,* 24th edition, Williams and Wilkins, 1982.

Stramler, J.H. *The Dictionary for Human Factors and Ergonomics,* CRC Press, 1993.

Szabo, R.M. and Gelberman, R.H. The pathophysiology of nerve entrapment syndromes. *The Journal of Hand Surgery,* 12A(5), 880-884, 1987.

Tanaka, S. and McGlothlin, J.D. A conceptual model to assess musculoskeletal stress of manual work for establishment of quantitative guidelines to prevent hand and wrist cumulative trauma disorders (CTDs), in *Advances in Industrial Ergonomics and Safety I,* Mital, A., editor, 419-426, Taylor & Francis, 1989.

Van de Graaff, K.M. and Rhees, R.W. *Human Anatomy and Physiology,* Schaum's Outline Series in Biology, McGraw-Hill Book Co., 1987.

Williams, H.J. and Ward, J.R. Musculoskeletal occupational syndromes. Chapter 29 in *Environmental and Occupational Medicine,* Little, Brown, & Co., 1983.

Yamaguchi, D.M., Lipscomb, P.R., and Soule, E.H. Carpal tunnel syndrome. *Minn Med,* 22-23, Jan. 1965.

## For Further Information

Although replete with medical terminology, an excellent reference for the biomechanical pathogenesis of cumulative trauma disorders of the upper extremity are chapters on the muscle–tendon unit and carpal tunnel syndrome in J.S. Moore's and A. Garg's *State of the Art Reviews: Ergonomics: Low- Back Pain, Carpal Tunnel Syndrome, and Upper Extremity Disorders in the Workplace.*

A good discussion of biomechanical models of the wrist is presented in Section 6.5.2 of *Occupational Biomechanics, 2nd Edition* by Don B. Chaffin and Gunnar B.J. Andersson. This book also has a chapter on handtool design (Chapter 5).

Basmajian's *Primary Anatomy, 8th edition* provides easy-to-read and well-illustrated depictions of the body's soft tissues (muscles, nerves, and tendons) and their structures and functions.

The *Clinical Mechanics of the Hand* by Paul W. Brand is particularly helpful in describing the function of specific muscles and tendons in the hand and forearm and how these interact to configure the hand in common pinch and grip postures.

The *Work-Related Musculoskeletal Disorders (WMSDs): A Reference Book for Prevention* (edited by Kuorinka and Forcier) is an excellent book that describes specific WMSDs and provides evidence for the association between work-related activity and each respective WMSD.

# 4

# Occupational Risk Factors for Shoulder Disorders

Eira Viikari-Juntura
*Finnish Institute of Occupational Health*

## 4.1  Introduction

This chapter deals with occupational risk factors of shoulder disorders. The shoulder is structurally and functionally intimately linked with the neck, and often neck–shoulder disorders have been dealt with as one group of disorders. In this chapter, the main emphasis is on the disorders of shoulder structures, although some overlap to the neck area cannot be avoided.

The shoulder is a complex system of bones, muscles, tendons, and ligaments that attach the upper extremity to the torso. The glenohumeral joint is the joint with the largest range of motion in the human body, allowing large mobility for the upper extremity and enabling the body to reach far in all directions. The primary function of the shoulder is to direct and support the hand in its activities.

Because of its supporting function, high forces are imposed on the shoulder, especially if the hand is holding a heavy object. Due to the long moment arm of the extended upper arm, fairly light objects, weighing about 1 kg, impose high mechanical stress on the shoulder. To be able to withstand such forces great stability is demanded of the structures of the shoulder. On the other hand, the long moment arm of the upper arm and the large mobility and relatively poor protection of the shoulder joint render various tissues liable to injuries associated with falls and other sudden movements. Such injuries may heal only partially and decrease the strength and stability of the structures permanently.

Physical load factors occurring at work associated with various shoulder disorders include manually strenuous activities, postural factors of the arm and torso, static work, repetitive work, lack of rest pauses, vibration from handheld tools, environmental factors, and work organizational factors. Only a small amount of data exist on which to base any reference values for acceptable load intensities, frequencies, and durations of such factors. Because of the great liability of the shoulder to acute injuries, nonoccupational

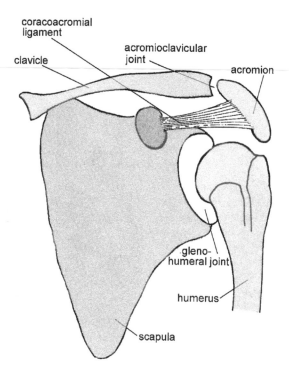

**FIGURE 4.1**   Bony structures in the shoulder, front view.

activities, especially hand-intensive sports, should be considered when investigating the etiology of shoulder pain in individual workers.

Due to the complexity of the shoulder, it has long been difficult to estimate the stresses to the different parts in the shoulder complex when exposed to various physical load factors. Recent advances in biomechanical modeling have markedly increased our knowledge of the stresses on the different structures of the shoulder and also of the capacity of shoulder structures.

## 4.2   Structure and Function of the Shoulder

### Bony Structures

The most important bones in the shoulder are the shoulder blade (scapula), the humerus, and the clavicle (Figure 4.1). The humerus is attached to the scapula by the glenohumeral joint at the lateral aspect. It is in this joint that the primary motion occurs during movements of the arm. The clavicle is attached to the acromion in the upper lateral aspect of the shoulder, and the other end attaches to the sternum. The scapula covers part of the dorsal aspect of the rib cage.

### Muscles and Tendons

The prime movers of the shoulder are the deltoid and four so-called rotator cuff muscles. The deltoid has its origin at the lateral part of the clavicle, the acromion process of the scapula, and the back of the scapula and inserts at the lateral aspect of the humerus. The rotator cuff muscles have their origin in the scapula and insert at the head of the humerus. Before the insertion, their tendons merge around the head of the humerus, thereby forming the rotator cuff (Figure 4.2). The trapezius is a flat muscle on the surface that has its origin at the occiput, neck, and thoracic vertebrae and inserts at the clavicle, acromion process of the scapula and spine of the scapula.

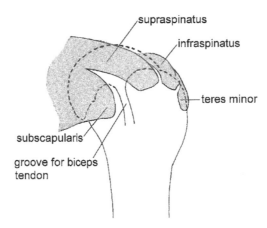

**FIGURE 4.2**   Insertion of rotator cuff tendons at the head of humerus, front view.

## Nerves and Vessels Around the Shoulder

In addition to the nerves and vessels that supply the shoulder muscles and bones, all vessels and nerves supplying the arm and hand pass by the shoulder, forming a neurovascular bundle at the so-called thoracic outlet area. Due to postural factors, structural anomalies, and tight bands of the muscles at the thoracic outlet, the course of these latter nerves and vessels may be interfered with.

## Movements of the Shoulder

The upper arm has a wide range of motion in all three planes: sagittal, transversal, and horizontal. The movement in the sagittal plane is the flexion–extension movement, the movement in the transversal plane is the abduction–adduction movement; and the movement in the horizontal plane is the horizontal abduction and adduction (Figure 4.3). Moreover, the arm has a wide range of rotation around its longitudinal axis. Most work activities demand varying degrees of flexion and abduction of the shoulder, often with the forearm in flexion at the elbow.

## Loading of Shoulder Muscles in Different Activities

Shoulder muscle loading may be estimated by recording the electrical activity of the muscles by electromyography and measuring the intramuscular pressure. The trapezius, deltoid, supraspinatus, and infraspinatus muscles have been measured most commonly. The abduction movement in the plane of the scapula creates more loading than the flexion movement, especially for the supraspinatus muscle. The intramuscular pressure shows a linear increase with increasing abduction or flexion angle. In the supraspinatus, a flexion or abduction angle of 30° is enough to raise the intramuscular pressure above the level where muscle blood flow is impeded. Adding hand load increases the intramuscular pressure markedly. The increase is somewhat higher in shoulder abduction than in flexion. Flexing the elbow by 90° reduces the intramuscular pressure by about 30% in shoulder abduction.

## Mechanisms of Injury in the Shoulder

Shoulder disorders may have their pathological process in the muscles, tendons, nerves, or joints. Commonly painful muscles are the trapezius, especially the descending part, supra- and infraspinatus, and the levator scapulae. *Tendinitis of the rotator cuff tendons* (e.g., *supraspinatous or infraspinatous tendinitis*) is the most common tendon disorder at the shoulder. In the *thoracic outlet syndrome*, the nerves and/or vessels of the neurovascular bundle are compressed at the thoracic outlet. The joints of the shoulder differ as to how prone they are to *degenerative changes*. In the glenohumeral joint, degenerative changes are uncommon, whereas in the acromioclavicular joint, such changes occur much more frequently.

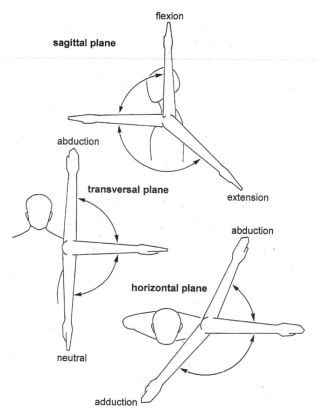

**FIGURE 4.3**   Movements of the shoulder in the three planes.

   As described above, flexion and abduction of the arm increase the intramuscular pressure, which in turn interferes with blood circulation and may produce local ischemia and cause muscle pain. Other possible mechanisms for muscle pain are disturbances in energy metabolism and mechanical failure, especially after heavy physical exercise.

   Rotator cuff tendinitis is usually associated with degenerative changes in the tendons of the rotator cuff. Such changes may be caused by impaired circulation in the tendon due to high tension in the muscle. When the hand is exposed to local low-frequency vibration, the muscles of the arm contract involuntarily as a consequence of the tonic vibration reflex. The tendons of the rotator cuff may also be mechanically injured due to compression under the acromion and the coracoacromial arch. The tendons may also be injured in falls and other accidents. A degenerated tendon is more likely to tear in an accident than is a healthy tendon.

## 4.3   Occupational Risk Factors

Knowledge of occupational risk factors for shoulder disorders is based on epidemiological studies in the field and experimental studies in the laboratory. Epidemiological studies have investigated the associations between physical load factors and clinically defined shoulder tendinitis, radiographically assessed degenerative joint disease, or reported shoulder pain. In many epidemiological studies, certain occupations have been selected to represent a certain type of work, e.g., welders for overhead work. In some other studies, exposure assessment has been based on self-assessment. Only rarely has the validity of such self-assessment been investigated. This means that the information on physical load factors in the epidemiological studies is usually crude and may have considerable inaccuracy and even bias. Only exceptionally have direct measurements been carried out.

In experimental studies, the intensity, frequency, and duration of the exposure can be determined and measured with high precision, but the outcome is different from that in the epidemiological studies. In addition to subjectively assessed discomfort, various physiological responses have been measured, such as myoelectrical activity of muscles, intramuscular pressure, and blood metabolites. Only some data exist on the associations between different types of physiological responses and the development of shoulder disorders. Based on epidemiological and experimental evidence, ten work-related risk factors may be recognized:

- Heavy physical work
- Manual handling
- Elevated postures of the arm
- Nonneutral trunk postures
- Static postures
- Repetitive work
- Lack of pauses
- Vibration
- Draft
- Work organizational factors

Each of these risk factors will be discussed below. A compilation of selected epidemiological studies (Table 4.1), studies in which various physiological responses have been measured during real or simulated work (Table 4.2), and studies that have measured various physiological responses during basic movements and loading situations of the shoulder in the laboratory (Table 4.3) serve as a source of original data for the text. In all studies, various physical load factors (exposure) have been the independent variables under study. The outcome or health effect has usually been more long term by nature in the epidemiologic studies and a short-term response in the experimental studies.

## Heavy Physical Work

Several studies have shown an association between heavy physical work and shoulder problems. The association has been found for shoulder tendinitis (Stenlund et al., 1993), radiographically defined acromioclavicular arthrosis (Stenlund et al., 1992), and reported shoulder pain (Viikari-Juntura et al., 1993). An increased risk of acromioclavicular arthrosis was observed for 10 to 28 years of manual work, and an even higher risk for more than 28 years of manual work, i.e., an exposure–response relationship was observed for cumulative exposure to manual work. Examples of occupational groups are rockblasters, bricklayers, and various jobs in forestry.

Heavy physical work may involve manual handling of heavy loads, nonneutral trunk postures, and elevated postures of the arm. Such work may also be associated with repeated minor traumas and a risk of major trauma.

## Manual Handling

Manual handling of loads has been associated with shoulder disorders in some studies. Wells et al. (1983) found a higher prevalence (13%) of recurrent shoulder pain for letter carriers than for meter readers (7%) and postal clerks (5%). The maximum bag weight of the letter carriers was 11.4 kg. Letter carriers whose maximum bag weight had been increased by 4.4 kg had a prevalence of 23% of recurrent shoulder pain. Stenlund et al. (1992) determined life-long lifting as lifted tonnes, and found an increased risk of radiographically assessed osteoarthrosis of the acromioclavicular joint for 710 to 26,000 tonnes, and an even higher risk for more than 26,000 tonnes. This means that an exposure–response relationship was also found for lifted load and acromioclavicular joint arthrosis. Unfortunately, it is not known whether intensive lifting for a short period is associated with a different risk than less intensive lifting for a longer period. It should also be noted that the data on lifting were based on questionnaires and interviews with no assessment of validity.

**TABLE 4.1**   Selection of Epidemiological Studies on Shoulder Disorders

| Study Population | Type of Study | Outcome | Exposure | Validity of Exposure Assessment Method | Adjustment for Confounders | Findings | Reference |
|---|---|---|---|---|---|---|---|
| 17 cases (all men) with shoulder pain for at least 3 months seeking medical advice, 34 age, sex, and workplace matched controls | Case-control | Prolonged shoulder pain (clinical signs of shoulder tendinitis in 12 cases) | Physical heaviness of work, shoulder load (work height of arms), based on interview of the cases and referents | Not reported | Not necessary (matched design) | No difference between cases and controls in physical heaviness of work 11/17 cases and 5/34 controls performed their work at or above acromion height | Bjelle et al., 1979 |
| 152 female assembly-line packers 133 female shop assistants | Cross-sectional | Humeral tendinitis (supraspinatous or bicipital tendinitis; clinically verified) | Job title (packer, shop assistant) used in analysis Assembly-line: repetitive motions up to 2,500/day; static muscle work; extreme work postures of the hands; varying amount of lifting | Not necessary | Not performed | Prevalence of humeral tendinitis 9.2% among the packers and 3.8% among the shop assistants | Luopajärvi et al, 1979 |
| 131 shipyard welders with more than 5 years of welding experience 56 male office clerks above 40 years of age | Cross-sectional + case-control among welders | Supraspinatus tendinitis (clinically verified) | Job title (welding work) Rating of shoulder load by expert | Not reported | Not performed in cross-sectional study Not necessary in case-control study (matched design) | Estimated prevalence of supraspinatus tendinitis 18.3% among welders. Uncertainty of true occurrence of disease because questionnaire and clinical study 1 year apart and only symptomatic were examined. In case-control study, tendency toward higher load among those with supraspinatus tendinitis | Herberts et al., 1981 |

| 92 letter carriers with 11.4 kg maximum weight, 104 letter carriers with 15.8 kg maximum weight (weight increase occurred on an average of 5 months prior to study), 76 gas meter readers, 127 postal clerks | Cross-sectional | Occurrence of and disability due to shoulder pain by telephone interview | Maximum weight carried at work: either 11.4 or 15.8 kg | Not necessary | Age, body mass index, years in current job, previous heavy work | Prevalence of "significant" shoulder problems 23, 13, 7, and 5% among letter carriers with weight increase, letter carriers without weight increase, meter readers, and postal clerks, respectively. Shoulder problems associated with weight carried. | Wells et al., 1983 |
| 96 female employees in the electronics industry | Cross-sectional | Neck, shoulder, and arm disorders (clinically verified) and symptoms (during preceding 12 months) | Observations from video taken from two projections (individually for each worker): work cycle time, number and duration of rest periods (>2 s) for arm, shoulder, and head, duration of different upper arm and head postures, frequency of changes between postural categories | Not reported | Age, anthropometry, maximum muscle strength and endurance, work history, leisure physical activity, hobbies, perceived psychological stress at work, work satisfaction, number of breaks and rest pauses at work, individual productivity | Long duration of employment in the electronics industry and % of work cycle time with upper arm abducted 0 to 30° associated with symptoms in the shoulder during preceding 12 months (slight or more severe). High stature and a high number of upper arm flexions per hour inversely related to shoulder symptoms. | Kilbom et al., 1986 |
| 54 bricklayers, 55 rock blasters, 98 construction foremen (all men) | Cross-sectional | Radiographical arthrosis of the acromioclavicular joints | Job title, life-long lifted load, life-long exposure to vibration, years of manual work based on questionnaire and interview | Not reported | Age, dexterity, smoking | Prevalence of osteoarthrosis in the right vs. left shoulder: rockblasters 61.8%; 56.4% bricklayers 59.3%; 40.7% foremen 36.7%; 23.4% Increased risk for years of manual work, cumulative lifted load and vibration (exposure-response relationship for all) | Stenlund et al., 1992 |

**TABLE 4.1 (continued)**     Selection of Epidemiological Studies on Shoulder Disorders

| Study Population | Type of Study | Outcome | Exposure | Validity of Exposure Assessment Method | Adjustment for Confounders | Findings | Reference |
|---|---|---|---|---|---|---|---|
| As above | As above | Shoulder tendinitis (rotator cuff tendinitis; clinically verified) | As above | Not reported | Age, dexterity, smoking, sports activities | Prevalence of rotator cuff tendinitis in the right vs. left shoulder: rockblasters 23.6%; 14.5% bricklayers 1.8%; 1.8% foremen 9.2%; 3.1% Increased risk for cumulative exposure to vibration | Stenlund et al, 1993 |
| 157 cases and 2,565 controls (incidence study) in a forest industry company, 352 cases and 122 controls (persistence study) | 1-year follow-up | Incidence of severe shoulder pain, persistence of severe shoulder pain | Trunk and shoulder postures and movements, physical heaviness of work, work characteristics (self-administered questionnaire) | Fair for most items | Sex, age, body mass index, smoking, sleep disorder | Frequent twisting of the torso associated with incidence of severe shoulder pain Physically heavy work and mental overload at work associated with persistently severe shoulder pain | Viikari-Juntura et al, 1993 |
| 52 female orchard farmers | Two cross-sections | Prevalence of neck and shoulder stiffness and pain; pain in joint motion and muscle tenderness in the neck and shoulder | Pear bagging (2,455/day) in the first cross-section, apple bagging (1,887/day) in the second cross-section. Upper arm flexion >90° for 75% of time in pear bagging and 41% of time in apple bagging (based on goniometer measurements of one worker) | Not reported | Not necessary, because the same subjects were investigated at two time points | Neck and shoulder stiffness, neck pain, shoulder muscle tenderness, and pain in movements of the neck more prevalent during pear bagging than apple bagging. It is not, however, clear whether the higher prevalence of pain could be attributed to higher work pace or higher work height, or both when bagging pears. | Sakakibara et al, 1995 |

**TABLE 4.2** Selected Studies on Physiological Responses During Real or Simulated Work Tasks Involving Activation of Shoulder Muscles

| Study Group | Task | Physiological Response | Muscles | Findings* | Reference |
|---|---|---|---|---|---|
| 41 female office workers (nonsymptomatic and symptomatic) | Various keyboard activities in the field: conventional typewriter, telex machine, visual display terminal, telephone exchanger, insurance calculating machine | Myoelectrical activity in forearm and shoulder muscles by surface electrodes | Trapezius, deltoid, rhomboideus, forearm extensors | The activities of the right trapezius ranged from 10 to 30% of maximal activity. Static activity in the trapezius measured also at times when the forearm was not active. | Onishi et al., 1982 |
| 25 assembly workers: 16 women and 9 men (nonsymptomatic and symptomatic) | Light assembly and soldering tasks in the field | Myoelectrical activity of shoulder muscles by surface electrodes | Anterior deltoid, upper trapezius, infraspinatus | High static activity levels of about 7–14% of maximal activity in all muscles; median activity levels high (about 16–20%) in infraspinatus and trapezius; acceptable maximum contraction levels (about 30%) in all muscles. | Christensen, 1986 |
| 6 healthy female secretaries | Wordprocessing: 5h, 3h, and 3h work period with 15 s pauses every 6th minute in the field | Myoelectrical activity by surface electrodes, discomfort in the eyes, neck, shoulders, elbows, hands, back | Upper trapezius | Median of static load 3.2% and 3.0%, of median load 7.3% and 5.5%, and of peak load 11.7% and 9.2% of maximal activity on right and left side, respectively. No difference in electrical activity between the three work periods. Discomfort ratings slightly lower after work with rest pauses than without rest pauses. | Hagberg and Sundelin, 1986 |
| 8 healthy female cashiers | Simulation of cash register operations in the laboratory using conventional keyboard, horizontal scanner, vertical scanner, and pen reader in the sitting posture. Horizontal scanner used also in the standing posture. | Myoelectrical activity of neck and shoulder muscles by surface electrodes | Upper trapezius, cervical erector spinae, levator scapulae, thoracal erector spinae/rhomboidei, infraspinatus | Static load between 2–5% of maximal activity; median load below 10%; peak load below 30% in all muscles. | Lannersten and Harms-Ringdahl, 1990 |

**TABLE 4.2 (continued)**   Selected Studies on Physiological Responses During Real or Simulated Work Tasks Involving Activation of Shoulder Muscles

| Study Group | Task | Physiological Response | Muscles | Findings* | Reference |
|---|---|---|---|---|---|
| 6 healthy female physiotherapy students | Work simulation in the laboratory of grasping a cylinder (weight 15 g) and releasing it through a hole in the table, paced as 2,466 cycles per hour (MTM-110). | Myoelectrical activity by surface electrodes | Lateral and cervical parts of the upper trapezius, infraspinatus | Static and peak loads were 4.4 and 31%, 17 and 37%, and 12 and 55% of the maximal activity for the lateral and cervical part of the upper trapezius, and the infraspinatus, respectively. Signs of fatigue seen in all subjects. The localization and time for these signs varied between different subjects. | Sundelin and Hagberg, 1992 |
| 12 healthy female secretaries | Wordprocessing task in the field with 3 different air velocities (0.96, 1.4, 1.96 m/s) | Myoelectrical activity by surface electrodes | Cervical and lateral part of upper trapezius, levator scapulae | Increased myoelectrical activity (increase in both mean power frequency and root-mean-square amplitude) in the lateral part of upper trapezius, possibly due to reflex recruitment of motor units. In the cervical part of upper trapezius, increase of root-mean-square amplitude and a decrease in mean power frequency, possibly due to cooling of the muscle. | Sundelin and Hagberg, 1992 |
| 12 healthy female physiotherapy students | 2-hour work simulation in the laboratory of grasping a cylinder and releasing it through a hole: one hour continuously (MTM-110), and another hour with 1-minute pause every 6th minute, consisting of lifting of a 2-kg box on a shelf slowly 5-6 times (MTM-132 for the repetitive task) | Myoelectrical activity by surface electrodes, rating of discomfort | Lateral and cervical parts of the upper trapezius, infraspinatus | Electromyographic signs of fatigue less pronounced with pause activities than without. Fatigue patterns lower during the second hour, indicating adaptation to work. Discomfort ratings did not differ between work with and without pauses and were higher during the second hour. | Sundelin, 1993 |

* Activity figures from different studies, given as % of maximal activity, can be compared only roughly, due to different calibration procedures in different studies.

**TABLE 4.3** Selected Studies on Physiological Responses Associated with Controlled Postures and Loading of the Shoulder Muscles in the Laboratory

| Study Group | Task | Physiological Response | Muscles | Findings* | Reference |
|---|---|---|---|---|---|
| 6 healthy female students | Repetitive shoulder flexions 0–90° with hand loads between 0 and 3.1 kg. Repetition rate 15/min, duration of each trial 60 min | Myoelectrical activity of shoulder muscles, tenderness of shoulder muscles, clinical tests for shoulder tendinitis, heart rate, rating of perceived exertion | Upper trapezius, anterior deltoid, biceps brachii | Peak load ranged between 13 and 60% of maximal activity. Ratings of perceived exertion increased more than heart rate, indicating the importance of local factors. After two days, tenderness in the descending part of the trapezius and supraspinatus in all subjects, signs of shoulder tendinitis in 2 subjects. | Hagberg, 1981 |
| 9 healthy men; 5 with welding experience | Laboratory experiment on arm posture and hand tool weight (upper arm flexion and abduction, elbow flexion, and upper arm rotation in 21 combinations. Hand load 0, 1, and 2 kg) | Myoelectrical activity of shoulder muscles by intramuscular electrodes | Anterior, middle, and posterior deltoid, supraspinatus, infraspinatus, upper trapezius | Degree of upper arm elevation most important determinant of shoulder muscle load. Muscles of the rotator cuff more hand-load dependent than the deltoid muscle. | Sigholm et al., 1984 |
| 10 healthy females (skilled in assembly work) | Laboratory experiment on different sitting postures and tilting the work object from the horizontal plane. Experimental work cycle of simulated soldering at an electronics plant. | Myoelectrical activity of neck and shoulder muscles | Cervical erector spinae/trapezius, upper trapezius, middle trapezius/supra-spinatus, thoracal erector spinae/rhomboidei, levator scapulae, sternocleidomastoid | Sitting posture with the spine slightly tilted backward and the cervical spine vertical was associated with the lowest activity in posterior neck and shoulder muscles. The posture with the whole spine straight and vertical resulted in a higher myoelectrical activity, and the posture with the whole spine flexed was associated with the highest myoelectrical activity. A backward inclination of the spine generally requires that the work object should be tilted from the horizontal plane. | Schüldt et al., 1986 |
| 10 healthy females (skilled in assembly work) | Laboratory experiment on the effects of arm support and suspension on myoelectrical activity of posterior neck and shoulder muscles in different sitting postures and tilting the work object from the horizontal plane. Simulated work cycle of soldering at an electronics plant. | Myoelectrical activity of neck and shoulder muscles | Cervical erector spinae/trapezius, upper trapezius, middle trapezius/supra-spinatus, thoracal erector spinae/rhomboidei, levator scapulae, sternocleidomastoid | Both arm support and suspension reduced the myoelectrical activity of the muscles in the whole-spine-straight-and-vertical posture. Activity values were in most cases below 10% of maximal activity. Arm support suspension might be more efficient than arm support in the posture with the trunk slightly inclined backward. Arm support might be more efficient than arm suspension in the whole-spine-flexed posture which even with support was associated with relatively high activities. | Schüldt et al., 1987 |

**TABLE 4.3 (continued)**    Selected Studies on Physiological Responses Associated with Controlled Postures and Loading of the Shoulder Muscles in the Laboratory

| Study Group | Task | Physiological Response | Muscles | Findings* | Reference |
|---|---|---|---|---|---|
| 12 healthy students | Laboratory experiment on arm posture and hand tool weight (upper arm flexion and abduction, elbow flexion in 15 combinations; hand load 0, 1, and 2 kg) | Intramuscular pressure | Supraspinatus | High intramuscular pressure in abductions ≥30° with and without elbow flexion and with and without shoulder load. Hand load increased intramuscular pressure in most postures. | Järvholm et al., 1988 |
| 6 healthy females | Continuous and intermittent holding of the arm horizontally at 60° to the sagittal plane (cycle time 10, 60, and 360 s and duty cycle 0.33, 0.50, 0.67, and 0.83) | Myoelectrical activity by surface electrodes, arterial blood pressure, heart rate, perceived fatigue during exercise; venous blood potassium, lactate, and ammonia pre- and postexercise; maximal voluntary contraction, pressure pain threshold, proprioceptive performance, and 1-minute arm holding at 25% MVC before and up to 4 h postexercise | Upper trapezius | Duty cycle influenced all variables, cycle time only blood pressure and fatigue perception. Cardiovascular and neuromuscular recovery incomplete for hours. Ranking of protocols differed according to the criterion variable | Mathiassen, 1993 |
| 72 healthy subjects with assembly work experience, 35 men and 37 women | Simulation of a repetitive assembly job by a work simulator, consisting of lifting and lowering of a tool handle and striking a metal pointer to a plate at the end of each excursion (in the sitting posture). Repetition rate (20, 24, 35 lifts/min), required force (10, 20, 30% of maximal activity), tool weight (2.1, 2.5, 3.0 kg) and reach height (109, 120, 131 cm) were varied. | Duration of work until given rating of perceived muscle discomfort was reached | Muscles in the neck, shoulder, or upper arm | Increase in force, repetition rate, tool weight, and height of upper target all reduced work trial duration, repetition rate and force having the largest effects. Interaction between repetition rate and force, so that an increase in each variable led to an attenuation of the other variable's effect. | Putz-Anderson and Galinsky, 1993 |

* Activity figures from different studies, given as % of maximal activity, can be compared only roughly, due to different calibration procedures in different studies.

Manual handling activities impose high loads on probably all shoulder structures of which the rotator cuff muscles and the deltoid have been the most investigated (Sigholm et al., 1984; Järvholm et al., 1988). Hand load has a strong effect on shoulder muscle activity, especially in the rotator cuff muscles, and on intramuscular pressure in the supraspinatus. Manual handling activities may also run a risk of trauma.

## Elevated Postures of the Arm

There is some epidemiological evidence to support an association between elevated postures of the arm and shoulder pain (Bjelle et al., 1979; Sakakibara et al., 1995) as well as supraspinatus tendinitis (Herberts et al., 1981). The occupations involved in the studies have been shipyard welders and orchard farmers. Experimental studies have shown that the activity of shoulder muscles increases with increasing elevation (flexion and abduction) of the arm (Sigholm et al., 1984). A flexion angle of $\geq 30°$ without hand load raises the intramuscular pressure at a level where blood circulation is disturbed (Sigholm et al., 1984; Järvholm et al., 1988). After longer periods of intensive shoulder muscle exercise, cardiovascular and neuromuscular recovery may be incomplete for hours (Mathiassen, 1993). Elevated arm postures may also be associated with mechanical irritation of the rotator cuff tendons under the acromion and cora-coacromial arch.

The loads imposed on shoulder structures in various activities with elevated arms may be decreased by suspending the arms. The results from the simulated work of welders showed that arm suspension reduced shoulder muscle load, but the intramuscular pressure of the supraspinatus remained at a level where muscle blood flow would still be compromised (Järvholm et al., 1991).

## Nonneutral Trunk Postures

In sedentary work, the workplace layout largely determines the posture of the torso, neck, and limbs. Schüldt et al. (1987) investigated the myoelectrical activity of several neck and shoulder muscles in different postures during simulated soldering work in the laboratory. Sitting with the spine slightly tilted backward and the cervical spine vertical was associated with the lowest activity. The posture with the whole spine straight and vertical resulted in a higher myoelectrical activity, and the posture with the whole spine flexed was associated with the highest activity. In the latter experiment, the work object was horizontal in the whole-spine-flexed posture, tilted 35° in the whole-spine-straight-and-vertical posture, and tilted 75° in the posture with the spine tilted backward and neck vertical. As stated by these authors, a backward inclination of the spine generally requires that the work object should be tilted from the horizontal plane.

## Static Work

According to epidemiological studies, shipyard welders (Herberts et al., 1981), orchard harvesters (Sakak-ibara et al., 1995), packers (Luopajärvi et al., 1979), garment workers (Punnett et al., 1985), workers in light assembly tasks (Kvarnström, 1983), and office workers with intensive use of the mouse (Hagberg and Karlqvist, 1994) have shown a high risk for shoulder disorders. Common to the tasks in these occupations is static exertion of shoulder muscles with or without elevation of the arm.

Measurements of myoelectrical activity in the field and simulated work in the laboratory have shown static activity levels ranging from 4 to 17% of maximal activity in different shoulder muscles (Christensen, 1986; Sundelin and Hagberg, 1992), the upper range being far above the 2 to 5% of maximal activity recommended by Jonsson (Jonsson, 1982). A simulation of cash register operation in the laboratory (Lannersten and Harms-Ringdahl, 1990) and word processing operations, both with spontaneous and forced pauses (Hagberg and Sundelin, 1986), was associated with a lower static load of 2 to 5% of maximal activity.

Measurements of various keyboard activities in the field showed activity in the trapezius also at times when no activity was performed by the forearm (Onishi et al., 1982). The importance of interruptions of activity was shown in a follow-up of workers at a chocolate manufacturing plant. The workers who had a smaller number of short unconscious interruptions of electromyographic activity (so-called emg-gaps) had a higher risk of contracting trapezius myalgia than those with a higher number of emg-gaps during a follow-up time of six months (Veiersted et al., 1993).

The load of the shoulder muscles may be reduced by arm support or arm suspension in activities involving static exertion of shoulder and arm muscles. A laboratory experiment of simulated soldering work in different postures of the trunk and neck suggested that arm suspension might be more efficient than arm support in the posture with the trunk slightly inclined backward. Arm support might be more effective than arm suspension in the whole-spine-flexed posture. This latter posture, however, showed generally high myoelectrical activities and should not be adopted for longer periods of work (Schüldt et al., 1987).

## Repetitive Work

A typical pattern of repetitive work is that the fingers perform quick movements while the shoulder muscles perform mostly static exertions to fulfil their primary task in supporting the arms. Word processing, packing, and light assembly tasks are examples of such repetitive work, and high risk for shoulder disorders has been shown in these tasks. In a simulation of assembly work in the laboratory, the introduction of a one-minute pause with dynamic lifting activity every 6th minute resulted in less pronounced myoelectrical signs of fatigue in the shoulder muscles, suggesting that dynamic activity might be effective in counteracting the effects of static loading of shoulder muscles (Sundelin, 1993). No difference was seen in discomfort ratings, however.

Certain work tasks may demand repetitive movements of the upper arm, but the health effects of repetitive arm flexions or abductions are not well known. In an experimental study in which six healthy students performed repetitive shoulder flexions with a frequency of 15/min for 60 minutes with loads varying from 0 to 3.1 kg, all subjects had tenderness in the descending part of the trapezius and supraspinatus, and two of them also had other signs of supraspinatus tendinitis after two days (Hagberg, 1981). A study among electronics assembly workers showed that the duration of mild upper arm flexion was associated with shoulder disorders, but the number of upper arm flexions was inversely related to shoulder disorders (Kilbom et al., 1986). This suggests that a more dynamic working style decreases the risk of shoulder disorders.

An experimental study used the psychophysical approach to investigate the effects of various repetition rates, forces, tool weights, and reach heights on work durations until a given degree of subjectively rated fatigue was achieved in repeated arm flexions. The repetition rate was the prime determinant for work duration, followed by force, height of upper target, and tool weight. Repetition rate and force showed an interaction, so that increases in each variable led to a slight attenuation of the other variable's effect (Putz-Anderson and Galinsky, 1993). This study is among the few sources of data on which reference values for the frequencies of shoulder elevations with given loads and elevation angles may be established.

## Work-Rest Schedule: Lack of Pauses

It is conceivable that the frequency, duration, and quality of pauses are crucial determinants for the development of fatigue in the muscles during forceful, repetitive, or static exertions. Work–rest schedules have been investigated in some studies in the field (Hagberg and Sundelin, 1986) and in the laboratory. In the aforementioned simulation of light industrial work in the laboratory, an MTM-110 pacing without pauses was compared with an MTM-132 pacing with one-minute pause every 6th minute, consisting of the dynamic lifting activity, the production rate of the task itself being equal in both schemes. As mentioned, electromyographic signs of fatigue were less pronounced with pause activities than without, despite the higher repetition rate and extra work of lifting during the pauses (Sundelin, 1993). In another experimental study, a range of physiological responses was measured when holding the arm continuously and intermittently in the horizontal plane at 60° to the sagittal plane. Duty cycle had a more pronounced effect on the various physiological responses than cycle time. A ranking of the protocols with different combinations of cycle time and duty cycle differed according to which physiological response was used (Mathiassen, 1993).

In conclusion, while the need for pauses is evident in tasks demanding forceful, repetitive, or static shoulder activities, only few data exist upon which to base any recommendation. Moreover, the recommendations will differ depending on what kind of physiological criterion is used. Sometimes objective and subjective criteria seem to contradict each other.

## Vibration

Vibration from hand-held tools has been shown to be associated with both radiographically assessed arthrosis of the acromioclavicular joint (Stenlund et al., 1992) and shoulder tendinitis (Stenlund et al., 1993). For both conditions, an exposure–response relationship between cumulative exposure to vibration and the disease have been observed. The assessment of cumulative exposure took into consideration the hours that each vibrating tool had been used and the energy emission from the tool.

## Draft

High air velocities in the work environment are perceived as draft and traditionally considered to increase neck and shoulder discomfort. Only some epidemiological evidence exists for the association of draft with neck and shoulder pain (Tola et al., 1988). The behavior of shoulder muscles was studied in an experiment with different air velocities in the office environment. The myoelectrical activity changes suggested increased recruitment of motor units in some muscles, and a possible cooling effect in others associated with increasing air velocity (Sundelin and Hagberg, 1992).

## Work Organizational Factors

Demands, control, and social support are work organizational factors that have been most often investigated in association with shoulder disorders. An association between high job demands such as time pressure, high concentration, high work load and shoulder disorders has been shown in many studies. Also low control and little autonomy has been associated with shoulder disorders, but the results concerning social support are conflicting (Bongers et al., 1993). In a follow-up study, mental overload (difficult phases at work, the need to hurry to get work done) was associated with the persistence of severe shoulder pain but not with the incidence of pain, indicating that work organizational factors might have a greater role in the prognosis than in the genesis of shoulder disorders (Viikari-Juntura et al., 1993).

## Summary of Occupational Risk Factors

The text above deals with individual occupational risk factors for shoulder disorders. In real work situations, many risk factors are present simultaneously and may have combined effects on the risk of shoulder disorders. In dynamic work, the overall risk of shoulder disorders is probably a function of arm posture, weight of load, and frequency of repetitions of arm movements. There is evidence that a high cumulative exposure of heavy work increases the risk of shoulder disorders, but whether a certain duration of heavy work per day could be tolerated for longer times is not known. Traumas to the shoulder may increase the risk of shoulder disorders in heavy work. There is convincing evidence of harmful effects of low-frequency vibration from handheld tools to the shoulder. Therefore, low-frequency vibration from tools should be eliminated or kept to the minimum.

In static work tasks with lower force demands, the elevation angle of the arm, the overall posture of the body, and the rest pauses largely determine the loading pattern of shoulder muscles. In this kind of work, optimal body and arm postures should be enabled by proper workplace layout and possibilities to support or suspend the arm according to the preference of the worker.

Work organizational factors may have an effect on the risk of shoulder disorders by influencing the intensity, frequency, or duration of physical load factors. They may also affect the reporting of the disorders or the recovery from them.

## References

Bjelle, A., Hagberg, M., and Michaelsson, G. 1979. Clinical and ergonomic factors in prolonged shoulder pain among industrial workers. *Scand. J. Work Environ. Health* 5(3):205-210.

Bongers, P.M., De Winter, C., Kompier, M.A.J., and Hildebrandt, V.H. 1993. Psychosocial factors at work and musculoskeletal disease. *Scand. J. Work Environ. Health* 19(5):297-312.

Christensen, H. 1986. Muscle activity and fatigue in the shoulder muscles of assembly-plant employees. *Scand. J. Work Environ. Health* 12(6):582-587.

Hagberg, M. 1981. Work load and fatigue in repetitive arm elevations. *Ergonomics* 24(7):543-555.

Hagberg, M. and Karlqvist, L. 1994. Symptoms and disorders related to keyboard and computer mouse use. International Conference on Occupational Disorders of the Upper Extremities, December 1-2, 1994, Miyako Hotel, San Francisco, California.

Hagberg, M. and Sundelin, G. 1986. Discomfort and load on the upper trapezius muscle when operating a wordprocessor. *Ergonomics* 29(12):1637-1645.

Herberts, P., Kadefors, R., Andersson, G., and Petersén, I. 1981. Shoulder pain in industry: An epidemiological study on welders. *Acta Orthop. Scand.* 52(3):299-306.

Jonsson, B. 1982. Measurement and evaluation of local muscular strain in the shoulder during constrained work. *J. Human Ergol.* 11(1):73-88.

Järvholm U., Palmerud, G., Kadefors, R., and Herberts, P. 1991. The effect of arm support on the supraspinatus muscle during simulated assembly work and welding. *Ergonomics* 34(1):57-66.

Järvholm, U., Palmerud, G., Styf, J., Herberts, P., and Kadefors, R. 1988. Intramuscular pressure in the supraspinatus muscle. *J. Orthop. Res.* 6(2):230-238.

Kilbom, Å., Persson, J., and Jonsson, B.G. 1986. Disorders of the cervicobrachial region among female workers in the electronics industry. *Int. J. Ind. Ergonomics* 1(1):37-47.

Kvarnström, S. 1983. Occurrence of musculoskeletal disorders in a manufacturing industry, with special attention to occupational shoulder disorders. *Scand. J. Rehab. Med.* (Suppl. 8):6-101.

Lannersten, L. and Harms-Ringdahl, K. 1990. Neck and shoulder muscle activity during work with different cash register systems. *Ergonomics* 33(1):49-65.

Luopajärvi, T., Kuorinka, I., Virolainen, M., and Holmberg, M. 1979. Prevalence of tenosynovitis and other injuries of the upper extremities in repetitive work. *Scand. J. Work Environ. Health* 5(Suppl. 3):48-55.

Mathiassen, S.-E. 1993. The influence of exercise/rest schedule on the physiological and psychophysical response to isometric shoulder-neck exercise. *Eur. J. Appl. Physiol.* 67(6):528-539.

Onishi, N., Sakai, K., and Kogi, K. 1982. Arm and shoulder muscle load in various keyboard operating jobs of women. *J. Human Ergol.* 11(1):89-97.

Punnett, L., Robins, J.M., Wegman, D.H., and Keyserling, W.M. 1985. Soft tissue disorders in the upper limbs of female garment workers. *Scand. J. Work Environ. Health* 11(6):417-425.

Putz-Anderson, V. and Galinsky, T.L. Psychophysically determined work durations for limiting shoulder girdle fatigue from elevated manual work. *Int. J. Ind. Ergon.* 11(1):19-28.

Sakakibara, H., Miyao, M., Kondo, T., and Yamada, S. 1995. Overhead work and shoulder-neck pain in orchard farmers harvesting pears and apples. *Ergonomics* 38(4):700-706.

Schüldt, K., Ekholm, J., Harms-Ringdahl, K., Németh, G., and Arborelius, U.P. 1986. Effects of changes in sitting work posture on static neck and shoulder muscle activity. *Ergonomics* 29(12):1525-1537.

Schüldt, K., Ekholm, J., Harms-Ringdahl, K., Németh, G., and Arborelius, U.P. 1987. Effects of arm support or suspension on neck and shoulder muscle activity during sedentary work. *Scand. J. Rehab. Med.* 19(2):77-84.

Sigholm, G., Herberts, P., Almström, C., and Kadefors, R. 1984. Electromyographic analysis of shoulder muscle load. *J. Orthop. Res.* 1(4):379-386.

Stenlund, B., Goldie, I., Hagberg, M., and Hogstedt, C. 1993. Shoulder tendinitis and its relation to heavy manual work and exposure to vibration. *Scand. J. Work Environ. Health* 19(1):43-49.

Stenlund, B., Goldie, I., Hagberg, M., Hogstedt, C., and Marions, O. 1992. Radiographic osteoarthrosis in the acromioclavicular joint resulting from manual work or exposure to vibration. *Br. J. Ind. Med.* 49(8):588-593.

Sundelin, G. 1993. Patterns of electromyographic shoulder muscle fatigue during MTM-paced repetitive arm work with and without pauses. *Int. Arch. Occup. Environ. Health* 64(7):485-493.

Sundelin, G. and Hagberg, M. 1992. Electromyographic signs of shoulder muscle fatigue in repetitive arm work paced by the Methods-Time-Measurement system. *Scand. J. Work Environ. Health* 18(4):262-268.

Sundelin, G., and Hagberg, M. Effects of exposure to excessive drafts on myoelectric activity in shoulder muscles. *J. Electromyogr. Kinesiol.* 2:36-41.

Tola, S., Riihimäki, H., Videman, T., Viikari-Juntura, E., and Hänninen, K. 1988. Neck and shoulder symptoms among men in machine operating, dynamic physical work and sedentary work. *Scand. J. Work Environ. Health* 14(5):299-305.

Veiersted, K.B., Westgaard, R., and Andersen, P. 1993. Electromyographic evaluation of muscular work pattern as a predictor of trapezius myalgia. *Scand. J. Work Environ. Health* 19:284-290.

Viikari-Juntura, E., Riihimäki, H., Takala, E.-P., Rauas, S., Leppänen, A., Malmivaara, A., Grönqvist, R., Härmä, M., Martikainen, R., Saarenmaa, K., and Kuosma, E. 1993. Niska-hartiaseudun ja yläraajan oireita ennustavat tekijät metsäteollisuudessa (Factors predicting pain in the neck, shoulders, and upper limbs in forestry work). *Työ ja ihminen* 7(4):233-253 (in Finnish with English summary).

Wells, J.A., Zipp, J.F., Schuette, P.T., and McEleney, J. 1983. Musculoskeletal disorders among letter carriers A comparison of weight carrying, walking & sedentary occupations. *J. Occup. Med.* 25(11):814-820.

## For Further Information

A recent scientific review on occupational risk factors of soft tissue disorders of the shoulder is presented by Sommerich, McGlothlin, and Marras in *Ergonomics* (1993, 36:697-717).

An extensive overview of the work-relatedness of shoulder disorders as well as other disorders of the neck and upper limbs with a practical approach to prevention is presented in Hagberg, Silverstein, Wells et al.'s *Work-Related Musculoskeletal Disorders (WMSDs): A Handbook for Prevention,* Taylor & Francis, 1995.

Results on extensive investigations using electromyographic recordings in simulated sedentary work in different postures have been presented in a supplement (No. 19, 1988) of *Scandinavian Journal of Rehabilitation Medicine* by Kristina Schüldt "On Neck Muscle Activity and Load Reduction in Sitting Postures."

Basic biomechanics of the shoulder are included in *Occupational Biomechanics* by Chaffin and Andersson (John Wiley & Sons, Inc., 1991).

# 5

# Hand Tools:
# Design and Evaluation

Robert G. Radwin
*University of Wisconsin–Madison*

## 5.1   Introduction

This chapter describes specific hand tool design features that help minimize physical stress and maximize task performance in jobs involving the continuous or repetitive use of hand tools. An important objective of ergonomics in the design, selection, installation, and use of hand tools is the reduction of muscle fatigue onset and the prevention of musculoskeletal disorders of the upper limb. It is not just the tool design, but how a tool is used for a specific task and workstation that imparts physical stress upon the tool operator. Consequently, there is no "ergonomic hand tool" *per se*. What makes sense in one situation can produce unnecessary stress in another.

It is generally agreed that physical stress, fatigue, and musculoskeletal disorders can be reduced and prevented by selecting the proper tool for the task. Tools used so that physical stress factors are minimized, such as reducing stress concentrations in the fingers and hands, producing low force demands on the operator, or minimizing shock, recoil, and vibration are usually the best tools for the job. Control of these factors depends on the tool and the specific tool application. Selection of tools should, therefore, be viewed within the context of the specific job being performed.

Tool selection should be based on (1) process engineering requirements, (2) human operator limitations, and (3) workstation and task factors. Some factors considered for each of these requirements are summarized in Table 5.1. A detailed description of each of these factors is contained in Radwin and Haney (1996). Manufacturing engineers often specify the process requirements with little regard for the operator and the workstation. Hand tool selection should therefore consider how the particular task and workstation relate to the capabilities and limitations of the human operator for a particular tool design. The process is not always simple and often involves an iterative approach, considering individual tool design features and their role in augmenting and mitigating physical stress. This chapter will describe some hand tool design features and the research leading to an understanding of how tool design can help reduce physical stress in hand tool operation.

**TABLE 5.1**　Requirements for Ergonomic Hand Tool Selection

| | |
|---|---|
| Process Engineering Requirements | Requirements specified in terms of the production process, such as how fast a drill bit should turn, or how much torque should be applied to a screw being tightened. |
| | Manufacturing process requirements are often based on the product design and parameters needed for accomplishing the task quickly and reliably at the desired level of quality. |
| Identify Human Operator Limitations | Consider how process and workstation requirements affect the tool operator's ability to perform the task. |
| | Human capabilities are limited by strength, fatigue, anthropometry, and manual dexterity. |
| Workstation and Task Factors | Consider the particular task and workstation where the tool is being used. |
| | Requirements may include work location, work orientation, tool shape, tool weight, gloves, frequency of operation, tool accessories, work methods and standards. |

## 5.2　Power Tool Triggers and Grip Force

Extended-length triggers (see Figure 5.1) that distribute force among two or more fingers are often suggested for minimizing stress concentrations at the volar aspects of the fingers (Lindquist et al., 1986; Putz-Anderson, 1988). The rationale is that the force for squeezing the trigger and grasping the handle will be distributed among several fingers to reduce the stress in the index finger. Following is a description of a study that investigated how this particular design feature affects the force in the hands.

In order to directly measure finger and hand force exerted during actual tool operation, an apparatus was constructed for simulating a functioning pistol grip pneumatic nutrunner (Oh and Radwin, 1993). Strain gages were installed in two aluminum bars that were used as the handle for measuring force exerted against the fingers and palm. The instrumented bars were constructed so they were insensitive to the point of force application and linearly summed force applied along the length of the handle (Pronk and Niesing, 1981; Radwin et al., 1991). This was accomplished by measuring shearing stress acting in the cross section of the beam. Strain gages were mounted on a thin web that was machined into the central longitudinal plane and aligned at 45° with respect to the long axis. The effect of bending stresses were completely removed from the strain gages by selecting a measurement point at the neutral axis of the beam, so that all the strain at the measurement point is strictly due to shear stress. Shear strain is totally independent of the point of application.

The strain gage instrumented handle was mounted on a rigid frame and attached perpendicular to a modified in-line pneumatic nutrunner motor in a configuration resembling a pistol-grip power tool (see Figure 5.2). The two dynamometers were mounted in parallel on a track so the handle span could be continuously adjusted. The apparatus was completely functional. The air motor contained an automatic air shut-off torque control mechanism and was operated at a 6.8 Nm target torque setting.

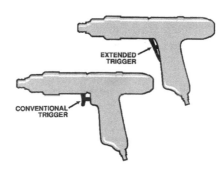

**FIGURE 5.1**　A conventional trigger and an extended-length trigger on pistol grip power hand tools. (Reprinted with permission from *Human Factors*, 35, 3, 1993. Copyright 1993 by the Human Factors and Ergonomics Society.)

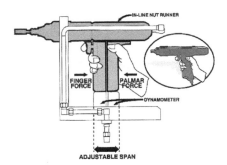

**FIGURE 5.2**   Dynamometer used for measuring finger and palm forces exerted when operating a completely functional simulation of a pistol-grip pneumatic power hand tool. (Reprinted with permission from *Human Factors*, 35, 3, 1993. Copyright 1993 by the Human Factors and Ergonomics Society.)

Plastic caps were formed and attached to each end of the dynamometer so the contours resembled a power tool handle. The handle circumference was 12 cm for a 4-cm span, measured between two points tangent to the handle contact surfaces. The handle circumference increased an additional 2 cm as the handle span was increased by 1 cm. A trigger was mounted on the finger side cap (see Figure 5.2), and a contact switch was installed inside the trigger. A leaf spring was used for controlling trigger tension. When the trigger was squeezed, the switch tripped a relay and a solenoid valve for supplying air to the pneumatic power tool motor.

Two different trigger types were tested. One was a conventional power tool trigger, activated using only the index finger. The second was longer than the conventional trigger and was activated using both the index and middle fingers (see Figure 5.1). The conventional trigger was 21 mm long and the extended trigger was 48 mm long. The conventional trigger required 8 N, and the extended trigger required 11 N for activation.

Use of the extended trigger was found beneficial for reducing grip force and exertion levels during tool operation. Average peak finger and palm forces were, respectively, 9% and 8% less for the extended trigger than for the conventional trigger. Eleven of eighteen subjects (61%) indicated that they preferred using the handle with the extended trigger after just an hour of use in the laboratory. The average finger and palmar holding force was 65% and 48%, respectively, less for the extended trigger, than for the conventional trigger. Since subjects spent 65% to 76% of the operating time holding the tool, using an extended trigger may have an important effect on reducing exposure to forceful exertions in the hand during power hand tool operation.

## 5.3   Handle Size and Grip Force

Research on handle design has typically focused on finding the optimal handle dimensions. Grip strength is affected through the biomechanics of grip from the relative position of the joints of the hand and by the position and length of the muscles involved. Consequently, grip strength is affected by the handle size. Recommendations for handle size are usually based on the span that maximizes grip strength, or the span that minimizes fatigue.

Hertzberg (1955), in an early Air Force study, reported that a handle span of 6.4 cm maximized power grip strength. Greenberg and Chaffin (1975) recommended that a tool handle span should be in the range between 6.4 cm and 8.9 cm in order to achieve high grip forces. Ayoub and Lo Presti (1971) found that a 3.8 cm diameter was optimum for a cylindrical handle. This was based on maximizing the ratio between strength and EMG activity, and on the number of work cycles before onset of fatigue. Another study by Petrofsky et al. (1980) showed that the greatest grip strength occurred at a handle span between 5 cm and 6 cm.

Grip strength is affected by hand size. Fitzhugh (1973) showed that the handle span resulting in maximum grip strength for a 95 percentile male hand length is larger than the handle span for a 50

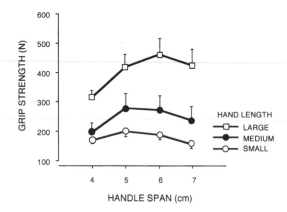

**FIGURE 5.3** Average grip strength plotted against handle span for three hand size categories. Error bars represent standard error of the mean. (Reprinted with permission from *Human Factors*, 35, 3, 1993. Copyright 1993 by the Human Factors and Ergonomics Society.)

percentile female. Consequently, a person with a small hand might benefit from using a smaller handle, and a person with a large hand might benefit from using a larger one.

Grip strength data often used for handle design are based on population measurements made using instruments like the Jamar or Smedley dynamometers (Schmidt and Toews, 1970; Young et al., 1989) rather than using handle dimensions representative of an actual tool. In most cases, only one dimension (handle span) has been controlled, while the other handle dimensions were not necessarily similar to a tool handle.

A power hand tool manufacturer considered offering a power hand tool that provided a handle that was adjustable in size. An investigation of grip strength using handle dimensions similar to power hand tool handles was conducted in order to explore the differences against published grip strength data (Oh and Radwin, 1993). Hand length up to 17 cm was classified as small, between 17 cm and 19 cm as medium, and greater than 19 cm as large. Average grip strength is plotted against handle span and hand size in Figure 5.3. Grip strength increased as hand length increased. Large hand subjects produced their maximum grip strength (mean = 463 N, SD = 128 N) for a handle span of 6 cm, while medium hand (mean = 280 N, SD = 122 N) and small hand (mean = 203 N, SD = 51 N) subjects produced their maximum strength for a handle span of 5 cm.

The span resulting in maximum grip strength agreed with the findings of previous strength studies. Hertzberg (1955) found that subjects exerted more force at a 6.4 cm span than among 3.8 cm, 6.4 cm, 10.2 cm, and 12.7 cm handle spans. Petrofsky et al. (1980) reported that on the average, subjects produced maximum grip force for a handle span between 5 cm and 6 cm.

Although the span resulting in maximum grip strength and the grip strength function agreed with previous findings, the maximum grip strength for both student and industrial worker subjects was markedly less than what has been previously reported in the literature. Schmidt and Toews (1970) collected grip strength data from 1,128 male and 80 female Kaiser Steel Corporation employee applicants, using a Jamar dynamometer. They reported for a handle span of 3.8 cm, an average of 499 N for the dominant male hand and 308 N for the dominant female hand. Swanson et al. (1970) measured the grip strength of 50 females and 50 males using a Jamar dynamometer. Among these subjects, 36 were light manual workers, 16 were sedentary workers, and 48 were manual workers. They reported for a handle span of 6.4 cm, 467 N for the male dominant hand and 241 N for the female dominant hand. These all exceeded the strength levels observed (see Figure 5.3).

A major difference between grip strength measured for tool handles by Oh and Radwin (1993) and previously reported strength data is in the handle dimensions. The Jamar and Smedely dynamometers have smaller circumferences and narrower widths than the tool handle used in this study. The tool handle curvature was also straight while the Jamar dynamometer has a curved surface at the grip center. The handle used in this study closely represented an actual tool handle in circumference and width. These

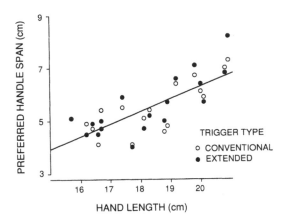

**FIGURE 5.4**   Preferred handle span plotted against hand length. (Reprinted with permission from *Human Factors*, 35, 3, 1993. Copyright 1993 by the Human Factors and Ergonomics Society.)

size and curvature differences can affect the position of the fingers and grip posture. These dimensional differences must be considered when designing handles based on strength using published grip strength data.

The investigation also found a difference in grip strength between student subjects and industrial workers. Grip strength, averaged over handle span, was 279 N (SD = 133 N) for the students and 327 N (SD = 90 N) for the workers. No significant grip strength differences, however, between the student and worker groups were observed within each hand size.

The underlying assumption in designing handles based on maximum strength is that the actual force exerted is independent of handle size. Exertion level is the ratio of the actual grip force used, to the maximum voluntary force generating capacity. If the grip force used during tool operation is the same for all handle sizes, then the handle span associated with the greatest grip strength should result in the lowest exertion level. If grip force, however is affected by handle size, then the handle span associated with the greatest grip strength may not be the handle span resulting in the minimum exertion level.

A series of experiments were performed using the pistol grip power hand tool with strain gage instrumented handles and an adjustable handle span as described above. Handle span affected peak finger and palmar force. Peak finger force increased 24% for a student subject group, and 30% for an industrial worker group, as handle span increased from 4 cm to 7 cm. Similarly, peak palmar force increased 21% for the student group and 22% for the worker group, as handle span increased from 4 cm to 7 cm. Handle span also influenced finger and palmar holding forces. Finger holding force increased 20%, and palmar holding force increased 16%, as handle span increased from 4 cm to 7 cm for the student subjects.

The study found that hand size was proportional to the handle span operators preferred when offered the opportunity to adjust the handle size to any size they desired. Operators with larger hand sizes reported they preferred using a tool with a larger handle. Preferred handle span is plotted against hand length in Figure 5.4 for both trigger types. There was no difference between the preferred handle span for the conventional trigger and the preferred handle span for the extended trigger. No anthropometric measurements, however, were related to the span resulting in the minimum peak exertion level. Exertion level when holding the tool was less for the large size hands than for the small size hands. Holding exertion level for the large hands was maximum for the 4 cm handle span, while holding exertion level for the small hands was maximum for the 7 cm handle span. In addition, the tendency for large hand subjects to prefer larger handle spans suggests that selectable size handles may be more desirable than having only a single size handle for power hand tools.

## 5.4   Static Hand Force

Safe power hand tool operation requires that an operator possess the ability to adequately support the tool in a particular position, apply the necessary forces, while reacting against the forces generated by the tool. Force demands that exceed an operator's strength capabilities can cause loss of control, resulting in an accident or an injury. Design and selection of power hand tools that minimize static grip and hand force will help reduce muscle fatigue and prevent upper limb disorders.

The force necessary for supporting a power hand tool depends on the tool weight, its center of gravity, the length of the tool, and air hose attachments. Power hand tools should be well balanced with all attachments installed. As a general rule, a hand tool center of gravity should be aligned with the center of the grasping hand so the hand does not have to overcome moments that cause the tool to rotate the operator's wrist and arm (Greenberg and Chaffin, 1977).

Psychophysical experiments have provided some insight into the load that power tool operators prefer. When experienced hand tool operators were asked to rate the mass of the power tools they operated, tools weighing 0.9 to 1.75 kg mass were rated "just right" (Armstrong et al. 1989). Other psychophysical experiments showed that perceived exertion for a tool mass of 1 kg was significantly less than for tools with a mass of 2 kg and 3 kg (Ulin and Armstrong, 1992).

There is a tradeoff between selecting a light tool and the benefit of the added weight for performing operations that require high feed force. The power available for a grinding task increases with increasing mass of the grinder. Reducing the weight of the grinder can increase the feed force the operator must provide and may increase the amount of time necessary for accomplishing the task, consequently subjecting the operator to more stressful work and greater vibration exposure. Heavy grinding tasks should be performed on horizontal surfaces so the weight of the tool does not have to be supported by the operator. Heavy power tools should be suspended using counterbalancing accessories.

In addition to supporting the tool load, power hand tool operators often have to exert push or feed force, or act against reaction forces. Feed force is necessary for starting a threaded fastener, advancing a bit, or keeping a bit or socket engaged during the securing cycle. Feed force is affected by the work material and design of the tool, bit, or fastener. Large feed forces are sometimes needed when operating power tools such as drills and screwdrivers. Repetitive or sustained exertions associated with these operations should be minimized. Drill feed force is affected by the drill power and speed, bit type, material, and diameter of the hole drilled. Power screwdriver feed force may be affected by the fastener head and screw tip used. Feed force for a slotted or Phillips head screw generally requires more feed force than for a torx head screw. Self-tapping screws require more force than screws tightened through pre-tapped holes. Material hardness is also a factor for self-tapping screws and drilling. Feed force requirements also increase as torque level increases for cross recess screws.

Power hand tools such as screwdrivers or nutrunners, used for tightening threaded fasteners, are commonly configured as (1) in-line, (2) pistol grip, and (3) right angle. A mechanical model of a nutrunner was developed for static equilibrium (no movement) conditions (Radwin et al., 1995). Hand force, reaction force from the workpiece, tool orientation, weight, and output torque were included in this model. This chapter will describe the model developed for pistol grip nutrunners.

The model uses a Cartesian coordinate system relative to the orientation of the handle grasped in the hand using a power grip. This coordinate system has the x-axis perpendicular to the axial direction of the handle; the y-axis is parallel to the long axis of the handle; and the z-axis is parallel to the tool spindle. The origin is the end of the tool bit or socket. Hand forces are described in relation to these coordinate axes. To simplify the model, an initial assumption is that orthogonal forces can be applied along the handle without producing coupling moments. This assumption allows force to be considered as having a single point of application. The resultant hand force $\mathbf{F}_H$ at the grip center is the vector sum of the three orthogonal force components

$$\mathbf{F}_H = F_{H_x}\mathbf{i} + F_{H_y}\mathbf{j} + F_{H_z}\mathbf{k} \tag{5.1}$$

where the hand force magnitude is:

$$\left|\mathbf{F}_H\right| = \sqrt{F_{H_x}^{\,2} + F_{H_y}^{\,2} + F_{H_z}^{\,2}} \tag{5.2}$$

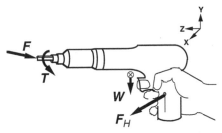

F = Workpiece Reaction Force
T = Spindle Torque
$F_H$ = Hand Force
W = Tool Weight

**FIGURE 5.5**   Free body diagram and orthogonal force components considered in the pistol grip force model.

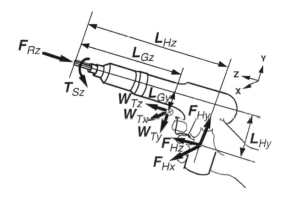

**FIGURE 5.6**   Power hand tool geometry and variables in the hand tool static force model.

and **i, j, k** are the unit vectors. The coordinates and respective force components are illustrated in Figure 5.5.

Consider the free-body diagram for the pistol grip nutrunner in Figure 5.6. The torque, $T_S$, acts in reaction to torque, $T$, applied by the tool to the fastener. The tool operator has to oppose this equal and opposite reaction torque in the counter-clockwise direction by producing a reaction force, $F_{Hx}$. That is not the only force, however, that the operator has to produce. A force acting in the $z$ direction, $F_{Hz}$, provides feed force and produces an equal and opposite reaction force, $F_{Rz}$. In addition, the operator has to react against the tool weight in order to support and position the tool by providing a vertical force component, $F_{Hy}$. The tool weight, $W_T$, and push force, $F_{Hz}$, tend to produce a clockwise moment about the tool spindle in the $yz$-plane which is countered by this vertical support force.

When a body is in static equilibrium, the sum of the external forces and the sum of the moments are equal to zero. Using that relationship, the following system of equations was developed for the pistol-grip nutrunner to describe these static forces:

$$
\begin{bmatrix} F_{Hx} \\ F_{Hy} \\ F_{Hz} \end{bmatrix} = \begin{bmatrix} -L_{Gz}/L_{Hz} & 0 & 0 & -1/L_{Hz} & 0 \\ 0 & -L_{Gz}/L_{Hz} & \left(-L_{Gy}-L_{Hy}\right)/L_{Hz} & 0 & -L_{Gy}/L_{Hz} \\ 0 & 0 & -1 & 0 & -1 \end{bmatrix} \cdot \begin{bmatrix} W_{Tx} \\ W_{Ty} \\ W_{Tz} \\ T_{Sz} \\ F_{Rz} \end{bmatrix} \quad (5.3)
$$

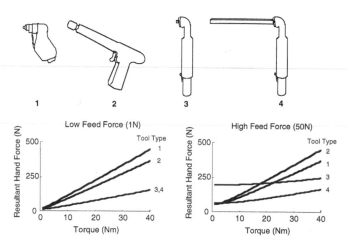

**FIGURE 5.7**  Comparison of predicted hand forces for four different tool configurations performing the same task for one-hand tool operation.

Assuming one-hand operation, resultant hand force magnitude was predicted using the model for the four different tools and plotted as a function of torque in Figure 5.7. Hand force was determined for both low feed force (1 N) and high feed force (50 N) conditions, when operating these tools against a vertical surface. When feed force was small, the resultant hand force was mostly affected by torque reaction force, which increased as torque increased for all four tools. Since the greatest force component in this case was torque reaction force, Tools 3 and 4 had the least resultant hand force since they both had the longest handles. Tool 3, however, had a considerably greater resultant hand force when feed force was high. This effect was not observed for Tool 4, which also had a similar handle, but contained a spindle extension shaft.

## 5.5  Dynamic Reaction Force

Whereas manual hand tools rely on the human operator for generating forces, power hand tools operate from an external energy source (i.e., electric, pneumatic, and hydraulic) for doing work. The tool operator provides static force for supporting the tool and for producing feed force, and must react against the forces generated by the power hand tool. Power hand tools such as nutrunners produce rapidly building torque reaction forces which the operator must react against in order to maintain full control of the tool.

Nutrunner reaction torque is produced by spindle rotation and is affected primarily by the spindle torque output and tool size. Nutrunner spindle torque can range from less than 0.8 Newton-meters (Nm) to more than 700 Nm. This torque is transmitted to the operator as a reaction force through the moment arm created by the tool and tool handle. A tool operator opposes reaction torque while supporting the tool and preventing it from losing control.

The three major operating modes for nutrunners include (1) mechanical clutch, (2) stall, or (3) automatic shut-off. When a stall tool is used, maximum reaction torque time is directly under operator control by releasing the throttle, which can last as long as several seconds. Stall tools tend to expose an operator to reaction torque the longest. Although clutch tools limit reaction torque exposure, ratcheting clutch tools can expose workers to significant levels of vibration if used frequently (Radwin and Armstrong, 1985). The speed of the shut-off mechanism controls exposure to peak reaction force for automatic shut-off tools. Consequently, automatic shut-off tools have the shortest torque reaction time because these tools cease operating immediately after the desired peak torque is achieved.

As torque is applied to a threaded fastener, it rotates at a relatively low spindle torque until the clamped pieces come into intimate contact. This torque can approach zero with free running nuts or can be rather significant as in the case where locking nuts, thread interference bolts, or thread-forming type fasteners are used. After the fastener brings the clamped members of the joint into initial intimate contact, it continues to draw the parts together until they form a solid joint. When the joint becomes solid, continued turning of the nut results in a proportionally increasing torque. This is the elastic portion of the cycle and is the time when reaction torque forces are produced. Torque build-up, and consequently torque reaction force, continues rising at a fixed rate until peak torque is achieved, which is the clamping force of the joint. Forearm muscle reflex responses when operating automatic air shut-off right angle nutrunners during the torque-reaction phase was more than four times greater than the muscle activity used for holding the tool and two times greater than the run-down phase (Radwin et al., 1989). Flexor EMG activity during the torque–reaction phase increased for tools having increasing peak spindle torque.

Threaded fastener joints are classified as "soft" or "hard" depending on the relationship between torque build-up and spindle angle. The International Standardization Organization (ISO) specifies that a hard joint has an angular displacement less than 30 degrees when torque increases from 50% to 100% of target torque, and a soft joint has an angular displacement greater than 360 degrees (ISO-6544).

Nutrunner torque reaction force is a function of several factors including target torque, spindle speed, joint hardness, and torque build-up time. Some of these factors are interdependent. Faster spindle speed results in shorter torque build-up time, and softer joints are related to longer build-up times. The duration of exertion is directly related to torque build-up time rather than just the speed of the tool or joint hardness.

Studies have shown that torque build-up time as well as the magnitude of torque reaction force has a significant influence on human operators during power nutrunner use. Kihlberg et al. (1995) studied right angle nutrunners having different shut-off mechanisms (fast, slow, and delayed) and found a strong correlation between perceived discomfort, handle displacement, and reaction forces. Radwin et al. (1989) investigated the effects of target torque and torque build-up time using right-angle pneumatic nutrunners and found that average flexor rms electromyography (EMG) activity scaled for grip force increased from 372 N for a low target torque (30 Nm) to 449 N for a high target torque (100 Nm), and that average grip force was 390 N for a long build-up time (2 s), and increased to 440 N for a medium build-up time (0.5 s). They also reported that EMG latency between tool torque onset and peak flexor rms EMG for the long torque build-up time (2 s) was 294 ms and decreased to 161 ms for the short build-up time (0.5 s). The findings suggested that torque reaction force can affect extrinsic hand muscles in the forearm, and hence grip exertions, by way of a reflex response. Johnson and Childress (1988) showed that low torque was associated with less muscular activity and reduced subjective evaluations of exertion.

Representative torque reaction force, handle kinematics, and EMG muscle activity are illustrated in Figure 5.8. Since torque builds up in a clockwise direction, the reaction torque has a tendency to rotate the tool counterclockwise with respect to the operator. When the operator has sufficient strength to react against the reaction torque, the tool remains stationary or rotates clockwise and the operator exerts concentric muscle contractions against the tool (positive work). However, when the tool overpowers the operator, it tends to move in a counterclockwise direction and the operator exerts eccentric muscle contractions against the tool (negative work). Therefore, measures of handle movement that occur (handle velocity and displacement) and the direction of rotation can indicate relative tool controllability. Handle movement direction was defined as positive when the handle moved in the direction of tool reaction torque (see Figure 5.1). If handle velocity increases after shutoff, it means that the tool and hand are unstable. The work done on the tool–hand system and the power involved in doing work during torque build-up were also assessed. If the operator has the capacity to successfully react against the torque build-up (positive work), then the tool is considered stable. This occurs when handle displacement and velocity were less than zero. If the handle become unstable and the net handle displacement occurs in the direction of torque reaction away from the operator, then work and power are negative.

A computer-controlled right angle nutrunner was used to study power hand tool reaction forces (Oh and Radwin, 1997). A torque transducer and an angle encoder were integrated into the tool spindle head which outputted analog torque and digital angular rotation signals. A threaded fastener joint simulator that could be oriented horizontally or vertically was mounted on a height adjustable platform. The longitudinal axis of the joint head was oriented perpendicular to the ground for the horizontal workstation setting, and oriented parallel to the ground for the vertical workstation setting.

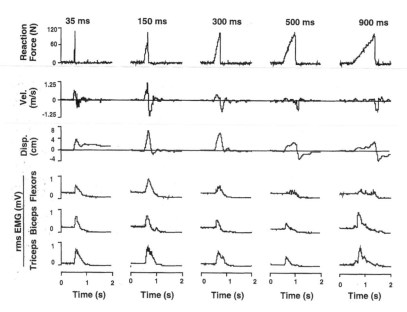

**FIGURE 5.8**    Representative torque reaction force, handle kinematics and EMG muscle activity for different torque build-up times.

The study showed that workstation orientation and tool dynamics (torque reaction force and torque build-up time) influenced operator muscular exertion and handle stability. In general, handle instability increased when the tool was operated on a vertical workstation (rather than a horizontal workstation), when torque reaction force was high (88.3 and 114.6 N), and for a 150 ms torque build-up time, regardless of torque reaction force.

As torque reaction force increased from 52.1 N to 114.6 N, peak hand velocity 89%, and peak hand displacement increased 113%. Peak hand velocity was greatest for a 150 ms build-up time and the least for a 900 ms build-up time. The effect of target torque was consistent with previous studies that showed that target torque was related to muscular exertion, subjective perceived exertion, and handle instability (Johnson and Childress, 1988; Lindqvist, 1993; Oh and Radwin, 1994; Radwin et al., 1989). As torque reaction force increased from 52.1 N to 114.6 N, the magnitude of negative work increased by 35%, and the magnitude of average power against the operator increased by 30%. Under these conditions, perceived exertion also increased from 2.7 to 4.3 (as rated on Borg's 10-point scale), and task acceptance rate decreased from 73% to 28%. When the tool was operated on a horizontal workstation, average finger flexor EMG was significantly influenced by torque reaction force. As torque reaction force increased from 52.1 N to 114.6 N, the average flexor EMG increased by 14%.

The effect of torque build-up time on power hand tool operators has been studied in terms of perceived exertion, muscular activity, and handle stability (Armstrong et al., 1994; Freivalds and Eklund, 1991; Lindqvist et al., 1986; Oh and Radwin, 1994; Radwin et al., 1989). Torque build-up time is a concern because it is directly related to assembly time and exertion duration. Increased duration may lead to earlier fatigue onset. Although longer build-up time results in longer duration exertions and increases the operation cycle time, it may provide an opportunity for better tool control since it gives the operator a longer time to react.

Peak hand velocity was 5.7% less for horizontal workstations (mean = 0.46 m/s, SD = 0.26 m/s) than for vertical workstations (mean = 0.67 m/s, SD = 0.34 m/s). A similar trend was observed for peak hand displacement. Peak hand displacement for horizontal workstations (mean = 4.0 cm, SD = 2.2 cm) was 90.2% less than peak hand displacement for vertical workstations (mean = 7.6 cm, SD = 4.6 cm). Previous findings agree that a horizontal workstation is preferable for right angle tool use. Ulin et al. (1992) showed that average subjective ratings of perceived exertion were significantly less when the tool was operated on horizontal workstations rather than vertical workstations. Also, 88% more negative work and 58% more power against the operator were recorded while the tool was operated on a vertical workstation. However, subjective ratings of perceived exertion and task acceptance rates did not differ between horizontal and vertical workstations. This might come from the fact that the torque levels in the current study were much greater than the torque level used for the Ulin et al. study (1992).

Although perceived exertion was less and task acceptance rate was greater for a 35 ms build-up time than for longer build-up times, the operator might not have sufficient time to voluntarily react against torque build-up with the 35 ms build-up time. On the average, the onset of the EMG burst occurred 40 ms after the onset of torque build-up for the 35 ms build-up time. This indicated that the muscles were not activated until a significant amount of torque had built up for the 35 ms build-up time. Lack of muscular contraction during torque build-up might explain why the peak handle velocity was higher for short build-up times. Without muscular contractions, the inertia of the tool and hand had to absorb all of the reaction force. Short exertion duration and lack of muscular contractions due to EMG latencies might contribute to lower subjective ratings of perceived exertion for the 35 ms build-up.

The larger torque variance that occurred for the 35 ms build-up time indicated that even though subjective perceived exertion was less, this condition might result in more target torque error. Also, the probability of increased handle instability after shutoff was significantly greater for the 35 ms build-up time. This suggests that even after shutoff, operators did not have sufficient capacity to control the tool reaction torque. Therefore, the 35 ms build-up time increased handle stability in terms of peak handle displacement and negative work, and reduced subjective perceived exertion, however, the lack of muscular contraction during torque build-up reduced tightening quality.

Methods for limiting reaction force include (1) use of torque reaction bars, (2) installing torque absorbing suspension balancers, (3) providing tool mounted nut holding devices, and (4) using tool support reaction arms. A torque reaction bar sometimes can be used to transfer loads back to the work piece. Tools that can be equipped with a stationary reaction bar adapted to a specific operation so reaction force can be absorbed by a convenient solid object can completely eliminate reaction torque from the operator's hand. These bars can be installed on in-line and pistol-grip tools. Right angle tools can react against a solid object instead of relying on the hand and arm. Reaction devices (1) remove reaction forces from the operator, (2) permit pistol-grip and in-line reaction bar tools to be operated using two hands, (3) free the operator from restricting postures, (4) provide weight improvements over right angle nutrunners, and (5) improve tool fastening performance. The disadvantages are that reaction bars must be custom made for each operation, and the combination of several attachments for one tool can be difficult. Torque reaction bars may also add weight to the tool and can make the tool more cumbersome to handle.

## 5.6   Vibration

Vibration can be a by-product of power hand tool operation, or it can even be the desired action as is the case with abrasive tools like sanders or grinders. Vibration levels depend on tool size, weight, method of propulsion, and the tool drive mechanism. It is affected by work material properties, disk abrasives, and abrasive surface area. Continuous vibration is inherent in reciprocating and rotary power tools. Impulsive vibration is produced by tools operating by shock and impact action, such as impact wrenches or chippers. The tool power source, such as air power, electricity, or hydraulics can also affect vibration. Vibration is also generated at the tool-material interface by cutting, grinding, drilling, or other actions.

Pneumatic hammer recoil was observed producing a stretch reflex and muscular contractions in the elbow and wrist flexors (Carlsöö and Mayr, 1974). Studies of the short-term neuromuscular effects of hand tool vibration have demonstrated that hand tool vibration can introduce disturbances in neuromuscular force control resulting in excessive grip exertions when holding a vibrating handle (Radwin et al., 1987). The results of these studies demonstrated that grip exertions increased with tool vibration. Average grip force increased for low frequencies (40 Hz) vibration but did not change for higher frequencies (160 Hz) vibration. Since forceful exertions are a commonly cited factor for chronic upper extremity muscle, tendon, and nerve disorders, vibrating hand tool operation may increase the risk of CTDs through increased grip force.

Vibration has also been shown to produce temporary sensory impairments (Streeter, 1970; Radwin et al., 1989). Recovery is exponential and can require more than 20 minutes (Kume et al., 1984). Workers often sand or grind surfaces and periodically inspect their work using tactile inspection to determine if the surface was sanded to the desired level of smoothness. Diminished tactility may result in a surface feeling smoother than it actually is, resulting in a rougher surface than is actually desired.

Vibration has not been shown to be significantly reduced by using resilient mounts on handles. Vibration isolation techniques have been generally unsuccessful for limiting vibration transmission from power tools to the hands and arms. Isolation has been particularly difficult for vibration frequencies less than 100 Hz. This is because attenuation only occurs when the vibration spectrum falls above the resonant frequency of the isolation system or material. When the vibration frequency is less than the resonant frequency of the isolating material, the handle acts as a rigid body and no vibration is attenuated. Grinding tools typically run at speeds near 6000 rpm (100 Hz) making it difficult to have a resilient vibration isolating handle. Furthermore, if the vibration frequency is approximately equivalent to the isolator resonant frequency, the system will actually intensify vibration levels. Weaker suspension systems have lower resonant frequencies, but are often impractical because such a system is usually too flexible for the heavily loaded handles of tools like grinders. Handles loaded with high forces must be very rigid.

## References

Armstrong, T.J., Bir, C., Finsen, L., Foulke, J., Martin, B., Sjøgaard, G., and Tseng, K. 1994. Muscle responses to torques of hand held power tools, *Journal of Biomechanics,* 26(6): 711-718.

Ayoub, M.M. and Lo Presti, P. 1971. The determination of an optimum size cylindrical handle by use of electromyography. *Ergonomics,* 14(4): 509-518.

Fitzhugh, F.E. 1973. *Dynamic Aspects of Grip Strength.* (Tech Report). Department of Industrial & Operations Engineering, Ann Arbor: The University of Michigan.

Freivalds, A. and Eklund, J. 1991. Subjective ratings of stress levels while using powered nutrunners, in W. Karwowski and J.W. Yates (Eds.), *Advances in Industrial Ergonomics and Safety III,* New York: Taylor & Francis, 379-386.

Greenberg, L. and Chaffin, D.B. 1975. *Workers and Their Tools: A Guide to the Ergonomic Design of Hand Tools and Small Presses.* Midland, MI: Pendell.

Hertzberg, H.T.E. 1955. Some contributions of applied physical anthropology to human engineering. *Annals of NY Academy of Science,* 63(4): 616-629.

International Organization for Standardization 1981. *Hand-held Pneumatic Assembly Tools for Installing Threaded Fasteners — Reaction Torque Reaction Force and Torque Reaction Force Impulse Measurements.* ISO-6544.

Johnson, S.L. and Childress, L.J. 1988. Powered screwdriver design and use: tool, task, and operator effects, *International Journal of Industrial Ergonomics,* 2: 183-191.

Kihlberg, S., Kjellberg, A., and Lindbeck, L. 1995. Discomfort from pneumatic tool torque reaction force reaction: acceptability limits, *International Journal of Industrial Ergonomics,* 15: 417-426.

Lindqvist, B. 1993. Torque reaction force reaction in angled nutrunners, *Applied Ergonomics,* 24(3): 174-180.

Lindquist, B., Ahlberg, E., and Skogsberg, L. 1986. *Ergonomic Tools in Our Time.* Atlas Copco Tools, Stockholm.

Oh, S. and Radwin, R.G. 1993. Pistol grip power tool handle and trigger size effects on grip exertions and operator preference, *Human Factors,* 35(3): 551-569.

Oh, S. and Radwin, R.G., 1994, Dynamics of power hand tools on operator hand and arm stability, in *Proceedings of the Human Factors and Ergonomics Society 38th Annual Meeting,* 602-606, Santa Monica, CA: Human Factors and Ergonomics Society.

Oh, S. and Radwin, R.G. 1998. The influence of target torque and torque build-up time on physical stress in right angle nutrunner operation, *Ergonomics,* 41(2): 188-206.

Petrofsky, J.S., Williams, C., Kamen, G., and Lind, A.R. 1980. The effect of handgrip span on isometric exercise performance, *Ergonomics,* 23(12): 1129-1135.

Pronk, C.N.A. and Niesing, R. 1981. Measuring hand grip force using an application of strain gages, *Medical, Biological Engineering and Computing,* 19: 127-128.

Putz-Anderson, V. 1988. *Cumulative Trauma Disorders.* New York: Taylor & Francis.

Radwin, R.G., Masters, G., and Lupton, F.W. 1991. A linear force summing hand dynamometer independent of point of application, *Applied Ergonomics,* 22(5): 339-345, 1991.

Radwin, R.G. and Haney, J.T. 1996. *An Ergonomics Guide to Hand Tools,* Fairfax, VA: American Industrial Hygiene Association.

Radwin, R.G., VanBergeijk, E., and Armstrong, T.J. 1989. Muscle response to pneumatic hand tool torque reaction force reaction forces, *Ergonomics,* 32(6): 655-673.

Radwin, R.G., Oh, S., and Fronczak. 1995. A mechanical model of hand force in power hand tool operation, *Proceedings of the Human Factors and Ergonomics Society 39th Annual Meeting,* Santa Monica: Human Factors and Ergonomics Society: 348-352.

Schmidt, R.T. and Toews, J.V. 1970. Grip strength as measured by the Jamar dynamometer, *Archives of Physical Medicine & Rehabilitation,* 51(6): 321-327.

Swanson, A.B., Matev, I.B., and Groot, G. 1970. The strength of the hand, *Bulletin of Prosthetics Research,* Fall: 145-153.

Ulin, S.S., Snook, S.H., Armstrong, T.J., and Herrin, G.D. 1992. Preferred tool shapes for various horizontal and vertical work locations, *Applied Occupational and Environmental Hygiene,* 7(5): 327-337.

Young, V.L., Pin, P., Kraemer, B.A., Gould, R.B., Nemergut, L., and Pellowski, M. 1989. Fluctuation in grip and pinch strength among normal subjects, *The Journal of Hand Surgery,* 14A(1): 125-129.

# 6

# Gloves

Ram Bishu
*University of Nebraska–Lincoln*

A. Muralidhar
*University of Nebraska–Lincoln*

## 6.1 Importance of the Hand

The hand is probably the most complex of all anatomical structures in the human body. Along with the brain, it is the most important organ for accomplishing the tasks of exploration, prehension, perception, and manipulation, unique to humans. The importance of the hand to human culture is emphasized by its depiction in art and sculpture, its reference frequency in vocabulary and phraseology, and its importance in communication and expression (Chao et al., 1989). The human hand is distinguished from that of the primates by the presence of a strong opposable thumb, which enables humans to accomplish tasks requiring precision and fine control. The hand provides humans with both mechanical and sensory capabilities.

## 6.2 Prehensile Capabilities of the Hand

Napier (1956) divides hand movements into two main groups — prehensile movements, in which an object is seized and held partly or wholly within the compass of the hand, and nonprehensile movements, where no grasping and seizing is involved but by which objects can be manipulated by pushing or lifting motions of the hand as a whole or of the digits individually.

Landsmeer (1962) further classifies human grasping capabilities as *power grip*, where a dynamic initial phase can be distinguished from a static terminal phase, and *precision handling*, where there is no static terminal phase. The dynamic phase as defined by Landsmeer includes the opening of the hand, positioning of the fingers, and the grasping of the object. Westling and Johansson (1984) state that the factors that influence force control during precision grip are friction, weight, and a safety margin factor related to the individual subject. They also found that in multiple trials, the frictional conditions during a previous trial could affect the grip force. They also showed that the grip employed when holding small objects stationary in space was critically balanced such that neither accidental slipping between the skin

**TABLE 6.1**   Comparison of Bare Hand–Gloved Hand Capabilities

| Indices | Bare Hand | Gloved Hand |
|---|---|---|
| Thermal Tolerance | Poor | Good |
| Tactile Perception | Excellent | Poor |
| Grip Strength | Good | Reduced |
| Range of Motion | Excellent | Poor |
| Manipulative Ability | Excellent | Reduced |
| Prehension | Excellent | Poor |
| Torque Capability | Poor | Improved |
| Vibration Tolerance | Poor | Good |
| Dexterity | Excellent | Reduced |
| Chemical Resistance | Poor | Excellent |
| Electrical Energy | Poor | Excellent |
| Radiation (all kinds) | Poor | Excellent |
| Biohazard Risk | Poor | Excellent |
| Abrasive Trauma | Poor | Improved |

and the object occurred, nor did the grip force reach exceedingly high values. This sense of critical balance as to the amount of force applied while gripping is important, as too firm a grip could result in the destruction of a fragile object, causing possible injury to the hand, or lead to muscle fatigue and interfere with further manipulative activity imposed upon the hand. Sensory perception in the hand is due to the presence of mechanoreceptors distributed all over the palmar area, especially at the tips of the fingers.

Thus, feedback from the hand is a critical component of the gripping task enabling the amount of force to be controlled. Anything that blocks the transmission of impulses from the hand interferes with the feedback cycle and affects grip force control. Gloves do affect the feedback cycle.

## 6.3   Need for Protection of the Hand

The hand, which provides humans with both mechanical and sensory capabilities, needs to be protected from the environment. Protection is needed from mechanical trauma (abrasions, cuts, pinches, punctures, crush injuries), thermal extremes (heat and cold), radiation (nuclear, ultraviolet, X-ray, and thermal), chemical hazards, blood-borne pathogens, electrical energy, and vibration.

There exist several forms of hand protection, which can be used as stand-alone protection, or in combination with other personal protective equipment. The commonly available hand protection are gloves, mittens, finger cots, and gauntlets made of several materials such as leather, cotton, rubber, nylon, latex, metal, and in combinations of the same, to provide maximum protection against the specific condition being guarded against. The use of gloves, although a necessity in many workplaces, has some associated disadvantages. Gloves have been found to affect hand performance adversely, and the performance parameters affected are dexterity, task time, grip strength, and range of motion. Table 6.1 provides a summary of bare hand and gloved hand capabilities.

Facilitation of these activities, with simultaneous protection from the hazards of the work environment, are often conflicting objectives of glove design. The conflicts associated with providing primary hand protection through the use of a glove while permitting adequate hand functioning has been widely recognized.

It will be relevant to give a brief description of a variety of gloves that are available today. It will also be relevant to discuss performance effects of gloves, before detailing the challenges of glove design.

## 6.4   Types of Gloves

There are a wide variety of gloves available today. Starting from a garden glove at 50 cents a pair from the local grocery store to the custom-fit shuttle gloves donned by astronauts for extra vehicular activities

(EVA), which cost a few hundred thousand dollars a pair, the variety among gloves can be so overwhelming as to defy easy categorization. Gloves can be categorized along a number of dimensions, such as materials, design, and location of use. According to the National Safety Council (1975, 1976), hand protection can be job-rated or general purpose. Job-rated hand protection is designed to protect against the hazards of specific operations, while general purpose gloves protect against many hazards. Materials used in gloves are cotton, nylon, duck, jersey, canvas, terry, flannel, lisle, leather, rubber, synthetic rubber, wire mesh, aluminized fabric, asbestos, plastic and synthetic coatings, impregnated fabrics, polyvinyl chloride, nitrile, neoprene, and many man-made fibers with identifiable brand names (Dionne, 1979; Riley and Cochran, 1988). Glove styles include liners, reversibles, open back, gloves or mittens with reinforced nubby palms and fingers, and double-thumb gloves. Certain tasks may need double or more gloves. For example, shuttle gloves are an assemblage of three layers of gloves, while latex-sensitive people in the medical community wear an inner liner with an outer shell. The length of glove may be wrist-, elbow-, or shoulder-length with exact dimensions depending on the manufacturer. In summary, the gloves range from easily available general purpose ones to highly task-specific and job-rated ones.

## 6.5   Glove Effect on Strength

*Grip strength:* Published evidence exists for glove effect on grip strength, grasp strength, pinch strength, grasp at submaximal levels of exertion, torque capabilities, and on endurance time. Reduction in grip and grasp force when gloves are donned has been reported by a number of investigators (Hertzberg, 1955; Lyman and Groth, 1958; Cochran et al. 1986; Wang et al. 1987; and Sudhakar et al. 1988). Hertzberg (1955), using a Smedley hand dynamometer, determined that grip strength was reduced by about 20% among gloved airplane pilots. Reduction in strength may be as much as 30% or more, according to Lyman and Groth (1958). Cochran et al. (1986) performed an experiment which examined the differences in grasp force degradation among five different types of commercially available gloves as compared to a bare-handed condition. The results indicated that the "no glove" condition was significantly higher in grasp force than any of the glove conditions. Wang et al. (1987) performed an experiment on strength decrements with three different types of gloves. The results of the study showed that there was a reduction in grip strength when comparing gloved performance to bare-handed performance. Bishu et al. (1995a, b) studied the effects of EVA gloves at different pressures on human hand capabilities. A factorial experiment was performed in which three types of EVA gloves were tested at five pressure differentials. The independent variables tested in this experiment were gender, glove type, pressure differential, and glove make. Six subjects participated in an experiment where a number of performance measures, namely grip strength, pinch strength, time to tie a rope, and the time to assemble a nut and bolt, were recorded. Tactile sensitivity was also measured through a two-point discrimination test. The salient results were that with EVA gloves strength is reduced by nearly 50%, and that performance decrements increase with increasing pressure differential. McMullin and Hallbeck (1991) studied the effect of wrist position, age, and glove type on the maximal power grasp force, and their findings indicate that a single-layer glove is better than several layers, as the bunching of glove material at the joints could cause strength decrement. More recently Muralidhar et al. (1999) evaluated two prototype gloves (contour and laminated) with a single layer and a double layered glove. Bare-hand performance was measured to assess the exact glove effect. Considerable reduction in grip strength with gloves was found. Figure 6.1 shows the effect of gloves on grip strength. Similar results were also reported by Bronkema and Bishu (1996).

In summary, most of the research evidence on gloves indicates that gloves reduce grip and grasp capabilities.

*Torque strength:* A number of studies have reported an increase in strength capabilities with gloves. Riley et al. (1985) examined forward handle pull, backward handle pull, maximum wrist flexion torque, and maximum wrist extension torque while using no-glove, one-glove, and double-glove conditions. The results of this study showed that the one-glove condition was superior to both the no-glove and two-glove conditions. Similar results have been reported by Adams and Peterson (1988), who investigated the effects of two types of gloves on torque capabilities. In this study, a two-layer work glove and a three-

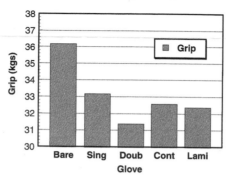

**FIGURE 6.1**   Gloves vs. grip strength.

layer chemical defense glove were found to enhance tightening performance, while only the work glove aided the loosening performance. Mital et al. (1994) have reported an increase in peak torquing exertion capabilities when gloves are donned, with the extent of increase being dependent on the type of gloves donned. In contrast, Cochran et al. (1988) found that gloves reduce torquing force. They had subjects perform a flexion torque task using four sizes of cylindrical handles (7 cm, 9 cm, 11 cm, and 13 cm) while wearing three types of gloves (cotton smooth leather, and suede leather). Using cotton gloves yielded the lowest torquing force, while bare-handed had the highest, and the two leather gloves were in between and were not significantly different from each other. These results are supported by Chen et al. (1989), who found the forces generated using cotton gloves of all sizes were significantly lower than the leather or deerskin gloves of different sizes in a similar torquing task. The effect of gloves on torque capabilities is far less clear. However, it is reasonable to assume that gloves would aid torquing tasks.

*Pinch strength:* As compared to grip or grasp capabilities, studies on glove effect on pinch strength are few and far between. Kamal et al. (1992) report that gloves do not affect lateral pinch capabilities. Hallbeck and McMullin (1991, 1993) found similar results for three jaw chuck pinch. Overall, gloves do not affect pinch strength.

*Endurance time:* Almost all activities with a gloved hand involve certain levels of hand exertions for periods of time. Therefore, two issues are relevant here: the extent of exertion and the time of exertion. Most of the published studies on gloves have addressed the issue of extent of exertion. Bishu et al. (1995b) addressed the question of how long a person can sustain a level of exertion in the gloved-hand condition. This deals with muscular fatigue and related issues. They reported that the endurance time at any exertion level depended, not on the glove, but just on the level of exertion expressed as a percentage of maximum exertion possible at that condition. There is, however, a glove effect for the maximum exertion. Figure 6.2 shows the plot of the exertion level effect on the endurance time, across all glove and pressure configurations. The endurance time is least at 100% exertion level, while it is greatest at 25% exertion level.

## 6.6   Glove Effect on Dexterity

Bradley (1969) showed that control operation time was affected while wearing gloves. Banks and Goehring (1979), while studying the effects of degraded visual and tactile information in diver performance, found that the use of gloves increased task time by 50 to 60%. McGinnis et al. (1973) investigated the effect of six different hand conditions on dexterity and torque capability. They used bare hand, leather glove, leather glove with inserts, impermeable glove, impermeable glove with inserts, and an impermeable glove with built-in insulation. They found that under dry conditions, the impermeable glove had the best torque capability, and that the bare-handed dexterity performance was superior to that of gloved-hand performance. Plummer et al. (1985) studied the effects of nine glove combinations (six double and three

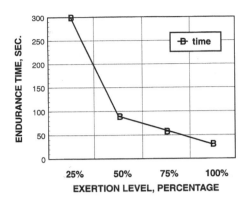

**FIGURE 6.2**  Endurance time.

single) on performance of the Bennett Hand Tool Dexterity Test apparatus. Results of the study indicated that subjects, with gloves donned, took longer to complete the task, with the double glove causing longer completion times. Cochran and Riley (1986) found that gloves generally reduce dexterity and force capability. Bensel (1993) conducted an experiment in which the effects of three thicknesses (0.18 mm, 0.36 mm, and 0.64 mm) of chemical protective gloves on five dexterity tests (the Minnesota rate of manipulation-turning; the O'Connor finger dexterity test; a cord and cylinder manipulation; the Bennet hand-tool dexterity test; and a rifle disassembly/assembly task) were investigated. Mean performance times were shortest for the bare-handed condition and longest for the thickest (0.64 mm) glove. Nelson and Mital (1995) found no appreciable differences in dexterity and tactility among latex gloves of five different thicknesses: 0.2083 mm; 0.5131 mm; 0.6452 mm; 0.7569 mm; and 0.8280 mm. The authors found the thickest latex glove (0.8280 mm) to be puncture resistant, with no loss in dexterity and tactility as compared to the thinner gloves. Bellinger and Slocum (1993) investigated the effect of protective gloves on hand movement and found that gloves decreased the range of motion in adduction/abduction and supination/pronation, while extension/flexion was not affected. Their findings suggest that there is an overall reduction in the kinematic abilities of the hand while wearing gloves. More recently Muralidhar et al. (1999) evaluated two prototype gloves (contour and laminated) with a single layer, and a double-layered glove. Bare-hand performance was measured to assess the exact glove effect. A battery of tests consisted of the Pennsylvania Bi-Manual Worksample Assembly Test (PBWAT), Minnesota Rate of Manipulation Test-Turning (MRMTT), a rope-tying task to evaluate dexterity for flexible object manipulation, and a manipulability test. Figure 6.3 shows the glove effect on MRMTT. Figure 6.4 shows the plot of the glove effect on PBWAT. Figure 6.5 shows the glove effect on the rope tying time, while Figure 6.6 shows the glove effect on the manipulation time. It is seen that gloves reduce dexterity. The reduction in gloved performance is seen consistently in all the measures.

Overall, gloves reduce finger dexterity, and manipulability.

## 6.7  Glove Effect on Tactility

Although intuitively most obvious, the effect of gloves on tactile sensitivity has not been well documented. The evidence on this matter is somewhat confusing mainly due to inadequacies of measures and inadequacies of instruments. The monofilament test (Weinstein, 1993) is by far the most popular to assess tactile sensitivity. Used in clinical testing, filaments with predetermined force are pressed against the fingers of the subjects by the experimenter until the sensation of touch is felt. The force is recorded as the tactile sensitivity. The two-point discrimination test used by O'Hara et al. (1988) and by Bishu and Klute (1995a) failed to give a clear indication of loss of tactile sensitivity. Bronkema et al. (1994) have used grasp force degradation at submaximal levels of exertion with gloves as a measure of the loss of

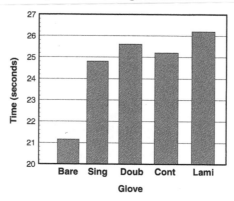

**FIGURE 6.3**    Gloves vs. pegboard time.

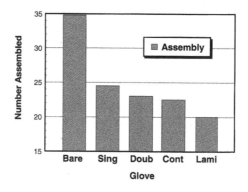

**FIGURE 6.4**    Gloves vs. number of assemblies.

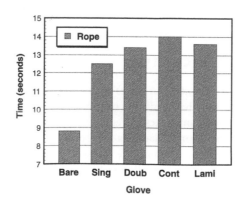

**FIGURE 6.5**    Gloves vs. rope knotting time.

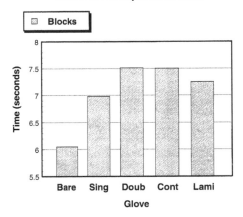

**FIGURE 6.6**  Gloves vs. blocks manipulation time.

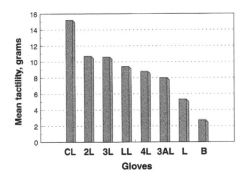

**FIGURE 6.7**  Glove effect on tactility.

tactility. Their results indicate that gloves do reduce tactile sensitivity. Desai and Konz (1983) studied the effect of gloves on tactile inspection performance and found that gloves had no significant effect on the inspection performance. In fact, they recommend that gloves be worn during tactile inspection tasks to protect the inspectors' hands from abrasion and to help in the detection of small surface irregularities. Nelson and Mital (1995) found no appreciable differences in dexterity and tactility among latex gloves of five different thicknesses. In summary, the effect of gloves on tactility has not been clearly understood and should be the focus of glove research in the future.

## 6.8  Liners

Today there is a growing trend toward the use of inner gloves or glove liners. For example, health care professionals often tend to use glove liners to prevent outer glove/skin interaction. Similarly, meat processors usually wear glove liners, while astronauts use multiple layers of liners. Almost all of the research efforts on gloves has focused on the outer glove, while liners have drawn little research attention. Using a standardized glove testing protocol, Bishu and Chin (1998) investigated the effect of a number of glove liners.

The study compared three types of liners: liners made from PTFE, cotton, and latex. A battery of evaluation tests, comprising some standardized tests and certain functional tests, was designed. The tests assessed the following capabilities: tactile sensitivity, dexterity, manipulability, strength, and effect of continuous use. The actual tests performed were the pinch test, finger strength test, monofilament test,

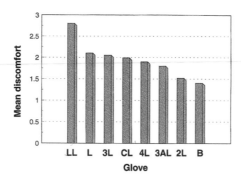

**FIGURE 6.8**   Glove effect on overall fatigue.

pain threshold test, rope-tying test, peg board test, and fatigue test. In the fatigue test the effect of continuous use of gloves was measured with Borg's RPE scale. Continuous use was simulated by making the subjects perform keyboard tasks and peg board tasks alternately for an hour. Discomfort measures were recorded every 10 minutes. Important findings were (a) liners made a distinct contribution to the performance decrements with gloves, and (b) liners had a significant effect on the overall comfort (discomfort) during extended periods of glove usage. Figure 6.7 shows the graph of effect of liners on tactility as measured with monofilament test. Figure 6.8 shows the liner effect on overall fatigue.

## 6.9   Glove Attributes

Bradley (1969) also investigated dexterity as a function of glove attributes such as snugness of fit, tenacity, and suppleness in a wide variety of 18 industrial gloves. The conclusions reached are that various glove attributes influence dexterity performance to varying extents. Bishu et al. (1987) found that glove attributes and the task performed had a significant effect on the force exertion. Wang et al. (1987) concluded that altered feedback from a gloved hand caused strength degradation. Batra et al. (1994) found grip strength reduction to be significantly correlated with glove thickness and subjective rating discomfort, and suggest that glove thickness should be minimized, while increasing the tenacity. In spite of these studies, no comprehensive model linking performance degradation with glove characteristics exists.

## 6.10   Challenges of Glove Design

In summary, gloves do reduce performance, but provide a vital protective function. Facilitation of performance, with simultaneous protection from the hazards of the work environment, are often conflicting objectives of glove design. The conflicts associated with providing primary hand protection through the use of a glove while permitting adequate hand functioning has been widely recognized. Looking at the glove attributes that cause performance differences, attributes or level of attributes that facilitate performance deteriorates safety function. This conflict poses certain challenges for the glove designer. Before attempting to design any kind of protection for the hand, it is necessary to first identify what it needs to be protected against. Human capability is limited to a narrow bandwidth of acceptable environmental conditions in which performance is not affected. There are a number of environmental hazards, often in combination, that are likely to pose a threat to the hands of workers interfacing with their workplace.

Glove material is often fixed by the environment. Environments that expose the worker to radiation, electrical, biological, fire, chemical, and extreme thermal hazards warrant that specific materials be incorporated into the hand protection, irrespective of the design. The hazard-specificity of such materials

**TABLE 6.2**   Glove Attributes as a Function of Design Parameters

| | |
|---|---|
| Prehension | Function (Design, Material) |
| Torque Capability | Function (Design, Material) |
| Dexterity | Function (Design, Material) |
| Tactile Feedback | Function (Material, User) |
| Grip Strength | Function (User, Material, Design) |
| Fit | Function (Design, Material) |
| Pressure-Pain Threshold | Function (User, Material, Design) |
| Range of Motion | Function (User, Design, Material) |
| Abrasion Resistance | Function (Material) |
| Puncture Resistance | Function(Material) |
| Cut & Tear Resistance | Function (Material) |
| Thermal Protection | Function (Material) |
| Chemical Resistance | Function (Material) |
| Biohazard Protection | Function (Material) |
| Radiation Protection | Function (Material) |
| Electrical Protection | Function (Material) |
| Vibration Protection | Function (Material, Design) |

also poses a problem for the glove designer, because the minimum thickness of the material required to provide adequate protection is usually a fixed value, limiting the designers' choice of variable parameters.

For example, a glove designed for use in an operating room or by a dental hygienist has to be capable of being sterilized either by steam or chemical disinfectants, impermeable to any potentially dangerous fluids, and of sufficient thickness and strength to maintain its integrity for a reasonable period of time.

However, in spite of the inflexibility in the materials' parameters, the glove is expected to enable the user to function without significant loss of desirable hand functions like grip strength, dexterity, range of motion, and tactile feedback. When these requirements are combined with multiple hazard condition protection requirements, glove design becomes a complex task. Table 6.2 shows the glove attributes as a function of design parameters. Muralidhar et al. (1999) suggest an ergonomic approach for glove design. Basing on published literature on force distribution in the hand during any task, they recommend that gloves should have variable thickness, with more thickness in regions where more force is exerted and less thickness in regions where force exertion is minimal. They argue that such an approach would yield gloves with minimal performance degradation and maximal protection.

## 6.11   Glove Evaluation Protocol

The question of how to evaluate a glove has always interested the designer, manufacturer, and the user of gloves. Standard evaluation protocols do not generally exist. Even in cases where they do exist, as in cases of rubber gloves used by utility people or fire fighters' gloves, the protocols are inadequate. It is recommended that a typical glove evaluation protocol include the following:

1. Strength tests including grip and pinch tests.
2. A battery of standardized tests to assess dexterity, tactility, and manipulability. Typical standard tests for these include Pennsylvania Bi-Manual Worksample Assembly Test (PBWAT), Minnesota Rate of Manipulation Test-Turning (MRMTT), Purdue peg board test, O'Connor dexterity test, and Monofilament test.
3. A battery of functional tests. This is what most of the existing glove evaluation protocols lack. Functional tests are task specific and should be appropriately designed to simulate actual tasks to be performed with the concerned gloves.

Evaluation protocols similar to one listed above have been used by O'Hara et al. (1988) and Bishu and Klute (1995a) in the evaluation of EVA gloves. Similar protocols have been used for evaluating liners by Bishu and Chin (1998).

## 6.12    Glove Standards

Existing glove standards are of three types. The standards generally describe protective requirements as in some of the U.S. Occupational Safety and Health Administration standards; or describe the protection the gloves must provide for safety as in gloves in the chemical industry or for the utility personnel; or specifically describe glove testing requirements.

## 6.13    Conclusion

In summary, the following statements can be made with regards to gloves.

1. Gloves protect the hand from the environment, but affect the performance.
2. Gloves range widely in size, type, and cost. They range from general purpose gloves to highly specialized task-specific gloves.
3. Gloves reduce grip or grasp strength capabilities, while they do not affect torque or pinch capabilities.
4. Glove reduce hand dexterity, tactility, and manipulability.
5. Providing protection without compromising performance is a continuous challenge for glove designers.

### References

Adams, S.K. and Peterson, P.J. (1988). Maximum voluntary hand grip torque for circular electrical connectors. *Human Factors*, 30 (6), pp. 733-745.

Banks, W.W. and Goehring, G.S. (1979). The effects of degraded visual and tactile information on diver work performance. *Human Factors*, 21 (4), pp. 409-415.

Batra, S., Bronkema, L.A., Wang, M., and Bishu, R.R. (1994). Glove attributes: can they predict performance? *International Journal of Industrial Ergonomics*, 14, pp. 201-209.

Bellingar, T.A. and Slocum, A.C. (1993). Effect of protective gloves on hand movement: An exploratory study. *Applied Ergonomics*, 24 (4), pp. 244-250.

Bensel, C.K. (1993). The effects of various thicknesses of chemical protective gloves on manual dexterity. *Ergonomics*, 36 (6), pp. 687-696.

Bishu, R.R., Batra, S., Cochran, D.J., and Riley, M.W. (1987). Glove effect on strength: an investigation of glove attributes. *Proceedings of the 31st Annual Meeting of the Human Factors Society*, pp. 901-905.

Bishu, R.R. and Chin, A. (1998). Inner gloves: How good are they? *Advances in Occupational Ergonomics and Safety 1998* (Editor S. Kumar) IDS Press, Ohio, 347-400.

Bishu, R.R. and Klute, G. (1995a). The effects of extra vehicular activity gloves on human performance. *International Journal of Industrial Ergonomics*, 16, pp. 165-174.

Bishu, R.R., Klute, G., and Kim, B. (1995b). Force endurance relationship: does it matter if gloves are donned? *Applied Ergonomics*, 26(3), pp. 179-185.

Bradley, J.V. (1969). Effect of gloves on control operation time. *Human Factors*, 11(1), pp. 13-20.

Bronkema, L. and Bishu R.R. (1996). The effects of glove frictional characteristics and load on grasp force and grasp control. *Proceedings of the 40th Annual Meeting of the Human Factors and Ergonomic Society*, Philadelphia, PA., pp. 702-706.

Bronkema, L., Bishu, R.R., Garcia, D., Klute, G., and Rajulu, S. (1994). Tactility as a function of grasp force: the effect of glove, pressure and load. *Advances in Ergonomics and Safety VI* (Editor: Aghazadeh), Taylor & Francis Ltd., London, pp. 627-632.

Chao, E.Y.S., An, K.N., Cooney, W.P., and Linschied, R.L. (1989). *Biomechanics of the Hand — A Basic Research Study*, World Scientific, Singapore.

Chen, Y., Cochran, D.J., Bishu, R.R., and Riley, M.W. (1989). Glove size and material effect on task performance. *Proceedings of the 33rd Annual Meeting of the Human Factors Society*, Denver, Colorado, pp. 708-712.

Cochran, D.J. and Riley, M. (1986). The effects of handle shape and size on exerted forces. *Human Factors*, 28 (3), pp. 253-265.

Cochran, D.J., Albin, T.J., Bishu, R.R., and Riley, M.W. (1986). An analysis of grasp force degradation with commercially available gloves. *Proceedings of the 30th Annual Meeting of the Human Factors Society*, pp. 852-855.

Cochran, D.J., Batra, S., Bishu, R.R., and Riley, M.W. (1988). The effects of gloves and handle size on maximum torque. *Proceedings of the 10th Congress of the International Ergonomics Association*, pp. 254-256.

Desai, S. and Konz, S. (1983). Tactile inspection performance with and without gloves. *Proceedings of the Human Factors Society*, pp. 782-785.

Dionne, E.D. (1979). How to select proper hand protection. *National Safety News*, 119, pp. 44-53.

Hallbeck, M.S. and McMullin, D.L. (1993). Maximal power grasp and three jaw chuck pinch as a function of wrist position, age and glove type. *International Journal of Industrial Ergonomics*, 11(3), pp. 195-206.

Hallbeck, M.S. and McMullin D.L. (1991). The effect of gloves, wrist position, and age on peak three-jaw chuck pinch force: a pilot study. *Proceedings of the Human Factors Society, 35th Annual Meeting*, pp. 753-757.

Hertzberg, T. (1955). Some contributions of applied physical anthropometry to human engineering. *Annals of New York Academy of Sciences*, 63, pp. 621-623.

Kamal, A.H., Moore, B.J., and Hallbeck, M.S. (1992). The effects of wrist position/glove type on maximal peak lateral pinch force. *Advances in Industrial Ergonomics and Safety IV* (Editor S. Kumar) London: Taylor & Francis, pp. 701-708.

Landsmeer, J.M.F. (1962). Power grip and precision handling. *Annals of Rheumatic Disease*, 21, pp. 164-169.

Lyman, J. and Groth, H. (1958). Prehension force as measure of psychomotor skill for bare and gloved hands. *Journal of Applied Psychology*, 42:1, pp. 18-21.

McGinnis, J.S., Bensel, C.K., and Lockhar, J.M. (1973). *Dexterity Afforded by CB Protective Gloves*. U.S. Army Natick Laboratories, Natick, Massachusetts, Report No. 73-35-PR.

McMullin, D.L. and Hallbeck, M.S. (1991). Maximal power grasp force as a function of wrist position, age, and glove type: A pilot study. *Proceedings of Human Factors Society 35th Annual Meeting*, pp. 733-737.

Mital, A., Kuo, T., and Faard, H.F. (1994). A quantitative evaluation of gloves used with non-powered hand tools in routine maintenance tasks. *Ergonomics*, 37, (2), pp. 333-343.

Muralidhar, A. and Bishu, R.R., (1994) Glove evaluation: a lesson from impaired hand testing. *Advances in Industrial Ergonomics and Safety VI*, Editor F. Aghazadeh. Taylor & Francis. pp. 619-625.

Muralidhar, A., Bishu, R.R., and Hallbeck, M.S. (1999). The development and evaluation of an ergonomic glove. *Applied Ergonomics*, 30:6, pp. 555-563.

National Safety Council, (1975). Programming personal protection: hands and fingers, *National Safety News*, 114, pp. 56-58.

National Safety Council, (1976). Keeping hands safe. *National Safety News*, 119, pp. 44-53.

Napier, J.R. (1956). The prehensile movements of the human hand. *The Journal of Bone and Joint Surgery*, 38B, pp. 902-913.

Nelson, J.B. and Mital, A., (1995). An ergonomic evaluation of dexterity and tactility with increase in examination/surgical glove thickness. *Ergonomics*, 38(4), pp. 723-733.

O'Hara, J.M., Briganti, M., Cleland, J., and Winfield, D. (1988). Extravehicular activities limitations study. Volume II: establishment of physiological and performance criteria for EVA gloves — final report (*Report number AS-EVALS-FR-8701, NASA Contract no NAS-9-17702*).

Plummer, R., Stobbe, T., Ronk, R., Myers, W., Kim, H., and Jaraiedi, M. (1985). Manual dexterity evaluation of gloves used in handling hazardous materials. *Proceedings of the 29th Annual Meeting of the Human Factors Society*, pp. 819-823.

Riley, M.W. and Cochran, D.J. (1988). Ergonomic aspects of gloves: design and use. *International Reviews of Ergonomics*, 2, Editor David J. Oborne, Taylor & Francis, pp. 233-250.

Riley, M.W., Cochran, D.J., and Schanbacher, C.A. (1985). Force capability differences due to gloves. *Ergonomics*, 28 (2) pp. 441-447.

Sudhakar, L.R., Schoenmarklin, R.W., Lavender, S.A., and Marras, W.S. (1988). The effects of gloves on muscle activity. *Proceedings of the 32nd Annual Meeting of the Human Factors Society*, pp. 647-650.

Wang, M.J., Bishu, R.R., and Rodgers, S.H. (1987). Grip strength changes when wearing three types of gloves. *Proceedings of the Fifth Symposium on Human Factors and Industrial Design in Consumer Products*, Interface 87, Rochester, NY.

Weinstein, S. (1993). Fifty years of somatosensory research: from the Semmes-Weinstein monofilaments to the enhanced sensory test. *Journal of Hand Therapy*, January-March, pp. 11-22.

Westling, G. and Johansson, R.S. (1984). Factors influencing the force control during precision grip. *Experimental Brain Research*, 53, pp. 277-284.

# 7

# Industrial Mats

Jung-Yong Kim
*Hanyang University*

## 7.1 Introduction

Prolonged standing on one's feet is very common in the workplace. Workers who are exposed to prolonged standing often experience fatigue, discomfort, and swelling of the legs and feet (Winkel, 1981; Rys and Konz, 1989). Ryan (1989) showed that supermarket cashiers, who stood 90% of their working hours, experienced discomfort mostly in the lower back area. Redfern and Chaffin (1988) reported a significant level of fatigue and discomfort in various areas of the body after prolonged standing.

Many studies have shown that the lack of venous return in the lower extremities increased discomfort during prolonged standing (Brantingham et al., 1970; Winkel and Jorgensen, 1986; Konz et al., 1990). Local muscle fatigue in the lower back area has also been observed after two hours of standing (Kim et al., 1994). Likewise, the cause of discomfort and fatigue can be different depending upon the related body parts.

In industry, floor mats have been widely distributed as a quick remedy to help reduce the discomfort and fatigue of workers. However, there is no documented guideline to choose the proper matting for an individual's working condition. This made it difficult for ergonomists or safety managers to objectively evaluate mats for their own workplaces. Therefore, in this chapter, various studies are introduced and compared to help readers understand different approaches and testing methods. Furthermore, a few tips or guidelines in the selection of a proper mat are summarized based upon the results of these studies.

## 7.2 Psychophysical Approach

A subjective rating technique for postural discomfort (Corlett and Bishop, 1976) has been employed to examine the level of discomfort in various body parts after prolonged standing. In this technique, workers can score the level of discomfort by using a body diagram, even though it only provides subjective opinions. Redfern and Chaffin (1988) used this technique to examine the overall body fatigue and leg fatigue in nine different floor conditions at the end of the workday. They asked about the discomfort level of the feet, ankle, shank, knee, thigh, hips, lower back, and upper back. They found that all the body parts except for the legs and hips indicated significant differences in a discomfort rating with a change in the floor conditions. In their study, the feet showed the highest discomfort rating followed by the ankle and shank. Regarding the floor conditions, the relatively soft mats consistently showed less discomfort than the concrete floor or the hard mat. However, the uneven soft surface showed a relatively higher rating of tiredness despite its softness. No quantitative data were reported to specify the proper compressibility of mats in this study. It was concluded that the different hardness and depth, as well as the viscoelastic property of the mat are the main factors in determining the effectiveness of the mat.

Konz et al. (1990) investigated three different mats and a concrete surface. Twenty college students stood on mats for 90 minutes, and discomfort levels were examined from the neck, shoulder, upper back, mid back, lower back, buttocks, upper leg, lower leg, ankle, hind foot, mid foot, and fore foot. As a result, all of the lower parts of the body from the buttocks down were significantly affected by the floor surfaces. All three mats in this study showed a significant reduction of discomfort compared to the concrete surface. Importantly, the compressibility of the mats was quantitatively reported based upon the technique developed by Konz and Subramanian (1989). In this study, the comfort level was inversely related to the mat compressibility. In other words, the harder the mat was, the more effective it was in reducing discomfort, which was the reverse outcome compared to the previous study by Redfern and Chaffin (1988).

Hinnen and Konz (1994) tested five different mats by using a compressibility measure (Konz and Subramanian, 1989). They tested 16 female subjects standing an entire shift for 2 days. Each hour they measured the discomfort levels of nine body regions. After they compared five mats in terms of discomfort level and compressibility, they found that there was an optimal range of thickness and compressibility of the mats to maximize the reduction of discomfort. They concluded that the most comfort can be provided when the mat is at least 1/2 inch thick and 3 to 4% compressible.

From the results of these studies, it was found that the discomfort level increased with time and appeared to be the greatest at the feet, and became progressively less and less from the feet up. Also, the surface type, thickness, and compressibility of the mat have been recognized as important factors in determining the anti-fatiguing effect during prolonged standing.

## 7.3  Physiological Approach

During a period of prolonged standing or sitting, the hampered venous and arterial circulation of the lower leg (Basmajian, 1979; Brantingham, 1970; Winkel and Jorgensen, 1986) can cause foot swelling or skin temperature change, which are signs of discomfort and tiredness of the leg.

Changes in foot dimension and skin temperature were measured by Rys and Konz (1989) and Konz et al. (1990). They examined the changes in foot length, width, thickness, and ankle thickness after standing for 90 minutes and found no significant differences with a variety of floor surfaces. The calf circumference was also measured by Kuorinka et al. (1978) after one hour of standing. The result showed 3.5 to 1 mm increase in calf circumference in an hour, but the increase somewhat leveled off toward the end of the session for half of the participants. Eventually, no difference in calf circumference was found on the three surfaces.

Skin temperature was measured from the calf and instep by Konz et al. (1990). They observed that the calf skin temperature had increased by 0.3° C for the concrete surface, while small or negative increases were recorded for the mat. Conversely, Rys and Konz (1989) reported that the calf skin temperature measured on the concrete was significantly lower, by 1.5 and 1.9° C, compared to the rubber mat after an hour of standing. No skin temperature change was found in the instep as the floor surface varied.

## 7.4  Postural Approach

The frequency of posture shifting was observed by a video recorder, and the center of gravity of body sway was measured by force platform by (Zhang et al., 1991). They found that the frequency of posture shift is a sensitive measure to show the effect of prolonged hours of standing, but it is a poor indicator of showing the difference in floor types. Kuorinka et al. (1978) also measured the frequency of a postural sway at the beginning and the end of one hour of standing. An increase of frequency was found at the end of the session, but no difference was found between the three different surfaces.

## 7.5 Biomechanical Approach

An electromyographic (EMG) study (Basmajian 1979) showed that the standing posture can be maintained by muscle activities in the solius, iliopsoas, sacrospinalis, and neck extensor. Kuorinka et al. (1978) examined the EMG signals of the solius muscle and showed a slight trend for the integrated EMG (IEMG) to rise on a concrete surface, although the initial IEMG was the lowest. Zhang et al. (1991) measured EMG from the tibialis anterior and gastrocnemius muscles during and after two hours of standing and found no differences in the fatigue level as the thickness of the floor changed.

Marras (1992) pointed out that the processed EMG used in previous studies simply indicated the level of muscle contraction that could be a very weak indicator of muscle fatigue during static standing. He suggested that the EMG power spectrum could be a more effective tool for detecting the localized muscle fatigue after prolonged standing than the processed EMG. LeVeau and Andersson (1992) used a spectral analysis to derive EMG power frequency distribution that is the second-order information of an IEMG signal. They showed that the shift of mean or median value of a frequency distribution could be used as a sign of local muscle fatigue.

Kim et al. (1994) studied muscle fatigue after a period of two hours of standing by using the EMG power spectrum. The erector spinae muscle as well as tibialis anterior and gastrocnemius muscle were examined. Three floor conditions including concrete, thin mat, and thick mat were tested. The compressibility of the mat was also quantified based on the method used by Konz and Subramanian (1989). The Kin/Com dynamometer was used to measure the 75% of maximum voluntary contraction (MVC) before and after prolonged standing. The IEMG immediately recorded after two hours of standing and the median frequency shift of the EMG power spectrum was computed. In the study, a more local muscle fatigue in the erector spinae muscle rather than the lower leg muscles was found. The significant increase of the processed IEMG signal was also observed in the leg muscles, however, that was not necessarily the sign of local muscle fatigue. It was stated that the decreased discomfort on the softer mat could be due to the active venous return following lower leg contraction on a soft surface.

## 7.6 Characteristics of Tested Mats

Various types of mats have been tested in different studies. Currently, individual mats cannot be quantitatively compared to each other because test protocols are not yet standardized for the studies. In spite of this, test results in different studies are summarized in Table 7.1. The comfort score in this table should not be used for direct comparison between mats.

## 7.7 A Standardized Protocol

A standardized protocol needs to be developed to quantitatively evaluate mats. The protocol is expected to specify the method and apparatus to measure the thickness and compressibility of material. The material can be further specified in terms of resiliency and elasticity if an apparatus such as an Instron machine is available. Moreover, surface type can be specified in terms of the shape and friction coefficient. In general, the standing period needs to be longer than an hour to see signs of discomfort. Since the standing posture may change the result greatly, the posture needs to be standardized and strictly instructed during the test.

### Standardized Compressibility Measure

Konz and Subramanian (1989) computed an average foot pressure based upon the average male body weight of 170 lb. (77.3 kg), which is 0.35 kg/cm². If one uses a 7 × 7 cm mat specimen, 17.15 kg of force

*Occupational Ergonomics: Engineering and Administrative Controls*

**TABLE 7.1**  Characteristics of Various Mats

| Mat Type | Description and Material | Thickness, mm (in) | Compression[1] (mm) | Compression[2] (%) | Source | Comfort score |
|---|---|---|---|---|---|---|
| Foam plastic | Dense surface (low resiliency, high elasticity) | 10 (0.4") | n/a | n/a | Kuorinka et al. (1978) | high |
| San-EZE | Resilient rubber | 22 (0.87") | 0.9 | 4.1 | Konz et al. (1990) Rys & Konz (1989) | high |
| Optimat | Honeycomb | 12 (0.5") | 0.7 | 5.8 | Konz et al. (1990) Rys & Konz (1990) | high |
| Traction mat | Rubber | 9.5 (0.37") | 0.7 | 7.4 | Rys & Konz (1989) | high |
| San-EZE II | n/a | 18 (0.7") | 1.7 | 9.4 | Konz et al. (1990) | intermediate |
| Footsaver | Thick solid surface | 12 (0.5") | n/a | n/a | Rys & Konz (1990) | low |
| Footsaver | n/a | 7 (0.3") | 1.3 | 18.6 | Konz et al. (1990) | low |
| Thin (Blue) | Thin hard rubber with padding | 8 (0.31") | 0.55 | 6.9 | Kim et al. (1994) | high |
| Thick (Black) | Thick hard rubber w/uneven surface | 22 (0.87") | 0.49 | 2.2 | Kim et al. (1994) | low |
| Thin | Rubber | 1.6 (0.06") | n/a | n/a | Redfern & Chaffin (1988) | low |
| Medium | Rubber | 6.4 (0.25") | n/a | n/a | Redfern & Chaffin (1988) | intermediate |
| Thick | Rubber | 9.5 (0.37") | n/a | n/a | Redfern & Chaffin (1988) | high |
| Hard w/trilaminate padding | Trilaminate padding | n/a | n/a | n/a | Redfern & Chaffin (1988) | high |
| Hard w/o trilaminate padding | No padding | n/a | n/a | n/a | Redfern & Chaffin (1988) | low |
| Viscoelastic mat | Viscoelastic material | n/a | n/a | n/a | Redfern & Chaffin (1988) | low |
| Uneven | Soft, uneven surface | n/a | n/a | n/a | Redfern & Chaffin (1988) | low |
| Workstation mat | PVC vinyl w/1/2" dia. hole | 22 (0.87") | n/a | 2.2 | Hinnen & Konz (1994) | high |
| Super sponge cushion matting | Rubber tile w/sponge base | 12.7 (0.5") | n/a | 3.3 | Hinnen & Konz (1994) | high |
| Comfort-EZE | Rubber w/5/16" dia. knobs | 9.5 (0.37") | n/a | 5.2 | Hinnen & Konz (1994) | low |
| Cushion-rib runner tread | Vinyl sponge/corrugated top | 9.5 (0.37") | n/a | 8.9 | Hinnen & Konz (1994) | low |
| Interlocking rubber matting | Rubber w/smooth top | 12.7 (0.5") | n/a | 3.5 | Hinnen & Konz (1994) | intermediate |

n/a: not available.
[1] The absolute amount of compressed part under the given pressure (Equation 1)
[2] The amount of compressed part relative to the original depth of mat

should be applied to generate the same average foot pressure on the specimen. In actual testing, 18 kg instead of 17.15 kg was used to record the compressibility of the mat. The duration of compression should be short enough (about a second) to simulate the actual foot-stepping situation. The equation computing the average adult foot pressure (Konz and Subramanian, 1989) is as follows:

$$\text{pressure} = 0.15 + 0.0026 \times \text{body weight (kg)} \tag{7.1}$$

A proper press machine should be used to control the pressure level precisely. The following are examples of devices used in previous studies:

1. Instron machine (Redfern and Chaffin, 1988), Instron model 1122 universal testing machine (Konz et al., 1990).
2. MTS Bionix 858 servo hydraulic materials testing system (Kim et al., 1994)

## 7.8   Foot Wear Conditions

The discomfort level can be greatly affected by footwear as well as the characteristics of the mat itself. That is, the standardized test also needs to be specified in terms of footwear conditions, and the results should be interpreted accordingly. The following are examples of the various footwear used in previous studies.

1. Working shoes (Redfern and Chaffin, 1988)
2. Thin-soled sneakers (Kuorinka et al., 1978)
3. Dress shoes with hard insole and sneaker with soft insole (Zhang et al., 1991)
4. Cotton socks and slippers (Konz et al., 1990)
5. Cotton socks without shoes (Kim et al., 1994)

Moreover, the final selection of mats should be made after considering the interactive effect between shoes and mats. The right combination of the two materials not only increases the comfort level but can also prevent slips and falls in the workplace. To quantitatively assess the interaction between shoes and mats, Redfern and Bidanda (1994) measured the friction level in terms of proper parameters including dynamic coefficient of friction. Leclercq et al. (1995) also suggested the use of reference conditions including lubricant, floor surface and footwear model to accurately measure the slipping resistance.

## 7.9   Suggestions for the Ergonomist

Selecting the best mat is not a simple task. Commercialized mats sometimes use materials with different compressibility for each foot. This type of mat may be able to stimulate the venous circulation of one leg better than the other. However, it has not been adequately studied to determine how effective those specially designed mats are. Presently, the simple and safe way of selecting a mat is to choose one that is "not too hard, not too soft." For example, workers who stand and walk around may need a good hard and even surface to minimize unnecessary ankle action to balance their posture. At the same time, the softness of the mat helps the blood pumping mechanism while standing quietly to reduce the discomfort level. Therefore, the combination of solid surface and a soft padding may meet the needs of workers who both walk and stand. In summary, the surface shape, softness, thickness, kind of padding, and material should be carefully considered in selecting the right mat to meet the specific requirements of an individual's workplace.

Further research should be conducted under the standardized protocol to acquire information on the anti-fatiguing and anti-slipping effect of various mats. Then, a quantitative ergonomic guideline can be developed based upon the standardized data. Eventually, such guidelines will help ergonomists and safety managers select the best mat for the individual worker.

# References

Basmajian, J.V. and DeLuca, C. 1979. Their functions revealed by electromyography, in *Muscle Alive*. Williams & Wilkins, Baltimore, MD.

Brantingham, C.R., Beekman, B.E., Moss, C.N., and Gorden, R.B. 1970. Enhanced venous blood pump activity as a result of standing on a varied terrain floor surface. *Journal of Occupational Medicine*, 12: 164-169.

Corlett, E.N. and Bishop, R.P. 1976. A technique for assessing postural discomfort. *Ergonomics*, 19(2): 175-182.

Hinnen, P. and Konz, S. 1994. Fatigue mats. *Advances in Industrial Ergonomics and Safety VI*, Taylor & Francis. 323-327.

Kim, J.Y., Stuart-Buttle, C., and Marras, W.S. 1994. The effects of mats on back and leg fatigue. *Applied Ergonomics*, 25(1): 29-34.

Konz, S., Bandla, V., Rys, M., and Sambasivan, J. 1990. Standing on concrete vs. floor mats. *Advances in Industrial Ergonomics and Safety II*, Taylor & Francis. 991-998.

Konz, S. and Subramanian, V. 1989. Footprints. *Advances in Industrial Ergonomics and Safety I*, Taylor & Francis. 203-205.

Kuorinka, I., Hakkanen, S., Nieminen, K., and Saari, J. 1978. Comparison of floor surfaces for standing work. *Biomechanics VI-B*, 207-211.

Leclercq, S., Tisserand, M., and Saulnier, H. 1995. Assessment of slipping resistance of footwear and floor surfaces. influence of manufacture and utilization of the products. *Ergonomics*, 38(2): 209-219.

LeVeau B. and Andersson, G. 1992. Output forms: data analysis and applications. Interpretation of electromyographic signals. *Selected Topics in Surface Electromyography for Use in the Occupational Setting: Expert Perspectives*. U.S. Department of Health and Human Services. 70-102.

Marras, W.S. 1992. Applications of electromyography in ergonomics. *Selected Topics in Surface Electromyography for Use in the Occupational Setting: Expert Perspectives*. U.S. Department of Health and Human Services. 122-143.

Redfern, M.S. and Bidanda, B. 1994. Slip resistance of the shoe-floor interface under biomechanically-relevant conditions. *Ergonomics*, 37(3): 511-524.

Redfern, M.S. and Chaffin, D.B. 1988. The effect of floor types on standing tolerance in industry. *Trends in Ergonomics/Human Factors V, (*Ed.) F. Aghazadeh, Elsevier Science Pub.

Ryan, G.A. 1989. The prevalence of musculoskeletal symptoms in supermarket workers. *Ergonomics*, 32: 359-371.

Rys, M.J. and Konz, S. 1989. Standing with one foot forward. *Advances in Industrial Ergonomics and Safety I, (*Ed.) A. Mital, Taylor & Francis.

Rys, M.J. and Konz, S. 1989. An evaluation of floor surfaces. *Proceedings of the Human Factors Society 33rd Annual Meeting*. 517-520.

Rys, M. and Konz, S. 1990. Floor surfaces. *Proceedings of the Human Factors Society 34th Annual Meeting*. 575-579.

Winkel, J. 1981. Swelling of the lower leg in sedentary work — pilot study. *Journal of Human Ergology*, 10: 139-149.

Winkel, J. and Jorgensen, K. 1986. Evaluation of foot swelling and lower-limb temperature in relation to leg activity during long-term seated office work. *Ergonomics*, 29(2): 313-328.

Zhang, L. Drury, C.G., and Woollet, S.M. 1991. Constrained standing: evaluation of the foot/floor interface. *Ergonomics*, 34: 175-192.

# Ergonomic Principles Applied to the Prevention of Injuries to the Lower Extremity

Steven A. Lavender

*Rush-Presbyterian-St. Luke's Medical Center*

Gunnar B.J. Andersson

*Rush-Presbyterian-St. Luke's Medical Center*

Most of the ergonomics literature dealing with the prevention and control of musculoskeletal disorders in the workplace has focused on the upper extremity and the back. Comparatively little attention has been given to lower extremity musculoskeletal disorders which occur in the workplace. One could argue that since the lower extremity problems are not well documented in ergonomic journals, the problems may not be of much practical significance. The first objective of this chapter is to review the current literature regarding occupational musculoskeletal disorders affecting the lower extremities and to demonstrate the significance of the problem. The second objective is to describe what types of intervention strategies are available to minimize the likelihood of future or recurrent injuries to the feet, ankles, knees, and hips.

## 8.1  Lower Extremity Injuries: Is There an Occupational Problem?

The sports medicine literature is full of lower extremity overuse injuries in athletes. All too often we have seen athletes relegated to the sidelines following some sort of soft tissue injury that is likely to be the effect of not just a single incidence, but rather a cumulative loading pattern during practice and competition. Luckily, in most occupational environments the intensity of the exercise is greatly diminished, however, the cumulative exposure problem still persists. Recent studies have begun to report the relationship between occupational factors and knee, hip, and foot trauma.

Lindberg and Axmacher (1988) reported the prevalence of coxarthosis in the hip to be greater in male farmers than in an age-matched group of urban dwellers. Vingard et al. (1991) classified blue-collar occupations as to whether static or dynamic forces could be expected to act on the lower extremity. The authors found that those employed in occupations that experienced greater loads on the lower extremity, namely farmers, construction workers, firefighters, grain mill workers, butchers, and meat preparation workers, had an increased risk of osteoarthrosis of the hip. Similarly, Vingard et al. (1992) found that disability pensions for hip osteoarthrosis were significantly more likely to be received by males employed as farmers, forest workers, and construction workers.

Lindberg and Montgomery (1987) reported that knee gonarthrosis (osteoarthritis), as defined by a "narrowing of the joint space with a loss of distance between the tibia and the femur in one compartment, of one-half or more of the distance in the other compartment of the same knee joint or the same compartment of the other knee, or less than 3 mm," was more common in those who had performed jobs that required heavy physical labor for a long time. Kohatsu and Schurman (1990) found that, relative to controls, the individuals with severe OA were two to three times more likely to have worked in occupations requiring moderate to heavy physical work. Anderson and Felson (1988) reported a relationship between the frequency of knee bending required in a respondent's occupation and osteoarthritis in the older working population (55 to 64 years). Moreover, these same authors have shown that the strength demands of the job were predictive of knee OA in the women from this older age group (Anderson and Felson, 1988). The authors suggest that the increased OA in those with long exposure to occupational tasks is indicative of the role of repetitive occupational exposure. Further supporting the link between material handling jobs and knee problems is the finding by McGlothlin (1996), who recently reported that beverage delivery personnel were experiencing discomfort in the knees, in addition to the anticipated discomfort in the back and shoulders. It should be recognized, though, that personal risk factors for osteoarthrosis of the knee include obesity and significant knee injury (Kohatsu and Schurman, 1990). These same authors found no relationship between leisure time activities and knee OA.

Torner et al. (1990) reported that chronic prepatellar bursitis was the predominant knee disorder in 120 fishermen who underwent an orthopedic physical examination. Forty-eight percent of the men examined showed this disorder. Interestingly, the finding was as common among younger men as in older men. The authors believe that this disorder is a secondary effect of the boat's motion. The knees are used to stabilize the body by pressing against gunwales or machinery as tasks are performed with the upper extremities. Furthermore, just standing in mild sea conditions (maximum roll angles of 8 degrees) has been shown to considerably elevate the moments at the knees as the motion in the lower extremities and the trunk are the primary means for counteracting a ship's motions (Torner et al., 1991).

The etiology of "beat knee" was described by Sharrard (1963). He reported on the examination of 579 coal miners. Forty percent of those examined were symptomatic or had previously experienced symptoms. Most of the injuries could be characterized as acute simple bursitis or chronic simple bursitis. The majority of the affected miners were colliers whose job requires constant kneeling at the mine face. There was a strong relationship between the coal seam height (directly related to roof height in a mine) and the incidence of beat knee. The incidence rates were much higher in mines with a roof height under four feet as compared with those with greater roof heights. Obviously, this factor greatly affects the work posture of the miners. With higher roof heights miners can alternate between stooped and kneeling postures, but when seams are one meter or less, the stooped posture is no longer an alternative. Gallagher and Unger (1990), for example, present recommendations for weight limits of handled materials in underground mines. Below 1.02 m these are based on miners in kneeling postures. Sharrard (1963) also speculated on the individual factors attributable to the disorder and found a higher incidence among younger men. However, this may be due to the "healthy worker effect" (Andersson, 1991) in which older miners with severe "beat knee" have left the mining occupation.

Tanaka et al. (1982) reported that the occupational morbidity ratios for workers' compensation claims of knee-joint inflammation among carpet installers was twice that found in tile setters and floor layers, and was over 13 times greater than that of carpenters, sheet metal workers, and tinsmiths. Others have shown the knees of those involved with carpet and flooring installation were more likely to have fluid collections in the superficial infrapatellar bursa, have a subcutaneous thickening in the anterior wall of the superficial infrapatellar bursa, and have an increased thickness in the subcutaneous prepatellar region (Myllymaki et al., 1993).

Thun et al. (1987) determined the incidence of repetitive knee trauma in the flooring installation professions. While all flooring installers spend a large amount of time kneeling, the authors divided the 154 survey respondents into two groups, "tilesetters" and "floor layers," based on their use of a "knee-kicker." This device is used to stretch the carpet during the installation process. These respondents were compared with a group of millwrights and brick layers whose jobs did not require extended kneeling and/or the use of a knee kicker. Of the 112 floor layers (those who used the knee kicker) the prevalence rate of bursitis was approximately twice that found in the 42 tilesetters, and over three times that found in the 243 millwrights and brick layers. However, the prevalence in both groups of flooring workers of having required needle aspiration of the knee was almost five times that of millwrights and bricklayers. These results suggest that long durations of occupational kneeling is related to fluid accumulation, yet the bursitis is due to the repetitive trauma endured by the floor layers using the knee kicker. Village et al. (1991) found that the peak impulse forces generated in the knees of carpet-layers when using the "knee-kicker" were on the order of 3000 N. The opposite knee which was supporting the body during this action had an average peak force of 893 N. Bhattacharya et al. (1985) reported knee impact forces of 2469 N (about three times body weight) for a light kick and 3019 N (or about four times body weight) for a hard kick. These light and hard kicks resulted in impact decelerations of 12.3 $g$ and 20 $g$, respectively. The authors observed that the knee kicking action during flooring installation occurred at a rate of 141 kicks per hour. Putting the knee injuries in perspective, pain was reported by 22% of questionnaire respondents in the tufting job at a carpet manufacturer. However, knees were only listed in 2.4% of the accident records. Thus, the knee is frequently the site of discomfort, although there may be few lost days associated with knee pain (Tellier and Montreuil, 1991).

Cumulative trauma injuries can take the form of stress or fatigue fractures. Linenger and Shwayhat (1992) reported training-related injuries to the foot occurred in military personnel undergoing basic training at a rate of three new injuries per 1,000 recruit days. These authors found that stress fractures to the foot, ankle sprains, and achilles tendinitis accounted for the bulk of the injuries. Anderson (1990) found the stress fractures to be most common in the distal second and third metatarsal bones but could occur in any of the bones in the foot. Giladi et al.'s (1985) findings indicated that 71% of the stress fractures in their sample of military recruits occurred in the tibia and 25% in the femoral shaft. Moreover, they found the fractures to occur later in the training process than reported by others. Jordaan and Schwellnus (1994) reported that overuse injuries, when normalized according to training hours per week, decreased from week 1 to week 4, showed a resurgence in week 5, and a large peak in the final week of training. The injury rates corresponded to the weeks in which there was increased marching and less field training.

With regard to the overuse injuries found in military recruits, some investigators have looked into aspects of lower limb morphology that may indicate which individuals will be more susceptible to injury while performing the tasks associated with military training. Giladi et al. (1991) reported the influence of individual factors on the incidence of fatigue fractures, specifically, they found that individuals with narrow tibiae, and/or a greater external rotation of the hip were more likely to experience fatigue fractures. Cowan and colleagues (1996) reported the relative risk of "overuse" injuries was significantly higher in military recruits with the most valgus knees. In addition, these authors showed that the "Q" angle, which defines the degree of deviation in the patellar tendon from the line of pull on the patella by the quadriceps muscles, was shown to be predictive of stress fractures.

In summary, several occupational risk factors have been identified which place an employee at increased risk for disorders in the lower extremity. The literature has shown that heavy physical labor and frequent knee bending are factors, especially in the older component of the work force, thereby suggesting an interaction between the age degenerative processes and cumulative work experience. In other occupations the risk of lower extremity disorders is increased through poor footing conditions. And clearly, the role of direct cumulative trauma in those employees who must maintain kneeling postures and use their knees to strike objects (knee kicker) cannot be overlooked when considering preventive measures.

## 8.2   Preventing Injury: Types of Ergonomic Controls

Several types of control mechanisms to prevent or accommodate lower extremity disorders are available. This section will focus on the techniques whereby the foot–floor interface can be optimized. This includes measures to prevent slips and falls, stress fractures, as well as improving circulation and comfort in the lower extremities for those who remain in relatively static work postures throughout the day.

### Floor Mats

Floor mats are often used for local slip protection. While inexpensive, they create a possible trip hazard, interfere with operations or cleanliness, and wear excessively (Andres et al., 1992). Several investigators have looked into the use of floor mats to reduce the fatigue effects observed in jobs that require prolonged standing. The subjects tested by Kuorinka et al. (1978) indicated through subjective ratings that they preferred to work on softer surfaces as opposed to harder surfaces. A foam plastic surface was rated the best and concrete the worst. These authors reported a moderate correlation between the subjective comfort ratings of the five surfaces tested and the order of surface hardness. However, integrated electromyographic (EMG) signals, median frequency of the EMG, measures of postural sway, and measures of calf circumference did not show any significant difference due to the floor covering. Hinnen and Konz (1994) asked employees in a distribution center to stand for two 8-hour shifts on each of five mats tested in the study. Approximately every hour the employees rated their comfort in several body regions including the upper leg, lower leg, ankle, and back. A scale of 0 (no discomfort at all) to 10 (extreme discomfort) was used. While these workers experienced relatively little discomfort, the mats with compressibility between 3 and 4% did best in the upper leg discomfort rankings as well as subject preference rankings. Marginally significant changes in the discomfort ratings were reported for the ankle. The ratings of lower leg and back discomfort showed no significant differences.

Rys and Konz (1989, 1990) reported on several anthropometric and physiological measures including changes in foot size and skin temperature at the instep and the calf. In general, the mats included in this study were significantly different from concrete in that there was greater skin temperature at both measured locations and greater comfort ratings. These authors report that the comfort was inversely related to mat compressibility.

Cook et al. (1993) used surface EMG to study the recruitment of the anterior tibialis and paraspinal muscles when standing on linoleum-covered concrete vs. an expanded vinyl 9.5 mm-thick surgical mat. After subjects stood for two sessions, of two-hour duration, on the mat and on the linoleum, it was concluded that there were no significant changes in the mean of the rectified EMG signals in either muscle due to the mats. As in the studies above, subjective data support the use of the mat.

Kim et al. (1994) tested two types of floor mats and a control condition in which subjects stood on concrete. While these authors observed muscular fatigue, as determined by a shift in the EMG median frequencies in the gastrocnemius and anterior tibialis muscle, the EMG median frequencies in these muscles were not affected by the use of floor mats. The median frequency shift in the erector spinae was reduced when subjects stood on the thinner and more compressible mat. The authors hypothesized that greater compressibility would have made for a less stable base of support, thereby, requiring more frequent postural changes in the trunk to overcome the destabilization associated with postural sway. Thus, the dynamic use of erector spinae muscles to correct for postural sway would facilitate the oxygen delivery and the removal of contractile by-products through increased blood flow. A further test of this hypothesis would evaluate whether this motion occurred only in the trunk, or if it occurred in the lower extremities which did not show the spectral shift due to the floor condition.

### Shoe Insoles

It is widely recognized that shoe design plays a critical role in the development of overuse syndromes in runners (Lehman, 1984; McKenzie et al., 1985; Pinshaw et al., 1984). Moreover, the role of the shoe in controlling lower extremity kinematics has been reviewed by Frederick (1986) and discussed by McKenzie et al. (1985). Similarly, the use of wedged insoles has been shown to alter the static posture of the lower extremity (Yasuda and Sasaki, 1987). Sasaki and Yasuda (1987) have shown the use of wedged insoles to be a good conservative treatment for medial osteoarthritis of the knee in the early stages. These authors reported that patients with early radiographic stages of osteoarthritis and who were provided a wedged insole had reduced pain and improved walking ability relative to controls without the insole.

Clearly, the lower extremity disorders reported by runners represent extreme overuse, however, the treatment and prevention mechanisms may be applicable to occupational settings where employees must stand, walk, run, or even jump during their normal work activities. Padded insoles have been investigated for the shock-abating effects on the skeletal system. Loy and Voloshin (1991) used lightweight accelerometers for measuring the shock waves as subjects walked, ascended and descended stairs, and jumped off platforms of a fixed height. The peak magnitude of the shock waves during jumping activities were approximately eight times that seen during normal walking. The results indicated that the insoles reduced the amplitude of the shock wave by between 9 and 41% depending upon the activity performed. The insoles were most effective at reducing heel strike impacts and had the largest effect with the jumping activities.

Milgrom et al. (1985) tested the effects of shock attenuation on the incidence of overuse injuries in infantry recruits. Earlier studies conducted by fixing accelerometers to the tibial tubercle showed that soldiers wearing modified basketball shoes had mean accelerations that were 19% less than soldiers wearing lightweight infantry boots. These authors also found that over the 14 weeks of basic training the modified basketball shoes reduced the metatarsal stress fractures. However, the tibial and femoral stress fractures were not affected by the shoes worn. Gardner et al. (1988) compared viscoelastic polymer insole and a standard mesh insole that were issued by platoon to over 3,000 marine recruits. While the polymer insole had good shock absorbing properties, the incidence of lower extremity stress injuries over the 12-week basic training program were unaffected by the insole used.

Several studies have been conducted to evaluate variations in insole materials. Leber and Evanski (1986) describe the characteristics of the following seven insole materials: Plastazote, Latex foam, Dynafoam, Ortho felt, Spenco, Molo, and PPT. These authors measured the plantar pressures in 26 patients with forefoot pain. All insole materials reduced the plantar pressure by between 28 and 53% relative to a control condition. However, PPT, Plastazote, and Spenco were the superior products. Viscolas and Poron were found to have the best shock absorbency of the five insole materials tested by Pratt et al. (1986). Maximum plantar pressures were found to be significantly reduced in the forefoot region with PPT, Spenco, and Viscolas, although the three materials were not significantly different (McPoil and Cornwall, 1992). In the rear foot region, however, McPoil and Cornwall (1992) report that only the PPT and the Spenco reduced the maximum plantar pressure relative to the barefoot condition. The plantar pressure in the rearfoot region was not significantly reduced with the Viscolas. Interestingly, based on the shock absorbency data from Pratt's (1988) 30-day durability test, the resilience of Viscolas, PPT, and Plastazote could be described as excellent, good, and poor, respectively. Sanfilippo et al. (1992) also reported the change in foot-to-ground contact area as a function of insole material. Plastazote, Spenco, and PPT led to a significantly greater contact area than the other materials tested.

In summary, insoles appear to be effective at modifying the lower extremity kinematics and reducing the peak plantar pressures, although their effectiveness is dependent upon the material used. Additional research is needed to clarify the effectiveness of insoles in controlling lower extremity stress injuries. Based on the previous discussion it should be clear that the effectiveness of this control strategy will be dependent upon shock absorbing capacity, the pressure dispersion, and the durability properties of the insole materials selected.

## The Foot–Floor Interface

Controlling slip and fall injuries requires a multifaceted approach. The foot–floor interface is analogous to the four-legged stool shown in Figure 8.1. To optimize the postural stability all four legs need to be

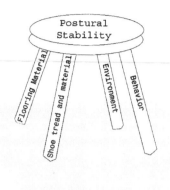

**FIGURE 8.1** The four-legged stool showing the interdependence of the factors affecting postural stability at the foot–floor interface.

in place and of equal length. The obvious legs are the flooring material and the shoe tread material and design. The environmental conditions represent the third leg as these affect the coefficient of friction between the shoe and the floor. And the fourth leg of the stool pertains to the behavior of the individual wearing the shoe. This behavioral component includes an individual's locomotion pattern, perception of environmental conditions, and allocation of the attentional resources necessary for adaptive behavior. If any of the four primary components are missing, at least to some degree, a leg of the stool is cut off by some random amount. For example, if the environmental conditions result in an oil film on the floor surface, the stool may still stand on the three remaining legs provided the shoe design and the floor material are adequate, and that the individual perceives the environmental conditions and adapts his or her behavior accordingly. Thus, the stool remains standing, although precariously, in spite of a shortened leg. If the individual did not attend to the environmental conditions the psychomotor leg would have been shortened, thereby making it unlikely that the stool will remain standing. In summary, the prevention of lower extremity injuries due to slips and falls requires attention be paid to each component or leg of the stool responsible for maintaining the body's stability.

In considering slip and trip prevention it is the dynamic friction of the interface, as opposed to static friction, which is considered more critical in determining slip potential as most slips occur when the heal initially contacts the ground (Redfern et al., 1992; Strandberg, 1983). Gronqvist et al. (1992) has quantified slip resistance by determining the coefficient of friction between the interacting surfaces and possible contaminants. These authors reported that the important counter measures against floor slipperiness are the microscopic porosity and roughness of floor coverings. Flooring materials which have rough, unglazed, raised patterns, or are made from porous ceramic tiles are best for reducing slipping hazards in areas which must maintain very high standards for hygiene. In environments where the hygiene standards can be relaxed, very rough epoxy or acrylic resin floor materials should be used. Floor surface issues become even more critical on ramps and other inclined work surfaces. Redfern and McVay (1993) reported that when walking down ramps, the *required* coefficient of friction increased in a nearly linear fashion as ramp angle increased. This was due to the high shear forces encountered during heal strike as an individual walks down a ramp.

Gronqvist and Hirvonen (1994) studied slip resistance properties of footwear on iced surfaces. They found that the shoe material significantly affected the slip resistance. Shoe heels and soles constructed from thermoplastic rubber with a large cleated area were best for dry ice conditions ($-10°$ C). Very few of the shoes tested functioned well on wet ice ($0°$ C) where there is a boundary layer of water on top of the ice. In fact, the shoes which worked best for dry ice conditions were among the worst when tested in the wet ice condition. Shoes with the sharpest cleats yielded the greatest friction readings under this condition. Hardness of the heel and sole material was not a significant factor on wet ice.

Little consensus is found in sole hardness data, and what differences exist are not of practical significance (Leclercq, 1994). In general, microcellular PU (polyurethane) heels and soles are recommended, although footwear showed less of an effect on the kinetic coefficient of friction than did variations in floor surfaces (Gronqvist et al., 1992). Relative to rubber soles and heels, Manning et al. (1985) reported that boots with microcellular polyurethane (PU) soles and heels better resisted polishing caused by smooth, wet, or oily floors; had a longer life; and had a greater average coefficient of friction. The COF of the PU improved over time, indicating that the initial smooth surface of the boots should be roughened prior to use. Tisserand (1985) recommends avoiding the use of "micro-treads" or small bumps. However, tread design and material are not independent factors when it comes to the coefficient of dynamic friction. Leclercq et al. (1994) stress that the ridges in the tread need to be as sharp as possible to wipe away contaminants on the surface. In addition, these authors point out that the tread design needs to provide channels for the surface contaminants to flow through. Slipping risk is greatest at heel-strike, and therefore, the heel of the shoe requires considerable attention. There is controversy regarding whether the use of a bevelled heel to increase contact surface area is superior to the use of a small contact surface that would result in high contact pressure (Leclercq et al., 1994).

Tisserand (1985) has promoted the concept of a mental model whereby an individual has constant input as to the friction available from the surface being walked upon. The model is updated as new information is received indicating a change in the surface conditions. Slips occur where there is a discrepancy between the mental model of the surface and the actual conditions. Thus, as Tisserand (1985) points out: "The risk of slipping lies more in the gradient of the friction coefficient of the surface than in its absolute value: such as when in a car, a small patch of ground with a low coefficient (an isolated patch of ice, for example, on a large surface with a high coefficient of friction) is more dangerous than a surface with a medium but constant coefficient of friction" (pp. 1039).

Swensen assessed the subjective judgments of surface slipperiness by having groups of iron workers and students walk across steel beams. The subjects were asked to rank the slipperiness of four types of steel coatings and four levels of surface contaminant: none, water, clay, plastic covering oil. Static coefficient of friction (COF) values ranged between .98 and .20. For the experienced and inexperienced subjects the correlation between the subjective surface ratings and the actual COF measured following the test were .75 and .90, respectively. The subjects exposed to the very slippery conditions (COF = .20) created by the oil and plastic compensated by shortening their stride length, thereby lessening the foot velocity and shear forces, and maintained the body's center of gravity within the smaller region of stability. Thus, people can detect the COF and adapt their gait accordingly. When this adaptation fails to take place, an individual is much more likely to slip and possibly fall.

Ideally, with the perfect shoe–floor interface, one with no environmental contaminants, the individual's behavior would be not be a factor. But in few cases would this exist, and then, even a small fluctuation in the environment would be enough to disrupt the balance of the now three-legged stool. Therefore, administrative control measures need to be considered as a means for maintaining the behavior necessary for stable work postures. Employees should be trained to recognize where slippery conditions are likely within a facility, and that they be alert to changing environmental conditions. Further, employees should be encouraged to report maintenance problems with machines that affect the flooring conditions.

## Stair Design

Pauls (1985) reports that only 6% of stair accidents entail slips, more often accidents are due to "overstepping." This author suggests that the overstepping occurs because the individual descending the stairs does not accurately perceive the stair width (also known as tread length). Thus, the foot is placed too far forward on the step. This scenario accounts for 19% of all stair accidents. This work highlights the interplay between engineering design factors (stair size) and behavioral factors. Research has suggested that stairs should have risers no higher than 178 mm (7 inches) and treads no shorter than 279 mm (11 inches).

A secondary issue in stair design is the complex effect of visual distractions. Pauls (1985) reported that when visual distractions were present people actually focused harder on descending the stairs. When no distractions were present people exhibited less caution. Similarly, patients commonly reported that their falls leading to femoral neck fractures were initiated with missing a step down, for example, unexpectedly stepping off a curb (Citron et al., 1985). This suggests that when the brain recognizes the potential distractions attention resources are allocated to the task at hand, whereas without some overt distraction the brain may allocate adequate attentional resources, or it may not. Archea (1985) has stressed that an older population that may not have the perceptual and motor capabilities found in younger individuals are more vulnerable to accidents on stairs. These findings suggest that the effects of distractions around stairs changes through the aging process.

## Help for Those in Kneeling Postures

Sharrard (1963) reported that there was no relationship between the type of knee pads used and the incidence of beat knee in miners. This author recorded peak pressures on the order 35.7 kg per square cm as simulated mining tasks were performed. These compression forces were shown to vary widely throughout the 2.5 second cycle time for a shoveling task. Unfortunately, the author had no instrumentation capable of determining the shear forces and the torsional moments placed on the knee during the simulated tasks. At the time of Sharrard's paper a "bursa pad" had been designed that allowed perspiration to escape, pushed coal particles away from the skin, and provided satisfactory cushioning. Although no control group was used, the author reported that of the 24 previously affected men selected to test the pad under working conditions only two reported a recurrence of beat knee after a 12-month period.

Ringen et al. (1995) reported on a new tool to reduce the knee and back trauma in those who tie rebar rods together in preparation for pouring concrete. No longer will concrete workers need to kneel or stoop for extended periods to interconnect the iron rods as this tool allows the operator to work in a standing posture.

Powered carpet stretching tools are available to remove the repeated trauma experienced by carpet layers. However, their widespread implementation depends upon educating flooring workers on the trade-offs between the additional time necessary to operate the tool and the knee disorders associated with the conventional technique.

## 8.3   Summary

Ergonomic texts historically have focused relatively little attention on the prevention of lower extremity disorders or the accommodation of individuals returning to work who have experienced a lower extremity disorder. In part this may be due to an underappreciation of the frequency and severity of occupational lower extremity disorders. Unlike many back or upper extremity disorders which have their origins in the repeated stresses placed on muscular, tendinous, and ligamentous tissues, many of the occupational lower extremity disorders occur through direct compression of the body tissues by a surface in the environment. As a result, the occupational lower extremity disorders often involve cartilaginous tissue and bone. Therefore, accommodation and prevention of these disorders occurs primarily through the optimizing the body's contact with surfaces in the environment. This chapter has illustrated some of the key ways in which this can be accomplished.

## References

Anderson, E.G. (1990). Fatigue fractures of the foot. *Injury*, 21, 275-279.

Anderson, J.J. and Felson, D.T. (1988). Factors associated with osteoarthritis of the knee in the first national health and nutrition examination survey (HANES I). *American J. of Epidemiology*, 128, 179-189.

Andersson, G.B.J. (1991). The epidemiology of spinal disorders, in J.W. Frymoyer (Ed.) *The Adult Spine: Principles and Practice*. New York: Raven, 107-146.

Andres, R.O., O'Conner, D., and Eng, T. (1992). A practical synthesis of biomechanical results to prevent slips and falls in the workplace, in S. Kumar (Ed.) *Advances in Industrial Ergonomics and Safety IV*, London: Taylor & Francis, 1001-1006.

Archea, J.C. (1985). Environmental factors associated with stair accidents by the elderly. *Clinics in Geriatric Medicine*, 1, 555-569.

Bhattacharya, A., Mueller, M., and Putz-Anderson, V. (1985). Traumatogenic factors affecting the knees of carpet installers. *Applied Ergonomics*, 16, 243-250.

Citron, N. (1985). Femoral neck fractures: are some preventable? *Ergonomics*, 28, 993-997.

Cook, J., Branch, T.P., Baranowski, T.J., and Hutton, W.C. (1993). The effect of surgical floor mats in prolonged standing: an EMG study of the lumbar paraspinal and anterior tibialis muscles. *J. Biomedical Engineering*, 15, 247-250.

Cowan, D.N., Jones, B.H., Frykman, P.N., Polly, D.W., Harman, E.A., Rosenstein, R.M., and Rosenstein, M.T. (1996). Lower limb morphology and risk of overuse injury among male infantry trainees. *Medicine and Science in Sports and Exercise*, 28, 945-952.

Frederick, E.C. (1986). Kinematically mediated effects of sport shoe design: a review. *Journal of Sports Sciences*, 4, 169-184.

Gallagher, S. and Unger, R.L. (1990). Lifting in four restricted lifting conditions. *Applied Ergonomics*, 21, 237-245.

Gardner, L.I., Dziados, J.E., Jones, B.H., and Brundage, J.F. (1988). Prevention of lower extremity stress fractures: a controlled trial of a shock absorbent insole. *American Journal of Public Health*, 78, 1663-1567.

Giladi, M., Ahronson, Z., Stein, M., Danon, Y.L., and Milgrom, C. (1985). Unusual distribution and onset of stress fractures in soldiers. *Clinical Orthopaedics & Related Research*, 192, 142-146.

Giladi, M., Milgrom, C., Simkin, A., and Danon, Y. (1991). Stress fractures. Identifiable risk factors. *American Journal of Sports Medicine*, 19, 647-652.

Gronqvist, R., Hirvonen M., and Skytta, E. (1992). Countermeasures against floor slipperiness in the food industry, in S. Kumar (Ed.) *Advances in Industrial Ergonomics and Safety IV*, London: Taylor & Francis, 989-996.

Gronqvist, R. and Hirvonen, M. (1994). Pedestrian safety on icy surfaces: Anti-slip properties of footwear, in F. Aghazadeh (Ed.) *Advances in Industrial Ergonomics and Safety VI*, London: Taylor & Francis, 315-322.

Hinnen, P. and Konz, S. (1994). Fatigue mats, in F. Aghazadeh (ed.) *Advances in Industrial Ergonomics and Safety VI*, London: Taylor & Francis, 323-327.

Jordaan, G. and Schwellnus, M.P. (1994). The incidence of overuse injuries in military recruits during basic military training. *Military Medicine*, 159, 421-426.

Kim, J.Y., Stuart-Buttle C., and Marras, W.S. (1994). The effects of mats on back and leg fatigue. *Applied Ergonomics*, 25, 29-34.

Kuorinka, I., Hakkanen, S., Nieminen, K., and Saari, J. (1978). Comparison of floor surfaces for standing work, in E. Asmussen and K. Jorgensen (Eds.) *Biomechanics VI-B, Proceedings of the Sixth International Congress of Biomechanics*, Baltimore, University Park Press, 207-211.

Kohatsu, N.D. and Schurman, D.J. (1990). Risk factors for the development of osteoarthrosis of the knee. *Clinical Orthopaedics and Related Research*, 261, 242-246.

Kuorinka, I, Hakkanen, S, Nieminen, K., and Saari, J. (1978). Comparison of floor surfaces for standing work, in E. Asmussen and K. Jorgensen (Eds.) *Biomechanics VI-B*, Baltimore: University Park Press, 207-210.

Leber, C. and Evanski, P.M. (1986). A comparison of shoe insole materials in plantar pressure relief. *Prosthetics and Orthotics International*, 10, 135-138.

Leclercq, S., Tisserand, M., and Saulnier, H. (1994). Slip resistant footwear: A means for the prevention of slipping, in F. Aghazadeh (Ed.) *Advances in Industrial Ergonomics and Safety VI*, London: Taylor & Francis, 329-337.

Lehman, W.L. (1984). Overuse syndromes in runners. *American Family Physician*, 29, 157-161.

Lindberg, H. and Axmacher, B. (1988) Coxarthrosis in farmers. *Acta Orthop. Scand.*, 59, 607.

Lindberg H. and Montgomery, F. (1987). Heavy labor and the occurrence of gonarthrosis. *Clinical Orthopaedics and Related Research*, 214, 235-236.

Linenger, J.M. and Shwayhat, A.F. (1992). Epidemiology of podiatric injuries in US Marine recruits undergoing basic training. *Journal of the American Podiatric Medical Association*, 82, 269-271.

Loy, D.J. and Voloshin, A.S. (1991). Biomechanics of stair walking and jumping. *Journal of Sports Sciences*, 9, 136-149.

Manning, D., Jones, C., and Bruce, M. (1985). Boots for oily surfaces. *Ergonomics*, 28, 1011-1019.

McKenzie, D.C., Clement, D.B., and Taunton, J.E. (1985). Running shoes, orthotics, and injuries. *Sports Medicine*, 2, 334-47.

McGlothlin, J.D. (1996). *Ergonomic Interventions for the Soft Drink Beverage Delivery Industry.* U.S. Department of Health and Human Services (NIOSH) Publication No. 96-109.

McPoil, T.G. and Cornwall, M.W. (1992). Effect of insole material on force and plantar pressures during walking. *Journal of the American Podiatric Medical Association*, 82, 412-416.

Milgrom, C., Finestone, A., Shlamkovitch, N., Wosk, J., Laor, A., Voloshin, A., and Eldad, A. (1992). Prevention of overuse injuries of the foot by improved shoe shock attenuation. A randomized prospective study. *Clinical Orthopaedics & Related Research*, 281, 189-192.

Myllymaki, T., Tikkakoski, T., Typpo, T., Kivimaki, J., and Suramo, I. (1993). Carpet layer's knee. An ultrasonographic study. *Acta Radiol.*, 34, 496-499.

Pinshaw, R., Atlas, V., and Noakes, T.D. (1984). The nature and response to therapy of 196 consecutive injuries seen at a runners' clinic. *South African Medical Journal*, 65, 291-298.

Pratt, D.J., Rees, P.H., and Rodgers, C. (1986). Assessment of some shock absorbing insoles. *Prosthetics & Orthotics International*, 19, 43-45.

Pratt, D.J. (1988). Medium term comparison of shock attenuating insoles using a spectral analysis technique. *Journal of Biomedical Engineering*, 10, 426-429.

Redfern, M.S. and Bidanda, B. (1992). The effects of shoe angle, velocity, and vertical force on shoe/floor slip resistance, in S. Kumar (Ed.) *Advances in Industrial Ergonomics and Safety IV*, London: Taylor & Francis, 997-1000.

Redfern, M.S. and McVay, E.J. (1993). Slip potentials on ramps, in *Proceedings of the Human Factors and Ergonomics Society 37th Annual Meeting*, 2, 701-703.

Ringen, K., Englund, A., and Seegal, J. (1995). Construction workers, in B.S. Levey, and D.H. Wegman (Eds.) *Occupational Health: Recognizing and Preventing Work-Related Disease*, Boston: Little, Brown, and Company, 685-701.

Rys, M. and Konz, S. (1990). Floor mats, in the *Proceedings of the Human Factors Society 34th Annual Meeting*, 1, 575-579.

Rys, M. and Konz, S. (1989). An evaluation of floor surfaces, in the *Proceedings of the Human Factors Society 33rd Annual Meeting*, 1, 517-520.

Sanfilippo, P.B., Stess, R.M., and Moss, K.M. (1992). Dynamic plantar pressure analysis. Comparing common insole materials. *Journal of the American Podiatric Medical Association*, 82, 502-513.

Sasaki. T. and Yasuda, K. (1987). Clinical evaluation of the treatment of osteoarthritic knees using a newly designed wedged insole. *Clinical Orthopaedics & Related Research*, 221, 181-187.

Sharrard, W.J.W. (1963). Aetiology and pathology of beat knee. *British Journal of Industrial Medicine*, 20, 24-31.

Strandberg, L. (1983). On accident analysis and slip-resistance measurement. *Ergonomics*, 26, 1983.

Swensen, E.E., Purswell, J.L., Schlegel, R.E., and Stanevich, R.L. (1992). Coefficient of friction and subjective assessment of slippery work surfaces. *Human Factors*, 34, 67-77.

Tanaka, S., Smith, A.B., Halperin, W., and Jensen, R. (1982). Carpet-layers knee. *The New England J. of Medicine*, 307, 1276-1277.

Tellier, C. and Montreuil, S. (1991) Pain felt by workers and musculoskeletal injuries: assessment relating to tufting shops in the carpet industry, in Y. Queinnec and F. Daniellou (Eds.) *Designing for Everyone*, 1, 287-289.

Thun, M., Tanaka, S., Smith, A.B., Haperin, W.E., Lee, S.T., Luggen, M.E., and Hess, E.V. (1987). Morbidity from repetitive knee trauma in carpet and floor layers. *British Journal of Medicine*, 44, 611-620.

Tisserand, M. (1985). Progress in the prevention of falls caused by slipping. *Ergonomics*, 28, 1027-1042.

Torner, M., Zetterberg, C., Hansson, T., and Lindell, V. (1990). Musculo-skeletal symptoms and signs and isometric strength among fishermen. *Ergonomics*, 33, 1155-1170.

Torner, M., Almstrom, C., Karlsson, R., and Kadefors, R. (1991). Biomechanical calculations of musculo-skeletal load caused by ship motions, in combination with work, on board a fishing vessel, in Y. Queinnec and F. Daniellou (Eds.) *Designing for Everyone*, 1, 293-295.

Village, J., Morrison, J.B., and Leyland, A. (1991). Carpetlayers and typesetters ergonomic analysis of work procedures and equipment, in Y. Queinnec and F. Daniellou (Eds.) *Designing for Everyone*, 1, 320-322.

Vingard, E., Alfredsson, L, Goldie, I., and Hogstedt, C. (1991). Occupation and osteoarthrosis of the hip and knee: A register-based cohort study. *International Journal of Epidemiology*, 20, 1025-1031.

Vingard, E., Alfredsson, L., Fellenius, E., and Christer, H. (1992). Disability pensions due to musculo-skeletal disorders among men in heavy occupations. *Scand. Journal Social Med.*, 20, 31-36.

Yasuda, K. and Sasaki, T. (1987). The mechanics of treatment of the osteoarthritic knee with a wedged insole. *Clinical Orthopaedics & Related Research*, 215, 162-172.

# 9

# Ergonomics of the Foot

Stephan Konz
*Kansas State University*

This chapter is divided into five sections. Section 1 (Foot/Leg) gives the anatomy, physiology, and dimensions of the foot. Section 2 (Activities) describes the activities of standing, walking, running, and stepping. Section 3 (Accidents) discusses falls, their causes and solutions. Section 4 (Fatigue/Comfort) discusses walking and standing. Section 5 (Foot Controls) briefly describes pedals and switches.

## 9.1 Foot/Leg

### Anatomy

Figure 9.1 shows the bones of the foot and ankle. The toes (foot fingers) are divided into *metatarsals* and three *phalanges* (except for the big toe, which only has two phalanges). In supporting the body, the *calcaneus* (heel) supports 50% of the weight, the 1st and 2nd metatarsal 25%, and the 3rd, 4th, and 5th metatarsal 25%. In between are two arches: (1) the medial arch (calcaneus, the talus, the navicular, the cuneiform bones, and the 1st, 2nd, and 3rd metatarsals) and (2) the lateral arch (calcaneus, talus, cuboid, and the 4th and 5th metatarsals).

Under the heel (calcaneus) is a very important shock absorber, the heel pad (about 1.8 cm thick). The bottom of the calcaneus is not spherical but has two small "mountains"; the pad reduces the pressure on these mountains, and thus on the ankle, knee, and back.

The foot is connected to the ankle with a *mortise and tenon* joint. The vertical leg of the mortise is short on the outside (*lateral* side); in addition, the ligaments holding the bottom of the fibula (lateral malleolus) to the talus and calcaneus are relatively weak. In contrast, the vertical leg of the inside (*medial*) mortise is longer, and the ligaments holding the bottom of the tibia (medial malleolus) to the talus are relatively strong.

Inward rotation (*inversion*) of the foot tends to pull the ligaments from the bone; with proper treatment, healing is usually complete in about 3 weeks. There is a danger that the injured person may not seek medical advice even with a complete tear of the ligaments (connecting either the malleolus and the talus or the tibia and fibula). Then there would be need for surgical repair and rigid fixation in a cast for 2 to 3 months.

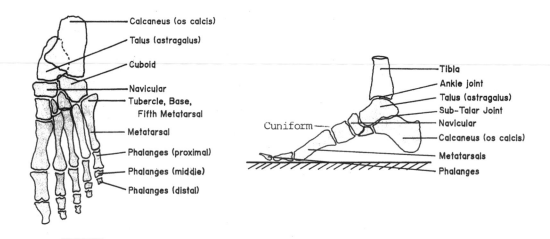

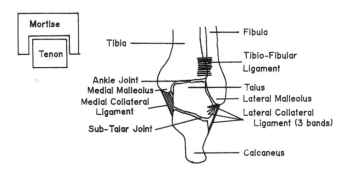

**FIGURE 9.1** The foot and ankle. The right foot is viewed from below (top left) and the outside (top right); the left ankle (bottom) is viewed from the front.

External rotation (*eversion*) of the foot tends to break one of the malleoli bones (vertical part of the mortise). These serious injuries tend to be recognized and the injured person goes to a physician.

Approximately 80% of all foot fractures involve the toes; almost all of them could be prevented by safety shoes since they lie within the area protected by the metal toe cap (Rowe, 1985).

Three venous systems drain the lower limbs: (1) a deep central system drains the muscles, (2) a superficial system drains the foot and the skin of the leg, and (3) a perforating system connects the deep and superficial systems.

## Physiology

The veins are the body's blood storage location. If the legs don't move, the blood from the heart tends to go down to the legs and stay there (*venous pooling*). This causes more work for the heart, as, for a constant supply of blood, when there is a lower ml of blood per beat, then there must be more beats. Venous pooling causes swelling of the legs (*edema*) and varicose veins. The foot swelling during stationary seated desk work can be overcome by modest leg activity (such as rolling the chair about the workstation) (Winkel and Jorgensen, 1986).

Venous pressure in the ankle of sedentary people is approximately equal to hydrostatic pressure from the right auricle. Pollack and Wood (1949) gave a mean ankle venous pressure of 56 mmHg for sitting and 87 for standing. Nodeland et al. (1983) gave 48 for sitting and 80 for standing. Pollack and Wood reported walking drops ankle venous pressure to about 23 mmHg (Nodeland et al. reported 21) in about

**TABLE 9.1** Selected Dimensions (cm) of Nude U.S. Adult Civilians

|  | Mean | | Std. Deviation | |
| --- | --- | --- | --- | --- |
|  | Females | Males | Females | Males |
| Stature | 162.9 (100%) | 175.6 (100%) | 6.4 | 6.7 |
| Crotch height | 74.1 (45%) | 83.7 (48%) | 4.4 | 4.6 |
| Knee height | 51.5 (32%) | 55.9 (32%) | 2.6 | 2.8 |
| Foot length | 24.4 (15%) | 27.0 (15%) | 1.2 | 1.3 |
| Foot breadth | 9.0 (6%) | 10.1 (6%) | .5 | .5 |

The percentages show the dimension as a percent of stature height. Shoes add 25 mm height for males and 15 mm for females. Shoes add .9 kg to body weight.

*Source:* Konz, S. 1995. *Work Design: Industrial Ergonomics*, 4th ed., p. 111. Publishing Horizons, Scottsdale, AZ. With permission.

ten steps. The fall occurs as the calf muscles contract in taking the next step before venous filling has been completed; thus additional blood is pumped out of the leg, causing a further drop in pressure when the calf muscles relax. The drop stabilizes in about ten steps when the incoming flow to the vein from the capillaries equals the flow out of the leg. Thus, walking can partially compensate for posture; for example, Nodeland et al. reported standing bench work (i.e., with occasional steps around the area) had ankle pressure approximately equal to sitting at a desk (48 mmHg).

Because of vasoconstriction, foot skin temperature (without shoes) usually is the lowest body skin temperature. Normal skin foot temperature = 33.3°C for males but 31.2° for females (Oleson and Fanger, 1973).

## Dimensions

Table 9.1 gives some dimensions for U.S. adults. A large portion of the variation in human stature is in leg length; the torso is relatively constant in height. Figure 9.2 shows the mean difference, when standing, between the inside of the two feet is about 107 mm. The distance between foot centerlines is about 107 + 90 = 197 mm (200 mm in round numbers). The distance between outside edges is 107 + 90 + 89 = 286 (300 mm in round numbers). Yet mean height for males is 1756 mm! Thus there is a base of only 200 to 300 mm for a structure of 1756 mm.

The mean pressure (Rys and Konz, 1994) on the feet can be estimated from:

$$MP = .15 + .002\,6\ WT \quad \left(r^2 = .49\right) \tag{9.1}$$

where
MP = mean pressure, kg/cm$^2$
WT = body weight, kg

Thus a 70 kg person would have an MP = .33 kg/cm$^2$. But the peak pressure could be much higher (say 10 kg/cm$^2$). Diabetics (who may have neuropathic feet) can have pressures of 20 to 30 kg/cm$^2$, leading to recurrent ulceration and eventual amputation (Boulton et al., 1984).

There is no significant difference between the left and right foot. However, for specific individuals, there often is considerable difference between the left and right foot — especially in width (see Figure 9.3.)

The technical name for differences in leg length in the same person is *leg length discrepancy* (LLD). Contreras et al. (1993), summarizing studies with N = 2377, reported that 40% of people had LLD ≤ 5 mm, 30% had LLD ≤ 9 mm, 20% had LLD ≤ 11 mm, and 10% had LLD ≤ 14 mm.

Weight of leg segments (Clauser et al., 1969), as a percent of body weight, are: 1.47 for foot, 4.35 for calf, and 10.27 for thigh; a total leg is 16.10 and both legs are 32.2. For example, the weight of both legs for a 70 kg person would average 70 (.322) = 22.5 kg.

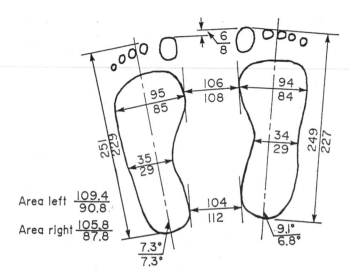

**FIGURE 9.2**   Footprint dimensions in mm (males above line; females below line); areas in mm²; angles in degrees (Rys and Konz, 1994). Toe area is 10% of contact area. They stood with the right foot slightly (6-8 mm) ahead of the left foot. The left foot (for males) averaged 7.3° to the left of the medial plane; the right foot averaged 9.1° to the right.

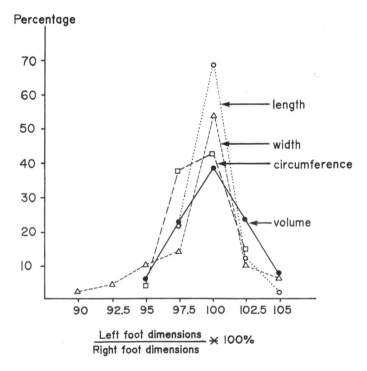

**FIGURE 9.3**   Distributions of left/right percentage of 84 Americans for foot length, width, circumference and volume (Rys and Konz, 1994). Although the mean of the left does not differ significantly from the right (i.e., the ratio does not differ from 100), for an individual the left can differ from the right.

When people stand at a work surface, there needs to be an indentation for their toes so they can stand close to the worksurface. Rys and Konz (1994) recommend a space at least 150 mm deep, 150 mm high, and 500 mm wide.

## 9.2 Activities of the Foot

### Standing

During standing, the legs will generally move occasionally. Satzler et al. (1993) recorded foot movements for 120 min of standing; people moved a foot approximately every 90 s.

### Walking

When walking, the activity of one leg has a shorter swing phase (when the foot is being passed forward) and a longer support (stepping, contact) phase (when the foot is on the ground). The support phase starts at heel strike and ends at toe-off; it has an early, passive section and a later active (propulsion) section (Davis, 1983).

At heel strike, the forward-moving heel hits the ground (causing deceleration). Continued forward motion of the body results in the forefoot contacting the ground; propulsion (acceleration) begins. The heel rises and the foot is pushed backward under the body. This tendency is resisted by friction under the sole; the body is propelled forward. The foot is everted, increasing forefoot contact area on the inner side, until only the skin around the big toe is in ground contact. Finally, contact ceases and the cycle repeats. At heel strike, horizontal velocity decreases from about 450 cm/s to 20 cm/s; heel angle to the floor changes from about 20° prior to heel contact to 0° at 100 ms after contact (Redfern and Rhoades, 1996). During a slip, instead of stopping, the heel continues to move and the leading foot moves out in front of the body.

Since the swing phase is shorter than the support phase, heel strike of the opposite limb occurs during the propulsion section of the support phase.

The length of stride (L) divided by stature height (h) varies linearly with velocity; L/h = .67 at v = .8 m/s and L/h = .9 at 1.7 m/s (Alexander, 1984).

### Running

Walking changes to running, for normal size adults, at about 2.5 m/s (6 miles/h) (Alexander, 1984) since it uses less energy (for the same speed). Running differs from walking in that both feet are off the ground for part of the stride. In addition, the heel strike should be renamed the foot strike, since the initial contact probably will be forward of the heel. Peak force is about 3 × body weight at about .1 s after contact. For walking, heel touchdown to toe push-off is .48 s, while, for running, the average contact duration is .29 s (Scanton and McMaster, 1976). After foot strike (usually on the outside edge of the foot), the foot rolls inward and flattens out (*pronation*). Then the foot rolls through the ball and rotates outward (*supination*).

### Stepping

Descending stairs demands a gait quite different from ascent (Templer, 1992). For descent, the leading foot swings forward over the nosing edge and stops its forward motion when it is directly over the tread below; the toe is pointed downward. Meanwhile the heel of the rear foot begins to rise, starting a controlled fall downward toward the tread. The heel of the forward foot then is lowered and the weight transferred to the forward foot. The rear foot then begins to swing forward. We tend to hold our center of gravity as far back as possible by leaning backward. Problems are overstepping the nosing with the forward foot, catching the toe of the forward foot, and snagging the heel of the rear foot on the nosing as it swings past. Falls tend to be down the stairs.

For ascent, the leading foot has a toe-off, swing, and first contact with the upper step. The foot is roughly horizontal. The ball of the foot is well forward on the tread; the heel may or may not be on the tread. The rear foot then rises on tiptoe, pushing down and back. The rear leg then begins the swing phase. The primary problem is catching the toe, foot, or heel of either foot on the stair nosing. Another problem is slipping by the rear foot when it pushes backward. Falls tend to be upward.

## 9.3  Accidents

### Falls

The annual death rate from falls in the United States is about 11,700; about 6,500 of these are in the home (especially affecting elderly women). About 15% of the population will have hospital treatment in their lifetime because of injuries from a stair accident (Pauls, 1985). Occupational exposures result in about 1,500 deaths and 300,000 injuries (Leamon, 1992). In industrial fatalities, falls account for about 12%, which is greater than the total for electrical current, fires, burns, and poisons of all types (Leamon, 1992). Of workers injured in falls from heights, about 20% die (Eisma, 1990).

Not all underfoot accidents result in falls, and not all falls result in a lost-time injury. Some falls result in no lost time, some result in sprains and strains, some in broken bones, and some in death. In addition, falls often are not recorded by accident recording procedures (Leamon, 1992). Thus the accident reports tend to drastically underestimate the number of falls. It needs to be emphasized that the risk of a fall varies very much with occupation; all people do not have a dry, level indoor floor. Construction workers, cleaning personnel, transportation workers, and restaurant serving personnel have higher risks (Chaffin et al., 1992).

Andersson and Lagerlof (1983) analyzed 121,000 occupational accidents resulting in injuries; 20,600 had a fall. Falls on the same level were 2/3 and falls between levels were 1/3. For the same level, the main pre-events were slipping (55%) and tripping (19%). (Manning et al. (1988) reported, for falls on the same level, 62% for slips and 17% for trips.) For falls to a lower level (e.g., from stairs, ladders, roofs, vehicles), the main pre-events were loss of support of underlying surface (28%), slipping (28%), and stepping-on-air (8%). The lower-level problem is focused in job trades such as roofers, painters, and maintenance workers. Since the fall has a greater distance, the body velocity and resulting deceleration become greater.

Falls which occur when the person is carrying something are especially dangerous. The object carried decreases stability as a function of the torque above the ankle (weight × object height above ankle). Other problems are that the arms cannot be used for balance (to prevent a fall), to grab a railing, or to break the fall impact.

### Causes of Falls

Falls can occur from slips (unexpected horizontal foot movement), trips (restriction of foot movement), and stepping-on-air (unexpected vertical foot movement). Loss of balance and falls can occur without a slip, trip, or stepping-on-air. Examples would be from alcohol or drugs, fainting, etc. Or, people who fall may have blood pressure problems or foot problems (Gabel et al., 1985).

The elderly may be more likely to fall because of deterioration in postural control mechanisms and decrements in visual acuity, strength, endurance, reaction time, and motor control. Furthermore, on falling, the elderly are more likely to sustain a bone fracture (due to osteoporosis) (Maki and Fernie, 1990).

#### Slips

Slips primarily occur during foot pushoff and heel strike. During pushoff, the person falls forward (less common and less dangerous). In addition, during pushoff, most of the weight has already been transferred to the other foot. If a slip occurs during heel strike, the person falls backward. The critical time is .05 to .1 s after heel strike. Leamon (1992) defines a microslip as a slip less than 2 cm, a slip as 8 to 10 cm, and a slide as uncontrolled movement of the heel. Microslips occur very often and normally are not perceived

**TABLE 9.2** Minimizing Slips and Slip Effects

1. Eliminate the lubricant.
   A. Avoid spilling of lubricant.
   B. Quickly clean up lubricant. Note that the lubricant (water, mud) could be on the shoe. "Lubricants" also can be solid objects such as coins, paper clips, hairpins, screws, and metal chips.
   C. Don't add lubricants during cleaning (e.g., don't use an oil mop to clean a waxed floor).
2. Choose good flooring.
   A. Carpet is best (high friction, low effect of lubricants)
   B. For hard surfaces, stainless steel and ceramic are worst as they are the smoothest. Grooved or porous floors reduce lubricant problems but are hard to clean.
   C. Use mats and duckboards (elevated slatted flooring) for local areas where wet floors are common. For example, building entrances often are wet due to tracking in of water and snow. Machines using oil or coolant can be a problem. Mats should have beveled edges to reduce tripping and holes to encourage drainage.
3. Choose good shoes. The heel is critical.
   A. Bevel the rear of heel (reduce the contact angle during heel strike to 0 from 10 to 15°).
   B. Have a tread to penetrate (squeeze out) the lubricant.
   C. Have soft material to increase contact area, and thus grip. Slip resistance of shoes increases after about 5 km of walking so tests on new shoes underestimate the slip resistance (Leclercq, Tisserand, and Saulnier, 1995).
4. Walk carefully (short slow steps).
   A. Keep body center of gravity within stride.
   B. Reduce heel angle at heel strike.
5. If there is a slip, eliminate the fall (e.g., handrails).
6. If there is a fall, decrease consequence of the fall.
   A. Less distance to fall.
   B. Lower impact force/pressure (use soft surfaces such as carpet; minimize sharp objects).

by the person. A slip is perceived and the person typically jerks the upper body, moves the arms, etc. — but does not fall. A slide involves loss of control and usually a fall.

During a slip, there normally is a "lubricant" (water, oil, grease, dust, ice, snow) either on the surface or on the shoe heel (Leclercq et al., 1994).

Slips can also occur, with stationary feet, during pushing and pulling. Although the feet slip, there does not tend to be a fall and injury.

Slips also can occur when the "ground" slopes (front to back or side to side). Examples are ramps and ladder rungs. When moving a cart up or down a ramp, stay above the cart to prevent injury from the cart if it "gets loose." Outdoor walkways often are sloped, have poor illumination, and have water and ice as lubricants.

In the special outdoor circumstances of snow and ice, slipping can be very common; in Finland, slipping outdoors is 10 times more common in winter (Gronqvist and Hirvonen, 1995). The most danger occurs when the ice is "wet" (i.e., close to the freezing point) as the water is a lubricant. For this situation, the best shoes have a soft heel/sole (Shore A hardness <60) made of thermoplastic rubber and exhibiting a large apparent contact area (good tread). Cleats (spikes, studs) are effective if they can penetrate the ice (i.e., if ice is close to the freezing point); if they cannot penetrate (i.e., ice is too hard (say at $-10°$ C), then shoes with spikes are very slippery. Strewing sand on ice is effective on wet ice (i.e., close to freezing) but has relatively little effect on dry ice. Adding salt to melt ice may work if the temperature is close to freezing; if the water then evaporates, this is good, but if the result is just a lubricant added to the ice, the result is bad.

Table 9.2 summarizes how to reduce slips.

## Trips

Trips occur during swing. As the foot swings forward, it hits an obstacle and the person falls forward. Thus, in contrast to slips where the problem is excessive horizontal leg movement, with trips the problem is lack of leg movement. Outdoor trips often occur from uneven surfaces (walkways, parking lots) which the person expects to be even. Indoor trips tend to be from objects on the floor or stairs. Usually there is a visual problem.

### Stepping-On-Air

Stepping-on-air occurs when the foot has unexpected vertical movement. This can occur on steps when the distance between stairs is unequal; it can occur when there is a hole in the ground; it can occur when there is no ground (i.e., "cliff," edge-of-scaffold, unexpected step, step on spiral stairs, unexpected curb or ramp). Very commonly, stepping-on-air occurs with "single steps" (small changes in elevation) such as curbs or one-step changes in floor level. Steps descending from large trucks and off-road vehicles often present problems. On steps, the fall usually occurs when descending; the fall can be for a considerable distance.

In some cases, the surface is there initially but breaks or moves (step breaks, floor mat slides, a chair used as a stepstool moves). Ladder "feet" need to be non-slip as slipping of the ladder base is the most common pre-event for portable ladder accidents (Alexsson and Carter, 1995). Often stepping-on-air has a visual cause.

## Solutions for Falls

The goals are to: (1) prevent the fall and (2) reduce the consequences of the fall.

### Prevent the Fall

Scaffolds and work platforms should have a waist-high (107 cm) guardrail as well as a 10 cm high toeboard (reduces slips over the edge as well as reducing falling objects). The top of the guardrail should discourage sitting.

Since the primary problem on stairs is overstepping (Pauls, 1985), for safe stairs: (1) have easily visible steps, (2) provide treads that are long enough, and (3) provide handrails that are both within reach and graspable.

Visual solutions consider both the quantity of light and the quality of light. Note that not everyone's vision is perfect. For example, consider people not wearing their glasses, vision of elderly, the "out of focus" of steps when people wearing bifocals descend stairs, etc.

The quantity of light typically is increased with fixed sources (ceiling lights, street lights, etc.). But lamps fail; the resulting lack of light is especially critical for stairs. One solution is to have two lamps illuminate critical areas. Portable sources such as flashlights are a temporary solution. Too much light causes glare; solutions include nonreflecting surfaces and glare shields for both natural and artificial illumination.

The quality of light is also important. Ideally, the light should give moderate shadows because shadows aid depth perception. Depth perception is improved by using multiple sources and considering the orientation (direction) of the light. Camouflage consists in obliterating contrasts; we are concerned with anticamouflage. Contrast is especially important on steps.

Because walking is automatic, attention must be drawn to steps, especially if the step is "camouflaged" so there is an "ambush." Do not distract attention from a stairs by providing a "view" as a person begins the descent. Call attention to steps by changing the color of the floor (e.g., red carpet on stairs vs. green carpet on approaches), having a handrail (especially for "one-step" stairs), changing wall color on stair walls (e.g., paint changes color and descends at the angle of the stairs), having the handrail color contrast with the wall and stair, and avoiding carpet patterns which confuse depth perception (e.g., narrow strips with strong contrast).

For stairs, the key for friction is the nosing, not the tread. Thus, have a high coefficient of friction for the nosing. Outdoor stairs often are lubricated by rain and snow. Such stairs should have a wash (slope of less than 1:60) to permit water to drain. Perhaps a roof is feasible — even if it doesn't give 100% protection. Prevent water (from the ground or a building) from draining onto the stairs.

Handrails on stairs help prevent falls. The handrail should permit a power grip (11 to 13 cm circumference), have a clearance from the wall of about 3.8 cm, and be 89 to 97 cm above the stair (Konz, 1994).

A handrail must be within reaching distance; at a minimum there should be a handrail on the right side descending. For detailed information on stair design, see Templer (1992) and NFPA (1991).

For mounting/dismounting vehicles, use the *three-contact rule* (at each phase of mounting/dismounting, at least three limbs should maintain contact with steps or handles at the same time).

Ramps for people should have a maximum angle of about 5°. If handtrucks are pushed up the ramp, there should be a landing at least every 3 m in elevation. Have a nonskid surface in the center of each lane. Ramps should have handrails and, if used by vehicles, heavy curbs. If vehicles use a ramp exposed to rain or snow, have a 60 cm strip of abrasive metal plate in the track of each wheel; attach it to the concrete with countersunk holes and flat-head expansion screws. Maximum ramp angles for vehicles are 3° for a power-operated hand truck, 7° for a powered platform truck, 10° for a low-lift pallet-skid truck, 10° for an electric fork truck, and 15° for a gasoline fork truck (Konz, 1994).

Don't use the hands to carry objects while on ladders and stairs.

If a person knows the surface is slippery, walking behavior can be changed. A short stride length reduces foot velocity, gives smaller foot shear force at the heel/ground interface, and keeps the body center of gravity between the feet. Leaning forward helps keep the center of gravity between the feet and, if you fall, it will be forward instead of backward. Sun et al. (1996) report that, when walking down a ramp, people over age 35 decrease stride length and steps/min.

## Reduce the Consequences of the Fall

The solution depends upon the task and environment. For example, for workers on a scaffold or roof, use a full-body harness attached to an anchorage point. Another choice is safety nets.

A fall down some stairs is comparable to falling into a hole with jagged rocks at the bottom. For falls on stairs or on the same level, carpet can reduce peak body deceleration on the hip by 20% over hard floors (Makie and Fernie, 1990). (Of course, in addition, carpets have high coefficients of friction and thus have very few slips.) Stair landings reduce the distance of a fall. To minimize impact injuries, stairs, handrails, and balustrading should be free from hostile elements such as projecting elements, sharp edges, and corners.

Box 1 discusses hard surface floors.

---

### Box 1
### Hard Surface Floors

Concrete floors can either be: (1) the wearing surface itself or (2) a base for other materials (e.g., terrazzo, plastic tiles or sheeting, or carpet).

If concrete is the wearing surface, it must be durable, have satisfactory slip resistance, and be easily cleaned.

Concrete typically is poured in slabs with joints. There are two types of joints: (1) contraction joints (5 to 10 mm wide) and (2) expansion (isolation) joints (about 20 mm wide, filled with compressible material). The expansion joints are not very necessary in a temperature-controlled building and are potential trip hazards as the slab shifts. Good design is to minimize contraction joints in a building. In addition, a horizontal reinforcing dowel between slabs will reduce slab tilting and thus trip hazards.

If tile or plastic is placed on concrete, the problem tends to be slips rather than trips.

If liquids are used or stored in an area, there will be spills. Install drains and slope the floor appropriately.

## 9.4   Fatigue/Comfort

The discussion is divided into walking and standing.

### Walking

The primary problem is the shock of heel strike being transmitted up the foot, leg and back. For shoe solutions, see Box 2.

The energy cost of walking depends upon the terrain, with a hard surface giving the minimum cost (Pandolf et al., 1976):

$$\text{WLKMET} = C \left( 2.7 + 3.2 \ (v - .7)^{1.65} \right) \tag{9.2}$$

where
    WLKMET= Walking metabolism, W/kg of body weight
    C = Terrain coefficient
      = 1.0 for treadmill, blacktop road
      = 1.1 for dirt road
      = 1.2 for light brush
      = 1.3 for hard-packed snow; C = 1.3 + .082 (foot depression, cm)
      = 1.5 for heavy brush
      = 1.8 for swamp
      = 2.1 for sand
    v = velocity, m/s (for v > .7 m/s (2.5 km/h))

---

### Box 2
### Shoes

Athletic shoes are divided into running shoes (designed for forward movement) and court shoes (designed for quick side to side movement).

For running, the main problem is the shock of foot strike, yielding arthritis of the knee and hip, Achilles tendonitis, low back pain, and shin splints. Overpronation gives knee pain.

For walking, heel strike is less forceful (1.5 × body weight) so cushioning is less critical (but still desirable); flexibility in the sole allows the normal heel-to-toe roll. For walking, deceleration properties of the cushioning material is important; for standing, time is not as critical and material stiffness and maximum compression is relevant to comfort (Goonetilleke and Himmelsbach, 1992).

At a given walking speed, the energy expenditure values increase by .7 to 1.0% per additional 100 g shoe weight; this increase in energy consumption is approximately 5 times higher than the same weight on the upper body (Smolander et al., 1989). Legg and Mahanty (1986) found it took 6.4 times as much energy to carry a kilogram on the feet as on the back.

Feet often swell so, for fit, buy shoes when your feet are swollen (late in the day). In addition, your left and right foot might vary slightly. Thus buy shoes with at least four pairs of eyelets as they increase the adjustment possibilities.

People with low arches (footprint has a broad connection of two areas) will be more comfortable with shoes with a straighter "inner line" (difficult to distinguish left from right shoe).

Boots support the ankle and calf as well as increasing insulation; they are especially useful for side support, such as when walking outdoors. Boots also protect against chemicals and animal products (such as fats and oils); some boot materials have a better life than others.

For impact protection, use a steel-toe shoe; in some industries (such as mining), metatarsal guards are also used.

## Standing

Problems can occur with floor temperature and static electricity. However, the primary problem is lack of circulation in the leg and static loading of the muscles.

When wearing normal shoes, 23° C is optimal comfort for floors for standing and walking people; use 25° C for sedentary people (ASHRAE, 1993). Heavy carpet will save about 1% of the total energy used to heat the building (Hager, 1977).

Static electricity is a problem in industries such as electronics. Some solutions are: (1) raise humidity in the air above 40%, (2) use conductive carpets (e.g., with carbon fibers), (3) use an antistatic floor mat, (4) connect the operator to ground with a static-bleed wrist strap, (5) use shoes with static-dissipating soles.

Teitelman et al. (1990) reported preterm births occurred more often (7.7%) when women had jobs with prolonged standing; the rate for sedentary jobs was 4.2% and for active jobs was 2.8%.

Avoid static standing by sitting, by walking, and by shifting posture while standing. If static standing is required, consider a cushioned floor or a footrest (Whistance et al., 1995).

### Sitting

Perhaps the person can sit instead of stand. Sitting does tend to restrict movement of the shoulders and thus reach distance. A compromise is a sit-stand seat, such as that used by post office workers sorting mail into boxes. Nijboer and Dul (1987) reported a sit-stand seat was beneficial for upholstery workers even though they could use the seat for only part of the work cycle. Another technique for supporting part of the body weight is to have something to lean on — typically a counter. Another possibility is to sit part of the time. For example, service personnel often are required by management to stand when serving customers; they should be provided chairs for times when there are no customers. Seats should be available for factory personnel during breaks. One firm used swing-down benches on the wall of an aisle; during work they were up against the wall but during worker breaks they were pivoted into sitting position.

### Walking

As mentioned earlier, blood circulation in the foot can be brought to normal in as few as ten steps. Thus design the job so there is occasional walking (such as to get supplies, dispose of materials).

### Shifting Posture

It also is possible to shift the posture while standing — remember the bar rail. Bar rails are designed to improve the comfort of those standing at the bar. Satzler et al. (1993) studied four conditions: (1) standing with one foot on a 100 mm high, flat platform, (2) standing with one foot on a 100 mm 15° angled platform, (3) standing with one foot on a 100 mm high, 50 mm diameter bar, and (4) standing with both feet flat on a concrete floor. The three standing aids were preferred over no aid; the two platforms were better than the bar. Note that bar patrons not only support their feet on bar rails but lean on the bar.

### Cushioned Floor

The entire floor can be cushioned (carpeted) or the floor can be cushioned locally with a mat. Mats can also be used to raise the feet off the floor (raise above liquid) and act as a frictional surface (avoid slips).

Brantigham et al. (1970) studied a "varied terrain" floor mat with nonuniform resilience density; each placement of the foot caused a slight change in horizontal angle of the foot during weight-bearing. The concept is that many foot problems are due to the "over flat" nature of the built environment. The varied terrain mat enhanced circulation in the lower extremities (reduced venous pressure) and increased skin temperature of the calf .3 to 1.0° C (improved circulation to the surface); about 2/3 of the subjects reported they were less tired when using the special mat.

Additional studies have been done on floor mats (Kuorinka et al., 1978; Rys and Konz, 1988; Redfern and Chaffin, 1988; Rys and Konz, 1989; Rys and Konz, 1990; Konz et al., 1990; Zhang et al., 1990; Stuart-Buttle et al., 1993; Redfern, 1995). Summarizing:

- Floor mats improve comfort (over hard-surfaced floors). Comfort may be increased in the back as well as the legs.
- Mats should compress but not too much. Optimum is about 6% under the feet of a 70 kg adult.
- Mats should have beveled edges to reduce tripping and falling.
- Mats should have a nonslip surface; drain holes may be useful to aid drainage of fluids. In addition, the mat should not slip on the floor.
- Mats may have to be cleaned periodically (e.g., in food service environments). In these cases, large mats are difficult to handle.
- If a raised work platform is used to stand on, the surface should be resilient rather than rigid (i.e., wood or plastic, not steel). The platform also should have a high ratio of surface to holes (i.e., you are not standing on "knives").

## 9.5   Foot Controls

Although most controls are operated by the hands, some controls are operated by the foot. The foot does not have the dexterity of the hand, but it is connected to the leg instead of the arm so it can exert more force. A leg has approximately 3 times the strength of an arm. A foot control also reduces use of the hand/arm.

Foot controls can be divided into pedals and switches.

### Pedals

Pedals can be used for power and control. Power generation can be continuous (bicycle) or discrete (nonpowered automobile brake pedal). For information on continuous power, see Whitt and Wilson (1982) and Brooks et al. (1986).

Discrete power generally is applied by one leg; there does not seem to be any advantage to using the left or right leg. Force using both feet is about 10% higher than using just one foot (Van Buseck, 1965).

A control example is an auto accelerator pedal.

### Switches

A foot switch can actuate a machine (such as a punch press). Generally the foot remains on the switch so the time and effort of moving the foot/leg is not important.

On–off controls (such as faucets, clamping fixtures) can be actuated by lateral motion of the knee as well as vertical motion of the foot. The knee should not have to move more than 75 to 100 mm; force requirements should be light. Hospitals use knee switches to actuate faucets to improve germ control on the hands.

Avoid foot pedals/switches which must be operated while standing, because they tend to distort posture and cause back problems.

### Defining Terms

**Calcaneus:**   The heel bone; see Figure 9.1.
**Edema:**   Swelling of legs due to fluid retention.
**Eversion:**   External rotation of the foot.
**Inversion:**   Inward rotation of the foot.
**Metatarsals:**   Bones in the foot; see Figure 9.1.
**Mortise and tenon joint:**   A type of joint; see Figure 9.1.
**Lateral:**   The outside (side farthest from the centerline).
**Leg length discrepancy:**   Differences in leg length (in the same person).
**Medial:**   The inside (side closest to the centerline).
**Phalanges:**   Bones in the foot; see Figure 9.1.
**Pronation:**   Rolling inward (toward the centerline) of the foot.

**Supination:** Rolling outward (away from centerline) of the foot.

**Three-contact-rule:** Rule used on ladders and steps. At least three limbs should be in contact with steps or handles at all times.

**Venous pooling:** Pooling of blood in the veins of the legs.

# References

Alexander, R. 1984. Stride length and speed for adults, children and fossil hominids. *American J. of Physical Anthropology,* 63: 23-27.

Alexsson, P. and Carter, N. 1995. Measures to prevent portable ladder accidents in the construction industry. *Ergonomics,* 38 (2): 250-259.

Andersson, R. and Lagerloff, E. 1983. Accident data in the new Swedish information system on occupational injuries. *Ergonomics,* 26 (1): 33-42.

ASHRAE, 1993. *Handbook of Fundamentals,* ed., Parsons, R. Am. Society of Heating, Refrigeration and Air Conditioning Engineers, Atlanta, GA.

Boulton, A., Franks, C., Betts, R., Duckworth, T., and Ward, J. 1984. Reduction of abnormal foot pressures in diabetic neuropathy using a new polymer insole material. *Diabetes Care,* 7(1): 42-46. Jan-Feb.

Brantigham, R., Beekman, B., Moss, C., and Gordon, R. 1970. Enhanced venous pump activity as a result of standing on a varied terrain floor surface. *J. of Occupational Medicine,* 12(5): 164-169.

Brooks, A., Abbott, A., and Wilson, D. 1986. Human-powered watercraft. *Scientific American,* 256 (12):120-130, December.

Chaffin, D., Woldstad, J., and Trujillo, A. 1992. Floor/shoe slip resistance measurement. *American Industrial Hygiene Association J.,* 53 (5): 283-289.

Clauser, C., McConnville, J., and Young, J. 1969. *Weight, Volume and Center of Mass of the Human Body, AMRL-TR-70,* Aerospace Medical Research Laboratory, Dayton, OH.

Contreras, R., Rys, M., and Konz, S. 1993. Leg length discrepancy, in *The Ergonomics of Manual Work,* eds. W. Marras, W. Karwowski, and L. Pacholski, 199-202, Taylor & Francis, London.

Davis, P. 1983. Human factors contributing to slips, trips and falls. *Ergonomics,* 26 (1): 51-59.

Eisma, T. 1990. Rules changes, worker training help simplify fall prevention. *Occupational Health and Safety,* 52-55, March.

Gabell, A., Simons, M., and Nayak, U. 1985. Falls in the healthy elderly: predisposing causes. *Ergonomics,* 28 (7): 965-975.

Goonetilleke, R. and Himmelsbach, J. 1992. Shoe cushioning and related material properties. *Proceedings of the Human Factors Society,* 519-522.

Grondqvist, R. and Hirvonen, M. 1995. Slipperiness of footwear and mechanisms of walking friction on icy surfaces. *Int. J. of Industrial Ergonomics,* 16:191-200.

Hager, N. 1977. Energy conservation and floor covering materials. *ASHRAE Journal,* 34-39, September.

Konz, S., Bandla, V., Rys, M., and Sambasivan, J. 1990. Standing on concrete vs. floor mats, in *Advances in Industrial Ergonomics and Safety II,* 526-529, ed., B. Das, Taylor & Francis, London.

Konz, S. 1994. Change-in-level. *Facility Design: Manufacturing Engineering,* 118-122, Publishing Horizons, Scottsdale, AZ.

Konz, S. 1995. *Work Design: Industrial Ergonomics,* Publishing Horizons, Scottsdale, AZ.

Kourinka, I., Haakanen, S., Nieminen, K., and Saari, J. 1978. Comparison of floor surfaces for standing work. *Biomechanics VI-B; International Series on Biomechanics, Proceedings of 6th Int. Congress on Biomechanics,* 207-211, University Park Press, Baltimore, MD.

Leamon, T. 1992. The reduction of slip and fall injuries: Part 1--Guidelines for the practitioner and Part II — The scientific basis (knowledge base) for the guide. *Int. J. of Industrial Ergonomics,* 10: 23-34.

Leclercq, S., Tisserand, M., and Saulnier, H. 1994. Assessment of the slip-resistance of floors in the laboratory and in the field: Two complementary methods for two applications. *Int. J. of Industrial Ergonomics,* 13: 297-305.

Leclercq, S., Tisserand, M., and Saulnier, H. 1995. Assessment of slipping resistance of footwear and floor surfaces. *Ergonomics,* 38 (2): 209-219.

Legg, S. and Mahanty, A. 1986. Energy cost of backpacking in heavy boots. *Ergonomics,* 29(3): 433-438.

Manning, D., Ayers, I., Jones, C., Bruce, M., and Cohen, K. 1988. The incidence of underfoot accidents during 1985 in a working population of 10,000 Merseyside people. *J. of Occupied Accidents,* 10: 121-130.

Maki, B. and Fernie, G. 1990. Impact attenuation of floor coverings in simulated falling accidents. *Applied Ergonomics,* 21(2): 107-114.

NFPA, *NFPA Life Safety Code 1991,* NFPA, 1 Batterymarch Park, Quincy, MA 02269.

Nijboer, I. and Dul, J. 1987. Introduction of standing aids in the furniture industry, in *Musculoskeletal Disorders at Work,* ed. P. Buckle, 227-233, Taylor & Francis, London.

Nodeland, H., Ingemansen, R., Reed, R., and Aukland, K. 1983. A telemetric technique for studies of venous pressure in the human leg during different positions and activities. *Clinical Physiology,* 3: 573-576.

Oleson, B. and Fanger, P. 1973. The skin temperature distribution for resting man in comfort. *Archives des Sciences Physiologigues,* 27(4): A385-93.

Pandolf, K., Haisman, M., and Goldman, R. 1976. Metabolic energy expenditure and terrain coefficients for walking on snow. *Ergonomics,* 19: 683-690.

Pauls, J. 1985. Review of stair safety research with an emphasis on Canadian studies. *Ergonomics,* 28(7): 999-1010.

Pollack, A. and Wood, E. 1949. Venous pressure in the saphenous vein at the ankle in man during exercise and changes in posture. *J. Applied Physiology,* 1: 649-662, March.

Redfern, M. and Chaffin, D. 1988. The effects of floor types on standing tolerance in industry, in *Trends in Ergonomics/Human Factors V,* ed., F. Aghazadeh, 401-405, Elsevier, Amsterdam.

Redfern, M. 1995. Influence of flooring on standing fatigue. *Human Factors,* 37 (3): 570-581.

Redfern, M. and Rhoades, T. 1996. Fall prevention in industry using slip resistance testing, in *Occupational Ergonomics,* Eds., Bhattacharya, A. and McGloughlin, J., 463-476, Marcel Dekker, New York.

Rowe, M. 1985. *Orthopaedic Problems at Work,* Perinton Press, Fairport, NY.

Rys, M. and Konz, S. 1988. Standing work: carpet vs. concrete, *Proceedings of the Human Factors Society,* 522-526.

Rys, M. and Konz, S. 1989. An evaluation of floor surfaces, *Proceedings of the Human Factors Society,* 517-520.

Rys, M. and Konz, S. 1990. Floor mats. *Proceedings of the Human Factors Society,* 575-579.

Rys, M. and Konz, S. 1994. Standing. *Ergonomics,* 37(4): 677-687.

Satzler, C., Satzler, L., and Konz, S. 1993. Standing aids. *Proceedings of the Ayoub Symposium,* 29-31, Texas Tech. University, Lubbock, TX.

Scanton, P. and McMaster, J. 1976. Momentary distribution of forces under the foot. *J. Biomechanics,* 9:45-48.

Smolander, J., Louhevarra, V., and Hakola, T. 1989. Cardiorespiratory strain during walking in snow with boots of different weights. *Ergonomics,* 32(1): 3-13.

Stuart-Buttle, C., Marras, W., and Kim, J. 1993. The influence of anti-fatigue mats on back and leg fatigue. *Proceedings of the Human Factors and Ergonomics Society,* 769-773.

Sun, J., Walters, W., Svensson, N., and Lloyd, D. 1996. The influence of surface slope on human gait characteristics: a study of urban pedestrians walking on an inclined surface. *Ergonomics,* 39(4): 677-692.

Templer, J. 1992. *The Staircase: Studies of Hazards, Falls, and Safer Design,* MIT Press, Cambridge, MA.

Teitelman, A., Welch, L., Hellenbrand, K., and Bracken, M. 1990. Effect of maternal work activity on preterm birth and low birth weight. *Am. J. Epidemiology,* 131: 104-113.

Van Buseck, C. 1965. Excerpts from maximal brake pedal forces produced by male and female drivers. Research Report EM-18, Warren, MI: GM, Jan.

Whistance, R., Adams, L., van Geems, B., and Bridger, R., 1995. Postural adaptations to workbench modifications in standing workers. *Ergonomics,* 38(12): 2485-2503.

Whitt, F. and Wilson, D. 1982. *Bicycling Science,* MIT Press, Cambridge MA.

Winkel, J. and Jorgensen, K. 1986. Evaluation of foot swelling and lower-limb temperature in relation to leg activity during long-term seated office work. *Ergonomics,* 29(2): 313-328.

Zhang, L., Drury, C., and Woollet, S. 1991. Constrained standing: evaluating the foot/floor interface. *Ergonomics,* 34: 175-192.

## For Further Information

Konz, S. 1995. *Work Design: Industrial Ergonomics,* 4th edition, Holcomb Hathaway, Scottsdale AZ. This popular textbook concisely summarizes many aspects of job design and gives detailed design guidelines.

Konz, S. 1994. *Facility Design: Manufacturing Engineering,* 2nd edition, Holcomb Hathaway, Scottsdale, AZ. Gives many details and design recommendations for design and arrangement of industrial facilities.

*Ergonomics.* This journal, published in England, publishes articles on ergonomics from authors around the world.

# Section II
## Low Back Disorders

# 10

# Epidemiology of Back Pain in Industry

Gunnar B.J. Andersson
*Rush-Presbyterian-St. Luke's Medical Center*

## 10.1  Introduction

Epidemiologic research in low back pain (LBP) has been, and still is, hampered by methodologic problems in definition, classification, and diagnosis. Objective evidence of existing low back pain is often lacking, and people's recall of previous episodes is poor. The intermittent nature of low back pain complicates prevalence studies, and studies of disability due to LBP are influenced by legal and socioeconomic factors. Methodologic problems also exist in the quantification of physical exposures that might be of etiologic importance.

In general, data about back pain may be obtained from official health registers or by retrospective, prospective, or cross-sectional surveys of general populations or of specific industrial populations. Such data are useful in defining the magnitude of the problem. Care must be taken when interpreting these data, however. As mentioned above there is no consensus on classification and diagnosis, making it difficult to rely on insurance and hospital data (Wood and Badley, 1987). Sickness absence and disability data are heavily influenced by work conditions and the legal and socioeconomic situation, and there is a poor correlation between tissue injury and disability.

Data from workers' compensation claims are affected by several inherent biases (Abenhaim and Suissa, 1987): (a) all workers are not covered by worker compensation programs; (b) the claims data are mainly administrative and therefore, while accurate on absence and cost, lack validity on symptoms and diagnosis; (c) all workers with back pain do not file a claim, and many do not stay away from work.

## 10.2  The Magnitude of the Problem

The magnitude of any health problem is measured by prevalence and incidence. In a prevalence study, the presence of LBP and other important variables is determined at one point in time (point prevalence) or during one period of time (period prevalence) for each member of the population studied or for a representative sample. Incidence may be defined as the number of people who develop LBP over a specified time period, such as their lifetimes (lifetime incidence, which is synonymous with lifetime prevalence) or in a single year (annual incidence). In short, prevalence means all cases of LBP, whereas incidence means all *new* cases of LBP.

## 10.3   National Studies

Information obtained from different countries is considered separately because the differing socioeconomic factors of these populations may influence the results. This is particularly true for disability data, which are significantly determined by local legal, social, and economic factors.

### United States

Between 10 and 17% of adults in the U.S. have a back pain episode in a given year (Cunningham and Kelsey, 1984; Deyo and Tsui-Wu, 1987; Praemer et al. 1992). In about one-third of these the pain is severe and chronic. In a large National Health Survey performed from 1985 through 1988, about 4.1 million persons per year reported a "disc disorder," and another 4.6 million a "back strain" (Praemer, 1992). Other epidemiologic data show that back pain is the most frequent cause of activity limitation in people below age 45, the second most common reason for patient visits (over 14% of new visits are for back pain), the fifth ranking reason for hospitalization, and the third most common reason for surgical procedures (Praemer et al., 1992; Taylor et al., 1994; Hart et al., 1995; Andersson, 1997). About 2% of the U.S. workforce (500,000 workers), report compensation back injuries each year (National Safety Council, 1991). The frequency of surgical procedures for back related conditions has risen dramatically in the U.S. over the past two decades. In fact, it has more than doubled from 1979 to 1990. In that later year, 279,000 back operations were performed on adults; 232,500 without fusion and 46,500 fusions (Taylor et al., 1994).

### United Kingdom

British surveys place low back pain at the top of the list of medical conditions as well. In 1992–93 there were 81 million certified back-related sick days, and about 7 million visits to general practitioners for back pain (National Back Pain Association, 1994). During the same period there were 33,000 work-related back injuries. Frank (1993) reported that back pain was the single largest cause of sick leave in 1988–89, responsible for 12.5% of all sickness absence days.

### Sweden

Swedish national insurance data show a consistent sickness absence in percent of all annual sickness absence from the early sixties to the late eighties. In the 1961 to 1971 period, the average absence was 12.5% or 1% of all workdays (Helander, 1973). In 1983 the percent was 10.9, and in 1987, 13.5% (Nachemson, 1991). Unfortunately, the number of sickness absence days rose dramatically during that period so that the percent of insured sick listed for back pain rose from 1% of the working population in 1970 to 8% in 1987, the number of days per absence rose similarly from 20 to 34 days per year, and the cost in terms of lost production increased 16-fold (Table 10.1). Retirement and disability pensions caused by back pain rose by 6000% from 1952 to 1987.

During 1983 and 1984, a prospective Swedish study analyzed all patients who were sicklisted for LBP in a district of Gothenburg containing 49,000 subjects from 20 to 65 years of age (Choler et al. 1985). A total of 7,526 sickness absence episodes for LBP were reported over an 18-month period. Fifty-seven percent of patients recovered in 1 week, 90% in 6 weeks, and 95% after 12 weeks. At the end of a year 1.2% remained work disabled. Those with sciatica were out of work for longer periods of time than were

**TABLE 10.1** Estimated Sicklisting for LBP and Associated Cost Due to Loss of Production in Sweden

| Year | % of Insured[1] | Number of Days | Cost of Loss of Production[2] ($ Million[3]) |
|------|-----------------|----------------|--------------------------------------|
| 1970 | 1 | 20 | 52 |
| 1975 | 3 | 22 | 179 |
| 1980 | 4 | 25 | 285 |
| 1987 | 8 | 34 | 806 |

[1]4.7 million
[2]Based on 1987 wages and social costs.
[3]Assuming $1 = 6 Sw. Cr.
Adapted from Nachemson AL (1991): Back Pain. Causes, diagnosis and treatment. The Swedish Council of Technology Assessment in Health Care. Stockholm.

patients who had back pain only. Recurrent pain and disability occurred in 12% over the 18-month period of observation.

## Canada

Lee et al. (1985) analyzed data on musculoskeletal complaints based on the 1978 to 1979 Canada Health Survey. A prevalence of 4.4% with "serious back and spine problems" was calculated. The total number of disability days exceeded 21 million, and the average sickness absence period was 21.4 days. There was no difference in prevalence between men and women.

## 10.4 Cross-Sectional Studies

A cross-sectional epidemiologic study is one in which a population is studied at a single point in time, or over a defined period, in an attempt to evaluate all members of that population. In the past decades several cross-sectional studies have been performed. Table 10.2 presents the prevalence and lifetime incidence of LBP, as determined by some of these studies. The prevalence rates vary from a low of 12.0% to a high of 35.0%. Some authors report a higher prevalence in females, but others found no difference. The lifetime incidence rates are higher and range from 48.8% to 69.9%.

### United States

In 1973, Nagi, Riley, and Newby determined the prevalence rates of persistent back pain of persons between 18 and 64 years residing in Columbus, Ohio (Nagi et al., 1973). A random sample of 1,135 subjects was studied, of whom 203 (18%) reported "often being bothered with pain in the back." Of those with back problems, 62% had had a spine radiograph; 26% had worn a back support; and 4% had had back operations.

Frymoyer et al. (1980, 1983) performed a retrospective and cross-sectional analysis of 1,221 males 18 to 55 years of age who had enrolled in a family practice facility from 1975 to 1978. Almost 70% had had LBP. When the data from that study were extrapolated to the 50 million working American males in the age group 18 to 55, it was calculated that 38.5 million workdays are lost annually. Patients with severe LBP had significantly more leg complaints, sought more medical care and treatment for LBP, and had lost more time from work for this reason when compared to subjects with no or moderate LBP. Sciatica-like symptoms were present in 28.9% of the males with moderate LBP and 54.5% of the males with severe LBP. Objective reports of numbness were present in 14.0% of the males with moderate LBP and 37.4% of those with severe LBP, while weakness was reported by 17.9% of those with moderate LBP and 44.0% with severe LBP.

**TABLE 10.2**  Prevalence and Lifetime Incidence of LBP in Difference Cross-Sectional Studies

| Lifetime Incidence (%) | Prevalence (%) | | Study Group | | | | |
| | Point | Period | N | Age | Sex | Comment | Reference |
| --- | --- | --- | --- | --- | --- | --- | --- |
| 62.6 | 12.0 | — | 449 | 30–60 | M | | Biering-Sorensen (1983) |
| 61.4 | 15.2 | — | 479 | 30–60 | F | | |
| 48.8 | — | — | 692 | 15–72 | F | | Hirsch, et al. (1969) |
| 60.0 | — | — | 1193 | 25–59 | M | Industrial Population | Hult (1954) |
| 69.9 | — | — | 1221 | 28–55 | M | | Frymoyer, et al. (1980) |
| — | 18.0 | — | 1135 | 18–64 | M/F | | Nagi, et al. (1973) |
| 61.0 | — | 31 | 716 | 40–47 | M | 1-month period | Svensson and Andersson (1983) |
| 67.0 | — | 35 | 1640 | 38–64 | F | 1-month period | Svensson, et al. (1988) |
| 51.4 | 22.2 | — | 3091 | 20+ | M | | Valkenburg and Haanen (1982) |
| 57.8 | 30.2 | — | 3493 | 20+ | F | | |
| — | 12.9 | — | 3316 | — | M/F | 8 work groups | Magora (1972) |
| — | — | 25 | — | 40–59 | M | 1-year period | Gyntelberg (1974) |
| — | 29.0 | — | 575 | 55– | M/F | | Bergenudd and Nilsson (1988) |
| 75.0 | — | 21 | 7217 | 30– | M/F | 1-month period | Heliovaara (1989) |
| 58.0 | — | 36 | 2667 | — | M/F | Adults, 1-year period | Walsh (1992) |
| 59.0 | 33.0 | — | 4000 | — | M/F | Adults | Skovrow, et al. (1993) |
| — | 18.0 | — | 4256 | — | F | Family care employees | Moens, et al. (1993) |

Andersson GBJ (1997): The epidemiology of spinal disorders, in *The Adult Spine: Principles and Practice*, 2nd edition, J.W. Frymoyer, Ed. Lippincott-Raven, Philadelphia, pp. 93-141.

Studies by Kelsey (1975a,b) and Kelsey and Hardy (1975) sampled 20- to 64-year-olds residing in the New Haven (Connecticut) area who had lumbar X-rays taken over a two-year period for suspected herniated nucleus pulposus. The researchers divided the sample into those with surgically confirmed herniated discs and those who had probable or possible herniated discs based on clinical signs and symptoms. She was able to define a variety of risk factors related to the diagnosis of herniated lumbar disc; including sedentary occupations, driving of motor vehicles, chronic cough and bronchitis, lack of physical exercise, participants in certain sports (baseball, golf, and bowling), suburban residence, and pregnancy.

Kelsey et al. (1984a,b) and Kelsey and Golden (1988) later performed another case-control study in Connecticut from 1979 to 1981 with minor methodologic modifications. The study population was 20- to 64-year-old women and men who had had X-rays and myelograms at various health centers in New Haven and Hartford. As in the previous study, they were divided into those with surgically confirmed disc herniations and those with probable or possible disc herniations. A control group of nonback patients admitted for in-hospital services was matched for sex and age. A number of possible risk factors were studied and odds ratios determined. Frequent lifting and twisting were both significant risk factors, as were driving and smoking.

The prevalence of low back pain in elderly people (over age 65) was determined in a survey of 3,097 persons living in rural parts of Iowa (Lavsky-Shulan et al., 1985). Twenty-four percent of the women and 18% of the men had low back pain in the year preceding the survey, and 40% had back pain at the time of the interview. Five percent of the population had been operated on.

## Scandinavia

A number of studies have been performed in the city of Göteborg, Sweden (about 450,000 inhabitants). Four are reviewed here. Hirsch, Jonsson, and Lewin (1969) interviewed 692 women (15 to 72 years of age), selected at random to represent the adult Swedish female population. The lifetime incidence of LBP was 48.8% and increased with age up to 55 years, after which no further increase was noted. Horal (1969) and Westrin (1970, 1973) studied a random sample of subjects who in 1964 had been sicklisted for LBP by physicians in Göteborg, Sweden. They were compared to a control group matched with respect to sex, age, and sickness benefit but not previously sicklisted for LBP. Of the total group, Horal studied 212 pairs of probands and controls, and

**TABLE 10.3**   Data on Prevalence and Use of Medical Services from
Two Retrospective Cross-Sectional Surveys in Göteborg, Sweden

|  | Men (Age 40–47) (N = 940) | Women (Age 38–64) (N = 1,760) |
| --- | --- | --- |
| Lifetime prevalence | 61 | 66 |
| One-Month prevalence | 31 | 35 |
| Chronic pain | 3.5–4 | |
| Physician visit | 40 | 38 |
| Radiograph | 23 | 30 |
| Hospitalized | 3.5 | 3.4 |
| Operation | 0.8 | 1.0 |

Adapted from Svensson HO, Andersson GBJ (1983): Low back pain in forty to forty-seven year old men: Work history and work environment factors. *Spine* 8:272-276. and Svensson HO, Andersson GBJ (1989): The relationship of low-back pain, work history, work environment, and stress: A retrospective cross-sectional study of 38- to 64-year-old women. *Spine* 14:517-522.

shortly thereafter Westrin studied 214 (78% of the base material). Ninety-five percent (95%) of the probands had had LBP in the preceding 3 to 4 years, and 52% had ongoing pain at the time of the interview. In the control group, the corresponding figures were 49% and 27%, respectively. This means that sickness absence statistics severely underestimate the true frequency of low back pain.

Svensson and Andersson (Andersson et al., 1983; Svensson, 1982; Svensson and Andersson, 1982, 1983; Svensson et al., 1983) studied a randomized sample of 940 40- to 47-year-old men in Göteborg, Sweden. Seven hundred and sixteen men were interviewed, and information about the remaining 234 was obtained from the Swedish National Health Insurance Office. Thirty-three percent of all sickness absence episodes experienced during their working life were spine related, constituting 47% of all sickness absence days; 3.6% were totally disabled and 4% had been off work more than 3 months because of LBP in the 3 years preceding the study. Forty percent had consulted a physician, 3.5% had been admitted to a hospital, and 0.8% had been operated on because of their LBP (Table 10.3).

The same study design was later used to survey 1,640 38- to 64-year-old women (Svensson et al., 1988; Svensson and Andersson, 1989). Of these, 19% had been off work because of LBP in the preceding three-year period, 3.5% for 3 months or longer. About 2.6% of 38- to 49-year-old women had significant work disability, whereas the corresponding percentage among 50- to 64-year-olds was 5.9.

Biering-Sorensen (1982) sampled 82% of all 30- to 60-year-old inhabitants in Glostrup, Denmark. There were 449 men and 479 women. An extensive questionnaire regarding low back problems was administered along with objective measurements of spine function. Twelve months after the examination 99% of the study population completed a follow-up questionnaire on LBP occurring in the intervening period. The lifetime prevalence/incidence of LBP appears in Figure 10.1 along with the one-year period and point prevalence data. In general, increasing age was associated with increasing episodes of LBP. Work absence at some time was reported by 22.5% of those who had LBP, 10% had needed some job adjustment, and 63% had changed their jobs because of back pain. Of those who had experienced LBP, 60% had consulted a physician, 25% a specialist, and 15% a chiropractor (Biering-Sorenson, 1983). About 30% had had radiographs taken of the lumbar spine, 4.5% had been admitted to a hospital, and 1% had been operated on because of LBP.

The prevalence rate of sciatica and its impact on Finnish society was estimated by Heliovaara 1988 (see also Heliovaara et al., 1987), based on a sample of 8,000 persons representative of the Finnish population aged 30 or over. Sciatica was present in 5.3% of men and 3.7% of women. In both genders the prevalence rates were highest in the 45- to 64-year-old group. The prevalence of definite herniated discs was 1.9% for men and 1.3% for women. Low back syndrome other than sciatica was present in 12.5% of men and 17% of women. Disability due to lumbar disc syndrome was estimated at 3.5% in men, 4.5% in women.

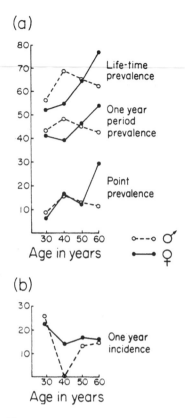

**FIGURE 10.1**   Top: Prevalence rates of "low back trouble" by age and sex. (Redrawn from Biering-Sorensen F (1982): Low back trouble in a general population of 30-, 40-, 50-, and 60-year-old men and women. Study design, representativeness, and basic results. *Dan Med Bull* 29:289.) Bottom: Incidence of "low back trouble" over one year in a general population. (From Biering-Sorensen F (1982): Low back trouble in a general population of 30-, 40-, 50-, and 60-year-old men and women. Study design, representativeness, and basic results. *Dan Med Bull* 29:289. With permission.)

## Israel

Magora and Taustein (1969) studies a random sample of 3,316 individuals taken at random. Present and past LBP was determined. Four hundred twenty-nine (12.9%) were found to suffer from LBP at the time of the survey (point prevalence), and 92% (394) of those had pain on and off from 6 months to 11 years or more before the investigation. The majority of the subjects with LBP did not take sick leave (57.8%) and, of those who did, 29.4% had absence periods from 1 to 10 days.

## The Netherlands

A study of 3,091 men and 3,493 women 20 years of age and older was performed by Valkenburg and Haanen (1982) in 1975 through 1978 in the Dutch city of Zoetermeer. A questionnaire, physical examination, and radiographs were obtained. The prevalence of LBP in men and women increased slightly with age up to 65 years, and thereafter decreased (Table 10.4). Disc prolapse, defined by clinical signs and symptoms, was found in 1.9% of men and 2.2% of women. Considerable disability was attributed to LBP: 85% had recurrences; 30% had LBP for more than 3 months, and 30% had become bedridden at some point by their symptoms. Nearly half of the men and one-third of the women reported that they had been unfit to work because of LBP at some time, and 8% of the men as well as 4% of the women had changed their jobs because of these complaints. Twenty-eight percent of the men and 42% of the women had consulted a physician for LBP.

**TABLE 10.4**   Low Back Complaints and Work Incapacity in the Zoetermeer Study

|                    | Men (%) | Rel % | Women (%) | Rel %* |
|--------------------|---------|-------|-----------|--------|
| Point-prevalence   | 22.2    |       | 30.2      |        |
| Lifetime Incidence | 51.4    |       | 57.8      |        |
| >3 Months          | 14.3    | 28    | 19.6      | 34     |
| Unfit for Work     | 24.3    | 47    | 19.5      | 34     |
| Work Change        | 4.2     | 8     | 2.4       | 4      |
| Recurrences        | —       | 85    | —         | 85     |

*Rel % refers to proportion among those with a lifetime history of low back pain.

**TABLE 10.5**   Rates of Selected Back Operations in the United States per 100,000 of General Population

| Procedure     | 1979 | 1981 | 1983 | 1985 | 1987 | % Increase |
|---------------|------|------|------|------|------|------------|
| Laminectomy   | 31   | 36   | 41   | 41   | 38   | 23         |
| Discectomy    | 59   | 57   | 81   | 96   | 103  | 75         |
| Lumbar Fusion | 5    | 9    | 10   | 18   | 15   | 200        |

Deyo RA, Cherwin D, Conrad D, Volinn E (1991): Cost, controversy, crisis: Low Back Pain and the Health of the Public. *Ann Rev Publ Health* 12:141-56.

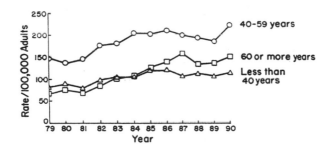

**FIGURE 10.2**   Low back surgery rates per 100,000 adults, by age, 1979-1990. (Adapted from Taylor VM, Deyo RA, Cherkin DC, Kreuter W (1994): Low back pain hospitalization. Recent United States Trends and Regional Variations. *Spine* 19:1207-1213.)

## Belgium

About 4,000 individuals were studied to explore the influence of sociocultural and employment variables on back pain (Skovron et al., 1994). The lifetime incidence of back pain was 59%, and 33% had ongoing back pain (point prevalence). Age (OR = odds ratio, 2.0) and female gender (OR = 1.42) were associated with an increased risk of a LBP history. Sociocultural factors influenced the risk of first-time back pain episodes, but not of disability or severity.

## Hospitalizations and Operations

Volinn et al. (1994) examined the National Hospital Discharge Survey for time trends (1979–1987), and for geographic variations (1987). The U.S. rate of lower back surgery increased 49% over the time period reviewed, while the rate of nonsurgical low back pain hospitalization decreased by 33%. Table 10.5 illustrates the dramatic increase in back operations in the U.S. and breaks it down into actual procedures (Deyo et al., 1991). A comparatively larger increase occurred in fusions than in laminectomies or discectomies. Volinn

**TABLE 10.6**   Hospitalizations in the U.S. for Back Conditions in 1988, Based on First Listed Diagnosis by ICD.9.CM.Code

| Diagnosis | # Hospitalizations | ICD.9.CM.Code |
|---|---|---|
| Intervertebral Disc Disorders | 417,000 | 722 |
| Other and Unspecified Back Disorders | 178,000 | 724 |
| Fracture of the Vertebral Column | 76,000 | 805 |
| Spondylosis and Allied Disorders | 75,000 | 721 |
| Sprains and Strains of Other and Unspecified Parts of Back | 55,000 | 847 |
| Sprains and Strains of Sacroiliac Region | 42,000 | 846 |

From National Hospital Discharge Survey, 1988.

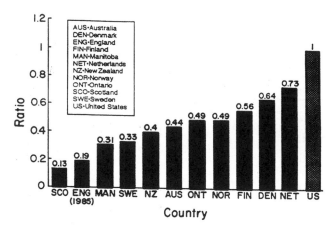

**FIGURE 10.3**   Ratios of back surgery rates in 11 countries or Canadian provinces compared to the back surgery rate in the U.S. (1988–1989). Adapted from Cherkin DC, Deyo RA, Loeser JD, Bush T, Waddell G (1994): An international comparison of back surgery rates. *Spine* 19:1201-1206.)

et al. (1994) also report large regional variation in hospitalization, surgical rates, and length of stay. They found that in 1987 the rate of surgery ranged from 77/100,000 adults in the Northeast region to 146/100,000 in the South region. Variations in surgical rates exist between areas of each region as well. Thus, among counties in the State of Washington rates of surgery for low back pain varied from 11.5/100,000 to 172/100,000, an almost 1.5-fold difference (Volinn, 1992). This indicates that cultural differences and practice patterns have a major influence on hospitalizations and procedures. Taylor et al. (1994) further analyzed the National Hospital Survey Data from 1979 through 1990. The increase in the number of surgical procedures continued from 1987 to 1990. Over the 11-year time period surveyed, adult low back operations increased by 55% from 147,500 (1979) to 279,000 (1990). This corresponds to an increase from 102 to 158 per 100,000 adults. The rise was particularly great for fusions which increased by 100% from 13 to 26/100,000. In 1990, there were 46,500 lumbar fusions and 232,500 low back pain operations without fusion. The upward trends occurred among all age groups, but were particularly great for patients aged 60 and older (Figure 10.2). Nonsurgical hospitalizations, on the other hand, decreased from 402/100,000 in 1979 to 150/100,000 in 1990. Table 10.6 shows the hospitalizations for back conditions in 1988.

There are marked international variations in rates of back surgery (Deyo et al., 1991, Cherkin et al., 1994). Nachemson (1991) has reported that the number of operative procedures for herniated nucleus pulposus (HNP) in Sweden has remained stable at 200/million inhabitants/year from the mid-1950s through the 1980s. The corresponding numbers in Finland (1967–1977) were 350 (Heliovaara, 1988) and in Great Britain 100 (Benn and Wood, 1975; Wood and Badley, 1987). The number of people operated on for HNP per million/year in the U.S. has been assessed in the past as ranging between 450 to 900

**TABLE 10.7**  Prevalence of Back and Spine Impairments in the United States, 1988

| Impairment by | | # in millions | per 1,000 population |
|---|---|---|---|
| Total | | 15.431 | 64.1 |
| Gender | Males | 6.701 | 57.4 |
| | Females | 8.730 | 70.3 |
| Race | White | 13.957 | 68.7 |
| | Black | 1.137 | 38.7 |
| | Other | 0.336 | — |
| Age | 0–17 | 0.714 | 11.2 |
| | 18–44 | 8.295 | 80.5 |
| | 45–64 | 4.105 | 90.1 |
| | 65–74 | 1.333 | 75.9 |
| | 75–84 | 0.780 | 87.2 |
| | 84– | 0.203 | 93.6 |

Adapted from Praemer A, Furnes S, Rice DP (1992): Musculoskeletal conditions in the United States. *Amer Acad Orthopaed Surg,* Park Ridge, IL pp. 1-199. Based on the National Health Interview Survey 1985–88.

(Kane, 1980; Frymoyer, 1988; Kelsey and White, 1980). In 1990, it was 1310 (Taylor et al. 1994). A recent study compares the back surgery rates of thirteen countries and provinces revealing that in the United States, it is at least 40% higher than in any other country, and more than five times higher than in Scotland and England (Figure 10.3). Differences in the underlying prevalence of back pain was not felt to explain the differences in surgical rates. Cultural differences, differences in practice patterns, and availability are more likely explanations.

## Chronic Back Pain

Limited data is available about the prevalence of chronic back pain. The NHANES II data suggest that 33.7% of those reporting back pain lasting for at least 2 weeks had pain for 6 months or longer. This percentage is similar to the one reported by Valkenburg and Haanen for a Dutch population, where 34% of those with back pain had a duration of 3 months or longer (14.3% of all males, 19.6% of all females). Praemer et al. (1992) used the 1988 National Health Interview Survey (1985–1988) to determine the frequency of impairment in the U.S. Impairment in the survey was defined as a chronic or permanent defect representing a decrease or loss of ability to perform various functions. Musculoskeletal impairment was the most prevalent impairment in people up to age 65, and back and spine impairments the most frequently reported subcategory of musculoskeletal (51.7%). The annual rates vary significantly by gender and age group (Table 10.7). Back and spine impairments were more common in females (70.3/1,000) than males (57.4/1,000), and more common among whites (68.7/1,000) than blacks (38.7/1,000). In 1988, back and spine impairments resulted in over 185 million restricted activity days (21.0/impairment), including 83 million bed days (5.4/impairment) (Table 10.8). About 56% of restricted activity days occurred among women.

Rossignol et al. (1988) (see also Abenhaim et al., 1988) followed a cohort of 2,341 cases, randomly sampled to represent occupational back compensated individuals in the state of Quebec (Canada) in 1981. From the initial injury date, 6.7% of compensated workers were still absent from work after 6 months, accounting for 68% of work days lost and 76% of the total compensation cost for low back pain. When the cumulative absence was calculated over the three years, 9.7% of workers were absent for 6 months or longer. This increase illustrates the recurrent nature of back pain. A logistic regression model was used to calculate risk factors associated with absences of 6 months or more. Age and site of symptoms were the two most important variables. An increase in age of 23 years doubled the odds of accumulating at least 6 months of absence and lumbar symptoms were 2.86 times more likely than thoracic symptoms

**TABLE 10.8**    Restricted Activity Days and Bed Days/Impairment for Back or Spine Disorders in 1988 by Major Population Subgroup

|        |        | Rescheduled Activity Days/Impairment | Bed Days/ Impairment |
|--------|--------|-------------------------------------|----------------------|
| Total  |        | 12.0                                | 5.4                  |
| Gender | Male   | 14.5                                | 5.9                  |
|        | Female | 10.1                                | 5.0                  |
| Race   | White  | 10.1                                | 4.2                  |
|        | Black  | 31.8                                | 15.4                 |
| Age    | >65    | 11.5                                | 5.4                  |
|        | 65–    | 15.0                                | 5.3                  |

Adapted from Praemer A, Furnes S, Rice DP (1992): Musculoskeletal conditions in the United States. *Amer Acad Orthopaed Surg*, Park Ridge, IL pp. 1-199.

to become chronic. The odds ratio for gender and occupation was not statistically significant. When occupation was eliminated, sex achieved statistical significance, however (OR 1.92; 95% CI = 1.19-3.11).

van Doorn (1995) studied a self-employed subset of Dutch dentists, veterinarians, physicians, and physical therapists. Twenty-three percent of claims lasted longer than 6 months or were considered chronic (van Doorn 1995). The risk for chronicity increased with advancing age. Using a predictive cox-regression model, van Doorn found that specific diagnosis, older age, previous back pain, and psycho-social problems were all factors negatively influencing recovery.

## 10.5   Occupational Risk Factors

The relationship between occupational factors and low back pain is difficult to determine because exposure is usually difficult and sometimes impossible to quantify. Using job titles alone is not appropriate. Healthy workers may well stay in the same occupation and job, while workers with low back pain may leave a job and move to a less taxing one. The result is a shift in prevalence of back pain from heavy to light jobs. Another problem is the definition of what is heavy and what is light. Traditionally, heavy physical jobs have been defined as jobs with high energy demand, and light as jobs with low energy demand. Many low energy jobs are static in nature, however, which as such can be a risk factor for LBP. Burdorf (1992) reviewed 81 original epidemiologic articles to determine how well they assessed exposure to postural loading. In 58% there was no exposure data, while in the remaining subjects, exposure was based on questionnaire in 33%, observation in 9%, and direct measurement in 5% only. Throughout, it was felt that the quality of exposure data was poor.

The recall of exposure is even worse than the recall of previous pain and disability, and is often influenced by the insurance system. Subjects will tend to relate LBP to an "injury" or particular exposure if this results in compensation. Kelsey (1975a) found, for example, that 70.9% of men receiving compensation associated the onset of their problem with a specific event, compared to 35.5% of men not receiving compensation. Swedish and Canadian injury statistics would suggest the same.

Further complicating the situation is the fact that exposure to several occupational risk factors often occurs in the same job. For example, a truck driver may have to load and unload his truck (lifting), sit for many hours in an unchanged posture (static loading), and be exposed to whole-body vibration.

The seven most frequently discussed occupational risk factors are listed in Table 10.9. Among them, the first six are all physical risk factors which have been experimentally associated with the development of injuries in spinal tissues. Because they often occur together, the relative importance of each is difficult to determine. The seventh, "psychological and psychosocial work factors," is probably more related to disability claims than to occurrence of a specific pathology.

**TABLE 10.9**   The Seven Most Frequently
Discussed Risk Factors for Lower Back Pain

Heavy Physical Work
Static Work Postures
Frequent Bending and Twisting
Lifting, Pushing and Pulling
Repetitive Work
Vibrations
Psychological and Psychosocial

## Heavy Physical Work

Several investigations indicate an increase in sickness absence due to low back pain, and also an increase in low back symptoms in individuals performing physically heavy work (Andersson, 1981, 1997; Snook, 1982). A few will be discussed here. Magora (1970) found prevalence to vary greatly between occupations. Uyttendaele et al. (1981) found that university and hospital employees with occupations demanding high physical strains were absent from work significantly more often due to LBP than those with light physical work. Unskilled laborers had the highest prevalence rate for disc prolapse and lumbago in the Dutch study by Valkenburg and Haanen (1982). Svensson and Andersson (1983) found heavy physical work to be strongly associated with the occurrence of low back pain; and the highest prevalence of low back pain in their first cross-sectional study, which included men only, was in men with physically heavy professions (Table 10.9). In a subsequent study in women, however, Svensson and Andersson (1989) found only psychological variables, such as dissatisfaction with the work environment and the work itself, and fatigue and worry at the end of the work day to be directly associated with low back pain. Forward bending, lifting, standing, and monotonous work correlated to LBP, but in the univariate analysis.

Klein et al. (1984) found the highest rates of back sprain/strains among workers in physically heavy industries and with physically heavy occupations, as did Behrens et al. (1994). Lloyd et al. (1986) report a higher lifetime prevalence and three-month prevalence in miners than in office workers. Finnish employees in the metal industry were followed over a 10-year period (Leino et al., 1987). Low back pain was more common among blue collar workers than among white collar workers. Among blue collar workers, however, only weak associations were found between morbidity and indices of physical work load.

Mitchell (1985) surveyed low back pain occurring in RAF (Royal Air Force) ground trades personnel. The male prevalence was 9.1% overall, and increased with the severity of the job grade. In all job grades, the prevalence was highest in the 40- to 49-year age group. Both sickness absence and frequency of employment restriction increased with the severity of the job grade. Herrin et al. (1986) analyzed musculoskeletal injury rates among 6,900 workers in 55 industrial jobs with almost 3,000 different manual tasks. The manual exertion requirements were determined using various job stress indices, and 2-year retrospective as well as 1-year prospective medical reports were gathered. Musculoskeletal problems were twice as common if predicted peak lumbosacral disc compression forces exceeded 6,800 N (1,500 lb.). Back problems were about 2.5 times higher in workers with high physical performance requirements. Using the U.S. "Quality of Employment Survey" for 1972–1973, Leigh and Sheetz (1989) found physically heavy work to be associated with back pain, particularly farming (OR = 5.17). A cross-sectional survey at a U.S. oil company employing 10,350 full-time regular employees found the relative risk for a low-back injury to be 1.57 in physically demanding jobs (Tsai et al., 1992).

Similar to LBP, sciatica also has an increased prevalence in physically heavy occupations. Wickstrom et al. (1978) found a higher prevalence of sciatica in concrete reinforcement workers than in computer technicians, and higher rates of HNP have been reported in physically demanding occupations by Hrubec and Nashold (1975), Kelsey et al. (1984), and Videman et al. (1984). Riihimaki (1985) found higher prevalances of sciatica among concrete reinforcement workers compared to house painters. Hrubec and Nashold (1975) found a negative association between HNP and clerical work, while craftsmen and

**TABLE 10.10**　Annual Prevalence of Back Sprains/Strains (BS) and Unspecified Back Injuries (BI) in Selected Studies

| Author & year | Prevalence (%) | Comment |
|---|---|---|
| Hult (1954) | 2.9 | BI. Worker samples |
| Blow/Jackson (1971) | 3.6 | BI. Dockworkers |
| Stubbs/Nicholson (1979) | 0.9 | BI. Construction workers |
| Klein et al. (1984) | 0.7 | BS. General population |
| Anderson (1986) | 4.5 | BI. Dockyard workers |
| Abenhaim/Suissa (1987) | 1.4 | BI. General population |

**TABLE 10.11**　Prevalence of LBP in Studies Comparing Physically Heavy and Light Work

| Reference | Physically heavy (%) | Physically light (%) | N | Comment |
|---|---|---|---|---|
| Hult (1954) | | | | |
| LBP | 64.2 | 52.7 | | |
| Severe LBP | 10.6 | 6.8 | | |
| Work absence | 43.5 | 22.5 | | |
| Lawrence (1955) | 41.0 | 29.0 | 362 | |
| Rowe (1963, 1969) | 47.0 | 35.0 | | Medical visits |
| Ikata (1965) | 22.4 | 5.2 | | Sciatica |
| Magora (1972) | 21.6 | 10.4 | | |
| Lloyd et al. (1986) | 69.0 | 58.0 | | |
| | 35.0 | 26.0 | | 3 months |

*Source:* Rowe ML (1963): Preliminary statistical study of low back pain. *J Occup Med* 5(7): 336-341; Rowe ML (1969): Low back pain in industry. A position paper. *J Occup Med* 11(4): 161-169.

foremen had a significantly higher than average risk. Heliovaara (1987), in a large Finnish study, found that men had a significantly higher probability of being hospitalized for HNP if they were blue collar workers or motor vehicle drivers. A lesser difference between occupations was present in women.

All these studies would seem to support the idea that heavy physical work increases the risk of low back pain. Confounding factors may exist, however, and the level at which physical work load becomes a risk factor is not determined. Other studies are less clear. Lockshin et al. (1969), Sairanen et al. (1981), Porter (1987), and Bigos and Battié (1992) did not find any differences in prevalence between heavy and light work. And occupational factors did not predict the incidence of LBP over a 1-year follow-up period in a Danish cross-sectional sample (Biering-Sorensen and Thomsen, 1986). It is not surprising in epidemiologic studies of this type that some studies will be negative. The association between work load and LBP is relevant in time order, strong based on the majority of studies, dose-related (Heliovaara et al., 1991; Burdorf et al., 1991; Kumar, 1990; Punnett et al., 1989), consistent, and biologically coherent. Table 10.10 summarizes the annual rates per 100 workers reported by different investigators among different worker samples, using injury reports, registry statistics, questionnaires, and interviews. Although differences in research methods make data difficult to compare directly, the average prevalence rates range from a low of 0.75 to a high of 4.5, i.e., large injury rates are reported everywhere. Prevalence rates are higher for workers in physically demanding jobs (Table 10.11). This reflects not only the risk, but also the difficulty of performing a heavy job with a painful lower back. It is obvious from the data, however, that back pain is very frequent in physically light jobs as well. A change in the work environment, therefore, cannot be expected to solve the back injury problem completely.

## Static Work Postures

Working in predominantly one posture, such as prolonged sitting, seems to carry an increased risk for back pain. But there is considerable disagreement. While many studies indicate an increased risk of low back pain in subjects with predominantly sitting work postures (Hult 1954; Kroemer and Robinette, 1969; Lawrence, 1955; Magora, 1972; Partridge, 1969), others do not (Bergquist-Ullman, 1977; Braun, 1969; Frymoyer et al., 1983; Damkot et al., 1984; Heliovaara, 1987; Riihimaki, 1989; Svensson and Andersson, 1983, 1989; Westrin, 1973). Kelsey (1975, 1978) and Kelsey and Hardy (1979) found that men who spend more than half their workday in a car have threefold increased risk of disc herniation. This could be due to the combined effect of sitting and vehicle vibration. Magora (1972) found that those who either sat or stood during most of the workday had an increased risk of low back pain. Frequent changes in posture were also found to increase the risk of back pain, however.

## Frequent Bending and Twisting

The association between low back symptoms and frequent bending and twisting is difficult to evaluate as a separate activity because lifting is usually also involved. A large number of studies report an association between these movements in general and low back pain (Bergquist-Ullman and Larsson, 1977; Brown, 1975; Damkot et al., 1984; Daniel et al., 1980; Frymoyer et al., 1983; Frymoyer et al., 1980; Lloyd et al., 1986; Maeda et al., 1980; Troup, 1984; Troup et al., 1970; Wickstrom et al., 1978).

Magora (1973) established a connection between both excessive bending and occasional bending on the one hand and low back pain on the other, and a similar finding was made by Chaffin and Park (1973). Keyserling et al. (1988) found low back pain to be related to asymmetric postures in an automobile assembly plant, and Riihimaki et al. (1989) report a relationship between sciatica and twisting and bending work postures not only in physically heavy jobs, but in office workers as well. Further analysis of the data from the automobile assembly plant (Punnett et al., 1991) revealed the following odds ratios (OR) and confidence intervals (CI); mild trunk flexion (OR 4.9, CI 1.4–17.4), severe trunk flexion (OR 5.7, CI 1.6–20.4), and trunk twist or lateral bend (OR 5.9, CI 1.6–21.4). The risk increased with exposure to multiple postures and with increasing duration of exposure. Thus, a combination of mild flexion and twisting produced an OR of 7.4 (CI 1.8–29.4).

## Lifting, Pushing, and Pulling

It has been clearly established that back pain can be triggered by lifting, but the frequency at which lifting is the main cause of back pain varies between studies (Bergquist-Ullman and Larsson, 1977; Bigos et al., 1986; Hult, 1954; Ikata, 1965; Kelsey, 1975; Klein et al., 1984; Lloyd et al., 1986; Magora, 1970, 1972). Sudden unexpected maximal efforts were found by Magora (1972) to be particularly harmful, and Glover (1960), Tichauer (1965), and Troup et al. (1970) express the same opinion about lifting in combination with lateral bending and twisting.

Chaffin and Park (1973) found that workers involved in heavy manual lifting had about eight times the number of lower back injuries as those with a more sedentary work situation. Svensson and Andersson (1983) found a direct association between occurrence of low back pain and frequent lifting, as did Frymoyer et al. (1980, 1983) and Hult (1954). Snook (1982) found that a worker was three times more susceptible to compensable low back injury if exposed to excessive manual handling tasks. The National Institute of Occupational Safety and Health (NIOSH) estimated in 1981 that one third of the U.S. workforce lifted in excess of what was considered acceptable, and that lifting was a major cause of low back pain. They also concluded that the severity of injury rate while lifting was proportional to the weight of the object, the bulk of the object, the location of the object at the start of the lift, and the frequency of lifting.

Kelsey (1975), on the other hand, in her first study found no indication that workers with herniated discs did more lifting on the job than workers without such symptoms. Further, there was no indication in that study that jobs requiring pushing, pulling, or carrying either increased or decreased the risk of herniated discs. In the second study, however, (Kelsey et al., 1984b) frequent lifting was identified as a risk factor for HNP, the risk increasing the heavier the weight lifted and the more frequent the lifts. Lifting while twisting increased the risk even further. The odds ratio for HNP in subjects performing frequent lifting of heavy weights while twisting was 3.4.

Troup et al. (1981) followed 802 workers over a 2-year period. Half of all episodes of LBP were associated with a "back injury" and of those, 1 of 3 occurred from manual material handling. A cross-sectional survey in a small market town in the south of England correlated the life-time occupational history of 545 adults with the prevalence of low back pain (Walsh et al., 1989). The strongest associations were for lifting and moving weights over 25Kg (RR 2.0, CI 1.1–3.7). When considering those individuals with severe unremitting back pain, the risk ratio for lifting increased to 5.3 among the men, and 2.9 among the women.

## Repetitive Work

Repetitive work increases, in general, the sickness absence rate. Low back pain seems to be no exception in this respect. This may explain, in part, why assembly line industries have a higher incidence of low back pain among their manual workers than among their office employees (Bergquist-Ullman and Larsson, 1977).

Duration of employment may also be associated with back pain, although the healthy worker effect makes this difficult to evaluate. Magora (1970) found an association with back pain among workers doing heavy work, and Astrand (1987) reports a direct relationship between back pain and time of employment among 391 male employees in a Swedish pulp and paper industry.

## Vibrations

There are several studies suggesting an increasing risk of low back pain in drivers of tractors (Christ, 1973, 1974; Christ and Dupuis, 1966; Christ and Dupuis, 1968; Damlund et al., 1986; Dupuis and Christ, 1966; Dupuis et al., 1972; Dupuis and Zerlett, 1986; Hulshof and van Zanten, 1987; Rosegger and Rosegger, 1960; Seidel et al., 1986; Seidel and Heide, 1986), of trucks (Gruber, 1976; Kelsey and Hardy, 1975; Kristen et al., 1981; Wilder et al., 1982; Behrens et al., 1994), of buses (Gruber and Ziperman, 1974; Kelsey and Hardy, 1975), and of airplanes (Fitzgerald and Crotty, 1972; Schulte-Wintrop and Knoche, 1978). These studies also suggest that low back pain occurs at an earlier age in subjects exposed to vibration.

Kelsey and Hardy (1975) found that truck driving increased the risk of disc herniation by a factor of four, while tractor driving and car commuting (20 miles or more per day) increased the risk by a factor of two. In a later study, the risk of HNP was related to the type of vehicle, indicating significant differences between different brands of cars (Kelsey et al., 1984a).

Hulshof and van Zanten (1987) have reviewed the epidemiologic data supporting a relationship between whole-body vibration and low back pain. They concluded that vibration was a probable risk factor in helicopter pilots, tractor drivers, construction machine operators, and transportation workers. They were critical of the data, however, concluding that none of the many studies reviewed was adequate in terms of the quality of exposure data, effect data, study design, and methodology. Most studies did not control for confounding variables, and only a few had control populations.

Dupuis and Zerlett (1986), in a 10-year prospective study (1961 to 1971), describe an increased incidence of backache reports from 47% to 58% among tractor drivers, while Hilfert et al. (1981) reported that 70% of 352 construction machine operators had periodic LBP compared to 54% in an unexposed control group. Gruber and Zipermann (1974) compared 1,448 male interstate bus drivers to three control groups. Experienced drivers had a higher prevalence of spinal disorders than controls. A significant correlation was found between prevalence rates and exposure level. In a later study Gruber (1976) found significantly higher back pain prevalence rates among 3,205 interstate truck drivers compared to 1,137 air traffic controllers. Behrens et al. (1994) found the highest prevalence estimates for back injuries among

U.S. occupational groups to occur among truck drivers. In a Danish study, 2,045 full-time male bus drivers in the three largest cities in Denmark were compared to 195 motormen (Netterstrom and Juel, 1989). The prevalence of low back pain was 57% vs. 40%. Burdorf and Zondervan (1990) found an odds ratio of 3.6 for low back pain among crane operators compared to controls.

Buckle et al. (1980), Frymoyer et al. (1980), Backman (1983), Damkot et al. (1984), Walsh et al. (1989), and Biering-Sorensen and Thomsen (1986) all report an association between automobile use and low back pain. The risk of being hospitalized because of HNP was high among motor vehicle drivers in the Finnish Study by Heliovaara (1987), and Pentinnen (1987) and Riihimaki et al. (1989) report an increased risk of sciatica with motor vehicle driving in other Finnish studies.

Studies of vibration-exposed populations have also indicated that radiographic changes occur in the spines of these subjects (Dupuis and Zerlett, 1986). These studies are retrospective and usually limited to selected subject groups. It is therefore difficult to make cause-effect conclusions. Further, the radiographic findings are diverse and cannot all be explained by mechanical theory. Nonetheless the prevalence rates of radiographic changes are very high.

## Psychological and Psychosocial Work Factors

Several psychological work factors, including monotony at work, work dissatisfaction, and poor relationship to co-workers have been found to increase the risk of complaining of low back pain and report workers compensation claims. (Astrand, 1987; Battie, 1989; Cunningham and Kelsey, 1984; Deyo and Tsui-Wu, 1987; Bergquist-Ullman and Larsson, 1977; Bigos et al., 1996; Bigos and Spengler et al., 1986; Damkot et al., 1984; Svensson and Andersson, 1983 1989, 1988). Monotony had a direct relationship to low back pain in the study by Svensson and Andersson (1983), while Bergquist-Ullman and Larsson (1977) found that workers with monotonous jobs, requiring less concentration, had a longer sickness absence following low back pain than the others. Diminished work satisfaction has also been found to be related to an increased risk of low back pain by Westrin (1970), Magora (1973), and Svensson et al. (1983). Bergenudd and Nilsson (1988) found that middle-aged workers had an increased prevalence of back pain if they had physically heavy jobs and that the association increased further when the workers were dissatisfied with their work. Individuals with back pain had been less successful in a childhood intelligence test, and on average had a shorter education. Kelsey and Golden (1988) point out that since most of the studies are retrospective it is difficult to determine whether psychological factors are antecedents or consequences of pain. Bigos et al. (1986) and Battie et al. (1992) prospective studies concluded, however, that psychological work factors were more important than physical work factors as risk indicators of low back pain.

## Acknowledgment

This manuscript is in part based on Chapter 7, The epidemiology of spinal disorders, by G.B.J. Andersson, in *The Adult Spine* (Ed. J. W. Frymoyer), Lippincott-Raven Publishers, Philadelphia, pp. 93-141, 1997.

## References

Abenhaim L, Suissa S, Rossignol M (1988): Risk of recurrence of occupational back pain over three year follow-up. *Brit J Ind Med* 45:829-833.

Abenhaim LL, Suissa S (1987): Importance and economic burden of occupational back pain: A study of 2500 cases representative of Quebec. *J Occup Med* 29:670-674.

Andersson GBJ (1997): The epidemiology of spinal disorders, in *The Adult Spine:* Principles and Practice, 2nd edition, J.W. Frymoyer, Ed. Lippincott-Raven, Philadelphia, pp. 93-141.

Andersson GBJ, Svensson HO, Oden A (1983): The intensity of work recovery in low back pain. *Spine* 8:880-884.

Andersson GBJ (1981): Epidemiologic aspects on low back pain in industry. *Spine* 6:53-60.

Astrand NE (1987): Medical, psychological, and social factors associated with back abnormalities and self reported back pain. *Brit J Ind Med* 44:327-336.

Backman AL (1983): Health survey of professional drivers. *Scand J Work Environ Health* 9:30-35.

Battié MC (1989): The reliability of physical factors as predictors of the occurrence of back pain reports. A prospective study within industry. Thesis, University of Goteborg, Goteborg, Sweden.

Battié MC, Bigos SJ, Fisher LD, Spengler DM, Hansson TH, Nachemson AL, Wortley D (1990): Anthropometric and clinical measurements as predictors of industrial back pain complaints: A prospective study. *J Spinal Disorders* 3:195-204.

Behrens V, Seligman P, Cameron L, Mathias CGT, Fine L (1994): The prevalence of back pain, hand discomfort and dermatitis in the U.S. working population. *Am J Public Health* 84:1780-1785.

Benn RT, Wood PH (1975): Pain in the back: an attempt to estimate the size of the problem. *Rheumatol Rehabil* 14:121-128.

Bergenudd H, Nilsson B (1988): Back pain in middle age; occupational work load and psychologic factors: an epidemiologic survey. *Spine* 13:58-60.

Bergquist-Ullman M, Larsson U (1977): Acute low back pain in industry. A controlled prospective study with special reference to therapy and confounding factors. *Acta Orthop Scand* (Suppl) (170):1-117.

Biering-Sorensen F, Thomsen C (1986): Medical, social and occupational history as risk indicators for low-back trouble in a general population. *Spine* 11:720-725.

Biering-Sorensen F (1982): Low back trouble in a general population of 30-, 40-, 50-, and 60-year-old men and women. Study design, representativeness, and basic results. *Dan Med Bull* 29:289.

Biering-Sorensen F (1983): A prospective study of low back pain in a general population, III: Medical service-work consequence. *Scand J Rehab Med* 15:89.

Bigos SJ, Battie MC (1992): Risk factors for industrial back problems, in *Seminars in Spine Surgery*, Vol. 4 (Ed. S.W. Wiesel), W.B. Saunders, Philadelphia, pp. 2-11.

Bigos SJ, Battie MC, Fisher LD, Fordyce WE, Hansson TH, Nachemson AL, Spengler DM (1996): A longitudinal, prospective study of acute industrial back problems: The influence of work perceptions and psychosocial factors. *Spine.*

Bigos SJ, Spengler DM, Martin NA, Zeh J, Fisher L, Nachemson A (1986): Back injuries in industry: A retrospective study. III. Employee-related factors. *Spine* 11:252-256.

Braun W (1969): Ursachen des lumbalen Bandscheiberverfalls. Die Wirbelsaule in Forschung und Praxis 43.

Brown JR (1975): Factors contributing to the development of low back pain in industrial workers. *Amer Industr Hyg Assoc J* 36:26-31.

Buckle PW, Kember PA, Wood AD, Wood SN (1980): Factors influencing occupational back pain in Bedfordshire. *Spine* 5:254-258.

Burdorf A, Govaert G, Elders L (1991): Postural load and back pain of workers in the manufacturing of prefabricated concrete elements. *Ergonomics* 34:909-18.

Burdorf A, Zondervan H (1990): An epidemiological study of low-back pain in crane operators. *Ergonomics* 33:981-987.

Burdorf A (1992): Exposure assessment of risk factors for disorders of the back in occupational epidemiology. *Scand J Work Environ Health* 18:1-9.

Chaffin DB, Park KS (1973): A longitudinal study of low-back pain as associated with occupational weight lifting factors. *Amer Ind Hyg Assoc J* 34:513-525.

Cherkin DC, Deyo RA, Loeser JD, Bush T, Waddell G (1994): An international comparison of back surgery rates. *Spine* 19:1201-1206.

Choler U, Larsson R, Nachemson A, Peterson LE (1985): Back pain. Spri report 188 (in Swedish):1-100.

Christ W (1973): Beanspruchung und Leistungsfahigkeit des Menschen bei underbrochener und Langzeit-Exposition mit stochastischen Schwingungen. Dissertation, Technical University, Darmstadt. VDI Ber 11:1-85.

Christ W (1974): Belastung durch mechanische Schwingungen und mogliche Gesunheitsschadigungen im Bereich der Wirbelsaule. *Fortschr Med* 92:705-708.

Christ W, Dupuis H (1968): Untersuchung der Moglichkeit von gesundheitlichen Schadigungen im Bereich der Wirbelsaule. *Med Welt* 36:1919-1920; 37:1967-1972.

Christ W, Dupuis H (1966): Uber die Beanspruchung der Wirbelsaule unter dem Einfluss sinusformiger und stochastischer Schwingungen. *Int Z Angew Physiol Einschl Arbeitsphysiol* 22:258-278.

Cunningham LS, Kelsey JL (1984): Epidemiology of musculoskeletal impairments and associated disability. *Am J Public Health* 74:574-579.

Cust G, Pearson JC, Mair A (1972): The prevalence of low back pain in nurses. *Int Nurs Rev* 19:169-179.

Damkot DK, Pope MH, Lord J, Frymoyer JW (1984): The relationship between work history, work environment and low-back pain in men. *Spine* 9:395-399.

Damlund M, Goth S, Hasle P, Munk K (1982): Low-back pain and early retirement among Danish semi-skilled construction workers. *Scand J Work Environ Health* (Suppl) 8:100-104.

Damlund M, Goth S, Hasle P, Munk K (1986): Low back strain in Danish semi-skilled construction work. *Applied Ergonomics* 17:31-39.

Daniel JW, Fairbank JC, Vale PT, O'Brien JP (1980): Low back pain in the steel industry: A clinical, economic and occupational analysis at a North Wales integrated steelworks of the British Steel Corporation. *J Soc Occup Med* 30:49-56.

Deyo RA, Cherwin D, Conrad D, Volinn E (1991): Cost, controversy, crisis: Low Back Pain and the Health of the Public. *Ann Rev Publ Health* 12:141-56.

Deyo RA, Tsui-Wu Y-J (1987): Descriptive epidemiology of low-back pain and its related medical care in the United States. *Spine* 12:264-268.

Dupuis H, Christ W (1966): Untersuchung der Moglichkeit von Gesundheits-schadigungen im Bereich der Wirbelsaule bei Schlepperfahrern. Research report, Max-Planck-Institute fur Landarbeit und Landtechnik, Bad Kreuznach.

Dupuis H, Zerlett G (1986): *The Effects of Whole-Body Vibration.* New York: Springer-Verlag.

Dupuis H, Hartung E, Louda L (1972): Vergleich regelloser Schwingungen eines berenzten Frequenzbereiches mit sinusformigen Schwingungen hinsichtlich der Einwirkung auf den Meschnen. *Ergonomics* 15:237-265.

Fitzgerald JG, Crotty J (1972): The incidence of backache among aircrew and groundcrew in the RAF.

Frank A (1993): Low Back Pain. *Brit Med Journal* 306:901-908.

Frymoyer JW (1988): Back pain and sciatica. *New Engl J Med* 318:291-300.

Frymoyer JW, Pope MH, Costanza MC, Rosen JC, Goggin JE, Wilder DG (1980): Epidemiologic studies of low-back pain. *Spine* 5:419-423.

Frymoyer JW, Pope MH, Clements JH et al. (1983): Risk factors in low back pain. *J Bone Joint Surg* 65:213.

Glover JR (1960): Back pain and hyperaesthesia. *Lancet* 1:1165-1169.

Gruber GJ (1976): Relationships between whole-body vibration and morbidity patterns among interstate truck drivers. U.S. Department of Health, Education and Welfare, DHEW (NIOSH) Publication No. 77-167.

Gruber GJ, Ziperman HH (1974): Relationship between whole-body vibration and morbidity patterns among motor coach operators. DHEW (NIOSH) Publication No. 75-104.

Hart LG, Deyo RA, Cherkin DC (1995): Physician office visits for low back pain. *Spine* 20:11-19.

Helander E (1973): Back pain and work disability, (in Swedish). *Socialmed Tidskr* 50:398.

Heliovaara M (1987): Occupation and risk of herniated lumbar intervertebral disc or sciatica leading to hospitalization. *J Chronic Dis* 40:259-264.

Heliovaara M (1988): Epidemiology of sciatica and herniated lumbar intervertebral disc. Helsinki: The Research Institute for Social Security, pp. 1-147.

Heliovaara M, Makela M, Knekt P, Impivaara O, Aromaa A (1991): Determinants of sciatica and low-back pain. *Spine* 16:608-14.

Heliovaara M, Knekt P, Aroma A (1987): Incidence and risk factors of herniated lumbar disc or sciatica leading to hospitalization. *J Chron Dis* 3:251-285.

Herrin GD, Jaraiedi M, Anderson CK (1986): Prediction of overexertion injuries using biomechanical and psychophysical models. *Am Industr Hygiene Assoc J* 47:322-330.

Hilfert R, Kohne G, Toussaint R, Zerlett G (1981): Probleme der Ganzkorperschwingungs-belastung von Erdbaumaschinenfuhrern. *Zentralblatt Arbeitsmedizin, Arbeitsschutz, Prophylaxe Ergonomie* 31:152-155.

Hirsch C, Jonsson B, Lewin T (1969): Low-back symptoms in a Swedish female population. *Clin Orthop* 63:171.

Horal J (1969): The clinical appearance of low back pain disorders in the city of Gothenburg, Sweden. Comparisons of incapacitated probands and matched controls. *Acta Orthop Scand* (suppl) 118:1.

Hrubec Z, Nashold BS Jr. (1975): Epidemiology of lumbar disc lesions in the military in World War II. *Am J Epidem* 102(5):367-376.

Hulshof C, van Zanten BV (1987): Whole body vibration and low back pain. A review of epidemiological studies. *Int Arch Occup Environ Health* 59:205-220.

Hult L (1954): Cervical, dorsal, and lumbar spinal syndromes. *Acta Orthop Scand* (Suppl) 17:1-102.

Ikata T (1965): Statistical and dynamic studies of lesions due to overloading on the spine. *Shikoku Acta Med* 40:262.

Kane WJ (1980): Worldwide incidence rates of laminectomy for lumbar disc herniations. Presented at the annual meeting of ISSLS, New Orleans, LA.

Kelsey JL, Golden AL (1988): Occupational and workplace factors associated with low-back pain. *Occup Med* 3:7-16.

Kelsey JL, Hardy RJ (1975): Driving of motor vehicles as a risk factor for acute herniated lumbar intervertebral disc. *Am J Epidemiol* 102(1):63-73.

Kelsey JL (1975a): An epidemiological study of the relationship between occupations and acute herniated lumbar intervertebral discs. *Int J Epidemiol* 4(3):197-205.

Kelsey JL (1975b): An epidemiological study of acute herniated lumbar intervertebral discs. *Rheumatol Rehabil* 14(3):144-159.

Kelsey JL, White AA III (1980): Epidemiology and impact on low back pain. *Spine* 5(2):133-142.

Keyserling WM, Punnett L, Fine LJ (1988): Trunk posture and back pain: Identification and control of occupational risk factors. *Appl Ind Hyg* 3:87-92.

Klaukka T, Sievers K, Takala J (1982): Epidemiology of rheumatic diseases in Finland in 1967-76. *Scand J Rheumatol* (suppl) 47:5-15.

Klein BP, Jensen RC, Sanderson LM (1984): Assessment of workers' compensation claims for back strains/sprains. *J Occup Med* 26:443-448.

Kristen H, Lukeschitsch G, Ramach W (1981): Untersuchung der Lendenwirbelsaule bei Kleinlasttransportarbeitern. *Arb Med Soz Med Prav Med* 61:226-229.

Kroemer KH, Robinette JC (1969): Ergonomics in the design of office furniture. *Industr Med Surg* 38:115-125.

Kumar S (1990): Cumulative load as a risk factor for back pain. *Spine* 15:1311-16.

Lavsky-Shulan M, Wallace RB, Kohout FJ et al. (1985): Prevalence and Functional Correlates of Low Back Pain in the Elderly: The Iowa 65+ Rural Health Study. *J Am Geriatrics Soc* 33:23-28.

Lawrence JS (1955): Rheumatism in coal miners, Part III. Occupational factors. *Br J Indust Med* 12:249-261.

Lee P, Helewa A, Smythe HA et al. (1985): Epidemiology of musculoskeletal disorders (complaints) and related disability in Canada. *J Rheumatol* 12:1169-1173.

Leigh JP, Sheetz RM (1989): Prevalence of back pain among full-time United States workers. *Brit J Industr Med* 4:651-657.

Leino P, Aro S, Hasan J (1987): Trunk muscle function and low back disorders: A ten-year follow-up study. *J Chronic Dis* 40:289-296.

Lloyd MH, Gauld S, Soutar CA (1986): Epidemiologic study of back pain in miners and office workers. *Spine* 11:136-140.

Lockshin MD, Higgins IT, Higgins MW, Dodge HJ, Canale N (1969): Rheumatism in mining communities in Marion County, West Virginia. *Am J Epidemiol* 90:17-29.

Maeda K, Harada N, Takamatsu M (1980): Factor analysis of complaints of occupational cervicobrachial disorder in assembly lines of a cigarette factory. *Kurume Med J* 27:253-261.

Magora A (1970): Investigation of the relation between low back pain and occupation. 2. Work history. *Industr Med Surg* 39(12):504-510.

Magora A (1973): Investigation of the relation between low back pain and occupation. 5. Psychological aspects. *Scand J Rehab Med* 5:191-196.

Magora A (1972): Investigation of the relation between low back pain and occupation. 3. Physical requirements: Sitting, standing and weight lifting. *Industr Med Surg* 41:5-9.

Magora A, Taustein I (1969): An investigation of the problem of sick-leave in the patient suffering from low back pain. *Ind Med Surg* 38:398.

Mitchell JN (1985): Low back pain and the prospects for employment. *J Soc Occup Med* 35:91-94.

Nachemson AL (1991): Back Pain. Causes, diagnosis and treatment. The Swedish Council of Technology Assessment in Health Care. Stockholm.

Nagi SZ, Riley LE, Newby LG (1973): A social epidemiology of back pain in a general population. *J Chron Dis* 26:769.

National Safety Council (1991). *Accident Facts*, Chicago.

Netterstrom B, Juel K (1989): Low back trouble among urban bus drivers in Denmark. *Scand J Soc Med* 17:203-206.

NIOSH (1981): Work practices guide for manual lifting. DHHS (NIOSH) Publication No 81-122.

Partridge REH, Anderson JAD (1969): Back pain in industrial workers. Proceedings of the International Rheumatology Congress, Prague, Czechoslovakia, Abstract 284.

Pentinnen J (1987): Back pain and sciatica in Finnish farmers. Helsinki: Publications of the Social Insurance Institution, ML:71.

Porter RW (1987): Does hard work prevent disc protrusion? *Clin Biomech* 2:196-198.

Praemer A, Furnes S, Rice DP (1992): Musculoskeletal conditions in the United States. *Amer Acad Orthopaed Surg*, Park Ridge, IL pp. 1-199.

Punnett L, Fine LJ, Keyserling WM, Herrin GO, Chaffin DB (1991): Back disorders and nonneutral trunk postures of automobile assembly workers. *Scand J Work Environ Health* 17:337-346.

Riihimaki H, Wickstrom G, Hanninen K, Luopajarvi T (1989): Predictors of sciatic pain among concrete reinforcement workers and house painters. A five year follow-up. *Scand J Work Environ Health* 15:415-423.

Riihimaki H (1985): Back pain and heavy physical work: A comparative study of concrete reinforcement workers and maintenance house painters. *Br J Ind Med* 42:226-232.

Rosegger R, Rosegger S (1960): Arbeitsmedizinische Erkenntnisse beim Schlepperfahren. *Arch Landtechn* 2:3-65.

Rossignol M, Suissa S, Abenhaim L (1988): Working disability due to occupational back pain: Three-year follow-up of 2300 compensated workers in Quebec. *J Occup Med* 30:502-505.

Sairanen E, Brushaber L, Kaskinen M (1981): Felling work, low back pain and osteoarthritis. *Scand J Work Environ Health* 7:18-30.

Schulte-Wintrop HC, Knoche H (1978): Backache in VH-ID helicopter crews. AGARD-CP-255.

Seidel H, Heide R (1986): Long-term effects of whole-body vibration: A critical survey of the literature. *Int Arch Occup Environ Health* 58:1-26.

Seidel H, Bluethner R, Hinz B (1986): Effects of sinusoidal whole-body vibration on the lumbar spine: The stress-strain relationship. *Int Arch Occup Environ Health* 57:207-223.

Skovron ML, Szpalski M, Nordin M, Melot C, Cukier D (1994): Sociocultural factors and back pain: a population-based study in Belgian adults. *Spine* 19:129-137.

Snook SH (1982): Low back pain in industry, in *Symposium on Idiopathic Low Back Pain*, AA White, SL Gordon, Eds. St. Louis: Mosby, pp. 23-28.

Svensson HO, Andersson GBJ (1983): Low back pain in forty to forty-seven year old men: Work history and work environment factors. *Spine* 8:272-276.

Svensson HO, Andersson GBJ, Johansson S, Wilhelmsson C, Vedin A (1988): A retrospective study of low back pain in 38- to 64-year-old women. Frequency and occurrence and impact on medical services. *Spine* 13:548-552.

Svensson HO, Andersson GBJ (1989): The relationship of low-back pain, work history, work environment, and stress: A retrospective cross-sectional study of 38- to 64-year-old women. *Spine* 14:517-522.

Svensson HO (1982): Low-back pain in 40-47 year old men: Some socioeconomic factors and previous sickness absence. *Scand J Rehabil Med* 14:54-59.

Svensson HO, Andersson GBJ (1982): Low back pain in 40-47 year old men. I: Frequency of occurrence and impact on medical services. *Scand J Rehabil Med* 14:47.

Svensson HO, Vedin A, Wilhelmsson C et al. (1983): Low back pain in relation to other diseases and cardiovascular risk factors. *Spine* 8:277.

Taylor VM, Deyo RA, Cherkin DC, Kreuter W (1994): Low back pain hospitalization. Recent United States trends and regional variations. *Spine* 19:1207-1213.

Tichauer ER (1965): The biomechanics of the arm-back aggregate under industrial working conditions. ASME Rep No 65-WA/HUE-1.

Troup JD (1984): Causes, prediction and prevention of back pain at work. *Scand J Work Environ Health* 10:419-428.

Troup JD, Martin JW, Lloyd DC (1981): Back pain in industry: A prospective study. *Spine* 6:61-69.

Troup JDG, Roantree WB, Archibald RM (1970): Survey of cases of lumbar spinal disability. A methodological study. Med Officers' Broadsheet, National Coal Board.

Tsai SP, Gilstrap EL, Cowles SR, Waddell Jr., LC, Ross CE (1992): Personal and job characteristics of musculoskeletal injuries in an industrial population. *Journal Occupational Medicine* 34:606-612.

Uyttendaele D, Vandendriessche G, Vercauteren M, DeGroote W (1981): Sicklisting due to low back pain at the Ghent State University and University Hospital. *Acta Orthop Belgica* 47:523-546.

Valkenburg HA, Haanen HCM (1982): The epidemiology of low back pain, in *Symposium on Idiopathic Low Back Pain,* AA White, SL Gordon, Eds. St. Louis: Mosby, pp. 9-22.

vanDoorn TWC (1995): Low back disability among self-employed dentists, veterinarians, physicians and physical therapists in The Netherlands. *Acta Orthop Scand* 66(suppl 263): 1-64.

Videman T, Numminen T, Tola S, Kuorinka I, Vanharanta H, Troup JDG (1984): Low back pain in nurses and some loading factors of work. *Spine* 9:400-404.

Volinn E, Turczyn KM, Loeser JD (1994): Patterns in low back pain hospitalizations: Implications for the treatment of low back pain in an era of health care reform. *The Clinical Journal of Pain* 10:64-70.

Walsh K, Varnes N, Osmond C, Styles R, Coggon D (1989): Occupational causes of low-back pain. *Scand J Environ Health* 15:54-59.

Westrin C-G (1973): Low back sick-listing. A nosological and medical insurance investigation. *Scand J Soc Med* (suppl) 7:1-116.

Westrin C-G (1970): Low back sick-listing. A nosological and medical insurance investigation. *Acta Soc Med Scand* 2:127-134.

Wickstrom G, Hanninen K, Lehtinen M, Riihimaki H (1978): Previous back syndromes and present back symptoms in concrete reinforcement workers. *Scand J Work Environ Health* (Suppl 4) 1:20-29.

Wilder DG, Woodworth BB, Frymoyer JW, Pope MH (1982): Vibration and the human spine. *Spine* 7:243-254.

Wood PHN, Badley EM (1987): Epidemiology of back pain, in *The Lumbar Spine and Back Pain.* M Jayson, Ed. London: Churchill Livingstone, pp. 1-15.

# 11

# Static Biomechanical Modeling in Manual Lifting

Don B. Chaffin
*The University of Michigan*

## 11.1   Introduction

Though most manual tasks in industry involve significant body motions, it continues to be very helpful to evaluate specific exertions within a manual task by performing a static biomechanical analysis. Such analyses are normally performed by combining the postural information (body angles) obtained from a stopped frame video image (or photograph) of a worker, and measured forces exerted at the hands. The latter is often obtained with a simple handheld force gauge.

What follows is a description of a computerized static biomechanical model which has been developed and used over the last 25 years to predict:

1. The percentage of men and women who would be capable of exerting specified hand forces in various work postures, and
2. The forces acting on various spinal motion segments.

Since these two different output predictions have specific criterion values referenced in the NIOSH Lifting Guideline, they are often used by professional ergonomists to determine the relative risk of injury associated with the performance of a manual exertion of interest (Chaffin, 1988). It also should be noted, that the prediction of the percent of the population capable of performing a specific exertion required on a job is often crucial to the determination of a job-specific strength test score for pre-employment and return to work purposes (Chaffin, 1996a). Finally, because the biomechanical population strengths and low back stresses are predicted by a computerized model which runs on common personal computer platforms, this has meant that job and product designers and engineers have been able to simulate various expected high exertion tasks during the early part of the design process, and thus avoid costly prototype evaluations and retrofits when the products and/or processes become operational (Chaffin, 1996b). It is this latter application of the biomechanical static strength prediction programs that provide perhaps the greatest potential benefit over other common job evaluation methods. Many other methods require a

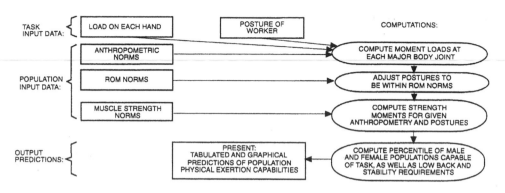

**FIGURE 11.1**   Biomechanical logic used to predict whole-body static exertion capabilities for given postures, hand force directions, and anthropometric groups.

person to be observed and measured, sometimes with expensive instrumentation. This precludes the use of these empirical methods for use in proscriptive job design, wherein the job exists only on paper or in a computer-rendered drawing of the workspace. By interfacing a computerized biomechanical model of a person (as described in the following) into a computerized rendering of the workspace, the designer can quickly perform a large number of simulated exertions to determine the human consequences of altering a proposed job design, much like that being done to accommodate various sized individuals using computerized anthropometric manikins.

What follows is a brief description of the development of a static biomechanical strength modeling technology, including illustrations of how it has been used to evaluate various manual lifting situations.

## 11.2   Development of Static Strength Prediction Programs

The general logic used to predict population static strengths in various jobs is depicted in Figure 11.1. In this model specific muscle group strength data and spinal vertebrae failure data are used as the limiting values for the reactive moments at various body joints created when a person of a designated stature and body weight attempts an exertion (i.e., lifts, pushes, or pulls in a specific direction with one or both hands while maintaining a known posture).

This logic has been well described for a sagittal, coplanar static strength analysis in Chaffin and Andersson (1991). When wishing to perform an analysis in three dimensions, the body is represented as a set of links with known mass, as depicted in Figure 11.2. The load moments $M_j$ are computed by the cross products of the unit distance vectors to each joint and body segment weights and hand forces.

The static moment equilibrium equations for the elbow and shoulder in this linkage can be defined as:

$$\overline{F}_{R\,HAND} = F_{RX}i + F_{RY}j + F_{RZ}k \tag{11.1}$$

where

$\overline{F}_{R\,HAND}$ is the Right Hand Force with X, Y, Z unit vector $(i, j, k)$ components for each,

$F_{RX}$: X component of R Hand Force,

$F_{RY}$: Y component of R Hand Force, and

$F_{RZ}$: Z component of R Hand Force.

$$\overline{V}_1 = \left(V_{1X}i + V_{1Y}j + V_{1Z}k\right) \bullet (\text{Lelink}) \tag{11.2}$$

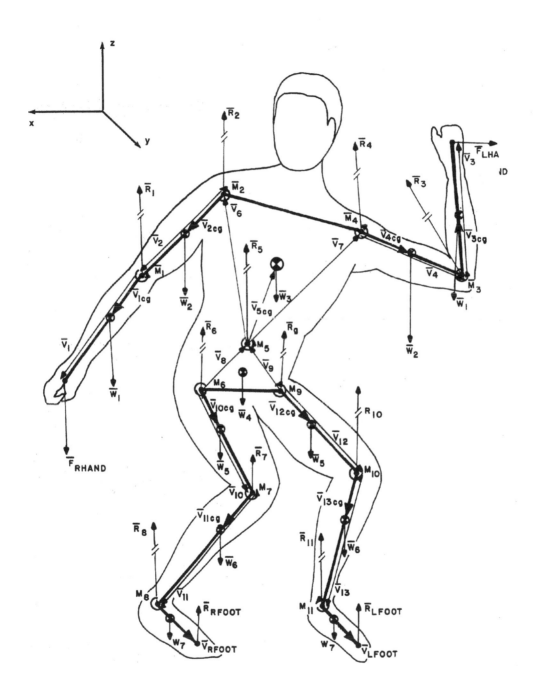

**FIGURE 11.2** Three-D distance, force, and moment vectors used in 12 link biomechanical model for strength prediction. From Chaffin, D.B. and Erig, M. 1991. Three-dimensional biomechanical static strength prediction model sensitivity to postural and anthropometric inaccuracies. *IIE Trans.* 23(3):216-227. With permission.)

where

$\overline{V_1}$ is the Forearm Length with Link unit Vectors,

$V_{1x}$: X component of Forearm Unit Vector,

$V_{1y}$: Y component of Forearm Unit Vector,

$V_{1z}$: Z component of Forearm Unit Vector, and

Lelink:    Magnitude of Forearm Length from Anthropometric Data.

$$\overline{V_{1cg}} = \left(V_{1X}i + V_{1Y}j + V_{1Z}k\right) \bullet \left(cglink\right) \tag{11.3}$$

where

$V_{1cg}$ is the cg distance from elbow to Forearm Center of Gravity Vector expressed in unit vector form, multiplied by the cglink, which is the magnitude of proximal distance to cg of forearm from anthropometric data.

$$\overline{M_1} = \left(M_{1X}i + M_{1Y}j + M_{1Z}k\right) \tag{11.4}$$

where

$\overline{M_1}$ is the Elbow Resultant Moment with X, Y, Z unit vector components,

$M_{1x}$: Elbow Moment about X axis,

$M_{1y}$: Elbow Moment about Y axis,

$M_{1z}$: Elbow Moment about Z axis,

and

$$\overline{M_1} = \overline{V_1} * \overline{F}_{R\,HAND} + \overline{V_{1cg}} * \overline{W_1} \tag{11.5}$$

where

$\overline{W_1} = 0i + 0j - W_{1Z}k,$ (which is Forearm Weight Vector),

and

$$\overline{R_1} = \left(R_{1X}i + R_{1Y}j + R_{1Z}k\right) \tag{11.6}$$

where

$\overline{R_1}$ is the Elbow Joint Reaction Force Vector with X, Y, Z unit vector components,

and

$$\overline{M_2} = \overline{M_1} + \overline{V_{2cg}} * \overline{W_2} + \overline{V_2} * \left(-\overline{R_1}\right) \tag{11.7}$$

where

$\overline{M}_2$ is the Right Shoulder Resultant Moment with:

$\overline{V_{2cg}}$:   Upperarm Center of Gravity Vector,

$\overline{V_2}$:   Upperarm Link Vector,

$\overline{W_2}$:   Upperarm Weight Vector.

and

$$\overline{M}_2 = M_{2X}i + M_{2Y}j + M_{2Z}k \qquad (11.8)$$

where

$\overline{M}_{2x}$: Shoulder Moment about X axis,

$M_{2y}$: Shoulder Moment about Y axis,

$M_{2z}$: Shoulder Moment about Z axis.

A recursive computational procedure is used to continue the analysis to compute external load moments and forces at the elbow and shoulder of the arm or arms doing the exertion, the lumbosacral joint, hip joints and knee and ankle joints.

The size and mass of the person (linkage) size is most often specified as a select strata of the population (i.e., a percentile of specific anthropometric dimensions is selected from population surveys). Thus, a small, medium, or large man or woman can be specified, or specific link anthropometry can be used if available. Link length-to-stature ratios from Drillis and Contini (1966) and link mass-to-bodyweight ratios from Dempster (1955) and Clauser et al. (1969) are used to simplify this procedure, if specific anthropometry is not available on a subject. Most often an average male or female anthropometry is chosen for assessing the strength requirements of a given task in industry.

The strength moment values used as population limit values in the program were measured by Stobbe (1982) for 25 men and 22 women employed in manual jobs in three different industries. These values have been combined with the earlier values from Chaffin and Baker (1970) and Schanne (1972) to form the statistical data for the population joint moment limits.

Once the size of the person has been specified or selected from a known anthropometric data source, the posture is entered with reference to either photographs or videos (or by manipulating a computer-generated hominoid) and then the hand forces of interest are keyed in. The program then computes the load moments at each joint of the linkage, and compares each to the corresponding strength moment capability obtained from the previously measured populations. This provides a prediction of the percent of the population that is capable of producing the necessary strength moments at each joint. The logic for computing the lumbar motion segment compression force is shown in Figure 11.3.

The logic assumes that once the lumbar moment is computed (as described in the preceding), torso muscles contract to stabilize the column. In the 2D Sagittal Plane logic a single torso muscle equivalent contraction force is included. When the necessary reactive torso muscle force is added to body segment weights and hand forces (with a minor adjustment for abdominal pressure effects) a prediction of the compression force on the L5/S1 disc results, as shown in Figure 11.4.

When an asymmetric exertion (e.g., one-handed force, or twisted or laterally bent torso) is being analyzed, many different torso muscle actions and passive supporting tissue reactions need to be considered. The first step in such a procedure requires that the position, orientation, cross-sectional size, and length of the various connective tissues be modeled at the lumbar spinal level. A geometric torso

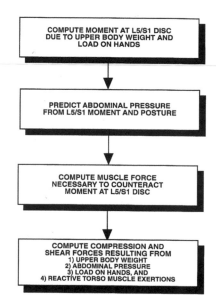

**FIGURE 11.3**   Logic for computing L5/S1 compression forces in 2D and 3D Static Strength Prediction Programs.

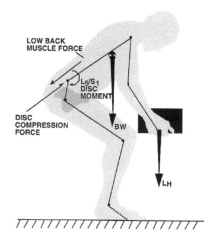

**FIGURE 11.4**   Simple low-back model of lifting for static coplanar lifting analyses. The load on the hands $L_H$ and torso and arm weights BW act to create moments at the L5/S1 disc of the spine. The moments are resisted primarily by the back muscles. The high muscle forces required in such a task cause high disc compression forces. (From Chaffin, D.B. and Andersson, G.B.J. 1991. *Occupational Biomechanics*, 2nd ed. John Wiley & Sons, Inc., NY. With permission.)

model proposed by Nussbaum and Chaffin (1996) for this purpose is shown in Figure 11.5. This model includes estimates of specific tissue geometry acquired from various CT scans (Tracy et al., 1989; Moga et al., 1993; Chaffin et al., 1990), along with passive tissue reaction forces estimated by McCully and Faulkner, (1983), Nachemson et al. (1979), Miller et al. (1986), and others.

The most important predictors of spinal column stress, however, are the muscle reaction forces required to stabilize the spine to external load moments. In the 3D torso models various approaches have been used to predict the required reactive muscle forces. Perhaps the most commonly cited torso biomechanical model for 3D Static Analysis is that developed by Schultz and Andersson (1981). It is depicted in Figure 11.6. A revised version of this model has been developed by Bean et al. (1988). This latter model

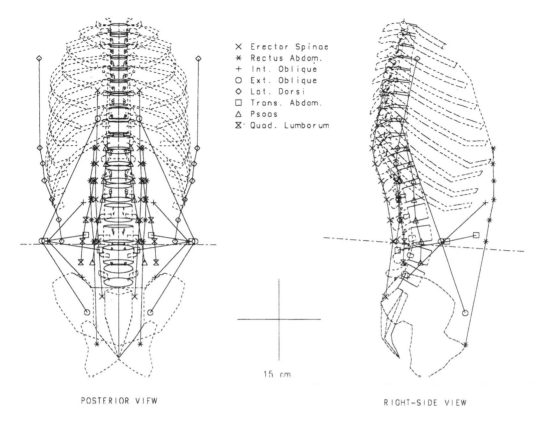

X Erector Spinae
\* Rectus Abdom.
+ Int. Oblique
O Ext. Oblique
◇ Lat. Dorsi
□ Trans. Abdom.
△ Psoas
⊠ Quad. Lumborum

15 cm

POSTERIOR VIEW

RIGHT-SIDE VIEW

**FIGURE 11.5** Muscle geometry illustrated for a 50th percentile male. Muscles are treated as pointwise connections from origin to insertion (see text). An imaginary cutting plane which bisects the $L_3/L_4$ motion segment is also shown. (From Nussbaum, M.A. and Chaffin, D.B. 1996. Development and evaluation of a scalable and deformable geometric model of the human torso. *Clin. Biomech.* 11(1):25-34. With permission.)

provides a more efficient computational method for solving the linear programs used to simultaneously minimize the torso muscle contraction intensities and motion segment compression forces. The present model predicts the minimum muscle force contractile intensities required to meet the moment equilibrium requirements about the three orthogonal axis-of-rotation of the motion segment. Given a set of optimal forces so computed, the model further seeks to minimize the disc compression forces. Because such an approach attempts to minimize *both* muscle intensity requirements and disc compression forces simultaneously, it is referred to as a "double linear optimization" approach.

More recently Hughes and Chaffin (1995) proposed that a nonlinear objective function be used as the basis for selecting the various muscle reaction forces during a given exertion. They referred to this as the sum of the cubed muscle intensity objective. Nussbaum et al. (1996) also have proposed a neural network model to predict torso muscle actions. And most recently, Raschke and Chaffin (1996) have proposed that the external moment is normally distributed about the torso, and activates several muscles simultaneously depending on the direction and magnitude of the external moment.

## 11.3 Computerization of Strength Prediction and Back Force Prediction Models

It should be clear from the preceding descriptions that the biomechanical models used for population strength and spinal motion segment force prediction are computationally intense, especially in the 3-dimensional form. For this reason a number of faculty, staff, and students associated with the Center

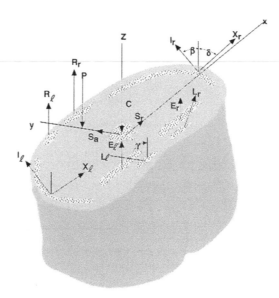

**FIGURE 11.6**    Schematic of 10 muscle model developed by Schultz and Andersson (1981).

for Ergonomics at the University of Michigan have worked to provide user friendly, computer programs of the models. These are referred to as the 2-dimensional and 3-dimensional Static Strength Prediction Programs™ (i.e., 2DSSPP™ and 3DSSPP™). Over 2,500 licenses for use of these programs have been granted by the University of Michigan's, IPO Software Group over the last 13 years.

The main screen of the 2DSSPP™ program is depicted in Figure 11.7. The input values, (i.e., body link angles, hand forces, and anthropometry) are shown in the upper left quadrant. A stick figure depicting the body posture, the hand location, and the hand force direction used as inputs is depicted in the upper right quadrant. The predicted percent of the male and female population having sufficient strength to perform the designated exertion (in this case lifting a 44-pound stock reel) is shown in tabular and graphical form in the lower left quadrant. The back compression force predictions for a man and women performing the 44-pound lift is shown in the lower right quadrant. From inspection of the percent capable predictions (left bottom) it is obvious that hip strengths are the most limiting muscle group strength (only 66% of women and only 87% of men have sufficient hip strength to lift the 44-pound reel). These values are below those recommended by NIOSH, which believes that jobs should accommodate 99% of men's and 75% of women's strength (or 90% of a mixed gender population). It also is shown in the right bottom quadrant that the L5/S1 compression forces of 924 and 845 pounds are above that recommended by NIOSH of 760 pounds.

The 3DSSPP program requires more input data than the 2D version depicted in Figure 11.7. Three-dimensional exertions often involve two hand forces which can act in any direction. Also, a model of the human body in 3D has 12 body links (some with three postural angles). Two types of input presentations are used to assist in assuring the correct input data are used for 3D analyses. One presentation has three orthogonal views of a stick figure, and the other presents a shaded, enfleshed hominoid, which can be viewed from any direction. Figure 11.8 depicts these two presentations. The use of the 3D hominoid was found by Beck and Chaffin (1992) to allow postures viewed on a video or photograph to be accurately represented in a computer. An inverse kinematic model with preferred postural prediction capability is included to allow the user to easily manipulate the figure into the posture to be analyzed. The output screens are similar to that shown in Figure 11.7 for the 2DSSPP. These depict both percent capable strength predictions (for 21 muscle functions) and lumbar compression forces; the latter can include 3D predictions of a large number of individual muscle and spinal forces.

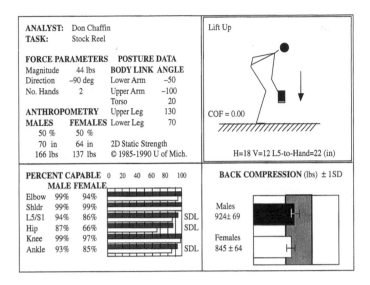

FIGURE 11.7   Main screen from the University of Michigan 2D Static Strength Prediction Program for personal computers. (From the University of Michigan, IPO Software, Ann Arbor, MI 48109.)

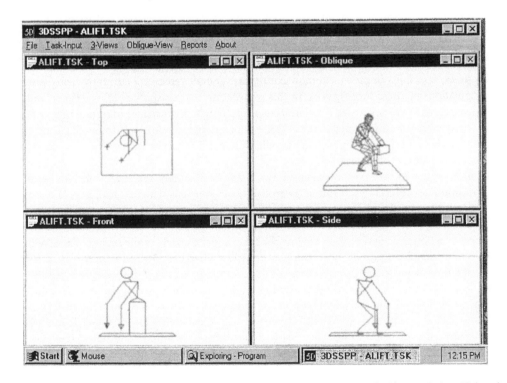

FIGURE 11.8   Input screen used in 3DSSPP™ to assure postural data are correct. (With permission, University of Michigan Regents.)

## 11.4 Validation of Strength and Back Force Prediction Models

The validation of the static strength predictions from the 2D and 3D strength models has been accomplished in three different studies. All three validations required using the models to simulate whole-body exertions and compare the percent capable predictions with the mean, 10, and 90 percentile strengths of a group of volunteers who performed the same tasks.

In the first validation Garg and Chaffin (1975) had 71 male Air Force personnel perform 38 different maximum arm exertions (i.e., lifts, pushes, pulls, etc.) in a variety of arm/torso postures while seated. They found the predicted strengths were highly correlated with the group strengths when performing the 38 upper body tasks (r = 0.93 to 0.97). Chaffin et al. (1987) simulated 15 different whole-body exertions in the sagittal plane which were also performed by both men and women from a variety of industries. In some of these tests over 1,000 people performed the exertions, though on average about 200 people performed each one. Comparison with the 2DSSPP program with the group strength data revealed a very high correlation (r = 0.92). This same study also included simulations with the 3DSSPP program of 72 different one-arm exertions performed by five male Army personnel. The correlations ranged from r = 0.71 to 0.83. Unfortunately, in this latter comparison, exact postural and bracing conditions were not available to use in the simulations. This may have contributed to the lower correlations.

The last validation involved simulations of 56 one- and two-handed, whole-body exertions in 14 different symmetric, bent, and twisted-torso postures (Chaffin and Erig, 1991). The simulation results were compared with the group strengths of 29 young males. Photographs from several views were available to assist in replicating the postures used by these subjects. The results indicated that if care is taken to assure that the postures used in the model simulation is the same as that chosen by people performing the exertions, the prediction error standard deviation will be less than 6%.

In conclusion, it appears that the strength prediction models and population norms used in the present models are accurate in predicting the percent of the population capable of performing a large variety of different types of maximal static exertions. One caution should be noted, however. At present the strength norms used as limits in the models are based on male and female populations who are relatively young (i.e., 18 to 49 years). To improve the models further, strength values are currently being gathered on older populations by these investigators. In this regard, one comparison involving 98 men and women with a mean age of 73 years, showed a major decrease in strength performance in certain muscle functions. When these decreases were included in the 3DSSPP population database, it was found that some exertions that could easily be performed by younger people were predicted to be impossible to perform by most older people (Chaffin et al., 1994).

Validation of the low back biomechanical model has been largely dependent on EMG estimates of muscle reactions in subjects performing controlled torso exertions. Hughes et al. (1994) discusses this procedure, and the results of comparisons with four different optimization procedures used to predict torso muscle responses to different external torso moment loads. Generally speaking, relatively high correlation (r > 0.8) is achieved when loading the torso approximately in the sagittal plane. With greater asymmetric or sudden loading, more complex muscle patterns result, sometimes with a 10 to 30% antagonistic type of muscle response. These complex responses are often not well predicted (r < 0.6) by existing models. Thus it is expected that existing models may underpredict the muscle-induced compression and shear forces on the spinal motion segments by as much as 30% during sudden (i.e., jerking) motions or lateral, asymmetric exertions. The newer neural network and/or geometric moment distribution models are yet to be thoroughly tested under complex loading conditions. They may be less sensitive to this cocontraction phenomenon than existing optimization models.

## 11.5 Final Comments

The development of biomechanically based, static strength and low back force prediction models provide powerful tools for performing ergonomic assessments of high physical effort jobs. With the advent of

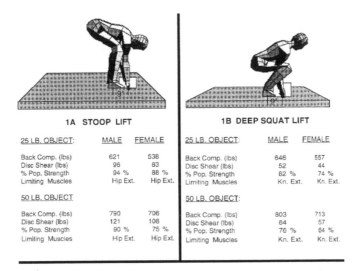

| 1A STOOP LIFT | | | 1B DEEP SQUAT LIFT | | |
|---|---|---|---|---|---|
| 25 LB. OBJECT: | MALE | FEMALE | 25 LB. OBJECT: | MALE | FEMALE |
| Back Comp. (lbs) | 621 | 538 | Back Comp. (lbs) | 646 | 557 |
| Disc Shear (lbs) | 96 | 83 | Disc Shear (lbs) | 52 | 44 |
| % Pop. Strength | 94 % | 88 % | % Pop. Strength | 82 % | 74 % |
| Limiting Muscles | Hip Ext. | Hip Ext. | Limiting Muscles | Kn. Ext. | Kn. Ext. |
| 50 LB. OBJECT | | | 50 LB. OBJECT: | | |
| Back Comp. (lbs) | 790 | 706 | Back Comp. (lbs) | 803 | 713 |
| Disc Shear (lbs) | 121 | 108 | Disc Shear (lbs) | 64 | 57 |
| % Pop. Strength | 90 % | 75 % | % Pop. Strength | 76 % | 64 % |
| Limiting Muscles | Hip Ext. | Hip Ext. | Limiting Muscles | Kn. Ext. | Kn. Ext. |

**FIGURE 11.9** Comparison of two different postures used to lift 25- and 50-pound objects from the floor close to feet using the Michigan 3DSSPP™.

faster personal computers these models are gaining popularity not only as a means to evaluate existing manual tasks, but also to aid in the design of new workplaces and equipment, as well as in the specification of personnel selection and training programs.

The inclusion of inverse kinematics, behaviorally based posture prediction methods, improved human form graphics, and direct human image video input data, have made the use of these models relatively easy for the ergonomics practitioner. By simply clicking and pointing to parts of the body, one can adjust postures and have a comprehensive biomechanical assessment of a specific exertion (such as depicted in Figure 11.9).

With the widely recognized need to "design workplaces right the first time," computerized biomechanical strength prediction models will become even more useful. It is hoped that this presentation assists those interested in knowing more about both the scientific basis for this rapidly expanding technology, and the potential benefits and limitations inherent in it.

## Acknowledgment

I wish to acknowledge NIH Grant AR-39599 for partial support for some of the work described in this presentation.

## References

Bean, J.C., Chaffin, D.B., and Schultz A.B. 1988. Biomechanical model calculation muscle contraction forces: a double linear programming method. *J. Biomech.* 21(1):59-66.

Beck, D.J. and Chaffin, D.B. 1992. An evaluation of inverse kinematics models for posture prediction. Proceedings of *CAES 1992*, Tampere, Finland. Center for Ergonomics, The University of Michigan, Ann Arbor, MI.

Chaffin, D.B. 1996a. Ergonomic basis for job-related strength testing, in *Disability Evaluations*, Eds. S.L. Demeter, G.B.J. Andersson, and G.M. Smith, p.159-167. American Medical Association, Mosby, St. Louis, MO.

Chaffin, D.B. 1997. Biomechanical aspects of workplace design, in *Handbook of Human Factors*, Ed. G. Salvendy. Wiley & Sons, Inc., New York, NY.

Chaffin, D.B. 1988. A biomechanical strength model for use in industry. *Appl. Ind. Hyg.* 3(3):79-86.

Chaffin, D.B. 1988. Biomechanical modelling of the low back during load lifting. *Ergonomics* 31(5):685-697.

Chaffin, D.B. and Andersson, G.B.J. 1991. *Occupational Biomechanics*, 2nd ed. John Wiley & Sons, Inc., NY.

Chaffin, D.B. and Baker, W.H. 1970. A biomechanical model for analysis of symmetric sagittal plane lifting. *AIIE Trans.* 2(1):16-27.

Chaffin, D.B. and Erig, M. 1991. Three-dimensional biomechanical static strength prediction model sensitivity to postural and anthropometric inaccuracies. *IIE Trans.* 23(3):216-227.

Chaffin, D.B., Redfern, M.S., Erig, M., and Goldstein, S.A. 1990. Lumbar muscle size and location measurements from CT scans of 96 older women. *Clin. Biomech.* 5(1):9-16.

Chaffin, D.B., Freivalds, A., and Evans, S.M. 1987. On the validity of an isometric biomechanical model of worker strengths. *IIE Trans.* 19(3):280-288.

Chaffin, D.B., Woolley, C.B., Buhr, T., and Verbrugge, L. 1994. Age effects in biomechanical modeling of static lifting strengths. Proceedings of Human Factors Society, Nashville, TN.

Clauser, C.E., McConville, J.T., and Young, J.W. 1969. Weight, volume and center of mass of segments of the human body. AMRL-TR-69-70, *Aerospace Med. Res. Labs.*, OH.

Dempster, W.T. 1955. Space requirements of the seated operator. WADC-TR-55-159, *Aerospace Med. Res. Lab.* Wright-Patterson AFB, OH.

Drillis, R. and Contini, R. 1966. *Body Segment Parameters*. BP174-945, Tech. Rep. No. 1166.03, S. of Engnrg. and Sci., NYU, NY.

Garg, A.D. and Chaffin, D.B. 1975. A biomechanic computerized simulation of human strength. *AIIE Trans.* 14:272-280.

Hughes, R.E. and Chaffin, D.B. 1995. The effect of strict muscle stress limits on abdominal muscle force predictions for combined torsion and extension loadings. *J. Biomech.* 28(5):527-533.

Hughes, R.E., Chaffin, D.B., Lavender, S.A., and Andersson, G.B.J. 1994. Evaluation of muscle force prediction models of the lumbar trunk using surface electromyography. *J. Orthop. Res.* 12:689-698.

Miller, J.A.A., Schultz, A.B., Warwick, D.N., and Spencer, D.L. 1986. Mechanical properties of lumbar spine motion segments under large loads. *J. Biomech.* 19:79-84.

Moga, P.J., Erig, M., Chaffin, D.B., and Nussbaum, M.A. 1993. Torso muscle moment arms at intervertebral levels T10 through L5 from CT scans on eleven male and eight female subjects. *Spine.* 18(15):2305-2309.

McCully, K.K. and Faulkner, J.A. 1983. Length-tension relationship of mammalian diaphragm muscles. *J. Appl. Physiol.* 54:1681-1686.

Nachemson, A.L., Schultz, A.B., and Berkson, M.H. 1979. Mechanical properties of human lumbar spine motion segments: influences of age, sex, disc level, and degeneration. *Spine.* 4:1-8.

Nussbaum, M.A. and Chaffin, D.B. 1996. Development and evaluation of a scalable and deformable geometric model of the human torso. *Clin. Biomech.* 11(1):25-34.

Nussbaum, M.A., Chaffin, D.B., and Martin, B.J. 1996. A back-propagation neural network model of lumbar muscle recruitment during moderate static exertions. *J. Biomech.* 28(9):1015-1024.

Raschke, U. and Chaffin, D.B. 1996. Trunk and hip muscle recruitment in response to external anterior lumbosacral shear and moment loads. *Clin. Biomech.* 11(3):145-152.

Schanne, F.T. 1972. Three dimensional hand force capability model for a seated person. An unpublished Ph.D. dissertation, University of Michigan, Ann Arbor, MI.

Schultz, A.B. and Andersson, B.J.G. 1981. Analysis of loads on the lumbar spine. *Spine.* 6(1):76-82.

Stobbe, T.J. 1982. The development of a practical strength testing program in industry. An unpublished Ph.D. dissertation, University of Michigan, Ann Arbor, MI.

Tracy, M.F., Gibson, M.J., Szypryt, E.P. et al. 1989. The geometry of the muscles of the lumbar spine determined by magnetic resonance imaging. *Spine.* 14:186-193.

# 12

# Dynamic Low Back Models: Theory and Relevance in Assisting the Ergonomist to Reduce the Risk of Low Back Injury

Stuart M. McGill
*University of Waterloo*

# 12.1   Introduction

This chapter will address some issues associated with "complex" models and highlight how ergonomists can take advantage of these special tools to reduce the risk of low back injury. Throughout the course of daily activity, the low back system is subjected to loading from external forces, and from forces produced by the internal tissues needed to create movement and maintain static postures. The fact that injury to the low back can only be caused by excessive loading of any given tissue cannot be easily dismissed. This overloading may occur during strenuous exertion or during prolonged postures; sitting may be such an example. On the other hand, too little tissue loading also leads to higher injury risk from atrophied tissues, detrained motor control systems and physiological impairment. Prevention of injury requires knowledge of the tissue loads during activity, which also enable testing of hypotheses designed to reduce the risk of injury. Because direct measurement of tissue loads *in vivo* is not feasible, the only tenable option for predicting tissue loads is to utilize modeling approaches.

Biomechanical models have been used to estimate loads in the low back tissues and identify high-risk jobs for approximately three decades. Several approaches to model development have been employed with each approach having a specific objective, and corresponding assets and liabilities. For example, some models were intended to reveal spine function and low back injury mechanisms which is background knowledge required to devise injury avoidance strategies in industry. Other model approaches were intended as simple tools for health and safety personnel to provide an approximate index of injury risk on the plant floor. The main issue boils down to the purpose, and necessary complexity, of the model. Complex models are required to reveal how tissues function inside a worker and to identify specific (and often subtle) injury mechanisms, while simpler models are needed to broaden utility of use and reduce the more overt risks on the plant floor. But the better "simple" models need the "complex" models to assess the many simplifying assumptions which affect accuracy and validity of output, depending on the type of application. Further, in most workplaces, the most blatant or overt ergonomic injury risks have been addressed, and only the more subtle and sometimes sublime risks remain. Ergonomists need "simple" models but also must be conversant with the more complex models which will assist in rectifying the more subtle injury risks, and assist in developing more effective intervention strategies.

The intent of this chapter is to first examine some issues associated with the development, interpretation, and application of "complex" models, then briefly describe a "complex" model, and finally apply the "complex" model to real occupational injury issues. Specifically, data and model output will be integrated into the formulation of a set of guidelines intended to reduce the risk of low back injury in a wide variety of occupational situations, ranging from heavy lifting to sedentary activities. Since some of these guidelines are based on recent research findings, they remain tentative until their efficacy under clinical trial is proven or disproven. While some may think it prudent to wait for the definitive study, ergonomists have not always the luxury of time to deal with injury issues, thus the intention in listing the tentative guidelines here is to provoke discussion and debate, and motivate the initiation of experiments to test their viability.

# 12.2   Issues Relevant for Complex Model Interpretation

Having made clear the need for "complex" models in the previous section, the ergonomist must understand the limitations, and conversely the most appropriate applications, associated with complex models. The issues of anatomical complexity and how tissue loading is determined will be discussed here. Simple and static models are described elsewhere in this handbook, together with a discussion of the static vs. dynamic modeling issue and when two-dimensional analysis or three-dimensional analysis is required (e.g., Chapter 13.).

## The Asset of Anatomical Complexity

There is no doubt that the most overt violations of biomechanical injury risk reduction principles can be addressed by quite "simple" models. However, if the purpose of a biomechanical model is to provide insight into the functional role of various tissues and how they become injured, it is necessary that the model represent the structure of the anatomy as closely as possible. Unfortunately, some researchers have worked beyond the anatomical limitations of the "simple" models and have made erroneous and unjustified conclusions about the best choice of injury risk reduction strategy. Examples of questionable anatomical/mechanical simplification are as follows:

1. Muscle areas used in some models were obtained from cadavers. Dimensions obtained in this way are difficult to justify for use in models of healthy, young, and working individuals when atrophy and distortion from fluids would greatly under-predict the force potential of the musculature. These underestimations of muscle area, and force-producing potential, have been pointed out with CT scan data of younger, ambulatory adults presented by Nemeth and Ohlsen (1986), Reid and Costigan (1985), and McGill et al. (1988, 1993, 1997). Models for ergonomic application must give proper credit to the musculature to most accurately estimate the risk of injury for a given work load.

2. Muscles in the trunk do not pull in straight lines as represented in many lumbar spine models. Several important muscles within the trunk act around pulley systems of bone, other muscle bulk, and pressurized viscera, which alters length, force, and vector direction properties together with the resultant joint loading (cf. McGill and Norman, 1987; McGill, 1996).

3. Models which assume the extensor musculature can be represented by a single equivalent force vector acting parallel to the compressive axis produce controversial output. Excellent work by Langenberg (1970) and Macintosh and Bogduk (1987) provides clear descriptions of the connections for the prime extensors of longissimus thoracic pars thoracic and pars lumborum, iliocostalis lumborum pars thoracic and pars lumborum, and multifidus. Very few of these fibers run parallel to the axis of spinal compression, demonstrating that they exert both compression and shear forces on the spine (Figure 12.1). In addition, the laminated architecture of the muscle fascicles provides for a much larger muscle cross-sectional area to contribute to extensor moment production than would be observed in a single transverse section of the torso. For this reason, an estimate of extensor moment potential from a transverse scan at a single level of the spine would result in large error as only a small portion of the musculature would be measured. The relatively large bulk of thoracic fibers (shown in Figure 12.2) is often neglected as an important contributor to extension as the fibers produce extensor forces over the full lumbar spine through a moment arm often exceeding 10 cm.

4. The extensor and torsional potential of latissimus dorsi is often neglected, yet it has the largest moment arm length of all of the posterior trunk muscles. Its association with the lumbodorsal fascia as a spine extensor is a contentious issue (cf. Tesh et al., 1987; McGill and Norman, 1988).

5. The passive force contributions of the supraspinous and interspinous ligaments are often modeled with a single equivalent element. However, interspinous fibers run obliquely to the supraspinous ligament, thus creating nonparallel forces (Figure 12.1a, vectors a,b,c). In fact, the interspinous acts to generate a shear force on the joint. However, the fiber direction of this ligament complex as depicted in most anatomical texts (e.g., *Gray's Anatomy* (1980)), is in error as pointed out by Heylings (1978). Instead, the interspinous ligament has been shown to contribute significant anterior not posterior, shear forces which increase, rather than reduce, facet load during large degrees of flexion (see Figure 12.1) (Shirazi-Adl and Drouin, 1987; McGill, 1988).

6. The diaphragm area and shape is fundamental to the calculation of the potential assistance provided by intra-abdominal pressure. It is suspected that the size of diaphragms that have been used in the past (up to 465 cm²) is a gross overestimate; the normal surface area on which pressure is exerted is probably closer to 243 cm² (McGill and Norman, 1987), 276 cm² (Schultz et al., 1982), or 299 cm² (Troup et al., 1983).

7. The psoas complex has long been considered and described in the textbooks as a flexor of the lumbar spine. Recent work has shown that while it is a flexor of the hip, its moment arm to flex the spine is extremely limited (Thorstensson et al., 1989; Santaguida and McGill, 1995), and its neural activation is not correlated with moments acting on the lumbar spine but rather with hip flexion moments (McGill et al., 1996).

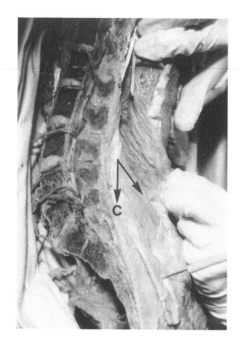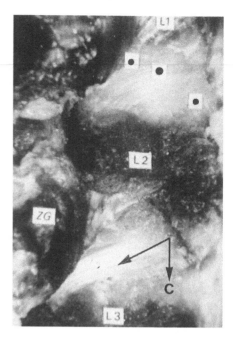

**FIGURE 12.1** (Left panel) Pars lumborum fibers of iliocostalis lumborum and longissimus thoracis create a posterior shear force on the superior vertebra compared to the compressive axis (C), (right panel) while the interspinous ligament imposes an anterior shear when strained in flexion. The general oblique line of action of the muscle and ligament which causes shear loading, is shown compared to the compressive axis (C). (From Heylings, D.J.A. (1978) Supraspinous and interspinous ligaments of the human lumbar spine. *J. Anat.* 123:127-131. With permission.)

## The Problem of Indeterminacy

Pain and disability have been documented to result from damage to ligaments, discs, vertebral end plates, vertebral bodies themselves, facet joints, various muscles, and to several other tissues. While investigation of these injuries requires a modeling approach that incorporates sufficient anatomical detail, a method is needed to solve for the inherent indeterminacy from so many unknown muscle, ligament, and bone forces. Because people differ in the way they perform work, causing only some of them to become injured, the modeling method should also be sensitive to the many different ways that individual people utilize their muscles and various ligaments to perform tasks. Historically, two basic approaches have been used to partition the supporting duties among the many components of the trunk musculature and ligamentous system — optimization approaches and biological approaches that utilize biological signals obtained from each subject (for example, measurements of muscle EMG and spine kinematics). The optimization approach attempts to satisfy the reaction moment requirements needed to support a posture by recruiting the various muscles based on an optimization criterion such as the minimization of joint compression and shear load (e.g., Gracovetsky et al., 1981), or first minimization of muscle contraction intensity and then spine compression load (e.g., Bean et al., 1988). Generally optimization assumes that the motor control system operates to fulfill objectives that can be mathematically defined and in so doing ignores individual variability and predicts the same tissue load distribution for all subjects performing a certain task. Furthermore, most currently reported optimization approaches are very poor at predicting patterns of muscle cocontraction so characteristic of three-dimensional spine motion (cf. Hughes et al., 1994, 1995) and for predicting muscle activity in other areas of the body (e.g., Collins, 1995). It may turn out in the future that more robust and comprehensive optimization "cost functions," yet to be defined, will prove more effective in predicting patterns of muscle activation and their resulting forces.

On the other hand, a modeling approach that uses biological signals obtained directly from each subject is inherently sensitive to the individual ways that people load their low back tissues and, hence

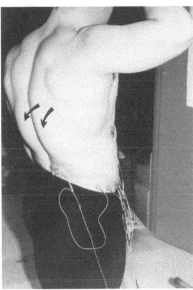

**FIGURE 12.2** A bundle of fibers of longissimus thoracis has been dissected and the tendon isolated to show the insertion on the sixth and seventh thoracic (T6 and T7) ribs and sacral origin (a). Hence, these muscles create an extensor moment over the full length of the lumbar spine and minimize the compressive penalty due to their mechanical advantage. Longissimus thoracis bulk in a developed weight lifter (b). (From McGill, 1990. With permission.)

are better suited to investigations of injury. For example, McGill and Norman (1986) and Marras and Sommerich (1991) have proposed dynamic models of the lumbar spine that attempted to determine the significant forces in many muscles in the low back based in part on their neural activation measured through calibrated surface EMG and in the passive structures based on estimates of strain from directly measured spine kinematics. The major asset of the biologically based approach is that muscle cocontraction is fully accommodated together with the approach being sensitive to the differences in the way individuals perform a movement. However, estimations of muscle force, based in part on myoelectric signals, are problematic because the force potential per muscle cross-sectional area must be assumed, together with other variables that are known to modulate muscle force production. Furthermore, one must rely on anatomical accuracy to satisfy the moment requirements about all three joint axes and about several joints simultaneously so that errors in achieving moment equilibrium remain. Nonetheless, the ability to predict measured moments from biologically driven models adds some degree of validity to the modeling approach. In addition, animal studies involving the direct measurement of tendon forces compared with predictions of tendon forces, from EMG-based models, appear encouraging (see the work of Komi, 1990; Gregor et al., 1987; Norman et al., 1988). Recent work by Cholewicki et al. (1995)

compared the assets and liabilities of an EMG and spine kinematics approach with an optimization approach to obtain tissue force distribution profiles and concluded that the biological–EMG-based approach was more suitable for investigation of injury mechanisms.

## Estimating Deep (and Inaccessible) Muscle Forces

A major drawback of the EMG-based approach using surface electrodes is the myoelectric inaccessibility of the deeper torso muscles (e.g., psoas, quadratus lumborum, three layers of the abdominal wall). In an attempt to address this criticism, recent work by McGill et al. (1996) utilized in-dwelling intramuscular electrodes with simultaneous surface electrode sites to evaluate the possibility, and validity, of using surface activity profiles as surrogates to activate deeper muscles over a wide variety of tasks and exercises common in industry and in rehabilitation programs (e.g., situps, curlups, leg raises, pushups, some spine extensor tasks, and lateral bending and twisting tasks). Prediction of these deeper muscles is possible from well-chosen surface electrodes within an error criterion of 15% of MVC (RMS difference), or less.

## Dynamic Models to Assess Movement Over Long Durations

Analysis of prolonged tasks requires assessment of tissue loads throughout the performance rather than trying to select a single event in time for analysis. Low back tissues are not always under static loading during static work, nor used in repeatable patterns during prolonged work (Potvin and Norman, 1992). For example, subtle shifts in tissue load distribution during the performance of lighter loading tasks, but of longer duration, can lead to situations where loads in a single tissue may rise to unreasonable levels (for example, during prolonged sitting or flexion where ligamentous creep transfers more load and strain to the posterior anulus). Once again, anatomical complexity is required to evaluate the interplay between muscle and passive tissues, but also a biologically driven model is required that is sensitive to subtle changes in spinal posture (and therefore sharing of ligament and disc loads) and sensitive to shifts in muscle activity (either between muscles or between muscles and passive tissues).

## A Brief Description of a Dynamic, 3-Dimensional, Anatomically Complex, Biologically Driven, Modeling Approach

While two groups (the Marras Group, e.g., Granata and Marras, 1993, and the McGill Group) have devoted much effort to the development of biologically driven models, the McGill model will be described here, given its familiarity to the author.

Individual tissue loads have been predicted from a laboratory technique and model developed over the past 14 years. The model is composed of two distinct parts. First, a three-dimensional linked segment representation of the body was constructed using the dynamic load in the hands as input, and working through the arm and trunk linkage, reaction forces and moments about a joint in the low back (usually L4/L5) are computed (previously described in McGill and Norman, 1985) (see Figure 12.3). Joint displacements are recorded on two or more video cameras at 30 Hz to reconstruct the joints and body segments in three dimensions. This first model produces the three reaction forces and corresponding moments about the three orthopedic axes of the low back (flexion–extension, lateral bend, axial twist). The second anatomically detailed model enables the partitioning of the reaction moment obtained from the link segment model into the substantial restorative moment components (supporting tissues) using an anatomically detailed three-dimensional representation of the skeleton, muscles, ligaments, nonlinear elastic intervertebral discs, etc. This part of the model was first described in McGill and Norman (1986), with full three-dimensional methods described in McGill (1992) and the most recent update provided by Cholewicki and McGill (1996) where 90 low back and torso muscles are represented in total. Very briefly, first the passive tissue forces are predicted by assuming stress–strain or load deformation relationships for the individual passive tissues and calibrated for the differences in flexibility of each subject by normalizing the stress–strain curves to the passive range of motion of the subject, which is detected by electromagnetic instrumentation which monitors the relative lumbar angles in three dimensions. Then

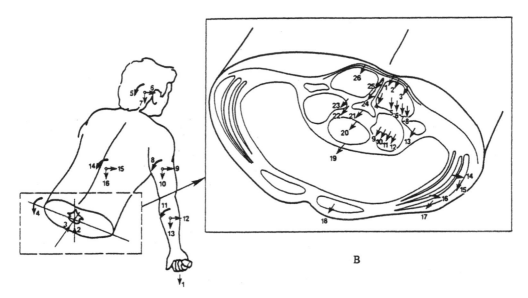

**FIGURE 12.3**   The tissue load prediction approach required two models: (A) the first is a dynamic, 3D linked segment model to obtain the three reaction moments about the low back; (B) the second model partitions the moments into tissue forces (muscle forces 1–18), ligaments 19–26, and moment contributions from deformed disc, gut, and skin in bending etc.).

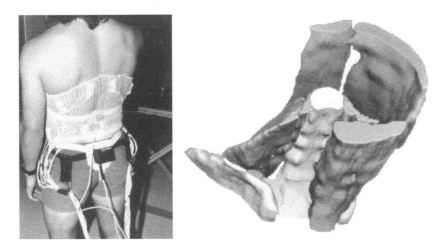

**FIGURE 12.4**   Subject monitored with EMG electrodes and electromagnetic instrumentation to directly measure 3D lumbar kinematics and muscle activity (left panel). The modeled spine (partially reconstructed for illustration purposes) moves in accordance with the subject's spine, whose muscles are activated by the subject.

the remaining moment is partitioned among the many laminae of muscle based on their myoelectric profile, their physiological cross-sectional area, and modulated with known relationships for instantaneous muscle length and of either shortening or lengthening velocity. Most recent improvements of the force velocity relationship have been described in Sutarno and McGill (1995). In this way, the modeled spine moves according to the movements of the subjects spine, and the virtual muscles are activated according to the activation measured directly from the subject (see Figure 12.4). This method of using biological signals to solve the indeterminacy of multiple load bearing tissues facilitates the assessment of the many ways that we choose to support loads, an objective that is necessary for evaluation of injury mechanisms and the formulation of injury avoidance initiatives.

## 12.3 Application of Complex Dynamic Models to Reduce the Risk of Low Back Injury

In the previous section a case was made, and hopefully justified, for the need for representing the musculature and ligaments as accurately as possible in spine models. Such models have enabled reassessment of many mechanical issues that pertain to spine function. The following section constitutes a discussion of some recent research findings as they relate to the issue of formulating guidelines for manual exertion for task assessment and implementation by the ergonomist. While not all issues were borne from the use of complex dynamic models, they were listed here to act as source material for the ergonomist.

### Should One Avoid End Range of Spine Motion During Exertion?

It is recognized that very few lifting tasks in industry can be accomplished by "bending the knees and not the back." Furthermore, most workers rarely adhere to this technique when repetitive lifts are required — a fact which is quite probably due to the increased physiologic cost of squatting compared with stooping (Garg and Herrin, 1979). However, a case can be formulated for the preservation of neutral lumbar spine curvature while lifting (specifically avoiding end range limits of spine motion about any of the three axes). This is a different concept than "trunk angle," as the posture of the lumbar spine can be maintained independent of thigh and trunk angles. Specifically, there has been much confusion in the literature between trunk angle or inclination, and the amount flexion in the lumbar spine. Bending over is accomplished by either hip flexion or spine flexion or both. It is the issue of specific lumbar spine flexion that is of importance here. Normal lordosis can be considered to be the curvature of the lumbar spine associated with the upright standing posture.

Using the tissue load distribution perspective, the following example demonstrates the shifts in tissue loading, predicted from our modeling approach, which has quite dramatic affects on shear loading of the intervertebral column and lends insight into the stoop–squat issue. First, the dominant direction of the pars lumborum fibers of longissimus thoracis and iliocostalis lumborum are noted to act obliquely to the compressive axis of the lumbar spine, producing a posterior shear force on the superior vertebra. In contrast, the interspinous ligament complex acts with the opposite obliquity to impose an anterior shear force on the superior vertebra (see Figure 12.1). Let's observe one example where spine posture determines the interplay between passive tissues and muscles which ultimately modulates the risk of several types of injury. For example, if a subject holds a load in the hands with the spine fully flexed sufficient to achieve myoelectric silence in the extensor muscles (reducing their tension), and with all joints held still so that the low back moment remains the same, then the recruited ligaments will add to the anterior shear to levels well over 1000 N, which is of great concern from an injury risk viewpoint (see Figure 12.5). However, when a more neutral lordotic posture is adopted, the extensor musculature is responsible for creating the extensor moment and at the same time will support the anterior shearing action of gravity on the upper body and handheld load. Disabling the ligaments by avoiding full flexion greatly reduced shear loading. Full flexion postures have injury implications on strained posterior tissues and also on structures affected by large shear loads (e.g., facet joints, neural arch, or conditions of spondylolisthesis).

Using knowledge of tissue loads, one could take the position that the important issue is not whether it is better to stoop lift or to squat lift but rather the emphasis should be to place the load close to the body to reduce the reaction moment (and the subsequent extensor forces and resultant compressive joint loading) and to avoid a fully flexed spine (i.e., maintain some lordosis) to minimize shear loading. In fact, sometimes it may be better to squat to achieve this, or in cases where the object is too large to fit between the knees, it may be better to stoop, flexing at the hips but always avoiding full lumbar flexion to minimize posterior ligamentous involvement. (For a more comprehensive discussion, see McGill and Norman, 1987; Potvin et al., 1991; McGill and Kippers, 1994.)

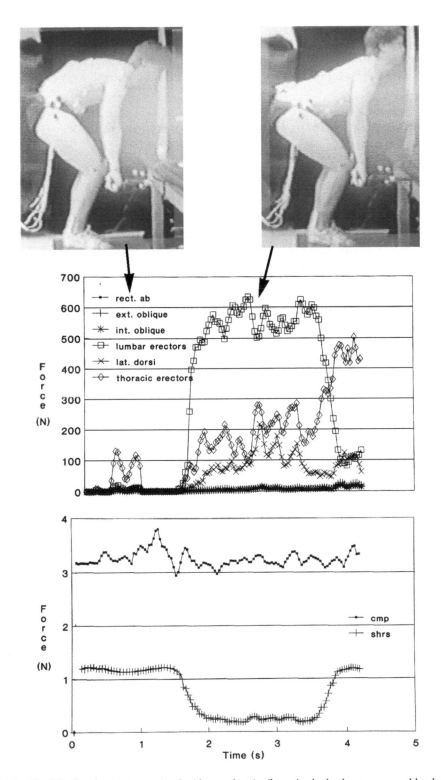

**FIGURE 12.5** The fully flexed spine is associated with myoelectric silence in the back extensors and loaded posterior passive tissues, and high shearing forces on the lumbar spine (these are computer digitized images). A more neutral posture recruits the shear supporting pars lumborum extensors, disables the interspinous ligaments minimizing their shear producing forces, and reduces the net shear (Shrs) on the spine. Compression (CMP) remains essentially unchanged.

The scientific evidence appears to point to the following positive effects if one lifts while avoiding full lumbar flexion: conscious control of lumbar musculature is retained, reducing shear load on the facet joints and providing neurological protection not present if one depends on passive tissue to support the load; the disc is loaded more uniformly permitting all of the annular fibers to share the stress rather than disabling some.

## Should One Lift or Perform Extreme Torso Bending Shortly After Rising from Bed?

The diurnal variation in spine length together with the ability to flex forward has been well documented. Losses in sitting height over a day have been measured up to 19 mm by Reilly et al. (1984), who also noted that approximately 54% of this loss occurred in the first 30 minutes after rising. Over the course of a day, hydrostatic pressures cause a net fluid outflow from the disc, resulting in the narrowing of the space between the vertebrae, which in turn reduces tension in the ligaments. When lying down at night, osmotic pressures exceed the hydrostatic pressure, causing the disc to expand. Adams et al. (1987) noted that the range of lumbar flexion increased by 5° throughout the day. The increased fluid content caused the lumbar spine to be more resistant to bending, while the musculature does not appear to compensate by restricting the bending range. They estimated that disc bending stresses increased by 300% and ligament stresses by 80% and concluded there is an increased risk of injury to these tissues when bending forward early in the morning.

## Should One Lift Immediately Following Prolonged Flexion Postures?

For a number of years, it has been proposed that the nucleus within the annulus "migrates" anteriorly during spinal extension and posteriorly during flexion (McKenzie, 1979). McKenzie's program of passive extension of the lumbar spine (which is presently popular in physical therapy) was based on the supposition that an anterior movement of the nucleus would decrease pressure on the posterior portions of the annulus, which is the most potentially problematic site of herniation. Due to viscous properties of the nuclear material, such repositioning of the nucleus is not immediate upon a postural change, but rather takes time. Krag et al. (1987) demonstrated anterior movement, albeit quite minute, from an elaborate experiment that placed radio-opaque markers in the nucleus of cadaveric lumbar motion segments. Whether this observation was due only to a redistribution of the centroid of the wedge-shaped nuclear cavity moving forward with flexion, or to a migration of the whole nucleus, remains to be seen. Nonetheless, hydraulic theory would suggest lower bulging forces on the posterior annulus if the nuclear centroid moved anteriorly during extension. If compressive forces are applied to a disc where the nuclear material is still posterior (as in lifting immediately after a prolonged period of flexion), then a concentration of stress will occur on the posterior annulus.

While this specific area of research needs more development, there does appear to be a time constant associated with this redistribution of nuclear material. If this is so, it would be unwise to lift an object immediately following prolonged flexion — such as sitting or stooping, as would a stooped gardener who may stand erect and lift a heavy object. Furthermore, it was suggested by Adams and Hutton (1988) that prolonged full flexion may cause the posterior ligaments to creep, which may allow damaging flexion postures to go unchecked if lordosis is not controlled during subsequent lifts. The data of McGill and Brown (1992), in a study of posterior passive tissue creep while slouching forward in a seated posture, showed that even after 2 minutes following 20 minutes of full flexion, subjects only regained half of their intervertebral joint stiffness, while even after 30 minutes of rest some residual joint laxity remained. This is of particular importance for those individuals whose work or movement patterns are characterized by cyclic bouts of full end range of motion postures followed by exertion.

These data suggest that the spine has a memory, since the mechanics of the joints are modulated by previous loading history. Before lifting, following a stooped posture, or after prolonged sitting, a case could be made for standing or even consciously extending the spine for a short period. Allowing the nuclear material to "equilibrate" or move anteriorly to a position associated with normal lordosis may decrease forces on the posterior nucleus. However, further research is required to assess this potentially important possibility.

## Should Intraabdominal Pressure Be Increased While Lifting?

It has been claimed for many years that intra-abdominal pressure (IAP) plays an important role in support of the lumbar spine, especially during strenuous lifting. This issue has been considered in lifting mechanics for years and, for some, has formed a cornerstone for prescription of abdominal belts to industrial workers and also has motivated various abdominal strengthening programs. Many have advocated the use of intra-abdominal pressure as a mechanism to reduce lumbar spine compression (Cyron et al., 1979; Troup et al., 1983; Thomson, 1988).

However, some have indicated that they believe the role of IAP in reducing spinal loads has been overemphasized (e.g., Bearn, 1961; Grew, 1980; Ekholm et al., 1982). In fact, some experimental evidence suggests that somehow, in the process of building up IAP, the net compressive load on the spine is increased! During squat lifts, it appears that the net effect of the involvement of the abdominal musculature and IAP is to increase compression rather than alleviate joint load (McGill and Norman, 1987). The size of the cross-sectional area of the diaphragm and the moment arm used to estimate force and moment produced by IAP had a major effect on conclusions reached about the role of IAP (see McGill and Norman, 1987). The diaphragm surface area was 243 cm$^2$ and the centroid of this area was 3.8 cm. anterior to the center of the T12 disc (compare with a 511 cm$^2$ pelvic floor and 465 cm$^2$ diaphragm together with moment arm distances [up to 11.4 cm] used in other studies). While these results were obtained with complex models, others have noted increased low back EMG activity with higher IAP during voluntary valsalva maneuvers (Krag and co-workers, 1986). Nachemson and Morris (1964) and Nachemson et al. (1986) showed an increase in intradiscal pressure during a valsalva maneuver indicating a net increase in spine compression with an increase in IAP, presumably a result of abdominal wall musculature activity.

The generation of appreciable IAP during load handling tasks is well documented; the role of IAP is not. Farfan (1973) has suggested that IAP creates a pressurized visceral cavity to maintain the hoop-like geometry of the abdominals. Recent work which measured the distance of the abdominals to the spine (their moment arms) was unable to confirm substantial changes in abdominal geometry when activated in a standing posture (McGill et al., 1996). However, the compression penalty of abdominal activity cannot be discounted. It appears that the spine prefers to sustain increased compression loads if intrinsic stability is increased. An unstabilized spine buckles under extremely low compressive load (e.g., approximately 20N, Lucas and Bresler, 1961). The geometry of the musculature suggests that individual components exert lateral and anterior–posterior forces on the spine, which perhaps can be thought of as guy wires on a mast to prevent bending and compressive buckling (Cholewicki et al., 1996). As well, activated abdominals create a rigid cylinder of the trunk resulting in a stiffer structure. Thus it appears that increased IAP, commonly observed during many activities including lifting, as well as in those people experiencing back pain, does not have a direct role to reduce spinal compression but rather is an agent used to stiffen the trunk and prevent tissue strain or failure from buckling.

## Should Abdominal Belts Be Prescribed to Manual Materials Handlers?

This topic is addressed in Chapter 33 of this volume.

## Should Workers Adopt a Lifting Strategy to Recruit the Lumbodorsal Fascia?

Studies have attributed various mechanical roles to the lumbodorsal fascia (LDF). In fact there have been some attempts to recommend lifting postures based on LDF hypotheses. Suggestions were originally made (Gracovetsky et al., 1981) that lateral forces generated by internal oblique and transverse abdominis are transmitted to the LDF via their attachments to the lateral border. This lateral tension was hypothesized to increase longitudinal tension, from Poisson's effect, pulling in the direction of the posterior midline of the lumbar spine, causing the posterior spinous processes to move together resulting in lumbar extension. This proposed sequence of events formed an attractive proposition because the LDF has the largest moment arm of all extensor tissues. As a result, any extensor forces within the LDF would impose the smallest compressive penalty to vertebral components of the spine.

However, this hypothesis was examined by three studies which used quite different methods, all published about the same time, and which, collectively, questioned its viability: Tesh et al. (1987), who performed mechanical tests on cadaveric material; Macintosh et al. (1987), who recognized anatomical inconsistencies with the abdominal activation; McGill and Norman (1988), who tested the viability of LDF involvement with latissimus dorsi as well as with the abdominals using a dynamic modeling approach. Regardless of the choice of LDF activation strategy, the LDF contribution to the restorative extension moment was negligible compared with the much larger low back reaction moment required to support the load in the hands. Although the LDF does not appear itself to be a significant active extensor of the spine, it is a strong tissue with a well-developed lattice of collagen fibers. Its function may be that of an extensor muscle retinaculum (Bogduk and Macintosh, 1984). The tendons of longissimus thoracis and iliocostalis lumborum pass under the LDF to their sacral and ilium attachments. Perhaps the LDF provides a form of "strapping" for the low back musculature. Hukins et al. (1990), on theoretical grounds only at this time, have proposed that the LDF acts to increase the force per unit cross-sectional area that muscle can produce by up to 30%. They suggest that it does this by constraining bulging of the muscles when they shorten. This contention remains to be proven. Tesh et al. (1987) have suggested that the LDF may be more important for supporting lateral bending. No doubt, this notion will be pursued in the future. Given the confused state of knowledge about the role, if any, of the LDF, the promotion of lifting strategies based on intentional LDF involvement cannot be justified at this time.

## Should the Trunk Musculature Be Cocontracted to Stabilize the Spine?

The ability of the joints of the lumbar spine to bend in any direction is accomplished with large amounts of muscle coactivation. Such coactivation patterns are counter productive to generating the torque necessary to support the applied load in a way that minimizes the load penalty imposed on the spine from muscle contraction. Several ideas have been postulated to explain muscular coactivation: the abdominals are involved in the generation of intra-abdominal pressure (Davis, 1959), or in providing support forces to the lumbar spine via the lumbodorsal fascia (e.g., Gracovetsky et al., 1981); however, these ideas have not been without opposition (see previous sections).

It appears that another explanation for muscular co-activation is tenable. As noted above, ligamentous spine (one in which muscles are removed) will fail under compressive loading in a buckling mode, at very low forces of only about 20N (Lucas and Bresler, 1961). The spine can be likened to a flexible rod where under compressive loading it will buckle. However, if the rod has guy wires connected to it, like the rigging on a ship's mast, more compression is ultimately experienced by the rod, but it is able to bear much more compressive load as it is stiffened and more resistant to buckling. The cocontracting musculature of the lumbar spine can perform the role of stabilizing guy wires to each lumbar vertebrae bracing against buckling. Work by Crisco and Panjabi (1990) and Cholewicki and McGill (1996) has begun to quantify the influence of muscle architecture and the necessary coactivation on stability of the lumbar spine. The architecture of the lumbar erector spinae is especially suited for this role (see Macintosh and Bogduk [1987] and McGill and Norman [1987]). In order to invoke this antibuckling and stabilizing mechanism when lifting, one could justify lightly cocontracting the musculature to both minimize the potential of buckling and remove the possibility of any tissue having to bear a surprise load.

## How Do People Hurt Their Backs Picking Up a Pencil?

While injury from large exertions is understandable, explanation of how people injure their backs performing rather benign-appearing tasks is more difficult — but the following is worth considering by

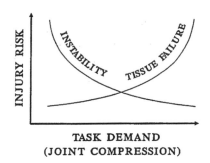

**FIGURE 12.6** While injury from high loading tasks is easier to rationalize, injury from low loading tasks appears to reduce spine stability and increase the possibility of injury from errors in motor control, and the resulting joint displacement and tissue overload. (From Cholewicki, J. and McGill, S.M. (1996) Mechanical stability of the *in vivo* lumbar spine: Implications for injury and chronic low back pain, *Clin. Biomech.* 11(1):1-15. With permission.)

the ergonomist. Continuing the considerations about stabilization from the previous section — a number of years ago, we were investigating the mechanics of power lifters' spines while they lifted extremely heavy loads, using video fluoroscopy for a sagittal view of the lumbar spine (Cholewicki and McGill, 1992). The range of motion of the power lifters' spines was calibrated and normalized to full flexion by first asking them to flex at the waist and support the upper body against gravity with no load in the hands. During their lifts, although they outwardly appeared to have a very flexed spine, in fact, the lumbar joints were all two to three degrees per joint from full flexion, explaining how they could lift such huge loads (up to 210 kg) without sustaining the injuries suspected to be linked with full lumbar flexion. However, during the execution of a lift, one lifter reported discomfort and pain. Upon examination of the video fluoroscopy records, one of the lumbar joints (specifically, the L4/L5 joint) reached the full flexion calibrated angle, while all other joints maintained their static position (2 to 3 degrees from full flexion). This is the first observation that we know of reported in the scientific literature documenting proportionately more rotation occurring at a single lumbar joint than at the other joints, and it would appear that this unique occurrence was due to an inappropriate sequencing of muscle forces (or a temporary loss of the normal motor control pattern). This motivated the work of our colleague, Dr. Jacek Cholewicki, to investigate and continuously quantify stability of the lumbar spine throughout a reasonably wide variety of loading tasks (Cholewicki and McGill, 1996). Generally speaking, it appears that the occurrence of a motor control error which results in a temporary reduction in activation to one of the intersegmental muscles, perhaps for example a laminae of multifidus, could allow rotation at just a single joint to the point where passive tissue or other muscle tissue could become irritated or injured. Dr. Cholewicki noted that the risk of such an event was greatest when there are high forces in the large muscles with simultaneous low forces in the small intersegmental muscles (a possibility with our power lifter) or when all muscle forces are low, such as during a low-level exertion. Thus, a mechanism is proposed, based on motor control error resulting in temporary inappropriate neural activation, that explains how injury might occur during extremely low load situations, for example, picking a pencil up from the floor following a long day at work performing a very demanding job (see Figure 12.6).

## Are Twisting Lifts Particularly Dangerous?

Twisting of the trunk has been identified as a factor in the incidence of occupational low back pain (Frymoyer et al., 1983; Troup et al., 1981), but the mechanisms of risk require some explanation. Some hypotheses have been based on an inertia argument in that twisting at speed will impose dangerous axial torques upon braking the axial rotation of the trunk at the end range of motion. Farfan and colleagues (1970) proposed that twisting of the disc is the only way to damage the collagenous fibers in the annulus leading to failure. They reported that distortions of the neural arch permitted such injurious rotations. More detailed analyses of the annulus under twist were conducted by Shirazi-Adl et al. (1986b) who supported Farfan's contention that twisting indeed can damage the annulus but also noted that twisting is not the sole mechanism of annulus failure. In contrast, some research has suggested that twisting *in vivo* is not dangerous to the disc as the facet in compression forms a mechanical stop to rotation well before the elastic limit of the disc is reached, and thus the facet is the first structure to sustain torsional failure (Adams and Hutton, 1981). Ligament involvement during twisting was studied by Ueno and Liu (1987), who concluded that the ligaments were under only negligible strain during a full physiological twist. However, an analysis of the L4/L5 joint by McGill and Hoodless (1990) suggested that posterior ligaments may become involved if the joint is fully flexed prior to twisting.

Certainly the mechanisms of injury from torsional loads applied under twisting conditions remain inconclusive. However, it is clear that the increase in compressive load on the spine is dramatic if a comparatively small amount of axial twist torque is required in addition to the dominant extensor torque. Using data from a combination of our previous studies, to support 50 N.m in extension imposes about 800 N of spinal compression, but 50 N.m in axial twist would impose over 2500 N, while 50 N.m of lateral bend imposes 1400 N of compression. These differences result from the difference in coactivation of the trunk musculature, combined with small moment arms in many cases, to generate the moments of force required. It appears that the joint pays dearly in order to support even small axial torques when extending during the lifting of a load.

## Is "Lifting Smoothly" and Not Jerking the Load Always Best?

We have all heard that a load should be lifted smoothly and not "jerked." This recommendation was most likely rationalized on the basis that accelerating a load upwards increases its effective mass by virtue of an additional inertial force acting downward together with the gravitational vector. However, this may not always be the case as it is possible to lift a load by transferring momentum from an already moving segment. The concept of momentum transfer during lifting has been referred to by Troup and Chapman (1969) and by Grieve (1975), who coined the term "kinetic lift." Later, McGill and Norman (1985) documented that smaller low back moments were possible in certain cases compared to moments during static lifts in the same posture using skillful transfer of momentum. For example, if a load is awkwardly placed, perhaps placed on a worktable at a distance of 75 cm from the worker, a slow-smooth lift would necessitate the generation of a large lumbar extensor torque for a lengthy duration of time — a situation that is most strenuous on the back. However, this load could be lifted with a very low lumbar extensor moment or quite possibly no moment at all. For example, if the worker leaned forward and placed his hands on the load, with bent elbows, the elbow extensors and shoulder musculature could thrust upward initiating upward motion of the trunk to create both linear and angular momentum in the upper body (note that the load has not yet moved). As the arms straighten, coupling takes place between the load and the large trunk mass (as the hands then start to apply upward force on the load) transferring some, or all, of the body momentum to the load causing it to be lifted with a jerk. This highly skilled "inertial" technique is observed quite frequently throughout industry and in some athletic events such as competitive weight lifting, but it must be stressed that such lifts are conducted by highly practiced and skilled individuals. In most cases, acceleration of loads to decrease low back stress in the manner described is not suitable for the "lay" individual when conducting the lifting chores of daily living.

The momentum — transfer technique is a skilled movement that requires practice, is only feasible for awkwardly placed lighter loads, and could not be justified for heavy lifts. However, there may be another mechanical variable to be integrated into the analysis of a dynamic technique. The tissue property of viscoelasticity enables tissues to sustain higher loads when loaded at rate (Burstein and Frankel, 1968). Troup (1977) suggested that the margin of safety for spine injury may be increased during a higher strain rate but cautioned that incorporation of this principle into lifting technique depends on the rate of increase in spinal stress, the magnitude of peak stress, and its duration. The lifting instruction to always lift a load "smoothly" may not invariably result in the least risk of injury. Indeed, it is possible to skillfully transfer momentum to an awkwardly placed object to position the load as close to the body quickly and minimize the extensor torque required to support the load. There can be no argument that reduction of the extensor moment required to support the hand load is paramount in reducing the risk of injury and that this is best accomplished by keeping the load as close to the body as possible.

### Is Sedentary-Seated Work Harmful?

Epidemiological evidence presented by Videman et al. (1990) documented the increased risk of disc herniation for those who perform sedentary jobs characterized by sitting. Known mechanical changes associated with the seated posture include the increase in intradiscal pressure when compared to standing postures (Andersson et al., 1975), increases in posterior annulus strain (Pope et al., 1977), creep in posterior passive tissues (McGill and Brown, 1992) which decreases anterior–posterior stiffness and increases shearing movement (Schultz et al., 1979), and posterior migration of the mechanical fulcrum (Wilder et al., 1988) which reduces the mechanical advantage of the extensor musculature (resulting in increased compressive loading). This has motivated occupational biomechanists to consider the duration of sitting as a risk factor when designing seated work in the interest of reducing the risk of injury. A recently proposed guideline has suggested a sitting limit of 50 minutes without a break, although this proposal will be tested and evaluated in the future.

## 12.4 Tentative Risk Reduction Guidelines for Occupational Injury

The following recommendations have been summarized from the biomechanical rationale developed in the previous section. Some are consistent with what has been advocated for years, while others contradict long-standing notions that were based on flawed, or unavailable, biomechanical understanding. They are more versatile and widely applicable than the commonly used instruction of "bend the knees, not the back" to reduce lifting stresses. However, they have not been subjected to rigorous scientific challenge and clinical trial. Due to the lack of conclusive evidence to support these recommendations, and acknowledging that the definitive studies will probably not occur in the near future, they are listed in the following form to provoke thought, stimulate discussion and generate research interest. Perhaps it is overly optimistic to expect that these recommendations will withstand time and remain intact. However, these recommendations may have the potential to reduce tissue loads during the performance of a wide range of industrial exertion tasks and they are able to accommodate all tasks including those outside the sagittal plane. Furthermore, the exact instructions issued to a specific worker should not be taken verbatim from the following list, but rather the biomechanical principle should be explained in a language and terminology which is familiar to the worker. In addition, often successful job incumbents have developed personal strategies for working that assist them in avoiding fatigue and injury. Their insights are the result of thousands of hours of performing the task, and they can be very perceptive — attempts should be made to accommodate them.

### Recommendations for Safer Work — A Tentative List

1. Avoid a fully flexed or bent spine and rotate trunk using hips.
   - Strain in the annular fibers is equalized (Hickey and Hukins, 1980; Shirazi-Adl et al., 1986a).
   - Posterior ligaments are not strained and cannot be injured (McGill, 1988).
   - Facets are in contact and can bear some load (e.g., Nachemson, 1960).
   - The anterior shearing effect from ligament involvement is minimized, and the posterior supporting shear of the musculature is maximized (McGill, 1989).
   - Compressive testing of lumbar motion units has shown increases in tolerance with partial flexion but decreased ability to withstand compressive load at full flexion (Adams and Hutton, 1988).
2. Choose a posture to minimize the reaction moment on low back so long as #1 is not compromised.
   - Neutral lordosis is still maintained but sometimes the load can be brought closer to the spine with bent knees (squat lift) or relatively straight knees (stoop lift). The key is to reduce the moment which has been shown to be a dominant risk factor (Marras et al., 1995).

3. Allow time for the disc nucleus to "equilibrate" and ligaments to regain stiffness after prolonged flexion.
   - After prolonged sitting or stooping, spend time standing to allow the nuclear material within the disc to equilibrate and equalize the stress on the annulus, and allow the ligaments to regain their rest length and provide protective stiffness to the lumbar spine (McGill and Brown, 1992).
4. Avoid lifting shortly after rising from bed.
   - Forward bending stresses on the disc and ligaments are higher in the early morning compared with later in the day (Adams et al., 1987).
5. Prestress system even during "light" tasks.
   - Lightly cocontract the stabilizing musculature to remove the slack from the system and stiffen the spine, even during "light" tasks such as picking up a pencil (Cholewicki and McGill, 1996).
   - Cocontraction and the corresponding increase in stability increases the margin of safety of material failure of the column under axial load (Crisco and Panjabi, 1990)
   - In this way, no tissue will have to bear a "surprise" load
6. Avoid twisting.
   - Twisting reduces the intrinsic strength of the annulus by disabling some of its supporting fibers while increasing the stress in the remaining fibers under load (Shirazi-Adl et al., 1986b)
   - Since there is no muscle designed to produce only axial torque, the collective ability of the muscles to resist axial torque is limited and may not be able to protect the spine in certain postures (McGill, 1991)
   - The additional compressive burden on the spine is substantial for even a low amount of axial torque production (McGill and Hoodless, 1990).
7. Exploit the acceleration profile of the load.
   - This is only for highly skilled individuals performing repetitive tasks.
   - Dangerous for heavy loads and should not be attempted.
   - It is possible that a transfer of momentum from the upper trunk to the load can start moving an awkwardly placed load without undue low back load (Grieve, 1975; Troup, 1977). Possibly, the viscoelastic property of biological material will safely absorb a momentary high load required to bring the load close to the trunk, which reduces the reaction moment.
8. Avoid prolonged sitting.
   - Prolonged sitting is associated with disc herniation (Videman et al., 1990).
   - When required to sit for long periods, adjust posture often, stand up, at least every 50 minutes, and extend spine and/or walk for a few minutes.
   - Organize work to break up bouts of prolonged sitting into shorter periods that are better tolerated by the spine.
9. Consider the best rest break strategies
   - Workers engaged in sedentary work would be best served by frequent, dynamic breaks to reduce tissue stress accumulation (Adams and Dolan, 1995) and migrate tissue load bearing (McGill, 1997).
   - Workers engaged in dynamic work may be better served with longer and more "restful" breaks.

## 12.5 Future Directions

The recent advances in low back research and ergonomic model sophistication has increased general understanding of lumbar mechanics, injury mechanisms, and planning injury avoidance strategies. However, the reader of this chapter becomes aware, quite quickly, that many issues pertaining to reducing the risk of industrial low back injury, and to the recommendation of injury risk reduction guidelines remain unsolved. The future of ergonomics and spinal biomechanics must address a range of issues that will demand the utmost effort in creating models that capture biological fidelity. Continued effort must be directed toward obtaining more sophisticated anatomy for model components, for it is the fine details that unlock the secrets of force generation, transmission, and sharing strategies among tissues. Tissue properties such as strength, viscoelasticity, and fatiguability must be better understood. The knowledge base of biomaterials is still in the developmental stage. Static behavior has been quite well documented for some tissues although not all. However, with the recent development of quite involved dynamic models, dynamic tissue behavior is desperately required. The property of viscoelasticity is paramount in the determination of dynamic tissue load due to its time and loading rate dependency. Examination of movement, particularly rapid movements of the trunk and limbs, is hindered by inadequate dynamic tissue information.

Muscle cocontraction and interplay with the passive tissues is far more prevalent under complex motion conditions which enforces the need to measure the activity of individual tissues. Both EMG techniques and optimization strategies to partition tissue loads require basic research to improve predictions of the cocontracting muscle forces. Perhaps major developments in artificial intelligence will provide the required interface with motor control in the future. However, at present only the mind is capable of such sophisticated processing, and research efforts would appear to be best directed toward the improvement of EMG techniques and the appropriate processing to obtain muscle force predictions. Work must continue to determine which muscles to monitor, where to place electrodes, and improve processing techniques to increase EMG reliability. Some international effort is ongoing, with the continual reporting of some quite impressive predictions of individual muscle force measures in animal preparations, from processed EMG.

Analysis of the single task, with no provision for repeated movements, has dominated the literature. However, most tasks in industry and those that are part of mundane daily events are repeated and prolonged. The effect of fatigue on the body system, intervertebral joints, muscles, and ligaments demand investigation. For example, at present the frequency of task repetition can only be recommended on the psychophysical criterion of what the individual "thinks" is appropriate. Data on repeated tissue loads is extremely scarce, although is appearing in the literature with greater frequency. Tissue fatigue must also be considered in the context of static holds that may occur in activities such as stooping for long periods of time while gardening. While ergonomic design of the workplace is of utmost importance in facilitating work postures that minimize joint loads, much basic research remains to test and refine low back hypotheses and related issues such as reducing the risk of low back injury for workers. Finally, there is a commonly held notion among ergonomists that minimal tissue loading is best. This is untrue. Biological tissues require repeated loading to be healthy. The trick is to develop a wise rest break strategy to facilitate optimal tissue adaption. The most healthy combination of work and rest will be achieved through first understanding the biomechanical, physiological, and psychological parameters of injury and human performance, followed with thoughtful application of this wisdom.

## Acknowledgment

The author wishes to acknowledge the many colleagues who have made intellectual contributions to the series of works documented in this chapter and who have made the journey such a joy: Jacek Cholewicki, Ph.D., John Sequin, M.D., Lina Santaguida, M.Sc., Chrisanto Sutarno, M.Sc., Daniel Juker, M.D., Michael Sharratt, Ph.D. and, in particular, Robert Norman, Ph.D.

## References

Adams, M.A. and Hutton, W.C. (1981) The relevance of torsion to the mechanical derangement of the lumbar spine, *Spine* 6:241-248.

Adams, M.A., Dolan, P., Hutton, W.C. (1987) Diurnal variations in the stresses on the lumbar spine, *Spine* 12(2):130:137.

Adams, M.A. and Hutton, W.C. (1988) Mechanics of the intervertebral disc, in *The Biology of the Intervertebral Disc* (Ed. P. Ghosh). CRC Press, Boca Raton.

Adams, M.A. and Dolan, P. (1995) Recent advances in lumbar spinal mechanics and their clinical significance, *Clin. Biomech.* 10(1):3-19.

Andersson, G.B.J., Ortengren, R., Nachemson, A., Elfstrom, G., Broman, M. (1975) The seated posture: an electromyographic and discometric study, *Orthop. Clin. N. Am.* 6:105-120.

Bean, J.C., Chaffin, D.B., Schultz, A.B. (1988) Biomechanical model calculation of muscle contraction forces: a double linear programming method, *J. Biomech.* 21:59-66.

Bearn, J.G. (1961) The significance of the activity of the abdominal muscles in weight lifting, *Acta Anat.* 45:83.

Bogduk, N. and Macintosh, J.E. (1984) The applied anatomy of the thoracolumbar fascia. *Spine*, 9:164-170.

Burstein, A.H. and Frankel, W.H. (1968) The viscoelastic properties of some biological material, *Ann. N.Y. Acad. Sci.*, 146:158-165.

Cholewicki, J. and McGill, S.M. (1992) Lumbar posterior ligament involvement during extremely heavy lifts estimated from fluoroscopic measurements, *J. Biomech.* 25(1):17-28.

Cholewicki, J., McGill, S.M., Norman, R.W. (1995) Comparison of muscle forces and joint load from an optimization and EMG assisted lumbar spine model: towards development of a hybrid approach, *J. Biomech.* 28(3):321-331.

Cholewicki, J. and McGill, S.M. (1996) Mechanical stability of the *in vivo* lumbar spine: Implications for injury and chronic low back pain, *Clin. Biomech.* 11(1):1-15.

Collins, J.J. (1995) The redundant nature of locomotor optimization laws, *J. Biomech.* 28(3):251-167.

Crisco, J. J. and Panjabi, M.M. (1990) Postural biomechanical stability and gross muscular architecture in the spine, in *Multiple Muscle Systems* (Eds. J. Winters, S. Woo), Springer-Verlag, New York. pp. 438-450.

Cyron, B.M., Hutton, W.C., Stott, J.R. (1979) The mechanical properties of the lumbar spine, *Mech. Eng.* 8(2):63-68.

Davis, P.R. (1959) The causation of herniae by weight-lifting, *Lancet*, 2:155-157.

Ekholm, J., Arborelius, U.P., Nemeth, G. (1982) The load on the lumbosacral joint and trunk muscle activity during lifting, *Ergonomics* 25(2):145-161.

Farfan, H.F. (1973) *Mechanical Disorders of the Low Back*, Lea and Febiger, Philadelphia.

Farfan, H.F., Cossette, J.W., Robertson, G.H., Wells, R.V., Kraus, H. (1970) The effects of torsion on the lumbar intervertebral joints: the role of torsion in the production of disc degeneration, *J. Bone Jt. Surg.* 52A (3):469-497

Frymoyer, J.W., Pope, M.H., Clements, J.H, Wilder, D.G., MacPherson, B., Ashikaga, T. (1983) Risk factors in low back pain, *J. Bone Jt. Surg.* 65A:213-218.

Garg, A. and Herrin, G. (1979) Stoop or squat: a biomechanical and metabolic evaluation, *AIIE. Trans.* 11:293-302.

Gracovetsky, S., Farfan, H.F., Lamy, C. (1981) Mechanism of the lumbar spine, *Spine* 6(1):249-262.

Granata, K.P. and Marras, W.S. (1993) An EMG-assisted model of loads on the lumbar spine during asymmetric, dynamic trunk extensions, *J. Biomech.* 26(12):1429-1438.

*Gray's Anatomy, Descriptive and Applied* (1980) 36th ed. (Eds. Warwick, R., and Williams, P.L.) Longmans, London.

Gregor, R.J., Komi, P.V., Jarvinen, M. (1987) Achilles tendon forces during cycling, *Int. J. Sports Med.* 8:9-14.

Grew, N.D. (1980) Intrabdominal pressure response to loads applied to the torso in normal subjects, *Spine* 5(2):149-154.

Grieve, D.W. (1975) Dynamic characteristics of man during crouch and stoop lifting, in *Biomechanics iv*, (Eds. Nelson, R.C. and Morehouse, C.A.) pp. 19-29, University Park Press, Baltimore.

Heylings, D.J.A. (1978) Supraspinous and interspinous ligaments of the human lumbar spine. *J. Anat.* 123:127-131.

Hickey, D.S. and Hukins, D.W.L. (1980). Relation between the structure of the annulus fibrosus and the function and failure of the intervertebral disc, *Spine* 5(20):106-116.

Hughes, R.E., Chaffin, D.B., Lavender, S.A., Anderson, G.B.J. (1994) Evaluation of muscle force prediction models of the lumbar trunk using surface electromyography, *J. Orthop. Res.* 12:689-698.

Hughes, R.E., Bean, J.C., Chaffin, D.B. (1995) Evaluating the effect of coordination in optimization models, *J. Biomech.* 28(7):875-878.

Hukins, D.W.L., Aspden, R.M., Hickey, D.S. (1990). Thoracolumbar fascia can increase the efficiency of the erector spinae muscles, *Clin. Biomech.* 5(1):30-34.

Komi, P.V. (1990) Relevance of in vivo force measurements to human biomechanics, *J. Biomech.* Suppl. 2:23-34.

Krag, M.H., Byrne, K.B., Gilbertson, L.G., Haugh, L.D. (1986) Failure of intraabdominal pressurization to reduce erector spinae loads during lifting tasks, in *Proceedings of the North American Congress on Biomechanics*, Montreal, 25-27 August, pp. 87-88.

Krag, M.H., Seroussi, R.E., Wilder, D.G., Pope, M.H. (1987) Internal displacement distribution from *in vitro* loading of human thoracic and lumbar spinal motion segments: Experimental results and theoretical predictions, *Spine* 12(10):1001-1007.

Langenberg, W. (1970) Morphologic, Physiologischer Querschnitt and Kraft des M. erector spinae in Lumbalbereich des Menshen, *Z. Anat. Entwickl.* 132:158-190.

Lucas, D. and Bresler, B. (1961) *Stability of the Ligamentous Spine*, Tech. Report No. 40, Biomechanics Laboratory, University of California, San Francisco.

Macintosh, J.E., Bogduk, N., Gracovetsky, S. (1987) The biomechanics of the thoracolumbar fascia. *Clin. Biomech.* 2:78-93.

Macintosh, J.E. and Bogduk, N. (1987) The morphology of the lumbar erector Spinae, *Spine*, 12(7):658-668.

Marras, W.S. and Sommerich, C.M. (1991) A three-dimensional motion model of loads on the lumbar spine I and II, *Human Factors*, 33(2):123-149.

Marras, W.S., Lavender, S.A., Leurgans, S.E., Fathallah, F.A., Ferguson, S.A., Allread, W.G., Rajulu, S.L. (1995) Biomechanical risk factors for occupationally-related low back disorders, *Ergonomics* 36(2):377-410.

McGill, S.M. and Norman, R.W. (1985) Dynamically and statically determined low back moments during lifting, *J. Biomech.* 18(12):877-885.

McGill, S.M. and Norman, R.W. (1986) Partitioning of the L4/L5 dynamic moment into disc, ligamentous and muscular components during lifting, *Spine* 11(7):666-678.

McGill, S.M. and Norman, R.W. (1987) Reassessment of the role of intraabdominal pressure in spinal compression, *Ergonomics* 30(11):1565-1588.

McGill, S.M. and Norman R.W. (1987) Effects of an anatomically detailed erector spinae model on L4/L5 disc compression and shear, *J. Biomech.* 20(6):591-600.

McGill, S.M. and Norman, R.W. (1988) The potential of lumbodorsal fascia forces to generate back extension moments during squat lifts, *J. Biomed Engng.* 10:312-318.

McGill, S.M., Patt, N., Norman, R.W. (1988) Measurement of the trunk musculature of active males using CT Scan radiography: duplications for force and moment generating capacity about the L4/L5 joint, *J. Biomech.* 21(4):329-341.

McGill, S.M. (1988) Estimation of force and extensor moment contributions of the disc and ligaments at L4/L5, *Spine* 13:1395-1402.

McGill, S.M. (1989) Loads of the lumbar spine and associated tissues, in *Biomechanics of the Spine — Clinical and Surgical Perspective*, (Eds. Goel, V.K. and Weinstein, J.N.) CRC Press, Boca Raton.

McGill, S.M. and Hoodless, K. (1990) Measured and modelled static and dynamic axial trunk torsion during twisting in males and females, *J. Biomed. Engng.* 12:403-409.

McGill, S.M. (1991) Electromyographic activity of the abdominal and low back musculature during the generation of isometric and dynamic axial trunk torque: implications for lumbar mechanics, *J. Orthop. Res.* 9:91-103.

McGill, S.M. (1991) The kinetic potential of the lumbar trunk musculature about three orthogonal orthopaedic axis in extreme postures, *Spine* 16(7):809-815.

McGill, S.M. (1992) The influence of lordosis on axial trunk torque and trunk muscle myoelectric activity, *Spine* 17(10):1187-1193.

McGill, S.M. (1992) A myoelectrically based dynamic 3-D model to predict loads on lumbar spine tissues during lateral bending, *J. Biomech.* 25(4):395-414.

McGill, S.M. and Brown, S. (1992) Creep response of the lumbar spine to prolonged lumber flexion, *Clin. Biomech.* 7:43-46.

McGill, S.M., Santaguida, L., Stevens, J. (1993) Measurement of the trunk musculature from T6 to L5 using MRI scans of 15 young males corrected for muscle fiber orientation, *Clin. Biomech.* 8:171-178.

McGill, S.M. and Kippers, V. (1994) Transfer of loads between lumbar tissues during the flexion relaxation phenomenon, *Spine* 19(19):2190-2196.

McGill, S.M. (1996) A revised anatomical model of the abdominal musculature for torso flexion efforts, *J. Biomech.* 29(7):973-977.

McGill, S.M., Juker, D., Axler, C. (1996) Correcting trunk muscle geometry obtained from MRI and CT scans of supine postures for use in standing postures, *J. Biomech.* 29(5):643-646.

McGill, S.M., Juker, D., Kropf, P. (1996) Appropriately placed surface EMG electrodes reflect deep muscle activities (psoas quadratus lumborum, abdominal wall) in the lumbar spine, *J. Biomech.* 29(11):1503-1507.

McGill, S.M. (1997) Biomechanics of low back injury: implications on current practice and the clinic. *J. Biomech.*, 30(5):465-475.

McKenzie, R.A. (1979) Prophylaxis in recurrent low back pain, *N.Z. Med. J.* 89:22.

Nachemson, A. (1960) Lumbar intradiscal pressure. *Acta Orthop. Scand.* suppl. 43.

Nachemson, A.L. and Morris, J.M. (1964) *In vivo* measurements of intradiscal pressure, *J. Bone Jt. Surg.* 46A:1077-1092.

Nachemson, A., Andersson, G.B.J., Schultz, A.B. (1986) Valsalva manoeuvre biomechanics: effects on lumbar trunk loads of elevated intra-abdominal pressure, *Spine* 11(5):476-479

Nemeth, G. and Ohlsen, H. (1986) Moment arm lengths of trunk muscles to the lumbosacral joint obtained *in vivo* with computer tomography, *Spine* 11(2):158

Norman, R.W., Gregor, R., Dowling, J. (1988) The prediction of cat tendon force from EMG in dynamic muscular contractions, in proceedings of the fifth biennial conference of the Canadian Society for Biomechanics 1988, O'Hara, August 16-19, pp. 120-121.

Pope, M.H., Hanley, E.N., Matteri, R.E., Wilder, D.G., Frymoyer, J.W. (1977) Measurement of intervertebral disc space height, *Spine* 2:282-286.

Potvin, J., Norman, R.W., McGill, S. (1991) Reduction in anterior shear forces on the L4/L5 disc by the lumbar musculature, *Clin. Biomech.* 6:88-96.

Potvin, J. and Norman, R.W. (1992) Can fatigue compromise lifting safety? Proceedings of the Second North American Congress on Biomechanics, Chicago, August 24-28, pp. 513-514.

Reid, J.G. and Costigan, P.A. (1985) Geometry of adult rectus abdominis and erector spinae muscles, *J. Orthop. Sports Phys. Ther.* 6:278-280.

Reilly, T., Tynell, A., Troup, J.D.G. (1984) Circadian variation in human stature, *Chronobiology Int.* 1:121-126

Santaguida, L. and McGill, S.M. (1995) The psoas major muscle: A three-dimensional mechanical modeling study with respect to the spine based on MRI measurement, *J. Biomech.* 28(3):339-345.

Schultz, A.B., Warwick, D.N., Berkson, M.H., Nachemson, A., (1979) Mechanical properties of the human lumbar spine motion segments — Part 1, Responses to flexion, extension, lateral bending and Torsion, *J. Biomech. Eng.* 101:46-52.

Schultz, A.B., Andersson, G.B.J., Ortengren, R., Haderspeck, K., Nachemson, A. (1982) Loads on the lumbar spine, *J. Bone Jt. Surg.* 64A(5):713-720.

Shirazi-Adl, A., Ahmed, A.M., Shrivastava, S.C. (1986a) A finite element study of a lumbar motion segment subjected to pure sagittal plane moments, *J. Biomech.* 19(4):331-350

Shirazi-Adl, A., Ahmed, A.M., Shrivastava, S.C. (1986b) Mechanical response of a lumbar motion segment in axial torque alone and combined with compression, *Spine* 11(9):914- 927

Shirazi-Adl, A. and Drouin, G. (1987) Load bearing role of the facets in the lumbar segment under sagittal plane Loadings *J. Biomech.* 20(6):601-603.

Sutarno, C. and McGill, S.M. (1995) Iso-velocity investigation of the lengthening behavior of the erector spinae muscles, *Eur. J. Appl. Physiol. Occup. Med.* 70(2):146-153.

Tesh, K.M., Dunn, J., Evans, J.H., (1987) The abdominal muscles and vertebral stability, *Spine* 12(5):501-508.

Thomson, K.D. (1988) On the bending moment capability of the pressurized abdominal cavity during human lifting activity, *Ergonomics* 31(5):817-828

Thorstensson, A., Andersson, E., Cresswell, A., (1989) Lumbar spine and psoas muscle geometry revised with magnetic resonance imaging, in *Proceedings of the International Society of Biomechanics,* June, Los Angeles, abstract #251.

Troup, J.D.G. and Chapman, A.E. (1969) The strength of the flexor and extensor muscles of the trunk, *J. Biomech.* 2:49-62.

Troup, J.D.G. (1977) Dynamic factors in the analysis of stoop and crouch lifting methods: a methodological approach to the development of safe materials handling standards, *Orthop. Clin. N. Am.* 8(1):201-209.

Troup, J.D.G., Martin, J.W., Lloyd, D.C. (1981) Back pain in industry: a prospective survey, *Spine* 6:61-69.

Troup, J.D.G., Leskinen, T.P.J., Stalhammear, H.R., Kuorinka, I.A. (1983) A comparison of intra-abdominal pressure increases, hip torque, and lumbar vertebral compression in different lifting techniques, *Human Factors* 25(5):517-525.

Ueno, K. and Liu, Y.K. (1987) A three-dimensional nonlinear finite element model of lumbar intervertebral joint in torsion, *J. Biomech. Eng.* 109:200-209.

Videman, T., Nurminen, J., Troup, J.D.G. (1990) Lumbar spinal pathology in cadaveric material in relation to history of back pain, occupation and physical loading, *Spine* 15(8):728-740.

Wilder, D.G., Pope, M.H., Frymoyer, J.W. (1988) The biomechanics of lumbar disc herniation and the effect of overload and instability, *J. Spine Disorders* 1(1):16-32.

# 13

# Selection of 2-D and 3-D Biomechanical Spine Models: Issues for Consideration by the Ergonomist

Robert W. Norman
*University of Waterloo*

Stuart M. McGill
*University of Waterloo*

## 13.1  Introduction

The usefulness and interpretation of complex biomechanical models have been explained in another chapter (McGill, 1999). Biomechanists continue to debate many contentious questions, such as whether there is a need for high levels of accuracy in the anatomical representation of spinal structures in a spine model, and whether the mathematical method used to solve large numbers of equations in these models reflects known physiological patterns of muscle activation in human movements. These authors believe that the answer to both of these questions is yes. However, the focus of this chapter is on simpler models intended for relatively easy use in workplace settings for the assessment of risk of back injury; but simplification raises another set of questions that ergonomists should ask before purchasing a biomechanical software package.

Some of the most pertinent questions are: How do biomechanists approach the assessment of level of risk of back injury in the workplace? Have biomechanical spine model estimates of risk factors (typically low back moments of force, spinal compression and shear forces are outputs from these models), been shown to be related to increased risk of injury? Are all spine models the same? Have spine models been validated, i.e., how close to reality are the model estimates of moments of force, spinal compression, and shear? How much anatomical detail in the model is necessary for adequate content validity? Is a model that will allow dynamic analysis of tasks always needed or are static models acceptable? Are single plane (2-D) assessments still useful or must 3-dimensional analyses always be performed? If a 3-dimensional model is needed, how does the model estimate forces on joints and how are the forces and moments interpreted? Are there any spine models that produce outputs that reflect the effects of prolonged loading?

This chapter will address these issues and will then describe a family of 2- and 3-dimensional models into which the authors have attempted to incorporate as many important biological features as they could within the constraints of producing a method to assess low back injury risk that is practical for workplace use by ergonomists. The chapter is not a critical, comparative review of commercially available spine models suitable for ergonomic use, although several interesting models have been described in the past decade by spinal biomechanists (e.g., Chaffin et al., 1989; Marras and Sommerich, 1991).

## 13.2   The Biomechanical Approach to Assessment and Reduction of Risk of Back Injury

Biomechanists generally operate under the concept that a tissue will become irritated or will fail when the forces to which it is exposed at a particular point in time are greater than the tolerance of that tissue to force at that point in time. Both the size of the force acting on a tissue and the tolerance of tissue to force change with time. The forces on tissues change with changing demands of a task or with the way in which the worker does the same task from time to time; the tolerance of the tissue to force (or other mechanical input) varies with factors such as state of repair from previous injury, conditioning or deconditioning, age, gender, disease, and inherited individual differences in factors such as the cross-sectional area of tissues.

Typically, the biomechanics approach to injury risk assessment is to represent the body of the worker as a system of individual body segments linked at the joints (linked segment model). An external force acting on the linkage, for example the force of a weight held in the hand, requires the production of forces in muscles and in other tissues at all joints in the linkage to prevent the linkage from collapsing or to cause it to move in a desired direction. At the level of the lumbar spine these "reaction forces" and "reaction moments of force" are produced by some combination of muscles, ligaments, discs, bones in contact with each other, and by other body tissues, depending upon what the posture is and where in the joint range of motion each body segment is at each instant in time throughout the movement. The reaction forces in the supporting tissues can become very large for even relatively light loads held in the hands because the weights of body parts are very large (head, arms, and trunk above the pelvis are about one half body weight), and the mechanical advantage of body tissues in producing moments of force is very poor. If the forces in tissues gets too large, the tissues are damaged.

In all biomechanical spine models, reaction moments are produced by simplified representations of the real anatomical structures that actually produce the reaction moments that have to support the weight of the force on the hands and of the body parts and any inertial forces caused by accelerations of these masses, for example by combining forces of some or many small muscles to make one force vector. Some models are much more simplified anatomically and physiologically than others leading to the question as to whether they are over-simplified and produce unrealistic estimates of exposure variables; the reverse question can also be asked. Are some models more complex than is necessary for ergonomic use?

The forces calculated in these supporting structures are the measures of exposure used to estimate injury risk. For spinal models, typically the reaction (or support) moment about the lumbar spine required to support the load moment, and the consequent compressive and shear forces acting on the lumbar spine at the L4/L5 or L5/S1 level are calculated. Quantifying the risk of injury usually involves comparing these estimates of supporting tissue forces with estimates of the failure tolerance of these tissues (e.g., compressive or shear strength of the spinal tissues) or reaction moments of forces with back extension strength under the knowledge that operating at or near a person's strength limits is getting close to tissue fatigue limits, if not failure limits. Simply, as a force applied to a tissue approaches its capacity to bear force, the risk of injury increases to the point where the limit of force tolerance is reached and injury (tissue damage) occurs. More recently, epidemiological studies have appeared in which injury or pain reporting odds ratios have been calculated for high vs. low exposure to biomechanical variables such as trunk angle (Punnett et al., 1991; Marras et al., 1993), back extension moment of force (Marras et al., 1993), and peak and prolonged spine compression and shear soon to be published from work of our own group and colleagues from the Institute for Work & Health (Toronto). The establishment of target values for these variables to reduce risk epidemiologically is promising.

The advantages of the biomechanical approach are several: from a good model, any task should be analyzable — lifting, lowering, pushing, pulling, while interacting with any material, tool or machine; the analysis should be individualizable for gender, height, weight, strength etc.; and, in principle, any region of the body may be considered. The critical issue for the ergonomist is the ability to analyze the demands of a job using reasonably simple methods, to assess the risk of injury, and to redesign jobs in a way that can be shown to actually reduce forces on vulnerable tissues to levels below injury threshold.

## 13.3 Have Biomechanical Spine Model Outputs Been Shown to Be Related to Injury Risk?

Both biomechanical and epidemiological evidence show clearly that high moments of force about the lumbar spine which, in turn, are highly related to large compressive or shear forces on lumbar motion units, increase risk of injury. Biomechanical models that output lumbar moments of force, compression and shear, therefore, are useful as tools for measuring exposure to risk of injury.

Many studies on cadavers and on animal tissue in the laboratory have shown that excessive compressive or shear forces on the spine will result in failure of various tissues including the cartilaginous end plate of the vertebral body, the laminae of the annulus fibrosis of the intervertebral discs, muscles, ligaments, facet joints, and other structures. Original data and excellent compilations of many years of experimental results on the compressive tolerance of spinal motion units (two vertebral bodies and a disc) can be found in papers by Brinckmann et al. (1989) and Jager et al. (1991). The compilation by Jager et al. shows that older spines and female spines are weaker in compressive loading than younger spines and, for the same age, male spines on average but with large individual differences in tolerance to load. Figure 13.1 is an adaptation of data from regression equations presented by Jager et al. (1991).

Cripton et al. (1994) have shown failure of spinal motion units exposed to anterior shear forces of about 2000 N. Troup and Chapman (1969) reported maximum trunk extension moments of force (back strength) of 485 N.m for 98 men aged 18 to 39 years and 302 N.m for 132 women aged 18 to 23 years. Although people can produce maximum effort muscular contractions without injury, particularly well-conditioned people, well-motivated maximum muscular contractions, particularly in eccentric (active muscle lengthening) efforts, are approaching tissue tolerance levels. It is also well known that maximum muscular efforts can be maintained for only a few seconds before fatigue results in continual drops in force with efforts to sustain or repeat the contraction. Muscular fatigue may be a protective mechanism for active tissue, but it causes changes in the ways people perform tasks, in some cases loading tissues that would normally not be used in that task and that should not be used because of a lower injury tolerance (Potvin and Norman, 1992).

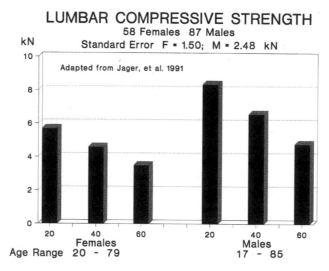

**FIGURE 13.1**  Lumbar spine compressive strength data compiled by Jager et al. (1991). Means by age and sex calculated from their regression equation.

As noted earlier, not only is there biomechanical justification for assessing risk using biomechanical spine models, there is an increasing body of epidemiological studies that support this notion. Marras et al. (1993) showed, in an epidemiological study of 403 lifting jobs in 40 different industries, that those workers exposed to high average or peak lumbar moments of force were 4 to 5 times more likely to be in a high injury risk group than those exposed to low moments. Punnett et al. (1991), in a study of 219 auto assemblers, showed that those who spent more than 10% of the cycle time with the trunk flexed more than 20 degrees were 6 times more likely to report a back injury than those who worked in a more upright posture. If they spent more than 10% of the time flexed more than 45 degrees, the risk was 9 times higher than those who could work in an upright posture with less than 20 degrees of flexion. A flexed trunk increases the moment of force because of the weight of the head, arms, and trunk alone, even without loads in the hands.

In a recent study (unpublished) from our own group we showed, in data from 235 auto workers, that the correlation of trunk moment of force with L4/L5 compression was 0.98 and with reaction shear was 0.88. The 105 auto workers who were in the back pain group had peak moments of force of 182 N.m compared with 141 N.m for 130 control workers who reported no back pain, peak spinal compression of 3402 N vs. 2740 N and peak spinal shear of 462 N vs. 353 N, respectively. All of these differences were highly significant statistically ($p<0.0001$). These values were obtained from video analysis of all workers in the study while they worked combined with a quasidynamic, two-dimensional, biomechanical spine model (2DWATBAK to be discussed later). Interestingly, the mean spinal compression forces of the back pain group were very close to the Action Limit for spinal compression (3434 N) proposed as a limit by NIOSH (1981), below which most people are expected to have nominal risk of back injury. As noted earlier, high compressive forces, in the absence of impacts, are produced primarily because of very large forces in torso muscles, particularly back extensor muscles, which have relatively poor mechanical advantage but are required to support or produce the sizable moments of force required to do some tasks. The independent odds ratios on three biomechanical variables, over the spread in observed responses for the 25% of random control subjects (no reports of back pain) who had the highest exposures compared to the 25% with the lowest exposures, showed that there was a 232% greater risk of being a low back pain case (those who reported back pain to plant medical personnel) if the shear forces on the spine were high, 189% greater risk if the "usual" force on the hands was high, and 170% greater risk if the average degrees of torso flexion were high. If all three of these variables were high, the risk of being a case was

750% higher than if the exposures were low. The univariable odds ratio for peak spinal compression was 1.86 and for accumulated compression over a shift, 1.33.

While it is infrequent to see spinal motion unit failure associated with reported occupational low back pain, high spinal compression almost invariably indicates high forces on spinal muscles and, if the torso is flexed, laterally bent or twisted to very near the extremes of the range of motion, on ligaments and other spinal structures.

## 13.4 Are All Spine Models the Same?

All spine models are not the same. Several different models which implicitly or explicitly contain different assumptions regarding function and dramatically different levels of anatomical realism, have been presented in the literature. An earlier chapter by McGill (1999) (Chapter 53) addressed generic functional differences among complex, EMG-assisted spine models (e.g., McGill and Norman, 1986; Marras and Sommerich, 1991), optimization models (e.g., Gracovetsky et al., 1981; Schultz et al., 1982; Bean et al., 1988), and single muscle equivalent models (e.g., Frievalds et al., 1984; Norman et al., 1994). Cholewicki et al. (1995) argue that outputs from current optimization spine models are not physiologically realistic in that they do not account for the simultaneous activation of "antagonistic" muscles (cocontraction), such as those of the low back and abdomen, that is observed when people handle loads. Cocontraction of trunk muscles is extreme when a trunk twist moment of force is present, for example when one bends forward and lifts a garage door or some other load with one hand. In this case, spinal compression would be considerably underestimated by optimization models or other types of models that do not account for cocontraction of antagonistic torso muscles.

Not only are there differences in the physiological behavior of different spine models because of differences in how the models are activated (e.g., EMG vs. optimization), there are structural differences in the anatomical representation of the models. It is well known that the low back muscles, when activated, produce force in a slightly backward as well as downward direction (with respect to the spine) during a lift. The backward (posterior) component of their force reduces the shear forces acting on the facet joints and other structures but not the compressive force (Potvin et al., 1991). Excessive shear is a risk factor. It is also well known that if the normal curvature (curve seen when standing upright) of the low back is lost during a lift, the low back muscles reduce their contribution to the support moment and with extreme spine flexion these muscles may shut off completely. This causes the moment to be supported entirely by passive structures such as ligaments, and the shear force reducing contribution of the muscles is lost. Ligament injury is more serious than muscle strain, and shear forces can irritate facet joints without causing tissue damage if the joints are sensitive or result in severe damage to facet joints and other parts of the vertebral body if the forces are excessive.

It is possible to build these structural and functional features of spinal mechanics into even relatively simple models. Biomechanical spine models intended for the assessment of back injury risk in industry should include these known phenomena to increase the number of known injury mechanisms to which they are sensitive. Not all of them do this; therefore all spine models are not the same.

## 13.5 Have Spine Models Been Validated?

It is important to ask the question as to how close to reality the spine model estimates of exposure measure are, the measures that will be compared with tolerance levels (the lumbar moments of force, spine compression, and shear). The fact is that none of the spine models referred to above has been directly validated, although one has come close. The authors of some of the models claim that their model has been validated, but their claims are, at best, based on correlations of estimates of muscle forces from the model output with EMG recordings. This is only an indirect and not entirely convincing validation.

To be assured that outputs from models are reasonably accurate, direct validation is necessary. Therefore, if a model is to predict a muscle force or spine compression force in Newtons (N), direct validation requires that measurement of muscle forces or spine compression (N) must be made against which to compare the predictions. Technically it is not possible to directly measure spine muscle forces in living, active humans. Therefore, direct validation of spinal models has not been reported, except for the work of Schultz et al. (1982).

Schultz et al. claimed validation of a 22-muscle optimization model on the grounds that the disc compression force (N) predicted by the model was close to the pressure (mmHg) that was directly measured from a pressure transducer inserted into the lumbar discs of humans while they performed quite light, static tasks. This is a reasonable approach to validating the disc compression prediction part of the model although even in this study, Newtons of force were plotted against mmHg of disc pressure.

These authors also recorded EMG amplitude activity observed during static efforts and correlated these static amplitudes with the muscle forces predicted by the model. The correlations were far from perfect but were encouraging. However, EMG is not a direct measure of muscle force and consequently the validation of the distribution of muscle forces in this, and in all other models, particularly during asymmetric tasks, is still in question.

As noted above, correlating predictions of muscle forces in Newtons with electrical activity from muscle in microvolts is, at best, indirect. If some in the scientific community consider this type of validation of spine, or other models to be acceptable, then the EMG-assisted models are inherently valid because they use the electrical activity on a trial-to-trial and instant-to-instant basis. They do not predict it.

Direct validation of predicted forces against directly measured forces acting on most spinal tissues is virtually impossible. Therefore, content validity in the form of realistic representation of anatomical structures together with "biological" recruitment of passive tissues and simultaneous activation (cocontraction) of both agonist and antagonist muscles is important. The ergonomist or other users of commercially available spine models should take claims of validation of these models skeptically, demand detailed descriptions of the anatomical structures built into the models, and clear statements of the implicit and explicit assumptions and limitations of the model before purchasing it.

## 13.6 How Much Anatomical Detail in the Model is Necessary for Adequate Content Validity?

As much biological realism as possible should be incorporated into simple/industrial models so that content validity is at least present, even if direct validation of predicted forces in low back tissues is not possible. For example, the architecture of the extensor muscles and ligaments of the low back, and their dramatic effect on low back shear loads were explained briefly above and in detail in Chapter 12 (McGill, 1999). These features should be incorporated into industrial low back models. Another issue concerns the representation of contraction of the many muscles acting about the torso. For example, cocontraction of antagonistic torso muscles is commonly observed even during sagittal plane load handling and is particularly dominant in 3-dimensional tasks. Cocontraction of antagonistic muscles results in higher muscle forces to satisfy the required moment, and in turn, higher spine compressive forces, than would be the case if there were no cocontractions (McGill and Norman, 1986; Marras and Granata, 1995). This effect can be easily incorporated into 2-dimensional models by simply reducing the effective moment arm lengths of single equivalent muscles from anatomical dimensions observed from CT and MRI (McGill et al., 1988, 1993) to more physiologically effective lengths (McGill, 1991, McGill and Norman, 1992). Table 13.1 shows this comparison. It is highly desirable to incorporate this phenomenon into any 2- or 3-dimensional model of the lumbar spine because the cocontraction can have a dramatic effect on the sizes of forces acting on other structures within the system.

However, the problem of incorporating the effects of muscular cocontractions into 3-dimensional spinal models is not trivial and is particularly problematic if a relatively simple model must be developed. In the opinion of the authors, optimization does not solve the problem. Indeed, one of the biggest

**TABLE 13.1** Approximate moment arms (cm) of equivalent force generators based on measurements of muscle geometry and also estimated from an anatomically detailed, myoelectrically driven model that attempted to accommodate measured muscular cocontraction. Cocontraction has the effect of decreasing the mechanical advantage of an equivalent force generator and imposes higher compressive loads on the spine.

| AXIS | Based on Muscle Geometry | Estimates from Model |
|---|---|---|
| Extension | 6.0–7.5 | 5.5–6.5 |
| Lateral bend | 5.0–6.0 | 3.0–4.0 |
| Axial twist | 7.0–10.0 | 1.0–3.0 |
| Flexion | 8.5–10 | 4.0–4.5 |

weaknesses in current spine optimization models is that they are poor at predicting patterns of cocontractions of antagonistic muscles (Hughes et al., 1995; Cholewicki et al., 1995) particularly in complex 3-D tasks. Anatomically complex models are available (and described in Chapter 53) that have biological and mechanical content validity and can produce estimates of forces on tissues during dynamic three-dimensional tasks (e.g., McGill and Norman, 1986; Marras and Sommerich, 1991; McGill, 1992). Unfortunately, these models require the use of multichannel electromyography and direct measurement of spine kinematics and, thus, are too complex for ready application in industry although they are extremely useful in laboratory experiments in formulating and testing hypotheses designed to reduce low back injury.

In order to facilitate industrial application, anatomically simplified 3-dimensional models have been reported in which the torso musculature is represented by a system of single equivalent force-moment generators (e.g., Schultz et al., 1982; Chaffin et al., 1989; Norman et al., 1994). During the generation of pure moment about any one axis, the single moment generators can be designed to quite closely predict the output from the detailed biomechanical models. The problem lies in tasks in which two or more moments must be satisfied. Should the three moment generators be activated separately, in spite of the fact that the human musculature does not work this way, or should they form a relationship that recognizes muscle coupling and allows the muscles to work together to satisfy the moments? One solution is to estimate spinal compression using a regression equation that relates compression, estimated from a complex EMG-assisted model, with the 3-dimensional low back moments that resulted in the compression. while subjects performed dynamic, asymmetrical, 3-dimensional tasks. An equation was proposed by McGill et al. (1996) that retained as much anatomical and physiological content validity as they could and, in particular, preserved the effects of antagonistic muscle cocontraction since it was based on subjects performing dynamic, asymmetrical, 3-dimensional tasks. This is described in the 3DWATBAK section of this chapter.

## 13.7 Is a Model Needed that Will Allow Dynamic Analysis of Tasks or Are Static Models Acceptable?

A model is needed that will at least allow accounting of the inertial forces on the hands (the live loads) during lifts and lowers, pushes and pulls, not just the static load weights of objects held in the hands. Some static models can provide this versatility and a "quasidynamic" analysis of human lifting tasks is often a reasonable approach to estimating the size of the moments of force, joint reaction forces, and lumbar compression and shear forces of tasks which, in reality, are dynamic. The extent to which an entirely static model is acceptable depends upon the size of the inertial forces in a dynamic lift compared with the static load and body segment weights which create forces and moments. For example, heavy loads may be analyzed statically in many cases, simply because the lifters are incapable of appreciably accelerating the load. An exception from the athletic world is the competitive "snatch" or "clean and jerk" of a heavy load, which contains periods of very high accelerations and decelerations and corresponding high and low inertial forces. While a static analysis would be inappropriate in these cases, most heavy loads in industry are not lifted with a technique characterized by high accelerations. Of course, to analyze pushes or pulls, knowledge of the size and direction of forces acting on the hands is essential.

Freivalds et al. (1984) estimated that accelerations of loads lifted by their subjects could increase the static load by as much as 40% of its weight. The loads lifted were maximum loads the subjects were willing to lift repeatedly according to their own feelings of exertion and fatigue (not single maximum lifts). They varied from about 10 kg to 66 kg depending upon the subject and conditions of the awkward lifts. The authors of this chapter reported that dynamic analysis of lifts of 20 kg masses resulted in lumbar moments 19% higher on average than those statically determined by zeroing all accelerations of the body segments and load in the hands. Individual trials were up to 50% higher, but one skilled subject showed slightly lower dynamic than static moments on two trials. He exploited trunk accelerations which were transferred to the load to assist in the lift. A subsequent reanalysis of the data, in which only the load weight acceleration but not the body segment accelerations were accounted for, showed that this "quasi-dynamic" moment overestimated the actual dynamic moment by about 25% (McGill and Norman, 1985).

These studies indicated that caution must be exercised in interpreting statically determined moments from simple industrial models and all other static models in the context of lifts which are highly dynamic. On the other hand, if one knows the actual dynamic load weight in the posture analyzed, from an instrumented handle on the load for example, using this instantaneous value (or "live" load) as a hand force, instead of the static load weight, will usually give a liberal estimate of the actual dynamic joint moment. This is an error on the "safe" side as far as estimating demands of the lift on the low back, or other joints is concerned, avoids the complexity of a fully dynamic analysis for the ergonomist, and allows some accounting for dynamic tasks in methods suitable for workplace use. To reiterate, a model which cannot account for varying sizes and directions of forces acting on the hands cannot be used to analyze pushes or pulls.

## 13.8   Are Single Plane (2-D) Models Useful or Must 3-Dimensional Analyses Always be Performed?

Many industrial tasks are dominated by joint loading in one plane even though there may be components of forces in other planes — the posture may be asymmetric but the loading is essentially planar and most often the dominant moment is in the flexion/extension (sagittal plane) of trunk movement. For many tasks, therefore, a 2-dimensional model is satisfactory and many load handling postures that appear to be 3-dimensional do not result in large twisting moments around the long axis of the trunk. On the other hand, some efforts that do not appear as trunk twists around its long axis in fact have a substantial moment about this axis or the lateral bend axis; handling loads with one hand but with no actual torso twist or lateral bend displacement are examples. Sensitivity analyses have demonstrated that joint moments and low back compression estimated from a 2-dimensional model are less than 10% different from those estimated from a 3-dimensional model for two-handed lifts in which the load is placed toward the side of the body up to 30 degrees away from the midline (Bone et al., 1990). This posture looks far more twisted than calculation of torsional moments reveals (Figure 13.2). This is a far smaller difference than the trigonometry would explain and is the result of movement toward the side of the body to reach the load that was accomplished by shifting the position of the legs, pelvis, and torso together, even though the feet were constrained to remain facing forward. These small errors can be further reduced by orienting the camera axis orthogonal to the dominant loading plane and a 2-dimensional analysis would not be out of the question. An example of the use of a 2- and 3-dimensional, quasidynamic spine model developed in our own laboratories and suitable for use in the workplace follows. Several models that have been reported in the literature by other authors were cited earlier.

## 13.9   A Simple 2-Dimensional, Quasidynamic Model for Risk Assessment and Job Design (2DWATBAK)

The objective of this model, called 2DWATBAK, is to facilitate rather routine injury risk assessment of industrial tasks, but in a way that incorporates as much biological detail and content validity as possible, consistent with a method that is usable in the workplace. This model uses joint position data, obtained

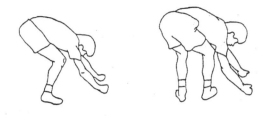

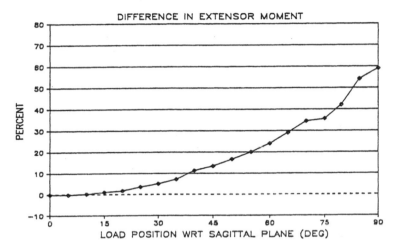

**FIGURE 13.2** (Top panel) Illustration of two of the lifting postures performed with the load positioned in the sagittal plane and 90° from the sagittal plane; (middle and bottom panels) Differences in L4/L5 extensor moment predicted by a 2-D model and the 3-D model, expressed as a percentage of the 3-D model prediction.

from a single frame of film, a slide or video, or from a manikin that the operator can move on the computer screen. The gender, body weight, and height of the worker and the sizes and directions of the static or dynamic forces acting on the hands must be input for the model to estimate moments of force about the elbows, shoulders, hips, knees, ankles, and the L4-L5 level of the low back. The compression and two shear forces, the reaction shear and the joint shear, acting at the L4/L5 level of the lumbar spine are then calculated.

To assess risk, the moments at all of the joints can be compared with strength data obtained from the literature or from the worker if measured, and the compression (e.g., Figure 13.1) and shear forces can be compared with tolerance data obtained on cadavers.

Several features are incorporated into 2DWATBAK in an attempt to increase biological content validity and to facilitate use:

1. Spinal posture (lordotic curvature) is monitored and entered by the operator to improve the accuracy of low back shear force predictions. Shear force injury risk has been underestimated in the past. For example, the size of the shear force supported by spinal structures depends upon whether the worker's lumbar spine is in a neutral posture (the curve observed during upright standing) or is flexed. A neutral posture requires erector spinae muscle activation to supply the supporting moment when the torso is inclined forward by rotating about the hips. When the posterior muscles are activated, they reduce the shear forces acting on the spinal motion unit joints (joint shear) below the level of anterior reaction shear. When the spine is severely flexed, the erector spinae muscles can be deactivated (flexion relaxation) and the moment has to be supported by ligaments and other passive tissues. When stretched, the interspinous ligament (Figure 13.3, force vectors 2a–d), described by Heylings (1978), adds undesirably to the anterior

shearing force produced by the weight of the head, arms, and trunk and forces in the hands; it does not offset this shearing force.

While most lumbar spines will tolerate 2000N of anterior shear (Cripton et al., 1994), one could argue for a workplace guideline of 1000N, half the average value, to provide a margin of safety for those in the bottom half of the shear strength distribution. Large individual differences typify tissue tolerance data.

2. Guidelines for interpreting low back compression values are improved with adjustment factors for gender differences and the effects of age (20, 40, 60 years) from data compiled by Jager et al. (1991).

3. Moments of force at each joint can be compared to joint strength data obtained from the literature (or from the worker) to identify joints at risk other than the low back (Figure 13.4).

4. The posture of a manikin can be altered on the computer screen by moving the body segment with the "arrow keys." 2DWATBAK then automatically recalculates the various exposure variable values to allow assessment of the redesigned task (Figure 13.5).

## 13.10 If a 3-Dimensional Model is Needed, How Does the Model Estimate Forces on Joints and How Are They Interpreted: An Example of a 3-Dimensional Industrial Risk Assessment Model (3DWATBAK)

While many occupational tasks are performed essentially in the sagittal plane and can be reasonably analyzed using two-dimensional biomechanical models, many other tasks require three-dimensional analysis for approximation of the actual demands. However, obtaining input data for a 3-dimensional model is a problem for industrial use in the absence of very expensive image reconstruction equipment or the need to use an often oversimplified look-up table of many possible 3-dimensional postures. 3DWATBAK will take digitized x,y,z coordinates if the user has a 3-dimensional image reconstruction system. An alternative to this expensive equipment for obtaining the necessary input coordinates of the body joints has been provided by means of a manipulable manikin on the computer screen.

First, a worker is captured in a frame of a video image or photograph. Then, 3DWATBAK provides three views (top, side, front) of a manikin which appears on the computer screen (Figure 13.6). The posture of the manikin is altered to approximate that of the worker by using a mouse. The body segments of any one of these views are moved one at a time. Movement of a segment in one view automatically moves the manikin in the other two views. When the posture of concern has been produced on the screen to the satisfaction of the analyst, the 3-dimensional coordinates of the end points (body joints) of the manikin segments are automatically generated, scaled, and stored.

As for 2DWATBAK, the model can be individualized for a particular worker by providing data on height, weight, gender, and user-defined or standard anthropometrics. Control of anthropometrics (e.g., mass of thigh or any other segment) allows alteration of a body segment mass or location of mass center if necessary, for example to add the mass of a tool belt or to accommodate analysis of an amputee. Size and direction of forces acting on each hand are defined as a push, pull, lift, lower, or as a load mass.

Anatomical detail has inherently been incorporated through the development of a regression equation to enhance biological fidelity in estimations of low back compressive load, and, in particular, attempt to account for the cocontracting musculature. This regression equation was based on the data of three subjects, performing a variety of tasks which produced various combinations of 3-dimensional moments about the low back. Compression was calculated using the anatomically detailed laboratory model which accommodated patterns of muscle/ligament interplay and muscle cocontraction measured in each subject-trial (see Chapter 12). The specific regression equation is as follows (R-square 0.936); the "shape" of the spinal compression/moment surface, from which the equation was derived, is seen in Figure 13.7.

WATBAK - 5.1

SPINE CURVATURE

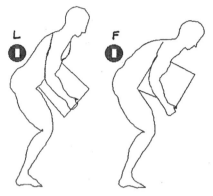

You can select the
the back posture that
best represents the
spinal curvature of
the subject. Choose 'L'
if the subject's back
is similar to the model
on the left.

←- posterior shear

←- anterior shear

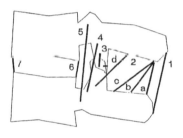

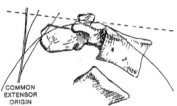

LUMBAR SPINE (L4/L5) PARAMETERS
----------------------------------

|  | SHEAR | | COMPRESSION | NIOSH | |
|---|---|---|---|---|---|
| MOMENT ARM | Reaction (N) | Joint (N) | (N) | AL (N) | MPL (N) |
| ERECTOR SPINAE  5cm: | 573. | 1699. | 5229. | 3433 | 6376 |
| ERECTOR SPINAE  6cm: | 573. | 1511. | 4396. | | |

```
ABDOMINAL PRESSURE IS:    9.8 kPa.
EFFECT OF ABDOMINAL PRESSURE ON COMPRESSION, SUBTRACT:  457.0 N.
SUBJECT'S HEIGHT (m):    1.75     WEIGHT (kg):    80    GENDER:   M
LOAD MASS (kg):          20       L4/L5 MOMENT (N.m):  250  .
FORCE ON RIGHT HAND (N): 98.1     ANGLE OF FORCE (DEG):  -90
FORCE ON LEFT HAND  (N): 98.1     ANGLE OF FORCE (DEG):  -90
```

**FIGURE 13.3** WATBAK allows the user to indicate the spine curvature (top panel), accommodates muscle and passive tissue load sharing (middle panel), and calculates the joint shear force — not only the reaction shear force (bottom panel). The joint shear force is the index of shear injury risk.

BODY JOINT STRENGTH DEMANDS(FEMALE)

| Calculated Joint Moments | | | | Selected Strength Data | | | | |
|---|---|---|---|---|---|---|---|---|
| JOINT | SIDE (R/L) | MOMENT (Nm) | | MOMENT (Nm) | S.D (+/-) | | ANGLE (DEG) | REFERENCE |
| ELBOW | R | 26.0 | Fl | 42 | 12 | Fl | 90 | Stobbe (1982) |
| ELBOW | L | 26.0 | Fl | | | | | |
| SHOULDER | R | 44.0 | Fl | 43 | 10 | Fl | 0 | Koski and McGill(1992) |
| SHOULDER | L | 44.0 | Fl | | | | | |
| ANKLE | R | 116.0 | Ex | 124 | 27 | Ex | DNA | Fugl-Meyer et al (1980) |
| ANKLE | L | 116.0 | Ex | | | | | |
| KNEE | R | 20.4 | Fl | 69 | 22 | Fl | 135 | Stobbe (1982) |
| KNEE | L | 20.4 | Fl | | | | | |
| HIP | R | 126.5 | Ex | 128 | 52 | Ex | 100 | Stobbe (1982) |
| HIP | L | 126.5 | Ex | | | | | |
| TORSO | | 205.8 | Ex | 244 | 53 | Ex | 180 | Troup and Chapman (1969) |

\* DNA: Data Not Available

**FIGURE 13.4**    Calculated joint moments from require by the task are compared with gender specific joint strength data from the literature. The average moment +/- 1 standard deviation at the joint angle reported in the studies cited are found under the "selected strength data" column. Data from a 50 kg woman holding a 20 kg load mass in front of her in a semi-squat with her torso inclined forward about 45°.

**FIGURE 13.5**    A manikin figure is moved into position in the sagittal plane, using arrow keys, which automatically creates posture data for joint loads. This method of data entry also facilitates rapid analysis of effects of redesign of tasks.

**FIGURE 13.6** Input to 3DWATBAK can be obtained by manually moving a 3-D manikin with a mouse into position to represent the posture of a worker seen on videotape (top panel). The manikin is automatically digitized and processed through the 3-D computations to produce 3-D output.

$$C = 1067.6 + 1.219F + 0.083F^2 - 0.0001F^3 + 3.299B + 0.119B^2$$
$$- 0.0001B^3 + 0.862T + 0.393T^2 - 0.0001T^3 \tag{13.1}$$

where

C = compression (N)

F = flexion–extension moment where negative values correspond to flexion (N.m)

B = lateral bending moment where bending to the right is positive (N.m)

T = axial twisting moment where CCW twist is positive (N.m)

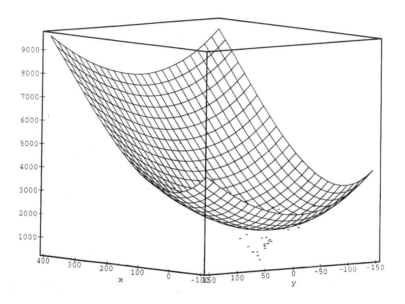

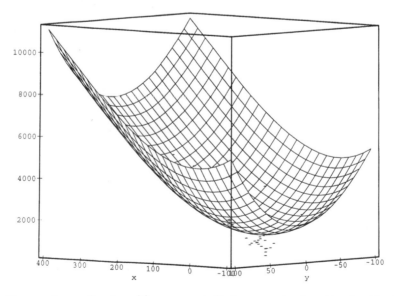

**FIGURE 13.7**   Compression as a function of three moments (flexion–extension, lateral bend, axial twist) is predicted using a 4-D regression equation. Since only three dimensions can be plotted at one time, two plots are shown, compression (top panel) as a function of flexion–extension moment (x) and lateral bending moment (y), compression (bottom panel) as a function of flexion–extension moment (x) and axial twist moment (y).

**TABLE 13.2**   Low back loads and some examples of 3-D joint moments produced during the barrel holding task shown in Figure 13.6. Typical moments about the x–y–z axes are automatically transformed into orthopedic orientations for each joint for functional interpretation.

| Segment | Proximal Joint | Orthopedic Moment (N.m) | Direction |
|---------|----------------|--------------------------|-----------|
| Right forearm | Right elbow | 2. 2 | Flexion |
| | | 0.6 | Pronation |
| | | 15. 2 | Adduction |
| Left forearm | Left elbow | 3.9 | Flexion |
| | | 0.6 | Pronation |
| | | 4.6 | Abduction |
| Right upperarm | Right shoulder | 1.6 | Flexion |
| | | 3.9 | Internal rotation |
| | | 27. 9 | Abduction |
| Left upperarm | Left shoulder | 21. 2 | Flexion |
| | | 4.6 | External rotation |
| | | 2.8 | Adduction |
| Thorax and abdomen | L4/L5 | 169 | Extension |
| | | 26.3 | Left axial twist |
| | | 6.9 | Left lateral bend |

Total compression: 3482 N

Anterior-posterior joint shear: 288 N

Anterior-posterior reaction shear: 534 N

Lateral reaction shear: 0 N

An example of the utility of 3DWATBAK is provided in Figure 13.6, where the real lifting task is observed and the ergonomist positions the manikin with the computer mouse to represent the posture, then 3DWATBAK uses linked segments to calculate the reaction moments (x, y, z), and then transforms the moments into specific joint axes (in this case flexion–extension, lateral bend, axial twist) and supports the moments with muscle and ligament forces (using the regression equation) to compute low back compression (Table 13.2).

It is interesting to compare the compressive cost of producing moments about the 3 lumbar axes. For example, a pure extensor moment of approximately 160 N.m results in a compression load close to 3000 N. On the other hand, an extensor moment of 100 N.m coupled with 50 N.m of left lateral bend and 50 Nm of CW axial twist produces the same level of compression. The spine pays dearly for supporting combinations of moments, particularly for those outside the sagittal plane.

In summary, the way in which 3DWATBAK represents the anatomy and biological muscle recruitment, represents a compromise between the relative simplicity that is needed in models for assessing the risk of injury in industry, and a rigorous and methodologically complex modeling approach that inherently contains quite high levels of biological fidelity in predicting low back loads but is impractical for industrial use. However, the equations within 3DWATBAK are based on model output that gives full credit to muscle cocontractions and also recognizes biological muscle coupling which occurs in three-dimensional spine loading. This method seems a reasonable way to examine risk of low back injury for three-dimensional occupational tasks.

## 13.11   Are There Any Spine Models that Reflect the Effects of Prolonged Loading?

Some experimental methods are under development but not yet routinely available to ergonomists. It is true that current biomechanical spine models provide an assessment of exposure to injury risk at a single

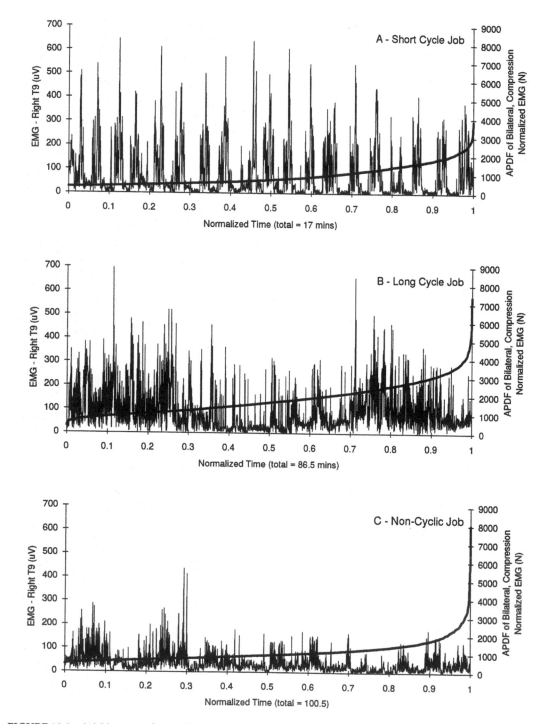

**FIGURE 13.8** (A) Linear envelope and accompanying APDF of short cycle job (1 min.); (B) Long cycle job (5 min.); (C) Noncyclic job illustrating conversion of extensor muscle EMG to Newtons of estimated compression during prolonged monitoring of jobs (Compression Normalized EMG). Interpretation in text.

instant in time. Our work with the complex, EMG-assisted model has shown a high correlation of back muscle EMG with spinal compression in load-handling tasks that do not have a dominant axial twist component. Modern data loggers permit the collection of EMG for long periods of time. The question then arose as to whether it was possible to estimate spinal compression continuously as a person worked by using an electronically processed version of the EMG signal calibrated against the spine compression observed when a worker held a load statically in a standard posture and the back muscle EMG recorded during that calibration effort. In other words, could we approximate the compression estimate that a biomechanical model would yield if it were applied 100 times per second for periods of work of several hours by normalizing the EMG to Newtons of compression from the calibration effort? Without going into detail, the answer was yes, as long as the job did not involve a large part of the total time in which the dominant torso moment was spinal twisting. The method did not work well for pure torso twisting.

Most jobs that we have observed in our field research do not have large periods of time in which torso twist predominates. Even in asymmetrical tasks, if the worker is bent over to some degree, the dominant support moment is trunk extension. This method seems to work well in this case. The most useful presentation of data obtained from this so-called "compression normalized EMG" is in the form of a (compression) amplitude probability distribution function. Figure 13.8 shows compression normalized APDF curves recorded from three different auto workers, one worker (A) who did a short cycle assembly job (1 minute), one worker (B) who did a longer job cycle (5 minutes), and a third worker (C) whose job was highly varied over a 2-hour period. These workers were monitored for 17, 86, and 100 minutes, respectively. If one uses the 90% level as the estimate of peak compressive loading (90% of the cycle time the compression was below this level and above this level for only 10% of the time), a range from about 1800 to 3000 N for the three workers is observed. The median level of compression, representing 50% of the work cycle time, ranged from about 800 to 1800N during the period of observation, and the compression never dropped below 400 N (level at 10% on the normalized time axis). Standing upright with no muscular activity the compression on the L4/L5 motion unit is about one half body weight.

This method, based upon spine model output, has considerable potential for monitoring spinal loading over prolonged periods of time, but the limitations of the method need further exploration. The percentages of the work cycle time at various levels of compression can be compared with data on spine tolerance to acute loading. It would be helpful if epidemiological data could be obtained on the risk associated with lower level accumulated loads as well. This would extend the utility of this type of monitoring of exposure to back injury risk.

# References

Bean, J.C., Chaffin, D.B., and Schulz, A.B. 1988. Biomechanical model calculation of muscle contraction forces: a double linear programming method. *J. Biomech.* 21:59-66.

Bone, B.C., Norman, R.W., McGill, S.M., and Ball, K.A. 1990. Comparison of 2D and 3D modelpredictions in analysing asymmetric lifting postures, in *Advances in Industrial Ergonomics and Safety II*, Taylor & Francis, pp. 543-550.

Brinckmann, P., Biggemann, M., and Hilweg, D. 1989. Prediction of the compressive strength of human lumbar vertebrae. *Clin. Biomech.* 4 (Suppl. 2).

Chaffin, D. Hughes, R., and Nussbaum, M. 1989. Towards a 3D biomechanical torso model for asymmetric loading. *Proceedings of the International Society for Biomechanics*, Los Angeles, abstract 132.

Cholewicki, J., McGill, S.M., and Norman, R.W. 1995. Comparison of muscle forces and joint load from an optimization and EMG assisted lumbar spine model: towards development of a hybrid approach. *J. Biomech.* 28(3):321-331.

Cripton, P., Berleman, U., Visarino, H., Begeman, P.C., Nolte, L.D., and Prasad, P. 1994. Response of the lumbar spine due to shear loading in: Injury prevention through biomechanics — *Symposium Proceedings*, May 4-5, Wayne State University.

Freivalds, A., Chaffin, D., Garg, A., and Lee, K.S. 1984. A dynamic biomechanical evaluation of lifting maximum acceptable loads. *J. Biomech.* 17(4):251-262.

Gracovetsky, S., Farfan, H.F., and Lamy, C. 1981. Mechanism of the lumbar spine. *Spine.* 6(1):249-262.

Heylings, D. 1978. Supraspinous and interspinous ligaments of the human lumbar spine. *J. Anat.* 123:127-131.

Hughes, R.E., Bean, J.C., and Chaffin, D.B. 1995. Evaluating the effect of cocontraction in optimization models. *J. Biomech.* 28(7):875-878.

Jager, M., Luttmann, A., and Laurig, W. 1991. Lumbar load during one-handed bricklaying. *Int. J. Ind. Ergonomics.* 8:261-277.

Marras, W.S. and Granata, K. 1995. A biomechanical assessment and model of axial twisting in the thoraco-lumbar spine. *Spine.* 20(13):1440-1451.

Marras, W. and Sommerich, C. 1991. A three-dimensional motion model of loads on the lumbar spine I: Model Structure. *Human Factors.* 33:123-137.

Marras, W.S., Lavender, S.A., Leurgans, S.E., Rajulu, S.L., Allread, W.G., Fathallah, F.A., and Ferguson, S.A. 1993. The role of dynamic three-dimensional trunk motion in occupationally-related low back disorders. *Spine.* 18(5):617-628.

McGill, S.M., and Norman R.W. 1985. Dynamically and statically determined low back moments during lifting. *J. Biomech.* 18(12):877-885.

McGill, S.M. 1999. Dynamic low back models: Theory and relevance in assisting the ergonomist to reduce the risk of low back injury, in *The Occupational Ergonomics Handbook* (eds. W. Karwowski, W. Marras) CRC Press, Boca Raton.

McGill, S.M. and Norman, R.W. 1986. Partitioning the L4-L5 dynamic moment into disc, ligamentous, and muscular components during lifting. *Spine.* 11(7):666-678.

McGill, S.M. and Norman, R.W. 1992. Loading of the low back during 3D moment generation. *Proceedings of the Human Factors Association of Canada*, Hamilton, October 25-18, pp. 73-79.

McGill, S.M. Norman, R.W., and Cholewicki, J. (1996). A Simple Polynomial that predicts low back compression during complex 3D tasks. *Ergonomics*, 39(9):1107-1118.

McGill, S.M., Patt, N., and Norman R.W. 1988. Measurement of the trunk musculature of active males using CT Scan radiography: Implications for force and moment generating capacity about the L4/L5 joint. *J. Biomech.* 21(4):329-341.

McGill, S.M., Santaguida, L., and Stevens, J. 1993. Measurement of the trunk musculature from T6 to L5 using MRI scans of 15 young males connected for muscle fibre orientation. *Clin. Biomech.* 8:171-178.

McGill, S.M. 1991. Lumbar loads from moments about three orthopaedic axes: Developing the architecture of a 3-D occupational low back model, *Proceedings of the XIIIth International Congress on Biomechanics*, Perth, Australia, December 9-13, pp. 545-547.

McGill, S.M. 1992. A myoelectrically based dynamic 3-D model to predict loads on lumbar spine tissue during lateral bending. *J. Biomech.* 25:395-414.

NIOSH–1981. National Institute for Occupational Safety and Health. *A Work Practices Guide for Manual Lifting*, Cincinnati; Taft Industries.

Norman, R., McGill, S., Lu. W., and Frazer, M. 1994. Improvements in biological realism in an industrial low back model: 3DWATBAK. *Proceedings of the 12th Triennial Congress of the International Ergonomics Association*, Toronto, Volume 2, pp. 299-301.

Potvin, J.R., Norman, R.W., and McGill, S.M. 1991. Reduction in anterior shear forces on the L4/L5 disc by the lumbar musculature. *Clin. Biomech.* 6:88-96.

Potvin, J.R. and Norman, R.W. 1992. Can fatigue compromise lifting safety? *Proceedings of the Second North American Congress on Biomechanics*, Chicago, August 24-28, pp. 513-514.

Punnett, L., Fine, L.J., Keyserling, W.M., Herrin, G.D., and Chaffin, D.B. 1991. Back disorders and nonneutral trunk postures of automobile assembly workers. *Scand. J. Work Environ. Health.* 17:337-346.

Schultz, A.B., Andersson, G.B.J., Ortengren, R., Haderspeck, K., and Nachemson, A. 1982. Loads on the lumbar spine. *J. Bone Jt. Surg.* 64A (5):713-720.

Troup, J.D.G. and Chapman, A.E. 1969. The strength of the flexor and extensor muscles of the trunk. *J. Biomech.* 2:49-62.

# 14

# Quantitative Assessment of Trunk Performance

**Mohamad Parnianpour**
*The Ohio State University*

**Aboulfazl Shirazi-Adl**
*École Polytechnique, Montreal*

## 14.1   Introduction

As early as 1700, Bernardino Ramazzini, one of the founders of occupational medicine, had associated certain physical activities with musculoskeletal disorders (MSD). He postulated that certain violent and irregular motions and unnatural postures of the body impair the internal structure (Snook et al., 1988; NIOSH, 1997). Presently, much effort is directed toward a better understanding of work-related musculoskeletal disorders involving the back, cervical spine, and upper extremities (Parnianpour et al., 1990). The World Health Organization (WHO) has defined occupational diseases as those work-related diseases where the relationship to specific causative factors at work has been fully established (WHO, 1985). Obtaining the occupational history is crucial to proper diagnosis and appropriate treatment of work-related disorders. The occupational physician must consider the conditions of both the workplace and the worker in the evaluation of the injured workers. Biomechanic and ergonomic evaluations have developed a series of techniques for quantification of the task demands and evaluation of the stresses in the workplace. Functional capacity evaluation has also been advanced to quantify the maximum performance capability of the workers. The motto of ergonomics is to avoid the mismatch between the task demand and the functional capacity of individuals. A multidisciplinary group of clinicians and engineers constitutes the rehabilitation team that will work together to implement the prevention measures. Through proper workplace design, workplace stressors and risk factors could be minimized. It is expected that one third of the compensable low back pain in industry could be prevented by proper ergonomic workplace or task design. In addition to reducing the probability of both the initial and recurring episodes, proper ergonomic design allows earlier return of injured workers by keeping the task demands at a lower level. Unfortunately, ergonomists are often asked to redesign the task or the workplace after a high incidence of injuries has already been experienced. The next preventive measure that has been suggested is preplacement of workers based on the medical history, strength, and physical examinations (Parnian-pour et al., 1987; Snook et al., 1988; Parnianpour and Engin, 1994).

Title I of the Americans with Disabilities Act (ADA, 1990) prohibits discrimination in regard to any aspect of the employment process. Thus, the development of preplacement tests has been impeded by the possibility of discrimination against individuals based on gender, age, or medical conditions. The ADA requires physical tests to simulate the "essential functions" of the task. In addition, one must be aware of "reasonable accommodations," such as lifting aids, that may make an otherwise infeasible task possible for a disabled applicant to perform. Health care providers who perform physical examinations and provide recommendations for job applicants must consider the rights of disabled applicants. It is extremely crucial to quantify the specific physical requirements of the job to be performed, and to examine an applicant's capabilities to perform those specific tasks, taking into account any reasonable accommo-dations that may be provided. Hence, task analysis and functional capacity assessment are truly intertwined.

This chapter is intended to illustrate the application of some principles and practices of human performance engineering, especially quantification of human performance. The problem of low back pain has been selected to illustrate a series of concepts that are essential to evaluation of both the worker and the workplace, while realizing the importance of the disorders of the neck and upper extremities. By inference and generalization, most of these concepts can be extended to these situations. Manual material handling tasks have been the focus of our attention due to their prevalence in industry. In addition, the trunk muscles and spine were selected for the most detailed investigation due to the observation that a large proportion of ADA cases involve low back disability (Khalaf et al., 1997a).

## 14.2  Principles

Assessment of function across various dimensions of performance (i.e., strength, speed, endurance, and coordination) has provided the basis for a rational approach to clinical assessment, rehabilitation strat-egies, and determination of return to work potential for injured employees (Kondraske, 1990). To understand the complex problem of trunk performance evaluation of (low back pain) LBP patients, the terminology of muscle exertion must first be defined. However, it should be noted that a number of excellent reviews of trunk muscle function have been performed (Andersson, 1991; Beimborn and Morrissey, 1988; Newton and Waddell, 1993; Pope, 1992). We do not intend to reproduce this extensive literature here, as our motive is to provide a critical analysis that will lead the reader toward an under-standing of the future of functional assessment techniques. A more extensive clinical application is presented elsewhere (Parnianpour and Tan, 1993; Parnianpour, 1995; Szpalski and Parnianpour, 1996).

### Impairment, Disability, and Handicap

The tremendous human suffering and economic costs of disability present a formidable medical, social, and political challenge in the midst of growing health care costs and scarcity of resources. The WHO (1980) distinguished among impairment, disability, and handicap. Impairment is any loss or abnormality of psychological, physiological, or anatomical structure or function — impairment reflects disturbances at the organ level. Disability is any restriction or lack (resulting from impairment) of ability to perform an activity in the manner or within the range considered normal for a human being — disability reflects disturbances at the level of person. Handicap is a disadvantage for a given individual, resulting from an impairment or a disability, that limits or prevents the fulfillment of a role that is normal (depending on age, sex, and social and cultural factors) for that individual. As disability is the objectification of an impairment, handicap represents the socialization of an impairment or disability. Despite the immense improvement presented by the International Classification of Impairments, Disabilities, and Handicaps (ICIDH), it is limited from an industrial medicine or rehabilitation perspective. The hierarchical organization lacks the specificity required for evaluating the functional state of an individual with respect to task demands.

Kondraske (1990) has suggested an alternative approach, using the principles of resource economics. The resource economics paradigm is reflective of the principal goal of ergonomics: fitting the demands of the task to the functional capability of the worker. The Elemental Resource Model (ERM) is based on the application of general performance theory that is to present a unified theory for measurement, analysis, and modeling of human performance across all different aspects of performance, across all human subsystems, and at any hierarchical level. This approach uses the same bases to describe both the fundamental dimensions of performance capacity and task demand (available and utilized resources) of each functional unit involved in performance of the high-level tasks. The elegance of the ERM is due to its hierarchical organization, allowing causal models to be generated based on assessment of the task demands and performance capabilities across the same dimensions of performance (Kondraske, 1990).

## Muscle Action and Performance Quantification

The details of the complex processes of muscle contraction in terms of the bioelectrical, biochemical, and biophysical interactions are under intense research. Muscle tension is a function of muscle length and its rate of change, and can be scaled by the level of neural excitation. These relationships are called the length–tension and velocity–tension relationships. From a physiological point of view, the measured force or torque applied at the interface is a function of: (a) the individual's motivation (magnitude of the neural drive for excitation and activation processes); (b) environmental conditions (muscle length, rate of change of muscle length, nature of the external load, metabolic conditions, pH level, temperature, and so forth); (c) prior history of activation (fatigue); (d) instruction and descriptions of the tasks given to the subject; (e) the control strategies and motor programs employed to satisfy the demands of the task; and (f) the biophysical state of the muscles and fitness (fiber composition, physiological cross-sectional area of the muscle, cardiovascular capability). It cannot be overemphasized that these processes are complex and interrelated (Kroemer et al., 1994). Other factors that may affect the performance of the patients are: misunderstanding of the degree of effort needed in maximal testing, test anxiety, depression, nociception, fear of pain and re-injury, as well as unconscious and conscious symptom magnification.

In the following sections we review some methods to quantify performance and lifting capability of isolated trunk muscles during a multilink coordinated manual material handling task. Relevant factors that influence the static and dynamic strength and endurance measures of trunk muscles will be addressed, and the clinical applications of these assessment techniques will be illustrated.

The central nervous system (CNS) appropriately excites the muscle, and the generated tension is transferred to the skeletal system by the tendon to cause motion, stabilize the joint, and/or resist the effect of external forces on the body. Hence, the functional evaluation of muscles cannot be performed without the characterization of the interfaced mechanical environment. The four fundamental types of muscle exertion or action are isometric, isokinetic, isotonic, and isoinertial. In isometric exertion, the muscle length is kept constant and there is no movement. Although mechanical work is not achieved, physiological work, i.e., static work, is performed and energy is consumed. When the internal force exerted by the muscle is greater than the external force offered by the resistance, then concentric, i.e., shortening, muscle action occurs; whereas if the muscle is already activated and the external force exceeds the internal force of the muscle, then eccentric, i.e., lengthening, muscle action occurs. When the muscle moves, either concentrically or eccentrically, dynamic work is performed. If the rate of shortening or lengthening of the muscle is constant, the exertion is called isokinetic. When the muscle acts on a constant inertial mass, the exertion is called isoinertial. Isotonic action occurs when the muscle tension is constant throughout the range of motion.

These definitions are very clear when dealing with isolated muscles during physiological investigations. However, terminologies employed in the literature of strength evaluation are imprecise. The terms are intended to refer to the state of muscles, but they actually refer to the state of the mechanical interface, i.e., the dynamometer. Isotonic exertion, as defined, is not as realizable physiologically because muscular tensions change as its lever arm changes despite the constancy of external loads. Special designs may vary the resistance level in order to account for changes in mechanical efficiency of the muscles. In addition, the rate of muscle length change may not remain constant even when the joint angular velocity is regulated by the dynamometer during isokinetic exertions. During isoinertial action, the net external resistance is not only a function of the mass (inertia) but is also a function of the acceleration. The acceleration, however, is a function of the input energy to the mass. Hence, to fully characterize the net external resistance we need to have both the acceleration and the inertial parameters (mass and moment of inertia) of the load and body parts. Future research should better quantify the inertial effects of the dynamometers particularly during nonisometric and nonisokinetic exertions.

For any joint or joint complex, muscle performance can be quantified in terms of the basic dimensions of performance: strength, speed, endurance, steadiness, and coordination. Muscle strength is the capacity to produce torque or work by voluntary activation of the muscles, whereas muscle endurance is the ability to maintain a predetermined level of motor output — e.g., torque, velocity, range of motion, work, or energy — over a period of time. Fatigue is considered to be a process under which the capability of muscles diminishes. However, neuromuscular adjustments take place to meet the task demands (i.e., increase in neural excitation) until there is final performance breakdown — endurance time. Coordination, in this context, is the temporal and spatial organization of movement and the recruitment patterns of the muscle synergies.

Despite the proliferation of various technologies for measurement, basic questions such as: "What needs to be measured and how can it best be measured?" are still being investigated. However, there is a consensus on the need to measure objectively the performance capability along the following dimensions: range of motion, strength, endurance, coordination, speed, acceleration, etc. Strength is one of the most fundamental dimensions of human performance and has been the focus of many investigations. Despite the general consensus about the abstract definition of strength, there is no direct method for measurement of muscle tension *in vivo*. Strength has often been measured at the interface of a joint (or joints) with the mechanical environment. A dynamometer, which is an external apparatus onto which the body exerts force, is used to measure strength indirectly.

Different modes of strength testing have evolved based on different levels of technological sophistication. The practical implication of contextual dependencies on the provided mechanical environment of the strength measures must be considered during the selection of the appropriate mode of measurement (Khalaf et al., 1997a). In this regard, equipment that can measure strength in different modes is more efficient in terms of both initial capital investment, required floor space in the clinics or laboratories, and the amount of time it takes to get the person in and out of the dynamometer.

## 14.3   Low Back Pain and Trunk Performance

The problem of LBP is selected to present important models that could be utilized by the entire multidisciplinary rehabilitation team for the measurement, modeling, and analysis of human performance (Kondraske, 1990). The inability to relate low back pain (LBP) to anatomical findings and the difficulties in quantifying pain have directed much effort toward quantification of spinal performance. The problem is made even more complex by the increasing demand of the health care system to quantify the level of impairment of patients reporting back pain without objective findings. Indeed some studies have indicated that a precise cause of nociception cannot be recognized in more than 80% of patients with low back pain complaints. However, work-related disorders of upper extremity, unlike low back disorders, can better be related to specific anatomical sites such as a tendon or compressed nerve. Examples of the growing number of cumulative trauma disorders of the upper extremities and the neck are: carpal tunnel syndrome (CTS), DeQuervain's disease, trigger finger, lateral epicondylitis (tennis elbow), rotator cuff tendinitis, thoracic outlet syndrome, and tension neck syndrome (Kroemer et al., 1994).

There are three basic impairment evaluation systems, each having its merits and shortcomings: (1) anatomic, based on physical examination findings; )2) diagnostic, based on pathology; and (3) functional, based on performance or work capacity (Luck and Florence, 1988). The earlier systems were anatomical, based on amputation and ankylosis. Although this approach may be more applicable to the hand, it is very inappropriate for the spine. The diagnostic-based systems suffer from lack of correspondence between the degree of impairment for a given diagnosis and the resultant disability and even more from the lack of a clear diagnosis. A large percentage of symptom-free individuals have anatomical findings detectable by the imaging technologies, while some LBP patients have no structural anomalies.

The function-based systems are more desirable from an occupational medicine perspective for the following reasons. They allow the rehabilitation team to rationally evaluate the prospect for return to light duty work and the type of "reasonable accommodations" needed (such as assistive devices) that could reduce the task demand below the functional capability of the individual. By focusing on remaining ability and transferable skills rather than the disability or structural impairment of the injured worker, the set of feasible jobs can be identified. These points are extremely important, given the natural history of work disability after a single low back pain episode causing loss of work time: 40% to 50% of workers return to work by 2 weeks, 60% to 80% return by 4 weeks, and 85% to 90% return by 12 weeks. The small portion of disabled workers who become chronic are responsible for the majority of the economic cost of LBP. It is therefore the primary goal of the rehabilitation team to prevent the LBP, which is self correcting in most cases no matter what kind of therapy is used, from becoming a chronic disabling predicament. Injured workers should neither be returned to work too early nor too late, as both could complicate the prognosis. The results of functional capacity evaluation and task demand quantification should guide the timing for returning to work. It is clear that psycho-socioeconomic factors become more important than physical factors as the disability progresses into "chronicity syndrome" and play a major role in defining the evolution of a low back disability claim. Future research should further establish the reliability and reproducibility of performance assessment tools to expedite their widespread use (Luck and Florence, 1988; Newton and Waddell, 1993; Parnianpour et al., 1989a,b).

## Maximal and Submaximal Protocols

Biomechanical strength models of the trunk are usually based on static maximal strength measurement. In real-life work situations, individuals rarely exert lengthy or maximum static effort. In most clinical situations, submaximal protocols are recommended, especially in patients with pain or with cardiovascular problems. Also, submaximal testing is less susceptible to fatigue and injury. The activity of daily living also has a great deal of submaximal efforts at the self-selected pace (Kim et al., 1996). Hence, it has been argued that testing at the preferred rate may be complementary to the maximal effort protocols. The preferred motion can be solicited by instructing the subject to perform repetitive movement at a pace and through the range of motion at which he/she feels the most comfortable. It has been shown that low back pain patients and normals have different resisted preferred flexion/extension motion characteristics. Having the subject perform against resistance is based on the hypothesis that, at higher resistance levels, the separation between the performance levels of patients and normal subjects becomes more evident. It has been shown, for example, that functional impairment of trunk extensors in LBP patients with respect to the normal population is larger at higher velocities during isokinetic trunk extension. However, the proponents of unconstrained testing have argued that separation of these groups can be performed based on the position, velocity, and acceleration profiles of the trunk during self-selected flexion/extension tasks. They have noted that pain and fear of re-injury may become the limiting factors. The sudden surge in acquiring performance measures of LBP patients during the initial rehabilitation process also underscores the validity of this concept.

## Static and Dynamic Strength Measurements of Isolated Trunk Muscles

Weakness of the trunk extensor and abdominal muscles in patients with low back pain was demonstrated using the cable tensiometer to measure isometric strength. The disadvantage of the cable tensiometer (which records applied force) is that it neglects to measure the lever arm distance from the center of trunk motion. It is also recommended that the cable tensiometer be used to determine peak isometric torques rather than the stable average torque exerted over a 3-second period. Dynamometers used for testing dynamic muscle performances contain either hydraulic or servo motor systems to provide constant velocity (e.g., isokinetic devices) or constant resistance (e.g., isoinertial devices) The isokinetic devices can be further categorized into passive and active types. The robotics-based dynamometers can actively apply force on the body and hence allow eccentric muscle performance assessments, while only concentric exertions can be measured by the passive devices. Eccentric muscle action can simulate the lowering phase of a manual material handling task. Based on sport medicine literature, eccentric action has been implicated for its significant role in the muscle injury mechanism. Using isokinetic dynamometers, the isometric and isokinetic strength of trunk extensor and abdominal muscles were shown to be weaker in low back pain patients compared to healthy individuals. Dedicated trunk testing systems have become the cornerstone of objective functional evaluation and have been incorporated in the rehabilitation programs in many centers.

Two issues of importance for future research are the role of pelvic restraints and the significance of using newly developed triaxial dynamometers as opposed to more traditional uniaxial dynamometers. Studies on healthy volunteers have shown that trunk motions occur in more than one plane — lateral bending accompanies the primary motion of axial rotation. Numerous attempts have been made to measure the segmental range of motion three dimensionally in the lumbar spine with the purpose of quantifying abnormal coupling and diagnosing instabilities.

The effect of posture on the maximum strength capability can be described based on the length–tension relationship of muscle action and the changes in the moment arm of trunk muscles (Parnianpour et al., 1991; Parnianpour et al., 1993; Khalaf et al., 1997a). Marras and Mirka (1989) studied the effect of trunk postural asymmetry, flexion angle, and trunk velocity (eccentric, isometric, and concentric) on maximal trunk torque production. It was shown that trunk torque decreased by about 8.5% of the maximum for every 15° of asymmetric trunk angle. At higher trunk flexion angles, extensor strength increased. Complex, significant interaction effects of velocity, asymmetry, and sagittal posture were detected. The range of velocity studies were more limited (±30°/second) than those used customarily in spinal evaluation.

Tan et al. (1993) tested 31 healthy males for the effects of standing trunk–flexion positions (0, 15, 35 degrees) on triaxial torques and electromyogram (EMG) of ten trunk muscles during isometric trunk extension at 30%, 50%, 70%, and 100% of maximum voluntary exertions (MVE). Trunk muscle strength was significantly increased at a more flexed position. But the accessory torques in the transverse and coronal planes were not affected by trunk postures. The recorded lateral bending and rotation accessory torques were less than 5% and 16% of the primary extension torque, respectively. The rectus abdominus were inactive during all the tests. The EMG of erector spinae varied linearly with higher values of MVE, while the latissimus dorsi had a nonlinear behavior. The obliques were coactivated only during 100% MVE. The neuromuscular efficiency ratio (NMER) was constructed as the ratio of the extension torque over the processed (RMS) EMG of the extensor muscles. It was hoped that NMER could be used in a clinical setting where generation of the maximum exertion is not indicated. However, the NMER proved to have a limited clinical utility because it was significantly affected by both exertion level and posture. The NMER of the extensor muscles increased at more flexed position. Studies that have combined the EMG activities and dynamometric evaluations have the potential of discovering the neuromuscular adaptation during different phases of injury and rehabilitation processes.

Khalaf et al. (1997a) measured concentric trunk isokinetic extension/flexion strength for 20 male and female healthy volunteers in the range of upright to 40 degrees of trunk flexion at six levels of angular velocity (10, 20, 40, 60, 80, and 100 degrees/sec). Results indicated that trunk strength is significantly influenced by trunk angular position, trunk angular velocity, gender, direction, as well as the interaction between trunk angular position and velocity. Three-dimensional surfaces of trunk strength in response to trunk angular position and velocity were constructed for each subject per direction. Such data presentation is more accurate and gives better insight about an individual's strength profile as compared to the traditional use of a single strength value (Figure 14.1). In addition, when comparing the task demands of a particular MMH task and maximum strength capacity of an individual subject, the contextual

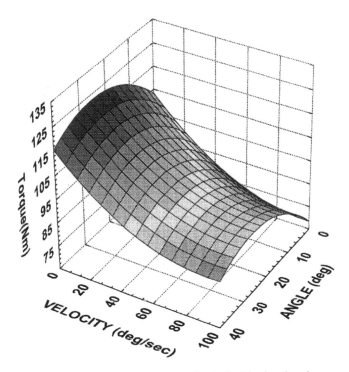

**FIGURE 14.1**   Trunk extension strength of a healthy female volunteer.

dependencies of fit or lack of it in terms of workplace design can better be illustrated. The workplace factors affect the required trunk speed, the muscular torque, and the range of motion of the worker. Concept of dynamic utilization ratio will be presented in the following section.

## Static and Dynamic Trunk Muscle Endurance

The high percentage of type I fibers in the back muscles, in addition to the better vascularization of these muscle groups, contributes to their superior endurance (Parnianpour et al., 1987a,b; Zhu et al., 1989). Physiological studies indicate that at higher muscle utilization ratios (relative muscle loads), fatigue is detected earlier. Isometric endurance tests have been used to compute the median frequency (MF) of the myoelectrical activities of trunk muscles in both normal and LBP populations. The expected decline of the median frequency with fatigue is parameterized by the intercept (initial MF) and the slope of the fall. It has been shown that trunk range of motion (ROM) and isometric strength suffered from lower specificity and sensitivity than spectral parameters. Trunk muscle endurance does differ between healthy subjects and those reporting LBP. During isometric endurance testing, trunk flexors develop fatigue faster than extensors in symptom-free subjects. The flexor fatiguability appeared significantly higher in patients with low back pain as compared to controls. Chronicity also influences trunk muscle endurance. Chronic LBP patients showed reduced abdominal as well as back muscle endurance as compared to the healthy controls, and lower back muscle endurance as compared to the intermittent LBP group. Individuals with a history of debilitating low back pain demonstrated less isometric trunk extensor endurance than either normal individuals or patients with history of lesser low back pain.

We have developed the following testing protocols for the quantification of trunk muscle endurance during diverse sets of fatiguing isometric submaximal tasks: (a) constant torque isometric exertions (Sparto et al., 1997b); (b) sustained varying torque isometric exertions (Sparto et al., 1997c; 1998). These isometric submaximal extension exertions are analyzed by novel modeling approaches to quantify the state of muscular fatigue using spectral analysis of trunk muscle activity (EMG). In addition, new EMG-driven models predicted the forces developed in the muscles during these tests so that the severity of

**TABLE 14.1**    Decline in Median Frequency of Power Spectrum of Erector Spinae (ES) at Five Different Locations During the Performance of a Static Endurance Test

| n = 10<br>Muscle Location | Median Frequency Decline (%/s)<br>Mean (S.D.) |
| --- | --- |
| Medial ES @ L1 | −0.36 (0.15) |
| Lateral ES @ L1 | −0.22 (0.11) |
| Medial ES @ L3 | −0.35 (0.13) |
| Lateral ES @ L3 | −0.30 (0.17) |
| Medial ES @ L5 | −0.39 (0.15) |

spinal loading could be quantified. The safety of the proposed protocols can be assessed by comparing the developed disc compression and shear forces with tissue tolerances in the literature. A detailed finite element model of the spine has been developed that quantifies the stress and strain in the highly innervated tissues such as the annulus fibers, ligaments, and vertebral end plates (Wang et al., 1997a,b,c; 1998).

While quantifying the trunk muscle recruitment during the isometric tasks, it was observed that as the subjects fatigued, muscles other than the primary trunk extensors became more active. The implication of this finding is that if a worker has a deficit in either the primary or *secondary* muscles, risk of injury may increase. We have found that more caudal and medial muscles fatigue the most (Table 14.1). Hence, if experimental resources are limited, based on our results, we have identified the medial erector spinae muscle at the level of L5 to be the location of choice for quantifying trunk fatigue (Sparto et al., 1997b).

Soft tissues subjected to repetitive loading, due to their viscoelastic properties, demonstrate creep and load relaxation (Wang et al. 1997b,c; 1998). The loss of precision, speed, and control of the neuromuscular system induced by fatigue reduces the ability of muscles to protect the weakened passive structure, which may explain many industrial, clinical, and recreational injury mechanisms. These results further indicate the necessity of relating clinical protocols to the job and show how short-duration maximal isometric testing alone cannot provide the complex functional interaction of strength, endurance, control, and coordination (Parnianpour et al., 1988; Sparto et al., 1997c).

Parnianpour et al. (1988) studied the effect of isoinertial fatiguing of flexion and extension trunk movements on the movement pattern (angular position and velocity profile) and the motor output (torque) of the trunk. They showed that, with fatigue, there is a reduction of the functional capacity in the main sagittal plane. There is also a loss of motor control, enabling a greater range of motion in the transverse and coronal planes while performing the primary sagittal task. Association of sagittal with coronal and transverse movements is considered more likely to induce back injuries; thus the effect of fatigue and reduction of motor control and coordination may be an important risk factor leading to injury-prone working postures. The endurance limit is a more useful predictor of incidence and recurrence of low back disorders than the absolute strength values. Although physiological criteria used in the National Institute for Occupational Safety and Health Lifting Guide (NIOSH, 1981) considered cardiovascular demands of dynamic repetitive lifting tasks, the limits of muscular endurance were not explicitly addressed. Future research should fill this gap, since the maximum strength measures should not guide the design decisions. Maximum level of performance can only be maintained for short periods of time, and muscular fatigue should be avoided to prevent the development of MSD. This caveat should be applied to all dimensions of performance capability (Kondraske, 1990; Parnianpour, 1995).

Other dynamic endurance protocols use repetitive isoinertial and isokinetic lifting tasks (Sparto et al., 1997c; 1998b). These two variations provide low stress to the musculoskeletal system and high stress to the cardiovascular system, and vice versa. Decreased knee and hip motion and increased spine flexion were evident in these studies, during repetitive sagittal lifting and lowering tasks, suggesting that the effects of fatigue may increase the loading of the passive tissues of the spine (Sparto et al., 1997c,d; 1998b). The changes in the phase angles of hip and spine motion between the rested and fatigued states during a repetitive lifting task are shown in Figure 14.2. The implications of this altered coordination should be investigated further in future research. Additional analyses have demonstrated a decrease in the postural

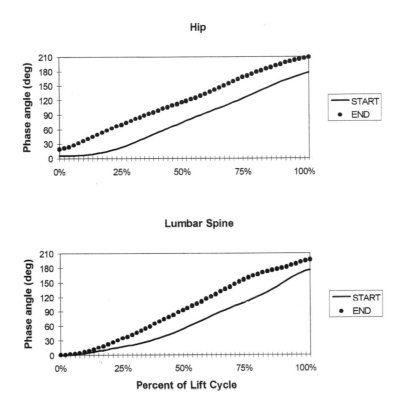

**FIGURE 14.2**  Change in the phase angles from the start to the end of the repetitive lifting test for one subject. The phase angles are shown for the hip and lumbar spine.

stability of subjects performing these tasks (Sparto et al., 1997d). The validation of the LIDOlift simulator was originally presented in Sparto et al. (1995).

Sparto and Parnianpour (1997) studied endurance of trunk extensors during repetitive submaximal trunk isokinetic extension. Trunk muscle electromyography (EMG) and torque output were measured in 16 healthy men [mean (s.d.) 23 (3) yrs, 178 (5) cm, 79 (9) kg] as they performed repetitive trunk extension exertions at 15 degrees/sec. A factorial design consisting of two load levels (35% and 70% of the maximum dynamic trunk extension torque) and two repetition rates (5 and 10 reps per minute) was implemented in order to cause different rates of fatigue. Each of the four load/rep combinations was tested during a single session, with at least one week in between. After the subjects stretched, bipolar surface electrodes were applied to the skin. Three bilateral trunk extensor muscles were sampled — latissimus dorsi (LAT), erector spinae (ERS), and posterior internal obliques (PINO) — as well as three trunk flexors — rectus abdomini (RAB), external obliques (EOB), and anterior internal obliques (AINO). The pelvis and lower extremities of the subjects were restrained within a trunk dynamometer. Maximal voluntary isometric exertions in each of the six cardinal directions were performed at 17.5 degrees of forward flexion to measure the maximum electromyographic activity of each muscle. The average of three dynamic maximal voluntary contractions (DMVC) in trunk extension was obtained, in order to measure the initial maximum torque-generating capability. During the dynamic exertions, subjects extended their trunks from a forward flexion posture of 35 degrees to upright standing, at a constant velocity of 15 degrees per second. The torque was measured using a load cell and knowing the distance from the point of force application to the center of rotation of the trunk. Two minutes of rest was provided between all maximum exertions.

During the repetitive endurance test, trunk extension was performed at 15 deg/sec while subjects controlled their torque output at 35% or 70% of the initial DMVC, by using visual feedback displayed on a computer monitor. The exertion rate, 5 or 10 reps per minute, was regulated by an audible tone.

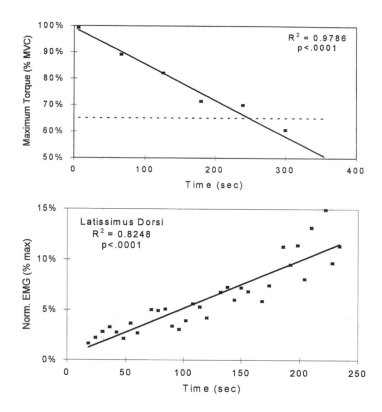

**FIGURE 14.3** (Top) Decline in maximum dynamic torque generating capability during an endurance test performed at 35% MVC, 10 reps/minute. A 35% decline in torque is indicated by the dashed line. (Bottom) The normalized EMG of the left latissimus dorsi during the same test.

Once per minute, a DMVC was performed, to quantify the reduction in the maximal torque-generating capability (fatigue). The exertions continued until a 35% reduction in the MVC occurred, the subject could no longer continue, or 30 minutes elapsed.

The raw EMG from each exertion was sampled at 1536 Hz, rectified, and low pass filtered at 3 Hz. After normalizing with the maximum EMG obtained from the isometric MVCs, the root-mean-square (rms) value of the EMG between the trunk angles of 25 and 10 degrees of flexion was computed. From the exertions whose rms torque was within 5% of the designated load level, the change in EMG with time, or SLOPE, was quantified using linear regression. In order to obtain an estimate of the slope for each muscle group, the slopes were averaged for the right and left muscle pairs. The effect of the load magnitude and repetition rate on the slope was tested using repeated measures multiple analysis of variance (MANOVA).

The effect of the repetitive endurance test on the decline in maximum torque-generating capability and change in muscle activation are shown in Figure 14.3. For this test, the maximum torque declined by 35% in about 4 minutes, during which the EMG of the left latissimus dorsi muscle increased by 10%. The endurance time (point at which the predicted maximum torque declined by 35%) for each of the load/reps combinations is shown in Table 14.2. The effect of load, repetition rate, and their interaction on the endurance time were all highly significant ($p < .001$), indicating the study design was effective at eliciting different rates of fatigue. It can be seen that doubling the repetition rate from 5 to 10 reps/min had a greater effect in causing fatigue than doubling the load magnitude.

The results of the MANOVA indicated that as a group, the slopes of all muscles were significantly affected by the effects of load, repetition rate, and their interaction ($p < .005$). Furthermore, Table 14.3 demonstrates that significant differences in slope were found in each of the trunk extensor muscles (LAT,

**TABLE 14.2**   Mean (s.d.) Endurance Time, in
Minutes, for the 4 Load/Reps Endurance Tests

|  | 35% DMVC | 70% DMVC |
| --- | --- | --- |
| 5 reps/min | 24.1 (8.1) | 9.8 (3.8) |
| 10 reps/min | 5.7 (3.9) | 2.4 (0.9) |

**TABLE 14.3**   P-values from the Univariate
ANOVA, Testing the Effect of Load, Repetition
Rate, and Interaction on the Slope

| SLOPE | Load (L) | Reps (R) | L * R |
| --- | --- | --- | --- |
| LAT | .0001 | .0001 | .0011 |
| ERS | .0001 | .0001 | .0001 |
| PINO | .0001 | .0001 | .0007 |
| RAB | .0209 | .0094 | .2441 |
| EXO | .0636 | .0836 | .2947 |
| AINO | .4040 | .1338 | .3949 |

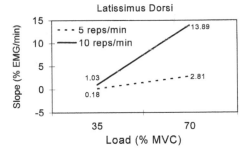

**FIGURE 14.4**   Increase in slope of the regression between the normalized EMG of the latissimus dorsi and time,
due to the interaction between the load and repetition rate.

ERS, PINO). Figure 14.4 shows the mean slope as a result of the interaction between load and repetition
rate for the latissimus dorsi. The other extensor muscles display a common pattern.

Because of the ambiguity in relating the amplitude of the processed EMG to muscle tension, one
cannot speculate if the increase in activation resulted in greater muscle force, or represented a decreased
efficiency in EMG/force production due to fatigue. Furthermore, it is not known if the function of the
increased activation of the secondary trunk extensors (LAT and PINO), is to provide greater stability, or
augment the torque generation. The increased and altered trunk extensor muscle activation due to fatigue
requires further investigation to determine its effect on the spinal loading. EMG-driven models accounting for muscle fatigue have been developed to address these issues (Sparto, 1998).

In addition, models are being developed that are attempting to predict the endurance time based on
parameters obtained from the noninvasive surface EMG of the trunk extensor muscles. Both short-time
Fourier transform and wavelet transform techniques are being used to process the EMG (Sparto et al., 1997a).
The use of the wavelet transform is more promising than the short-time Fourier transform because it does
not depend on stationarity of the signal, the bases of analysis are not restricted to sinusoids, and it is not
limited by the time-frequency resolution tradeoff. Furthermore, the analysis techniques can be incorporated
into digital signal processing hardware that can be used to monitor the extent of back muscle fatigue in real
time. The real time displays will allow the biofeedback application as well in the retraining of muscle
endurance so essential in the conservative treatment of low back pain patients with spinal instability.

## Lifting Strength Testing

The National Institute for Occupational Safety and Health (NIOSH, 1981) recommended static, i.e., isometric, strength measurement as its standard for lifting tasks. This was based on the evidence that associated low back pain with inadequate isometric strength. The incidence of an individual's sustaining an on-the-job back injury increases threefold when the task-lifting requirements approach or exceed the individual's strength capacity. However, lifting strength is not a true measure of trunk function but is a global measure taking into account arm, shoulder, and leg strength as well as the individual's lifting technique and overall fitness. It has been shown that strength tests were more valid and predictive of risk of low back disorders if they simulated the demands of the job. The clinicians must be aided with easy-to-use and validated instruments or questionnaires to gather information about the task demands in order to decide what testing protocol best simulates the applicant's spinal loading conditions.

Static strength measurements have been reported to underestimate significantly the loads on the spine during dynamic lifts. Comparing static and dynamic biomechanical models of the trunk, the predicted spinal loads under static conditions were 33% to 60% less than those under dynamic conditions, depending on the lifting technique. The recruitment patterns of trunk muscles (and thus the internal loading of the spine) are significantly different under isometric and dynamic conditions. General manual material handling tasks require a coordinated multilink activity which can be simulated using classical psychophysical techniques or the robotics-based lift task simulators. Various lifting tests, including static, dynamic, maximal, and submaximal, are currently available. The experimental results of correlational studies have confirmed the theoretical prediction that strength will be dependent upon the measurement technique (Balague et al., 1993; Parnianpour et al., 1987b; Parnianpour et al., 1993). Since muscle action requires external resistance, the effect of muscle action will depend on the nature of the resistance. These results refute the implicit assumption that a generic strength test exists that can be used for preplacing workers (pre-employment) and predicting the risk of injury or future occurrence of low back pain. The psychometric properties of isokinetic and isoresistive modes of strength testing were recently addressed. The quantification of the surface response of strength as function of joint angle and velocity was only possible for isokinetic testing, while isoresistive tests yielded a very sparse data set (Parnianpour, 1995).

The widely conflicting results found in the literature regarding the relationship of an individual's strength to the risk of developing LBP may be due to inappropriate modes of strength measurements, i.e., lack of job specificity (Parnianpour, 1995; Szpalski and Parnianpour, 1996; Parnianpour et al., 1977b). Isometric strength testing of the trunk is still widely used, especially in large-scale industrial or epidemiological studies, because it has been standardized and studied prospectively in industry. Compared to trunk dynamic strength testing protocols, the trunk isometric strength testing protocols are simpler and less expensive.

One outstanding issue during dynamic testing is the unresolved problem of how the wealth of information can be presented in a succinct and informative fashion. One approach has been to compare the statistical features of the data with the existing normal databases. This is particularly crucial because we do not have the option of comparing the results to the "contralateral asymptomatic joint," as we have with lower or upper extremity joints. Given the large differences between individuals, we recommend comparison be made to job-specific databases (Parnianpour et al., 1991; Parnianpour et al., 1994). For example, it is more appropriate for the trunk strength of an injured construction worker to be compared to age- and gender-controlled healthy construction workers than to data from healthy college graduate students or office workers. However, given the scarcity of such data, we argue for comparison of performance capacity with job demand based on task analysis. The performance capacity evaluation is once again linked to task demand quantification (Parnianpour, 1995).

In evaluating any manual material handling (MMH) tasks such as lifting, the quantification of the various generated parameters (kinetic, kinematic, and electromyographic) is essential to assessing functional capacity and development of a biomechanical profile of task demands. We have developed a methodology for representing and quantifying performance data variability of the kinematic and kinetic motion profiles as a function of the different lift characteristics (load, mode, and speed) during MMH

**TABLE 14.4**    P-values of MANOVA to Test the Effect of Load
(L), Mode (M), Speed (S), and Their Interactions on the
Extracted Features (Coefficients) of the Joint Torque Profiles

| N = 20 | L | M | S | L*M | L*S | M*S |
|---|---|---|---|---|---|---|
| Ankle | .032 | .001 | .033 | NS | NS | NS |
| Knee | .018 | .001 | NS | NS | NS | NS |
| L5/S1 | .003 | .003 | .011 | .001 | NS | NS |
| Shoulder | .001 | .001 | .046 | .005 | NS | NS |
| Elbow | .045 | .001 | .039 | NS | NS | NS |

NS: Not Significant

tasks (Khalaf et al., 1997b,c). Using a database of motion profiles from a manual lifting experiment, the Karhunen–Loeve Expansion (KLE) was shown to be quite effective for representing the various motion profiles, where the number of basis vectors (eigenvectors) and coefficients needed to accurately represent the data were substantially smaller than the original data set resulting in lower order space or dimension. The factorial lifting experiment required subjects to lift two masses (6.8 and 13.6 kg), at three different speeds (2, 4, 6 seconds per lifting/lowering cycle), using three modes of lift (preferred, straight leg, bent leg). Table 14.4 demonstrates how the lift characteristics affect the extracted features of the joint torque profiles. We can see that the main effects of load (L), mode (M), and speed (S) significantly affect the torque profiles (Khalaf et al., 1997). Further inspection of the data revealed that the 13.6 kg load resulted in greater peak lumbosacral (L5/S1) torques; the greater speeds of lift resulted in greater peak lumbosacral torque; and that the preferred mode of lift was a compromise between the straight leg and bent leg modes, in terms of the peak lumbosacral torque. The following analysis also revealed that the patterns of joint movement was invariant with respect to the level of load and speed tested in this experiment (Khalaf et al., 1997c).

The effects of lift characteristics were also investigated using analysis of variance techniques which recognize the vectorial constitution of the waveforms as opposed to the traditional descriptive statistics representing the data over the lifting cycle (Khalaf et al., 1997). The application of these techniques will enhance the ability to document the effect of intervening measures such as education or physical training/exercise on the kinematic and kinetic patterns of performance (Parnianpour et al., 1987). In addition, the ADA-related applications are significant since this will aid in the assessment of the feasibility of task performance as a function of existing impairments, and assessment of limitations on functional capacity, allowing appropriate task assignment based on a worker's capabilities. Furthermore, the differential influence of lift characteristics on the variability of performance during different phases of lifting and lowering may provide added resolution in the analysis of MMH tasks. The most effective workplace modifications can be recommended based on the preceding results.

## Inverse and Direct Dynamics

A major task of biomechanics has been to estimate the internal loading of musculoskeletal structure and establish the physiological loading during various daily activities. Kinematics studies deal with joint movement with no emphasis on the forces involved. However, kinetic studies address the effect of forces that generate such movements. Using sophisticated experimental and theoretical stress/strain analyses, hazardous/failure levels of loads have been determined. The estimated forces and stresses are used to estimate the level of deformation in the tissues. This technique allows us to assess the risk of over-exertion injury associated with any physical activity. Given repetitive motions and exertion levels much lower than the ultimate strength of the tissues, an alternative injury mechanism, the cumulative trauma model, has been used to describe much of the musculoskeletal disorders of the upper extremities (Wang et al., 1997a,b,c).

The experimental data on the joint trajectories are differentiated to obtain the angular velocity and acceleration. Appropriate inertial properties of the limb segments are used to compute the net external moments about each joint. This mapping from joint kinematics to net moments is called inverse dynamics. Direct dynamics refers to studies that simulate the motion based on known actuator torques at each joint. The key issue in these investigations is understanding the control strategies underlying the trajectory planning and performance of purposeful motion. A highly multidisciplinary field has emerged to address these unsolved questions [see Berme and Cappozzo (1990) for a comprehensive treatment of these issues].

It should be pointed out that determination of the external moments about different joints during manual material handling tasks is based on the well-established laws of physics. However, the determination of human performance and assessment of functional capacity are based on other disciplines, e.g., psychophysics, that are not as exact or well developed. We can describe the job demand, in terms of the required moments about each joint, easily by analyzing the workers performing the tasks. However, we are unable to predict the ability to perform an arbitrary task based on our incomplete knowledge of functional capacities at the joint levels. A task is easily decomposed to its demands at the joint level, however, we cannot compose (construct) the set of feasible tasks based on our functional capacity knowledge. The mapping from high-level task demands to the joint-level functional capacity for a given performance trial is unique. However, the mapping from joint-level functional capacity to the high-level task demand is one to many (not unique). The challenge to the human performance research community is to establish this missing link. Much of the integration of ergonomics and functional analysis depends upon the removal of this obstacle. The question of whether a subject can perform a task based on the knowledge of his/her functional capacity at the joint level remains an area of open research. When ergonomists or occupational physicians evaluate the fitness of task demands and worker capability, the following clinical questions will be presented: (1) Which space should be explored for determining normalcy, fit, or equivalence? (2) Should we consider the performance of the multilink system in the joint space or end-effector (cartesian work space)? These issues have profound effects on both the development of new technologies and evaluation of trunk or lifting performance (Sparto et al., 1995).

The enormous degree of freedom existing in the neuromusculoskeletal system provides the control centers for both the kinematic and actuator redundancies. The redundancies provide optimization possibility. Since we can lift an object from point A to point B with infinite postural possibilities, it can be suggested that certain physical parameters may be optimized for the learned movements. The possible candidates for objective function to be optimized are: movement time, energy, smoothness, muscular activities, etc. This approach, though still in its early stage, may be very important for spine functional assessment. We could compare the given performance to the optimal performance that is predicted by the model. This approach provides specific goals and gives biofeedback with respect to the individual's performance.

## Biomechanical Models

To complement the trunk strength and endurance testing protocols, a number of biomechanical simulation models have been developed which will be briefly discussed in the following section: (a) simulation of trunk strength based on anatomical cross-sectional area of muscles, line of action, and their lever arms (Parnianpour et al., 1997b); (b) simulation of trunk motion for healthy and impaired low back patients (Parnianpour et al., 1997a); (c) simulation of lifting tasks considering the strength and range of motion of each joint based on the load characteristics: initial and final load position, load magnitude, duration of lift (Khalaf et al., 1996).

Figure 14.5 represents the three-dimensional trunk strength boundaries based on the anatomical data of Hughes et al. (1995). Any task that requires moments that are interior to the surface are considered feasible, while those on the outside of the boundary are infeasible. These models can utilize the readily available CT scan or MRI data that the patients may already have. Any conceivable MMH task can be reduced to the three components of moment that it requires from the trunk muscles to keep it in equilibrium. Hence, the model is not limited to any specific task (pushing, lifting etc.), which is highly desirable. The preceding strength model can quickly identify the tasks that cannot be performed without any ergonomic modifications. It can easily be interfaced with the databases that may have been developed based on biomechanical analyses of the task in question (Parnianpour, 1995; Khalaf et al., 1997a).

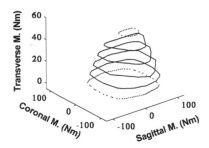

**FIGURE 14.5** Three-dimensional boundaries for trunk strength, assuming muscle geometry from Hughes et al. (1995).

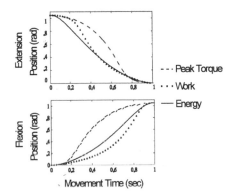

**FIGURE 14.6** Effect of minimizing different cost functions (energy, work, peak torque) on the optimized trunk trajectory during simulation of trunk extension and flexion movements. The movement time has been specified at 1 second.

Figure 14.6 shows the simulated trajectory of flexion and extension of the trunk using a number of candidate cost functions in the range of upright to 60 degrees of flexion. The simulation models require specification of initial and final angular position, velocity and acceleration of the trunk and the subject's trunk flexion/extension strength. The movement time could be specified or kept as another free parameter. This model can indicate whether a task is feasible within the desired movement time or could provide a physical basis for real-time feedback to individual during the training. To simulate the impairment, we have limited the trunk extension strength to 200 Nm, and report the simulated trajectories for flexion task in Figure 14.7. It is clear that the imposed limitation on strength reduces the variability of predicted trunk trajectory (Parnianpour et al., 1997a). Although we may not be able to ever verify the cost function being used or whether CNS uses the optimization algorithms, the ability to simulate trajectories of movement is still invaluable. The numerical simulations may allow us to estimate the bounds on the required functional capacity or changes in the workplace to make the task feasible.

Biomechanical simulation models provide a time- and cost-effective tool for answering "what if" types of questions. In the light of the ADA, this is of great value in predicting the consequences of task modifications and/or workstation alterations without subjecting an injured worker or a disabled individual to unnecessary testing. Consequently, a computer-based simulation program of multilink coordinated lifting that predicts the optimum motion pattern(s) required to perform a wide range of lifting tasks subjected to constraints based on experimental strength profiles has been developed (Khalaf et al., 1996). The model uses nonlinear optimization techniques to investigate the feasibility of task performance as a function of existing impairments, and limitations on functional capacity such as the range of motion, strength, and speed of lifting.

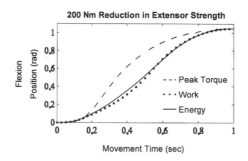

**FIGURE 14.7**   Effect of limiting trunk extension strength to 200 Nm on the optimized trunk flexion trajectory. The movement time has been specified at 1 second to allow comparison with unconstrained strength simulation presented in Figure 14.6.

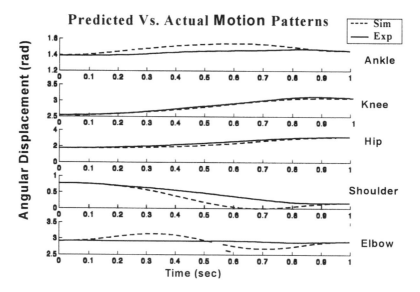

**FIGURE 14.8**   Comparison between the simulated and experimental motion patterns for the five joints during a lifting simulation (6.8 kg mass).

   The model's sensitivity to various types of strength constraints was studied such as upper and lower bounds of joint strength, joint strength as a function of joint position, and dynamic joint strength as a function of joint position and velocity (based on the previously obtained experimental strength profiles). The simulation was validated by comparing the predicted motion patterns with the experimental data generated for a similar lifting task (Figure 14.8). The results could be used as a biofeedback tool for training injured workers during rehabilitation, return to work assessment, as well as workplace modifications or "reasonable accommodations" as dictated by the ADA. The proper choice of a cost function remains an open area of research. Present work includes further experimental validation for a diverse set of loading conditions. In addition, we are in the process of implementing a nonlinear feedback control scheme to enhance the model by providing stability.

## Comparison of Task Demands and Performance Capacity

The regression analysis was used to model the dynamic torque, velocity, and power output as a function of resistance level during isoresistive flexion and extension using the B-200 Isostation (Parnianpour et al., 1990). Results indicated that the measured torque was not a good discriminator of the 10th, 50th, and 90th percentile populations. However, velocity and power were shown to effectively discriminate the three populations. Based on these data, it was suggested that during clinical testing, sagittal plane resistance should not be set at higher than about 80 Nm in order to minimize the internal loading of spine while taxing trunk functional capacity. This presentation of data may be useful to the physician or ergonomist in evaluating the functional capacity requirements of workplace manual material handling tasks. For example, a manual material handling task that requires about 80 Nm (61 ft-lbs) of trunk extensor strength could be performed by 90% of the population in the normal database if the required average trunk velocity does not exceed 40°/sec, while only 50% could perform the task if the velocity requirement exceeds 70°/sec; but more importantly, only the top 10th percentile population could perform the task if the velocity requirement approaches 105°/sec (Parnianpour et al., 1990). A few versions of lumbar motion monitors that can record the triaxial motion in the workplace have been used to provide the trunk movement requirements. The following example also illustrates the importance of having the same bases for evaluation of both task and the functional capability of the worker.

### Lifting Utilization Ratio

The utilization ratio provides a joint-specific unified scalar quantity representing the task demand normalized by an individual's maximum capacity. It would indicate whether a subject is capable of performing the task and how much of his/her maximum capacity is taxed by a given physical activity such as lifting. To illustrate this concept, Khalaf et al. (1997a) combined the trunk strength surface responses and the results of the task demand profiles of a lifting task performed by a typical female subject (Khalaf et al., 1997b). Very different results were predicted depending on whether the denominator of the utilization ratio was based on a single isometric strength value measured in upright position, posture dependent, or a function of both posture and trunk velocity. One of the reasons for the poor results of many epidemiological investigations for prediction of future low back injury based on isometric strength is shown here. To increase the predictive power, it has been suggested that the measured strength must simulate the essential functions of the task as closely as possible. The results of this analysis indicated that accurate estimation of the utilization ratio requires both the dynamic measurements of the external moment and strength in the range of joint positions being experienced during the performance of the lifting task. Moreover, the dynamic characteristics of the task should be considered. It is clear that the higher magnitudes of the dynamics components increase the task demand and reduce the maximum functional capacity, hence increasing the utilization ratio. The dynamic utilization ratio is computed considering the strength dependence on both the angular position and velocity. At the faster rate of lift, the numerator of the utilization ratio (task demand) is increased due to the increased acceleration. On the other hand, the denominator (maximum strength capacity) is decreased since the tension is inversely proportional to a muscle's speed of shortening (Hill's tension–velocity relationship). Hence, the utilization ratio approaches unity even for lifts that are only a fraction of the maximum lifting capacity. Moreover, the higher muscular activities (including higher coactivation) experienced at the faster lifting tasks may augment the risk of injury due to higher spinal loading (Ross et al., 1993).

## 14.4  Clinical Applications

Clinical studies have utilized quantitative human performance, i.e., strength and endurance measures, to predict the first incidence or recurrence of LBP, disability outcome, and also as a prognosis measure during the rehabilitation process. Training programs to enhance the endurance and strength of workers have been implemented in some industries. A prospective randomized study among employees in a geriatric hospital showed that exercising during work hours to improve back muscle strength, endurance, and coordination proved cost effective in preventing back symptoms and absence from work (Gundewall et al., 1993). Every hour spent by the physiotherapist on the exercise group reduced the work absence by 1.3 days. In this study both training and testing equipment were very modest. More studies on effectiveness of these programs are needed (Nordin et al., 1987; Kahanovitz et al., 1987). It can be hypothesized that these programs complement the stress management programs to enhance both worker satisfaction and coping strategies with regard to physical and nonphysical stressors at the workplace.

Functional-based impairment evaluation schemes have traditionally used spinal mobility (Buchalter et al., 1998; Elnaggar et al., 1991). Given the poor reliability of range of motion (ROM), its large variability among individuals, and the static psychometric nature of ROM, the use of continuous dynamic profiles of motion with the higher order derivatives has been suggested (Marras et al., 1993). Dynamic performances of 281 consecutive patients from the Impairment Evaluation Center at Mayo Clinic were used (Sharafeddin et al., 1996). As part of the comprehensive physical and psychological evaluation, 281 consecutive LBP patients underwent isometric and dynamic trunk testing using the B200 Isostation. Feature Extraction and Cluster Analysis techniques were used to find the main profiles in dynamic patient performances. The middle three cycles of movements were interpolated and averaged into 128 data points; thus the data were normalized with respect to cycle time. This allowed for comparison between individuals. The LBP patients in this study were shown to be heterogeneous with respect to their movement profile (Sharafeddin et al., 1996). Uniform treatment of these patients is questionable, and rehabilitation programs should consider their specific impairments. Future research should incorporate the clinical profiles with these movement profiles to further delineate the heterogeneity of low back patients (Parnianpour et al., 1994; Szpalski and Parnianpour, 1996).

Marras et al. (1993) used similar feature extraction techniques to characterize the movement profiles of 510 subjects belonging to normal (N = 339) and ten LBP patient groups (N = 171). Subjects were asked to perform flexion/extension trunk movement at five levels of asymmetry, while the three-dimensional movement of the spine was monitored by the Lumbar Motion Monitor (an exoskeleton goniometer developed at the Biodynamics Lab of The Ohio State University). Trunk motions were performed against no resistance, and no pelvic stabilization was required. The quadratic discriminant analysis was able to correctly classify over 80% of the subjects. The same technology was used to develop logistic regression models to identify the high-risk jobs in industrial workplaces. Hence, principles of human performance can successfully be applied to the worker and the task to avoid the mismatch between performance capability and task demand. A more detailed clinical implication of quantitative assessment of trunk performance is presented in Szpalski and Parnianpour (1996).

## 14.5 Conclusions

The outcome of trunk performance is affected by the many neural, mechanical, and environmental factors which must be considered during quantitative assessment. The objective evaluation of the critical dimensions of functional capacity and its comparison with the task demands is crucial to the decision-making processes in the different stages of the ergonomic prevention and rehabilitation process. Knowing the tissue tolerance limits from biomechanical studies, task demands from ergonomic analysis, and function capacities from performance evaluation, the rehabilitation team will optimize the changes to the workplace or the task that will maximize the functional reserves (unutilized resources) to reduce the occurrence of fatigue or overexertion. This will enhance worker satisfaction and productivity, while reducing the risk of MSD.

Based on ergonomic and motor control literature, the testing protocols that best simulate the loading conditions of the task will yield more valid results and better predictive ability. Ergonomic principles indicate that the ratio of the functional capacity to the task demand (utilization ratio) is critical to the development of muscular fatigue which may lead to more injurious muscle recruitment patterns and movement profiles due to loss of motor control and coordination. However, large prospective studies are still needed to verify this. The most promising application for these quantitative measures is to be used as a benchmark for the safe return to work of injured workers, given the enormous variations within the normal population.

The ability to identify subgroups of patients or high-risk individuals based on their functional performance will remain an open area of research to interested biomedical engineers within the multidisciplinary group of experts addressing neuromusculoskeletal occupational disorders. With the advent of technologies to monitor trunk performance in the workplace, we can obtain estimates of the injurious levels of task demands (kinematics and kinetic parameters) which can be used to guide our preplacement and rehabilitation strategies. The more functional the clinical tests become, the more clinicians need a complex interpretation scheme. An increasingly complex interpretation scheme opens the possibility of using mathematical modeling with intelligent computer interfaces.

The integration of complex biomechanical modeling of viscoelastic elements of spine, EMG-driven models quantifying the muscular activity, and realistic anatomical models with the well-controlled experimental designs motivated by the task demand analyses in the workplace is crucial to further understanding of low back disorders and their work-related risk factors. We need to accelerate our research efforts toward understanding the stabilization role of muscles in spine and the consequences of various neuro-control strategies by applying detailed biomechanical models to quantify the tissue stresses and strains (Parnianpour, 1991; Parnianpour et al., 1997b; Shirazi-Adl and Parnianpour, 1993, 1996a,b, 1998; Kiefer et al., 1997) during physical exertions.

## Acknowledgment

The authors acknowledge the support from OSURF and NIDRR H133E30009. The authors would like to thank, for their invaluable comments and contributions, George V. Kondraske, Margareta Nordin, Victor H. Frankel, Elen Ross, Jackson Tan, Robert R. Crowell, William Marras, Sheldon R. Simon, Ali Sheikhzadeh, Jung Yong Kim, Patrick Sparto, Kinda Khalaf, Jaw L. Wang, Alexander Kiefer, Marek Szpalski, and Sue Ferguson.

## References

Andersson GBJ. 1991. Evaluation of muscle function, in *The Adult Spine: Principles and Practice*, Ed. J.W. Frymoyer, p. 241. Raven Press Ltd., New York, NY.

Buchalter D, Parnianpour M, Viola K, Nordin M, Kahanovitz N. 1988. Three-dimensional spinal motion measurements. Part 1: A technique for examining posture and functional spinal motion. *Journal of Spinal Disorders*, 1(4), 279-83.

Balague F, Damidot P, Nordin M, Parnianpour M, Waldburger M. 1993. Cross-sectional study of isokinetic muscle trunk strength among school children. *Spine*, 18, 1199-1205.

Beimborn DS and Morrissey, MC. 1988. A review of literature related to trunk muscle performance. *Spine*, 13(6): 655-660.

Berme N and Cappozzo, A. 1990. *Biomechanics of Human Movement: Applications in Rehabilitation, Sports and Ergonomics*. Bertec Corporation, Worthington, OH.

Elnaggar IM, Nordin M, Sheikhzadeh A, Parnianpour M, Kahanovitz N. 1991. Effects of spinal flexion and extension exercises on low-back pain and spinal mobility in chronic mechanical low-back pain patients. *Spine*, 16(8), 967-72.

Gundewall B, Liljeqvist M, Hansson T. 1993. Primary prevention of back symptoms and absence from work. *Spine*, 18(5): 587-594.

Hughes RE, Bean JC, Chaffin DB. 1995. Evaluating the effect of coactivation in optimization models. *J. Biomechanics*, 28(7): 875-878.

Kahanovitz N, Nordin M, Verderame R, Yabut S, Parnianpour M, Viola K, Mulvihill M. 1987. Normal trunk muscle strength and endurance in women and the effect of exercises and electrical stimulation. Part 2: Comparative analysis of electrical stimulation and exercises to increase trunk muscle strength and endurance. *Spine*, 12(2), 112-118.

Khalaf KA, Parnianpour M, Wade L, Sparto PJ, Simon SR. 1996. Biomechanical simulation of manual multi-link coordinated lifting. *Proceedings of the 15th Southern Biomedical Engineering Conference*, Dayton, Ohio, 197-198.

Khalaf KA, Parnianpour M, Wade L. 1997. Feature extraction and modeling of the variability of performance in terms of biomechanical motion patterns during MMH tasks, in *Biomedical Sciences Instrumentation*, Eds. Benghuzzi, HA, Bajpai PK, Instrument Society of America, Vol. 33, 35-40.

Khalaf KA, Parnianpour M, Sparto PJ, Simon SR. 1997a. Modeling of functional trunk muscle performance: Interfacing ergonomics and spine rehabilitation in response to the ADA, *Journal of Rehabilitation Research and Development*, 34(4), 459-469.

Khalaf KA, Parnianpour M, Sparto PJ, Barin K, and Simon, SR. 1997b. Feature extraction and quantification of the variability of dynamic performance profiles at different sagittal lift characteristics. *IEEE Transactions on Rehabilitation Engineering*, in review.

Khalaf KA, Parnianpour M, Sparto PJ, Barin K. 1997c. The determination of the effect of lift characteristics on dynamic performance profiles during manual material handling (MMH) tasks. *Ergonomics*, in press.

Kiefer A, Shirazi-Adl A, Parnianpour M. 1997. On the stability of human spine in neutral postures. *European Spine Journal*, 6(1), 45-53.

Kim JY, Parnianpour M, Marras WS. 1996. Quantitative assessment of the control capability of the trunk muscles during oscillatory bending motion under a new experimental protocol. *Clinical Biomechanics*, 11(7), 385-391

Kondraske GV. 1990. Quantitative measurement and assessment of performance, in Smith RV and Leslie JH (Eds.) *Rehabilitation Engineering*, CRC Press, Boca Raton, FL.

Kroemer KE, Kroemer H, Kroemer-Elbert K. 1994. *Ergonomics: How to Design for Ease & Efficiency*. Prentice Hall, Inc., Englewood Cliffs, New Jersey.

Luck JV and Florence DW. 1988. A brief history and comparative analysis of disability systems and impairment evaluation guides. *Office Practice*, 19: 839-844.

Marras WS and Mirka GA. 1989. Trunk strength during asymmetric trunk motion. *Human Factors*, 31(6): 667-677.

Marras WS, Parnianpour M, Ferguson SA et al. 1993. Quantification and classification of low back disorders based on trunk motion. *European Journal of Medical Rehabilitation*, 3(6): 218-235.

National Institute for Occupational Safety and Health (NIOSH). 1981. Work practices guide for manual lifting (DHHS Publication No. 81122). Washington, D.C.: U.S. Government Printing Office.

National Institute for Occupational Safety and Health (NIOSH). 1997. Musculoskeletal Disorders and Workplace Factors (DHHS Publication No. 97-141). Cincinnati, Ohio: NIOSH Publication Dissemination.

Newton M and Waddell G. 1993. Trunk strength testing with iso-machines, Part 1: Review of a decade of scientific evidence. *Spine*, 18(7): 801-811.

Nordin M, Kahanovitz N, Verderame R, Parnianpour M, Yabut S, Viola K, Greenidge N, Mulvihill M. 1987. Normal trunk muscle strength and endurance in women and the effect of exercises and electrical stimulation. Part 1: Normal endurance and trunk muscle strength in 101 women. *Spine*, 12(2), 105-111.

Parnianpour M. 1991. Modeling of trunk muscles recruitment during isometric exertions. *IEEE Engineering in Medicine & Biology*, 10(2), 51-54.

Parnianpour M. 1995. Applications of quantitative assessment of human performance in occupational medicine, in Bronzino J. (Ed.). *Handbook of Biomedical Engineering*. CRC Press, Boca Raton, 1230-1239.

Parnianpour M, Bejjani FJ, Pavlidis L. 1987. Worker training: the fallacy of a single, correct lifting technique. *Ergonomics*, 30 (2), 331-334

Parnianpour M, Schecter S, Moritz U, Nordin M. 1987a. Back muscle endurance in response to external load. *Proceedings of the American Society of Biomechanics*, University of California Davis, pp. 41-42, 1987.

Parnianpour M, Nordin M, Moritz U et al. 1987b. Correlation between different tests of trunk strength. Buckle P. (Ed.): *Musculoskeletal Disorders at Work*. Taylor & Francis, London, pp. 234-238.

Parnianpour M, Nordin M, Kahanovitz N, Frankel VH. 1988. The triaxial coupling of torque generation of trunk muscles during isometric exertions and the effect of fatiguing isoinertial movements on the motor output and movement patterns. *Spine*, 13: 982-992.

Parnianpour M, Li F, Nordin M, Frankel VH. 1989. Reproducibility of trunk isoinertial performances in the sagittal, coronal, and transverse planes. *Bulletin of the Hospital for Joint Diseases Orthopaedic Institute*, 49(2), 148-154.

Parnianpour M, Li F, Nordin M, Kahanovitz N. 1989. A database of isoinertial trunk strength tests against three resistance levels in sagittal, frontal, and transverse planes in normal male subjects. *Spine*, 14(4), 409-411.

Parnianpour M, Nordin M, Skovron ML, Frankel VH. 1990. Environmentally induced disorders of the musculoskeletal system. *Medical Clinics of North America*, 74(2), 347-59.

Parnianpour M, Nordin M, Sheikhzadeh A. 1990. The relationship of torque, velocity and power with constant resistive load during sagittal trunk movement. *Spine*, 15: 639-643.

Parnianpour M, Campello M, Sheikhzadeh A. 1991. The effect of posture on triaxial trunk strength in different directions: Its biomechanical consideration with respect to incidence of low-back problem in construction industry. *International Journal of Industrial Ergonomics*, 8(3), 279-288.

Parnianpour M, Hasselquest L, Aaron A, Fagan L. 1993. The intercorrelation among isometric, isokinetic and isoinertial muscle performance during multi-joint coordinated exertions and isolated joint trunk exertion. *European Journal of Physical Medicine and Rehabilitation*, 3, 114-122.

Parnianpour M, Tan JC. 1993. Objective quantification of trunk performance, in D'Orazio B. (Ed.) *Back Pain Rehabilitation*. Andover Medical Publishers, Boston, 205-237.

Parnianpour M, Davoodi M, Forman M, Rose D. 1994. The normative database for the quantitative trunk performance of female dancers: Isometric and isoinertial strength and endurance. *Medical Problems of Performing Artists*, 9, 50-57.

Parnianpour M, Hanson T, Goldman S, Madson T, Sparto P. 1994a. The variability of trunk muscle performance in three distinct groups of females: impairment evaluation center, preplacement, and normal volunteers. *Proceedings of North American Spine Society Annual Meeting*, Minneapolis, MN, p. 165.

Parnianpour M, Engin AE. 1994. A more quantitative approach to classification of impairments, disabilities, and handicaps. *Journal of Rheum. Med. Rehab.*, 5(1), 52-64.

Parnianpour M, Wang JL, Shirazi-Adl A, Khayatian B, and Lafferriere G. 1997a. A computational method for simulation of trunk motion: Toward a theoretical based quantitative assessment of trunk performance, *IEEE Transactions on Rehabilitation Engineering*, in review.

Parnianpour M, Wang JL, Shirazi-Adl A, Wilke HJ. 1997b. The effect of trunk models in predicting muscle strength and spinal loading. *Journal of Musculoskeletal Research*, 1(1), 55-69.

Pope MH. 1992. A critical evaluation of functional muscle testing, in *Clinical Efficacy and Outcome in the Diagnosis and Treatment of Low Back Pain*, Ed. J.N. Weinstein, p. 101, Raven Press Ltd., New York, NY.

Ross EC, Parnianpour M, Martin D. 1993. The effects of resistance level on muscle coordination patterns and movement profile during trunk extension. *Spine*, 18, 1829-1839.

Sharafeddin H, Parnianpour M, Hemami H, Hanson T, Goldman S, Madson T. 1996. Computer aided diagnosis of low back disorders using the motion profile. *15th Southern Biomedical Engineering Conference*, Dayton, Ohio, pp. 431-432.

Shirazi-Adl A, Parnianpour M. 1993. Nonlinear response analysis of the human ligamentous lumbar spine in compression: On mechanisms affecting the postural stability. *Spine*, 18, 147-158.

Shirazi-Adl A, Parnianpour M. 1996a. Stabilizing role of moments and pelvic rotation on the human spine in compression. *ASME Journal of Biomechanical Engineering*, 118(1), 26-31.

Shirazi-Adl A, Parnianpour M. 1996b. Role of posture in mechanics of the lumbar spine. *Journal of Spinal Disorders*, 9, 277-286.

Shirazi-Adl A, Parnianpour M (1998) Finite element model studies in lumbar spine biomechanics, in Leondes C (Ed.) *Biomechanics Systems Techniques and Application*. Gordon and Breach Publishing Group, Newark, 127-145.

Snook SH, Fine LJ, Silverstein BA. 1988. Musculoskeletal disorders, in *Occupational Health: Recognizing and Preventing Work-related Disease*, B.S. Levy and D.H. Wegman (Eds.), p. 345-370. Little, Brown and Co., Boston/Toronto.

Sparto PJ. 1998. Trunk muscle electromyographic responses, wavelet detection of fatigue, and spinal loading during fatiguing repetitive trunk extension. Doctoral Dissertation, The Ohio State University, Columbus, Ohio.

Sparto PJ, Parnianpour M, Khalaf KA, Simon SR. 1995. The reliability and validity of a lift simulator and its functional equivalence with free weight lifting tasks, *IEEE Transactions on Rehabilitation Engineering*, 3(2), 155-165.

Sparto PJ, Parnianpour M. 1997. Changes in muscle recruitment patterns during fatiguing dynamic trunk extension exertions. *American Society of Biomechanics*, Clemson, South Carolina, September 24-27, 23-24.

Sparto PJ, Jagadeesh JM, Parnianpour M. 1997a. Real time wavelet analysis of electromyography for back muscle fatigue detection during dynamic work, in Benghuzzi HA, Bajpai PK (Eds.) Biomedical Sciences Instrumentation, Instrument Society of America, Vol. 33, 82-87.

Sparto PJ, Parnianpour M, Reinsel TE, Simon SR. 1997b. Spectral and temporal responses of trunk extensor EMG to an isometric endurance test, *Spine*, 22(4), 418-425.

Sparto PJ, Parnianpour M, Marras WS, Granata KP, Reinsel TE, Simon SR. 1997c. Neuromuscular trunk performance and spinal loading during a fatiguing isometric trunk extension with varying torque requirements, *Journal of Spinal Disorders*, 10(2), 145-156.

Sparto PJ, Parnianpour M, Reinsel TE, Simon SR. 1997c. The effect of fatigue on multi-joint kinematics and load sharing during a repetitive lifting test, *Spine*, 22(22), 2647-2654.

Sparto PJ, Parnianpour M, Reinsel TE, Simon SR. 1997d. The effect of fatigue on multi-joint kinematics, coordination, and postural stability during a repetitive lifting test, *Journal of Orthopaedic and Sports Physical Therapy*, 25(1), 3-12.

Sparto PJ, Parnianpour M, Marras WS, Granata KP, Reinsel TE, Simon SR. 1998a. The effect of EMG-force relationships and method of gain estimation on the predictions of an EMG-driven model of spinal loading, *Spine*, 23(4), 423-429.

Sparto PJ, Parnianpour M, Reinsel TE, Simon SR. 1998b. The effect of lifting belt use on multi-joint motion and load bearing during repetitive and asymmetric lifting, *Journal of Spinal Disorders*, 11(1), 57-64.

Szpalski M, Parnianpour M. 1996. Trunk performance, strength and endurance: measurement techniques and application. In Weisel S., & Weinstein J. (Eds.): *The Lumbar Spine*, second edition. WB Saunders, Philadelphia, 1074-1105.

Tan JC, Parnianpour M, Nordin M, et al. 1993. Isometric maximal and submaximal trunk exertion at different flexed positions in standing: Triaxial torque output and EMG. *Spine*, 18(16), 2480-1011.

Wang JL, Parnianpour M, Shirazi-Adl A, Engin AE. 1997a. The review and evaluation of viscoelastic models for collagen fiber during constant strain rate loading. *Biomedical Engineering, Application, Basis, Communication*, 9 (1), 5-19.

Wang JL, Parnianpour M, Shirazi-Adl A, Engin AE. 1997b. The simulation of viscoelastic behaviors under experimental controlled loading and the failure criterion of collagen fiber. *Journal of Theoretical and Applied Fracture Mechanics*, 27(1), 1-12.

Wang JL, Parnianpour M, Shirazi-Adl A, Engin AE. 1997c. Development and validation of a viscoelastic finite element model of an L2-L3 motion segment. *Journal of Theoretical and Applied Fracture Mechanics*, 28(1), 81-93.

Wang JL, Parnianpour M, Shirazi-Adl A, Engin AE. 1998. The dynamic response of L2/L3 motion segment to cyclic axial compressive loading. *Clinical Biomechanics*, 13, 516-525.

World Health Organization. 1980. *International Classification of Impairments, Disabilities, and Handicaps*. Geneva.

World Health Organization. 1985. *Identification and Control of Work-related Diseases*. Technical report no. 174. Geneva.

Zhu XZ, Parnianpour M, Nordin M, Kahanovitz N. 1989. Histochemistry and morphology of erector spinae muscle in lumbar disc herniation. *Spine*, 14(4), 391-397.

# 15

# Perspective on Industrial Low Back Pain

Malcolm H. Pope
*University of Iowa*

Donald R. McIntyre
*Interlogics*

## 15.1   Introduction

The first recorded case of occupational low back pain (LBP) was during the building of the pyramids. Although there is an extensive literature supporting a positive relationship between LBP and workplace factors (see reviews by Andersson et al., 1991; Garg and Moore, 1992; Kelsey and Hochberg, 1988; Riihimaki, 1995), recent publications have claimed that psychosocial factors are much more important (Nachemson, 1991; Bigos et al., 1991; Hadler, 1994). This paper will address the important ergonomic principles.

The industrial risk factors that have been identified are:

1. Manual materials handling (both frequent and heavy lifting)
2. Awkward postures
3. Fixed postures (including sitting)
4. Whole body vibration
5. Slipping and tripping
6. Psychosocial factors

There is a lengthy epidemiological literature dealing with the relationship of these risk factors with LBP. The epidemiological studies have several generic limitations:

1. Exposure data is generally limited (job title alone reduces risk toward null effects)
2. Confounding factors are often present
3. A definitive diagnosis is often absent

### Manual Materials Handling (MMH)

Numerous studies have found a positive relationship between MMH and LBP. Heliövaara (1987) found that the odds ration (OR) for herniated *nucleus pulposus* (HNP) for heavy MMH was 2.5 to 4.6. Riihimäki

(1995) summarized reviews of the relationship of LBP and MMH. Of 26 studies since 1950 only two found no relationship. However, all of the studies were cross sectional rather than prospective and used job title as a surrogate of exposure.

*Biomechanical* studies, as summarized by Pope et al. (1991), show that a large load or one held at a distance from the body leads to extremely high forces in the disc due to contraction of the erector spinae. Likewise, flexion of the trunk increases compression forces. A simple means of addressing the issue of the allowable lift is to use the NIOSH lifting formula (Waters et al., 1993).

1. Recommended Weight Limit (RWL)

$$RWL = LC \times HM \times VM \times DM \times AM \times CM \times FM \tag{15.1}$$

where
LC = Load constant
HM = Horizontal multiplier
VM = Vertical multiplier
AM = Asymmetric multiplier
CM = Coupling multiplier
FM = Frequency multiplier

The six multipliers are the penalties for deviating from the ideal lifting situation. The ideal lifting situation is defined as lifting from a 30 in. (75 cm) height, to a distance of up to 10 in. (25 cm), with hands close to the body (up to 10 in. [25 cm] from the ankles), with no twisting, with good grasp, a lifting frequency of once every 5 minutes or less, and with the lifting duration not exceeding 1 hour.

Thus, the load constant (LC) is the maximum weight allowed to be lifted. Since the six multipliers are penalties, none of them can be more than 1. In other words, under ideal conditions, when all six multipliers are 1, the RWL = LC.

2. Lifting Index (LI)

The LI gives the ratio of the weight to be lifted to the recommended weight limit.

$$LI = \frac{\text{Load Weight (lbs) or (kg)}}{\text{Recommended Weight Limit (lbs) or (kg)}} \tag{15.2}$$

A convenient means of assessing the NIOSH lifting risk is via a tool such as Video Works (see example). The data can be digitized directly from the video into the NIOSH equations and the effects of ergonomic changes assessed.

Control of the risk can include reducing the:

1. Postural demands
2. Burden
3. Size of the load
4. Repetitions
5. Twisting

Other control strategies can include the use of:

1. Lifting equipment
2. Parts delivered at waist height
3. Two-man lift

A logical control technique would appear to be matching the worker to the task. Chaffin (1974) found that a mismatch increased the injury risk. Liles et al. (1984) found that the incidence of injury was related to the job severity index (JSI). The JSI is the weight of a lift divided by the acceptable weight of a lift as defined by the worker. A JSI of 0.6 increasing to 2.0 increases the risk of injury by 4.5 times. Repetitive lifting of greater than 5 kg increases the risk twofold.

Testing of workers to assess lifting ability is limited in the United States under the provisions of the ADA unless the lifting represents an essential function of the job.

## Postural Demands

The literature contains numerous papers relating severe postural demands to LBP and HNP. These postural demands are often linked together with MMH. An exception is the static awkward postures normally associated with sedentary work. Punnett (1991), while studying auto workers, found that LBP was associated with awkward postures. Mild flexion has an OR = 4.9, high flexion 5.7 and twist or lateral bend 5.9. Pope et al. (1986) showed there is a higher level of antagonistic muscle activity in twisted postures, leading to high muscle and disc forces. This was supported by Kelsey et al. (1984) who found an OR of 3.0 between lifting while twisting and HNP. Adams and Hutton (1985), in laboratory studies, showed that loading a cadaver disc in a flexed, twisted orientation leads to an HNP.

Controls mean a redesign of the workplace to prevent high kinematic demands such as reaching, bending, and twisting. Conveyor belts, adjustable tables, and delivery of goods at waist height are some of the solutions that have been employed.

## Whole Body Vibration (WBV)

There is considerable literature relating exposure to WBV through driving vehicles to LBP and HND. Hulshof and Van Zanton (1987) found OR to range from 1.6 to 7.0. Kelsey and Hardy (1975) found that spending over half the day in a car gives an OR of 3.0 for HND.

Studies of the occupational environment reveal that many vehicles subject the worker to levels of vibration greater than that recommended by the International Standards Organization (ISO, 1978). The human spinal system has a characteristic response to vibration in a seated posture. The first resonance occurs within a band of 4.5 to 5.5 Hz. Similar, but less tightly defined resonances, are also in the 9.4 to 13.1 Hz range (Pope et al., 1986, 1987, 1989). The resonance is markedly affected by the pelvis–buttocks system. The response of the human is due to a combination of a vertical subsystem and a rotational subsystem. The latter is by rocking of the pelvis.

The back muscle response with respect to the vibration stimulus was studied by Seroussi et al. (1989). The muscles are not able to protect the spine from adverse loads. At many frequencies, the muscle response is so far out of phase that muscle forces are added to those of the stimulus. The fatigue that was found in muscles, after vehicular vibration, is indicative of the loads in the muscles. Exposure of the seated subject to whole body vibrations of 5 Hz, in a position that ensured back muscle activity, increased the rate of development of fatigue in the erector spinae muscles, as compared to the same conditions without vibrations. After WBV, there is an increased latency in muscle recruitment (Wilder et al., 1996). This may be due to the fatigue inherent in WBV exposure. This suggests that if a worker unloads a vehicle after exposure to WBV this could present a problem for the back.

Many ergonomic interventions are possible. WBV should be attenuated at the cab or seat to comply with ISO 2631. The layout of the vehicle cab should be optimized to prevent awkward postures. This is particularly true for tractor and forklift drivers. Excessive driving shifts should be avoided and frequent stretch breaks encouraged. Lifting directly after driving should be avoided. The psychological environment should not be ignored and good health (i.e., aerobic strength, trunk strength, and endurance) encouraged.

## Slipping and Tripping

It is reported in some studies that as much as 50% of LBP injuries occur in conjunction with slipping and tripping (Pope et al., 1991). In the absence of blunt trauma, the slip may cause a sudden load to be applied to the ligamentous spine or may cause the muscles to contract with high force (Wilder et al., 1996).

## Psychosocial factors

Many studies support a positive relationship between LBP report or disability and psychosocial stressors. These stressors include:

1. Monotonous work
2. Poor social support
3. High work pressure
4. Lack of control
5. Job dissatisfaction
6. Mutual dislike of boss and/or co-workers

In the Boeing study, Bigos et al. (1991) found the OR for LBP disability was 1.7. Magnusson et al. (1996) found that mechanical stressors lead to LBP, but psychosocial factors are responsible for low back pain disability.

The challenge would seem to be to separate the issues of LBP and disability since the solutions may be different. Clearly, a systematic method needs to be employed to deal with this difficult problem.

## 15.2   An Ergonomics and Injury Evaluation System

With this in mind, we developed a systems approach to the ergonomics and injury evaluation. The overall system is shown in Figure 15.1. It will be noted that the usually disparate functions of the industrial and medical facilities now overlap, leading to vastly improved communications and efficiencies. The system can be thought of as having both baselining, or pre-injury phases and post-injury components. With the emphasis on evaluating and treating both the worker and workplace, the baselining or pre-injury phase includes task analysis, task screening, and training, baseline testing of the worker, ergonomics intervention/reengineering, while the post-injury phase includes testing of both the worker and workplace. The system is operated by Interlogics, Hillsborough, NC.

## Baselining or Pre-Injury Phase

Software has been developed to download information from the standard OSHA 2000 or other customized databases in order to determine the incidence, nature, and severity of musculoskeletal-related injuries. These data are transferred to a central processing facility to identify jobs, activities, or sub populations with an increased injury risk (high-priority jobs). A typical data collection screen for a given injured worker is shown in Figure 15.2.

### Task Analysis

The system for ergonomics assessment and intervention includes task analysis. Once the high-priority jobs have been identified, trained technicians travel to the workplace where they use novel portable measurement devices for data collection. Data are initially collected at the workplace using a pentop computer (Figure 15.3). The collection system is based on a time and motion analysis and identifies the essential functions of the job and selects tasks and workplace hazards leading to a higher risk of injury.

This process also provides detailed job descriptions of the jobs analyzed. Specific job data include:

- Posture data, including sustained postures, improper sitting positions, and documentation of the orientation of the ankles, hips, lower back, shoulders, wrists, neck, and elbows
- Lifting hazards data

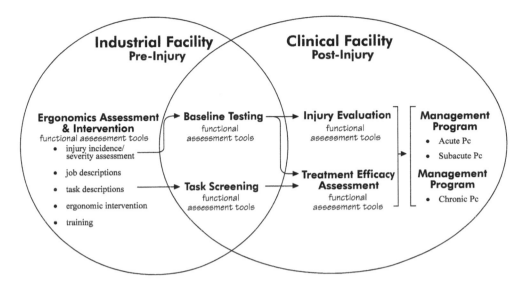

**FIGURE 15.1** Overview of the system used to evaluate, reduce, and manage job-related injuries.

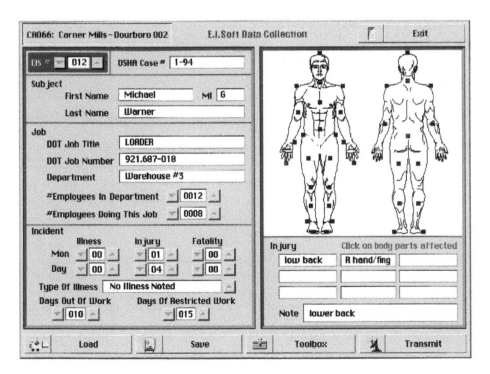

**FIGURE 15.2** Injury incidence and severity data collection screen.

- Pushing/pulling hazards data
- Repetition hazards data, including constant work cycles, lack of stretching, and heavy duty cycles
- Fatigue hazards data, including abnormal speech patterns and worker behavior changes, such as looking for resting places and rubbing body parts used in performing a task
- Environmental hazards data, including flagging for hot or cold extremes, wetness, slippery conditions, sharp edges, noise, and obstructions

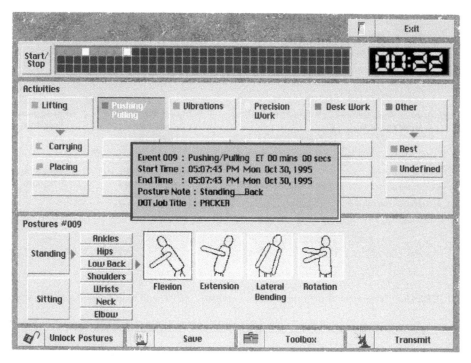

**FIGURE 15.3** Example screen used during the process of collecting job description and job hazard information.

- Tool/equipment hazards data
- Desk job hazard data, including ergonomic risks for workstations, excessive reaching for tools and/or equipment, VDT height and distance violations, excessive low or high positions of chairs, and cluttering of desktop work spaces
- Time-based summary data for the *Dictionary of Occupational Titles* (1987) jobs performed
- Time-based summary data for lifting, carrying, placing, pushing/pulling, vibrations, precision work, desk work, locomotion (walking), and rest
- Correlation data of perceived pain vs. actual joint stresses observed, including Genaidy and Karwowski (1993) Joint Stress Tables, Nordic Questionnaire (Dickenson et al., 1992), and extreme posture flags

Following this the data are transmitted to the centralized facility where professional ergonomists will make recommendations and generate reports focusing on potentially injurious task characteristics.

At this stage trained technicians would then be instructed to quantify the potentially hazardous aspects of a given job. If the job is not repetitive or stereotyped, this may mean that measurements should occur throughout the working day.

For example, InterLogics' B-Tracker™ is used to measure the functional characteristics of low-back movement. At the central processing facility, professional ergonomists analyze the data to precisely define awkward and prolonged trunk postures and repetitive trunk movements. (An illustration of this application of the B-Tracker™ in the workplace is shown in Figure 15.4.)

The report generated by the professional ergonomist with the help of the technicians would then make recommendations for workplace modification to reduce ergonomic risk or to hasten return to work for an injured worker.

### Task Screening and Training

The same data transmitted to the central processing facility to define potentially injurious task characteristics can also be used to generate a task-specific protocol to test job candidates. Candidates can be

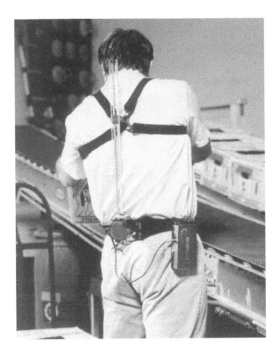

**FIGURE 15.4** Example of a proprietary measurement tool used to collect task specific functional information at the worksite.

screened "post-offer and pre-placement," using the task-specific protocol to determine whether they can perform the extreme characteristics required of the task. This screening process, which is based on the actual demands of a particular job, complies with the Americans with Disabilities Act of 1990.

A typical screening test, allowable under the ADA, would be to have the prospective worker lift a load repetitively from the floor to a shelf across the room.

Task analysis information can also be used to create task-specific training programs. The actions of the "expert" worker can often be used as a model for a videotape. Risks of injury can therefore be reduced by ergonomic intervention and training program implementation, for example, by correcting hazardous lifting techniques, encouraging other productive noninjurious workplace behaviors, and teaching workers how to recognize and respond to potentially hazardous situations. To determine the effectiveness of ergonomics intervention and training, workers are subsequently analyzed performing their tasks.

## Baseline Testing

Baseline testing of the worker, if economically feasible, can be a very useful tool for following a worker if he or she subsequently becomes injured. However, typically once injury incidence and severity data have been collected and analyzed, regardless of whether task-specific screening has occurred, testing can be conducted to generate baseline data for the identified "at risk" population. For instance, low-back baseline assessments are accomplished by using the B-Tracker, a portable stabilization system, and a standardized protocol consisting of software directed movements in all cardinal planes, circumductions, and stimulated sagittal plane lifts. (The testing set-up is shown in Figure 15.5.) These baseline data serve as a record of the functional characteristics of the healthy worker's low back. Each worker's data are subsequently compared to healthy and low-back problem population profiles, and a certainty of group membership in either of those populations is assigned. Current work employs a neural network to refine this population.

Baseline data are gathered in the event a worker later becomes injured. The comparison of baseline and post-injury clinical data enables a determination of the precise nature and extent of changes in the worker's functional status, which in turn leads to the assignment of an appropriate treatment path.

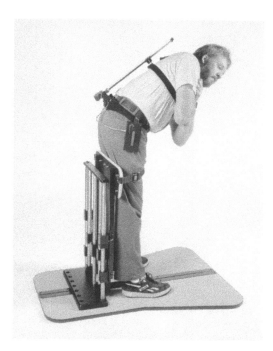

**FIGURE 15.5**  Testing set-up used for baselining healthy workers and for injury evaluations. (Photo by Dan Crawford ©1996.)

Baseline data also inform the clinician when a worker has recovered to his or her pre-injury functional status.

## Post-Injury Analysis

### The Worker

After an injury, the worker's functional status is assessed using the same protocols and devices used at baseline. In the case of a low-back injury, for instance, additional information is also collected, including a pain discomfort rating and a measure of nonorganic physical signs. This and the protocol data are transmitted to the central processing facility for analysis and report generation. The report identifies the functional status of the injured worker with respect to healthy and low-back problem populations, and it assigns a degree of certainty for the injured worker's membership in the low-back problem population. Work is under way to develop an expert system using neural networks that can learn to improve and refine its characterization. These data, especially the changes with time, will be helpful to the clinician in optimizing return to work and early activation while avoiding chronicity.

Comparing a worker's healthy baseline functional status with that derived from the clinical evaluation shows the extent of loss of functional ability. Subsequent testing of the injured worker can be used to track the recovery process.

Before release to return-to-work status, the injured worker can be screened for functional ability to perform specific task characteristics of his or her job. This is accomplished by using the task-specific screening protocol previously created from data collected at the work site. If the injured worker is found to be functionally unable to perform the specific task characteristics, the analyzed data are used to assign different job tasks suitable to the injured worker based on an analysis of the essential functions of alternative jobs.

In an effort to return injured workers to the workplace as soon as possible, separate programs are implemented for managing acute or chronic injuries. The acute low-back program is based on the clinical practice guidelines for acute low-back problems in adults as described by the Agency for Health Care

Policy and Research (1994). The chronic low-back pain management program was created and validated by the Iowa Spinal Research Center. An objective of the software-driven program for managing acutely injured patients is to begin the appropriate treatment regimens as early as possible in an effort to prevent any injuries from becoming chronic. The treatment costs are contained by selecting appropriate treatment protocols, maintaining contact with the injured worker, accurately tracking recovery progress, and objectively documenting the worker's return to pre-injury status.

## 15.3   Case Study

In many cases it would be helpful to evaluate the workplace (see task analysis) to assess whether ergonomic changes can be made to hasten return to work and prevent reinjury.

### Injury Incidence/Severity Analysis

Bandy Boxing Group's OSHA Logs for the 6-month period Jan-Jun 1996 were input into the OCM software. The resulting injury incidence and severity report shows the Stacker job has the highest incidence/severity rate. (Table 15.1) The incidence rate was found to be 19%. This represents the number of injuries per 100 full-time workers. The industry incidence rate for the same time period is 6%. The high incidence rate at Bandy Boxing Group compared to the industry incidence rate indicates further investigation is warranted.

Severity is the number of average lost workdays per total lost workday cases. On average injured workers missed 5 days of work due to injury. Industry severity for the same time period is 5.4. In general it is desirable to reduce the number of days of missed work, i.e., assign alternate duties to keep workers at work.

### Job Analysis

An ergonomist does an on-site analysis of the tasks associated with the Stacker job, using a pentop computer running TCM software. The resulting Job Analysis report pinpoints for further analysis the specific task that puts the Stacker at risk: the lifting task. The ergonomist's analysis of the Stacker job and related jobs at Bandy Boxing provides the data composing several job description reports (Tables 15.2A-C).

### Task Analysis

A videotaped performance of the Stacker's lifting task generates a VideoWorks report, documenting the task's high lifting index. Simultaneous use of the B-Tracker provides data for the B-Tracker Job Analysis report, which shows the Stacker repetitively using extreme flexion and extreme lateral bending postures (Tables 15.2A-E).

### Recommendations

All reports are analyzed by the ergonomic experts at InterLogics' Central Processing Facility, yielding an Ergonomic Recommendations report with the following recommendations:

- Engineering Re-design
- Ergonomic Intervention
- Education and Training
- Screening
- Baseline Testing

(See Table 15.4.)

**Injury Incidence/Severity Report**

Bandy Boxing Group
July 14, 1996

## Company Data

| | |
|---|---|
| Company #: | 39-0004-02 |
| Address: | 2912 Sousa Blvd. |
| | Muskegon, MI 49444 |
| Contact: | Beth Fingers |
| Phone: | (616) 733-3800 |

## Session Data

| | |
|---|---|
| Period: | Jan 95 - Jun 95 |
| # Job Titles: | 15 |
| Data Entry: | Tracy Marker |
| Phone: | (616) 733-3800 |

## Incident Summary

| Job Title | DOT Job Code | Injury | Lost WD | Rstrctd WD | Illness | Lost WD | Rstrctd WD | Death |
|---|---|---|---|---|---|---|---|---|
| Stacker | 929.687-030 | 5 | 12 | 11 | 0 | 0 | 0 | 0 |
| Gluer | 795.687-014 | 4 | 10 | 10 | 0 | 0 | 0 | 0 |
| Air-Drier-Machine Op | 534.682-010 | 3 | 9 | 8 | 0 | 0 | 0 | 0 |
| Rotary-Cutter Feeder | 640.686-010 | 3 | 8 | 3 | 0 | 0 | 0 | 0 |
| Tractor-Trailer-Truck Dr | 904.383-010 | 2 | 3 | 1 | 0 | 0 | 0 | 0 |
| Maintenance Mechan | 638.281-014 | 2 | 1 | 1 | 0 | 0 | 0 | 0 |
| Air Hammer Stripper | 794.687-050 | 2 | 1 | 0 | 0 | 0 | 0 | 0 |
| Flexographic-Press Op | 651.682-010 | 2 | 1 | 0 | 0 | 0 | 0 | 0 |
| Laminator | 554.685-030 | 1 | 2 | 1 | 0 | 0 | 0 | 0 |
| Packer | 929.684-010 | 1 | 1 | 0 | 0 | 0 | 0 | 0 |
| General Supervisor | 183.167-018 | 1 | 0 | 1 | 0 | 0 | 0 | 0 |
| Supplies Packer | 919.687-022 | 1 | 0 | 1 | 0 | 0 | 0 | 0 |
| Corrugator Operator | 641.562-010 | 1 | 0 | 0 | 0 | 0 | 0 | 0 |
| Printer-Slotter Helper | 659.686-014 | 1 | 0 | 0 | 0 | 0 | 0 | 0 |
| Quality-Control Tester | 559.367-010 | 1 | 0 | 0 | 0 | 0 | 0 | 0 |
| **Totals** | | **30** | **48** | **37** | **0** | **0** | **0** | **0** |

TABLE 15.1

## Baseline Testing

All stackers at Bandy Boxing are baseline tested to provide benchmark data about the current functional status of their low backs. The resulting Baseline Testing reports indicate that all but one of the Stackers achieved scores that place them in the healthy low-back population (Tables 15.5, 15.6A-B, and 15.7A-D).

## Clinical Evaluation

An injury to the Stacker Bobby Carrera immediately initiates a B-Tracker clinical evaluation, precisely the same protocol that was used during baseline testing. The Clinical Evaluation report score places Bobby Carrera in the problem low-back population, and also provides information about the relative extent of the injury. The attending physician prescribes a rehabilitation program and returns Bobby to work with alternative work duties, assigned on the basis of the Job Description reports previously compiled. Treatment and a subsequent Clinical Evaluation report show Bobby Carrera's improved functional status.

| Job Analysis Report | |
|---|---|
| | Bandy Boxing Group |
| | July 17, 1996 |

## Company Data

| | |
|---|---|
| Company #: | 39-0004-02 |
| Address: | 2912 Sousa Blvd. |
| | Muskegon, MI 49444 |
| Contact: | Beth Fingers |
| Phone: | (616) 733-3800 |
| Ergonomist: | Tracy Marker |

## Job/Task Data

| | |
|---|---|
| Job Title: | Stacker |
| Job Code: | 929.687-030 |
| Dept.: | Outgoing |

## Demographic Data

| | |
|---|---|
| Worker Name: | Bobby Carrera |
| Worker ID: | 123-45-6789 |
| Height: | 6'1" |
| Weight: | 170# |
| Hand Dominance: | Right |
| Time on Job: | 2 yrs |

## Job/Task Summary

| Event | %Time | Time | #Events | Avg Duration |
|---|---|---|---|---|
| All | 100 | 04:33 | 98 | 00:03 |
| Lifting | 75 | 03:24 | 68 | 00:03 |
| Pushing/ Pulling | 2 | 00:05 | 2 | 00:03 |
| Precision Work | 19 | 00:52 | 23 | 00:02 |
| Passive Work | 1 | 00:04 | 2 | 00:02 |
| Locomotion (Walking) | 2 | 00:05 | 2 | 00:03 |
| Resting | 1 | 00:03 | 1 | 00:03 |

**TABLE 15.2A**

At this point the physician, in conjunction with InterLogics, has Bobby return to work at full capacity as a Stacker. A later Baseline Testing report shows that Bobby Carrera is at or near pre-injury low-back functional status (Table 15.8).

## 15.4 Conclusions

The cost and disability associated with LBP remain very high. The vast majority of studies show causality between mechanical loading factors and LBP. It is evident that psychosocial factors have an important role in the disability from LBP.

The ergonomics and injury evaluation system works well in practice and represents a novel, comprehensive system for health care cost containment. This program links the industrial and clinical arenas to

## Job Analysis Report

### Pre-Analysis Pain Locations[†]

Low Back

### Relative Joint Stress Ratings[*]

| | Neck | Shoulder | Elbow | Wrist | Back | Hip | Ankle |
|---|---|---|---|---|---|---|---|
| **Lifting** | 68 | 204 | 136 | 0 | 340 | 0 | 0 |
| **Pushing/Pulling** | 0 | 0 | 0 | 0 | 0 | 0 | 0 |
| **Precision Work** | 69 | 46 | 115 | 23 | 69 | 0 | 0 |
| **Passive Work** | 0 | 0 | 0 | 0 | 0 | 0 | 0 |
| **Locomotion (Walking)** | 0 | 0 | 0 | 0 | 0 | 0 | 0 |
| **Totals** | 137 | 250 | 251 | 23 | 409 | 0 | 0 |

### Observed Task Risks

Awkward positions
Sustained postures
Extreme temperatures
Inadequate rest period

### Job Hazard Summary

Worker's pre-analysis pain location is matched with high stress rating at same joint (low back), suggesting further investigation warranted (VideoWorks, B-Tracker). Observed task risks suggest ergonomic intervention, administrative controls, or engineering re-design warranted.

**Report Reviewed by:** _____ **Date:** _____

Tracy Marker, Ergonomist

[†] from Nordic Musculoskeletal Questionnaire developed by Dickenson et al., 1992
[*] based on research by Genaidy & Karwowski, 1993

TABLE 15.2B

promote an integrated injury prevention and injury management program which effectively reduces injuries and more quickly returns injured workers to their jobs, all with the benefit of reduced expense. Injury prevention proceeds by initially identifying the causes of industry's job-related injuries and the at-risk worker populations; this then leads to baselining and screening procedures, to ergonomic intervention, or to training programs. On the clinical side, injury management proceeds by use of the information gathered at the workplace and by further evaluation with the same portable assessment tools. By use of such functional capability information, the injured worker can be placed in the chronic or acute treatment program, and his or her rehabilitation efficiently managed to expedite a speedy return-to- workplacement in a functionally appropriate job.

| **Job Description** | |
|---|---|
| | Bandy Boxing Group |
| | July 14, 1996 |

### Job/Company Data

Company #:  39-0004-02
Address:    2912 Sousa Blvd.
              Muskegon, MI 49444
Job Title:   Stacker
DOT Code:  929.687-030
Industry:   Packaging

| **Essential Job Functions** | **Essential Requirements** |
|---|---|
| 1. Re-stack paper stock onto pallet by flipping it over and evening it out in preparation for entry into printing press | 1. a) Sufficient hand grip strength to grasp stack of paper stock<br>b) Ability to lift, control, and flip large paper stock stacks up to 60 lbs each<br>c) Work in noisy environment |
| 2. Move re-stacked pallet with hand truck onto platform of printing press | 2. a) Ability to push and maneuver hand truck in close quarters<br>b) Properly position pallet in printing press |

| **Secondary Job Functions** | **Secondary Requirements** |
|---|---|
| 1. Maintain appropriate flow of material into press | 1. a) Verbally communicate with press operator<br>b) Assist other stacker in flipping and restacking large sheets of paper stock |
| 2. Assist printer operator as needed. | 2. a) Fold cleaning rags<br>b) Watch printing operation and signal to operator when problem arises |

### Strength Rating

```
          ▼
0   10  20        50           100
Sed. Light   Medium      Heavy   Very Heavy
```

DOT Physical Demand Strength Rating (PDSR) Scale
for Occasional Weight

| **Job Rotation Options** | **DOT Job Code** | **PDSR** |
|---|---|---|
| Scrapper | 794.687-050 | Heavy |
| Hand Packager | 920.587-018 | Medium |

Job rotation reduces the continuous performance of a single task, preventing the hazards associated with work that is repetitive, static, or awkward in nature.

**Report Reviewed by:** _____ **Date:** _____

Tracy Marker, Ergonomist

**TABLE 15.2C**

The cost of injuries in the workplace will continue to be expensive, but the costs can be effectively contained. Integrating injury prevention and injury management programs reduces health care costs by reducing initial injuries, speeding recovery, and leveraging professional expertise via a central processing facility.

## Defining Terms

**Accelerometer:** Instrument that measures acceleration.
**Biomechanics:** The application of principles of mechanics to the human body.

---

| VideoWorks Report: Individual |
|---|

Bandy Boxing Group
July 17, 1996

## Company/Task Data

| | |
|---|---|
| Company #: | 39-0004-02 |
| Address: | 2912 Sousa Blvd. |
| | Muskegon, MI 49444 |
| Job Title: | Stacker |
| Job Code: | 929.687-030 |
| PDSR: | Heavy Work |

## Performance Statement

On July 17, 1996, 3 lifting tasks associated with the job of Stacker at Bandy Boxing Group were performed by Bobby. An analysis of the performance was conducted with VideoWorks, yielding results based on the Revised NIOSH Lifting Equation (1991). **The greatest Lifting Index (LI) associated with the tasks performed was 4.46, indicating a high risk of injury**, i.e., this specific task was highly stressful and likely to put nearly all workers at increased risk of lifting-related injury. See the Recommended Adjustments section for specific ways in which to reduce the lifting index of the individual tasks performed.

## Performance Description

| | Name | Duration | Frequency (lifts/min) |
|---|---|---|---|
| **Lift #1** | Lift 1 | Moder | 4.30 |
| **Lift #2** | Lift 2 | Moder | 4.30 |
| **Lift #3** | Lift 3 | Moder | 4.30 |

## Performance Evaluation

| Name | L (lbs) | RWL(lbs) | FILI | LI(origin) | LI(dest) | Risk Level |
|---|---|---|---|---|---|---|
| **Lift 1** | 50.00 | 11.20 | 3.05 | 2.19 | 4.46 | high |
| **Lift 2** | 50.00 | 12.26 | 2.79 | 1.95 | 4.08 | high |
| **Lift 3** | 50.00 | 13.71 | 2.49 | 2.01 | 3.65 | high |

**Report Reviewed by:** _____ **Date:** _____
Tracy Marker, Ergonomist

TABLE 15.3A

**Electromyograph:** Recording of electrical output of the contraction of a muscle.
**Erector spinae:** Back muscles that extend the spine.
**Ergonomics:** Science that seeks to adapt working conditions to suit the worker.
**Extensor:** Any muscle that performs extension.
**Flexor:** Any muscle that flexes a joint.
**Intervertebral:** Situated between two adjacent vertebrae of the spine.
**Isometric:** A type of muscle contraction in which the length of the muscle remains constant.
**Kinematics:** Science of motion, including movements of body.
**Nucleus pulposus:** The central, more viscous portion of the intervertebral disc.

| VideoWorks Report: Individual | | | | | |
|---|---|---|---|---|---|

Bandy Boxing Group
July 17, 1996       Page 2

## Recommended Adjustments

| | | | **Ideal** | **Compromises** | |
|---|---|---|---|---|---|
| | **Actual** | **NIOSH** | **100%** | **50%** | **25%** |
| H | 24.07 | **10.0″** | -14.07 | -7.03 | -3.52 |
| A | 0.00 | **0°** | 0.00 | 0.00 | 0.00 |
| V | 41.51 | **30.0″** | -11.51 | -5.76 | -2.88 |
| D | 15.04 | **10.0″** | -5.04 | -2.52 | -1.26 |
| C | poor | **good** | good | fair | poor |
| F | 4.30 | **0.2** | -4.10 | -2.05 | -1.03 |
| | | | | | |
| H | 22.97 | **10.0** | -12.97 | -6.49 | -3.24 |
| A | 0.00 | **0** | 0.00 | 0.00 | 0.00 |
| V | 42.33 | **30.0** | -12.33 | -6.17 | -3.08 |
| D | 10.66 | **10.0** | -0.66 | -0.33 | -0.17 |
| C | poor | **good** | good | fair | poor |
| F | 4.30 | **0.2** | -4.10 | -2.05 | -1.03 |
| | | | | | |
| H | 20.50 | **10.0** | -10.50 | -5.25 | -2.63 |
| A | 0.00 | **0** | 0.00 | 0.00 | 0.00 |
| V | 43.69 | **30.0** | -13.69 | -6.84 | -3.42 |
| D | 10.12 | **10.0** | -0.12 | -0.06 | -0.03 |
| C | poor | **good** | good | fair | poor |
| F | 4.30 | **0.2** | -4.10 | -2.05 | -1.03 |

## Adjustment Results

| | **Posture Analyzed** | **Ideal** 100% | **Compromises** 50% | 25% |
|---|---|---|---|---|
| **Lifting Indexes #1** | 4.46 (end) | 0.98 | 2.19 | 2.54 |
| **Lifting Indexes #2** | 4.08 (end) | 0.98 | 2.06 | 2.72 |
| **Lifting Indexes #3** | 3.65 (end) | 0.98 | 1.91 | 2.47 |

### Variable Key

| | | | |
|---|---|---|---|
| **H** | (distance of load from subject) | **D** | (distance load moves vertically) |
| **A** | (angular rotation while lifting) | **C** | (coupling; good/fair/poor load handhold) |
| **V** | (distance of handhold from floor) | **F** | (frequency of lifts/minute) |
| | | **L** | (load weight) |

TABLE 15.3B

**Static Dynamic Task:** Characterized by both static and nonrepetitive low-back movements which show no obvious repeated pattern.

**Repetitive/Excursion Task:** Characterized by cyclical movement pattern where no posture is held continuously and the changes in trunk orientation are predictable.

**Excursion:** A continuous movement in a single direction with defined position end points (its length in degrees defines a range of motion for one or more axial movements).

**Oscillation Angle:** The angle which describes the symmetry of the excursion with respect to the subjects's neutral posture.

**Repetitive/Excursion Parameters:** Number of excursions, movement range (range of motion), velocity, oscillation angle, successive excursions (repetitions).

---

## VideoWorks Report: Comparative

Bandy Boxing Group
July 17, 1996

---

### Company/Task Data

| | |
|---|---|
| Job Title: | Stacker |
| Job Code: | 929.687-030 |
| PDSR: | Heavy Work |

### Comparison Statement

This report is based on a single task analysis session from July 17, 1996, when lifting tasks for the position of Stacker at Bandy Boxing Group were performed by Bobby. Task #1 of the 3 tasks analyzed at that time has had parameters modified in order to make it more nearly comply with NIOSH guidelines. The results are presented below.

### Selected Task with Modifications

| | Actual | NIOSH | What If? |
|---|---|---|---|
| L | 50.00 | 50.00# | 40.00 |
| H | 10.00 | 10.0" | 10.00 |
| V | 56.56 | 30.0" | 56.56 |
| D | 15.04 | 10.0" | 20.05 |
| A | 10.00 | 0° | 10.00 |
| F | 4.30 | 0.2 | 3.30 |
| C | poor | good | poor |

| | Actual | NIOSH | What If? |
|---|---|---|---|
| L | 50.00 | 50.00 | 40.00 |
| H | 24.07 | 10.0 | 19.07 |
| V | 41.51 | 30.0 | 36.51 |
| D | 15.04 | 10.0 | 20.05 |
| A | 10.00 | 0 | 0.00 |
| F | 4.30 | 0.2 | 3.30 |
| C | poor | good | poor |

### Modification Results

| | Actual | NIOSH | What If? |
|---|---|---|---|
| RWL | 11.20 | 51.0 | 16.02 |
| LI | 4.46 (end) | 0.98 | 2.50 (end) |
| FILI | 3.05 | 0.98 | 1.92 |

---

### Variable Key

| | | | | |
|---|---|---|---|---|
| | | **D** | (distance load moves vertically) | |
| **H** | (distance of load from subject) | **C** | (coupling; good/fair/poor load handhold) | |
| **A** | (angular rotation while lifting) | **F** | (frequency of lifts/minute) | |
| **V** | (distance of handhold from floor) | **L** | (load weight) | |

---

TABLE 15.3C

**Graphical Regions:** Class 1: denotes normal velocities, oscillation angles, and ranges of motion for which no ergonomic/safety action is required (within 31% of mean values for each direction for each axis). Class 2: denotes velocities, oscillation angles, and ranges of motion for which postures need consideration in the near future (between 31% and 62% of mean values for each direction for each axis). Class 3: denotes velocities, oscillation angles, and ranges of motion for which postures need consideration immediately and job-redesign may be required (beyond 62% of mean values for each direction for each axis).

| B-Tracker Task Analysis Report |
|---|

Bobby Carrera, 123-45-6789
July 17, 1996

## Company Data

Company: Bandy Boxing Group
Company #: 39-0004-02
Address: 2912 Sousa Blvd.
Muskegon, MI 49444

## Job/Task Data

Job Title: Stacker
Job Code: 929.687-030
Dept.: Outgoing
PDSR: Heavy

## Session Summary

Duration: 0:04:09
Task Name: Stacking
Task Type: Repetitive/Excursion

## Analysis Conclusions

This task involves repetitive trunk movements with biased flexion and lateral bending. Excursion analysis of the task reveals that the worker moved rapidly through large ranges with off-center orientations. These results suggest training, ergonomic intervention, engineering re-design or job-specific screening may be warranted.

## Ergonomic/Safety Concerns

Average percentage of oscillation angles, velocities, and ranges of motion within each class for all axes.

**Oscillation Angles:** Class 1: 44
Class 2: 55
Class 3: 1

**Velocities** Class 1: 48
Class 2: 51
Class 3: 1

**Ranges of Motion** Class 1: 46
Class 2: 53
Class 3: 1

**Report Reviewed by:** _____ **Date:** _____

Tracy Marker, Ergonomist

TABLE 15.3D

**B-Tracker Task Analysis Report**

### Demographic Data

| | |
|---|---|
| Worker Name: | Bobby Carrera |
| Worker ID: | 123-45-6789 |
| Height: | 6'1" |
| Weight: | 170# |
| Hand Dominance: | Right |
| Time on Job: | 2 yrs |

### Oscillation Angles (by axis)

Flex/Ext

| 0 | 0 | 0 | 14 | 84 | 2 |
|---|---|---|----|----|---|

Lateral

| 1 | 67 | 30 | 1 | 1 | 0 |
|---|----|----|---|---|---|

Rotation

| 0 | 6 | 46 | 42 | 6 | 0 |
|---|---|----|----|---|---|

Percentage of oscillation angles in each class. Each axis is shown comprising two directions of movement, e.g. right and left lateral bending.

### Velocities (by axis)

Flex/Ext

| 56 | 44 | 0 |
|----|----|---|

Lateral

| 25 | 75 | 0 |
|----|----|---|

Rotation

| 62 | 37 | 1 |
|----|----|---|

Percentage of velocities in each class.

### Ranges of Motion (by axis)

Flex/Ext

| 35 | 64 | 1 |
|----|----|---|

Lateral

| 28 | 72 | 0 |
|----|----|---|

Rotation

| 75 | 25 | 0 |
|----|----|---|

Percentage of ranges of motion in each class.

TABLE 15.3E

| **Ergonomic Recommendations** | |
|---|---|
| | Bandy Boxing Group<br>July 19, 1996 |

### Re-Engineering

Eliminate the need to manually re-stack paper stock for printing press by re-engineering the previous operation to include mechanical stacking.

### Ergonomic Intervention

Use a weight-dependent, adjustable pallet which maintains a constant height as stock is removed or added. Modify the orientation of the pallets with respect to each other to decrease awkward postures, e.g., butt pallets side by side to enable the worker to flip stock without lifting it completely off the pallet.

### Education and Training

The trainer, a safety professional or a supervisor with knowledge of ergonomics, introduces the worker to the stacking job both verbally and by demonstration, and works side-by-side with worker for one to two days. The worker is monitored while performing job for compliance with job description, productivity, and safety/ergonomic guidelines (use B-Tracker or BackWorks Targeting/Avoidance software where appropriate). The worker is tested using a written scrambled job description. The worker is also tested practically by walking the trainer through the stacking job. The trainer calculates a learning curve and revisits training as needed.

### Screening

BackWorks generates a protocol that can be used to test job candidates "post-offer and pre-placement." The protocol is based on the actual physical movements of the stacking task and complies with the Americans with Disabilities Act of 1990.

### Baseline Testing

The B-Tracker is used to gather baseline or benchmark data on the current functional status of the low back. Baseline data directs managed care towards a known pre-injury functional status.

**Report Reviewed by:**_____ **Date:**_____

Tracy Marker, Ergonomist

**TABLE 15.4**

| Backability Baseline Report | Bobby Carrera, 123-45-6789<br>August 1, 1996 | Level III |

### Demographic Data

Company: Bandy Boxing Group
Job Title: Stacker
Job Code: 929.687-030
Division: Packaging
Dept: Outgoing
Physical Demand Strength Rating:  Heavy
Non-Work Activity Level:  Heavy
Gender:  Male
Age:  30
Birth Date:  February 19, 1966
Height:  6 ft. 1 in.
Weight:  170 lbs.
Previous History of LBP:  No
Date of last LBP Incident:  NA

### B-Tracker Analysis: 27

0 _____ 100

27

The lower the B-Tracker Analysis Performance Score, the greater the certainty that this individual is a member of the healthy back population.

### Conclusions

On August 1, 1996, Bobby Carrera participated in a functional low-back analysis.  Results indicate that

• a low-back exercise program is recommended
• correct lifting techniques should be used; heavy lifting and prolonged and/or awkward trunk postures should be avoided
• performance of job tasks and work environment may be evaluated for potential improvement
• baseline testing should be repeated in 12 months

### Test Site:
### Administered by:
### Title:

Account Charged:  INT001
System Ref#:  Oct3096  121428

**TABLE 15.5**

**Low-Back Clinical Report**

Bobby Carrera, 123-45-6789
August 29, 1996

**Demographic Data**

DOT Job Title: Stacker
DOT Job Code: 929.687-030

**Conclusions**

On August 29, 1996, Bobby Carrera participated in a functional low-back analysis. The test results confirm membership in a low-back problem population. Bobby's Waddell score was consistent with an organic etiology.

**B-Tracker Analysis: 87**

0 ——————————————————————— 100
                                                87

The higher the B-Tracker Analysis Performance Score, the greater the certainty that the patient is a member of the low-back problem population. Scores between 0 and 5 indicate that the individual is a member of the healthy low-back population.

**VAS Test Results: 65**

0 ——————————————————————— 100
                                          65

Visual Analog Scale (VAS) prior to the B-Tracker test: "Mark on the scale from 0 to 100 your level of pain discomfort with 0 being None and 100 being Unbearable."

**Waddell's Signs: 0**

| Tenderness | Simulation | Distraction | Regional | Overreaction |
|------------|-----------|-------------|----------|--------------|

0 - 2 of 5 Waddell's signs indicate a negative result;
3 - 5 of 5 Waddell's signs indicate a positive result.

**Test Site** Stewart PT--Burlington, NC #3954-1AE4
**Signature** _____
**Title** _____

**TABLE 15.6A**

**Low-Back Clinical Report**

Bobby Carrera, 123-45-6789
September 18, 1996

### Demographic Data

DOT Job Title: Stacker
DOT Job Code: 929.687-030

### Conclusions

On September 18, 1996, Bobby Carrera participated in a functional low-back analysis. The test
results confirm membership in a low-back problem
population. Bobby's Waddell score was consistent
with an organic etiology.

### B-Tracker Analysis: 35

0 _____▲_____ 100
                      35

The higher the B-Tracker Analysis Performance
Score, the greater the certainty that the patient is a
member of the low-back problem population.
Scores between 0 and 5 indicate that the individual
is a member of the healthy low-back population.

### VAS Test Results: 11

0 ____▲_____ 100
     11

Visual Analog Scale (VAS) prior to the B-Tracker test:
"Mark on the scale from 0 to 100 your level of pain
discomfort with 0 being None and 100 being Unbearable."

### Waddell's Signs: 0

Tenderness  Simulation  Distraction  Regional  Overreaction
                                                    ▲

0 - 2 of 5 Waddell's signs indicate a negative result;
3 - 5 of 5 Waddell's signs indicate a positive result.

### Test Site
### Signature
### Title

Stewart PT--Burlington, NC  #3954-1AE4
_____
_____

**TABLE 15.6B**

| Backability Baseline Report | Bobby Carrera, 123-45-6789<br>September 20, 1996 | Level III |

### Demographic Data

Company: Bandy Boxing Group
Job Title: Stacker
Job Code: 929.687-030
Division: Packaging
Dept: Outgoing
Physical Demand Strength Rating: Heavy
Non-Work Activity Level: Heavy
Gender: Male
Age: 30
Birth Date: February 19, 1966
Height: 6 ft. 1 in.
Weight: 170 lbs.
Previous History of LBP: No
Date of last LBP Incident: NA

### B-Tracker Analysis: 31

0 _____ 100
▲
31

The lower the B-Tracker Analysis Performance Score, the greater the certainty that this individual is a member of the healthy back population.

### Conclusions

On September 20, 1996, Bobby Carrera participated in a functional low-back analysis. Results indicate that

• a low-back exercise program is recommended
• correct lifting techniques should be used; heavy lifting and prolonged and/or awkward trunk postures should be avoided
• performance of job tasks and work environment may be evaluated for potential improvement
• baseline testing should be repeated in 12 months

### Test Site:
### Administered by:
### Title:

Account Charged: INT001
System Ref#: Oct3096_121428

**TABLE 15.7A**

## BackAbility Baseline Summary

Bandy Boxing Group
August 1, 1996

### Baseline Testing Summary

On August 1, 1996, Bandy Boxing Group had 18 individuals with 3 different job titles participate in the BackAbility Baseline Program.

### Baseline Summary Data

| | Job Code | # | IV | III | II | I |
|---|---|---|---|---|---|---|
| | | | **Level** | | | |
| Stacker | 929.687-030 | 11 | 4 | 6 | 0 | 1 |
| Gluer | 795.687-014 | 4 | 1 | 3 | 0 | 0 |
| Air-Drier-Machine Op | 534.682-010 | 3 | 0 | 3 | 0 | 0 |
| | | **18** | **5** | **12** | **0** | **1** |

### Baseline Level Descriptions

**Level I**
- a special medical intervention is necessary to determine if a low-back problem exists

**Level II**
- a low-back exercise program is strongly recommended
- correct lifting techniques should be used; heavy lifting and prolonged and/or awkward trunk postures should be avoided
- performance of job tasks and work environment may be evaluated for potential improvement
- baseline testing should be repeated in 6 months

**Level III**
- a low-back exercise program is recommended
- correct lifting techniques should be used; heavy lifting and prolonged and/or awkward trunk postures should be avoided
- performance of job tasks and work environment may be evaluated for potential improvement
- baseline testing should be repeated in 12 months

**Level II**
- current low-back exercise should be maintained
- correct lifting techniques should be continued; avoidance of heavy lifting and prolonged and/or awkward trunk postures should persist

Account Charged: INT001
System Ref#: Oct3096_121428

TABLE 15.7B

| **BackAbility Baseline Summary** | | |
|---|---|---|

Bandy Boxing Group
August 1, 1996                                                                          page 2

## Individual Baseline Scores

**Level I**

**Stacker (1)**

Philbert, Yoakum          98

**Level II**

**NA**

**Level III**

**Stacker (6)**

| Carrera, Bobby | 69 | Barley, Filbert | 57 | Oliver, Clem | 51 |
|---|---|---|---|---|---|
| Philbert, Yoakum | 49 | Melbourne, Dan | 49 | Carrera, Bobby | 22 |

**Gluer (3)**

| Philbert, Yoakum | 67 | Laxaline, Martha | 15 | Bertram, Randy | 7 |
|---|---|---|---|---|---|

**Air-Drier-Machine (3)**

| Ballinger, Raymond | 57 | Mayonnaire, Lakisha | 54 | Blessingame, Chad | 6 |
|---|---|---|---|---|---|

**Level IV**

**Stacker (4)**

| Phillips, Tanya | 5 | Crosby, Willie | 2 | Boyles, Darlene | 2 |
|---|---|---|---|---|---|
| Fargo, Janelle | 0 | | | | |

**Gluer (1)**

Reynolds, Billy          0

TABLE 15.7C

## BackAbility Baseline Protocol

An Explanation of the Protocol and the Report

The B-Tracker BackAbility Baseline Protocol is used for testing subjects who have no low-back problem. Specific directions are provided about placement of the B-Tracker and use of a stabilization platform. The formalized protocol requires that the clinician demonstrate each baseline movement immediately preceding its performance. The following three movements compose the protocol:

Flexibility
    two repetitions of movements for each axis (flexion/extension, lateral bending, rotation). Subjects are instructed to move as far as possible.

Free Dynamic Movement
    seven repetitions of movements for each axis (flexion/extension, lateral bending, rotation). Subjects are instructed to move as far as feels comfortable at their preferred pace.

Circumduction
    four circumductions to the right, then four to the left. Subjects are instructed to move as far as is comfortable at their preferred pace.

### The Baseline Score

The purpose of the baseline score is to confirm—by degrees of certainty—the subject's low-back health and to establish for future reference a quantifiable benchmark. There are four levels to which individuals are assigned on the basis of their B-Tracker baseline score:

| | |
|---|---|
| Level I | 95-100 |
| Level II | 70-94 |
| Level III | 6-69 |
| Level IV | 1-5 |

### Use of the Baseline Report

In the event of injury, the subject's baseline data will serve as the goal to which recovery should be directed in order to achieve pre-injury functional status and return-to-work readiness. Categorization by level also enables recommendations which can assist individuals improve or maintain their low-back status. Of course, the recommendations should be tailored towards the functional demands placed on an individual's low back due to specific job requirements.

Level I
* a special medical intervention is necessary to determine if a low-back problem exists

Level II
* a low-back exercise program is strongly recommended
* correct lifting techniques should be used; heavy lifting and prolonged and/or awkward trunk postures should be avoided
* performance of job tasks and work environment may be evaluated for potential improvement
* baseline testing should be repeated in 6 months

Level III
* a low-back exercise program is recommended
* correct lifting techniques should be used; heavy lifting and prolonged and/or awkward trunk postures should be avoided
* performance of job tasks and work environment may be evaluated for potential improvement
* baseline testing should be repeated in 12 months

Level IV
* current low-back exercise should be maintained
* correct lifting techniques should be continued; avoidance of heavy lifting and prolonged and/or awkward trunk postures should persist

**TABLE 15.7D**

### Low-Back Clinical Protocol

An Explanation of the Protocol and the Report

The B-Tracker Low-Back Clinical protocol is used when the patient asserts he/she has a low-back problem. The protocol comprises the following three parts:

- Waddell Signs Test
- VAS (Visual Analog Scale) Test
- B-Tracker movement exercises.

The Waddell Signs Test is administered in accordance with procedures documented by Waddell, et al. in *Spine* (5:2, 1980), "Nonorganic Physical Signs in Low-Back Pain."

The VAS requires patients to rate their pain just prior to beginning the B-Tracker movement exercises. As patients view a scale whose only markings are '0' and 'None' at one end and '100' and 'Unbearable' at the other end, they are asked to respond to the following question: "Mark on the scale from 0 to 100 your level of pain discomfort, with 0 being None and 100 being Unbearable."

The B-Tracker movement exercises are performed in strict order with specific directions. Further assuring consistent testing are specific directions about placement of the B-Tracker and use of a stabilization platform. The protocol also requires that the clinician demonstrate each exercise immediately preceding its performance. The following three exercises compose the protocol's movement exercises:

Flexibility
two repetitions of movements for each axis (flexion/extension, lateral bending, rotation). Patients are instructed to move as far as possible.

Free Dynamic Movement
seven repetitions of movements for each axis (flexion/extension, lateral bending, rotation). Patients are instructed to move as far as feels comfortable at their preferred pace.

Circumduction
four circumductions to the right, then four to the left. Patients are instructed to move as far as is comfortable at their preferred pace.

## The Clinical Report Score

The purpose of the resulting B-Tracker score is to confirm—by degrees of certainty—the patient's assertion that he/she is experiencing a low-back problem without a specific diagnosis. To this end, there is a range of scores (6 - 100) which confirms a patient's low-back problem. A score falling in the range 0 - 5 suggests the patient's claim of a low-back problem should be further tested.

The Clinical Report score provides the means to make decisions regarding treatment paths. Consider the following two scenarios: confirmation can be used to initiate a treatment path involving exercise and other non-surgical procedures; or, confirmation can be used to initiate surgical intervention. In the first case, the treatment is conservative and in the second, aggressive. When considering an aggressive treatment path, greater certainty is desired.

An example may serve to illustrate: Consider a patient who obtains a score of 20. The score is sufficient to initiate a treatment path involving rehabilitation/exercise. However, the score may not be sufficient as the basis for consideration of surgical intervention due to the consequences associated with treatment if, in fact, the individual was not a member of the low-back problem population.

## Use of the Low-Back Clinical Report

The B-Tracker Low-Back Clinical Report, when compared to a patient's earlier Low-Back Baseline Report, objectively documents the precise extent of low-back functional loss. The B-Tracker Low-Back Clinical Report further provides means to ensure a patient is directed to an appropriate patient management program. Subsequent assessments with the Low-Back Clinical protocol objectively indicate the patient's level of functional improvement, when a patient has returned to his/her pre-injury functional status, and the patient's return-to-work readiness.

**TABLE 15.8**

# References

Adams, M.A., Hutton, W.C. 1985. Gradual disc prolapse. *Spine.* 10:524-531.

Andersson, G.B.J., Chaffin, D.B., Pope, M.H. 1991. Occupational biomechanics of the lumbar spine, in *Occupational Low Back Pain*, Mosby-Year Book, Inc., St. Louis, MO.

Bigos, S.J., Battié, M.C., Spengler, D.M. et al. 1991. A prospective study of work perceptions and psycho-social factors affecting the report of back injury. *Spine.* 16:1-6.

Dickenson, C.E. et al. 1992. Questionnaire development: an examination of the Nordic musculoskeletal questionnaire. *Applied Ergonomics.* 23:197-201.

*Dictionary of Occupational Titles.* 1977. U.S. Department of Labor Employment and Training Administration, Washington, D.C.

Chaffin, D.B. 1974. Human strength capability and low-back pain. *J. Occup. Med.* 16(4).

Garg A., Moore, S.J. 1992. Epidemiology of low-back pain in industry, in *Occupational Medicine, State of the Art Reviews: Ergonomics: Low-back Pain, Carpal Tunnel Syndrome, and Upper Extremity Disorders in the Workplace*, Eds. J. Moore and A. Garg, p. 599-608. Hanley & Belfus, Inc., Philadelphia, PA.

Genaidy, A.M., Karwowski, W. 1993. The effects of neutral posture deviations on perceived joint discomfort ratings in sitting and standing postures. *Ergonomics.* 36:785-792.

Hadler, N.M. 1994. Backache and work incapacity in Japan. *J. Occup. Med.* 36(10):1110-1114.

Heliövaara, M., Knekt, P., Aromaa, A. 1987. Incidence and risk factors of herniated lumbar intervertebral disc or sciatica leading to hospitalization. *J. Chronic Dis.* 40:251-258.

Hulsof, C., Veldhuizen van Zantan, B. 1987. Whole-body vibration and low-back pain: A review of epidemiologic studies. *International Archives of Occupational and Environmental Health.* 59:205-220.

ISO: Evaluation of human response to whole body vibration. *Intl. Org. for Standardization* Ref. No: 2631 (E), 1978.

Kelsey, J.L., Hochberg, M.C. 1988. Epidemiology of chronic musculoskeletal disorders. *Ann. Rev. Public Health.* 9:379-401.

Kelsey, J.L., Githens, P.B., White, A.A. III et al. 1984. An epidemiologic study of lifting and twisting on the job and risk factors for acute prolapsed lumbar intervertebral disc. *J. Orthop. Research.* 2:61.

Kelsey, J.L., Hardy, R.J. 1975. Driving of motor vehicles as a risk factor for acute herniated lumbar intervertebral disc. *Am. J. Epidemiology.* 102(1):63-73.

Liles, D.H., Deivanayagam, S., Ayoub, M.M., Mahajan, P. 1984. A job severity index for the evaluation and control of lifting injury. *Human Factors.* 26:683-694.

Magnusson, M.L., Pope, M.H., Wilder, D.G., Areskoug, B. 1996. Are occupational drivers at an increased risk for developing musculo-skeletal disorders? *Spine.* 21(6):710-717.

Nachemson, A.L. 1991. Spinal disorders: overall impact in society and the need for orthopaedic resources. *Acta Orthop. Scand. Suppl.* 241:17-22.

Pope, M.H., Frymoyer, J.W., Andersson, G.B.J., Chaffin, D. 1991. *Occupational Low Back Pain: Assessment, Treatment and Prevention*, 2nd ed. Mosby Press, St. Louis, MO.

Pope, M.H., Andersson, G.B.J., Broman, H., Svensson, M., Zetterberg, C. 1986. Electromyographic studies of the lumbar trunk musculature during the development of axial torques. *J. Ortho. Res.* 4(3):288-297.

Pope, M.H., Svensson, M., Broman, H., Andersson, G.B.J. 1986. Mounting of the transducer in measurements of sequential motion of the spine. *J. Biomech.* 19(8):675-677.

Pope, M.H., Wilder, D.G., Jorneus, L., Broman, H., Svensson, M., Andersson, G.B.J. 1987. The response of the seated human to sinusoidal vibration and impact. *J. Biomech. Eng.* 109:279-284.

Pope, M.H., Broman, H., Hansson, T. 1989. The dynamic response of a subject seated on various cushions. *Ergonomics.* 32(10):1155-1166.

Punnett, L., Fine, L.T., Keyserling, W.M. et al. 1991. Back disorders and non-neutral postures of automobile assembly workers. *Scandinavian J. of Work, Environ. and Health.* 17:337-346.

Riihimäki, H. 1995. Back and limb disorders, in *Epidemiology of Work Related Diseases,* Ed. J. C. McDonald, Chapter 10. BMJ Publishing Group, London, England.

Seroussi, R., Wilder, D., Pope, M.H. 1989. Trunk muscle electromyography and whole body vibration. *J. Biomech.* 22(3):219-229.

Waters, T.R., Putz-Anderson, V., Garg, A. et al. 1993. Revised NIOSH lifting equation for the design and evaluation of manual lifting tasks. *Ergonomics.* 36:749-776.

Wilder, D.G., Aleksiev, A., Magnusson, M., Pope, M.H., Spratt, K., Goel, V.K. 1996. Muscular response to sudden load: a tool to evaluate fatigue and rehabilitation. *Spine,* in press.

## For Further Information

Contact Interlogics, 328 Elizabeth Brady Rd., Hillsborough, NC 27278.

# 16

# Revised NIOSH Lifting Equation

Thomas R. Waters
*National Institute for Occupational
Safety and Health*

Vern Putz-Anderson
*National Institute for Occupational
Safety and Health*

## 16.1    Introduction

This chapter provides information about a revised equation for assessing the physical demands of certain two-handed manual lifting tasks that was developed by the National Institute for Occupational Safety and Health (NIOSH) and described earlier in an article by Waters et al.[1] We discuss what factors need to be measured, how they should be measured, what procedures should be used, and how the results can be used to ergonomically design new jobs or make decisions about redesigning existing jobs that may be hazardous. We define all pertinent terms and present the mathematical formulas and procedures needed to properly apply the NIOSH lifting equation. Several example problems are also provided to demonstrate how the equations should be used. An expanded version of this chapter is contained in a NIOSH document.[2]

Historically, NIOSH has recognized the problem of work-related back injuries and published the *Work Practices Guide for Manual Lifting* (WPG) in 1981.[3] The WPG contained a summary of the lifting-related literature up to 1981; analytical procedures and a lifting equation for calculating a recommended weight for specified two-handed, symmetrical lifting tasks; and an approach for controlling the hazards of low back injury from manual lifting. The approach to hazard control was coupled to the *action limit* (AL), a term that denoted the recommended weight derived from the lifting equation.

In 1985, NIOSH convened an *ad hoc* committee of experts who reviewed the current literature on lifting, including the NIOSH WPG.[*] The literature review was summarized in a document containing updated information on the physiological, biomechanical, psychophysical, and epidemiological aspects of manual lifting.[4] Based on the results of the literature review, the *ad hoc* committee recommended criteria for defining the lifting capacity of healthy workers. The committee used the criteria to formulate the revised lifting equation.[**] Subsequently, NIOSH staff developed the documentation for the equation and played a prominent role in recommending methods for interpreting the results of the lifting equation. The revised lifting equation reflects new findings and provides methods for evaluating asymmetrical lifting tasks, and lifts of objects with less than optimal couplings between the object and the worker's hands. The revised lifting equation also provides guidelines for a more diverse range of lifting tasks than the earlier equation.[3]

The rationale and criterion for the development of the revised NIOSH lifting equation are provided in a journal article by Waters et al.[1] We suggest that those users who wish to achieve a better understanding of the data and decisions that were made in formulating the revised equation consult that article. It provides an explanation of the selection of the biomechanical, physiological, and psychophysical criterion as well as a description of the derivation of the individual components of the revised lifting equation. For those individuals, however, who are primarily concerned with the use and application of the revised lifting equation, this chapter provides a more complete description of the method and limitations.

Although the revised lifting equation has not been fully validated, the recommended weight limits derived from the revised equation are consistent with, or lower than, those generally reported in the literature. Moreover, the proper application of the revised equation is more likely to protect healthy workers for a wider variety of lifting tasks than methods that rely only on a single task factor or single criterion.

Finally, it should be stressed that the NIOSH lifting equation is only one tool in a comprehensive effort to prevent work-related low back pain and disability. Some examples of other approaches are described elsewhere.[5] Moreover, lifting is only one of the causes of work-related low back pain and disability. Other causes that have been hypothesized or established as risk factors include whole body vibration, static postures, prolonged sitting, and direct trauma to the back. Psychosocial factors, appropriate medical treatment, and job demands may also be particularly important in influencing the transition of acute low back pain to chronic disabling pain.

## 16.2　Definition of Terms

This section provides the basic technical information needed to properly use the revised lifting equation to evaluate a variety of two-handed manual lifting tasks. Definitions and data requirements for the revised lifting equation are also provided.

### Recommended Weight Limit (RWL)

The *recommended weight limit* (RWL) is the principal product of the revised NIOSH lifting equation. The RWL is defined for a specific set of task conditions as the weight of the load that nearly all healthy workers could perform over a substantial period of time (e.g., up to 8 h) without an increased risk of developing lifting-related low back pain (LBP). By "healthy workers" we mean workers who are free of adverse health conditions that would increase their risk of musculoskeletal injury.

---

[*]The *ad hoc* 1991 NIOSH Lifting Committee members included: M.M. Ayoub, Donald B. Chaffin, Colin G. Drury, Arun Garg, and Suzanne Rodgers. NIOSH representatives included Vern Putz-Anderson and Thomas R. Waters.

[**]For this document, the revised 1991 NIOSH lifting equation will be identified simply as "the revised lifting equation."[1,2] The abbreviation WPG will continue to be used as the reference to the earlier NIOSH lifting equation, which was documented in a publication entitled *Work Practices Guide for Manual Lifting*.[3]

The concept behind the revised NIOSH lifting equation is to start with a recommended weight that is considered safe for an "ideal" lift (i.e., load constant equal to 51 lb) and then reduce the weight as the task becomes more stressful (i.e., as the task-related factors become less favorable). The precise formulation of the revised lifting equation for calculating the RWL is based on a multiplicative model that provides a weighting (multiplier) for each of six task variables:

1. Horizontal distance of the load from the worker ($H$)
2. Vertical height of the lift ($V$)
3. Vertical displacement during the lift ($D$)
4. Angle of asymmetry ($A$)
5. Frequency ($F$) and duration of lifting
6. Quality of the hand-to-object coupling ($C$)

The weightings are expressed as coefficients that serve to decrease the load constant, which represents the maximum recommended load weight to be lifted under ideal conditions. For example, as the horizontal distance between the load and the worker increases from 10 in, the recommended weight limit for that task would be reduced from the ideal starting weight.

The *recommended weight limit* (RWL) is defined as

$$RWL = LC \times HM \times VM \times DM \times AM \times FM \times CM$$

where

|      |   |                       |   | METRIC | U.S. CUSTOMARY |
|------|---|-----------------------|---|--------|----------------|
| LC   | = | Load Constant         | = | 23 kg  | 51 lb          |
| HM   | = | Horizontal Multiplier | = | (25/H) | (10/H)         |
| VM   | = | Vertical Multiplier   | = | 1–(.003 \|V-75\|) | 1–(.0075 \|V-30\|) |
| DM   | = | Distance Multiplier   | = | .82 + (4.5/D) | .82 + (1.8/D) |
| AM   | = | Asymmetric Multiplier | = | 1–(.0032A) | 1–(.0032A) |
| FM   | = | Frequency Multiplier  | = | From Table 16.5 | From Table 16.5 |
| CM   | = | Coupling Multiplier   | = | From Table 16.7 | From Table 16.7 |

The term *task variables* refers to the measurable task-related measurements that are used as input data for the formula (i.e., H, V, D, A, F, and C); whereas, the term *multipliers* refers to the reduction coefficients in the equation (i.e., HM, VM, DM, AM, FM, and CM).

## Measurement Requirements

The following list briefly describes the measurements required to use the revised NIOSH lifting equation. Details for each of the variables are presented later in this chapter.

$H$ = horizontal location of hands from midpoint between the inner ankle bones. Measure at the origin and the destination of the lift (cm or in).

$V$ = vertical location of the hands from the floor. Measure at the origin and destination of the lift (cm or in).

$D$ = vertical travel distance between the origin and the destination of the lift (cm or in).

$A$ = angle of asymmetry or angular displacement of the load from the worker's sagittal plane. Measure at the origin and destination of the lift (degrees).

$F$ = average frequency rate of lifting measured in lifts/min. Duration is defined to be ≤ 1 h; ≤ 2 h; or ≤ 8 h assuming appropriate recovery allowances (see Table 16.5).

$C$ = quality of hand-to-object coupling (quality of interface between the worker and the load being lifted). The quality of the coupling is categorized as good, fair, or poor, depending upon the type and location of the coupling, the physical characteristics of load, and the vertical height of the lift.

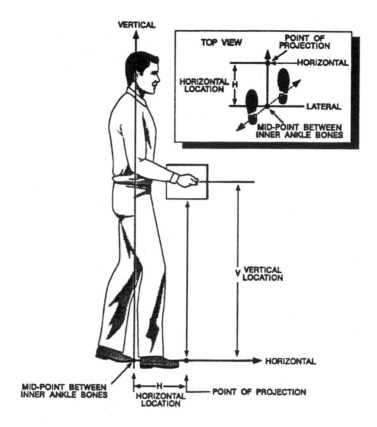

**FIGURE 16.1**   Graphic representation of hand location.

## Lifting Index (LI)

The *lifting index* (LI) is a term that provides a relative estimate of the level of physical stress associated with a particular manual lifting task. The estimate of the level of physical stress is defined by the relationship between the weight of the load lifted and the recommended weight limit. The LI is defined by the equation

$$LI = \frac{Load\ Weight}{Recommended\ Weight\ Limit} = \frac{L}{RWL}$$

where Load Weight (L) = weight of the object lifted (lbs or kg).

## Miscellaneous Terms

*Lifting task* The act of manually grasping an object of definable size and mass with two hands, and vertically moving the object without mechanical assistance.

*Load weight* (L) Weight of the object to be lifted, in pounds or kilograms, including the container.

*Horizontal location* (H) Distance of the hands away from the midpoint between the ankles, in inches or centimeters (measure at the origin and destination of lift). See Figure 16.1.

*Vertical location* (V) Distance of the hands above the floor, in inches or centimeters (measure at the origin and destination of lift). See Figure 16.1.

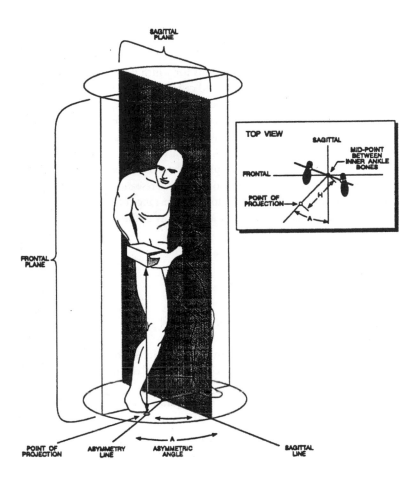

**FIGURE 16.2** Graphic representation of angle of asymmetry (A).

*Vertical travel distance* (*D*) Absolute value of the difference between the vertical heights at the destination and origin of the lift, in inches or centimeters.

*Angle of asymmetry* (*A*) Angular measure of how far the *object* is displaced from the front (midsagittal plane) of the worker's body at the beginning or end of the lift, in degrees (measure at the origin and destination of lift). See Figure 16.2. The asymmetry angle is defined by the location of the load relative to the worker's midsagittal plane, as defined by the neutral body posture, rather than the position of the feet or the extent of body twist.

*Neutral body position* Position of the body when the hands are directly in front of the body and there is minimal twisting at the legs, torso, or shoulders.

*Frequency of lifting* (*F*) Average number of lifts per minute over a 15-minute period.

*Duration of lifting* Three-tiered classification of lifting duration specified by the distribution of work-time and recovery-time (work pattern). Duration is classified as either short (1 h), moderate (1 to 2 h), or long (2 to 8 h), depending on the work pattern.

*Coupling classification* Classification of the quality of the hand-to-object coupling (e.g., handle, cut-out, or grip). Coupling quality is classified as good, fair, or poor.

*Significant control* A condition requiring "precision placement" of the load at the destination of the lift. This is usually the case when (1) the worker has to regrasp the load near the destination of the lift, or (2) the worker has to momentarily hold the object at the destination, or (3) the worker has to carefully position or guide the load at the destination.

## 16.3   Limitations of Equation

The lifting equation is a tool for assessing the physical stress of two-handed manual lifting tasks. As with any tool, its application is limited to those conditions for which it was designed. Specifically, the lifting equation was designed to meet specific lifting-related criteria that encompass biomechanical, physiological, and psychophysical assumptions and data used to develop the equation. To the extent that a given lifting task accurately reflects these underlying conditions and criteria, this lifting equation may be appropriately applied.

The following list identifies a set of work conditions in which the application of the lifting equation could either under- or overestimate the extent of physical stress associated with a particular work-related activity. Each of the following task limitations also highlights research topics in need of further research to extend the application of the lifting equation to a greater range of real-world lifting tasks.

The revised NIOSH Lifting Equation does not apply if any of the following occur:

Lifting/lowering with one hand
Lifting/lowering for over 8 hours
Lifting/lowering while seated or kneeling
Lifting/lowering in a restricted work space
Lifting/lowering unstable objects
Lifting/lowering while carrying, pushing, or pulling
Lifting/lowering with wheelbarrows or shovels
Lifting/lowering with "high speed" motion (faster than about 30 in./second)
Lifting/lowering with unreasonable foot/floor coupling (<0.4 coefficient of friction between the sole and the floor)
Lifting/lowering in an unfavorable environment (temperature significantly outside 66 to 79°F [19 to 26°C] range; relative humidity outside 35 to 50% range)

## 16.4   Obtaining and Using the Data

### Horizontal Component

#### Definition and Measurement

Horizontal location ($H$) is measured from the midpoint of the line joining the inner ankle bones to a point projected on the floor directly below the midpoint of the hand grasps (i.e., load center), as defined by the large middle knuckle of the hand (Figure 16.1). Typically, the worker's feet are not aligned with the midsagittal plane, as shown in Figure 16.1, but may be rotated inward or outward. If this is the case, then the midsagittal plane is defined by the worker's neutral body posture as defined above. If significant control is required at the destination (i.e., precision placement), then $H$ should be measured at both the origin and destination of the lift.

Horizontal distance ($H$) should be measured. In those situations where the $H$ value cannot be measured, then $H$ may be approximated from the following equations:

| Metric<br>(All distances in cm) | U.S. Customary<br>(All distances in inches) |
| --- | --- |
| $H = 20 + W/2$<br>for $V \geq 25$ cm | $H = 8 + W/2$<br>for $V \geq 10$ in. |
| $H = 25 + W/2$<br>for $V < 25$ cm | $H = 10 + W/2$<br>for $V < 10$ in. |

where $W$ is the width of the container in the sagittal plane and $V$ is the vertical location of the hands from the floor.

**TABLE 16.1**  Horizontal Multiplier

| in | Hm | cm | Hm |
|-----|------|------|------|
| ≤10 | 1.00 | ≤25 | 1.00 |
| 11 | .91 | 28 | .89 |
| 12 | .83 | 30 | .83 |
| 13 | .77 | 32 | .78 |
| 14 | .71 | 34 | .74 |
| 15 | .67 | 36 | .69 |
| 16 | .63 | 38 | .66 |
| 17 | .59 | 40 | .63 |
| 18 | .56 | 42 | .60 |
| 19 | .53 | 44 | .57 |
| 20 | .50 | 46 | .54 |
| 21 | .48 | 48 | .52 |
| 22 | .46 | 50 | .50 |
| 23 | .44 | 52 | .48 |
| 24 | .42 | 54 | .46 |
| 25 | .40 | 56 | .45 |
| >25 | .00 | 58 | .43 |
|  |  | 60 | .42 |
|  |  | 63 | .40 |
|  |  | >63 | .00 |

## Horizontal Restrictions

If the horizontal distance is less than 10 in. (25 cm), then $H$ is set to 10 in. (25 cm). Although objects can be carried or held closer than 10 in. from the ankles, most objects that are closer than this cannot be lifted without encountering interference from the abdomen or hyperextending the shoulders. Although 25 in. (63 cm) was chosen as the maximum value for $H$, it is probably too great a distance for shorter workers, particularly when lifting asymmetrically. Furthermore, objects at a distance of more than 25 in. from the ankles normally cannot be lifted vertically without some loss of balance.

## Horizontal Multiplier

The horizontal multiplier (HM) is $10/H$ for $H$ measured in inches and $25/H$ for $H$ measured in centimeters. If $H$ is less than or equal to 10 in. (25 cm), the multiplier is 1.0. HM decreases with an increase in $H$ value. The multiplier for $H$ is reduced to 0.4 when $H$ is 25 in. (63 cm). If $H$ is greater than 25 in., then HM = 0. The HM value can be computed directly or determined from Table 16.1.

# Vertical Component

## Definition and Measurement

Vertical location ($V$) is defined as the vertical height of the hands above the floor. $V$ is measured vertically from the floor to the midpoint between the hand grasps, as defined by the large middle knuckle. The coordinate system is illustrated in Figure 16.1.

## Vertical Restrictions

The vertical location ($V$) is limited by the floor surface and the upper limit of vertical reach for lifting (i.e., 70 in. or 175 cm). The vertical location should be measured at the origin and the destination of the lift.

## Vertical Multiplier

To determine the vertical multiplier (VM), the absolute value or deviation of $V$ from an optimum height of 30 in. (75 cm) is calculated. A height of 30 in. above floor level is considered "knuckle height" for a

**TABLE 16.2**   Vertical Multiplier

| V | | V | |
|---|---|---|---|
| in | VM | cm | VM |
| 0 | .78 | 0 | .78 |
| 5 | .81 | 10 | .81 |
| 10 | .85 | 20 | .84 |
| 15 | .89 | 30 | .87 |
| 20 | .93 | 40 | .90 |
| 25 | .96 | 50 | .93 |
| 30 | 1.00 | 60 | .96 |
| 35 | .96 | 70 | .99 |
| 40 | .93 | 80 | .99 |
| 45 | .89 | 90 | .96 |
| 50 | .85 | 100 | .93 |
| 55 | .81 | 110 | .90 |
| 60 | .78 | 120 | .87 |
| 65 | .74 | 130 | .84 |
| 70 | .70 | 140 | .81 |
| >70 | .00 | 150 | .78 |
| | | 160 | .75 |
| | | 170 | .75 |
| | | 175 | .70 |
| | | >175 | .00 |

worker of average height (66 in. or 165 cm). The vertical multiplier (VM) is $(1-(.0075|V\text{-}30|))$ for $V$ measured in inches, and VM is $(1-(.003|V\text{-}75|))$, for $V$ measured in centimeters.

When $V$ is at 30 in. (75 cm), the vertical multiplier (VM) is 1.0. The value of VM decreases linearly with an increase or decrease in height from this position. At floor level, VM = 0.78, and at 70 in. (175 cm) height, VM = 0.7. If $V$ is greater than 70 in., then VM = 0. The VM value can be computed directly or determined from Table 16.2.

## Distance Component

### Definition and Measurement

The distance variable ($D$) is defined as the vertical travel distance of the hands between the origin and destination of the lift. For lifting, $D$ can be computed by subtracting the vertical location ($V$) at the origin of the lift from the corresponding $V$ at the destination of the lift (i.e., $D$ is equal to $V$ at the destination minus $V$ at the origin). For a lowering task, $D$ is equal to $V$ at the origin minus $V$ at the destination.

### Distance Restrictions

The distance variable ($D$) is assumed to be at least 10 in. (25 cm), and no greater than 70 in. (175 cm). If the vertical travel distance is less than 10 in. (25 cm), then $D$ should be set to the minimum distance of 10 in. (25 cm).

### Distance Multiplier

The distance multiplier (DM) is $(.82 + (1.8/D))$ for $D$ measured in inches and $(.82 + (4.5/D))$ for $D$ measured in centimeters. For $D$ less than 10 in. (25 cm) $D$ is assumed to be 10 in. (25 cm), and DM is 1.0. The distance multiplier, therefore, decreases gradually with an increase in travel distance. The DM = 1.0 when $D$ is set at 10 in., (25 cm); DM = 0.85 when $D$ is 70 in. (175 cm). Thus, DM ranges from 1.0 to 0.85 as the $D$ varies from 0 in. (0 cm) to 70 in. (175 cm). The DM value can be computed directly or determined from Table 16.3.

**TABLE 16.3** Distance Multiplier

| D | | D | |
|---|---|---|---|
| ≤10 | DM | cm | DM |
| ≤10 | 1.00 | ≤25 | 1.00 |
| 15 | .94 | 40 | .93 |
| 20 | .91 | 55 | .90 |
| 25 | .89 | 70 | .88 |
| 30 | .88 | 85 | .87 |
| 35 | .87 | 100 | .87 |
| 40 | .87 | 115 | .86 |
| 45 | .86 | 130 | .86 |
| 50 | .86 | 145 | .85 |
| 55 | .85 | 160 | .85 |
| 60 | .85 | 175 | .85 |
| 70 | .85 | >175 | .00 |
| >70 | .00 | | |

## Asymmetry Component

### Definition and Measurement

Asymmetry refers to a lift that begins or ends outside the midsagittal plane (See Figure 16.2). In general, asymmetric lifting should be avoided. If asymmetric lifting cannot be avoided, however, the recommended weight limits are significantly less than those used for symmetrical lifting.[*]

An asymmetric lift may be required under the following task or workplace conditions:

1. The origin and destination of the lift are oriented at an angle to each other.
2. The lifting motion is across the body, such as occurs in swinging bags or boxes from one location to another.
3. The lifting is done to maintain body balance in obstructed workplaces, on rough terrain, or on littered floors.
4. Productivity standards require reduced time per lift.

The asymmetric angle ($A$), which is depicted graphically in Figure 16.2, is operationally defined as the angle between the asymmetry line and the midsagittal line. The *asymmetry line* is defined as the line that joins the midpoint between the inner ankle bones and the point projected on the floor directly below the midpoint of the hand grasps, as defined by the large middle knuckle. The *sagittal line* is defined as the line passing through the midpoint between the inner ankle bones and lying in the midsagittal plane, as defined by the neutral body position (i.e., hands directly in front of the body, with no twisting at the legs, torso, or shoulders). *Note*: The asymmetry angle is not defined by foot position or the angle of torso twist, but by the location of the load relative to the worker's midsagittal plane.

In many cases of asymmetric lifting, the worker will pivot or use a step turn to complete the lift. Because this may vary significantly between workers and between lifts, we have assumed that no pivoting or stepping occurs. Although this assumption may overestimate the reduction in acceptable load weight, it will provide the greatest protection for the worker.

---

[*]It may not always be clear if asymmetry is an intrinsic element of the task or just a personal characteristic of the worker's lifting style. Regardless of the reason for the asymmetry, any observed asymmetric lifting should be considered an intrinsic element of the job design and should be considered in the assessment and subsequent redesign. Moreover, the design of the task should not rely on worker compliance, but rather the design should discourage or eliminate the need for asymmetric lifting.

**TABLE 16.4**  Asymmetric Multiplier

| A deg | AM |
|---|---|
| 0 | 1.00 |
| 15 | .95 |
| 30 | .90 |
| 45 | .86 |
| 60 | .81 |
| 75 | .76 |
| 90 | .71 |
| 105 | .66 |
| 120 | .62 |
| 135 | .57 |
| >135 | .00 |

The asymmetry angle ($A$) must always be measured at the origin of the lift. If significant control is required at the destination, however, then angle $A$ should be measured at both the origin and the destination of the lift.

### Asymmetry Restrictions

The angle $A$ is limited to the range from 0° to 135°. If $A > 135$°, then AM is set equal to zero, which results in an RWL of zero, or no load.

### Asymmetric Multiplier

The asymmetric multiplier (AM) is $1-(.0032A)$. AM has a maximum value of 1.0 when the load is lifted directly in front of the body and decreases linearly as the angle of asymmetry ($A$) increases. The range is from a value of 0.57 at 135° of asymmetry to a value of 1.0 at 0° of asymmetry (i.e., symmetric lift). If $A$ is greater than 135°, then AM = 0, and the RWL = 0.0. The AM value can be computed directly or determined from Table 16.4.

## Frequency Component

### Definition and Measurement

The frequency multiplier is defined by (1) the number of lifts per minute (frequency), (2) the amount of time engaged in the lifting activity (duration), and (3) the vertical height of the lift from the floor. Lifting frequency ($F$) refers to the average number of lifts made per minute, as measured over a 15-minute period. Because of the potential variation in work patterns, analysts may have difficulty obtaining an accurate or representative 15-minute work sample for computing $F$. If significant variation exists in the frequency of lifting over the course of the day, analysts should employ standard work sampling techniques to obtain a representative work sample for determining the number of lifts per minute. For those jobs where the frequency varies from session to session, each session should be analyzed separately, but the overall work pattern must still be considered. For more information, most standard industrial engineering or ergonomics texts provide guidance for establishing a representative job sampling strategy (e.g., Eastman Kodak Company).[6]

### Lifting Duration

Lifting duration is classified into three categories based on the pattern of continuous work time and recovery time (i.e., light work) periods. A continuous work time (WT) period is defined as a period of uninterrupted work. Recovery time (RT) is defined as the duration of light work activity following a period of continuous lifting. Examples of light work include activities such as sitting at a desk or table, monitoring operations, light assembly work, etc. The three categories are short duration, moderate duration, and long duration.

**TABLE 16.5**  Frequency Multiplier (FM)

| Frequency[‡] Lifts/min (F) | ≤1 Hour | | >1 but ≤ 2 Hours | | >2 but ≤ 8 Hours | |
|---|---|---|---|---|---|---|
| | V[†] < 30 | V ≥ 30 | V < 30 | V ≥ 30 | V < 30 | V ≥ 30 |
| ≤0.2 | 1.00 | 1.00 | .95 | .95 | .85 | .85 |
| 0.5 | .97 | .97 | .92 | .92 | .81 | .81 |
| 1 | .94 | .94 | .88 | .88 | .75 | .75 |
| 2 | .91 | .91 | .84 | .84 | .65 | .65 |
| 3 | .88 | .88 | .79 | .79 | .55 | .55 |
| 4 | .84 | .84 | .72 | .72 | .45 | .45 |
| 5 | .80 | .80 | .60 | .60 | .35 | .35 |
| 6 | .75 | .75 | .50 | .50 | .27 | .27 |
| 7 | .70 | .70 | .42 | .42 | .22 | .22 |
| 8 | .60 | .60 | .35 | .35 | .18 | .18 |
| 9 | .52 | .52 | .30 | .30 | .00 | .15 |
| 10 | .45 | .45 | .26 | .26 | .00 | .13 |
| 11 | .41 | .41 | .00 | .23 | .00 | .00 |
| 12 | .37 | .37 | .00 | .21 | .00 | .00 |
| 13 | .00 | .34 | .00 | .00 | .00 | .00 |
| 14 | .00 | .31 | .00 | .00 | .00 | .00 |
| 15 | .00 | .28 | .00 | .00 | .00 | .00 |
| >15 | .00 | .00 | .00 | .00 | .00 | .00 |

[†] Values of V are in inches.

[‡] For lifting less frequently than once per 5 minutes, set F = .2 lifts/minute.

*Short Duration.* Short duration lifting tasks are those that have a work duration of 1 h or less, followed by a recovery time equal to 1.2 times the work time (i.e., at least a 1.2 recovery-time to work time ratio [RT/WT]). For example, to be classified as short-duration, a 45-minute lifting job must be followed by at least a 54-minute recovery period prior to initiating a subsequent lifting session. If the required recovery time is not met for a job of one hour or less, and a subsequent lifting session is required, then the total lifting time must be combined to correctly determine the duration category. Moreover, if the recovery period does not meet the time requirement, it is disregarded for purposes of determining the appropriate duration category.

As another example, assume a worker lifts continuously for 30 min, then performs a light work task for 10 min, and then lifts for an additional 45-minute period. In this case, the recovery time between lifting sessions (10 min) is less than 1.2 times the initial 30-minute work time (36 min). Thus, the two work times (30 min and 45 min) must be added together to determine the duration. Since the total work time (75 min) exceeds 1 hour, the job is classified as moderate duration. On the other hand, if the recovery period between lifting sessions was increased to 36 min, then the short duration category would apply, which would result in a larger FM value.

A special procedure has been developed for determining the appropriate lifting frequency (F) for certain repetitive lifting tasks in which workers do not lift continuously during the 15-minute sampling period. This occurs when the work pattern is such that the worker lifts repetitively for a short time and then performs light work for a short time before starting another cycle. For work patterns such as this, F may be determined as follows, as long as the actual lifting frequency does not exceed 15 lifts/min:

1. Compute the total number of lifts performed for the 15-minute period (i.e., lift rate times work time).
2. Divide the total number of lifts by 15.
3. Use the resulting value as the frequency (F) to determine the frequency multiplier (FM) from Table 16.5.

For example, if the work pattern for a job consists of a series of cyclic sessions requiring 8 min of lifting followed by 7 min of light work, and the lifting rate during the work sessions is 10 lifts/min, then the frequency rate (*F*) that is used to determine the frequency multiplier for this job is equal to (10 × 8)/15 or 5.33 lifts/min. If the worker lifted continuously for more than 15 min, however, then the actual lifting frequency (10 lifts/min) would be used.

When using this special procedure, the duration category is based on the magnitude of the recovery periods *between* work sessions, not *within* work sessions. In other words, if the work pattern is intermittent and the special procedure applies, then the intermittent recovery periods that occur during the 15-minute sampling period are *not* considered as recovery periods for purposes of determining the duration category. For example, if the work pattern for a manual lifting job was composed of repetitive cycles consisting of 1 min of continuous lifting at a rate of 10 lifts/min, followed by 2 minutes of recovery, the correct procedure would be to adjust the frequency according to the special procedure [i.e., *F* = (10 lifts/min × 5 min)/15 min = 50/15 = 3.4 lifts/min]. The 2-minute recovery periods would not count toward the RT/WT ratio, however, and additional recovery periods would have to be provided as described above.

*Moderate Duration.* Moderate duration lifting tasks are those that have a duration of more than 1 h but not more than 2 h, followed by a recovery period of at least 0.3 times the work time [i.e., at least a 0.3 recovery time to work time ratio (RT/WT)].

For example, a worker who continuously lifts for 2 h would need at least a 36-minute recovery period before initiating a subsequent lifting session. If the recovery time requirement is not met, and a subsequent lifting session is required, then the total work time must be added together. If the total work time exceeds 2 h, then the job must be classified as a long duration lifting task.

*Long Duration.* Long duration lifting tasks are those that have a duration of between 2 and 8 h, with standard industrial rest allowances (e.g., morning, lunch, and afternoon rest breaks). *Note*: No weight limits are provided for more than 8 h of work.

The difference in the required RT/WT ratio for the short (<1 hour) duration category, which is 1.2, and the moderate (1 to 2 h) duration category, which is 0.3, is due to the difference in the magnitudes of the frequency multiplier values associated with each of the duration categories. Since the moderate category results in larger reductions in the RWL than the short category, there is less need for a recovery period between sessions than for the short duration category. In other words, the short duration category would result in higher weight limits than the moderate duration category, so larger recovery periods would be needed.

### Frequency Restrictions

Lifting frequency (*F*) for repetitive lifting may range from 0.2 lifts/min to a maximum frequency that is dependent on the vertical location of the object (*V*) and the duration of lifting (Table 16.5). Lifting above the maximum frequency results in an RWL of 0.0 (except for the special case of discontinuous lifting discussed above, where the maximum frequency is 15 lifts/min).

### Frequency Multiplier

The FM value depends upon the average number of lifts/min (*F*), the vertical location (*V*) of the hands at the origin, and the duration of continuous lifting. For lifting tasks with a frequency less than 0.2 lifts/min, set the frequency equal to 0.2 lifts/minute. Otherwise, the FM is determined from Table 16.5.

## Coupling Component

### Definition and Measurement

The nature of the hand-to-object coupling or gripping method can affect not only the maximum force a worker can or must exert on the object, but also the vertical location of the hands during the lift. A "good" coupling will reduce the maximum grasp forces required and increase the acceptable weight for lifting, while a "poor" coupling will generally require higher maximum grasp forces and decrease the acceptable weight for lifting.

The effectiveness of the coupling is not static, but may vary with the distance of the object from the ground, so that a good coupling could become a poor coupling during a single lift. The entire range of

**TABLE 16.6**  Hand-to-Container Coupling Classification

| Good | Fair | Poor |
|---|---|---|
| 1. For containers of optimal design, such as some boxes, crates, etc., a "Good" hand-to-object coupling would be defined as handles or hand-hold cut-outs of optimal design [see notes 1 to 3 below]. | 1. For containers of optimal design, a "Fair" hand-to-object coupling would be defined as handles or hand-hold cut-outs of less than optimal design [see notes 1 to 4 below]. | 1. Containers of less than optimal design or loose parts or irregular objects that are bulky, hard to handle, or have sharp edges [see note 5 below]. |
| 2. For loose parts or irregular objects, which are not usually containerized, such as castings, stock, and supply materials, a "Good" hand-to-object coupling would be defined as a comfortable grip in which the hand can be easily wrapped around the object [see note 6 below]. | 2. For containers of optimal design with no handles or hand-hold cut-outs or for loose parts or irregular objects, a "Fair" hand-to-object coupling is defined as a grip in which the hand can be flexed about 90 degrees [see note 4 below]. | 2. Lifting non-rigid bags (i.e., bags that sag in the middle). |

1. An optimal handle design has .75 to 1.5 inches (1.9 to 3.8 cm) diameter, ≥4.5 inches (11.5 cm) length, 2 inches (5 cm) clearance, cylindrical shape, and a smooth, nonslip surface.
2. An optimal hand-hold cut-out has the following approximate characteristics: ≥1.5 inch (3.8 cm) height, 4.5 inch (11.5 cm) length, semi-oval shape, ≥2 inch (5 cm) clearance, smooth nonslip surface, and ≥ 0.25 inches (0.60 cm) container thickness (e.g., double thickness cardboard).
3. An optimal container design has ≤16 inches (40 cm) frontal length, ≤12 inches (30 cm) height, and a smooth nonslip surface.
4. A worker should be capable of clamping the fingers at nearly 90° under the container, such as required when lifting a cardboard box from the floor.
5. A container is considered less than optimal if it has a frontal length >16 inches (40 cm), height >12 inches (30 cm), rough or slippery surfaces, sharp edges, asymmetric center of mass, unstable contents, or requires the use of gloves.
6. A worker should be able to comfortably wrap the hand around the object without causing excessive wrist deviations or awkward postures, and the grip should not require excessive force.

the lift should be considered when classifying hand-to-object couplings, with classification based on overall effectiveness. The analyst must classify the coupling as good, fair, or poor. The three categories are defined in Table 16.6. If there is any doubt about classifying a particular coupling design, the more stressful classification should be selected.

The decision tree shown in Figure 16.3 may be helpful in classifying the hand-to-object coupling.

### Limitations

There are no limitations for classifying the coupling, but both hands must be observed when assessing the coupling. If one hand is predominately used to lift the load, and the fingers are flexed at 90° under the load, then the coupling should be rated as fair, regardless of the position of the other hand.

### Coupling Multiplier

Based on the coupling classification and vertical location of the lift, the Coupling Multiplier (CM) is determined from Table 16.7.

## 16.5  Procedures

Prior to data collection, the analyst must decide (1) if the job should be analyzed as a single-task or multitask manual lifting job, and (2) if significant control is required at the destination of the lift. This is necessary because the procedures differ according to the type of analysis required.

A manual lifting job may be analyzed as a single-task job if the task variables do not differ from task to task, or if only one task is of interest (e.g., single most stressful task). This may be the case if one of the tasks clearly has a dominant effect on strength demands, localized muscle fatigue, or whole-body

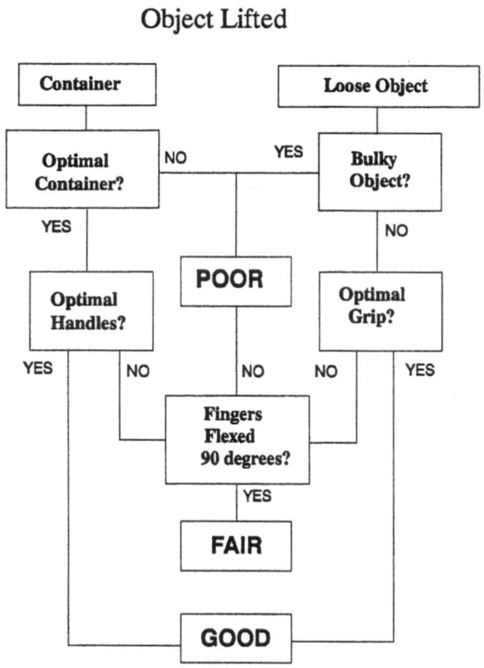

## Object Lifted

**FIGURE 16.3**   Decision tree for coupling quality.

**TABLE 16.7**   Coupling Multiplier

| | Coupling Multiplier | |
| --- | --- | --- |
| Coupling Type | V < 30 inches (75 cm) | V ≥ 30 inches (75 cm) |
| Good | 1.00 | 1.00 |
| Fair | 0.95 | 1.00 |
| Poor | 0.90 | 0.90 |

fatigue. On the other hand, if the task variables differ significantly between tasks, it may be more appropriate to analyze a job as a multitask manual lifting job. A multitask analysis is more difficult to perform than a single-task analysis because additional data and computations are required. The multitask approach, however, will provide more detailed information about specific strength and physiological demands.

For many lifting jobs, it may be acceptable to use either the single- or multitask approach. The single-task analysis should be used when possible, but when a job consists of more than one task and detailed information is needed to specify engineering modifications, then the multitask approach provides a reasonable method of assessing the overall physical demands. The multitask procedure is more compli-cated than the single-task procedure, and requires a greater understanding of assessment terminology and mathematical concepts. Therefore, the decision to use the single- or multitask approach should be based on: (1) the need for detailed information about all facets of the multitask lifting job, (2) the need for accuracy and completeness of data regarding assessment of the physiological demands of the task, and (3) the analyst's level of understanding of the assessment procedures.

The decision about control at the destination is important because the physical demands on the worker may be greater at the destination of the lift than at the origin, especially when significant control is required. When significant control is required at the destination, for example, the physical stress is increased because the load will have to be accelerated upward to slow down its descent. This acceleration

## JOB ANALYSIS WORKSHEET

**DEPARTMENT** _____   **JOB DESCRIPTION**
**JOB TITLE** _____   _____
**ANALYST'S NAME** _____   _____
**DATE** _____

### STEP 1. Measure and record task variables

| Object Weight (lbs) | | Hand Location (in) | | | | Vertical Distance (in) | Asymmetric Angle (degrees) | | Frequency Rate lifts/min | Duration (HRS) | Object Coupling |
| | | Origin | | Dest. | | | Origin | Destination | | | |
| L (AVG.) | L (Max.) | H | V | H | V | D | A | A | F | | C |
| | | | | | | | | | | | |

### STEP 2. Determine the multipliers and compute the RWL's

$$RWL = LC \times HM \times VM \times DM \times AM \times FM \times CM$$

**ORIGIN**      RWL = 51 × ☐ × ☐ × ☐ × ☐ × ☐ × ☐ = [        ] **Lbs**

**DESTINATION**  RWL = 51 × ☐ × ☐ × ☐ × ☐ × ☐ × ☐ = [        ] **Lbs**

### STEP 3. Compute the LIFTING INDEX

**ORIGIN**      LIFTING INDEX = $\dfrac{\text{OBJECT WEIGHT (L)}}{\text{RWL}}$ = ———— = [     ]

**DESTINATION**  LIFTING INDEX = $\dfrac{\text{OBJECT WEIGHT (L)}}{\text{RWL}}$ = ———— = [     ]

**FIGURE 16.4**   Single task job analysis worksheet.

may be as great as the acceleration at the origin of the lift and may create high loads on the spine. Therefore, if significant control is required, then the RWL and LI should be determined at both locations and the lower of the two values will specify the overall level of physical demand.

To perform a lifting analysis using the revised lifting equation, two steps are undertaken: (1) data are collected at the worksite as described in step 1 below; and, (2) the recommended weight limit and lifting index values are computed using the single-task or multitask analysis procedure described in step 2 below.

## Step 1: Collect Data

The relevant task variables must be carefully measured and clearly recorded in a concise format. As mentioned previously, these variables include the horizontal location of the hands ($H$), vertical location of the hands ($V$), vertical displacement ($D$), asymmetric angle ($A$), lifting frequency ($F$), and coupling quality ($C$). A job analysis worksheet, as shown in Figure 16.4 for single-task jobs or Figure 16.5 for multitask jobs, provides a simple form for recording the task variables and the data needed to calculate the RWL and the LI values. A thorough job analysis is required to identify and catalog each independent lifting task in the worker's complete job. For multitask jobs, data must be collected for each task.

## Step 2: Single-Task Procedure

For a single-task analysis, step 2 consists of computing the recommended weight limit (RWL) and the lifting index (LI). This is accomplished as follows.

Calculate the RWL at the origin for each lift. For lifting tasks that require significant control at the destination, calculate the RWL at *both* the origin and the destination of the lift. The latter procedure is required if (1) the worker has to regrasp the load near the destination of the lift, (2) the worker has to momentarily hold the object at the destination, or (3) the worker has to position or guide the load at

# MULTI-TASK JOB ANALYSIS WORKSHEET

DEPARTMENT _____    JOB DESCRIPTION _____

JOB TITLE _____

ANALYST'S NAME _____

DATE _____

## STEP 1. Measure and Record Task Variable Data

| Task No. | Object Weight (lbs) | | Hand Location (in) | | | | Vertical Distance (in) | Asymmetry Angle (degs) | | Frequency Rate lifts/min | Duration Hrs | Coupling |
|---|---|---|---|---|---|---|---|---|---|---|---|---|
| | L (Avg.) | L (Max.) | Origin H | V | Dest. H | V | D | Origin A | Dest. A | F | | C |
| | | | | | | | | | | | | |
| | | | | | | | | | | | | |
| | | | | | | | | | | | | |
| | | | | | | | | | | | | |
| | | | | | | | | | | | | |

## STEP 2. Compute multipliers and FIRWL, STRWL, FILI, and STLI for Each Task

| Task No. | LC x HM x VM x DM x AM x CM | FIRWL x FM | STRWL | FILI = L/FIRWL | STLI = L/STRWL | New Task No. | F |
|---|---|---|---|---|---|---|---|
| 51 | | | | | | | |
| 51 | | | | | | | |
| 51 | | | | | | | |
| 51 | | | | | | | |
| 51 | | | | | | | |

## STEP 3. Compute the Composite Lifting Index for the Job    (After renumbering tasks)

CLI = STLI$_1$ + $\triangle$FILI$_2$ + $\triangle$FILI$_3$ + $\triangle$FILI$_4$ + $\triangle$FILI$_5$

| FILI$_2$(1/FM$_{1,2}$ - 1/FM$_1$) | FILI$_3$(1/FM$_{1,2,3}$ - 1/FM$_{1,2}$) | FILI$_4$(1/FM$_{1,2,3,4}$ - 1/FM$_{1,2,3}$) | FILI$_5$(1/FM$_{1,2,3,4,5}$ - 1/FM$_{1,2,3,4}$) |
|---|---|---|---|
| | | | |

CLI =

FIGURE 16.5    Multitask job analysis worksheet.

the destination. The purpose of calculating the RWL at both the origin and destination of the lift is to identify the most stressful location of the lift. Therefore, the lower of the RWL values at the origin or destination should be used to compute the LI for the task, as this value would represent the limiting set of conditions.

The assessment is completed on the single-task worksheet by determining the LI for the task of interest. This is accomplished by comparing the actual weight of the load (*L*) lifted with the RWL value obtained from the lifting equation.

## Step 2: Multitask Procedure

For a multitask analysis, step 2 comprises three steps:

1. Compute the frequency-independent recommended weight limit (FIRWL) and single-task recommended weight limit (STRWL) for each task.
2. Compute the frequency-independent lifting index (FILI) and single-task lifting index (STLI) for each task.
3. Compute the composite lifting index (CLI) for the overall job.

*Compute the Frequency-Independent Recommended Weight Limits (FIRWLs).* Compute the FIRWL value for each task by using the respective task variables and setting the frequency multiplier (FM) to a value of 1.0. The FIRWL for each task reflects the compressive force and muscle strength demands for a single repetition of that task. If significant control is required at the destination for any individual task, the FIRWL must be computed at both the origin and the destination of the lift, as described above for a single-task analysis.

*Compute the Single-Task Recommended Weight Limit (STRWL).* Compute the STRWL for each task by multiplying its FIRWL by its appropriate FM value. The STRWL for a task reflects the overall demands of that task, assuming it was the only task being performed. *Note*: This value does not reflect the overall demands of the task when the other tasks are considered. Nevertheless, it is helpful in determining the extent of excessive physical stress for an individual task.

*Compute the Frequency-Independent Lifting Index (FILI).* The FILI is computed for each task by dividing the *maximum* load weight (*L*) for that task by the respective FIRWL. The maximum weight is used to compute the FILI because the maximum weight determines the maximum biomechanical loads to which the body will be exposed, regardless of the frequency of occurrence. Thus, the FILI can identify individual tasks with potential strength problems for infrequent lifts. If any of the FILI values exceed a value of 1.0, then job design changes may be needed to decrease the strength demands.

*Compute the Single-Task Lifting Index (STLI).* The STLI is computed for each task by dividing the *average* load weight (*L*) for that task by the respective STRWL. The average weight is used to compute the STLI because the average weight provides a better representation of the metabolic demands, which are distributed across the tasks, rather than being dependent on individual tasks. The STLI can be used to identify individual tasks with excessive physical demands (i.e., tasks that would result in fatigue). The STLI values do not indicate the relative stress of the individual tasks in the context of the whole job, but they can be used to prioritize the individual tasks according to the magnitude of their physical stress. Thus, if any of the STLI values exceed a value of 1.0, then ergonomic changes may be needed to decrease the overall physical demands of the task. *Note*: It may be possible to have a job in which all of the individual tasks have an STLI less than 1.0 and yet is physically demanding due to the combined demands of the tasks. In cases where the FILI exceeds the STLI for any task, the maximum weights may represent a significant problem and careful evaluation is necessary.

*Compute the Composite Lifting Index (CLI).* The assessment is completed on the multitask worksheet by determining the composite lifting index (CLI) for the overall job. The CLI is computed as follows:

1. The tasks are renumbered in order of decreasing physical stress, from the task with the greatest STLI down to the task with the smallest STLI. The tasks are renumbered in this way so that the more difficult tasks are considered first.

**TABLE 16.8**   Computations from Multitask Example

| Task # | Load Weight (L) | Task Frequency (F) | FIRWL | FM | STRWL | FILI | STLI | New Task # |
|--------|-----------------|--------------------|-------|-----|-------|------|------|------------|
| 1 | 30 | 1 | 20 | .94 | 18.8 | 1.5 | 1.6 | 1 |
| 2 | 20 | 2 | 20 | .91 | 18.2 | 1.0 | 1.1 | 2 |
| 3 | 10 | 4 | 15 | .84 | 12.6 | .67 | .8 | 3 |

2. The CLI for the job is then computed according to the following formula:

$$CLI = STLI_1 + \sum \Delta LI$$

where:

$$\sum \Delta LI = \left( FILI_2 \ X \left( \frac{1}{FM_{1,2}} - \frac{1}{FM_1} \right) \right)$$

$$+ \left( FILI_3 \ X \left( \frac{1}{FM_{1,2,3}} - \frac{1}{FM_{1,2}} \right) \right)$$

$$+ \left( FILI_4 \ X \left( \frac{1}{FM_{1,2,3,4}} - \frac{1}{FM_{1,2,3}} \right) \right)$$

$$\cdot$$
$$\cdot$$
$$\cdot$$

$$+ \left( FILI_n \ X \left( \frac{1}{FM_{1,2,3,4,...,n}} - \frac{1}{FM_{1,2,3,...,(n-1)}} \right) \right)$$

*Note:* (1) The numbers in the subscripts refer to the new task numbers; and, (2) the FM values are determined from Table 16.5, based on the sum of the frequencies for the tasks listed in the subscripts.

*An Example.* The following example is provided to demonstrate this step of the multi- task procedure. Assume that an analysis of a typical three-task job provided the results shown in Table 16.8.

To compute the Composite Lifting Index (CLI) for this job, the tasks are renumbered in order of decreasing physical stress, beginning with the task with the greatest STLI. In this case, as shown in Table 16.8, the task numbers do not change. Next, the CLI is computed according to the formula shown above. The task with the greatest CLI is Task 1 (STLI = 1.6). The sum of the frequencies for Tasks 1 and 2 is 1+2, or 3, and the sum of the frequencies for Tasks 1, 2 and 3 is 1+2+4, or 7. Then, from Table 16.5, $FM_1$ is 0.94, $FM_{1,2}$ is 0.88, and $FM_{1,2,3}$ is 0.70. Finally, the CLI = 1.6 + 1.0(1/.88 − 1/.94)+.67(1/.70 − 1/.88) = 1.6 + .07 + .20 = 1.9. *Note:* The FM values were based on the sum of the frequencies for the subscripts, the vertical height, and the duration of lifting.

# 16.6 Applying the Equations

## Using the RWL and LI to Guide Ergonomic Design

The recommended weight limit (RWL) and lifting index (LI) can be used to guide ergonomic design in several ways.

1. The individual multipliers can be used to identify specific job-related problems. The relative magnitude of each multiplier indicates the relative contribution of each task factor (e.g., horizontal, vertical, frequency, etc.).
2. The RWL can be used to guide the redesign of existing manual lifting jobs or to design new manual lifting jobs. For example, if the task variables are fixed, then the maximum weight of the load could be selected so the RWL is not exceeded; if the weight is fixed, then the task variables could be optimized so that the weight does not exceed the RWL.
3. The LI can be used to estimate the relative magnitude of physical stress for a task or job. The greater the LI, the smaller the fraction of workers capable of safely sustaining the level of activity. Thus, two or more job designs could be compared.
4. The LI can be used to prioritize ergonomic redesign. For example, a series of suspected hazardous jobs could be rank ordered according to the LI and a control strategy could be developed according to the rank ordering (i.e., jobs with lifting indices above 1.0 or higher would benefit the most from redesign).

## Rationale and Limitations for LI

The NIOSH recommended weight limit (RWL) and lifting index (LI) equations are based on the concept that the risk of lifting-related low back pain increases as the demands of the lifting task increase. In other words, as the magnitude of the LI increases, (1) the level of the risk for a given worker would be increased and (2) a greater percentage of the workforce is likely to be at risk for developing lifting-related low back pain. The shape of the risk function, however, is not known. Without additional data showing the relationship between low back pain and the LI, it is impossible to predict the magnitude of the risk for a given individual or the exact percent of the work population who would be at an elevated risk for low back pain.

To gain a better understanding of the rationale for the development of the RWL and LI, consult Waters et al.,[1] which provides a discussion of the criteria underlying the lifting equation and of the individual multipliers and identifies both the assumptions and uncertainties in the scientific studies that associate manual lifting and low back injuries.

## Job-Related Intervention Strategy

The lifting index may be used to identify potentially hazardous lifting jobs or to compare the relative severity of two jobs for the purpose of evaluating and redesigning them. From the NIOSH perspective, it is likely that lifting tasks with an LI > 1.0 pose an increased risk for lifting-related low back pain for some fraction of the workforce.[1] Hence, to the extent possible, lifting jobs should be designed to achieve an LI of 1.0 or less.

Some experts believe, however, that worker selection criteria may be used to identify workers who can perform potentially stressful lifting tasks (i.e., lifting tasks that would exceed an LI of 1.0) without significantly increasing their risk of work-related injury above the baseline level.[6,7] Those who endorse the use of selection criteria believe that the criteria must be based on research studies, empirical observations, or theoretical considerations that include job-related strength testing and/or aerobic capacity testing. Even these experts agree, however, that many workers will be at a significant risk of a work-related injury when performing highly stressful lifting tasks (i.e., lifting tasks that would exceed an LI of

**TABLE 16.9**    General Design/Redesign Suggestions

| | |
|---|---|
| If HM is less than 1.0 | Bring the load closer to the worker by removing any horizontal barriers or reducing the size of the object. Lifts near the floor should be avoided; if unavoidable, the object should fit easily between the legs. |
| If VM is less than 1.0 | Raise/lower the origin/destination of the lift. Avoid lifting near the floor or above the shoulders. |
| If DM is less than 1.0 | Reduce the vertical distance between the origin and the destination of the lift. |
| If AM is less than 1.0 | Move the origin and destination of the lift closer together to reduce the angle of twist, or move the origin and destination farther apart to force the worker to turn the feet and step, rather than twist the body. |
| If FM is less than 1.0 | Reduce the lifting frequency rate, reduce the lifting duration, or provide longer recovery periods (i.e., light work period). |
| If CM is less than 1.0 | Improve the hand-to-object coupling by providing optimal containers with handles or hand-hold cut-outs, or improve the hand-holds for irregular objects. |
| If the RWL at the destination is less than at the origin | Eliminate the need for significant control of the object at the destination by redesigning the job or modifying the container/object characteristics. (See requirements for significant control in text.) |

3.0). Also, "informal" or "natural" selection of workers may occur in many jobs that require repetitive lifting tasks. According to some experts, this may result in a unique workforce that may be able to work above a lifting index of 1.0, at least in theory, without substantially increasing their risk of low back injuries above the baseline rate of injury.

## Example Problems

Two example problems are provided to demonstrate the proper application of the lifting equation and procedures. The procedures provide a method for determining the level of physical stress associated with a specific set of lifting conditions and assist in identifying the contribution of each job-related factor. The examples also provide guidance in developing an ergonomic redesign strategy. Specifically, for each example, a job description, job analysis, hazard assessment, redesign suggestion, illustration, and completed worksheet are provided.

A series of general design/redesign suggestions for each job-related risk factor are provided in Table 16.9. These suggestions can be used to develop a practical ergonomic design/redesign strategy.

### Loading Bags Into A Hopper, Example 1

*Job Description.* The worker positions himself midway between the handtruck and the mixing hopper, as illustrated in Figure 16.6. Without moving his feet, he twists to the right and picks up a bag off the handtruck. In one continuous motion he then twists to his left to place the bag on the rim of the hopper. A sharp-edged blade within the hopper cuts open the bag to allow the contents to fall into the hopper. This task is done infrequently (i.e., 1 to 12 times per shift) with large recovery periods between lifts (i.e., > 1.2 Recovery Time/Work Time ratio). In observing the worker perform the job, it was determined that the nonlifting activities could be disregarded because they require minimal force and energy expenditure. Significant control is not required at the destination, but the worker twists at the origin and destination of the lift. Although several bags are stacked on the hand truck, the highest risk of overexertion injury is associated with the bag on the bottom of the stack; therefore, only the lifting of the bottom bag will be examined. Note, however, that the frequency multiplier is based on the overall frequency of lifting for all of the bags.

*Job Analysis.* The task variable data are measured and recorded on the job analysis worksheet (Figure 16.7). The vertical location of the hands is 15 inches at the origin and 36 inches at the destination. The horizontal location of the hands is 18 inches at the origin and 10 inches at the destination. The asymmetric angle is 45° at the origin and 45° at the destination of the lift, and the frequency is less than .2 lifts/min for less than 1 hour (see Table 16.5).

Using Table 16.6, the coupling is classified as fair because the worker can flex the fingers about 90° and the bags are semi-rigid (i.e., they do not sag in the middle). Significant control of the object is not

**FIGURE 16.6**  Loading bags into hopper, example 1.

## JOB ANALYSIS WORKSHEET

**DEPARTMENT** Manufacturing
**JOB TITLE** Batch Processor
**ANALYST'S NAME**
**DATE**

**JOB DESCRIPTION**
Dumping bags into mixing hopper
Example 1

### STEP 1. Measure and record task variables

| Object Weight (lbs) | | Hand Location (in) | | | | Vertical Distance (in) | Asymmetric Angle (degrees) | | Frequency Rate lifts/min | Duration (HRS) | Object Coupling |
|---|---|---|---|---|---|---|---|---|---|---|---|
| | | Origin | | Dest. | | | Origin | Destination | | | |
| L (AVG.) | L (Max.) | H | V | H | V | D | A | A | F | | C |
| 40 | 40 | 18 | 15 | 10 | 36 | 21 | 45 | 45 | <.2 | <1 | Fair |

### STEP 2. Determine the multipliers and compute the RWL's

RWL = LC × HM × VM × DM × AM × FM × CM

**ORIGIN** RWL = $\boxed{51}$ × $\boxed{.56}$ × $\boxed{.89}$ × $\boxed{.91}$ × $\boxed{.86}$ × $\boxed{1.0}$ × $\boxed{.95}$ = $\boxed{18.9}$ **Lbs**

**DESTINATION** RWL = $\boxed{51}$ × $\boxed{\phantom{x}}$ × $\boxed{\phantom{x}}$ × $\boxed{\phantom{x}}$ × $\boxed{\phantom{x}}$ × $\boxed{\phantom{x}}$ = **Lbs**

### STEP 3. Compute the LIFTING INDEX

**ORIGIN** LIFTING INDEX = $\dfrac{\text{OBJECT WEIGHT (L)}}{\text{RWL}}$ = $\dfrac{40}{18.9}$ = $\boxed{2.1}$

**DESTINATION** LIFTING INDEX = $\dfrac{\text{OBJECT WEIGHT (L)}}{\text{RWL}}$ = —— =

**FIGURE 16.7**  Job analysis worksheet, example 1.

required at the destination of the lift so the RWL is computed only at the origin. The multipliers are computed from the lifting equation or determined from the multiplier tables (Tables 16.1 to 16.5, and Table 16.7). As shown in Figure 16.7, the RWL for this activity is 18.9 lbs.

*Hazard Assessment.* The weight to be lifted (40 lbs) is greater than the RWL (18.9 lbs). Therefore, the LI is 40/18.9 or 2.1. This job would be physically stressful for many industrial workers.

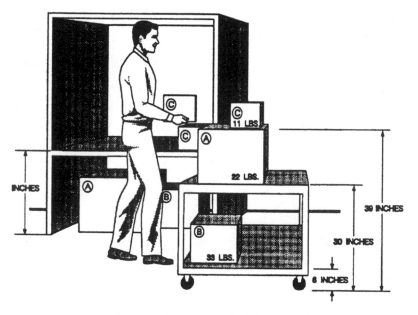

**FIGURE 16.8**   Warehouse order filling, example 2.

*Redesign Suggestions.* The worksheet shows that the smallest multipliers (i.e., the greatest penalties) are .56 for the HM, .86 for the AM, and .89 for the VM. Using Table 16.8, the following job modifications are suggested:

1. Bring the load closer to the worker to increase the HM.
2. Reduce the angle of asymmetry to increase AM. This could be accomplished either by moving the origin and destination points closer together or farther apart.
3. Raise the height at the origin to increase the VM.

If the worker could get closer to the bag before lifting, the H value could be decreased to 10 inches, which would increase the HM to 1.0, the RWL would be increased to 33.7 lbs, and the LI would be decreased to 1.2 (i.e., 40/33.7).

*Comments.* This example demonstrates that certain lifting jobs may be evaluated as a single-task or multitask job. In this case, only the most stressful component of the job was evaluated. For repetitive lifting jobs, the multitask approach may be more appropriate.

## Warehouse Order Filling, Example 2

*Job Description.* A worker lifts cartons of various sizes from supply shelves onto a cart as illustrated in Figure 16.8. There are three box sizes (i.e., A, B, and C) of various weights. These lifting tasks are typical in warehousing, shipping, and receiving activities in which loads of varying weights and sizes are lifted at different frequencies. Assume that the following observations were made: (1) control of the load is not required at the destination of any lift; (2) the worker does not twist when picking up and putting down the cartons; (3) the worker can get close to each carton; and, (4) walking and carrying are minimized by keeping the cart close to the shelves.

*Job Analysis.* Since the job consists of more than one distinct task and the task variables often change, the multitask lifting analysis procedure should be used. This job can be divided into three tasks represented by cartons A, B, and C. The following measurements were made and recorded on the job analysis worksheet (Figure 16.9):

1. The horizontal locations (H) for each task at the origin and destination are as follows: Box A, 16 inches; Box B, 12 inches; and, Box C, 8 inches.
2. The vertical locations (V) at the origin are taken to be the position of the hands under the cartons as follows: Box A, 0 inches; Box B, 0 inches; and, Box C, 30 inches.

## MULTI-TASK JOB ANALYSIS WORKSHEET

**DEPARTMENT** Warehouse
**JOB TITLE** Shipping Clerk
**ANALYST'S NAME**
**DATE**

**JOB DESCRIPTION**
Selecting an order for shipment
Warehouse order filling
Example 2

### STEP 1. Measure and Record Task Variable Data

| Task No. | Object Weight (lbs) L (Avg.) | L (Max.) | Hand Location (in) Origin H | V | Dest. H | V | Vertical Distance (in) D | Asymmetry Angle (degs) Origin A | Dest. A | Frequency Rate lifts/min F | Duration Hrs | Coupling C |
|---|---|---|---|---|---|---|---|---|---|---|---|---|
| 1 (A) | 22 | 33 | 16 | 0 | 16 | 30 | 30 | 0 | 0 | 1 | 8 | Fair |
| 2 (B) | 33 | 44 | 12 | 0 | 12 | 6 | 6 | 0 | 0 | 2 | 8 | Fair |
| 3 (C) | 11 | 22 | 8 | 30 | 8 | 39 | 9 | 0 | 0 | 5 | 8 | Fair |
| | | | | | | | | | | | | |
| | | | | | | | | | | | | |

### STEP 2. Compute multipliers and FIRWL, STRWL, FILI, and STLI for Each Task

| Task No. | LC | HM | VM | DM | AM | CM | FIRWL | FM | STRWL | FILI = L/FIRWL | STLI = L/STRWL | New Task No. | F |
|---|---|---|---|---|---|---|---|---|---|---|---|---|---|
| 1 | 51 | .63 | .78 | .88 | 1.0 | .95 | 21.0 | .75 | 15.8 | 1.6 | 1.4 | 2 | 1 |
| 2 | 51 | .83 | .78 | 1.0 | 1.0 | .95 | 31.4 | .65 | 20.4 | 1.4 | 1.6 | 1 | 2 |
| 3 | 51 | 1.0 | 1.0 | 1.0 | 1.0 | 1.0 | 51.0 | .35 | 17.8 | .4 | .6 | 3 | 5 |
| | 51 | | | | | | | | | | | | |
| | 51 | | | | | | | | | | | | |

### STEP 3. Compute the Composite Lifting Index for the Job (After renumbering tasks)

$$CLI = STLI_1 + \Delta FILI_2 + \Delta FILI_3 + \Delta FILI_4 + \Delta FILI_5$$

| | $FILI_2(1/FM_{1,2} - 1/FM_1)$ | $FILI_3(1/FM_{1,2,3} - 1/FM_{1,2})$ | $FILI_4(1/FM_{1,2,3,4} - 1/FM_{1,2,3})$ | $FILI_5(1/FM_{1,2,3,4,5} - 1/FM_{1,2,3,4})$ | |
|---|---|---|---|---|---|
| | 1.6(1/.55-1/.65) | .4(1/.18-1/.55) | | | |
| CLI = 1.6 | .45 | 1.5 | | | 3.6 |

**FIGURE 16.9** Job analysis worksheet, example 2.

3. The vertical locations (V) at the destination are the vertical position on the cart as follows: Box A, 30 inches; Box B, 6 inches; and, Box C, 39 inches.
4. The average weights lifted for each task are as follows: Box A, 22 lbs; Box B, 33 lbs; and, Box C, 11 lbs.
5. The maximum weights lifted for each task are as follows: Box A, 33 lbs; Box B, 44 lbs; and, Box C, 22 lbs.
6. No asymmetric lifting is involved (i.e., A = 0).
7. The lifting frequency rates for each task are as follows: Box A, 1 lift/min; Box B, 2 lifts/min; and Box C, 5 lifts/min.
8. The lifting duration for the job is 8 hours, however, the maximum weights are lifted infrequently (i.e., less than or equal to once every 5 minutes for 8 hours).
9. Using Table 16.6, the couplings are classified as fair.

The multitask lifting analysis consists of the following three steps:

1. Compute the frequency-independent-RWL (FIRWL) and frequency-independent-lifting index (FILI) values for each task using a default FM of 1.0.
2. Compute the single-task-RWL (STRWL) and single-task-lifting index (STLI) for each task.
3. Renumber the tasks in order of decreasing physical stress, as determined from the STLI value, starting with the task with the largest STLI.

**Step 1** — Compute the FIRWL and FILI values for each task using a default FM of 1.0. The other multipliers are computed from the lifting equation or determined from the multiplier tables (Tables 16.1 to 16.5, and Table 16.7). Recall, that the FILI is computed for each task by dividing the *maximum* weight of that task by its FIRWL.

|        |   | FIRWL    | FILI |
|--------|---|----------|------|
| Task 1 | = | 21.0 lbs | 1.6  |
| Task 2 | = | 31.4 lbs | 1.4  |
| Task 3 | = | 51.0 lbs | .4   |

These results indicate that *some of the tasks requires excessive strength*. Remember, however, that these results do not take the frequency of lifting into consideration.

**Step 2** — Compute the STRWL and STLI values for each task, where the STRWL for a task is equivalent to the product of the FIRWL and the FM for that task. Recall, that the STILI is computed for each task by dividing the *average* weight of that task by its STRWL. The appropriate FM values are determined from Table 16.5.

|        |   | STRWL    | STLI |
|--------|---|----------|------|
| Task 1 | = | 15.8 lbs | 1.4  |
| Task 2 | = | 20.4 lbs | 1.6  |
| Task 3 | = | 17.8 lbs | .6   |

These results indicate that Tasks 1 and 2 *would be stressful* for some workers, if performed individually. Note, however, that these values do not consider the combined effects of all of the tasks.

**Step 3** — Renumber the tasks, starting with the task with the largest STLI value, and ending with the task with the smallest STLI value. If more than one task has the same STLI value, assign the lower task number to the task with the highest frequency.

*Hazard Assessment.* Compute the composite-lifting index (CLI) using the renumbered tasks. As shown in Figure 16.9, the CLI for this job is 3.6, which indicates that this job would be physically stressful for nearly all workers. Analysis of the results suggests that the combined effects of the three tasks are significantly more stressful than any individual task.

*Redesign Suggestions.* Developing a redesign strategy for a job depends on tangible and intangible factors that may be difficult to evaluate, including costs/benefits, feasibility, and practicality. No preferred procedure has been developed and tested. Therefore, the following suggestions represent only one approach to ergonomic job modification.

In this example, the magnitude of the FILI, STLI, and CLI values indicate that both strength and endurance would be a problem for many workers. Therefore, the redesign should attempt to decrease the physical demands by modifying the job layout and decrease the physiological demands by reducing the frequency rate or duration of continuous lifting. If the maximum weights were eliminated from the job, then the CLI would be significantly reduced, the job would be less stressful, and more workers could perform the job than before.

Those lifts with strength problems should be evaluated for specific engineering changes, such as (1) decreasing carton size or removing barriers to reduce the horizontal distance; (2) raising or lowering the origin of the lift; (3) reducing the vertical distance of the lift; improving carton couplings, and (4) decreasing the weight to be lifted. The redesign priority for this example is based on identifying interventions that provide the largest increase in the FIRWL for each task (Step 2 on worksheet). For example, the maximum weight lifted for carton A is unacceptable; however, if the carton at the origin were on the upper shelf, then the FIRWL for Task 1 would increase from 21.0 lbs to 27.0 lbs. The maximum weight lifted still exceeds the FIRWL, but lifts of average weight are now below the FIRWL. Additionally, providing handles, decreasing box size, or reducing the load to be lifted will decrease the stress of manual lifting.

*Comments.* This example demonstrates the complexity of analyzing multitask lifting jobs. Errors resulting from averaging, and errors introduced by ignoring other factors (e.g., walking, carrying, holding, pushing and pulling activities, and environmental stressors), can only be resolved with detailed biomechanical, metabolic, cardiovascular, and psychophysical evaluations.

Several important application principles are illustrated in this example:

1. The horizontal distance (H) for Task 3 was less than the 10.0 inches minimum. Therefore, H was set equal to 10 inches (i.e., multipliers must be less than or equal to 1.0).
2. The vertical travel distance (D) in Task 2 was less than the 10 inches minimum. Therefore, D was set equal to 10 inches.

## References

1. Waters, T.R., Putz-Anderson, V., Garg, A., and Fine, L.J., 1993, Revised NIOSH equation for the design and evaluation of manual lifting tasks. *Ergonomics.* Vol. 36(7), 749-776.
2. Waters, T.R., Putz-Anderson, V., and Garg, A., 1994, *Applications Manual for the Revised NIOSH Lifting Equation.* National Institute for Occupational Safety and Health, Technical Report. DHHS(NIOSH) Pub. No. 94-110. Available from the National Technical Information Service (NTIS). NTIS document number PB94-176930 (1-800-553-NTIS).
3. NIOSH, 1981, *Work Practices Guide for Manual Lifting*, NIOSH Technical Report No. 81-122 (U.S. Department of Health and Human Services, National Institute for Occupational Safety and Health, Cincinnati, OH).
4. NIOSH, 1991, *Scientific Support Documentation for the Revised 1991 NIOSH Lifting Equation: Technical Contract Reports, May 8, 1991* (U.S. Department of Health and Human Services, National Institute for Occupational Safety and Health, Cincinnati, OH). Available from the National Technical Information Service (NTIS No. PB-91-226-274).
5. ASPH/NIOSH, 1986, *Proposed National Strategies for the Prevention of Leading Work-Related Diseases and Injuries: Part 1.* Published by the Association of Schools of Public Health under a cooperative agreement with the National Institute for Occupational Safety and Health.
6. Eastman Kodak, 1986, *Ergonomic Design for People at Work, Vol. 2*, Van Nostrand Reinhold, New York.
7. Ayoub, M.M. and Mital, A. 1989, *Manual Materials Handling*, Taylor & Francis, London.
8. Chaffin, D.B. and Andersson, G.B.J. 1984, *Occupational Biomechanics*, John Wiley & Sons, New York.

# 17

# A Population-Based Load Threshold Limit (LTL) for Manual Lifting Tasks Performed by Males and Females

Waldemar Karwowski
*University of Louisville*

Paul Gaddie
*University of Louisville*

Renliu Jang
*University of Louisville*

Wook Gee Lee
*University of Louisville*

## 17.1 Introduction

In order to reduce the risk of low back pain and injuries due to manual lifting tasks, we need guidelines for load limits with respect to different percentiles of capable industrial population that are simple to understand by the practitioners, and easy to apply in practical settings. Unfortunately, the revised NIOSH (1991) method, which aims to protect 99% of males and 75% of females of the U.S. population, is quite cumbersome to use, and by design very restrictive in setting the acceptable load limits. This chapter introduces population-based load threshold limits (LTL) for evaluation of manual lifting tasks performed by males and females. The proposed limits are based on the simulation of the revised (1991) NIOSH lifting equation, and application of the threshold lifting index (scaling factors) developed in reference to the comprehensive database for manual lifting tasks reported by Snook and Ciriello (1991).

As described in Chapter 16 of this handbook, the National Institute for Occupational Safety and Health proposed the recommended weight limit (RWL) as the guide for evaluating manual tasks in industry (Waters et al., 1993). The revised lifting equation (RLE) and its predecessor (NIOSH, 1981) aim to simultaneously satisfy the biomechanical, physiological, and psychophysical criteria for setting limits in manual lifting tasks (Waters et al., 1993; Dempsey, 1998). The 1991 equation expands the previous guidelines, and it is claimed that it can be applied to a larger percentage of lifting tasks (Waters et al., 1993).

Because the RWL is intended to protect 90% of the mixed industrial working population (99% males and 75% females), it is designed for the least physically capable percent of the industrial workforce. Given that more capable populations may be available at the workplace to perform physically demanding lifting tasks (for example strong young males and females), the application of the RWL-based limits that accommodate the majority of mixed male/female population may result in some work inefficiency. As discussed by Waters et al. (1993), a critical point concerning the 1991 RLE concept is that the NIOSH perspective was very conservative in nature. Furthermore, as demonstrated by Karwowski (1992), an application of the 1991 RLE model leads to a fairly narrow range of RWL values.

According to Waters and Putz-Anderson (1998), the lifting index (LI) may be used to identify potentially hazardous lifting jobs. NIOSH believes that jobs with LI smaller than 1.0 would pose an increased risk for lifting-related injury to less than 25% of female and 1% of male industrial workers. However, it was also recognized, that worker selection criteria may be used to identify workers who can perform lifting tasks that would exceed an LI = 1.0 without significantly increasing their risk of work-related injury above the baseline level. It was also noted that the said selection criteria must be based on research studies, job-related strength testing, and/or aerobic capacity (Waters and Putz-Anderson, 1998). Although the risk function representing the relationship between the lifting index and probability of injury is mostly unknown today (for some preliminary data see Karwowski et al., 1994), some experts hypothesize that many workers would be at significant risk of a work-related injury when performing lifting jobs with the LI values above 3.0, which are classified as highly stressful tasks (Waters et al., 1993; Waters and Putz-Anderson, 1998). Such assumption is yet to be validated in the field.

## 17.2    Threshold RWL Values

Given the above discussion, until comprehensive epidemiological data that relate the LI values to the risk of lifting-related injury are developed, the paradigm of protecting most of the population represented by the RLE concept seems a plausible and desirable strategy. There is also, however, a pressing need, at least for some of the industrial applications, to identify lifting limits and the population of capable workers (both males and females), who are able to handle physically demanding jobs with a lifting index above 1.0 (but below 3.0). Such data could be developed with respect to the simulated threshold RWL (TRWL) values corresponding to different levels of lifting index.

The threshold RWL values were defined by Karwowski and Gaddie (1995) as those limiting values of RWL (corresponding to LI = 1.0) that would be expected to represent at least 99.5% of all of the potential industrial jobs. In other words, these values mean that only in 0.5% of all possible cases (lifting tasks), the expected recommended weight limit would exceed the specific threshold RWL value. From the practical standpoint that means that for the majority of industrial jobs (99.5%), the expected RWL would be at or below the specific TRWL value. This implies that when LI is set to 1.0 for task design or evaluation purposes, it can be expected that only half of one percent of the plausible industrial lifting tasks would have the recommended weight greater than the specific TRWL value. Assuming that the 0.5% is the additional safety factor, the TRWL indicates the maximum values of RWL that one would expect to get when using the RLE model for evaluating any of the real lifting jobs encountered in industry.

## 17.3    The 1991 Revised NIOSH Lifting Equation

Detailed description of the relevant terms utilized by the RLE (1991) is provided in Chapter 16. For the purpose of this discussion, it is noted that the recommended weight limit (RWL) is the product of the load constant (LC) and six multipliers (M) which account for seven risk factors, i.e., RWL (Kg) = LC *

HM * VM * DM * AM * FM * CM. The multipliers are defined in terms of the related risk factors, including the horizontal location (H), vertical location (V), vertical travel distance (D), frequency of lift (F), asymmetry angle (A), and coupling (C). The multipliers for frequency and coupling are defined using relevant tables. In addition to lifting frequency, the work duration and vertical distance factors are used to compute the frequency multiplier.

## 17.4 Computer Simulation of the NIOSH (1991) Lifting Equation

One way to investigate the practical implications of the RLE (1991) for industry is to determine the likely results of the equation when applying a realistic and practical range of values for the risk factors (Karwowski, 1992). Karwowski and Gaddie (1995) introduced a computer simulation modeling approach to derive the threshold RWL values for 90% of the mixed male/female population (i.e., 99% of male and 75% of female industrial workers). Shortly, a computer simulation was written in SLAM II, a Simulation Language for Alternative Modeling (Pritsker, 1986). The NIOSH 1991 Lifting Equation was modeled as the product of factor multipliers represented as attributes of an entity flowing through the network. The developed model is a sequence of operations which draws random values from each of the relevant factor distributions and tables, and then calculates the corresponding RWL values.

## 17.5 Summary of Simulated Characteristics of Lifting Factors and Multipliers

As much as possible, the probability distributions for the risk factors utilized by the RLE (1991) were designed to be representative of the real industrial workplaces (Ciriello et al., 1990; Brokaw, 1992; Karwowski and Brokaw, 1992; Marras et al., 1995). Except for the vertical travel distance factor, coupling and asymmetry multipliers, all factors were defined using lognormal or normal distributions. For all the factors defined as having lognormal distributions, the following procedure was used to adjust for the required range of real values whenever necessary. After the given distribution was generated, it was then adjusted for the range, i.e., truncated for values below the lower limit and above the upper limit. Specifically, the values outside the range were removed, and the new values that would fall within the desired range were generated. This process normalized the derived distributions to the area of 1.0.

### Horizontal Factor

The horizontal factor (H) represents the distance of the hands from the midpoint of the ankles, measured in centimeters at the origin and the destination of lift. This horizontal distance was assumed to be the maximum of the two positions, and that no intermediate position was greater than this maximum. The range of horizontal distance as specified by NIOSH (Waters et al., 1993) is from 25 cm to 63 cm, and is used to define the horizontal multiplier (HM). A lognormal distribution for H (mean value of 44 cm, standard deviation of 6 cm) was used to generate random values within the required range. These values were chosen because H = 44 cm is the midpoint of the range from 25 cm to 63 cm, and the standard deviation of 6 cm allows almost all the values generated to fall within this range. It should be noted that the lognormal distribution gives greater weight to the values which are located near the mean.

### Vertical Height Factor

The vertical height factor (V) indicates the vertical distance of the hands from the floor, with the range from the "floor position" (of 0 cm at a minimum) to the upper limit of vertical reach of 175 cm (Waters et al., 1993). This factor was defined using a lognormal distribution (mean = 100 cm, std = 25 cm). It should be noted that the vertical height factor is also used as an independent variable in the definition of vertical multiplier (VM), the coupling multiplier (CM), and the frequency multiplier (FM).

## Distance Multiplier

The vertical travel distance factor (D), i.e., the difference between the origin and destination of the lift, is used by NIOSH (Waters et al., 1993) to define the distance multiplier (DM). For the purpose of this simulation, the vertical travel distance (D) was assumed uniform over the interval of (0, [175-V]). Since the vertical distance traveled is the difference between the origin and the destination heights, this distance can be no greater than the one defined by the origin of lift and the maximum vertical reach limit. A variable called the maximum vertical distance (MAVED) was created based on the vertical factor (V). The travel distance factor (D) was then drawn from a uniform distribution ranging from 0 to MAVED. The MAVED variable was then modified in order to reset all values of D which were less than 25 cm to 25 cm. This procedure guaranteed that the DM was set at 1.0 for such D values.

## Asymmetry Factor

The asymmetry factor is defined for the (0 – 135 degrees) range of the upper body twisting angle. This factor was modeled using a lognormal distribution (45, 15), i.e., the mean of 45° and a standard deviation of 15°. This mean and standard deviation provide a natural range from 0° to 90° of twist for the underlying distribution. The mean of 45 degrees was chosen based on the conservative assumption that in most of the industrial lifting tasks the upper body twisting does not exceed 90 degrees. The standard deviation was chosen to ensure that almost all the lifting posture asymmetry values will fall within the NIOSH-specified range limit.

## Work Duration Factor

The work duration factor is used to define the frequency multiplier (FM). For the purpose of this simulation, the work duration was conservatively modeled as a normal variable with the mean of 2.0 hours and standard deviation of 2.0 hours. This distribution was then adjusted for the range, i.e., truncated for values of work duration of zero or less, and those which were greater than 8 hours. For the purpose of simulation, the values outside the range of work duration were removed, and the new values that would fall within the desired range were generated. This process normalized the derived distribution to the area of 1.0. These random values were then used in the selection of the frequency multiplier from the NIOSH table for FM (Waters et al., 1993).

## Frequency Factor

The frequency variable F (lifts/minute), for the range of lifting frequency from 0.2 lifts/min to 15 lifts/min, is used in the definition of the frequency multiplier (FM). The frequency multipliers (FMs) were selected from the NIOSH table based on randomization of three factors: (1) the frequency of lift, (2) work duration, and (3) vertical height (V). Due to the discrete nature of the values for frequency of lift, the probability distribution function for this factor was generated using normal distribution. The parameters for this distribution were defined as follows: a mean of 4.7 lifts/min, and a standard deviation 1.75 lifts/min. These values were derived based on the information provided by results of industrial surveys of manual lifting tasks. For example, Ciriello et al. (1990) reported that 94% of industrial lifting tasks are performed at frequencies of 4.3 lifts/min or slower. Brokaw (1992) analyzed 31 industrial lifting tasks and reported that in about 80% of these tasks the observed frequency was lower than 5 lifts/min. For the purpose of this simulation, however, it was decided that better representation of the higher lifting frequencies is warranted by the fact that such frequencies of lifting tasks are common today in the service industry, and by the attention given to such tasks by NIOSH (1993), as reflected by their inclusion in the frequency multiplier table. The randomly generated values of the frequency of lift which were less than or equal to zero were eliminated, and the truncated normal distribution was renormalized to equate the area under the curve to 1.0. The resulting distribution was used to generate the probability density function values for entries from the frequency of lift table.

## Coupling Multiplier

Coupling describes the quality of coupling between the hand and the load. This quality has three linguistic values, i.e., *good*, *fair* and *poor*. The discrete values for the coupling multiplier (CM) were chosen based on the results of an industrial survey of 31 manual lifting tasks performed in three different companies reported by Brokaw (1992). This survey showed that slightly less than 10% of the observed lifting tasks had good couplings, about 64% had fair couplings, while about 26% had poor couplings. Due to the limited sample size of the above study, it was decided to allow for more liberal coupling assumption for the simulation purposes. Therefore, the probability of good hand-to-container coupling was set at 0.3, for fair coupling at 0.5, and the probability for poor coupling was set at 0.2.

## 17.6 Results of Simulation: Threshold Values of RWL

The results showed that overall, in conditions studied (represented by 100,000 randomly generated lifting task scenarios), the threshold RWL values (the RWL values which account for 99.5% of all simulated lifting tasks) were equal to or lower than: (1) 13.0 kg (or 28.6 lbs) for up to 1 hour of lifting, (2) 12.5 kg (or 26.4 lbs) for less than 2 hours of exposure, and (3) 10.5 kg (or 23.1 lbs) for lifting over an 8-hour shift. The above results indicate that under most of the examined lifting conditions (99.5% of the simulated cases), one can reasonably expect that an implementation of the RLE (1991) model at the level of Lifting Index of 1.0, which is designed to protect 90% of the mixed industrial working population, would necessitate redesign of those manual lifting tasks for which the load lifted exceeds the TRWL values reported above. From a practical point of view, these results define the threshold RWL values that can be used by practitioners for the purpose of immediate risk assessment of manual lifting tasks performed in industry.

## 17.7 Development of the Population-Based Design Load Threshold Limits

The next step in development of the population-based design load threshold limits (LTL) for manual lifting tasks was to define the threshold lifting index (TLI) values for different percentiles of capable male and female populations. This was done by using the scaling factors derived from the psychophysical data reported by Snook and Ciriello (1991), which account for different percentiles of the capable U.S. population. The scaling factors were defined in reference to 90% of male and 75% of female populations, at which their values were equated to 1.0 (e.g., LI = 1.0). These scaling factors (see Table 17.1) were derived based on the ratios between the 10th, 50th, and 90th percentile values of the maximum acceptable weights of lift (MAWL) under frequency of 1 lift per minute (box width

TABLE 17.1  Values of Threshold Lifting Index (TLI) for Male and Female Populations

| Percent capable | Males | | | Females | | |
|---|---|---|---|---|---|---|
| | V < 75 | V > 75 | SR | V < 75 | V > 75 | SR |
| 90% | 1.00 | 1.00 | 1.00 | 0.80 | 0.86 | 0.86 |
| 80% | 1.33 | 1.24 | 1.22 | 0.95 | 0.96 | 0.96 |
| 70% | 1.56 | 1.40 | 1.37 | 1.05 | 1.03 | 1.03 |
| 60% | 1.76 | 1.54 | 1.50 | 1.13 | 1.10 | 1.09 |
| 50% | 1.95 | 1.67 | 1.62 | 1.22 | 1.16 | 1.15 |
| 40% | 2.14 | 1.81 | 1.75 | 1.30 | 1.22 | 1.21 |
| 30% | 2.34 | 1.95 | 1.88 | 1.38 | 1.28 | 1.27 |
| 20% | 2.57 | 2.11 | 2.03 | 1.48 | 1.35 | 1.34 |
| 10% | 2.90 | 2.35 | 2.25 | 1.62 | 1.44 | 1.44 |

V = vertical height of lift (cm)
SR = shoulder-to-reach height

of 34 cm), assuming normal distribution of these values. Three lifting conditions with vertical heights of lift, i.e., (1) V < 75 cm, (2) V>75 cm, and the shoulder-to-reach, were adopted for the purpose of calculating the scaling factors for males and females based on the appropriate percentile distribution values (Snook and Ciriello, 1991).

The derived scaling factors were defined as the threshold lifting index (TLI) values (see Table 17.1). It should be noted that all of the TLI values are less than 3. For comparison purposes, the results of the field study of 24 lifting jobs (Karwowski et al., 1994) revealed the value of lifting index of 4.1 (at lift destination), which corresponds to the lumbar compressive strength of 4100 N (with one standard deviation) for 40-year-old males, as proposed by Jager and Luttman (1992).

The population-based load threshold limits (LTL) corresponding to different percentiles of capable workers were defined as follows:

$$LTL = TRWL \times TLI \ [kg]$$

The derived distributions of load threshold limits are illustrated in Figures 17.1 and 17.2, for males and females, respectively. Furthermore, Tables 17.2 and 17.3 show the LTL values as the function of work exposure and vertical lifting height. These values range from the most capable population of workers (10% value) to the least capable (90% value), and are expressed as a function of: (1) task exposure (up to one hour, one-to-two hours, and eight hours of work), and (2) vertical location of the load, i.e., the vertical height of the hands above the floor (V < 75 cm and V > 75 cm). In addition, the LTL values for lifting from shoulder-to-reach (SR) are also given. It should be remembered that the LTL values represent the threshold RWLs which apply to 99.5% of all possible simulated cases based on the NIOSH (1991) RLE model, and, therefore, represent the largest variation of all possible lifting tasks that can be encountered in industry today.

**TABLE 17.2**   Load Threshold Limits (LTL) for Male Population (kg) at Different Work Exposures

| Percent capable | V < 75 | | | V > 75 | | | Shoulder-to-reach | | |
|---|---|---|---|---|---|---|---|---|---|
| | 1 hr | 2 hrs | 8 hrs | 1 h | 2 hrs | 8 hrs | 1 hr | 2 hrs | 8 hrs |
| 90% | 13.00 | 12.50 | 10.50 | 13.00 | 12.50 | 10.50 | 13.00 | 12.50 | 10.50 |
| 80% | 17.29 | 16.63 | 13.97 | 16.12 | 15.50 | 13.02 | 15.86 | 15.25 | 12.81 |
| 70% | 20.28 | 19.50 | 16.38 | 18.20 | 17.50 | 14.70 | 17.81 | 17.13 | 14.39 |
| 60% | 22.88 | 22.00 | 18.48 | 20.02 | 19.25 | 16.17 | 19.50 | 18.75 | 15.75 |
| 50% | 25.35 | 24.38 | 20.48 | 21.71 | 20.88 | 17.54 | 21.06 | 20.25 | 17.01 |
| 40% | 27.82 | 26.75 | 22.47 | 23.53 | 22.63 | 19.01 | 22.75 | 21.88 | 18.38 |
| 30% | 30.42 | 29.25 | 24.57 | 25.35 | 24.38 | 20.48 | 24.44 | 23.50 | 19.74 |
| 20% | 33.41 | 32.13 | 26.99 | 27.43 | 26.38 | 22.16 | 26.39 | 25.38 | 21.32 |
| 10% | 37.70 | 36.25 | 30.45 | 30.55 | 29.38 | 24.68 | 29.25 | 28.13 | 23.63 |

V = vertical height of lift (cm)

**TABLE 17.3**   Load Threshold Limits (LTL) for Female Population (kg) at Different Work Exposures

| Percent capable | V < 75 | | | V > 75 | | | Shoulder-to-reach | | |
|---|---|---|---|---|---|---|---|---|---|
| | 1 hr | 2 hrs | 8 hrs | 1 h | 2 hrs | 8 hrs | 1 hr | 2 hrs | 8 hrs |
| 90% | 10.40 | 10.00 | 8.40 | 11.18 | 10.75 | 9.03 | 11.18 | 10.75 | 9.03 |
| 80% | 12.35 | 11.88 | 9.98 | 12.48 | 12.00 | 10.08 | 12.48 | 12.00 | 10.08 |
| 70% | 13.65 | 13.13 | 11.03 | 13.39 | 12.88 | 10.82 | 13.39 | 12.88 | 10.82 |
| 60% | 14.69 | 14.13 | 11.87 | 14.30 | 13.75 | 11.55 | 14.17 | 13.63 | 11.45 |
| 50% | 15.86 | 15.25 | 12.81 | 15.08 | 14.50 | 12.18 | 14.95 | 14.38 | 12.08 |
| 40% | 16.90 | 16.25 | 13.65 | 15.86 | 15.25 | 12.81 | 15.73 | 15.13 | 12.71 |
| 30% | 17.94 | 17.25 | 14.49 | 16.64 | 16.00 | 13.44 | 16.51 | 15.88 | 13.34 |
| 20% | 19.24 | 18.50 | 15.54 | 17.55 | 16.88 | 14.18 | 17.42 | 16.75 | 14.07 |
| 10% | 21.06 | 20.25 | 17.01 | 18.72 | 18.00 | 15.12 | 18.72 | 18.00 | 15.12 |

V = vertical height of lift (cm)

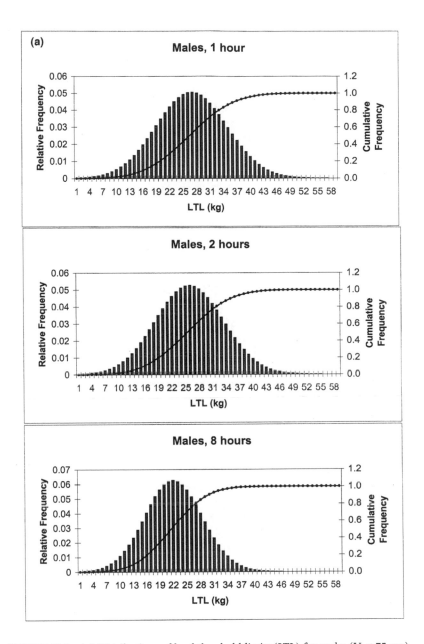

**FIGURE 17.1**    (a) Distributions of load threshold limits (LTL) for males (V < 75 cm).

## 17.8    The LTL Model Application

The population-based load threshold limits could be very useful for immediate risk assessment of manual lifting tasks performed in industry at large. Such values of lifted loads, if exceeded, provide an indication of the need for a more thorough examination of the manual lifting tasks performed in a given industrial setting, and, if appropriate, evaluation of the physical capacity of the exposed workers. It should be noted that the developed model does not imply that the LTL values are safe for any given individual performing a specific manual lifting task. Rather, these data should be used a guide for design and evaluation of manual lifting tasks performed in different types of industry by a given percentile of capable workers,

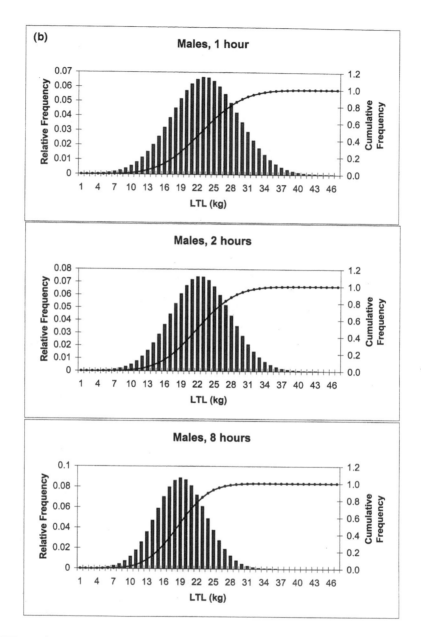

**FIGURE 17.1** (continued)    (b) Distributions of load threshold limits (LTL) for males (V > 75 cm).

until more objective epidemiological evidence regarding the causal relationship between the loads lifted on the job and related risk of low-back injury is available (Dempsey, 1998; Karwowski et al., 1998).

It should be remembered that the LTL data apply to 99.5% of the expected cases (i.e., potential lifting tasks that can be expected to occur in real industrial environments), and would be somewhat greater for 0.5% of the remaining cases. Since the TRWL concept, which integrates all aspects of the NIOSH revised lifting equation (1991), was used as the basis for development of the population-based design limits, the proposed LTL approach is applicable to a large variety of possible lifting task conditions as it integrates all lifting factors considered by the RLE model. Finally, the developed threshold load limit approach extends application of the original RWL concept to different percentiles of capable male and female populations, making it much more relevant and applicable to the real life situations.

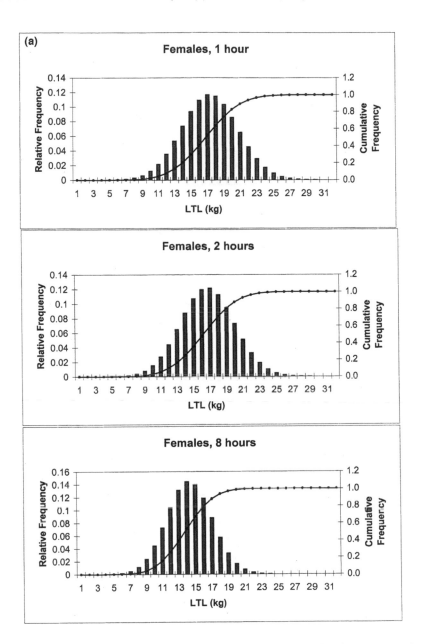

**FIGURE 17.2**    (a) Distributions of load threshold limits (LTL) for females (V < 75 cm).

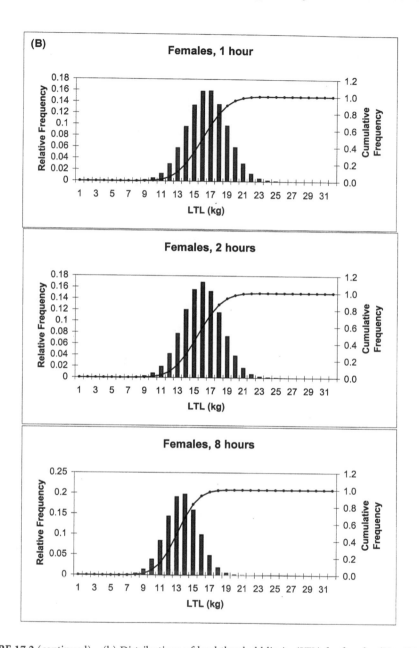

**FIGURE 17.2** (continued)     (b) Distributions of load threshold limits (LTL) for females (V > 75 cm).

# References

Ayoub, M. M., Dempsey, P. G., and Karwowski, W. 1997, Manual materials handling, in *Handbook of Human Factors & Ergonomics*, G. Salvendy, Ed., John Wiley & Sons, New York, pp. 1085-1123.

Brokaw, N. 1992, Implications of the Revised NIOSH Lifting Guide of 1991: A Field Study. Unpublished M.S. Thesis, Department of Industrial Engineering, University of Louisville, Louisville, Kentucky.

Dempsey, P. 1998, A critical review of biomechanical, epidemiological, physiological and psychophysical criteria for designing manual materials handling tasks. *Ergonomics*, 41 (1) 73-88.

Jager, M. and Luttman, A. 1992, The load on the lumbar spine during asymmetrical bi-manual materials handling. *Ergonomics*, 35 (7/8), 783-805.

Karwowski, W. and Brokaw, N. 1992, Implications of the Proposed Revisions in a Draft of the Revised NIOSH Lifting Guide (1991) for Job Redesign: A Field Study, *Proceedings of the 36th Annual Meeting of the Human Factors Society*, Santa Monica, California, pp. 659-663.

Karwowski, W. 1992, Comments on the assumption of multiplicity of risk factors in the draft revisions to NIOSH Lifting Guide, in S. Kumar (Ed.), *Advances in Industrial Ergonomics and Safety IV*, Taylor & Francis, London, pp. 905-910.

Karwowski, W., Caldwell, M., and Gaddie, P. 1994, Relationship between the NIOSH lifting index, compressive and shear forces on the lumbosacral joint, and low back injury incidence rate based on industrial field study, in *Proceedings of the 36th Annual Meeting of the Human Factors Society*, Santa Monica, California, pp. 654-659.

Karwowski, W., Lee, W. G., Jamaldin, B., Gaddie, P., and Jang, R. 1998, Beyond psychophysics: a need for cognitive modeling approach to setting limits in manual lifting tasks, *Ergonomics*, in press.

Marras, W. S., Lavender, S. A., Leurgans, S. E., Fathallah, F. A., Ferguson, S. A., Allread, W. G., and Rajulu, S. L. 1995, Biomechanical risk factors for occupationally related low back disorders. *Ergonomics*, 38, 377-410.

Snook, S. H. and Ciriello, V. M. 1991, The design of manual handling tasks: Revised tables of maximum acceptable weights and forces. *Ergonomics*, 34 (9), 1197-1213.

Waters, T. R., Putz-Anderson, V., Garg, A., and Fine, L.J. 1993, Revised NIOSH equation for the design and evaluation of manual lifting tasks, *Ergonomics*, 36, 749-776.

Waters, T. R. and Putz-Anderson, V. 1998, Revised NIOSH Lifting Equation, in W. Karwowski and W. S. Marras (Eds.), *Handbook of Occupational Ergonomics*, CRC Press, Boca Raton.

# 18

# Occupational Low Back Disorder Risk Assessment Using the Lumbar Motion Monitor

William S. Marras
*The Ohio State University*

W. Gary Allread
*The Ohio State University*

Richard G. Ried
*The Ohio State University*

## 18.1 Background

### Occupational Back Injuries

Low back disorders (LBD) are among the most common occupationally related injuries. According to Andersson (1997), LBDs affect an estimated 80% of the population during their working career. The National Center for Health Statistics (1977) has documented that LBDs are the prime reason for activity limitation for those under 45 years of age. Cats-Baril (1996) has shown that LBDs cost society up to 100 billion dollars annually. Despite the prevalence and cost of these injuries, there are relatively few accurate methods available to predict the risk of occupationally related LBDs.

### Tools for Analyzing Low Back Injury Risk

There is a host of tools available to evaluate LBD risk in industrial jobs. These tools vary widely in their assumptions and complexity. When choosing an analysis technique it is important to match the capabilities of tool to the characteristics of the job being evaluated.

Two- and three-dimensional static models are easy to apply but assume that motion is not a significant factor in LBD risk. The 1981 and 1991 NIOSH lifting equations are also easy to apply but assume that all movements are slow and smooth. Within their limitations, these methods can provide insight on LBD risk. However, recent research has suggested that motion may play a role in LBD risk.

Numerous epidemiologic studies have specifically indicated that the risk of LBD increases when dynamic lifting occurs. Data from a retrospective study of 4,645 injuries by Bigos et al. (1986) suggest that there were greater reports of LBD with dynamic tasks relative to awkward static tasks. Magora (1973) concluded that lateral bending and twisting were only significant risk factors when they occurred simultaneously with sudden (quick) movements. Punnett et al. (1991) studied non-neutral postures in automobile assembly plants and reported that postural stress to the back was more dynamic than static. All of these studies have indicated that the risk of LBD increases with dynamic activity, especially when the body moves asymmetrically. Therefore, if the job has a dynamic component it is important to choose an analysis method that accounts for the additional risk that motion imparts.

Video-based motion analysis systems offer one way to study the dynamic component of a job, but they have several drawbacks when used in an industrial setting. Video assessments must take place in a calibrated space of usually no more than 2 to 3 cubic meters. Cameras must be carefully placed to obtain data for all three planes of motion, and time-consuming analysis is necessary to obtain usable data. In industrial environments, tasks often involve movement outside the calibrated space, work areas limit camera placement, and a great deal of labor must be expended to analyze a few minutes of usable data. These limitations often make video-based systems impractical for routinely evaluating a large number of industrial jobs.

## 18.2   Development of the Lumbar Motion Monitor

### Physical Description

The Lumbar Motion Monitor (LMM) (Figure 18.1) was developed in the Biodynamics Laboratory at The Ohio State University in response to the need for a practical method of assessing the dynamic component of occupationally related LBD risk in industrial settings. The patented LMM is a triaxial electrogoniometer that acts as a lightweight exoskeleton of the lumbar spine. It is positioned on the back of a subject directly in line with the spine and attached by harnesses at the pelvis and thorax. Four potentiometers at the base of the LMM measure the instantaneous position of the spine (as a unit) in three-dimensional space relative to the pelvis. Position data from the potentiometers are recorded at 60 Hz, transmitted to an analog-to-digital (A/D) converter, and then recorded on a microcomputer. The data are then processed to calculate the position, velocity, and acceleration of the spine in each of the three planes of motion as a function of time.

### Calibration and Validation

During data collection and analysis, each LMM is used with its matching calibration file. Before its first use, each LMM is calibrated individually, using a specially designed reference frame (Figure 18.2). During the calibration process the LMM is positioned on the calibration frame at 225 different positions in three-dimensional space and the voltage outputs recorded. The calibration process eliminates any individual variability in each LMM.

The LMM was validated to ensure its accuracy and sensitivity with a video-based motion analysis system (Marras et al., 1992). During the validation process the predicted velocities and accelerations of the LMM were compared relatively (to position accuracy) against the predicted values of the motion analysis system. The results of the validation process (Table 18.1) show high correlation coefficient values and significance levels ($r > .95$, $p < 0.0001$) for all three planes of motion. An independent group has also determined that the reproducibility of the LMM is suitably high for range of motion and velocity for the device to be used for evaluation in a clinical and research setting (Gill and Callaghan, 1996).

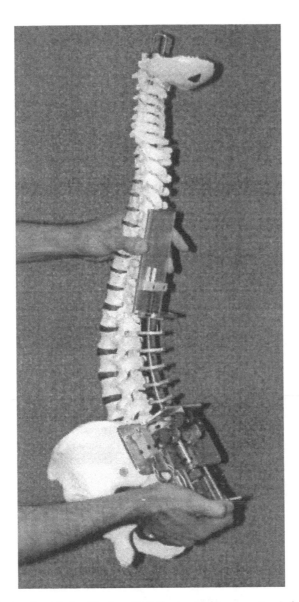

**FIGURE 18.1**  The Lumbar Motion Monitor (LMM) compared to an anatomical model of the spine.

## Development of the LBD Risk Model

An *in vivo* study was undertaken to determine quantitatively whether dynamic trunk motions in combination with workplace and environmental factors may better describe the risk of LBD in repetitive manual materials handling (MMH) tasks.

### Approach

The study involved an industrial surveillance of the trunk motions and workplace factors involved in high- and low-risk repetitive MMH jobs. The approach used in this phase of the project was to (1) identify industries involved with repetitive MMH work; (2) examine the company medical records as well as the health and safety records to identify those repetitive MMH jobs that were associated, historically, with either a high or low risk of occupationally related LBD; (3) quantitatively monitor the trunk motions and workplace factors associated with each of these jobs; and (4) evaluate the data to determine which combination of trunk motion and workplace factors was most closely associated with LBD risk.

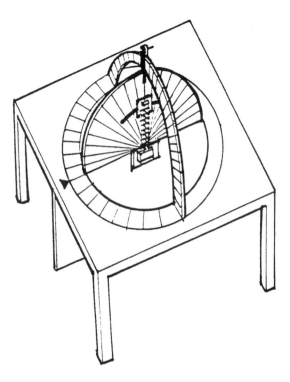

**FIGURE 18.2**    The three-dimensional reference frame.

**TABLE 18.1**    Correlations of the Velocities and
Accelerations of the Motion Analysis System in
the Three Planes of Motion with the Velocities
and Accelerations of the LMM

|          | Correlation* | |
|----------|----------|--------------|
| Plane    | Velocity | Acceleration |
| Lateral  | 0.95     | 0.95         |
| Sagittal | 0.99     | 0.96         |
| Twisting | 0.99     | 0.99         |

$^*\ p < 0.0001$

## Study Design

This study was a cross-sectional study of 403 industrial jobs from 48 manufacturing companies through-out the midwestern United States. Only repetitive jobs without job rotation were examined in this study. This was necessary to prevent the confounding effects created by alternate jobs. Jobs examined in this study were divided into two groups, high and low risk of LBD, based on examination of the injury and medical records. Whenever possible, company medical reports were used to categorize risk. In some cases only injury logs (OSHA 200 logs) were available. All medical reports, injury records and logs were scrutinized to ensure they were as accurate as possible. The outcome measure (LBD risk) derived from these medical and injury records consisted of the normalized rate of reported occupationally related LBD. Incidence of reported LBD was considered regardless of whether there was any restricted or lost time associated with the incident.

Thus, the dependent variable in this study consisted of two levels of job-related LBD risk categories. Low-risk group jobs were defined as those jobs with at least 3 years of records showing no injuries and no turnover. Turnover is defined as the average number of employees who left a job per year. High-risk group jobs were those jobs associated with at least 12 injuries per 200,000 hours of exposure (average of 26/200,000 hours). The high-risk group category incidence rate corresponds to the 75th percentile value of risk for the 403 jobs examined. Of the 403 jobs examined, 124 of the jobs were categorized as low-risk and 111 were categorized as high-risk. The remainder of the jobs (168) were categorized as medium-risk and were not used in this particular analysis.

The independent variables in this study consisted of workplace, individual, and trunk motion characteristics that were indicative of each job. The workplace and individual characteristics consisted of variables typically considered in current workplace guidelines for materials handling (NIOSH, 1981; Putz-Anderson and Waters, 1991). Specifically, these variables were: (1) the maximum horizontal distance of the load from the spine; (2) the weight of the object lifted; (3) the height of the load at the origin of the lift; (4) the height of the load at the destination of the lift; (5) the frequency of lifting (lift rate); (6) the asymmetric angle of the lift (as defined by NIOSH, 1981); (7) employee anthropometry (12 measures); (8) employee injury history; (9) employee satisfaction; and (10) trunk motion. Trunk motion characteristics were those variables obtained using the LMM. These variables consisted of the trunk angular position, velocity, and acceleration characteristics (i.e., means, ranges, maximums, minimums, etc.) in each of the cardinal planes. Selected trunk motion factors along with selected workplace factors were used to develop a quantitative model of occupational risk factors.

## Data Collection

Initially, data about employee health, employment history, and anthropometry were collected. Next, the employee was fitted with an LMM. A baseline reading from the LMM was then taken, while the individual stood erect and rigid. The employee then was asked to return to work and wore the LMM for at least ten job cycles. Thus, the length of time the employee wore the monitor depended upon the cycle time of the job. Monitoring of back motion was initiated as the employee began the MMH task and concluded when the employee completed the task. Extraneous activities not involving MMH were not monitored. Signals from the LMM were sampled at 60 Hz via an analog-to-digital converter and stored on a portable microcomputer. The data were further processed in the laboratory to determine position, velocity, and acceleration of the trunk as a function of time in the sagittal, lateral, and axial twisting planes of the body.

## Analysis

The data were initially examined to determine whether the trunk motions were repeatable. This analysis indicated that more than half of the variation was attributable to the job. Hence, trunk motions were dictated largely by the design of the task, and repetitive trials resulted in motions that were fairly similar.

The various personal, environmental, and workplace factors from the database were analyzed using logistic regression techniques to determine if any single factor could distinguish jobs associated with high-risk group membership from those that were not. The most powerful single variable was maximum moment, which yielded an odds ratio of 5.17. Overall, however, the odds ratios were low, indicating that few of the individual variables discriminate well between high- and low-risk jobs. Of the trunk motion factors, the velocity variables generally produced greater odds ratios than maximum or minimum position, range of motion, or acceleration. Table 18.2 shows the descriptive statistics of the workplace and trunk motion factors for the high- and low-risk groups.

Next, multiple logistic regression was used to predict the probability of high-risk group membership as a function of the values of several workplace and trunk motion factors. A five-variable model incorporating the trunk motion and workplace factors was developed and further refined after examining a series of stepwise logistic regression models (containing different variables, e.g., velocity, acceleration) fitted to several intermediate data sets. A combination of five variables (moment, frequency of lift, sagittal flexion, twisting velocity, and lateral velocity) was found to have the greatest odds of predicting high-risk group membership. This combination of workplace and trunk motion factors forms the basis of the LBD Risk Model. The model was selected for the statistical importance of the predictors and for biomechanical plausibility. The model variables remained consistent when tested with the various intermediate

**TABLE 18.2**   Descriptive Statistics of the Workplace and Trunk Motion Factors in Each of the Risk Groups

| Factors | High Risk (N = 111) | | | | Low Risk (N = 124) | | | |
|---|---|---|---|---|---|---|---|---|
| | Mean | SD | Min | Max | Mean | SD | Min | Max |
| WORKPLACE FACTORS | | | | | | | | |
| Lift Rate (lifts/hr) | 175.89 | 8.65 | 15.30 | 900.00 | 118.83 | 169.09 | 5.40 | 1500.00 |
| Vertical load location at origin (m) | 1.00 | 0.21 | 0.38 | 1.80 | 1.05 | 0.27 | 0.18 | 2.18 |
| Vertical load location at destination (m) | 1.04 | 0.22 | 0.55 | 1.79 | 1.15 | 0.26 | 0.25 | 1.88 |
| Vertical distance traveled by load (m) | 0.23 | 0.17 | 0.00 | 0.76 | 0.25 | 0.22 | 0.00 | 1.04 |
| Average weight handled (N) | 84.74 | 79.39 | 0.45 | 423.61 | 29.30 | 48.87 | 0.45 | 280.92 |
| Maximum weight handled (N) | 104.36 | 88.81 | 0.45 | 423.61 | 37.15 | 60.83 | 0.45 | 325.51 |
| Average horizontal distance between load and $L_5/S_1$ (N) | 0.66 | 0.12 | 0.30 | 0.99 | 0.61 | 0.14 | 0.33 | 1.12 |
| Maximum horizontal distance between load and $L_5/S_1$ (N) | 0.76 | 0.17 | 0.38 | 1.24 | 0.67 | 0.19 | 0.33 | 1.17 |
| Average moment (Nm) | 55.26 | 51.41 | 0.16 | 258.23 | 17.70 | 29.18 | 0.17 | 150.72 |
| Maximum moment (Nm) | 73.65 | 60.65 | 0.19 | 275.90 | 23.64 | 38.62 | 0.17 | 198.21 |
| Job satisfaction | 5.96 | 2.26 | 1.00 | 10.00 | 7.28 | 1.95 | 1.00 | 10.00 |
| TRUNK MOTION FACTORS | | | | | | | | |
| Sagittal Plane | | | | | | | | |
| Maximum extension position (°) | −8.30 | 9.10 | −30.82 | 18.96 | −10.19 | 10.58 | −30.00 | 33.12 |
| Maximum flexion position (°) | 17.85 | 16.63 | −13.96 | 45.00 | 10.37 | 16.02 | −25.23 | 45.00 |
| Range of motion (°) | 31.50 | 15.67 | 7.50 | 75.00 | 23.82 | 14.22 | 399.00 | 67.74 |
| Average velocity (°/sec) | 11.74 | 8.14 | 3.27 | 48.88 | 6.55 | 4.28 | 1.40 | 35.73 |
| Maximum velocity (°/sec) | 55.00 | 38.23 | 14.20 | 207.55 | 38.69 | 26.52 | 9.02 | 193.29 |
| Maximum acceleration (°/sec²) | 316.73 | 224.57 | 80.61 | 1341.92 | 226.04 | 173.88 | 59.10 | 1120.10 |
| Maximum deceleration (°/sec²) | −92.45 | 63.55 | −514.08 | −18.45 | −83.32 | 47.71 | −227.12 | −4.57 |
| Lateral Plane | | | | | | | | |
| Maximum left bend (°) | −1.47 | 6.02 | −16.80 | 24.49 | −2.54 | 5.46 | −23.80 | 13.96 |
| Maximum right bend (°) | 15.60 | 7.61 | 3.65 | 43.11 | 13.24 | 6.32 | 0.34 | 34.14 |
| Range of motion (°) | 24.44 | 9.77 | 7.10 | 47.54 | 21.59 | 10.34 | 5.42 | 62.41 |
| Average velocity (°/sec) | 10.28 | 4.54 | 3.12 | 33.11 | 7.15 | 3.16 | 2.13 | 18.86 |
| Maximum velocity (t/sec) | 46.36 | 19.12 | 13.51 | 119.94 | 35.45 | 12.88 | 11.97 | 76.25 |
| Maximum acceleration (°/sec²) | 301.41 | 166.69 | 82.64 | 1030.29 | 229.29 | 90.90 | 66.72 | 495.88 |
| Maximum deceleration (°/sec²) | −103.65 | 60.31 | −376.75 | 0.00 | −106.20 | 58.27 | −294.83 | 0.00 |
| Twisting Plane | | | | | | | | |
| Maximum left twist (°) | 1.21 | 9.08 | −27.56 | 29.54 | −1.92 | 5.36 | −30.00 | 11.44 |
| Maximum right twist (°) | 13.95 | 8.69 | −13.45 | 30.00 | 10.83 | 6.08 | −11.20 | 30.00 |
| Range of motion (°) | 20.71 | 10.61 | 3.28 | 53.30 | 17.08 | 8.13 | 1.74 | 38.59 |
| Average velocity (°/sec) | 8.71 | 6.61 | 1.02 | 34.77 | 5.44 | 3.19 | 0.66 | 17.44 |
| Maximum velocity (°/sec) | 46.36 | 25.61 | 8.06 | 136.72 | 38.04 | 17.51 | 5.93 | 91.97 |
| Maximum acceleration (°/sec²) | 304.55 | 175.31 | 54.48 | 853.93 | 269.49 | 146.65 | 44.17 | 940.27 |
| Maximum deceleration (°/sec²) | −88.52 | 70.30 | −428.94 | −5.84 | −100.32 | 72.40 | −325.93 | −2.74 |

data sets. The empirical stability of the model was checked by predicting the classification of 100 jobs based on the preliminary model. This model resulted in an odds ratio of 10.6.

By averaging individual probability values for moment, frequency of lift, sagittal flexion, twisting velocity, and lateral velocity, the LBD Risk Model is able to predict the probability of high-risk group membership (LBD risk) for any repetitive job (Figure 18.3). It is important to understand that the predictive power of the model is a result of the interaction of these five variables. Individually, each of these factors is unable to reliably distinguish between a high-risk and a low-risk situation, but when they are considered in combination the predictive power increases tenfold.

The LBD Risk Model is currently being validated in a longitudinal study. Jobs are prospectively tracked (over time) to determine if changes in injury rates after the implementation of ergonomic changes correspond with predicted value changes in the LBD Risk Model. Preliminary results indicate the changes

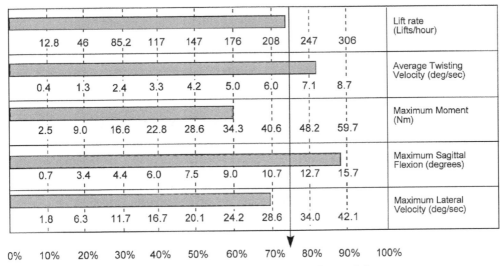

**FIGURE 18.3**   The LBD Risk Model showing the five risk variables scaled relative to risk. The arrow points to the overall probability of high risk of LBD group membership for a particular job.

in LMM probability of risk corresponds well with an observed change in actual injury rates after changes have been made to the work environment.

## 18.3   Benefits of the LMM and the LBD Risk Model

Use of the LMM and its computation of risk provides several advantages to the ergonomist studying injury risk for material handling activities. First, the LMM allows one to determine the instantaneous three-dimensional position of the trunk while individuals perform their actual job tasks, not work simulated instead in a laboratory. This ability eliminates the question of whether a material handling study conducted in an artificial setting can be generalized to the workplace. Also, the LMM data are gathered objectively. Resulting calculations of injury risk are determined irrespective of an investigator's (perhaps unintentionally biased) view of the work.

A second advantage of using the LMM system is that material handling jobs can be assessed relative to a large database covering jobs from diverse manufacturing environments and with different levels of risk. This enables the ergonomist to determine if a job has a high probability of injuring employees who perform that job. It also allows one to rank several jobs, based on their risk values, and to study solutions for those having the greatest chance of producing injury.

Third, the probability model enables the ergonomist to quantitatively assess and compare each task within a job. Specific factors that contribute to a task's probability are identified, as are the tasks that most contribute to the job's overall risk. This information pinpoints the specific tasks and the factors therein that must be addressed during job redesign to reduce the job's injury risk potential. This method of identifying and changing only those components that contribute to risk eliminates the need to change the entire job. This quantitative assessment permits the ergonomist to make decisions about what level of risk is acceptable for a job. It is not possible to totally reduce a job's risk to zero. However, the LBD Risk Model allows one to determine if a job's risk is above a criteria of acceptance set by the company or the ergonomist.

A fourth benefit of this LBD Risk Model is the assistance it can provide to the ergonomic intervention process. Modified jobs can be re-monitored using the LMM, and the effects of those changes can be quantified and compared with those values determined prior to the intervention. Traditionally, the effects of job changes on the numbers of related musculoskeletal strains (the job's incident rate) may take several years to appear. The LMM can produce more timely feedback to the ergonomist regarding anticipated returns on the redesign investment — actually, just the time needed to analyze the data. For jobs that produce minimal reductions in risk due to redesign efforts, further improvements can be attempted sooner. In other words, this risk model can assist in determining whether or not an ergonomic intervention has produced *enough* of an improvement in the job.

## 18.4    Applications: How to Use the LMM and LBD Risk Model

### Recommended Equipment

It is important to make sure that all equipment used in LMM data collection is available before going to a job site. The following list of items represents the essential equipment needed.

### Lumbar Motion Monitors

Two sizes of LMMs are available commercially, large and small. These LMMs differ only in their length; the large LMM is suitable for taller individuals or individuals of moderate height who must bend forward extremely far in their work. The small LMM is more appropriate for shorter individuals or those of moderate height who do little forward bending.

### LMM Harnesses

Elastic waist and shoulder harnesses are supplied with the commercial LMMs and consist of three sizes: large, medium, and small. The large harnesses are for taller or heavier people, the medium harnesses are for those with average body builds, and the small harnesses are for shorter individuals or those with slighter builds. All harnesses have adjustable Velcro straps to allow for individual differences. The shoulder harness is fitted over one's torso before the waist harness and LMM are attached. The waist harness attaches to the LMM before it is placed on the individual.

### Laptop/Notebook Computer

The LMM software requires a PC-based operating system. (The software will be discussed in a later section.) The computer should have, as a minimum, a 486 processor. However, the faster the processing speed of the computer used, the more efficiently data will be collected and stored. Most industrial environments will have 110-volt outlets for computers, though battery-operated laptops may make data collection more flexible. Be aware of the length of battery life before recharge, however.

### LMM/Computer Connections

A cable that connects the LMMs to the computer is supplied with the LMMs. This cable transmits information regarding trunk positions from the LMM to the computer for storage. It is important that this cable be in good working order, since data collection in an industrial environment can often put heavy wear on equipment. Also available is a digital telemetry system. This "wireless" system consists of a transmitter and battery pack, worn on the waist harness, and a receiver unit, placed alongside the computer. Both systems (the cable and the telemetry unit) provide the same information. The telemetry system, however, allows data to be collected without the cable connection between the monitored individual and the computer, providing greater freedom of movement.

### Other Equipment

The LMM provides data regarding the trunk motions required of a job. However, other information is important and needs to be collected at the job site to fully assess the job's risk. Listed below is additional equipment that should be part of the LMM gear.

    *Scale and Push/Pull Gauge* — It is critical to record the weights of items handled by individuals during data collection. A heavy-duty scale is needed, one that is capable of reading object weights from 1 to 100 pounds. Occasionally, during material handing, objects are not lifted but are pushed or pulled instead. A gauge capable of recording these forces also is important to have. Both devices are available commercially.

*Tape Measure* — A heavy-duty tape measure is needed to record the reach distance and the vertical and horizontal locations of objects that are handled at the job site. The size of the objects themselves also may be of importance and can be measured using a tape.

*Extension Cord/Outlet Strip* — Power outlets in industry are not always conveniently located near where data collection is to take place. An extension cord permits greater flexibility in locating the computer closer to the job site and in full view of the individual being monitored. An outlet strip also allows the use of other electronic devices such as video recorders and battery chargers.

*Wheeled Cart* — Collapsible carts are useful to move equipment into the job site, and from one site to another, if multiple sites are to be monitored.

## Selecting the Job(s) to Monitor

The LMM can be used to monitor the positions, velocities, and accelerations of the trunk in the three cardinal planes for any job in which trunk movements are required. This type of information can be important when one needs to describe the motion characteristics required of a job or activity. The LBD Risk Model is composed of five workplace factors that, when combined, assess a job's probability of having a high number of LBDs. This model is also useful for establishing relative comparisons between several tasks that comprise a job.

The industrial jobs used to develop the risk model had MMH frequencies ranging from approximately 6 to 1500 lifting tasks required per hour. Thus, the tasks to be assessed using this risk model should fall within this scope of lifting rates. Most MMH jobs fit this profile. For example, in automobile assembly, job cycles are repeated every one to two minutes, and the parts themselves often are standard in size and weight. Palletizing jobs may require very different types of objects to be handled, but the task of continual lifting from one storage area to another remains the same. These types of jobs are very consistent with those used to develop the risk model database, and the job's LBD risk can be easily assessed with the LMM's software.

Some jobs are less repetitive or have more job variation than assembly or palletizing tasks. However, the LBD risk model still can be used to make relative comparisons between the tasks that comprise the job. For jobs that require a larger number of tasks, the risk model is helpful in comparing the factors that make up the model. This will allow for the ergonomist to assess trade-offs between such factors as lifting frequency, object weights handled, and the trunk motions required for the different tasks. It should be noted that there is inherent variability in the way an individual may perform the same task repeatedly. Because of this difference, a minimum of five repetitions should be collected of any given task.

Another issue to be considered is job rotation. Employers often use a variety of rotation schemes for job processes. If the job to be monitored requires no rotation (employees perform the same job every day/week/month), then the risk assessment can be directly related to the tasks observed. Jobs in which individuals rotate regularly between a few work areas also may be used in assessing LBD risk, if this rotation schedule is fixed. When the job rotation requires employees to do completely different jobs on an hourly or weekly basis, it becomes difficult to relate a task's risk values to the overall job risk, since many tasks could contribute to the risk assessment. This issue is key to determining a job's suitability for LBD risk assessment. That is, does the job's work structure enable one to define the job in terms of a few repeatable, consistently performed tasks?

There may be some jobs that fit within the LBD Risk Model profile but still should not be monitored. Seated jobs may require repetitive activities, but they usually are not ones that require significant material handling. In any event, the LMM may rub against a chair's back or the waist harness will shift from its position on the hips during seated work, and erroneous LMM output will result. Also, jobs that require close contact of the LMM with a finished product could produce scratches on the product, and the employer may not want to risk product damage. Finally, exposure to water or other liquids may damage the LMM or its components.

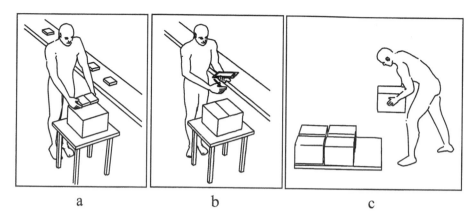

a                    b                    c

**FIGURE 18.4**   The three tasks of the frozen food packager job: (a) place two packages in box; (b) record date/time of packaging; (c) load full box onto pallet.

**TABLE 18.3**   Task Analysis of One Cycle of the Frozen Food Packager Job

| Task | Length of Task | Notes |
|------|------|------|
| Place two food packages in box | 42 seconds | Task time is seven seconds per two boxes packaged; twelve packages fit into each box |
| Record date/time of packaging | 10 seconds | No material handling required |
| Load full box onto pallet | 8 seconds | Full pallet contains seven layers of boxes |
| *Total time of job tasks* | *60 seconds* | |

## Defining the Major Components of the Job through a Task Analysis

It is very important to properly define all relevant tasks that make up the job under investigation. These tasks should encompass the range of materials handling work that is required of the job — especially those that may present a risk of LBD. The tasks also should be defined so they are meaningful to those who will be interested in the LMM results. One way to define these job tasks is through a task analysis. A task analysis defines the discrete events of a job.

Take, for example, a job performed in a food processing plant (shown in Figure 18.4). An employee places twelve frozen food packages, two at a time, into a box (a), records the date and time of the packaging (b), then loads the packed box onto a pallet (c). A simple task analysis of this job is shown in Table 18.3. Two of the three tasks identified would qualify as material handling tasks for this job — loading the individual packages and placing the full box onto the pallet. Recording the date and time of the packaging would not be considered relevant material handling since low forces and exertions are required. Any trunk movement related to this task would not be considered in further analyses. However, trunk kinematics and the probability of risk could be determined for the two relevant MMH tasks identified.

It may be convenient to subdivide and redefine a task that is similar in all work dimensions except for one. For instance, in the above example, a fully loaded pallet may contain many boxes stacked several layers high. Tasks could be defined separately as "Place box on Layer 1," "Place box on Layer 2," etc. This may assist data interpretation. That is, differences in trunk kinematics could be interpreted not only as a function of the defined job tasks, but further in terms of workplace factors such as a box's location on a pallet.

## Collecting and Recording Workplace Data for Risk Assessment

The trunk motion information that is used in the LBD Risk Model is automatically stored in the data collection software. Trunk motions include the position, the velocity of movement, and the related acceleration of the trunk in all three directions of motion during a task — the lateral (side-bending) plane, the sagittal (forward bending) plane, and the rotational (twisting) plane. Two other components, lift rate and maximum moment, must be determined and input manually.

## Lift Rate

Lift rate is defined as the *total* number of MMH actions that a job requires per hour. This number is related to the lifts for all tasks combined; it does not change from one task to another in a job. As the task analysis in Table 18.2 found, on average, one box is fully prepared and loaded onto the pallet per minute. Assuming this rate represents an average job cycle across the work shift, the packaging task alone would require 360 lifts per hour, since the product is placed in each of the 60 boxes six times (two packages per lift). For the palletizing task, 60 lifts are required per hour, for each fully packed box. Both tasks combine for a total of 420 lifts required per hour for this job. The non-MMH date/time recording task would not influence the lifting frequency of the job.

Because the lift rate value is directly input into the LBD Risk Model, it is very important to get an accurate estimate of the lifting frequency. It may be necessary to confirm the task analysis results by questioning employees familiar with the job or the job supervisor.

## Maximum Moment

Maximum moment is defined as the external moment generated about the spine. A moment is composed of two factors, the weight of the object being handled and the horizontal distance from the spine at which it is handled.

*Weight* — Each object handled on a job must be weighed and recorded in the software or noted manually and input during data analysis. If objects of varying weights are handled, then each must be weighed individually and recorded. This often occurs in mail and freight delivery operations. In the aforementioned food processing example, the weights are constant for each task. The combined weight of the two food packages lifted together needs to be recorded, as should the weight of a fully packed box that is palletized.

*Horizontal Distance* — A tape measure is needed to determine how far from the spine objects are being handled for each task. With the tape measure held *horizontally*, one must measure the distance from the spine at the lumbosacral joint (near the top of the hips) to the center of the hands when the task is being performed. Obviously, as individuals handle objects, this distance changes as the object is moved. It is important to determine at what point during the task the distance is the greatest (i.e., generating the greatest moment about the spine) and to record this length. An ergonomist correctly measuring the horizontal distance is shown in Figure 18.5a. Here, the tape measure is kept level to determine the length from the individual's $L_5/S_1$ joint to the center of the hands. The *incorrect* approach is being used in Figure 18.5b, since the distance being measured is not horizontal.

For some MMH jobs, perhaps those on an assembly line, work requirements can be very repetitive and the employees' actions and movements rather consistent. In these cases, the horizontal distances may not vary much as identical objects are handled on each cycle. For other jobs, such as when pallets are loaded or unloaded, each cycle can produce very different trunk motions, since objects are being handled from different areas. This will likely change the horizontal distance at which an employee handles the load. Because the maximum moment value, a combination of an object's weight and its horizontal distance from the spine, is directly input into the LBD Risk Model, it is important that these distances be accurately measured and that changes in the distances for each task cycle being monitored by the LMM are recorded.

The measurement of these horizontal distances should not interfere with the work being done by the employee. The ergonomist who measures these distances should stand close enough to the individual to get accurate readings, but far enough away to not disturb the work. It is important that the individual be able to move naturally at the job site. A monitored employee being crowded by an investigator will move differently, change his/her trunk motions, and give erroneous information via the LMM.

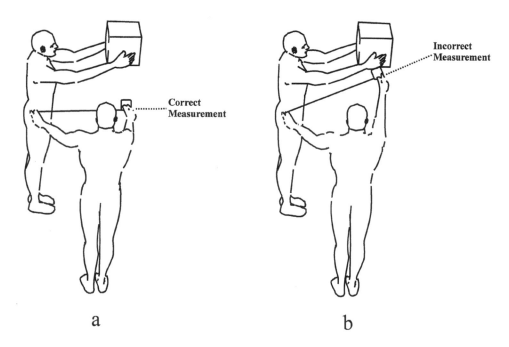

a                                    b

**FIGURE 18.5**     The correct (a) and incorrect (b) methods of measuring the distance a load is held from the $L_5/S_1$ joint.

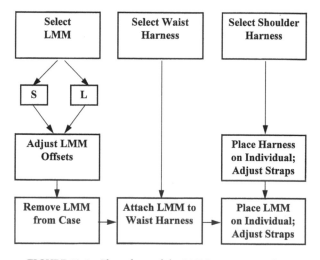

**FIGURE 18.6**     Flow chart of the LMM set-up procedure.

## Setting Up the LMM for Data Collection

After the job to be monitored has been selected, the tasks identified, and basic workplace information has been gathered, the data collection process can start. To begin, the proper LMM and harnesses need to be chosen and prepared for use. This process is depicted in a flow chart in Figure 18.6 and is described below.

### Selecting the Correct LMM

The standing height of the individual to be monitored is usually the best indicator of which LMM to use. As shown in Figure 18.7a, a small LMM placed on a taller individual will restrict his/her range of motion. A large LMM put on a shorter person will cause the LMM to buckle and give erroneous readings

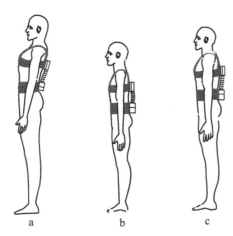

**FIGURE 18.7** LMM sizing- (a) the small LMM incorrectly placed on a tall individual; (b) a large LMM incorrectly put on a small individual; (c) properly sized LMM for the individual.

(Figure 18.7b). Both situations could damage the LMM. A properly sized LMM (Figure 18.7c) will enable an individual to stand upright without the LMM pulling him/her backward or buckling due to the waist and shoulder harnesses being too close to one another. The shoulder harness also can be adjusted to create a proper fit.

## Adjusting the LMM

LMMs are individually calibrated during their manufacture, and there is a specific position designated as its "neutral" position. During normal usage, this neutral position may change slightly and must be readjusted (through the software) before every subsequent use. To adjust the LMM, choose which LMM is to be used, but keep it in the LMM case. Attach the cable from the computer to the LMM (or use the telemetry device), and adjust the voltage offsets using the supplied LMM software. Every LMM should be adjusted in this manner before it is placed on an individual. Without disconnecting the LMM, use the software to monitor the LMM's three motion traces. Take the LMM out of the case and move the top of the LMM while keeping the bottom in place. The three traces on the computer screen should show realistic position values.

## Putting the LMM on the Employee

After the LMM has been checked to be working properly, it is ready to be placed on an employee. When handling the LMM, note that it should never be bent into extreme angles; damage to the LMM could result. The following steps, diagrammed in Figure 18.6, will ensure proper placement.

*Select the Shoulder Harness* — The shoulder harness should fit snugly when worn but still enable the individual to breath normally. It's important that most of the Velcro strap that crosses the chest be overlapped during wear, so that it and the LMM remain in place during data collection. Place the harness on the individual to be monitored.

*Attach LMM on Waist Harness* — Three sets of harnesses are available. Select the size that will fit on the individual with the greatest amount of Velcro that will overlap when worn. Place the base of the LMM evenly on the waist harness and secure tightly.

*Place Waist Harness/LMM on Individual* — First slide the top of the LMM into place on the shoulder harness. Then, fit the waist harness over the hips and momentarily secure the harness with the Velcro strap. The base of the LMM (at the lowest t-section) should be aligned with the lumbosacral joint ($L_5/S_1$) of the trunk. This position can be found by first locating the tops of the hips (the iliac crest) with the fingers, then placing the thumbs about an inch lower than the fingers. See Figure 18.8 for proper LMM alignment on an individual. When the LMM base has been positioned, check that the LMM's t-sections are aligned vertically. If not, one or both of the following adjustments can be made:

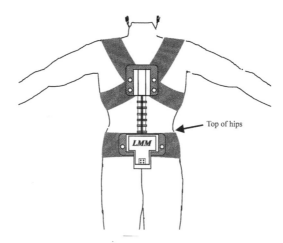

**FIGURE 18.8** Proper placement of the LMM on the torso. The base of the LMM is located slightly below the top of the hips, and the LMM t-sections are aligned vertically.

1. The waist harness can be moved to the left or right.
2. The shoulder harness can be loosened and shifted. It can be adjusted by shortening or lengthening the Velcro straps that cross over the shoulders.

Once the LMM is properly positioned, securely tighten the Velcro strap on the waist harness, and insert each leg strap through the buckle. It is important to always use these leg straps. They prevent the LMM from moving up on the hips, which otherwise would move the base of the LMM from its $L_5/S_1$ location and result in erroneous data.

## Collecting the Data

After the LMM is properly placed on an individual and is correctly hooked to the computer, the investigator is ready to gather job data. It is suggested that the individual be given some time (a few job cycles, for example) to become accustomed to the LMM before collecting data.

If not already done, the company, job, task, and employee data need to be entered into the LMM software. Follow the software instructions for this procedure. Examples of the job profile set-up screens for the Chattanooga™ DOS and Windows software are shown in Appendix A-18.1. After the task data are input and the data collection screen is open and showing the three traces (for the lateral, sagittal, and twisting planes of motion), data collection can begin.

The main goal of data collection is to gather information on trunk motions that is representative of all the work required of the job. For example, if the job requires handling objects of widely varying weights or from different locations (on a pallet, for example), then the data should be gathered of each of these tasks. The more data that are collected, the more it is likely to represent the requirements of the job. The amount of data collected should be reflective of the job and how it is structured. For example, if 80% of the job requires doing Task A and only 20% doing Task B, then a majority of the data should be of the employee performing Task A. In other words, data collection should be proportional to the job being studied. Example LMM data collection screens for DOS and Windows software packages are shown in Appendix A-18.2. Notice that only the motions associated with the designated tasks are marked using the computer keyboard's function keys.

There is no set number of job cycles that indicates when "enough" data have been collected. It is dependent on the nature of the tasks. Jobs that are highly consistent in their activities usually require fewer numbers of collected cycles than those with more widely varying requirements. The authors suggest, though, that a minimum of seven to ten cycles of any task be collected.

**TABLE 18.4** Example Data Collection Form

| Company: | | | | | | |
|---|---|---|---|---|---|---|
| Job: | | | | | | |
| Employee: | | | | | | |
| Date: | | | | | | |
| Trial # | Task | Object Weight | Horizontal Distance | Start Height | Finish Height | Comments |
| | | | | | | |
| | | | | | | |
| | | | | | | |
| | | | | | | |
| | | | | | | |
| | | | | | | |
| | | | | | | |
| | | | | | | |

It's extremely important to keep a record of the weights handled and the measured horizontal lifting distances for *each task cycle collected*, so that these data are matched with the corresponding trunk motion data for that cycle. An example form is shown in Table 18.4. The ergonomist may wish to develop a customized data collection form beforehand to keep track of this information.

## Analyzing and Interpreting the LMM Data

By following the guidelines in the previous sections of this chapter, data analysis and interpretation will be made easier. Tasks of a job that are carefully chosen will assist in the job's interpretation of its risk probability and lead to results having more practical significance. The software used to collect and analyze LMM data provides trunk kinematic information and risk probability charts for each task defined for a job and for each employee who was monitored. It is beyond the scope of this chapter to provide LMM software documentation. Instead, explanations of how these software outputs can be interpreted will be discussed.

### Trunk Kinematic Information

Information on the specific trunk motions produced by employees during each task can be obtained from the software. This includes the trunk positions and the velocities and accelerations produced by the trunk for each plane of motion for those cycles of a task chosen by the ergonomist. This information can be useful for general descriptions of the material handling or for comparisons with other tasks or jobs. Use of these data can be valuable for investigators who have formed hypotheses about, for example, what tasks require more trunk motions than others.

Output from the frozen foods example job is shown in Table 18.5. As the header shows, these data include trials of data collected for the food packaging task of this job. Information includes minimum and maximum trunk positions, average and maximum velocities, and maximum accelerations and decelerations for each plane of motion. From these data, average motions over the six trials could be computed, as could the range of movement required and the amount of variation for each kinematic parameter.

### Probability of High Risk Group Membership (LBD Risk)

The software will produce information that compares the job tasks of interest with a database of jobs previously determined to be at a "high" or "low" risk in terms of low back injury. The interpretation of data is best described through continued use of the food packager example.

**TABLE 18.5**  Output Data Showing Low-Back Kinematics for Six Trials of One Job Task

Company: Frosty Foods
Job: Frozen Food Packager
Employee: Allen Shaffer
Task: Place food packages in box
Date: 10/26/96
Run: 1

| Trial | Time | Side-Bending (Lateral) | | | | | | Forward-Bending (Sagittal) | | | | | | Twisting (Rotational) | | | | | |
|---|---|---|---|---|---|---|---|---|---|---|---|---|---|---|---|---|---|---|---|
| | | Min. Pos. | Max. Pos. | Avg. Vel. | Max. Vel. | Max. Acc. | Max. Dec. | Min. Pos. | Max. Pos. | Avg. Vel. | Max. Vel. | Max. Acc. | Max. Dec. | Min. Pos. | Max. Pos. | Avg. Vel. | Max. Vel. | Max. Acc. | Max. Dec. |
| 1 | 09:11:33 | -1.4 | 7.3 | 12.4 | 32.7 | 66.7 | -228.5 | -8.5 | 4.5 | 4.5 | 45.3 | 215.4 | -97.4 | -2.4 | 12.3 | 6.3 | 35.4 | 276.4 | -101.2 |
| 2 | 09:11:42 | -0.7 | 12.9 | 8.7 | 19.9 | 77.3 | -197.3 | -10.3 | 9.6 | 6.8 | 34.6 | 227.5 | -101.2 | -1.2 | 10.2 | 2.3 | 29.5 | 256.2 | -98.6 |
| 3 | 09:11:49 | -3.2 | 4.2 | 13.8 | 39.5 | 86.3 | -207.8 | -5.4 | 8.8 | 3.9 | 41.2 | 196.8 | -125.5 | -5.7 | 8.6 | 6.1 | 40.4 | 295.3 | -112.5 |
| 4 | 09:11:59 | -1.1 | 6.8 | 21.0 | 38.4 | 84.1 | -216.7 | -11.3 | 7.3 | 9.8 | 29.7 | 246.8 | -89.4 | -3.3 | 4.3 | 5.4 | 41.2 | 299.5 | -95.2 |
| 5 | 09:12:06 | 0.2 | 11.0 | 17.7 | 29.0 | 56.8 | -156.4 | -9.5 | 9.0 | 7.2 | 39.4 | 209.5 | -112.2 | -6.7 | 9.4 | 5.8 | 36.5 | 255.4 | -110.4 |
| 6 | 09:12:15 | -3.8 | 3.2 | 22.5 | 35.7 | 80.0 | -164.9 | -4.3 | 6.7 | 5.5 | 42.1 | 198.5 | -102.3 | -0.2 | 10.0 | 4.8 | 32.1 | 267.4 | -105.6 |

In the food packaging job, two tasks were defined. The first task involved placing individual frozen food packages into a box (12 in all), and the second task involved loading the filled box of packages onto a pallet. Risk probability charts are shown for each task in Figure 18.9. In Figure 18.9a, The package loading task was found to have a risk probability of 40%. (Probability values are calculated by averaging the individual logits from each of the five risk factors — in Figure 18.9a, (99%+30%+3%+23%+45%)/5 = 40%.) The risk charts for this first task that were produced using the DOS and Windows LMM software are shown in Appendix A-18.3. The probability value for the palletizing task (Figure 18.9b) was calculated to be 64%, clearly the task more similar to those considered "high risk." These two probability values are to be used only for comparison purposes. It is the chart in Figure 18.9c that reflects the true risk value for the entire job. This chart summarizes the largest values for each risk factor across both tasks making up this job. The value shown on this chart, and thus the risk for this material handling job, is 66%.

A closer examination of the top two charts in Figure 18.9 shows which factors most contributed to the job summary values in Figure 18.9c. Each of the five factors is discussed separately.

*Lift Rate* — The lifting frequency for the entire job was 420 lifts/hour. Because this variable is composed of the total number of lifts from both tasks, this value is shown to be the same on all charts in Figure 18.9. As indicated by the length of the lift rate bar on the charts, this rate is very rapid and is comparable to some of the highest frequency material handling jobs found in industry.

*Average Twisting Velocity* — The amount of twisting velocity required for package loading was fairly low, but it was moderately high for the palletizing task, as indicated by the length of the bars on these charts in Figure 18.9. The greater value of the two is used in the job summary chart in Figure 18.9c, which was taken from the palletizing task.

*Maximum Moment* — As shown on the charts in Figure 18.9, the moment values were very low for both tasks comprising this job. The low weight of the individual packages (each at one pound) and the fully packed box being palletized (12 pounds) generated low maximum moment values. The greater moment value from the palletizing task was used in the job summary chart.

*Maximum Sagittal Flexion* — The package loading task was performed while employees were in relatively upright postures. This is reflected in Figure 18.9a by a short sagittal flexion bar on the chart. However, during box palletizing, those boxes placed on the lower layers of the stack required much forward trunk flexion. Figure 18.9b depicts these higher angles by the long bar for this factor. Subsequently, this higher value of the two tasks resulted in it being used in the job summary chart in Figure 18.9c.

*Maximum Lateral Velocity* — The LMM determined that lateral velocities generated during the package loading task actually were higher than found during box palletizing. While the values for the previous factors all were larger during handling of the full box, and thus were used in the job summary chart, it is the lateral velocity value from the package loading task that must be used in the job summary, since it is the greater of the two tasks analyzed.

It is the job summary value of 66% (taken from the chart in Figure 18.9c) that represents the LBD risk for this example food processing job. The value indicates that, on the continuum of low-risk jobs (0%) to high-risk jobs (100%), this particular job has a 66% likelihood of being considered "high risk." As stated earlier in this chapter, a high-risk job was defined as one having twelve or more (with an average of 26.4) low back strains per 200,000 hours (or 100 workers/year) of employee exposure. Results here could be interpreted as indicating that this particular job has a moderately high chance of producing a large number of low back strain injuries among individuals who do this job.

The primary goal of the LMM analysis could be simply to determine whether or not a job presents a risk of low back strain to its employees. This can be determined by calculating the LBD risk value from a job summary. As Figure 18.10 shows, there is some overlap in risk values among jobs defined as low and high risk. However, our analyses (Marras et al., 1993) found that those jobs with fairly low probability values (below 30%, for example) were much more likely to be low-risk jobs. Similarly, very few jobs having a risk value over 50% were defined as "low risk," and no low-risk jobs had risk values over 70%. That is, jobs with probability values above 60% are virtually assured to have some low back injury risk associated with them.

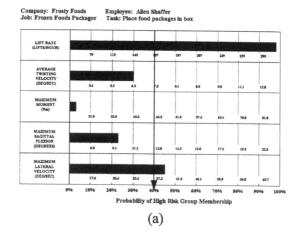

(a)

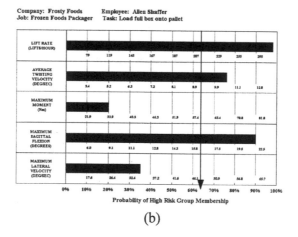

(b)

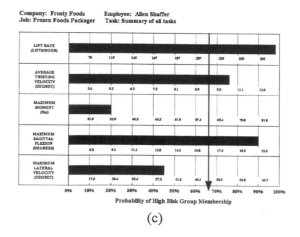

(c)

**FIGURE 18.9**   Risk value charts for the frozen food packager job: (a) "Place Two Food Packages in Box" task; (b) "Load Full Box onto Pallet" task; and (c) summary of all job tasks.

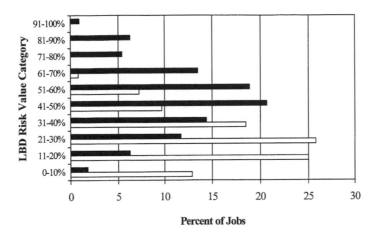

**FIGURE 18.10**   Distributions of the LBD risk values for jobs defined as "low risk" and those defined as "high risk" in Marras et al. (1993).

A second goal of the analysis may be to compare one job task with another, to determine which one(s) require more back motions and external moments about the trunk. This exercise can assist in learning which task(s) should be the focus of redesign efforts. Introducing ergonomic modifications to tasks already found to have low risk values probably will have little real impact on improving the job overall. However, making changes to tasks whose individual factors contribute most to the job's summary risk probability will reduce the probability for the entire job.

A third goal of the analysis may be to determine, for specific tasks, which individual components are most responsible for its composite probability value. This type of analysis provides direct information regarding how the job's requirements may affect those factors used in this probability model. Examples of ergonomic improvements that can be made on a job to affect each component are listed below. These examples should not be considered an exhaustive list of possible changes that can be implemented to improve working conditions.

*Lift Rate* — The total number of material handling actions required of the job affects this risk factor. Thus, reducing this rate will reduce the overall risk value. This type of change is not usually favored by management, since it will likely reduce productivity. However, there are ways to redesign jobs to reduce lifting frequency:

1. Rearrange Job Tasks — For jobs in which a number of tasks comprise the job, one method is to rearrange tasks with those of other jobs. This may more evenly distribute the lifting frequencies of several jobs, some of which may be considerably lower than the job monitored.

2. Rotate Jobs — Rotating employees between a job having a high lifting frequency with one having a much lower frequency also will reduce the rate for the job of interest. The effects of the job(s) into which employees are rotated must also be considered, however. It should be cautioned that this and the previous approach are most beneficial if the jobs that are rearranged or included in the rotation allow employees to use different muscle groups to perform the job. Rotating individuals into jobs that require the same muscles be used likely will have either no benefit or could produce greater musculoskeletal stress.

3. Add Employees — Dividing the job so that added personnel are available to perform the job will distribute the work across more people and lower the job's required lift rate. Of course, the cost of the additional employee(s) must be compared with the benefits of reduced low back strains and their related costs.

4. Automate — It may be possible that some job tasks can be automated through new equipment or robotics. This method of assisting the material handler will undoubtedly reduce the lifting frequency of the job and the overall job requirements.

*Average Twisting Velocity* — Rapid twisting of the trunk can result from a number of situations. If a work area is designed such that material transfer from point A to point B is difficult or if these two locations are not convenient to one another, a high twisting velocity may result. High velocities often result because work areas do not allow employees to move their feet to handle goods. As a result, a turning action an employee would normally do that included movement of the trunk, hips, legs, and ankles is more concentrated in the trunk, and higher twisting velocities can occur. It would be difficult to reduce the speed at which individuals twist simply by instructing them to "slow down." However, engineering controls that can reduce twisting velocity include:

1. Place Work in Front of Material Handler — Move the locations of point A relative to point B, so that they are more convenient to one another. In other words, create a workplace in which the material transfer requires moving in as few planes of motion as possible, thus allowing the employee to remain in a more neutral posture

2. Spread Out Congested Work Areas — Material handling areas that allow employees to walk or take at least one step between handling points A and B often reduces the twisting velocity of the job. This occurs because the added movement allows one to get into a position in which the entire body can assist in the transfer, rather than just the trunk.

3. Raise Working Heights — Depending on the location of goods to be handled, lower levels of work that require a great deal of sagittal flexion also may require additional trunk twisting. This is often the case when the work requires asymmetric lifting. By raising the work heights, not only will forward flexion be reduced, but twisting velocity can decrease because the handling distances are more appropriately located.

*Maximum Moment* — Because an external moment is the product of an object's weight and the distance from the body at which it is handled, reducing either of these two factors will reduce the moment. Various examples are described below.

1. Reduce Weight Requirements — Some work situations allow employees to handle as many units or as much weight as they feel is acceptable. However, to work faster, some employees may handle more goods at one time than is physically safe. A limit on the numbers/weights of objects handled in any given time can reduce Maximum Moment values. In some environments, raw materials handled are received from a supplier in bulk quantities. The weights of these materials may produce excessive moment values. By working with these suppliers, an arrangement may be possible to package materials in smaller, lower-weight containers. The weight changes just described likely will *increase* the lifting frequency of the job, however, so this trade-off in the risk model should be examined.

2. Install Material Handing Aids — For goods that are of uniform shape or size, several types of lifting aids are commercially available to provide handling assistance. These can be adapted to a wide range of work environments. Handling aids, such as lifting hoists, when incorporated successfully, greatly reduce the loading forces on the spine and result in much lower moment values. The device should be considered carefully, since a handling aid that is difficult to use or greatly slows the job process will likely be abandoned by the employees. In addition, some material handling aids can reduce the distance at which objects are handled. For example, lift/tilt tables are available commercially and able to be adapted to specific needs. These devices can raise or angle objects or bins of goods so that they can be more easily accessed. This can result in reduced horizontal reach distances.

3. Evaluate the Transfer Locations — The distance from the body at which individuals handle goods often is greatest during the initial or final contact with the product. This is often true during palletizing operations, in which cases need to be placed properly on a skid to ensure the load's stability. During carrying, for example, people tend to bring the product closer to their bodies. An evaluation of these locations may detect workplace arrangements that cause individuals to reach farther than is necessary to handle objects.

*Maximum Sagittal Flexion* — The more that individuals work in upright positions during material handling, the lower their trunk flexion will be. This reduces subsequent risk probabilities for jobs. Reducing sagittal flexion can be accomplished by eliminating tasks that, for example, require loads below knee level to be handled. Several interventions can prevent these situations.

1. Raise the Heights of Loads Placed near the Floor — Objects can be raised from the floor during material handling work a number of different ways. For palletizing tasks, stacking a skid underneath the pallet will raise the height of the bottom-most objects. Lift tables and self-raising devices placed underneath objects also will raise objects higher off the floor and reduce sagittal flexion. These changes should be evaluated for safety considerations and to ensure that objects at the top of the pallet can still be accessed. Other tables are available commercially that tilt or swivel objects, such as those on pallets, that bring the loads closer to the material handlers.

2. Adjust Working Heights Relative to an Individual's Standing Height — Work areas that are adjustable to accommodate those performing the work also can reduce trunk flexion. Work tables can be constructed or purchased that move vertically to a position most comfortable to the user. On assembly lines, conveyors can be adjusted that raise or lower objects depending on the work being performed at a specific location. Alternately, work areas alongside the lines, under the employees themselves, can be raised or lowered to produce the correct working height.

3. Train Employees on Proper Lifting Techniques — For some work situations, vertical adjustability may not be technically or economically feasible. In these cases, employees can be educated on proper lifting techniques aimed to reduce back strain and reduce the amount of sagittal flexion required.

*Maximum Lateral Velocity* — High lateral velocity values on the risk charts indicate that the material handling work requires rapid sideways bending. This motion may be difficult to visualize, but it usually indicates that work is not being performed in front of one's body, but asymmetrically instead. Workplace modifications that more conveniently locate or raise the work relative to the material handler (as already noted above) can assist in reducing this factor on the probability charts. A case study conducted by Stuart-Buttle (1995) found that reductions in lateral velocities and sagittal flexion can be achieved through workplace interventions in MMH tasks using lift tables. This paper also illustrated the importance of testing the impact of workplace modifications. The initial installation of this table actually produced higher LMM risk values and more employee dissatisfaction. The LMM analysis identified the problems with the new system and provided feedback about how the workplace needed to be further changed to produce actual risk reduction.

From the example job modifications just discussed, it is important to understand that these five workplace factors are interrelated. None of these factors responds independently from the others. For instance, adding a lift table to palletizing work may reduce sagittal flexion, because the load is being raised. However, is also can lower the maximum moment and average twisting velocity values, because the load may be held closer to the body during handling and be more easily accessed. If the work is self-paced, lift rate actually could *increase* since the work may be less physically demanding and those affected may be capable and willing to handle more material. This example illustrates the trade-offs that must be considered when evaluating the probability of risk for a job and implementing ergonomic interventions.

## Interpreting Results from Several Individuals

Casual observation of any material handling activity will show that different people usually perform the same job slightly differently. Employees inherently vary in how they do manual work, and this will likely produce different trunk motions. To account for these differences, the ergonomist may wish to use the LMM to monitor several employees who perform the same job and then analyze and interpret the results in terms of average probabilities across these individuals. The authors know of no literature that specifies how many employees must be monitored to account for all variability in a job. This process is task-dependent, and the ergonomist should study the amount of variability of the job tasks and work layout before deciding how many employees to monitor. Assembly-line or machine-paced operations may limit the freedom employees have in performing job tasks in comparison with self-paced material handling work. The ergonomist may want to adjust the number of individuals monitored accordingly.

**TABLE 18.6**  Computational Example for Assessing LBD Risk
from Three Employees Performing the Frozen Packager Job

Task 1: Place Two Food Packages in Box

| Risk Factor | Emp. 1 | Emp. 2 | Emp. 3 | Average |
|---|---|---|---|---|
| Lift Rate | 99.0% | 99.0% | 99.0% | 99.0% |
| Average Twisting Velocity | 30.0% | 21.0% | 42.0% | 31.0% |
| Maximum Moment | 3.0% | 2.0% | 4.0% | 3.0% |
| Maximum Sagittal Flexion | 23.0% | 26.0% | 17.0% | 22.0% |
| Maximum Lateral Velocity | 45.0% | 32.0% | 43.0% | 40.0% |
| **Average** | **40.0%** | **36.0%** | **41.0%** | **39.0%** |

Task 2: Load Full Box onto Pallet

| Risk Factor | Emp. 1 | Emp. 2 | Emp. 3 | Average |
|---|---|---|---|---|
| Lift Rate | 99.0% | 99.0% | 99.0% | 99.0% |
| Average Twisting Velocity | 76.0% | 75.0% | 80.0% | 77.0% |
| Maximum Moment | 20.0% | 22.0% | 24.0% | 22.0% |
| Maximum Sagittal Flexion | 90.0% | 94.0% | 89.0% | 91.0% |
| Maximum Lateral Velocity | 35.0% | 30.0% | 43.0% | 36.0% |
| **Average** | **64.0%** | **64.0%** | **67.0%** | **65.0%** |

Job Summary

| Risk Factor | Emp. 1 | Emp. 2 | Emp. 3 | Average |
|---|---|---|---|---|
| Lift Rate | 99.0% | 99.0% | 99.0% | 99.0% |
| Average Twisting Velocity | 76.0% | 75.0% | 80.0% | 77.0% |
| Maximum Moment | 20.0% | 22.0% | 24.0% | 22.0% |
| Maximum Sagittal Flexion | 90.0% | 94.0% | 89.0% | 91.0% |
| Maximum Lateral Velocity | 45.0% | 32.0% | 43.0% | 40.0% |
| **Average** | **66.0%** | **64.4%** | **67.0%** | **65.8%** |

If the LMM data collection and analysis software does not assess the average risk probability across employees doing the same tasks, then these values can be calculated manually, by following the steps listed below. A computational example of the results from three employees monitored who performed the frozen food packager job is shown in Table 18.6.

1. For Task 1, the average probabilities range from 36.0% to 41.0% for these three individuals. For Task 2, the range is 64.0% to 67.0%. To find the probability average for each of these tasks, average the values of *each* factor used in the model across all employees. That is, add the lift rate, average twisting velocity, maximum moment, maximum sagittal flexion, and maximum lateral velocity values separately and divide that number by the total number of employees monitored. Do this calculation for each job task. In Table 18.6, the right-most column shows the average values for each risk factor for the two material handling tasks of this job.

2. Compute the average probability for each task. To do this, sum the individual factor *probabilities* and divide by five. This value will be the average probability across employees. In the right-most column, Table 18.6 shows that the average probability for Task 1 is 39.0%, and it is 65% for Task 2. Also note that, due to individual differences in how the job is done, risk values vary from employee to employee.

3. Determine the *job* risk probability by reviewing all averaged task probabilities (again, those in the right-most column of Table 6 for Task 1 and Task 2). Find the highest probability value for each of the five factors and record it. This must be done for each employee and overall, to find the average risk for the job. As Table 18.6 shows, the largest average values for all five factors but maximum lateral velocity come from Task 2. These are shown in the bottom section of this table. Averaging these five values generates a job risk of 65.8%. This represents the risk for the entire job, averaged across all employees monitored.

# References

Andersson, GBJ, *The Adult Spine. Principles and Practice,* 2nd ed. Raven, New York, 1997.

Andersson GBJ, Epidemiologic aspects on low back pain in industry. *Spine,* 6(1):53-60, 1981.

Bigos SJ, Spengler DM, Martin NA, Zeh J, Fisher L, Nachemson A, and Wang MH, Back injuries in industry: A retrospective study. II. Injury factors. *Spine,* 11(3):246-251, 1986.

Cats-Baril WL and Frymoyer JW, The economics of spinal disorders, in Frymoyer JW, Ducker TB, Hadlet NM, Kostuik JP, Weinstein JW, and TS Whitecloud III (Eds.): *The Adult Spine: Principles and Practice.* Raven Press, New York, 85-105, 1991.

Gill, KP, and Callaghan MJ, Intratester and intertester reproducibility of the lumbar motion monitor as a measure of range, velocity and acceleration of the thoracolumbar spine. *Clinical Biomechanics,* 11(7):418-421, 1996.

Magora A, Investigation of the relation between low back pain and occupation. 4. Physical requirements: Bending, rotation, reaching and sudden maximal effort. *Scand J Rehab Med,* 5(4):186-190, 1973.

Marras, WS, Fathallah FA, Miller RJ, Davis SW, and Mirka GA, Accuracy of a three dimensional lumbar motion monitor for recording dynamic trunk motion characteristics. *International Journal of Industrial Ergonomics,* 9(1):75-87, 1992.

Marras WS, Lavender SA, Leurgans SE, Rajulu SL, Allread WG, Fathallah FA, and Ferguson SA, The role of dynamic three-dimensional motion in occupationally-related low back disorders. *Spine,* 18(5):617-628, 1993.

National Center for Health Statistics, *Prevalence of Selected Impairments,* United States Government Printing Office, Series 10, No. 134, 1977.

National Institute for Occupational Safety and Health (NIOSH): Work practices guide for manual lifting. Department of Health and Human Services (DHHS), National Institute for Occupational Safety and Health (NIOSH), Publication No. 81-122, 1981.

Putz-Anderson V and Waters TR. Revisions in NIOSH guide to manual lifting. Paper presented at national conference entitled "A national strategy for occupational musculoskeletal injury prevention — Implementation issues and research needs." Ann Arbor, MI, University of Michigan, 1991.

Punnett L, Fine LJ, Keyserling WM, Herrin GD, and Chaffin DB. Back disorders and nonneutral trunk postures of automobile assembly workers. *Scand J Work Environ Health,* 17:337-346, 1991.

Stuart-Buttle C, A case study of factors influencing the effectiveness of scissor lifts for box palletizing. *American Industrial Hygiene Association Journal,* 56:1127-1132, 1995.

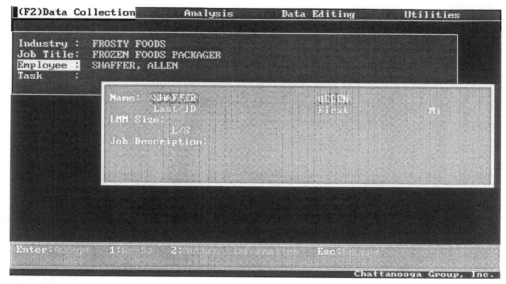

(a)

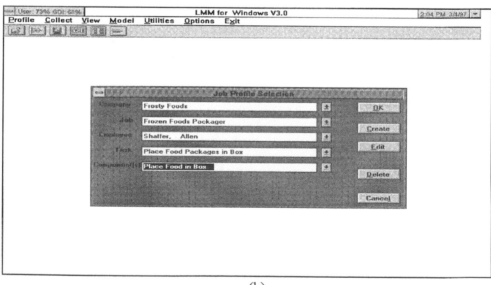

(b)

**APPENDIX A-18.1**    Example job profile selection screens for Chattanooga software; (a) DOS version; (b) Windows version.

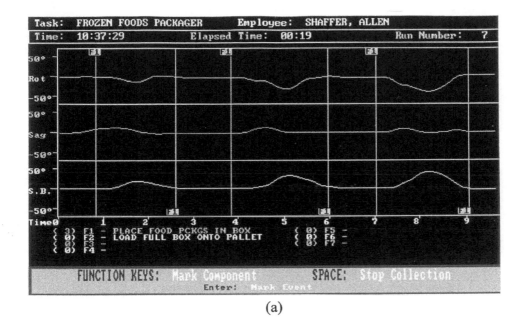

(a)

(b)

**APPENDIX A-18.2** Example data collection screens for Chattanooga software; (a) DOS version; (b) Windows version.

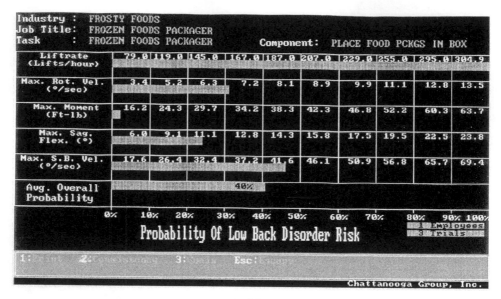

(a)

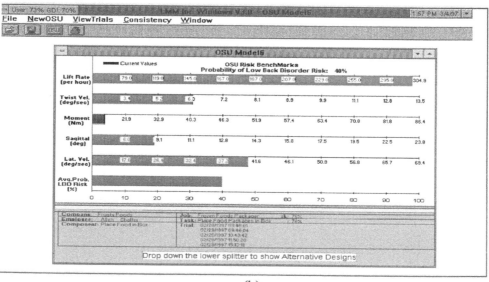

(b)

**APPENDIX A-18.3**    Example OSU Model screens for Chattanooga software; (a) DOS version; (b) Windows version.

# 19

# Prevention of Musculoskeletal Disorders: Psychophysical Basis

Patrick G. Dempsey
*Liberty Mutual Research Center for*
*Safety & Health*

## 19.1   Introduction

Occupational ergonomics is a discipline that seeks to maximize the performance of systems found in diverse settings such as production work, service industries, and health care. In this context, the system consists of the personnel, the equipment, the organizational and ambient environments, and the tasks required to produce a good or service. System performance refers to the ratio between system outputs (goods and services) and inputs to the system (personnel, capital, etc.). Not only is productivity at individual workstations important, but so is preventing injuries and illnesses associated with the tasks performed. A detriment to system performance in many settings in which manual work is performed is musculoskeletal disorders.

The direct costs (medical and indemnity payments) associated with work-related musculoskeletal disorders (WRMSDs) represent a significant source of financial losses. These losses have a direct economic impact on employers and insurers, which in turn are passed on to the customer. For example, customers who buy goods or services from an organization must bear the additional costs of the WRMSDs associated with producing those goods or providing those services.

An often overlooked aspect of WRMSDs is the negative impact of indirect costs (lost production time, training new workers, litigation, etc.) on system performance. For WRMSDs that develop over a period of time, there may be a gradual reduction in individual productivity. Once a disorder (whether it be the result of cumulative trauma or the direct result of a single overexertion) progresses to the point that the individual cannot continue to perform a job, there are additional indirect costs associated with lost production, replacing the individual, as well as potential performance decrements until the replacement worker becomes proficient at the task. The extent of indirect costs is partially a function of the injured individual's function in the system.

One approach to the prevention of WRMSDs is the psychophysical approach. This approach seeks to provide limits and guidelines for manual work that represent "maximum acceptable" work loads that minimize the injury potential of the work. To some extent, it is unknown if the psychophysical approach achieves this goal, as will be discussed in more detail later. In general, the extent to which many ergonomic criteria represent "optimal" limits with respect to optimizing the performance of systems comprised of manual work is unknown. Work loads and physical stresses have to be at or below levels which protect workers, but at the same time permit output levels which are economically sound. The psychophysical approach is an approach that seeks a balance between productivity and health and safety concerns.

## Chapter Goal and Outline

The goal of this chapter is to provide the reader with information concerning basic theory behind the psychophysical approach, the availability of data for designing manual tasks, the methods of applying the data, and the limitations of the data. The primary focus will be on the application of psychophysical techniques rather than on empirical methodologies or a literature review of the theoretical underpinnings of the psychophysical approach. Where necessary, theoretical and empirical results will be used to justify specific application techniques or to explain caveats of psychophysical data. Manual materials handling (MMH) and upper-extremity intensive (UEI) tasks have been the focus of psychophysical research, and each will be considered separately due to the disparity in application methodologies.

Readers interested in the empirical and theoretical aspects of the psychophysical approach to MMH task design are referred to Snook[1,2] and Ayoub and Mital[3] for further reading. For specific information on the comparisons of the psychophysical approach to the biomechanical and physiological approaches to MMH task design, the reader should consult Ayoub,[4] Mital et al.,[5] and Nicholson.[6] Less thorough information on the empirical and theoretical aspects of the psychophysical approach to UEI task design is available. However, there are useful discussions in Kim et al.[7] and Fernandez et al.[8]

## Introduction to Psychophysics

Psychophysics is a branch of psychology dealing with the relationships between stimuli and sensations. These relationships can be best described by the psychophysical power law. The psychological magnitude (sensation) $\psi$ grows as a power function of the physical magnitude $\phi$ (stimulus) in the following manner:[9]

$$\psi = k\phi^n$$

The value of the constant $k$ depends on the units of measure, while the exponent $n$ has a value that varies for different sensations. The value of $n$ may be lower than 1 for stimuli such as smell and brightness, or as high as 3.5 for electric shock.[9] Ljungberg et al.[10] found a value of $n = 1.86$ for a simulated brewery lifting task, whereas Gamberale et al.[11] found a value of 2.43 for a similar task. Gamberale et al. attributed the larger value in the latter study to more demanding lifting cycles.

In MMH experiments, the subject adjusts the magnitude of the stimulus (weight, force, or frequency) to correspond to a sensation which is "dictated" by the instructions given by the experimenter, i.e., "without straining yourself, or becoming unusually tired, weakened, overheated, or out of breath."[12] Adaptations of these instructions have been used to study UIE tasks. For UEI tasks, the instructions are directed at having subjects select work loads that do not result in "unusual discomfort in the hands, wrists, or forearms."[13] Subjects are also monitored during experimentation for signs of soreness, stiffness, and numbness.

## The Current State of Psychophysical Data

Psychophysical data are currently available for designing MMH tasks as well as UEI tasks. For MMH tasks, there are fairly extensive data available for maximum acceptable weights and forces for lifting, lowering, pushing, pulling, holding, and carrying tasks. Research of the psychophysical approach to MMH task design has spanned approximately three decades, and current work continues to expand the range of task conditions for which psychophysical data are available.

For UEI tasks, data are available for maximum acceptable frequencies and forces for a variety of tasks. It should be noted that there are considerably more data available for designing MMH tasks than UEI tasks. The increased attention to cumulative trauma disorders (CTDs) in the past decade or so led researchers to adapt psychophysical techniques used in MMH research to the study of UEI tasks. In part, this was done because of the lack of quantitative guidelines for forces, durations, and postures associated with UEI work. Although there are not extensive data, there are data that can be used to design manual work involving the upper extremities.

## 19.2 The Psychophysical Approach to Designing Manual Materials Handling Tasks

One of the first approaches to the control of MMH injuries through specifying task limits was the psychophysical approach. Applications in the military which relied on subjective estimates of load handling limits[14,15] were followed by the psychophysical approach being utilized to set industrial materials handling limits.[16] The methodologies and database developed by Snook and his colleagues[1,16-18,19] at the Liberty Mutual Research Center have been used by researchers and practitioners for the past several decades. This has resulted in the availability of a wide range of data available for designing MMH tasks (e.g., References 5, 19–22).

## Setting Weight and Force Limits

The application of the psychophysical approach to MMH task design is performed by using databases in the literature which provide limits specific to task conditions such as frequency, pushing or pulling distance, and load dimensions. Figure 19.1 presents a general model of the procedures associated with applying psychophysical data.

To use a database, the user records the relevant task parameters, then finds the value in the database applicable to the specific task. Since it is impossible to collect data for all combinations of tasks, one can either use interpolation to find the appropriate value, or one can use the closest value. In the latter case, the user should use the lower of the values in between which the task parameters fall. Next, the analyst determines if a task is acceptable. At a minimum, an MMH task should be designed to accommodate 75% of the population:[1] however, one should strive to accommodate at least 90% of the population whenever possible. If females perform the task, then the design should be based upon accommodating the female workers.

When a task does not accommodate at least 75% of the population, or if the task does accommodate 75% of the population and minor changes can be made to increase the acceptability of the task at little or no cost, then the task and/or workplace should be redesigned to accommodate at least 90% of the population. Options to increase the acceptability of a task will be described in more detail later. If a task cannot be redesigned to accommodate at least 75% of the population, then preplacement testing should be considered.

### Psychophysical Data

Most psychophysical databases present maximum acceptable weights or forces. There is limited information available on maximum acceptable frequencies.[23-25] The data described in detail here are all related to force and weight data.

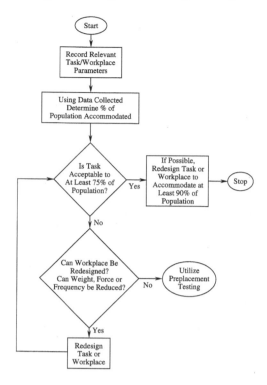

**FIGURE 19.1**    General model of the psychophysical approach to analyzing manual tasks.

The largest and most comprehensive single database for designing MMH tasks is that of Snook and Ciriello.[19] The data were collected from industrial subjects. The database includes maximum acceptable weights for lifting and lowering tasks, maximum acceptable initial and sustained forces for pushing and pulling tasks, and maximum acceptable weights for carrying tasks. The database provides values for males and females, as well as values that accommodate 10, 25, 50, 75, and 90% of the respective populations. An expanded computerized version of the database which covers a wider range of task conditions is available which performs analyses for both single- and multiple-component jobs (CompuTask™ *).

For lifting and lowering tasks, Snook and Ciriello's[19] database accommodates frequencies between one lift/lower every eight hours and 12 lifts/lowers per minute. Box dimensions in the sagittal plane between 34 and 75 cm are accommodated, as are lifts and lowers with vertical distances between 25 and 76 cm. Finally, data for floor to knuckle height, knuckle to shoulder height, and shoulder height to arm reach ranges of lift/lower are available.

Snook and Ciriello[19] presented maximum acceptable forces for pushing and pulling tasks, including forces required to start an object in motion (initial force) and forces required to keep an object in motion (sustained forces). Vertical handle heights between 57 and 135 cm are accommodated by the data. Frequencies for pushing/pulling are the same as those for lifting. The push/pull distances range between 2.1 m and 61.0 m.

Finally, Snook and Ciriello's[19] database also contains maximum acceptable weight of carry values for the same frequencies as the other MMH tasks mentioned above. Carry distances of 2.1 m, 4.3 m, and 8.5 m are available with vertical heights of 79 and 111 cm for males and 72 and 105 cm for females. Snook and Ciriello's[19] data are presented in Tables 19.1 through 19.9.

### Modified Psychophysical Databases

Mital et al.[5] modified Snook and Ciriello's[19] database to conform with biomechanical, physiological, and epidemiological criteria. Mital et al. also present data from other sources for additional types of MMH

---

*Registered trademark of Liberty Mutual Insurance Group, Boston, Massachusetts

tasks. Snook and Ciriello's[19] database was modified to satisfy the following criteria: spinal compression limits of 2689 and 3920 N for females and males, respectively, in consideration of the results of Liles et al.;[26] an intra-abdominal pressure limit of 90 mmHg; and an energy expenditure limit of 21 to 23% of treadmill aerobic capacity or 28 to 29% of bicycle aerobic capacity. Mital et al.[5] also present maximum acceptable frequencies of lift for one-handed lifting, acceptable holding time data, and maximum acceptable forces and weights of lift for MMH tasks performed in nonstandard postures such as lying down and kneeling. In order to extend the range of applicability of the modified data, Mital et al.[5] presented seven multipliers to adjust the data for the following factors: work duration, limited headroom, asymmetrical lifting, load asymmetry, couplings, load placement clearance, and heat stress.

Mital[27] found that psychophysical data collected in short periods (i.e., 20 to 25 min.) assuming a longer work period (8 to 12 hours) should be reduced. Subsequently, Mital[21] presented psychophysical data for males and females performing lifting tasks for 8-hour work shifts based on the adjustments determined in the earlier study. The data were collected from 37 males and 37 females experienced in manual lifting. Mital[21] also presented a modified database representing the combined data from his study, Snook's[1] data, and data collected by Ayoub et al.[20] Although the modified database only accommodates lifting tasks, the combined sample size is considerable. Similarly, Mital[22] presented a psychophysical database for lifting tasks for males and females working 12-hour shifts. The database represents values valid for 12 hours based on adjustments of 8-hour data.

## Data for Nonstandard MMH Tasks

One advantage of the psychophysical approach is that it allows for the realistic simulation of many types of materials handling tasks. Several such examples will be provided to illustrate how psychophysics has been used to develop guidelines for specific applications.

Smith et al.[28] presented a psychophysical database for evaluating MMH tasks performed in unusual postures. Maximum acceptable weight data for 99 different tasks were presented, including data for one- and two-handed lifting and lowering tasks performed in postures such as lifting on one knee, lifting on two knees, lifting while lying down, etc. These data are particularly appropriate for occasional maintenance tasks which impose postural constraints on the operator.

Like maintenance, mining is comprised of many activities for which standard psychophysical data are not applicable. Mining tasks are often performed under postural constraints, such as limited headroom. Gallagher[29] collected psychophysical data to address tasks performed under restricted headroom conditions and provided guidelines for tasks performed in low-seam coal mines requiring lifting while kneeling. Mining also requires handling of a variety of materials. Gallagher and Hamrick[30] provided psychophysical guidelines for the handling of rock dust bags, ventilation stopping blocks, and crib blocks.

## Assessment of Multiple Component MMH Tasks

More often than not, workers who perform MMH tasks perform a number of tasks, often in sequence. For example, a common combination task in industry is where a worker lifts material, carries it for some distance, then lowers the material. In such situations, all of the tasks should be evaluated.

Snook and Ciriello[19] recommend using the weight or force limit for the task with lowest percentage of the population accommodated as the design criterion for multiple component tasks. This recommendation was based upon the findings of Ciriello et al.[17,18] Thus, for a combination where the worker lifts materials, carries materials, and then lowers material, the limiting component task would be the lift. Snook and Ciriello[19] do caution that this method of analysis may result in violation of recommended energy expenditure criteria for some multiple component tasks.

Straker et al.[31] disagree with the multiple component methodology described in the preceding paragraph, stating that "combination tasks should probably be assessed as whole entities and not separated into components for analysis." However, Straker et al.[31] recommend no alternative methodology to assess multiple component tasks as "whole entities." Ciriello et al.[17,18] did evaluate combination tasks as whole entities using psychophysics, and as stated above, there is typically one component that is the limiting component. In general, the design and evaluation of multiple component MMH jobs is one of the more underdeveloped areas of MMH research and practice.

**TABLE 19.1** Maximum acceptable weight of lift for males (kg).

| Width‡ | Distance§ | Percent¶ | Floor level to knuckle height One lift every | | | | | | | | Knuckle height to shoulder height One lift every | | | | | | | | Shoulder height to arm reach One lift every | | | | | | | |
|---|---|---|---|---|---|---|---|---|---|---|---|---|---|---|---|---|---|---|---|---|---|---|---|---|---|---|
| | | | 5 | 9 | 14 | 1 | 2 | 5 | 30 | 8 | 5 | 9 | 14 | 1 | 2 | 5 | 30 | 8 | 5 | 9 | 14 | 1 | 2 | 5 | 30 | 8 |
| | | | s | s | s | min | min | min | min | h | s | s | s | min | min | min | min | h | s | s | s | min | min | min | min | h |
| | 76 | 90 | 6 | 7 | 9 | 11 | 13 | 14 | 14 | 17 | 8 | 10 | 12 | 13 | 14 | 14 | 16 | 17 | 6 | 8 | 9 | 10 | 10 | 11 | 12 | 13 |
| | | 75 | 9 | 11 | 13 | 16 | 19 | 20 | 21 | 24 | 10 | 14 | 16 | 18 | 18 | 19 | 21 | 23 | 8 | 10 | 12 | 14 | 14 | 14 | 16 | 17 |
| | | 50 | 12 | 15 | 17 | 22 | 25 | 27 | 28 | 32 | 13 | 17 | 20 | 22 | 23 | 24 | 26 | 29 | 10 | 13 | 15 | 17 | 17 | 18 | 20 | 22 |
| | | 25 | 15 | 18 | 21 | 28 | 31 | 34 | 35 | 41 | 16 | 21 | 24 | 27 | 27 | 28 | 32 | 35 | 11 | 16 | 18 | 21 | 21 | 22 | 24 | 27 |
| | | 10 | 18 | 22 | 25 | 33 | 37 | 40 | 41 | 48 | 19 | 24 | 28 | 31 | 32 | 33 | 37 | 40 | 14 | 18 | 21 | 24 | 24 | 25 | 28 | 31 |
| 75 | 51 | 90 | 6 | 8 | 9 | 12 | 13 | 15 | 15 | 17 | 8 | 11 | 13 | 15 | 15 | 16 | 18 | 19 | 6 | 8 | 9 | 12 | 12 | 12 | 14 | 15 |
| | | 75 | 9 | 11 | 13 | 17 | 19 | 21 | 22 | 25 | 11 | 15 | 17 | 20 | 20 | 21 | 23 | 25 | 8 | 11 | 12 | 15 | 15 | 16 | 18 | 20 |
| | | 50 | 13 | 15 | 18 | 23 | 26 | 28 | 29 | 34 | 14 | 19 | 21 | 25 | 25 | 26 | 29 | 32 | 10 | 14 | 16 | 19 | 20 | 20 | 23 | 25 |
| | | 25 | 16 | 19 | 22 | 29 | 33 | 35 | 36 | 42 | 17 | 23 | 26 | 30 | 31 | 32 | 36 | 39 | 13 | 17 | 19 | 23 | 24 | 25 | 27 | 30 |
| | | 10 | 19 | 22 | 26 | 34 | 38 | 42 | 43 | 50 | 20 | 26 | 30 | 35 | 36 | 37 | 41 | 45 | 15 | 19 | 22 | 27 | 27 | 29 | 32 | 35 |
| | 25 | 90 | 8 | 9 | 11 | 13 | 15 | 16 | 17 | 20 | 10 | 13 | 15 | 18 | 18 | 19 | 21 | 23 | 7 | 10 | 11 | 14 | 14 | 14 | 16 | 18 |
| | | 75 | 11 | 13 | 15 | 19 | 22 | 24 | 24 | 28 | 13 | 17 | 20 | 23 | 24 | 25 | 27 | 30 | 10 | 13 | 15 | 18 | 18 | 19 | 21 | 23 |
| | | 50 | 15 | 18 | 21 | 26 | 29 | 32 | 33 | 38 | 17 | 22 | 25 | 30 | 30 | 31 | 35 | 38 | 12 | 16 | 19 | 23 | 23 | 24 | 27 | 29 |
| | | 25 | 18 | 22 | 26 | 33 | 37 | 40 | 41 | 48 | 20 | 27 | 30 | 36 | 36 | 38 | 42 | 46 | 15 | 20 | 22 | 28 | 28 | 29 | 32 | 35 |
| | | 10 | 22 | 26 | 31 | 38 | 44 | 47 | 49 | 57 | 23 | 31 | 35 | 42 | 42 | 44 | 49 | 53 | 17 | 23 | 26 | 32 | 32 | 34 | 38 | 41 |
| | 76 | 90 | 7 | 8 | 10 | 13 | 15 | 16 | 17 | 20 | 8 | 10 | 12 | 13 | 14 | 14 | 16 | 17 | 7 | 9 | 10 | 12 | 12 | 13 | 14 | 16 |
| | | 75 | 10 | 12 | 14 | 19 | 22 | 24 | 24 | 28 | 10 | 14 | 16 | 18 | 18 | 19 | 21 | 23 | 9 | 11 | 13 | 16 | 16 | 17 | 19 | 21 |
| | | 50 | 14 | 16 | 19 | 26 | 29 | 32 | 33 | 38 | 13 | 17 | 20 | 22 | 23 | 24 | 26 | 29 | 11 | 15 | 17 | 20 | 21 | 21 | 24 | 26 |
| | | 25 | 17 | 20 | 24 | 33 | 37 | 40 | 41 | 48 | 16 | 21 | 24 | 27 | 27 | 28 | 32 | 35 | 13 | 18 | 20 | 25 | 25 | 26 | 29 | 31 |
| | | 10 | 20 | 24 | 28 | 38 | 43 | 47 | 48 | 57 | 19 | 24 | 28 | 31 | 32 | 33 | 37 | 40 | 15 | 21 | 23 | 28 | 29 | 30 | 33 | 36 |

| | | | | | | | | | | | | | | | | | | | | | | | | | | |
|---|---|---|---|---|---|---|---|---|---|---|---|---|---|---|---|---|---|---|---|---|---|---|---|---|---|---|
| | | 90 | 7 | 9 | 10 | 14 | 16 | 17 | 18 | 20 | 8 | 11 | 13 | 15 | 15 | 16 | 13 | 19 | 7 | 9 | 11 | 14 | 14 | 14 | 16 | 18 |
| | | 75 | 10 | 13 | 15 | 20 | 23 | 25 | 25 | 30 | 11 | 15 | 17 | 20 | 20 | 21 | 23 | 25 | 9 | 12 | 14 | 18 | 18 | 19 | 21 | 23 |
| 49 | 51 | 50 | *14* | 17 | 20 | 27 | 30 | 33 | 34 | 40 | 14 | 19 | 21 | 25 | 25 | 26 | 29 | 32 | 12 | 15 | 18 | 23 | 23 | 24 | 27 | 29 |
| | | 25 | *18* | *21* | 25 | 34 | 38 | 42 | 43 | 50 | 17 | 23 | 26 | 30 | 31 | 32 | 36 | 39 | 14 | 19 | 21 | 28 | 28 | 29 | 32 | 35 |
| | | 10 | *21* | *25* | *29* | 40 | 45 | 49 | 50 | 59 | *20* | *26* | *30* | 35 | 36 | 37 | 41 | 45 | *16* | *22* | 25 | 32 | 32 | 34 | 37 | 41 |
| | | 90 | 8 | 10 | 12 | 16 | 18 | 19 | 20 | 23 | 10 | 13 | 15 | 18 | 18 | 19 | 21 | 23 | 9 | 11 | 12 | 16 | 16 | 17 | 19 | 21 |
| | | 75 | 12 | 15 | 17 | 23 | 26 | 28 | 29 | 33 | 13 | 17 | 20 | 23 | 24 | 25 | 27 | 30 | 11 | 14 | 16 | 21 | 21 | 22 | 25 | 27 |
| | 25 | 50 | *16* | 20 | 23 | 30 | 34 | 37 | 38 | 45 | 17 | 22 | 25 | 30 | 30 | 31 | 35 | 38 | 14 | 18 | 21 | 27 | 27 | 28 | 32 | 35 |
| | | 25 | *21* | *25* | 29 | 38 | 43 | 47 | 48 | 56 | *20* | *27* | 30 | 36 | 36 | 38 | 42 | 46 | 16 | 22 | 25 | 33 | 33 | 34 | 38 | 42 |
| | | 10 | *24* | *29* | *34* | 45 | 51 | 56 | 57 | 67 | *23* | *31* | *35* | 42 | 42 | 44 | 49 | 53 | *19* | *25* | *29* | 38 | 38 | 40 | 44 | 48 |
| | | 90 | 8 | 10 | 11 | 15 | 17 | 19 | 19 | 23 | 8 | 11 | 13 | 15 | 15 | 16 | 18 | 19 | 8 | 10 | 12 | 14 | 14 | 15 | 16 | 18 |
| | | 75 | *12* | 14 | 17 | 22 | 25 | 28 | 28 | 33 | 11 | 15 | 17 | 20 | 20 | 21 | 23 | 25 | 10 | 14 | 16 | 18 | 19 | 19 | 22 | 24 |
| | 76 | 50 | *16* | *19* | 22 | 30 | 34 | 37 | 38 | 44 | 14 | 19 | 21 | 25 | 25 | 26 | 29 | 32 | 13 | 17 | 20 | 23 | 24 | 25 | 27 | 30 |
| | | 25 | *20* | *24* | *28* | 37 | 42 | 47 | 47 | 55 | 17 | *23* | 26 | 30 | 31 | 32 | 36 | 39 | *16* | *21* | 24 | 28 | 29 | 30 | 33 | 36 |
| | | 10 | *24* | *29* | *33* | 44 | 50 | 54 | 56 | 65 | *20* | *26* | *30* | 35 | 36 | 37 | 41 | 45 | *18* | *24* | *28* | 33 | 33 | 34 | 38 | 42 |
| | | 90 | 9 | 10 | 12 | 16 | 18 | 20 | 20 | 24 | 9 | 12 | 14 | 17 | 17 | 18 | 20 | 22 | 8 | 11 | 13 | 16 | 16 | 17 | 18 | 20 |
| | | 75 | *12* | 15 | 18 | 23 | 26 | 28 | 29 | 34 | 12 | 16 | 18 | 22 | 23 | 23 | 26 | 29 | 11 | 14 | 17 | 21 | 21 | 22 | 24 | 26 |
| 34 | 51 | 50 | *17* | 20 | 24 | 31 | 35 | 38 | 39 | 46 | 15 | 20 | 23 | 28 | 29 | 30 | 33 | 36 | 14 | 18 | 21 | 26 | 27 | 28 | 31 | 34 |
| | | 25 | *21* | *25* | *30* | 39 | 44 | 48 | 49 | 57 | 18 | 24 | 27 | 34 | 35 | 36 | 40 | 44 | *17* | 22 | 25 | 32 | 32 | 33 | 37 | 41 |
| | | 10 | *25* | *30* | *35* | 46 | 52 | 57 | 58 | 68 | *21* | *28* | *32* | 40 | 40 | 42 | 46 | 51 | *19* | *26* | *29* | 37 | 37 | 39 | 43 | 47 |
| | | 90 | 10 | 12 | 14 | 18 | 20 | 22 | 23 | 27 | 11 | 14 | 16 | 20 | 20 | 21 | 23 | 26 | 10 | 13 | 15 | 19 | 19 | 19 | 22 | 24 |
| | | 75 | *15* | 18 | 21 | 26 | 30 | 32 | 33 | 38 | 14 | 18 | 21 | 26 | 27 | 28 | 31 | 34 | 13 | 17 | 20 | 24 | 25 | 26 | 29 | 31 |
| | 25 | 50 | *20* | 24 | 28 | 35 | 40 | 43 | 44 | 52 | 18 | 23 | 27 | 33 | 34 | 35 | 39 | 43 | 16 | 22 | 25 | 31 | 31 | 33 | 36 | 40 |
| | | 25 | *26* | *30* | *35* | 44 | 50 | 54 | 55 | 65 | *21* | *28* | 32 | 40 | 41 | 42 | 47 | 52 | *20* | *26* | *30* | 37 | 38 | 39 | 44 | 46 |
| | | 10 | *29* | *35* | *41* | 52 | 59 | 64 | 66 | 76 | *25* | *33* | *37* | 47 | 47 | 49 | 55 | 60 | *23* | *30* | *35* | 43 | 44 | 45 | 51 | 55 |

‡Box width (the dimension away from the body) (cm).

§Vertical distance of lift (cm).

¶Percentage of industrial population.

Italicized values exceed 8 h physiological criteria.

From Snook, S. H. and Ciriello, V. M., The design of manual handling tasks: revised tables of maximum acceptable weights and forces, *Ergonomics*, 34, 1197, 1991. With permission.

**TABLE 19.2** Maximum acceptable weight of lift for females (kg).

| Width† | Distance§ | Percent¶ | Floor level to knuckle height — One lift every | | | | | | | | Knuckle height to shoulder height — One lift every | | | | | | | | Shoulder height to arm reach — One lift every | | | | | | | |
|---|---|---|---|---|---|---|---|---|---|---|---|---|---|---|---|---|---|---|---|---|---|---|---|---|---|---|
| | | | 5 (s) | 9 (s) | 14 (s) | 1 (min) | 2 (min) | 5 (min) | 30 (min) | 8 (h) | 5 (s) | 9 (s) | 14 (s) | 1 (min) | 2 (min) | 5 (min) | 30 (min) | 8 (h) | 5 (s) | 9 (s) | 14 (s) | 1 (min) | 2 (min) | 5 (min) | 30 (min) | 8 (h) |
| 75 | 76 | 90 | 5 | 6 | 7 | 7 | 8 | 8 | 9 | 12 | 5 | 6 | 7 | 9 | 9 | 9 | 10 | 12 | 4 | 5 | 5 | 6 | 7 | 7 | 7 | 8 |
| | | 75 | 7 | 8 | 9 | 9 | 10 | 10 | 11 | 14 | 6 | 7 | 8 | 10 | 11 | 11 | 12 | 14 | 5 | 6 | 6 | 7 | 8 | 8 | 8 | 10 |
| | | 50 | 8 | 10 | 10 | 11 | 12 | 12 | 13 | 17 | 7 | 8 | 9 | 11 | 12 | 12 | 13 | 16 | 6 | 7 | 7 | 8 | 9 | 9 | 10 | 11 |
| | | 25 | 9 | 11 | 12 | 13 | 14 | 14 | 15 | 21 | 8 | 9 | 10 | 13 | 14 | 14 | 15 | 18 | 7 | 7 | 8 | 9 | 10 | 10 | 11 | 13 |
| | | 10 | 11 | 13 | 14 | 14 | 15 | 16 | 17 | 23 | 9 | 10 | 11 | 14 | 15 | 15 | 17 | 20 | 7 | 8 | 9 | 10 | 11 | 11 | 12 | 14 |
| | 51 | 90 | 6 | 7 | 8 | 8 | 9 | 9 | 10 | 14 | 6 | 7 | 8 | 9 | 10 | 10 | 11 | 13 | 5 | 6 | 7 | 7 | 7 | 7 | 8 | 9 |
| | | 75 | 7 | 9 | 9 | 10 | 11 | 11 | 13 | 17 | 7 | 8 | 9 | 11 | 12 | 12 | 13 | 15 | 6 | 7 | 8 | 8 | 8 | 9 | 9 | 11 |
| | | 50 | 9 | 10 | 11 | 12 | 13 | 14 | 15 | 21 | 9 | 9 | 11 | 13 | 14 | 14 | 15 | 17 | 7 | 8 | 9 | 9 | 10 | 10 | 11 | 13 |
| | | 25 | 10 | 12 | 13 | 15 | 16 | 16 | 18 | 24 | 10 | 11 | 12 | 14 | 16 | 16 | 17 | 20 | 8 | 9 | 10 | 10 | 11 | 11 | 12 | 14 |
| | | 10 | 11 | 14 | 15 | 17 | 18 | 18 | 20 | 27 | 11 | 12 | 14 | 16 | 17 | 17 | 19 | 22 | 9 | 10 | 11 | 12 | 13 | 13 | 14 | 16 |
| | 25 | 90 | 6 | 8 | 8 | 9 | 9 | 9 | 11 | 14 | 6 | 7 | 8 | 10 | 11 | 11 | 12 | 14 | 5 | 6 | 7 | 8 | 8 | 8 | 9 | 10 |
| | | 75 | 8 | 10 | 11 | 11 | 12 | 12 | 13 | 18 | 7 | 8 | 9 | 12 | 13 | 13 | 14 | 17 | 6 | 7 | 8 | 9 | 9 | 9 | 10 | 12 |
| | | 50 | 10 | 12 | 13 | 13 | 14 | 14 | 16 | 21 | 9 | 10 | 11 | 14 | 15 | 15 | 16 | 19 | 7 | 8 | 9 | 10 | 11 | 11 | 12 | 14 |
| | | 25 | 11 | 14 | 15 | 15 | 16 | 17 | 19 | 25 | 10 | 11 | 12 | 16 | 17 | 17 | 19 | 22 | 8 | 9 | 10 | 12 | 12 | 12 | 14 | 16 |
| | | 10 | 13 | 16 | 17 | 17 | 19 | 19 | 21 | 29 | 11 | 12 | 14 | 18 | 19 | 19 | 21 | 24 | 9 | 10 | 11 | 13 | 14 | 14 | 15 | 17 |
| 76 | 76 | 90 | 5 | 6 | 7 | 8 | 8 | 8 | 9 | 13 | 5 | 6 | 7 | 9 | 9 | 9 | 10 | 12 | 4 | 5 | 5 | 7 | 7 | 7 | 8 | 9 |
| | | 75 | 7 | 8 | 9 | 10 | 10 | 10 | 12 | 16 | 6 | 7 | 8 | 10 | 11 | 11 | 12 | 14 | 5 | 6 | 6 | 8 | 8 | 8 | 9 | 11 |
| | | 50 | 8 | 10 | 10 | 12 | 12 | 13 | 14 | 19 | 7 | 8 | 9 | 11 | 12 | 12 | 13 | 16 | 6 | 7 | 7 | 9 | 10 | 10 | 11 | 12 |
| | | 25 | 9 | 11 | 12 | 14 | 15 | 15 | 17 | 22 | 8 | 9 | 10 | 13 | 14 | 14 | 15 | 18 | 7 | 7 | 8 | 10 | 11 | 11 | 12 | 14 |
| | | 10 | 11 | 13 | 14 | 15 | 17 | 17 | 19 | 25 | 9 | 10 | 11 | 14 | 15 | 15 | 17 | 20 | 7 | 8 | 9 | 11 | 12 | 12 | 13 | 15 |

‡Box width (the dimension away from the body) (cm).

§Vertical distance of lift (cm).

¶Percentage of industrial population.

Italicized values exceed 8 h physiological criteria.

From Snook, S. H. and Ciriello, V. M., The design of manual handling tasks: revised tables of maximum acceptable weights and forces, *Ergonomics*, 34, 1197, 1991. With permission.

**TABLE 19.3** Maximum acceptable weight of lower for males (kg).

| Width‡ | Distance§ | Percent¶ | Knuckle height to floor level One lower every | | | | | | | | Shoulder height to knuckle height One lower every | | | | | | | | Arm reach to shoulder height One lower every | | | | | | | |
|---|---|---|---|---|---|---|---|---|---|---|---|---|---|---|---|---|---|---|---|---|---|---|---|---|---|---|
| | | | 5 | 9 | 14 | 1 | 2 | 5 | 30 | 8 | 5 | 9 | 14 | 1 | 2 | 5 | 30 | 8 | 5 | 9 | 14 | 1 | 2 | 5 | 30 | 8 |
| | | | s | s | | min | min | min | | h | s | s | | min | min | min | | h | s | s | | min | min | min | | h |
| 75 | 76 | 90 | 7 | 9 | 10 | 12 | 14 | 15 | 16 | 20 | 10 | 11 | 14 | 14 | 15 | 15 | 16 | 19 | 6 | 7 | 9 | 9 | 10 | 10 | 11 | 13 |
| | | 75 | 10 | 13 | 14 | 18 | 20 | 22 | 22 | 29 | 13 | 16 | 18 | 18 | 21 | 21 | 21 | 26 | 9 | 10 | 12 | 12 | 14 | 14 | 14 | 18 |
| | | 50 | *14* | 17 | 19 | 23 | 27 | 29 | 30 | 38 | 18 | 20 | 24 | 24 | 27 | 27 | 28 | 34 | 11 | 13 | 15 | 16 | 18 | 18 | 19 | 23 |
| | | 25 | *17* | *21* | 24 | 29 | 33 | 36 | 37 | 47 | 21 | 25 | 29 | 29 | 34 | 34 | 34 | 42 | 14 | 16 | 19 | 20 | 23 | 23 | 23 | 28 |
| | | 10 | *20* | *25* | *28* | 34 | 39 | 42 | 44 | 56 | *25* | 29 | 34 | 34 | 39 | 39 | 39 | 49 | 16 | 19 | 22 | 23 | 26 | 26 | 27 | 33 |
| | 51 | 90 | 8 | 10 | 11 | 13 | 15 | 16 | 17 | 21 | 11 | 12 | 14 | 15 | 17 | 17 | 18 | 22 | 7 | 8 | 9 | 10 | 12 | 12 | 12 | 15 |
| | | 75 | 11 | 14 | 15 | 18 | 21 | 23 | 23 | 30 | 14 | 17 | 20 | 21 | 24 | 24 | 24 | 30 | 9 | 11 | 13 | 14 | 16 | 16 | 16 | 20 |
| | | 50 | *14* | *18* | 20 | 24 | 28 | 30 | 31 | 40 | 19 | 21 | 25 | 27 | 31 | 31 | 31 | 38 | 12 | 14 | 16 | 18 | 21 | 21 | 21 | 26 |
| | | 25 | *18* | *22* | 25 | 30 | 34 | 37 | 39 | 49 | 23 | 26 | 31 | 33 | 38 | 38 | 38 | 47 | 15 | 17 | 20 | 22 | 25 | 25 | 26 | 32 |
| | | 10 | *21* | *26* | 29 | 36 | 41 | 44 | 46 | 58 | *27* | 31 | 36 | 38 | 44 | 44 | 44 | 55 | 17 | 20 | 24 | 26 | 30 | 30 | 30 | 37 |
| | 25 | 90 | 9 | 11 | 12 | 15 | 17 | 18 | 19 | 24 | 12 | 14 | 17 | 18 | 21 | 21 | 21 | 26 | 8 | 9 | 11 | 12 | 14 | 14 | 14 | 17 |
| | | 75 | 13 | 16 | 17 | 21 | 24 | 25 | 26 | 34 | 17 | 20 | 23 | 24 | 28 | 28 | 28 | 35 | 11 | 13 | 15 | 16 | 19 | 19 | 19 | 24 |
| | | 50 | *17* | *21* | 23 | 27 | 31 | 34 | 35 | 45 | 22 | 25 | 30 | 32 | 36 | 36 | 37 | 45 | 14 | 16 | 19 | 21 | 24 | 24 | 25 | 31 |
| | | 25 | *21* | *26* | 29 | 34 | 39 | 42 | 44 | 56 | 27 | 31 | 37 | 39 | 44 | 44 | 45 | 56 | 17 | 20 | 24 | 26 | 30 | 30 | 30 | 38 |
| | | 10 | *24* | *31* | 34 | 40 | 46 | 49 | 51 | 66 | *31* | 36 | 43 | 45 | 52 | 52 | 52 | 65 | 20 | 23 | 28 | 30 | 35 | 35 | 35 | 44 |
| 25 | 76 | 90 | 8 | 10 | 11 | 15 | 17 | 18 | 19 | 24 | 10 | 11 | 14 | 14 | 15 | 15 | 16 | 19 | 7 | 8 | 10 | 11 | 12 | 12 | 12 | 15 |
| | | 75 | 12 | 15 | 16 | 21 | 24 | 26 | 26 | 34 | 13 | 16 | 18 | 18 | 21 | 21 | 21 | 26 | 10 | 11 | 14 | 15 | 17 | 17 | 17 | 21 |
| | | 50 | *15* | *19* | 21 | 27 | 31 | 34 | 35 | 45 | 18 | 20 | 24 | 24 | 27 | 27 | 28 | 34 | 13 | 15 | 17 | 19 | 22 | 22 | 22 | 27 |
| | | 25 | *19* | *24* | 26 | 34 | 39 | 42 | 44 | 56 | 21 | 25 | 29 | 29 | 34 | 34 | 34 | 42 | 16 | 18 | 21 | 23 | 27 | 27 | 27 | 33 |
| | | 10 | *25* | *28* | *31* | 40 | 46 | 49 | 51 | 65 | 25 | 29 | 34 | 34 | 39 | 39 | 39 | 49 | 18 | 21 | 25 | 27 | 31 | 31 | 31 | 39 |

| Box width ‡ (cm) | Vertical distance § (cm) | Percent ¶ | | | | | | | | | | | | | | | | | | | | | | | | |
|---|---|---|---|---|---|---|---|---|---|---|---|---|---|---|---|---|---|---|---|---|---|---|---|---|---|---|
| 49 | 51 | 90 | 17 | 14 | 14 | 14 | 12 | 10 | 9 | 8 | 22 | 18 | 17 | 17 | 15 | 14 | 12 | 11 | 25 | 19 | 19 | 17 | 15 | 12 | 11 |
| | | 75 | 24 | 19 | 19 | 19 | 16 | 14 | 12 | 10 | 30 | 24 | 24 | 24 | 21 | 20 | 17 | 15 | 35 | 28 | 26 | 25 | 22 | 17 | 14 |
| | | 50 | 31 | 25 | 24 | 24 | 21 | 18 | 16 | 14 | 38 | 31 | 31 | 31 | 27 | 24 | 21 | 19 | 47 | 37 | 35 | 33 | 29 | 21 | 19 |
| | | 25 | 37 | 30 | 30 | *35* | 26 | 23 | 19 | 17 | 47 | 38 | 38 | 38 | 33 | 31 | 26 | 23 | 58 | 46 | 44 | 41 | 36 | 26 | 23 |
| | | 10 | 44 | 35 | 35 | 35 | 30 | 26 | 22 | 19 | 55 | 44 | 44 | 44 | 38 | 35 | 31 | 27 | 68 | 54 | 51 | 48 | 42 | 31 | 27 |
| | 25 | 90 | 20 | 16 | 16 | 16 | 14 | 12 | 10 | 9 | 26 | 21 | 21 | 21 | 18 | 17 | 14 | 12 | 28 | 22 | 22 | 21 | 19 | 14 | 12 |
| | | 75 | 28 | 22 | 22 | 22 | 19 | 17 | 14 | 12 | 35 | 28 | 28 | 28 | 24 | 23 | 20 | 17 | 40 | 31 | 31 | 30 | 25 | 20 | 17 |
| | | 50 | 36 | 29 | *29* | 29 | 25 | 22 | 18 | 16 | 45 | 37 | 36 | 36 | 32 | 30 | 25 | 22 | 54 | 41 | 41 | 39 | 34 | 25 | 21 |
| | | 25 | 44 | *23* | 35 | 35 | 31 | 27 | 23 | 20 | 56 | 45 | 44 | 44 | 39 | 37 | 31 | 27 | 65 | 51 | 49 | 46 | 40 | 31 | 26 |
| | | 10 | 52 | *27* | *34* | *37* | 36 | 31 | 26 | 23 | 65 | 52 | 52 | 52 | 45 | 43 | 38 | 33 | 77 | 60 | 58 | 54 | 47 | 36 | 31 |
| 34 | 76 | 90 | 18 | 14 | 14 | 14 | 12 | 10 | 9 | 9 | 22 | 18 | 17 | 17 | 15 | 14 | 13 | 11 | 27 | 21 | 21 | 19 | 17 | 13 | 11 |
| | | 75 | 24 | *14* | *17* | 19 | 17 | 13 | 13 | 12 | 30 | 24 | 24 | 24 | 21 | 20 | 18 | 17 | 39 | 30 | 29 | 27 | 24 | 18 | 17 |
| | | 50 | 31 | *18* | *23* | 25 | 22 | 19 | 17 | 15 | 38 | 31 | 31 | 31 | 27 | 25 | 25 | 21 | 51 | 40 | 39 | 36 | 31 | 25 | 21 |
| | | 25 | 38 | *23* | *29* | 31 | 27 | 23 | 21 | 19 | 47 | 38 | 38 | 38 | 33 | 31 | 31 | 26 | 64 | 50 | 48 | 45 | 39 | 31 | 25 |
| | | 10 | 45 | *27* | *34* | *37* | 31 | 27 | 25 | 22 | 55 | 44 | 44 | 44 | 38 | 36 | 36 | 31 | 75 | 59 | 57 | 53 | 46 | 36 | 30 |
| | 51 | 90 | 20 | 16 | 16 | 16 | 14 | 12 | 10 | 9 | 24 | 20 | 20 | 20 | 17 | 15 | 14 | 11 | 29 | 22 | 22 | 20 | 18 | 14 | 12 |
| | | 75 | 27 | 22 | 22 | 22 | 19 | 17 | 14 | 12 | 33 | 27 | 27 | 27 | 23 | 21 | 19 | 15 | 40 | 32 | 30 | 28 | 24 | 19 | 17 |
| | | 50 | 35 | 28 | 28 | 28 | 24 | 21 | 19 | 16 | 43 | 35 | 35 | 35 | 30 | 27 | 24 | 20 | 53 | 42 | 40 | 37 | 33 | 24 | 22 |
| | | 25 | 43 | *24* | *30* | 35 | 30 | 27 | 23 | 20 | 53 | 43 | 43 | 42 | 37 | 31 | 30 | 23 | 67 | 52 | 50 | 47 | 42 | 30 | 24 |
| | | 10 | 50 | *28* | *35* | *38* | 35 | 31 | 27 | 23 | 62 | 50 | 50 | 49 | 43 | 35 | 35 | 27 | 78 | 62 | 59 | 55 | 50 | 35 | 28 |
| | 25 | 90 | 23 | 19 | 19 | 19 | 16 | 15 | 12 | 11 | 23 | 23 | 23 | 23 | 20 | 18 | 16 | 13 | 32 | 25 | 24 | 23 | 19 | 16 | 15 |
| | | 75 | 32 | *17* | 26 | 26 | 22 | 20 | 17 | 15 | 39 | 31 | 31 | 31 | 27 | 25 | 22 | 18 | 46 | 36 | 34 | 32 | 27 | 21 | 20 |
| | | 50 | 41 | *23* | 33 | 33 | 29 | 26 | 22 | 19 | 51 | 41 | 41 | 41 | 35 | 32 | 29 | 23 | 60 | 47 | 46 | 42 | 38 | 26 | 26 |
| | | 25 | 51 | *28* | 42 | 42 | 35 | 32 | 27 | 23 | 63 | 50 | 50 | 50 | 42 | 39 | 35 | 27 | 75 | 59 | 57 | 53 | 46 | 35 | 32 |
| | | 10 | 59 | *33* | *41* | *45* | 41 | 37 | 31 | 27 | 73 | 59 | 58 | 58 | 51 | 46 | 41 | 33 | 89 | 70 | 67 | 63 | 57 | 41 | 37 |

‡Box width (the dimension away from the body) (cm).

§Vertical distance of lower (cm).

¶Percentage of industrial population.

Italicized values exceed 8 h physiological criteria.

From Snook, S. H. and Ciriello, V. M., The design of manual handling tasks: revised tables of maximum acceptable weights and forces, *Ergonomics*, 34, 1197, 1991. With permission.

**TABLE 19.4** Maximum acceptable weight of lower for females (kg).

| Width‡ | Distance§ | Percent¶ | Knuckle height to floor level — One lower every | | | | | | | | Shoulder height to knuckle height — One lower every | | | | | | | | Arm reach to shoulder height — One lower every | | | | | | | |
|---|---|---|---|---|---|---|---|---|---|---|---|---|---|---|---|---|---|---|---|---|---|---|---|---|---|---|
| | | | 5 (s) | 9 (s) | 14 | 1 | 2 | 5 (min) | 30 | 8 (h) | 5 (s) | 9 (s) | 14 | 1 | 2 | 5 (min) | 30 | 8 (h) | 5 (s) | 9 (s) | 14 | 1 | 2 | 5 (min) | 30 | 8 (h) |
| | 76 | 90 | 5 | 6 | 7 | 7 | 8 | 8 | 9 | 12 | 6 | 6 | 7 | 8 | 9 | 10 | 10 | 13 | 5 | 5 | 5 | 6 | 7 | 7 | 7 | 9 |
| | | 75 | 6 | 8 | 8 | 9 | 10 | 10 | 11 | 14 | 7 | 8 | 8 | 11 | 11 | 12 | 12 | 15 | 5 | 6 | 6 | 7 | 8 | 8 | 9 | 11 |
| | | 50 | 7 | 9 | 10 | 11 | 12 | 12 | 13 | 17 | 8 | 9 | 10 | 12 | 13 | 14 | 14 | 18 | 7 | 8 | 8 | 8 | 10 | 10 | 10 | 13 |
| | | 25 | *9* | 11 | 12 | 12 | 14 | 14 | 15 | 20 | *9* | 11 | 11 | 13 | 15 | 17 | 17 | 21 | *8* | 9 | 9 | 10 | 11 | 12 | 12 | 15 |
| | | 10 | *10* | 13 | 13 | 14 | 15 | 16 | 17 | 23 | *11* | 12 | 13 | 15 | 17 | 19 | 19 | 24 | *9* | *10* | 10 | 11 | 12 | 14 | 14 | 17 |
| 75 | 51 | 90 | 6 | 7 | 7 | 8 | 9 | 10 | 10 | 14 | 7 | 8 | 8 | 9 | 10 | 11 | 11 | 14 | 5 | 6 | 6 | 6 | 7 | 8 | 8 | 10 |
| | | 75 | 7 | 8 | 9 | 10 | 11 | 12 | 13 | 17 | 8 | 9 | 9 | 11 | 12 | 13 | 13 | 17 | 7 | 7 | 8 | 8 | 9 | 10 | 10 | 12 |
| | | 50 | *8* | 10 | 11 | 12 | 14 | 14 | 15 | 20 | 10 | 11 | 11 | 13 | 15 | 16 | 16 | 20 | *8* | 9 | 9 | 9 | 11 | 12 | 12 | 15 |
| | | 25 | *10* | 12 | 13 | 14 | 16 | 17 | 18 | 24 | 11 | 13 | 13 | 15 | 17 | 19 | 19 | 23 | *9* | *10* | 11 | 11 | 12 | 13 | 13 | 17 |
| | | 10 | *11* | *13* | 14 | 16 | 18 | 19 | 20 | 27 | *13* | 15 | 15 | 17 | 19 | 21 | 21 | 26 | *10* | *12* | 12 | 12 | 14 | 15 | 15 | 19 |
| | 25 | 90 | 6 | 8 | 8 | 9 | 10 | 10 | 11 | 14 | 7 | 8 | 8 | 10 | 11 | 12 | 12 | 15 | 5 | 6 | 6 | 7 | 8 | 9 | 9 | 11 |
| | | 75 | 8 | 10 | 10 | 11 | 12 | 12 | 13 | 17 | 8 | 9 | 9 | 12 | 13 | 15 | 15 | 19 | 7 | 7 | 8 | 9 | 10 | 11 | 11 | 13 |
| | | 50 | *9* | 11 | 12 | 13 | 14 | 15 | 16 | 21 | 10 | 11 | 11 | 14 | 16 | 18 | 18 | 22 | 8 | 9 | 9 | 10 | 12 | 13 | 13 | 16 |
| | | 25 | *11* | 13 | 14 | 15 | 17 | 17 | 19 | 25 | 11 | 13 | 13 | 16 | 19 | 20 | 20 | 26 | 9 | 10 | 11 | 12 | 13 | 15 | 15 | 19 |
| | | 10 | *12* | 15 | 16 | 17 | 19 | 20 | 21 | 28 | *13* | 15 | 15 | 19 | 21 | 23 | 23 | 29 | 10 | 12 | 12 | 13 | 15 | 17 | 17 | 21 |
| | 76 | 90 | 5 | 6 | 7 | 8 | 8 | 9 | 10 | 13 | 6 | 6 | 7 | 8 | 9 | 10 | 10 | 13 | 5 | 5 | 5 | 6 | 7 | 8 | 8 | 10 |
| | | 75 | 6 | 8 | 8 | 9 | 10 | 11 | 12 | 16 | 7 | 8 | 8 | 10 | 11 | 12 | 12 | 15 | 5 | 6 | 6 | 8 | 9 | 9 | 9 | 12 |
| | | 50 | 8 | 9 | 10 | 11 | 13 | 13 | 14 | 19 | 8 | 9 | 10 | 12 | 13 | 14 | 14 | 18 | 7 | 8 | 8 | 9 | 11 | 11 | 11 | 14 |
| | | 25 | 9 | 11 | 11 | 13 | 15 | 16 | 17 | 22 | 9 | 11 | 11 | 13 | 15 | 17 | 17 | 21 | *8* | *9* | 9 | 11 | 12 | 13 | 13 | 16 |
| | | 10 | *10* | 13 | 13 | 15 | 17 | 18 | 19 | 25 | *11* | 12 | 13 | 15 | 17 | 19 | 19 | 24 | *9* | *10* | 10 | 12 | 13 | 15 | 15 | 19 |

| Box width ‡ | Vert dist § | % ¶ | | | | | | | | | | | | | | | | | | | | | | | | | | | | |
|---|---|---|---|---|---|---|---|---|---|---|---|---|---|---|---|---|---|---|---|---|---|---|---|---|---|---|---|---|---|---|
| 49 | 51 | 90 | 6 | 6 | 7 | 7 | 9 | 10 | 10 | 11 | 11 | 15 | 11 | 8 | 7 | 5 | 6 | 7 | 8 | 9 | 9 | 8 | 11 | 9 | 9 | 8 | 7 | 6 | 5 |
| | | 75 | 7 | 8 | 9 | 8 | 11 | 12 | 13 | 13 | 14 | 18 | 14 | 9 | 8 | 7 | 7 | 8 | 9 | 10 | 10 | 9 | 13 | 10 | 10 | 10 | 8 | 7 | 7 |
| | | 50 | *8* | *10* | *11* | 10 | 13 | 15 | 16 | 16 | 16 | 22 | 16 | 11 | 10 | 8 | 8 | 9 | 11 | 13 | 13 | 11 | 16 | 13 | 13 | 11 | 10 | 9 | 8 |
| | | 25 | *10* | *12* | *13* | 12 | 15 | 17 | 18 | 19 | 19 | 25 | 19 | 13 | 12 | 9 | 10 | 10 | 13 | 15 | 15 | 13 | 18 | 15 | 15 | 13 | 11 | 10 | 9 |
| | | 10 | *11* | *13* | 13 | 14 | 17 | 19 | 20 | 21 | 21 | 29 | 22 | 15 | 13 | *10* | *12* | 12 | 15 | 16 | 16 | 15 | 21 | 16 | 16 | 15 | 13 | 12 | 11 |
| | 25 | 90 | 6 | 6 | 8 | 8 | 10 | 11 | 12 | 13 | 10 | 15 | 12 | 9 | 7 | 5 | 6 | 8 | 9 | 9 | 9 | 8 | 12 | 9 | 9 | 8 | 8 | 6 | 6 |
| | | 75 | 8 | 8 | 10 | 10 | 13 | 13 | 15 | 15 | 12 | 19 | 14 | 10 | 8 | 7 | 7 | 9 | 10 | 12 | 12 | 9 | 14 | 12 | 12 | 10 | 9 | 8 | 7 |
| | | 50 | 9 | 9 | 11 | 12 | 15 | 15 | 16 | 16 | 14 | 23 | 17 | 11 | 9 | 8 | 8 | 11 | 13 | 14 | 14 | 11 | 17 | 14 | 14 | 13 | 11 | 9 | 8 |
| | | 25 | *11* | *11* | 13 | 14 | 18 | 18 | 20 | 20 | 16 | 27 | 20 | 13 | 11 | 9 | 9 | 13 | 15 | 16 | 16 | 13 | 20 | 16 | 16 | 15 | 13 | 11 | 9 |
| | | 10 | *12* | *12* | 15 | 16 | 20 | 20 | 23 | 23 | 19 | 30 | 25 | 15 | 13 | *10* | *10* | 15 | 16 | 18 | 18 | 15 | 23 | 18 | 18 | 16 | 15 | 13 | 11 |
| 76 | 90 | 90 | 6 | 6 | 8 | 8 | 9 | 10 | 11 | 11 | 9 | 15 | 11 | 8 | 7 | 5 | 6 | 7 | 8 | 9 | 9 | 9 | 12 | 9 | 9 | 8 | 7 | 6 | 5 |
| | 75 | 75 | 8 | 8 | 10 | 10 | 11 | 13 | 13 | 13 | 11 | 19 | 14 | 9 | 8 | 7 | 8 | 9 | 10 | 11 | 11 | 9 | 14 | 11 | 11 | 10 | 8 | 7 | 7 |
| | 50 | 50 | *10* | *10* | 12 | 12 | 13 | 15 | 16 | 17 | 13 | 23 | 17 | 11 | 10 | 8 | 9 | 11 | 13 | 13 | 14 | 11 | 17 | 13 | 13 | 11 | 11 | 9 | 8 |
| | 25 | 25 | *11* | *11* | 14 | 14 | 15 | 18 | 19 | 20 | 15 | 27 | 20 | 13 | 11 | *9* | *11* | 13 | 15 | 15 | 16 | 13 | 20 | 16 | 16 | 13 | 13 | 11 | 9 |
| | 10 | 10 | *13* | *13* | 16 | 16 | 17 | 20 | 21 | 23 | 17 | 30 | 23 | 14 | 13 | *11* | *12* | 14 | 16 | 16 | 18 | 15 | 23 | 18 | 18 | 16 | 14 | 13 | 11 |
| 34 | 51 | 90 | 7 | 7 | 9 | 9 | 11 | 12 | 13 | 14 | 9 | 18 | 14 | 8 | 7 | 7 | 8 | 8 | 9 | 11 | 11 | 8 | 13 | 11 | 11 | 10 | 8 | 8 | 7 |
| | | 75 | 9 | 9 | 11 | 11 | 13 | 15 | 15 | 17 | 11 | 22 | 17 | 9 | 8 | 8 | 9 | 9 | 10 | 13 | 13 | 9 | 16 | 13 | 13 | 12 | 10 | 9 | 8 |
| | | 50 | *10* | *10* | 13 | 14 | 16 | 18 | 18 | 20 | 13 | 27 | 20 | 11 | 10 | *10* | *11* | 11 | 13 | 14 | 15 | 11 | 19 | 15 | 15 | 14 | 11 | 11 | 10 |
| | | 25 | *12* | *12* | 15 | 16 | 19 | 21 | 21 | 24 | 15 | 31 | 26 | 13 | 12 | *11* | *13* | 13 | 14 | 16 | 18 | 13 | 22 | 18 | 18 | 16 | 14 | 13 | 11 |
| | | 10 | *14* | *14* | 17 | 18 | 21 | 24 | 25 | 27 | 18 | 35 | 29 | 15 | 13 | *13* | *15* | 15 | 16 | 18 | 20 | 15 | 25 | 20 | 20 | 18 | 16 | 15 | 13 |
| | 25 | 90 | 8 | 8 | 10 | 11 | 12 | 13 | 13 | 14 | 9 | 19 | 13 | 9 | 8 | 7 | 7 | 9 | 11 | 12 | 12 | 9 | 15 | 12 | 12 | 11 | 9 | 8 | 8 |
| | | 75 | *10* | *10* | 12 | 13 | 15 | 16 | 17 | 17 | 11 | 23 | 16 | 11 | 9 | 8 | 8 | 11 | 13 | 14 | 14 | 11 | 18 | 14 | 14 | 13 | 11 | 10 | 9 |
| | | 50 | *12* | *12* | 14 | 15 | 19 | 20 | 21 | 20 | 13 | 28 | 20 | 13 | 11 | *10* | *11* | 13 | 15 | 17 | 17 | 13 | 21 | 17 | 17 | 15 | 13 | 11 | 10 |
| | | 25 | *14* | *14* | 17 | 18 | 22 | 23 | 24 | 23 | 15 | 33 | 23 | 16 | 13 | *11* | *13* | 15 | 18 | 19 | 19 | 15 | 24 | 19 | 19 | 18 | 15 | 13 | 11 |
| | | 10 | *15* | *15* | 18 | 20 | 25 | 26 | 28 | 28 | 17 | 37 | 26 | 18 | 15 | *13* | *15* | 17 | 20 | 22 | 22 | 15 | 28 | 22 | 22 | 20 | 18 | 15 | 13 |

‡Box width (the dimension away from the body) (cm).
§Vertical distance of lower (cm).
¶Percentage of industrial population.
Italicized values exceed 8 h physiological criteria.

From Snook, S. H. and Ciriello, V. M., The design of manual handling tasks: revised tables of maximum acceptable weights and forces, *Ergonomics*, 34, 1197, 1991. With permission.

**TABLE 19.5**   Maximum acceptable forces of push for males (kg).

**Initial forces¶**

| Height† | Percent§ | 2.1 m push 6 s | 12 s | 1 min | 2 min | 5 min | 30 min | 8 h | 7.6 m push 15 s | 22 s | 1 min | 2 min | 5 min | 30 min | 8 h | 15.2 m push 25 s | 35 s | 1 min | 2 min | 5 min | 30 min | 8 h | 30.5 m push 1 min | 2 min | 5 min | 30 min | 8 h | 45.7 m push 1 min | 2 min | 5 min | 30 min | 8 h | 61.0 m push 1 min | 2 min | 5 min | 30 min | 8 h |
|---|---|---|---|---|---|---|---|---|---|---|---|---|---|---|---|---|---|---|---|---|---|---|---|---|---|---|---|---|---|---|---|---|---|---|---|---|---|
| 144 | 90 | 20 | 22 | 25 | 25 | 26 | 26 | 31 | 14 | 16 | 21 | 21 | 22 | 22 | 26 | 18 | 19 | 19 | 19 | 20 | 21 | 25 | 15 | 16 | 19 | 19 | 24 | 13 | 14 | 16 | 16 | 20 | 12 | 14 | 14 | 14 | 18 |
|  | 75 | 26 | 29 | 32 | 32 | 34 | 34 | 41 | 18 | 20 | 27 | 27 | 28 | 28 | 34 | 23 | 25 | 25 | 25 | 26 | 27 | 32 | 19 | 21 | 25 | 25 | 31 | 16 | 18 | 21 | 21 | 26 | 16 | 18 | 18 | 18 | 23 |
|  | 50 | 32 | 36 | 40 | 40 | 42 | 42 | 51 | 23 | 25 | 33 | 33 | 35 | 35 | 42 | 29 | 31 | 31 | 31 | 33 | 33 | 40 | 24 | 27 | 31 | 31 | 38 | 20 | 23 | 26 | 26 | 33 | 20 | 22 | 22 | 22 | 28 |
|  | 25 | 38 | 43 | 47 | 47 | 50 | 51 | 61 | 27 | 31 | 40 | 40 | 42 | 42 | 51 | 35 | 37 | 37 | 37 | 40 | 40 | 48 | 28 | 32 | 37 | 37 | 46 | 24 | 27 | 32 | 32 | 39 | 23 | 27 | 27 | 27 | 34 |
|  | 10 | 44 | 49 | 55 | 55 | 58 | 58 | 70 | 31 | 35 | 46 | 46 | 48 | 48 | 58 | 40 | 43 | 43 | 43 | 45 | 46 | 55 | 32 | 37 | 42 | 42 | 53 | 28 | 31 | 36 | 36 | 45 | 27 | 31 | 31 | 31 | 39 |
| 95 | 90 | 21 | 24 | 26 | 26 | 28 | 28 | 34 | 16 | 18 | 23 | 23 | 25 | 25 | 30 | 21 | 22 | 22 | 22 | 23 | 24 | 28 | 17 | 19 | 22 | 22 | 27 | 14 | 16 | 19 | 19 | 23 | 14 | 16 | 16 | 16 | 20 |
|  | 75 | 28 | 31 | 34 | 34 | 36 | 36 | 44 | 21 | 23 | 30 | 30 | 32 | 32 | 39 | 27 | 28 | 28 | 28 | 30 | 30 | 36 | 21 | 24 | 28 | 28 | 35 | 18 | 21 | 24 | 24 | 30 | 18 | 21 | 21 | 20 | 26 |
|  | 50 | 34 | 38 | 43 | 43 | 45 | 45 | 54 | 26 | 29 | 38 | 38 | 40 | 40 | 48 | 33 | 35 | 35 | 35 | 37 | 38 | 45 | 27 | 30 | 35 | 35 | 44 | 23 | 26 | 30 | 30 | 37 | 22 | 26 | 26 | 26 | 32 |
|  | 25 | 41 | 46 | 51 | 51 | 54 | 55 | 65 | 31 | 35 | 45 | 45 | 48 | 48 | 58 | 40 | 42 | 42 | 42 | 45 | 45 | 54 | 32 | 36 | 42 | 42 | 54 | 27 | 31 | 36 | 36 | 45 | 27 | 31 | 31 | 31 | 38 |
|  | 10 | 47 | 53 | 59 | 59 | 62 | 63 | 75 | 35 | 40 | 52 | 52 | 55 | 56 | 66 | 46 | 49 | 49 | 49 | 52 | 52 | 62 | 37 | 41 | 48 | 48 | 60 | 32 | 36 | 41 | 41 | 52 | 31 | 35 | 35 | 35 | 44 |
| 64 | 90 | 19 | 22 | 24 | 24 | 25 | 26 | 31 | 13 | 14 | 20 | 20 | 21 | 21 | 26 | 17 | 19 | 19 | 19 | 20 | 20 | 24 | 14 | 16 | 19 | 19 | 23 | 12 | 14 | 16 | 16 | 20 | 12 | 14 | 14 | 14 | 17 |
|  | 75 | 25 | 28 | 31 | 31 | 33 | 33 | 40 | 16 | 19 | 26 | 26 | 27 | 28 | 33 | 21 | 24 | 24 | 24 | 26 | 26 | 31 | 18 | 21 | 24 | 24 | 30 | 16 | 18 | 21 | 21 | 26 | 15 | 18 | 18 | 18 | 22 |
|  | 50 | 31 | 35 | 39 | 39 | 41 | 41 | 50 | 20 | 23 | 32 | 32 | 34 | 35 | 41 | 27 | 30 | 30 | 30 | 32 | 33 | 39 | 23 | 26 | 30 | 30 | 37 | 20 | 22 | 26 | 26 | 32 | 19 | 22 | 22 | 22 | 28 |
|  | 25 | 38 | 42 | 46 | 46 | 49 | 50 | 59 | 25 | 28 | 39 | 39 | 41 | 41 | 50 | 32 | 36 | 36 | 36 | 39 | 39 | 47 | 28 | 31 | 36 | 36 | 45 | 24 | 27 | 31 | 31 | 39 | 23 | 26 | 26 | 26 | 33 |
|  | 10 | 43 | 48 | 53 | 53 | 57 | 57 | 68 | 28 | 32 | 45 | 45 | 47 | 48 | 57 | 37 | 42 | 42 | 42 | 44 | 45 | 54 | 32 | 36 | 41 | 41 | 52 | 27 | 31 | 36 | 36 | 44 | 26 | 30 | 30 | 30 | 38 |

**Sustained forces***

| Height† | Percent§ | 2.1 m push 6 s | 12 s | 1 min | 2 min | 5 min | 30 min | 8 h | 7.6 m push 15 s | 22 s | 1 min | 2 min | 5 min | 30 min | 8 h | 15.2 m push 25 s | 35 s | 1 min | 2 min | 5 min | 30 min | 8 h | 30.5 m push 1 min | 2 min | 5 min | 30 min | 8 h | 45.7 m push 1 min | 2 min | 5 min | 30 min | 8 h | 61.0 m push 1 min | 2 min | 5 min | 30 min | 8 h |
|---|---|---|---|---|---|---|---|---|---|---|---|---|---|---|---|---|---|---|---|---|---|---|---|---|---|---|---|---|---|---|---|---|---|---|---|---|---|
| 144 | 90 | 10 | 13 | 15 | 15 | 16 | 18 | 22 | 8 | 9 | 13 | 13 | 15 | 16 | 18 | 9 | 11 | 11 | 11 | 12 | 13 | 16 | 8 | 10 | 12 | 13 | 16 | 7 | 8 | 10 | 11 | 13 | 7 | 8 | 9 | 11 | 13 |
|  | 75 | *13* | *17* | *21* | *18* | *24* | 25 | 30 | *10* | *13* | *17* | *18* | 20 | 21 | 25 | *11* | *13* | *15* | *16* | *18* | 18 | 21 | *11* | *13* | *16* | 18 | 21 | *9* | *11* | 13 | 15 | 18 | *9* | *11* | 11 | 13 | 15 |
|  | 50 | *17* | *22* | *27* | *28* | *31* | 32 | 38 | *13* | *16* | *22* | *23* | *26* | 27 | 32 | *14* | *17* | *20* | *21* | *24* | 24 | 28 | *15* | *17* | *20* | 23 | 28 | *12* | *14* | *17* | 19 | 23 | *12* | *14* | *13* | 16 | 19 |
|  | 25 | *21* | *27* | *33* | *34* | *38* | 40 | 47 | *16* | *20* | *28* | *29* | *32* | 33 | 39 | *17* | *20* | *24* | *25* | *28* | 29 | 34 | *18* | *21* | *25* | 28 | 34 | *15* | *18* | *21* | 24 | 28 | *15* | *17* | *16* | 20 | 24 |
|  | 10 | *25* | *31* | *38* | *40* | *45* | 46 | 54 | *19* | *23* | *32* | *33* | *38* | 39 | 46 | *20* | *24* | *28* | *28* | *33* | 34 | 40 | *21* | *25* | *29* | 33 | 40 | *18* | *21* | *24* | 28 | 33 | *19* | *20* | *20* | 23 | 28 |
| 95 | 90 | 10 | 13 | 16 | 16 | 17 | 19 | 23 | 8 | 10 | 13 | 13 | 15 | 15 | 18 | 8 | 10 | 11 | 11 | 12 | 13 | 16 | 8 | 10 | 12 | 13 | 16 | 7 | 8 | 9 | 11 | 13 | 7 | 8 | 9 | 11 | 13 |
|  | 75 | *14* | *18* | *22* | *22* | 25 | 26 | 31 | *8* | *11* | *13* | *17* | 20 | 21 | 25 | *11* | *13* | *15* | *16* | 18 | 18 | 21 | *11* | *13* | *16* | 18 | 21 | *9* | *11* | 13 | 15 | 18 | *9* | *11* | 11 | 12 | 15 |
|  | 50 | *18* | *23* | *28* | *28* | 29 | 34 | 40 | *14* | *17* | *19* | *23* | 26 | 27 | 32 | *15* | *17* | *19* | *21* | 23 | 23 | 28 | *15* | *17* | *20* | 23 | 28 | *12* | *14* | 17 | 19 | 23 | *12* | *14* | 14 | 16 | 19 |
|  | 25 | *22* | *28* | *34* | *34* | 35 | 41 | 49 | *17* | *21* | *21* | *29* | 32 | 33 | 39 | *18* | *21* | *24* | *24* | 28 | 28 | 34 | *18* | *21* | *25* | 28 | 34 | *15* | *18* | 21 | 24 | 28 | *15* | *17* | 17 | 20 | 23 |
|  | 10 | *26* | *33* | *40* | *40* | 41 | 48 | 57 | *20* | *25* | *27* | *32* | 37 | 38 | 45 | *20* | *25* | *28* | *28* | 32 | 33 | 39 | *21* | *25* | *29* | 33 | 40 | *17* | *20* | 24 | 27 | 32 | *17* | *20* | 20 | 23 | 27 |
| 64 | 90 | 10 | 13 | 16 | 16 | 18 | 19 | 23 | 8 | 10 | 12 | 13 | 14 | 15 | 18 | 8 | 10 | 11 | 11 | 11 | 13 | 15 | 8 | 9 | 11 | 13 | 15 | 7 | 8 | 9 | 11 | 13 | 7 | 8 | 9 | 11 | 13 |
|  | 75 | *14* | *18* | *21* | *22* | 25 | 26 | 31 | *11* | *13* | *17* | *17* | 19 | 20 | 24 | *11* | *13* | *14* | *15* | 17 | 17 | 21 | *11* | *13* | *15* | 17 | 20 | *9* | *11* | *12* | 14 | 17 | *9* | *10* | 10 | 12 | 14 |
|  | 50 | *18* | *23* | *29* | *29* | 32 | 33 | 39 | *14* | *17* | *21* | *21* | 25 | 26 | 31 | *14* | *17* | *19* | *19* | 22 | 22 | 27 | *14* | *16* | *19* | 22 | 26 | *12* | *14* | *16* | 18 | 22 | *12* | *14* | 14 | 15 | 18 |
|  | 25 | *22* | *28* | *34* | *35* | 39 | 41 | 48 | *17* | *21* | *26* | *27* | 31 | 32 | 37 | *18* | *21* | *23* | *23* | 27 | 28 | 33 | *17* | *20* | *24* | 27 | 32 | *14* | *17* | *20* | 23 | 27 | *14* | *17* | 17 | 19 | 22 |
|  | 10 | *26* | *32* | *39* | *41* | 46 | 48 | 56 | *20* | *25* | *30* | *32* | 36 | 37 | 44 | *21* | *25* | *27* | *27* | 31 | 32 | 38 | *20* | *24* | *28* | 32 | 37 | *17* | *20* | *23* | 26 | 31 | *16* | *19* | 19 | 22 | 26 |

† Vertical distance from floor to hands (cm).
§ Percentage of industrial population.
¶ The force required to get an object in motion.
* The force required to keep an object in motion.
Italicized values exceed 8 h physiological criteria.

From Snook, S. H. and Ciriello, V. M., The design of manual handling tasks: revised tables of maximum acceptable weights and forces, *Ergonomics*, 34, 1197, 1991. With permission.

**TABLE 19.6**   Maximum acceptable forces of push for females (kg).

**Initial forces¶**

| Height‡ | Percent§ | 2.1m 6s | 2.1m 12s | 2.1m 1 | 2.1m 2 | 2.1m 5 | 2.1m 30 | 2.1m 8h | 7.6m 15s | 7.6m 22s | 7.6m 1 | 7.6m 2 | 7.6m 5 | 7.6m 30 | 7.6m 8h | 15.2m 25s | 15.2m 35s | 15.2m 1 | 15.2m 2 | 15.2m 5 | 15.2m 30 | 15.2m 8h | 30.5m 1 | 30.5m 2 | 30.5m 5 | 30.5m 30 | 30.5m 8h | 45.7m 1 | 45.7m 2 | 45.7m 5 | 45.7m 30 | 45.7m 8h | 61.0m 2 | 61.0m 5 | 61.0m 30 | 61.0m 8h |
|---|---|---|---|---|---|---|---|---|---|---|---|---|---|---|---|---|---|---|---|---|---|---|---|---|---|---|---|---|---|---|---|---|---|---|---|---|
| 135 | 90 | 14 | 15 | 17 | 18 | 20 | 21 | 22 | 15 | 16 | 16 | 16 | 18 | 19 | 20 | 12 | 14 | 14 | 14 | 15 | 16 | 17 | 12 | 13 | 14 | 15 | 17 | 12 | 13 | 14 | 15 | 17 | 12 | 12 | 14 | 15 |
| | 75 | 17 | 18 | 21 | 22 | 24 | 25 | 27 | 18 | 19 | 19 | 20 | 22 | 23 | 24 | 15 | 17 | 17 | 17 | 18 | 20 | 21 | 15 | 16 | 17 | 19 | 21 | 15 | 16 | 17 | 19 | 21 | 14 | 15 | 17 | 19 |
| | 50 | 20 | 22 | 25 | 26 | 29 | 30 | 32 | 21 | 23 | 23 | 24 | 26 | 27 | 29 | 18 | 20 | 20 | 20 | 22 | 23 | 25 | 18 | 19 | 21 | 22 | 25 | 18 | 19 | 21 | 22 | 25 | 17 | 18 | 20 | 22 |
| | 25 | 24 | 25 | 29 | 30 | 33 | 35 | 37 | 25 | 26 | 27 | 28 | 31 | 32 | 34 | 20 | 23 | 23 | 24 | 26 | 27 | 29 | 20 | 22 | 24 | 26 | 29 | 20 | 22 | 24 | 26 | 29 | 20 | 21 | 23 | 26 |
| | 10 | 26 | 28 | 33 | 34 | 38 | 39 | 41 | 28 | 30 | 30 | 31 | 34 | 36 | 38 | 23 | 26 | 26 | 26 | 29 | 31 | 32 | 23 | 25 | 27 | 30 | 33 | 23 | 25 | 27 | 30 | 33 | 22 | 24 | 26 | 29 |
| 89 | 90 | 14 | 15 | 17 | 18 | 20 | 21 | 22 | 14 | 15 | 16 | 17 | 19 | 19 | 21 | 11 | 13 | 14 | 14 | 16 | 16 | 17 | 12 | 14 | 15 | 16 | 18 | 12 | 14 | 15 | 16 | 18 | 12 | 13 | 14 | 16 |
| | 75 | 17 | 18 | 21 | 22 | 24 | 25 | 27 | 17 | 18 | 20 | 20 | 22 | 23 | 25 | 14 | 16 | 17 | 17 | 19 | 20 | 21 | 15 | 16 | 18 | 19 | 21 | 15 | 16 | 18 | 19 | 21 | 15 | 16 | 17 | 19 |
| | 50 | 20 | 22 | 26 | 26 | 29 | 30 | 32 | 20 | 21 | 24 | 24 | 27 | 28 | 30 | 16 | 19 | 21 | 21 | 24 | 24 | 25 | 18 | 20 | 21 | 23 | 26 | 18 | 20 | 21 | 23 | 26 | 18 | 19 | 20 | 23 |
| | 25 | 24 | 25 | 30 | 30 | 33 | 35 | 37 | 23 | 25 | 28 | 28 | 31 | 33 | 34 | 19 | 22 | 23 | 24 | 27 | 28 | 30 | 21 | 23 | 24 | 26 | 30 | 21 | 23 | 24 | 26 | 30 | 20 | 22 | 24 | 27 |
| | 10 | 26 | 28 | 34 | 34 | 38 | 39 | 41 | 28 | 28 | 31 | 32 | 35 | 37 | 39 | 22 | 24 | 26 | 27 | 30 | 31 | 34 | 24 | 26 | 28 | 30 | 33 | 24 | 26 | 28 | 30 | 33 | 23 | 25 | 26 | 30 |
| 57 | 90 | 11 | 12 | 14 | 14 | 16 | 17 | 18 | 11 | 12 | 14 | 14 | 16 | 16 | 17 | 9 | 11 | 12 | 12 | 13 | 14 | 15 | 11 | 12 | 12 | 13 | 15 | 11 | 12 | 12 | 13 | 15 | 10 | 11 | 12 | 13 |
| | 75 | 14 | 15 | 17 | 17 | 19 | 20 | 21 | 14 | 15 | 17 | 17 | 19 | 20 | 21 | 11 | 13 | 14 | 15 | 16 | 17 | 18 | 13 | 14 | 15 | 16 | 18 | 13 | 14 | 15 | 16 | 18 | 12 | 13 | 14 | 16 |
| | 50 | 16 | 17 | 20 | 21 | 23 | 24 | 25 | 16 | 18 | 20 | 21 | 23 | 24 | 25 | 14 | 15 | 17 | 18 | 19 | 20 | 21 | 15 | 17 | 18 | 19 | 22 | 15 | 17 | 18 | 19 | 22 | 15 | 16 | 17 | 19 |
| | 25 | 19 | 20 | 24 | 24 | 27 | 28 | 30 | 19 | 21 | 24 | 24 | 27 | 28 | 29 | 16 | 18 | 20 | 20 | 23 | 24 | 25 | 18 | 19 | 21 | 22 | 25 | 18 | 19 | 21 | 22 | 25 | 17 | 19 | 20 | 23 |
| | 10 | 21 | 23 | 26 | 27 | 30 | 31 | 33 | 22 | 23 | 26 | 27 | 30 | 31 | 33 | 18 | 20 | 22 | 23 | 25 | 26 | 28 | 20 | 22 | 23 | 25 | 28 | 20 | 22 | 23 | 25 | 28 | 19 | 21 | 23 | 25 |

*(Column time units: s for the first columns of 2.1/7.6/15.2 m sections; min for the intermediate columns; h for the 8-hour columns.)*

**Sustained forces\***

| Height‡ | Percent§ | 2.1m 6s | 2.1m 12s | 2.1m 1 | 2.1m 2 | 2.1m 5 | 2.1m 30 | 2.1m 8h | 7.6m 15s | 7.6m 22s | 7.6m 1 | 7.6m 2 | 7.6m 5 | 7.6m 30 | 7.6m 8h | 15.2m 25s | 15.2m 35s | 15.2m 1 | 15.2m 2 | 15.2m 5 | 15.2m 30 | 15.2m 8h | 30.5m 1 | 30.5m 2 | 30.5m 5 | 30.5m 30 | 30.5m 8h | 45.7m 1 | 45.7m 2 | 45.7m 5 | 45.7m 30 | 45.7m 8h | 61.0m 2 | 61.0m 5 | 61.0m 30 | 61.0m 8h |
|---|---|---|---|---|---|---|---|---|---|---|---|---|---|---|---|---|---|---|---|---|---|---|---|---|---|---|---|---|---|---|---|---|---|---|---|---|
| 135 | 90 | 6 | 8 | 10 | 10 | 11 | 12 | 14 | 6 | 7 | 7 | 8 | 8 | 9 | 11 | 5 | 6 | 6 | 6 | 6 | 7 | 9 | 5 | 6 | 6 | 6 | 8 | 5 | 5 | 5 | 6 | 8 | 4 | 4 | 4 | 6 |
| | 75 | 9 | 12 | 14 | 14 | 16 | 17 | 21 | 9 | 10 | 11 | 11 | 12 | 13 | 16 | 7 | 8 | 9 | 9 | 10 | 11 | 13 | 7 | 8 | 9 | 9 | 12 | 7 | 7 | 8 | 8 | 11 | 6 | 6 | 6 | 9 |
| | 50 | 12 | 16 | 19 | 20 | 21 | 23 | 28 | 12 | 14 | 15 | 15 | 16 | 17 | 21 | 9 | 11 | 12 | 13 | 13 | 14 | 17 | 10 | 11 | 12 | 12 | 16 | 10 | 11 | 11 | 11 | 15 | 8 | 9 | 9 | 12 |
| | 25 | 16 | 20 | 24 | 25 | 27 | 29 | 36 | 15 | 17 | 18 | 19 | 20 | 22 | 27 | 12 | 14 | 15 | 16 | 16 | 18 | 22 | 13 | 15 | 15 | 16 | 20 | 13 | 14 | 14 | 14 | 19 | 11 | 11 | 12 | 15 |
| | 10 | 18 | 23 | 28 | 29 | 32 | 34 | 42 | 18 | 20 | 22 | 22 | 24 | 26 | 32 | 14 | 16 | 17 | 18 | 18 | 20 | 25 | 15 | 17 | 17 | 18 | 25 | 14 | 16 | 16 | 17 | 22 | 13 | 13 | 14 | 17 |
| 89 | 90 | 6 | 7 | 9 | 9 | 10 | 11 | 13 | 6 | 7 | 8 | 8 | 9 | 9 | 11 | 5 | 6 | 6 | 7 | 7 | 8 | 10 | 5 | 6 | 6 | 7 | 9 | 5 | 6 | 6 | 7 | 9 | 4 | 4 | 5 | 6 |
| | 75 | 8 | 11 | 13 | 13 | 15 | 16 | 19 | 8 | 9 | 11 | 11 | 12 | 13 | 16 | 7 | 8 | 9 | 10 | 10 | 11 | 14 | 8 | 9 | 9 | 11 | 13 | 7 | 8 | 8 | 9 | 12 | 6 | 6 | 7 | 9 |
| | 50 | 11 | 15 | 18 | 18 | 20 | 21 | 26 | 11 | 13 | 15 | 15 | 17 | 18 | 22 | 9 | 11 | 13 | 13 | 14 | 15 | 19 | 10 | 12 | 12 | 13 | 17 | 10 | 11 | 11 | 12 | 16 | 8 | 8 | 9 | 12 |
| | 25 | 14 | 18 | 22 | 23 | 25 | 27 | 33 | 15 | 17 | 19 | 19 | 21 | 23 | 28 | 12 | 14 | 16 | 16 | 17 | 19 | 24 | 13 | 15 | 15 | 16 | 22 | 12 | 14 | 14 | 14 | 20 | 11 | 11 | 12 | 15 |
| | 10 | 17 | 22 | 26 | 27 | 30 | 32 | 39 | 18 | 20 | 22 | 23 | 25 | 27 | 33 | 15 | 17 | 18 | 19 | 21 | 23 | 28 | 16 | 18 | 19 | 19 | 26 | 14 | 16 | 17 | 17 | 24 | 13 | 13 | 14 | 18 |
| 57 | 90 | 5 | 6 | 8 | 8 | 9 | 10 | 12 | 5 | 6 | 7 | 7 | 7 | 7 | 9 | 5 | 5 | 6 | 6 | 6 | 6 | 8 | 5 | 6 | 6 | 6 | 7 | 4 | 5 | 5 | 6 | 7 | 4 | 4 | 4 | 6 |
| | 75 | 7 | 9 | 11 | 12 | 13 | 14 | 17 | 7 | 8 | 8 | 8 | 8 | 10 | 11 | 7 | 8 | 8 | 9 | 9 | 10 | 12 | 7 | 8 | 8 | 8 | 9 | 6 | 6 | 7 | 7 | 9 | 5 | 6 | 6 | 8 |
| | 50 | 10 | 13 | 15 | 16 | 18 | 18 | 23 | 10 | 11 | 12 | 13 | 13 | 14 | 16 | 9 | 10 | 11 | 12 | 12 | 13 | 15 | 9 | 11 | 11 | 12 | 14 | 9 | 10 | 10 | 11 | 13 | 8 | 8 | 9 | 11 |
| | 25 | 12 | 16 | 19 | 20 | 22 | 23 | 29 | 13 | 14 | 15 | 16 | 17 | 18 | 21 | 11 | 13 | 14 | 15 | 15 | 16 | 19 | 12 | 14 | 14 | 14 | 18 | 11 | 13 | 13 | 14 | 17 | 10 | 10 | 11 | 14 |
| | 10 | 15 | 19 | 23 | 23 | 26 | 28 | 34 | 15 | 17 | 18 | 20 | 21 | 23 | 26 | 14 | 16 | 17 | 18 | 18 | 20 | 23 | 15 | 16 | 16 | 17 | 22 | 13 | 15 | 15 | 16 | 21 | 12 | 12 | 13 | 17 |

‡Vertical distance from floor to hands (cm).

§Percentage of industrial population.

¶The force required to get an object in motion.

\*The force required to keep an object in motion.

*Italicized values exceed 8 h physiological criteria.*

From Snook, S. H. and Ciriello, V. M., The design of manual handling tasks: revised tables of maximum acceptable weights and forces, *Ergonomics*, 34, 1197, 1991. With permission.

**TABLE 19.7** Maximum acceptable forces of pull for males (kg).

| Height† | Percent§ | 2.1 m pull — One pull every | | | | | | | 7.6 m pull — One pull every | | | | | | | 15.2 m pull — One pull every | | | | | | | 30.5 m pull — One pull every | | | | | 45.7 m pull — One pull every | | | | | 61.0 m pull — One pull every | | | |
|---|---|---|---|---|---|---|---|---|---|---|---|---|---|---|---|---|---|---|---|---|---|---|---|---|---|---|---|---|---|---|---|---|---|---|---|---|
| | | 6 | 12 | 1 | 2 | 5 | 30 | 8 | 15 | 22 | 1 | 2 | 5 | 30 | 8 | 25 | 35 | 1 | 2 | 5 | 30 | 8 | 1 | 2 | 5 | 30 | 8 | 1 | 2 | 5 | 30 | 8 | 2 | 5 | 30 | 8 |
| | | s | s | min | min | min | min | h | s | s | min | min | min | min | h | s | s | min | min | min | min | h | min | min | min | min | h | min | min | min | min | h | min | min | min | h |
| **Initial forces¶** | | | | | | | | | | | | | | | | | | | | | | | | | | | | | | | | | | | | | |
| 144 | 90 | 16 | 18 | 18 | 19 | 19 | 19 | 23 | 11 | 13 | 16 | 16 | 17 | 18 | 23 | 13 | 15 | 15 | 15 | 16 | 17 | 21 | 12 | 13 | 15 | 15 | 19 | 10 | 11 | 13 | 13 | 16 | 10 | 11 | 11 | 14 |
| | 75 | 19 | 22 | 22 | 23 | 23 | 24 | 28 | 14 | 15 | 20 | 20 | 21 | 21 | 28 | 16 | 18 | 19 | 19 | 20 | 20 | 26 | 14 | 16 | 19 | 19 | 23 | 12 | 14 | 16 | 16 | 20 | 12 | 14 | 14 | 17 |
| | 50 | 23 | 26 | 26 | 28 | 28 | 29 | 33 | 16 | 18 | 24 | 24 | 25 | 26 | 31 | 19 | 21 | 22 | 22 | 24 | 24 | 31 | 17 | 19 | 22 | 22 | 27 | 15 | 16 | 19 | 19 | 24 | 14 | 16 | 16 | 20 |
| | 25 | 27 | 31 | 31 | 32 | 33 | 33 | 39 | 19 | 21 | 28 | 28 | 29 | 30 | 39 | 22 | 25 | 26 | 26 | 28 | 28 | 36 | 20 | 22 | 26 | 26 | 32 | 17 | 19 | 22 | 22 | 28 | 16 | 19 | 19 | 24 |
| | 10 | 30 | 34 | 34 | 36 | 37 | 37 | 44 | 21 | 24 | 31 | 31 | 33 | 33 | 44 | 24 | 28 | 29 | 29 | 31 | 31 | 40 | 22 | 25 | 29 | 29 | 37 | 20 | 22 | 25 | 25 | 31 | 18 | 21 | 21 | 27 |
| 95 | 90 | 22 | 25 | 25 | 27 | 27 | 27 | 32 | 15 | 18 | 23 | 23 | 24 | 24 | 32 | 18 | 20 | 21 | 21 | 23 | 23 | 29 | 16 | 18 | 21 | 21 | 26 | 14 | 16 | 18 | 19 | 23 | 13 | 16 | 16 | 19 |
| | 75 | 27 | 30 | 31 | 31 | 32 | 33 | 39 | 19 | 21 | 28 | 28 | 29 | 30 | 39 | 22 | 25 | 26 | 26 | 28 | 28 | 36 | 20 | 22 | 26 | 26 | 32 | 17 | 19 | 22 | 22 | 28 | 16 | 19 | 19 | 24 |
| | 50 | 32 | 36 | 36 | 39 | 39 | 39 | 47 | 23 | 26 | 33 | 33 | 35 | 35 | 47 | 26 | 29 | 31 | 31 | 33 | 33 | 42 | 24 | 27 | 31 | 31 | 38 | 20 | 23 | 27 | 27 | 33 | 20 | 23 | 23 | 28 |
| | 25 | 37 | 42 | 42 | 45 | 45 | 51 | 54 | 26 | 30 | 39 | 39 | 41 | 41 | 54 | 30 | 34 | 36 | 36 | 38 | 39 | 49 | 27 | 31 | 36 | 36 | 45 | 24 | 27 | 31 | 31 | 38 | 23 | 26 | 26 | 33 |
| | 10 | 42 | 48 | 48 | 51 | 51 | 51 | 61 | 30 | 33 | 43 | 43 | 43 | 46 | 61 | 33 | 38 | 41 | 41 | 43 | 44 | 56 | 31 | 35 | 40 | 40 | 50 | 27 | 30 | 35 | 35 | 43 | 26 | 30 | 30 | 37 |
| 64 | 90 | 25 | 28 | 28 | 30 | 30 | 30 | 36 | 18 | 20 | 26 | 26 | 27 | 27 | 36 | 20 | 23 | 24 | 24 | 26 | 26 | 33 | 18 | 21 | 24 | 24 | 30 | 16 | 18 | 21 | 21 | 26 | 15 | 18 | 18 | 22 |
| | 75 | 30 | 34 | 34 | 37 | 37 | 37 | 44 | 21 | 24 | 31 | 31 | 33 | 34 | 44 | 24 | 28 | 29 | 29 | 31 | 32 | 40 | 22 | 26 | 31 | 31 | 38 | 19 | 22 | 25 | 25 | 32 | 19 | 21 | 21 | 27 |
| | 50 | 36 | 41 | 41 | 44 | 44 | 44 | 53 | 25 | 29 | 37 | 37 | 40 | 40 | 53 | 29 | 33 | 35 | 35 | 37 | 38 | 48 | 27 | 30 | 35 | 35 | 45 | 23 | 26 | 30 | 30 | 37 | 22 | 26 | 26 | 32 |
| | 25 | 42 | 48 | 48 | 51 | 51 | 51 | 61 | 30 | 34 | 44 | 44 | 46 | 47 | 61 | 34 | 39 | 41 | 41 | 43 | 44 | 56 | 31 | 35 | 41 | 41 | 52 | 27 | 30 | 35 | 35 | 43 | 26 | 30 | 30 | 37 |
| | 10 | 48 | 54 | 54 | 57 | 58 | 58 | 69 | 33 | 38 | 49 | 49 | 52 | 53 | 69 | 38 | 43 | 46 | 46 | 49 | 49 | 63 | 35 | 39 | 46 | 46 | 57 | 30 | 34 | 39 | 39 | 49 | 29 | 34 | 34 | 42 |
| **Sustained forces\*** | | | | | | | | | | | | | | | | | | | | | | | | | | | | | | | | | | | | |
| 144 | 90 | 8 | 10 | 12 | 13 | 15 | 15 | 18 | 8 | 10 | 10 | 11 | 12 | 12 | 15 | 7 | 8 | 9 | 9 | 9 | 11 | 15 | 7 | 8 | 9 | 11 | 13 | 6 | 7 | 9 | 11 | 13 | 6 | 6 | 7 | 9 |
| | 75 | 10 | 13 | 16 | 17 | 19 | 20 | 23 | 10 | 13 | 13 | 14 | 16 | 16 | 19 | 9 | 10 | 12 | 12 | 14 | 14 | 19 | 9 | 10 | 12 | 14 | 16 | 7 | 9 | 11 | 14 | 16 | 7 | 8 | 10 | 11 |
| | 50 | *13* | *15* | *20* | *21* | *23* | *24* | *28* | *13* | *15* | *16* | *17* | *19* | *20* | *23* | *11* | *13* | *14* | *15* | *17* | *17* | *23* | *11* | *13* | *15* | *17* | *20* | *9* | *11* | *13* | *14* | *18* | *8* | *10* | *12* | *14* |
| | 25 | *15* | *17* | *23* | *25* | *27* | *29* | *32* | *15* | *17* | *18* | *20* | *23* | *24* | *28* | *13* | *15* | *17* | *18* | *20* | *21* | *28* | *13* | *15* | *17* | *20* | *24* | *11* | *13* | *16* | *18* | *23* | *10* | *12* | *14* | *17* |
| | 10 | *17* | *19* | *26* | *28* | *30* | *33* | *36* | *17* | *19* | *22* | *23* | *27* | *27* | *32* | *15* | *17* | *19* | *20* | *23* | *24* | *32* | *15* | *17* | *20* | *24* | *28* | *12* | *15* | *18* | *20* | *27* | *11* | *14* | *16* | *19* |
| 95 | 90 | 10 | 13 | 16 | 17 | 19 | 20 | 24 | 10 | 13 | 13 | 14 | 16 | 16 | 19 | 9 | 10 | 12 | 12 | 14 | 14 | 19 | 9 | 10 | 12 | 14 | 17 | 7 | 9 | 10 | 13 | 18 | 7 | 9 | 10 | 12 |
| | 75 | *13* | *17* | *21* | *22* | *25* | *26* | *30* | *13* | *17* | *17* | *18* | *20* | *21* | *25* | *11* | *14* | *15* | *15* | *18* | *18* | *25* | *12* | *13* | *15* | *18* | *21* | *10* | *12* | *14* | *17* | *22* | *9* | *11* | *13* | *15* |
| | 50 | *16* | *21* | *26* | *27* | *31* | *32* | *37* | *16* | *20* | *21* | *22* | *25* | *26* | *31* | *14* | *17* | *19* | *19* | *22* | *23* | *31* | *14* | *17* | *19* | *22* | *26* | *13* | *15* | *18* | *21* | *26* | *12* | *12* | *14* | *18* |
| | 25 | *19* | *26* | *31* | *32* | *35* | *36* | *38* | *20* | *23* | *26* | *27* | *30* | *31* | *37* | *17* | *20* | *22* | *24* | *26* | *27* | *37* | *17* | *20* | *23* | *27* | *32* | *15* | *18* | *21* | *24* | *30* | *14* | *16* | *19* | *22* |
| | 10 | *22* | *29* | *35* | *38* | *41* | *43* | *39* | *23* | *26* | *29* | *31* | *34* | *36* | *42* | *19* | *23* | *26* | *27* | *30* | *31* | *42* | *19* | *23* | *27* | *31* | *36* | *17* | *20* | *24* | *27* | *38* | *16* | *19* | *21* | *25* |
| 64 | 90 | 11 | 14 | 17 | 18 | 20 | 20 | 25 | 11 | 14 | 15 | 16 | 17 | 17 | 20 | 10 | 12 | 13 | 13 | 15 | 15 | 20 | 9 | 11 | 13 | 15 | 18 | 8 | 9 | 11 | 12 | 15 | 8 | 9 | 10 | 12 |
| | 75 | *14* | *19* | *23* | *23* | *26* | *27* | *32* | *14* | *19* | *19* | *20* | *22* | *23* | *26* | *12* | *16* | *16* | *17* | *19* | *20* | *25* | *12* | *14* | *17* | *19* | *23* | *10* | *12* | *14* | *16* | *19* | *10* | *12* | *13* | *16* |
| | 50 | *17* | *23* | *28* | *29* | *32* | *34* | *40* | *17* | *23* | *24* | *24* | *27* | *28* | *33* | *15* | *18* | *20* | *21* | *23* | *24* | *33* | *15* | *18* | *20* | *24* | *28* | *12* | *15* | *17* | *20* | *23* | *12* | *14* | *16* | *20* |
| | 25 | *20* | *27* | *32* | *33* | *37* | *38* | *48* | *20* | *27* | *27* | *28* | *32* | *33* | *39* | *18* | *21* | *24* | *25* | *28* | *29* | *39* | *18* | *21* | *25* | *28* | *33* | *15* | *18* | *21* | *24* | *28* | *15* | *17* | *20* | *23* |
| | 10 | *23* | *31* | *38* | *40* | *44* | *46* | *54* | *23* | *31* | *31* | *32* | *37* | *38* | *45* | *20* | *24* | *27* | *28* | *32* | *33* | *45* | *21* | *24* | *28* | *32* | *38* | *17* | *20* | *24* | *27* | *32* | *17* | *20* | *23* | *27* |

† Vertical distance from floor to hands (cm).
§ Percentage of industrial population.
¶ The force required to get an object in motion.
\* The force required to keep an object in motion.
Italicized values exceed 8 h physiological criteria.

From Snook, S. H. and Ciriello, V. M., The design of manual handling tasks: revised tables of maximum acceptable weights and forces, Ergonomics, 34, 1197, 1991. With permission.

**TABLE 19.8**  Maximum acceptable forces of pull for females (kg).

**Initial forces¶**

| Height‡ | Percent§ | 2.1 m pull — 6 s | 12 s | 1 min | 2 min | 5 min | 30 min | 8 h | 7.6 m pull — 15 s | 22 s | 1 min | 2 min | 5 min | 30 min | 8 h | 15.2 m pull — 25 s | 35 s | 1 min | 2 min | 5 min | 30 min | 8 h | 30.5 m pull — 1 min | 2 min | 5 min | 30 min | 8 h | 45.7 m pull — 1 min | 2 min | 5 min | 30 min | 8 h | 61.0 m pull — 2 min | 5 min | 30 min | 8 h |
|---|---|---|---|---|---|---|---|---|---|---|---|---|---|---|---|---|---|---|---|---|---|---|---|---|---|---|---|---|---|---|---|---|---|---|---|---|
| 135 | 90 | 13 | 16 | 17 | 18 | 20 | 21 | 22 | 14 | 16 | 16 | 18 | 19 | 20 | 23 | 10 | 12 | 13 | 14 | 15 | 16 | 17 | 12 | 13 | 14 | 15 | 17 | 12 | 13 | 14 | 15 | 17 | 12 | 13 | 14 | 15 |
| | 75 | 16 | 19 | 20 | 21 | 24 | 25 | 26 | 17 | 19 | 19 | 21 | 22 | 23 | 26 | 12 | 14 | 16 | 16 | 18 | 19 | 20 | 14 | 16 | 17 | 18 | 20 | 14 | 16 | 17 | 18 | 20 | 14 | 15 | 16 | 18 |
| | 50 | 19 | 22 | 24 | 25 | 28 | 29 | 31 | 20 | 22 | 23 | 25 | 26 | 29 | 31 | 14 | 16 | 19 | 19 | 21 | 22 | 24 | 17 | 18 | 20 | 21 | 24 | 17 | 18 | 20 | 21 | 24 | 16 | 18 | 19 | 21 |
| | 25 | 21 | 25 | 28 | 29 | 32 | 33 | 35 | 23 | 25 | 26 | 29 | 30 | 33 | 35 | 16 | 19 | 21 | 22 | 25 | 26 | 27 | 19 | 21 | 23 | 24 | 27 | 19 | 21 | 24 | 24 | 27 | 19 | 20 | 22 | 24 |
| | 10 | 24 | 28 | 31 | 32 | 36 | 37 | 39 | 26 | 29 | 29 | 32 | 34 | 36 | 39 | 18 | 21 | 24 | 25 | 27 | 30 | 30 | 22 | 24 | 25 | 27 | 31 | 22 | 24 | 25 | 27 | 31 | 21 | 23 | 24 | 27 |
| 89 | 90 | 14 | 16 | 18 | 19 | 20 | 23 | 23 | 15 | 16 | 17 | 19 | 20 | 21 | 24 | 10 | 12 | 14 | 14 | 15 | 17 | 18 | 13 | 14 | 15 | 16 | 18 | 13 | 14 | 15 | 16 | 18 | 12 | 13 | 14 | 16 |
| | 75 | 16 | 19 | 21 | 22 | 25 | 26 | 27 | 16 | 18 | 20 | 22 | 22 | 23 | 27 | 12 | 15 | 17 | 17 | 19 | 20 | 21 | 15 | 16 | 18 | 19 | 21 | 15 | 16 | 18 | 19 | 21 | 15 | 16 | 17 | 19 |
| | 50 | 19 | 23 | 25 | 26 | 29 | 30 | 32 | 20 | 22 | 23 | 25 | 26 | 29 | 31 | 14 | 18 | 20 | 20 | 22 | 23 | 26 | 18 | 19 | 20 | 22 | 25 | 18 | 19 | 21 | 22 | 25 | 17 | 18 | 20 | 22 |
| | 25 | 22 | 26 | 29 | 30 | 33 | 35 | 37 | 23 | 25 | 27 | 28 | 30 | 32 | 35 | 17 | 20 | 23 | 24 | 26 | 28 | 30 | 20 | 22 | 23 | 25 | 27 | 21 | 23 | 25 | 26 | 30 | 20 | 21 | 23 | 26 |
| | 10 | 25 | 29 | 32 | 33 | 37 | 39 | 41 | 26 | 29 | 30 | 32 | 34 | 36 | 39 | 20 | 24 | 27 | 28 | 30 | 32 | 34 | 23 | 25 | 27 | 28 | 31 | 24 | 26 | 28 | 30 | 34 | 22 | 24 | 25 | 29 |
| 57 | 90 | 15 | 17 | 19 | 19 | 22 | 23 | 24 | 15 | 17 | 18 | 20 | 20 | 21 | 24 | 11 | 13 | 15 | 15 | 17 | 18 | 19 | 13 | 14 | 15 | 17 | 19 | 13 | 14 | 15 | 17 | 19 | 13 | 14 | 15 | 17 |
| | 75 | 17 | 20 | 22 | 23 | 26 | 27 | 28 | 17 | 19 | 21 | 21 | 23 | 24 | 28 | 13 | 15 | 17 | 18 | 20 | 21 | 22 | 15 | 16 | 18 | 20 | 22 | 16 | 17 | 18 | 20 | 22 | 15 | 16 | 18 | 20 |
| | 50 | 20 | 24 | 26 | 27 | 30 | 32 | 33 | 20 | 22 | 24 | 25 | 26 | 29 | 33 | 15 | 18 | 20 | 21 | 23 | 24 | 26 | 18 | 20 | 22 | 23 | 26 | 18 | 20 | 22 | 23 | 26 | 18 | 19 | 21 | 23 |
| | 25 | 23 | 27 | 30 | 31 | 35 | 36 | 38 | 23 | 25 | 27 | 29 | 30 | 32 | 35 | 17 | 21 | 24 | 24 | 27 | 28 | 30 | 21 | 23 | 25 | 27 | 30 | 21 | 23 | 25 | 27 | 30 | 21 | 22 | 24 | 27 |
| | 10 | 26 | 31 | 34 | 35 | 39 | 40 | 43 | 26 | 29 | 31 | 32 | 35 | 37 | 39 | 19 | 23 | 26 | 27 | 30 | 33 | 34 | 24 | 26 | 28 | 30 | 34 | 24 | 26 | 28 | 30 | 34 | 23 | 25 | 27 | 30 |

**Sustained forces***

| Height‡ | Percent§ | 2.1 m pull — 6 s | 12 s | 1 min | 2 min | 5 min | 30 min | 8 h | 7.6 m pull — 15 s | 22 s | 1 min | 2 min | 5 min | 30 min | 8 h | 15.2 m pull — 25 s | 35 s | 1 min | 2 min | 5 min | 30 min | 8 h | 30.5 m pull — 1 min | 2 min | 5 min | 30 min | 8 h | 45.7 m pull — 1 min | 2 min | 5 min | 30 min | 8 h | 61.0 m pull — 2 min | 5 min | 30 min | 8 h |
|---|---|---|---|---|---|---|---|---|---|---|---|---|---|---|---|---|---|---|---|---|---|---|---|---|---|---|---|---|---|---|---|---|---|---|---|---|
| 135 | 90 | *9* | *9* | 10 | 10 | 11 | 12 | 13 | 6 | *9* | *9* | 10 | 10 | 11 | 13 | 6 | 7 | 7 | 8 | 8 | 9 | 11 | 6 | 7 | 7 | 8 | 10 | 6 | 6 | 7 | 7 | 9 | *5* | 5 | 5 | 7 |
| | 75 | *8* | *12* | 13 | 14 | 15 | 16 | 18 | 8 | *11* | *12* | 12 | 13 | 14 | 18 | 7 | 9 | 10 | 10 | 11 | 12 | 15 | 8 | 9 | 10 | 10 | 14 | 8 | 9 | 9 | 9 | 12 | *7* | 7 | 7 | 10 |
| | 50 | *10* | *16* | 17 | 18 | 19 | 21 | 22 | *10* | *15* | 15 | 16 | 18 | 18 | 22 | 9 | 11 | 13 | 13 | 14 | 15 | 19 | *11* | 12 | 13 | 13 | 17 | *10* | 11 | 11 | 12 | 16 | *8* | 9 | 9 | 12 |
| | 25 | *13* | *19* | 21 | 21 | 23 | 25 | 27 | *13* | *18* | 18 | 19 | 21 | 22 | 27 | *11* | 14 | 15 | 16 | 16 | 18 | 23 | *13* | 15 | 15 | 16 | 21 | *12* | 13 | 14 | 14 | 19 | *10* | 11 | 11 | 15 |
| | 10 | *15* | *22* | 24 | 25 | 27 | 29 | 32 | *15* | *21* | 21 | 22 | 24 | 26 | 32 | *13* | 16 | 18 | 17 | 20 | 20 | 27 | *15* | 17 | 17 | 18 | 25 | *14* | 15 | 16 | 17 | 23 | *12* | 12 | 13 | 17 |
| 89 | 90 | *6* | *9* | 10 | 10 | 11 | 12 | 13 | *7* | *8* | *9* | 9 | 10 | 10 | 13 | *5* | 6 | 7 | 7 | 8 | 9 | 11 | *5* | 6 | 7 | 7 | 10 | *5* | 6 | 6 | 7 | 9 | *5* | 5 | 5 | 7 |
| | 75 | *8* | *12* | 13 | 13 | 15 | 16 | 17 | *7* | *10* | 11 | 12 | 13 | 13 | 17 | *7* | 8 | 9 | 10 | 11 | 11 | 14 | *7* | 8 | 9 | 9 | 13 | *7* | 8 | 8 | 8 | 12 | *6* | 6 | 7 | 9 |
| | 50 | *10* | *15* | 16 | 17 | 18 | 20 | 22 | *9* | *13* | 15 | 15 | 16 | 17 | 22 | *9* | 11 | 13 | 13 | 15 | 15 | 18 | *9* | 10 | 12 | 12 | 17 | *9* | 9 | 11 | 12 | 15 | *8* | 8 | 8 | 12 |
| | 25 | *12* | *18* | 20 | 20 | 22 | 24 | 25 | *11* | *16* | 16 | 18 | 18 | 18 | 22 | *11* | 13 | 15 | 15 | 17 | 17 | 22 | *11* | 13 | 13 | 14 | 21 | *11* | 11 | 13 | 13 | 19 | *8* | 10 | 10 | 15 |
| | 10 | *14* | *21* | 23 | 23 | 26 | 28 | 31 | *14* | *18* | 18 | 21 | 23 | 23 | 31 | *13* | 15 | 17 | 18 | 18 | 20 | 26 | *13* | 15 | 15 | 16 | 24 | *13* | 13 | 15 | 16 | 22 | *10* | 10 | 11 | 17 |
| 57 | 90 | *5* | *8* | 9 | 9 | 11 | 11 | 12 | *6* | *7* | 8 | 8 | 9 | 9 | 13 | *5* | 6 | 7 | 7 | 7 | 8 | 10 | *5* | 6 | 6 | 7 | 9 | *5* | 6 | 6 | 6 | 8 | *4* | 5 | 5 | 6 |
| | 75 | *7* | *11* | 12 | 13 | 14 | 14 | 16 | *8* | *9* | 11 | 11 | 12 | 13 | 18 | *7* | 8 | 9 | 10 | 10 | 11 | 13 | *7* | 8 | 8 | 9 | 11 | *7* | 8 | 8 | 8 | 11 | *6* | 6 | 6 | 9 |
| | 50 | *9* | *14* | 15 | 16 | 17 | 18 | 20 | *10* | *12* | 13 | 14 | 15 | 16 | 22 | *8* | 10 | 11 | 11 | 13 | 13 | 17 | *9* | 11 | 11 | 12 | 16 | *9* | 10 | 10 | 10 | 14 | *8* | 8 | 8 | 11 |
| | 25 | *11* | *17* | 18 | 19 | 21 | 22 | 24 | *13* | *15* | 16 | 17 | 19 | 19 | 27 | *10* | 12 | 14 | 14 | 16 | 16 | 21 | *11* | 13 | 13 | 14 | 21 | *11* | 12 | 12 | 13 | 17 | *8* | 8 | 8 | 13 |
| | 10 | *13* | *20* | 21 | 22 | 24 | 26 | 28 | *15* | *17* | 19 | 20 | 22 | 23 | 32 | *12* | 14 | 16 | 16 | 18 | 19 | 24 | *13* | 15 | 16 | 16 | 28 | *12* | 14 | 14 | 15 | 20 | *11* | 11 | 12 | 16 |

‡Vertical distance from floor to hands (cm).
§Percentage of industrial population.
¶The force required to get an object in motion.
*The force required to keep an object in motion.
Italicized values exceed 8 h physiological criteria.

From Snook, S. H. and Ciriello, V. M., The design of manual handling tasks: revised tables of maximum acceptable weights and forces, *Ergonomics*, 34, 1197, 1991. With permission.

**TABLE 19.9** Maximum acceptable weight of carry (kg).

| Height‡ (cm) | Percent§ | 2.1 m carry — One carry every | | | | | | | 4.3 m carry — One carry every | | | | | | | 8.5 m carry — One carry every | | | | | | |
|---|---|---|---|---|---|---|---|---|---|---|---|---|---|---|---|---|---|---|---|---|---|---|
| | | 6 s | 12 s | 1 min | 2 min | 5 min | 30 min | 8 h | 10 s | 16 s | 1 min | 2 min | 5 min | 30 min | 8 h | 18 s | 24 s | 1 min | 2 min | 5 min | 30 min | 8 h |
| **Males** | | | | | | | | | | | | | | | | | | | | | | |
| 111 | 90 | 10 | 14 | 17 | 17 | 19 | 21 | 25 | 9 | 11 | 15 | 15 | 17 | 19 | 22 | 10 | 11 | 13 | 13 | 15 | 17 | 20 |
| | 75 | 14 | 19 | 23 | 23 | 26 | 29 | 34 | 13 | 16 | 21 | 21 | 23 | 26 | 30 | 13 | 15 | 18 | 18 | 20 | 23 | 27 |
| | 50 | *19* | *25* | 30 | 30 | 33 | 38 | 44 | 17 | 20 | 27 | 27 | 30 | 34 | 39 | 17 | 19 | 23 | 24 | 26 | 29 | 35 |
| | 25 | *23* | *30* | 37 | 37 | 41 | 46 | 54 | *20* | *25* | 33 | 33 | 37 | 41 | 48 | *21* | *24* | 29 | 29 | 32 | 36 | 43 |
| | 10 | *27* | *35* | 43 | 43 | 48 | 54 | 63 | *24* | *29* | 38 | 39 | 43 | 48 | 57 | *24* | *28* | 34 | 34 | 38 | 42 | 50 |
| 79 | 90 | 13 | 17 | 21 | 21 | 23 | 26 | 31 | 11 | 14 | 18 | 19 | 21 | 23 | 27 | 13 | 14 | 17 | 18 | 20 | 22 | 26 |
| | 75 | 18 | 23 | 28 | 29 | 32 | 36 | 42 | 16 | 18 | 25 | 25 | 28 | 32 | 37 | 17 | 18 | 24 | 24 | 27 | 30 | 35 |
| | 50 | *23* | *30* | 37 | 37 | 41 | 46 | 54 | *25* | *30* | 32 | 33 | 36 | 41 | 48 | *22* | *24* | 31 | 31 | 35 | 39 | 46 |
| | 25 | *28* | *37* | 45 | 46 | 51 | 57 | 67 | *30* | *37* | 40 | 40 | 45 | 50 | 59 | *27* | *30* | 38 | 38 | 42 | 48 | 56 |
| | 10 | *33* | *43* | 53 | 53 | 59 | 66 | 78 | *35* | *43* | 47 | 47 | 52 | 59 | 69 | *32* | *35* | 44 | 45 | 50 | 56 | 65 |
| **Females** | | | | | | | | | | | | | | | | | | | | | | |
| 105 | 90 | *11* | *12* | 13 | 13 | 13 | 13 | 18 | *9* | *12* | 13 | 13 | 13 | 13 | 18 | *10* | *11* | 12 | 12 | 12 | 12 | 16 |
| | 75 | *13* | *14* | 15 | 15 | 16 | 16 | 21 | *11* | *14* | 15 | 15 | 16 | 16 | 21 | *12* | *13* | 14 | 14 | 14 | 14 | 19 |
| | 50 | *15* | *16* | 18 | 18 | 18 | 18 | 25 | *12* | *15* | 18 | 18 | 18 | 18 | 24 | *14* | *15* | 16 | 16 | 16 | 16 | 22 |
| | 25 | *17* | *18* | 20 | 20 | 21 | 21 | 28 | *14* | *17* | 20 | 20 | 21 | 21 | 28 | *15* | *17* | 18 | 18 | 19 | 19 | 25 |
| | 10 | *19* | *20* | 22 | 22 | 23 | 23 | 31 | *16* | *19* | 22 | 22 | 23 | 23 | 31 | *17* | *19* | 20 | 20 | 21 | 21 | 28 |
| 72 | 90 | *14* | *14* | 14 | 14 | 16 | 16 | 22 | *10* | *11* | 14 | 14 | 16 | 16 | 20 | *12* | *12* | 14 | 14 | 14 | 14 | 19 |
| | 75 | *15* | *16* | 16 | 16 | 18 | 18 | 25 | *11* | *13* | 16 | 16 | 18 | 18 | 23 | *14* | *15* | 16 | 16 | 17 | 17 | 23 |
| | 50 | *17* | *18* | 19 | 19 | 21 | 21 | 29 | *13* | *15* | 19 | 19 | 21 | 21 | 26 | *16* | *17* | 19 | 19 | 20 | 20 | 26 |
| | 25 | *20* | *21* | 21 | 22 | 24 | 24 | 33 | *15* | *17* | 21 | 22 | 24 | 24 | 30 | *18* | *19* | 21 | 22 | 22 | 22 | 30 |
| | 10 | *22* | *24* | 24 | 24 | 27 | 27 | 37 | *17* | *19* | 24 | 24 | 27 | 27 | 33 | *20* | *21* | 24 | 24 | 25 | 25 | 33 |

‡Vertical distance from floor to hands (cm).
§Percentage of industrial population.
Italicized values exceed 8 h physiological criteria (see text).

From Snook, S. H. and Ciriello, V. M., The design of manual handling tasks: revised tables of maximum acceptable weights and forces, *Ergonomics*, 34, 1197, 1991. With permission.

The only other method for multitask assessment that incorporates MMH tasks in addition to lifting that the author was able to find was the method presented by Mital.[32] The data used with this methodology are from Mital et al.,[5] which are modified psychophysical data as described earlier. The method is similar to that developed by Jiang and Mital,[33] except that capacity is predicted using more contemporary data.

In general, this method requires that each MMH task is analyzed, and data regarding work duration, etc., are also needed. The analyst then determines the percentage of the population that the design should accommodate, which should be 75 or 90%. The next step involves calculating the recommended work rate (kg-m/min) for the percentage of the population being analyzed using the Mital et al.[5] data. The actual work rate is divided by the recommended work rate, which yields the risk potential. Any risk potential values greater than 1 signal the need for task redesign. This method focuses on the individual components that are unacceptable, as with the Snook and Ciriello[19] method.

### Example

An example of Snook and Ciriello's[19] multicomponent task assessment will be used to illustrate how psychophysical data are used to analyze MMH tasks. The analysis was performed with the CompuTask computer program. The set of tasks is fairly simple and includes a worker bending over and lifting a box, carrying the box 5 feet, and lowering the box to the floor. The set of tasks is performed three times per minute for 8 hours. The relevant data that need to be collected as well as the analysis are shown in Figure 19.2. Aside from psychophysical results, physiological analyses and NIOSH lifting equation computations (STRWL = single task recommended weight limit, FIRWL = frequency independent recommended weight limit, STLI = single task lifting index, FILI = frequency independent lifting index) are provided by the software.

The task with the lowest percentage of the population accommodated is the lifting task. This task accommodates 75% of the male population and <10% of the female population. Thus, the set of tasks is marginally acceptable for males and unacceptable for females. Also, the overall physiological evaluation shows that the task is not acceptable for eight hours. As was discussed earlier, the method of analysis being used may result in violation of energy expenditure criterion. Redesign efforts would be focused on eliminating the tasks through materials handling devices or a conveyor, or eliminating the need to lift and lower the boxes by increasing the vertical origin and destination of the lifting and lowering tasks, respectively.

## Task and Workplace Design

An often overlooked use of psychophysical data is workplace design. In situations where unacceptable tasks cannot be eliminated, or loads, frequencies, or forces cannot be reduced to "acceptable" levels, psychophysical results can be used to suggest workplace design changes to increase the percentage of the population that a task or job will accommodate. For example, a common problem is that the dimensions and/or weight of the load cannot be reduced. In situations such as this, the task and/or workplace can be redesigned to decrease the physical demands of the task.

The following list provides several examples of task and workplace redesign principles to reduce physical demands when altering the material being handled is infeasible or mechanical aids cannot alleviate the need to handle materials manually. Application of these principles will increase the percentage of the population a task will accommodate.

1. For lifting or lowering tasks, bring the load closer to the body.
2. When lifting low loads, bring the vertical origin of the load as close to knuckle height as possible. Alternately, when lowering loads, the destination should be as close to knuckle height as possible. This principle will reduce bending. In general, try to avoid lifting or lowering to or from high and low locations.
3. Decrease the vertical distance that loads must be lifted/lowered and the distance which loads must be pushed, pulled, or carried.
4. Decrease the frequency of the task or increase the number or workers performing the task.

**Shipping Department**

## Evaluation Results

LIBERTY MUTUAL®

| COMPONENT | FREQUENCY ONE EVERY | FORCE (lbs) | HAND HEIGHT AT START (in) | HAND HEIGHT AT END (in) | DISTANCE MOVED | HAND DISTANCE FROM BODY (in) | BODY MOTION: TWIST | BODY MOTION: REACH | BODY MOTION: BEND | POPULATION PERCENTAGES: MALE | POPULATION PERCENTAGES: FEMALE | SUGGESTED MAXIMUM DURATION (hrs) MALE / FEMALE |
|---|---|---|---|---|---|---|---|---|---|---|---|---|
| Lift | 20 sec. | 34 | 5 | 33 | 28 in. | 10 | Yes | Moderate | Considerable | 75 | <10 | 8 / EL |
| Carry | 20 sec. | 34 | 33 | | 5 ft. | | No | | None | >90 | 77 | 8 / 8 |
| Lower | 20 sec. | 34 | 33 | 5 | 28 in. | 10 | No | Moderate | Considerable | 83 | <10 | 8 / 8 |
| EVALUATION FOR ENTIRE TASK | | | | | | | Yes | Moderate | Considerable | 75 | <10 | EL / EL |

EL —> Exceeds energy expenditure limit for the task duration specified

"POPULATION PERCENTAGES" are the percentages of the male and female population that can be expected to perform the task without excessive stress or excessive fatigue.

"SUGGESTED MAXIMUM DURATION" is the recommended continuous time the job can be performed during an 8-hour workday before exceeding the Energy Expenditure (kcal/Min) guidelines for males and females.

| | STRWL | FIRWL | STLI | FILI |
|---|---|---|---|---|
| Component # 1 - Lift | 11.6 | 21.1 | 2.9 | 1.6 |
| Component # 3 - Lower | 15.2 | 27.6 | 2.2 | 1.2 |

**FIGURE 19.2** Example of psychophysical analysis of a multiple-component MMH job using CompuTask™.

5. For pushing and pulling tasks, provide equipment that provides the least resistance so that initial forces required to overcome inertia are as low as possible. Maintenance of mechanical assists is very important with regard to this principle.
6. For all MMH tasks, provide good hand-to-object coupling when possible, i.e., tote boxes with handles, carts with handle bars, etc.
7. Decrease the duration over which the task is performed.
8. Change pulling tasks to pushing tasks.

## 19.3 The Psychophysical Approach to Designing Upper Extremity Tasks

The primary risk factors for WRMSDs of the upper extremity are fairly well known.[34] Task-related risk factors include posture, force, and repetition. Vibration and cold are task-related risk factors for some disorders such as carpal tunnel syndrome. Duration of the task and rest periods are also important since these factors affect the acceptability of task. Altering work–rest relationships can alter the acceptability of a particular combination of posture, force, and repetition.

There are few quantitative guidelines for limits of posture, force, and repetition. Although general guidelines suggest maintaining a neutral wrist posture and reducing the force requirements and frequency of a task, these guidelines do not indicate acceptable levels of the variables. Once ergonomic analyses and task redesign are done, a decision as to the acceptability of a task is difficult.

The application of the psychophysical approach to the design of UEI tasks was a response to the need for establishing quantitative guidelines with which to assess tasks. Currently, quantitative dose–response relationships developed with epidemiological techniques that provide relationships between individual risk factors and their interactions and the risk of upper-extremity WRMSDs do not exist. In the absence of such relationships, psychophysical data will continue to be one option of setting task limits. The remainder of this section will provide an overview of the current state of psychophysical data as well as discussion of how these data are applied in the workplace.

One advantage of the psychophysical approach is that data can be developed that incorporate force, posture, and repetition into the development of data for different durations. This is important in that this approach allows for trade-offs between variables, i.e., for some tasks, it is not always possible to modify all factors.

### Setting Acceptable Force and Frequency Limits

Fernandez and his colleagues have collected maximum acceptable frequency data for several types of tasks include drilling,[35-38] riveting,[39,40] and tasks requiring pinch and power grasps.[41,42] In these studies, factors such as wrist posture and duration were incorporated into the experimental protocol to provide frequency limits for a variety of UEI tasks. In a similar study, Abu-Ali et al.[43] had subjects control the length of rest periods for task combinations of varying wrist postures, exertion periods, and power grip forces.

In order to use these data, one would record relevant task parameters, find the data relevant for a particular task in the database, and determine if the task is acceptable to the majority of the population, just as with MMH data. If the task is not acceptable, then the frequency would need to be reduced, the duration would need to be reduced, or factors such as wrist deviation would need to be modified to increase the acceptability of a task.

While Fernandez and his colleagues chose frequency as the variable that subjects manipulate, Snook et al.[13] chose force as the manipulated variable. Also, Snook et al.[13] used a 7-hour adjustment period which was much longer than the shorter (20 to 25 min.) period used in the studies cited above. Snook et al.[13] studied tasks requiring wrist flexion with a power grasp, wrist flexion with a pinch grip, and extension with a power grasp. Frequencies between 2 and 20 repetitive motions per minute were studied. Female subjects adjusted wrist torque during the experiment by manipulating the resistance offered by

**TABLE 19.10**   Maximum acceptable forces for female wrist flexion (power grip) (N).

| Percent of population | Repetition rate | | | | |
|---|---|---|---|---|---|
| | 2/min | 5/min | 10/min | 15/min | 20/min |
| 90 | 14.9 | 14.9 | 13.5 | 12.0 | 10.2 |
| 75 | 23.2 | 23.2 | 20.9 | 18.6 | 15.8 |
| 50 | 32.3 | 32.3 | 29.0 | 26.0 | 22.1 |
| 25 | 41.5 | 41.5 | 37.2 | 33.5 | 28.4 |
| 10 | 49.8 | 49.8 | 44.6 | 40.1 | 34.0 |

From Snook, S. H., Vaillancourt, D. R., Ciriello, V. M., and Webster, B. S., Psychophysical studies of repetitive wrist flexion and extension, *Ergonomics*, 38, 1488, 1995. With permission.

**TABLE 19.11**   Maximum acceptable forces for female wrist flexion (pinch grip) (N).

| Percent of population | Repetition rate | | | | |
|---|---|---|---|---|---|
| | 2/min | 5/min | 10/min | 15/min | 20/min |
| 90 | 9.2 | 8.5 | 7.4 | 7.4 | 6.0 |
| 75 | 14.2 | 13.2 | 11.5 | 11.5 | 9.3 |
| 75 | 19.8 | 18.4 | 16.0 | 16.0 | 12.9 |
| 75 | 25.4 | 23.6 | 20.6 | 20.6 | 16.6 |
| 10 | 30.5 | 28.2 | 24.6 | 24.6 | 19.8 |

From Snook, S. H., Vaillancourt, D. R., Ciriello, V. M., and Webster, B. S., Psychophysical studies of repetitive wrist flexion and extension, *Ergonomics*, 38, 1488, 1995. With permission.

**TABLE 19.12**   Maximum acceptable forces for female wrist extension (power grip) (N).

| Percent of population | Repetition rate | | | | |
|---|---|---|---|---|---|
| | 2/min | 5/min | 10/min | 15/min | 20/min |
| 90 | 8.8 | 8.8 | 7.8 | 6.9 | 5.4 |
| 75 | 13.6 | 13.6 | 12.1 | 10.9 | 8.5 |
| 75 | 18.9 | 18.9 | 16.8 | 15.1 | 11.9 |
| 75 | 24.2 | 24.2 | 21.5 | 19.3 | 15.2 |
| 10 | 29.0 | 29.0 | 25.8 | 23.2 | 18.3 |

From Snook, S. H., Vaillancourt, D. R., Ciriello, V. M., and Webster, B. S., Psychophysical studies of repetitive wrist flexion and extension, *Ergonomics*, 38, 1488, 1995. With permission.

a magnetic particle brake. Aside from reporting the torques, forces were also reported which were computed by dividing the torques by the moment arms. The forces are reported in Tables 19.10 through 19.12. Currently, studies are being conducted that address other motions such as ulnar deviation, etc.

Snook et al.'s[13] data would be applied in a manner similar to that described earlier in this section. However, these data are more generic than some of the data collected by Fernandez and his colleagues. For example, the data collected by Snook et al.[13] do not apply only to specific tasks such as drilling, as do some of the data from other studies mentioned.[35-38]

Krawczyk et al.[14] presented preferred weights for manual transfer tasks for transfer distances of 0.5 and 1.0 m. and frequencies between 10 and 30 transfers per minute for an 8-hour work duration. Thus, depending on the situation, one could adjust frequency or transfer distance for a particular weight of object being transferred.

## Tool and Workplace Design

Up to this point, the psychophysical results discussed have all involved experimentation where subjects control a variable. Another approach to the study of manual work using psychophysics is to elicit perceived exertion or discomfort ratings from subjects performing specific tasks. The results can then be used to select the task conditions with the lowest perceived exertion or discomfort.

Ulin et al.[45-47] utilized perceived exertions to study tool masses, tool shapes, and horizontal and vertical work locations. A very general summary of the results indicates that 114 cm was the preferred vertical height for driving screws with a variety of tools. At that height, a pistol-shaped tool was the most preferred. The ratings of perceived exertion with respect to the horizontal distance of the workpiece from the body indicated that the workpiece should not be greater than 38 cm from the front of the worker. The optimal workpiece location was a vertical height of 114 cm and a horizontal distance of 13 cm. Likewise, increasing the tool mass increased the ratings, as would be expected.

In general, the following recommendations are examples of means to increase the psychophysical acceptability of a UEI task:

1. Maintain a neutral wrist posture.
2. Decrease the force requirements of a task.
3. Decrease the duration over which a task is performed. Worker rotation can be very helpful.
4. Decrease the frequency at which the task is performed. This will increase recovery time between exertions and help to prevent fatigue and possible CTDs.
5. Reduce the horizontal distance between the workpiece and the worker.
6. For work with hand tools performed while the operator is standing, position the workpiece at a vertical height of approximately 114 cm. Ideally, the location of the workpiece should be adjustable to accommodate different operators.

# 19.4 Advantages and Disadvantages of the Psychophysical Approach

Like any approach to setting limits for manual work, there are advantages and disadvantages of the psychophysical approach. This approach is only one of several approaches available for designing MMH and UEI tasks, and the advantages and disadvantages of each approach should be examined to determine the best fit for a particular situation. The advantages and disadvantages of the psychophysical approach to the design of MMH and UEI tasks are given below. When necessary, a distinction is made if the advantage or disadvantage is specific to MMH or UEI tasks.

The advantages of the psychophysical approach include:

- Psychophysics allows for the realistic simulation of industrial work.[2]
- Currently, there is a considerable amount of psychophysical data for MMH tasks available that were collected from industrial workers. Many physiological models were developed from limited samples of university students. Likewise, the representativeness of cadaver data used to set certain biomechanical criteria such as lumbosacral compression limits is questionable.
- Psychophysical results are consistent with the industrial engineering concept of a "fair day's work for a fair day's pay."[2]
- Psychophysics can be used to study intermittent MMH tasks which are common in industry.[2] Such tasks are not amenable to physiological analyses.
- Psychophysical results are very reproducible.[2]
- For MMH tasks, psychophysical judgments take into account the whole job, and integrate biomechanical and physiological factors.[48,49]
- Psychophysical results for MMH tasks appear to be related to low-back pain.[1,2,26]

- For MMH tasks, psychophysical data apply to a wider array of tasks than either the biomechanical or physiological approach.
- For MMH tasks that must necessarily be performed under postural restrictions (i.e., maintenance work and mining), psychophysics is one technique that can be used to develop handling limits specific to the tasks being examined.
- The psychophysical approach is less costly and time consuming to apply in industry than many of the biomechanical and physiological techniques.
- Currently, psychophysical data represent one of the only quantitative guides for the design of force limits for UEI work. In the absence of objective biomechanical or physiological criteria, psychophysics may be used to elicit acceptable task parameters for UEI work.[7]

The disadvantages and limitations of the psychophysical approach include:

- Psychophysics is a subjective method.[2]
- The assumption that the subjective work loads selected by subjects are below the threshold for injury has not been validated.[50] There is not extensive epidemiological support for psychophysical data for MMH tasks and no epidemiological support for using psychophysical data for the design of UEI data. However, the same is true of most of the other criteria currently in use for designing manual work.
- Psychophysical results for high-frequency MMH tasks exceed energy expenditure criteria.[2]
- Some psychophysical values for MMH tasks may violate the biomechanical spinal compression criterion of 3400 N.[51] However, this assumes that the spinal compression criterion of 3400 N is correct, for which there is not much support.
- Psychophysics does not appear to be sensitive to bending and twisting while performing MMH tasks, both of which have been related to compensable low-back pain cases.[2]
- The range of data for designing UEI tasks is somewhat limited at this time.

## 19.5   Conclusions

Psychophysical data are one option available to the ergonomist for designing manual tasks and assessing whether or not a task or set of tasks needs to be redesigned. These data have been applied in the workplace for many years with considerable success. As with any assessment tool, there are advantages and limitations associated with psychophysics as discussed in the previous section. When used properly, psychophysical data provide the analyst with a tool applicable to a diverse set of tasks involving manual work.

## References

1. Snook, S. H., The design of manual handling tasks, *Ergonomics*, 21, 963, 1978.
2. Snook, S. H., Psychophysical considerations in permissible loads, *Ergonomics*, 28, 327, 1985.
3. Ayoub, M. M. and Mital, A., *Manual Materials Handling*, Taylor & Francis, London, 1989.
4. Ayoub, M. M., Problems and solutions in manual materials handling: the state of the art, *Ergonomics*, 35, 713, 1992.
5. Mital, A., Nicholson, A. S., and Ayoub, M. M., *A Guide to Manual Materials Handling*, Taylor & Francis, London, 1993.
6. Nicholson, L. M., A comparative study of methods for establishing load handling capabilities, *Ergonomics*, 32, 1125, 1989.
7. Kim, C. H., Marley, R. J., Fernandez, J. E., and Klein, M. G., Acceptable work limits for the upper extremities with the psychophysical approach, in *Proceedings of the 3rd Pan Pacific Conference on Occupational Ergonomics*, 1994, 312.

8. Fernandez, J. E., Fredericks, T. K., and Marley, R. J., The psychophysical approach in upper extremities work, in *Contemporary Ergonomics 1995*, Robertson, S. A., Ed., Taylor & Francis, London, 1995, 456.

9. Stevens, S. S., The psychophysics of sensory function, *American Scientist*, 48, 226, 1960.

10. Ljungberg, A. S., Gamberale, F., and Kilbom, Å., Horizontal lifting — physiological and psychological responses, *Ergonomics*, 25, 741, 1982.

11. Gamberale, F., Ljungberg, A. S., Annwall, G., and Kilbom, Å., An experimental evaluation of psychophysical criteria for repetitive lifting work, *Applied Ergonomics*, 18, 311, 1987.

12. Snook, S. H. and Ciriello, V. M., Maximum weights and work loads acceptable to female workers, *Journal of Occupational Medicine*, 16, 527, 1974.

13. Snook, S. H., Vaillancourt, D. R., Ciriello, V. M., and Webster, B. S., Psychophysical studies of repetitive wrist flexion and extension, *Ergonomics*, 38, 1488, 1995.

14. Emanuel, I., Chaffee, J. W., and Wing, J., A study of human weight lifting capabilities for loading ammunition into the F-86H aircraft, WADC Technical Report 56-367, Wright Air Development Center, Wright-Patterson Air Force Base, Ohio, 1956.

15. Switzer, S. A., Weight-Lifting Capabilities of a Selected Sample of Human Subjects, Technical Document Report No. MRL-TDR-62-57, Aerospace Medical Research Laboratories, Wright-Patterson Air Force Base, Ohio, 1962.

16. Snook, S. H. and Irvine, C. H., The evaluation of physical tasks in industry, *American Industrial Hygiene Association Journal*, 27, 228, 1966.

17. Ciriello, V. M., Snook, S. H., and Hughes, G. J., Further studies of psychophysically determined maximum acceptable weights and forces, *Human Factors*, 35, 175, 1993.

18. Ciriello, V. M., Snook, S. H., Blick, A. C., and Wilkinson, P. L., The effects of task duration on psychophysically-determined maximum acceptable weights and forces, *Ergonomics*, 33, 187, 1990.

19. Snook, S. H. and Ciriello, V. M., The design of manual handling tasks: revised tables of maximum acceptable weights and forces, *Ergonomics*, 34, 1197, 1991.

20. Ayoub, M. M., Bethea, N., Deivanayagam, S., Asfour, S., Bakken, G., Liles, D., Selan, J., and Sherif, M., Determination and modeling of lifting capacity, Final Report, HEW (NIOSH) Grant No. 5R010H00545-02, 1978.

21. Mital, A., Comprehensive maximum acceptable weight of lift database for regular 8-hour work shifts, *Ergonomics*, 27, 1127, 1984a.

22. Mital, A., Maximum weights of lift acceptable to male and female industrial workers for extended work shifts, *Ergonomics*, 27, 1115, 1984b.

23. Fox, R. R., A psychophysical study of high-frequency lifting, Unpublished doctoral dissertation, Texas Tech University, Lubbock, Texas, 1993.

24. Nicholson, L. M. and Legg, S. J., A psychophysical study of the effects of load and frequency upon selection of work load in repetitive lifting, *Ergonomics*, 29, 903, 1986.

25. Snook, S. H. and Irvine, C. H., Maximum frequency of lift acceptable to male industrial workers, *American Industrial Hygiene Association Journal*, 29, 531, 1968.

26. Liles, D. H., Deivanayagam, S., Ayoub, M. M., and Mahajan, P., A job severity index for the evaluation and control of lifting injury, *Human Factors*, 26, 683, 1984.

27. Mital, A., The psychophysical approach in manual lifting — a verification study, *Human Factors*, 25, 485, 1983.

28. Smith, J. L., Ayoub, M. M., and McDaniel, J. W., Manual materials handling capabilities in non-standard postures, *Ergonomics*, 35, 807, 1992.

29. Gallagher, S., Acceptable weights and physiological costs of performing combined manual handling tasks in restricted postures, *Ergonomics*, 34, 939, 1991.

30. Gallagher, S. and Hamrick, C. A., Acceptable work loads for three common mining materials, *Ergonomics*, 35, 1013, 1992.

31. Straker, L. M., Stevenson, M. G., and Twomey, L. T., A comparison of risk assessment of single and combination manual handling tasks: 1. Maximum acceptable weight measures, *Ergonomics*, 39, 128, 1996.

32. Mital, A., Using "A Guide to Manual Materials Handling" for designing/evaluating multiple activity manual materials handling tasks, in *Proceedings of the IEA World Conference on Ergonomic Design, Interfaces, Products, and Information*, 1995, 550.

33. Jiang, B. C. and Mital, A., A Procedure for designing/evaluating manual materials handling tasks, *Int J Prod Res*, 24, 913, 1986.

34. Putz-Anderson, V. (Ed.), *Cumulative Trauma Disorders: A Manual for Musculoskeletal Disorders of the Upper Limb*, Taylor & Francis, London, 1988.

35. Davis, P. J. and Fernandez, J. E., Maximum acceptable frequencies for females performing a drilling task in different wrist postures, *J Human Ergol*, 23, 81, 1994.

36. Fernandez, J. E., Dahalan, J. B., and Klein, M. G., Using the psychophysical approach in hand-wrist work, in *Proceedings of the M.M. Ayoub Occupational Ergonomics Symposium*, Institute for Ergonomics Research, Lubbock, Texas, 1993, 63.

37. Kim, C. H. and Fernandez, J. E., Psychophysical frequency for a drilling task, *International Journal of Industrial Ergonomics*, 12, 209, 1993.

38. Vaidyanathan, V. and Fernandez, J. E., MAF for males performing drilling tasks, in *Proceedings of the Human Factors Society 36th Annual Meeting*, Human Factors Society, Santa Monica, 1992, 692.

39. Fredericks, T. K., The effect of vibration on maximum acceptable frequency for a riveting task, Unpublished doctoral dissertation, The Wichita State University, Wichita, Kansas, 1995.

40. Fredericks, T. K. and Fernandez, J. E., The effect of vibration on maximum acceptable frequency for a riveting task, in *Proceedings of the Konz/Purswell Occupational Ergonomics Symposium*, Institute for Ergonomics Research, Lubbock, Texas, 1995, 27.

41. Dahalan, J. B. and Fernandez, J. E., Psychophysical frequency for a gripping task, *International Journal of Industrial Ergonomics*, 12, 219, 1993.

42. Klein, M. G. and Fernandez, J. E., The effects of posture, duration, and force on pinching frequency, *International Journal of Industrial Ergonomics*, 20, 267, 1997.

43. Abu-Ali, M., Purswell, J. L., and Schlegel, R. E., Psychophysically determined work-cycle parameters for repetitive hand gripping, *International Journal of Industrial Ergonomics*, 17, 35, 1996.

44. Krawczyk, S., Armstrong, T. J., and Snook, S. H., Preferred weights for hand transfer tasks for an eight hour day, in *Proceedings of International Scientific Conference on Prevention of Work-related Musculoskeletal Disorders*, Hagberg, M., and Kilbom, Å., Eds., 1992, 157.

45. Ulin, S. S., Armstrong, T. J., Snook, S. H., and Franzblau, A., Effect of tool shape and work location on perceived exertion for work on horizontal surfaces, *Am Ind Hyg Assoc J*, 54, 383, 1993a.

46. Ulin, S. S., Armstrong, T. J., Snook, S. H., and Keyserling, W. M., Examination of the effect of tool mass and work postures on perceived exertion for a screw driving task, *International Journal of Industrial Ergonomics*, 12, 105, 1993b.

47. Ulin, S. S., Snook, S. H., Armstrong, T. J., and Herrin, G. D., Preferred tool shapes for various horizontal and vertical work locations, *Appl Occup Environ Hyg*, 7, 327, 1992.

48. Haslegrave, C. M. and Corlett, E. N., Evaluating work conditions for risk of injury — techniques for field surveys, in *Evaluation of Human Work*, 2nd. ed., Wilson, J. R., and Corlett, E. N., Eds., Taylor & Francis, London, 1995, 892.

49. Karwowski, W. and Ayoub, M. M., Fuzzy modelling of stresses in manual lifting tasks, *Ergonomics*, 27, 641, 1984.

50. Gamberale, F. and Kilbom, Å., An experimental evaluation of psychophysically determined maximum acceptable work load for repetitive lifting work, in *Ergonomics International 88*, A. S. Adams, R. R. Hall, B. J. McPhee, and Oxenburgh, M. S., Eds., Ergonomics Society of Australia, Sydney, 1988, 233.

51. Chaffin, D. B. and Page, G. B., Postural effects on biomechanical and psychophysical weight-lifting limits, *Ergonomics*, 37, 663, 1994.

# 20

# The Relative Importance of Biomechanical and Psychosocial Factors in Low Back Injuries

A. Kim Burton
*University of Huddersfield*

Michele C. Battié
*University of Alberta*

Chris J. Main
*University of Manchester*

## 20.1 Introduction

That low back trouble (LBT) is an increasing problem in industrialized society has become obvious. The realization that this increase is occurring despite the best efforts of ergonomists, clinicians, and legislators has not gone unnoticed (Nachemson, 1996; Waddell, 1992; Bigos and Battié, 1990; Winkel and Westgaard, 1996). Arguably the field of ergonomics has been thrown a little off balance by developments in the clinical arena over recent years. On the one hand, ergonomists and biomechanists strive to reduce physical stress at the workplace with the intent of lowering the risk of musculoskeletal problems. On the other hand, the clinicians and psychologists are suggesting that rehabilitation of the back-injured worker should involve not only activity but physical challenges to the musculoskeletal system (Kohles et al., 1990; Bendix et al., 1995). This apparent dichotomy cannot readily be dismissed; to understand how it has arisen requires an exploration of reports from the various fields of endeavor involved.

This chapter will attempt to bring together, from a variety of perspectives, evidence on the development, recurrence, and persistence of work-related low back trouble. Some of the issues raised are covered in detail elsewhere in this book; the intention here is to provide a rationale which links the considerations of ergonomics/biomechanics with those psychosocial aspects discussed in the two following chapters. Rather than present a systematic review, or attempt a meta-analysis, the starting point is the commonly held tenet that physically demanding work is detrimental to the back, i.e., it will cause injury leading to low back pain and consequent disability. From this it would follow logically that management of work-related low back disorders can best be achieved by reduction of occupational spinal stressors.

The basic "injury/damage" model can be expressed as one where exposure to mechanical overload, whether a single event or cumulative stress, results in some form of damage to spinal tissues, and that further exposure leads to further damage and/or lack of recovery, which in turn leads to disabling consequences. On the face of it, this model would appear to be intuitively reasonable and valid. What follows is a selective exploration of the literature focusing on evidence which challenges the basic premise that ergonomic intervention is the key to reducing the burden of occupational back pain, and highlights the emergent role of psychosocial influences.

## 20.2   Background

There is no shortage of epidemiological reports that link heavy, strenuous work (physical stress) with back pain (Suadicani et al., 1994; Riihimaki, 1989), but this link is not universally reported (Burton et al., 1989; Spengler et al., 1986; Riihimaki et al., 1994); seemingly much depends on definitions for back pain and workload. Similarly, reports of an association between heavy work and absenteeism (Riihimaki et al., 1989; Andersson et al., 1983), are not entirely consistent (Lindstrom et al., 1994). Physical spinal stress from whole body vibration (WBV) also has been associated with an increased risk of occupational back injury (Wikstrom et al., 1994), and has been linked specifically with symptoms (Burton and Sandover, 1987), but a dose–response relationship has not been confirmed. There is certainly some experimental biomechanical evidence suggesting that strenuous work environments are likely to be detrimental; *in vitro* experiments, simulating physiological occupational loads, can result in fatigue damage of numerous spinal soft tissues (Adams and Dolan, 1995) and to the end plates of the vertebral bodies (Brinckmann et al., 1988).

The injury model suggests that if occupational loads are reduced there should be a concomitant reduction in occupational back trouble. It is highly likely that occupational physical stressors on workers' spines have progressively reduced over the last 20 years or so, due to the combined effects of increasing mechanization and ergonomics-driven legislative procedures. However, there is no evidence that back pain has decreased, and in fact the disability due to LBT continues to grow exponentially (Clinical Standards Advisory Group, 1994).

It might, at this point, be helpful to put the work-relatedness of LBT into perspective. Occupational low back trouble must be viewed in the light of a high prevalence of LBT in the general population, at least in the industrialized nations. The background prevalence is often quoted as 80% or more, but a more realistic figure for what might be termed "notable" back trouble (i.e., that which results in care-seeking or a period of disability, as opposed to the transient twinges suffered by most) is probably an average of around 60% (Papageorgiou et al., 1995). The lifetime prevalence is similar among males and females, but does seem to vary with age. Few adults can recall trouble before the age of 18 years (Burton, 1987), but then the prevalence rises through working years (from ~50% to ~65%), but does not increase thereafter (Papageorgiou et al., 1995). The 1-month prevalence rate amounts to 39% on average among adults (Papageorgiou et al., 1995), while the point prevalence rate is of the order of 14 to 30% (Clinical Standards Advisory Group. 1994). Such figures are derived from cross-sectional studies and are at the mercy of reporting errors; for instance it has been found that some will deny ever having back trouble despite the fact that previously they have been sick listed for the condition (Svennson and Andersson, 1982). Longitudinal studies of school children and adolescents have reported a surprisingly high lifetime prevalence of over 50% by 16 years of age (Burton et al., 1996b; Balagué et al., 1995). Undoubtedly back pain is a common life experience virtually irrespective of age, but adolescent back pain differs markedly from the adult experience in that, while a recurrent phenomenon, it is rarely disabling (Burton et al., 1996b).

Recent epidemiological studies have revealed that LBT can be as prevalent among sedentary workers as among manual workers (Hemmingway et al., 1994) but heavy jobs do seem to be associated with an increased work loss due to LBT (Riihimaki et al., 1989), though that increase may be due to longer spells rather than more spells (Andersson et al., 1983). However, the association between heavy work and absence rates is not a consistent finding (Lindstrom et al., 1994). A study of construction workers did find a dose–response relationship between "severe" low back pain and stooping and kneeling, but back pain in general was associated more with a range of psychosocial parameters (Holmstrom et al., 1992).

The identification of risk factors for a condition with such a high background prevalence has long been recognized as problematic (Troup and Edwards, 1985). The etiology is clearly multifactorial (Troup and Videman, 1989) so it is exceedingly difficult to be certain that a particular job is involved in causation in individual cases. The problem may have arisen even if the individual had not been employed, there may have been a non-work etiology, or it may have resulted from a previous job. The high prevalence in adolescents, coupled with the fact that only 3 to 10% of episodes in adults are brought to medical attention (Wells, 1985), is a fair indication that much back pain is relatively benign and can be considered a normal life experience.

It should be remembered that LBT usually occurs simply as a complaint of pain (in back and/or leg, with or without sensory symptoms); very rarely is it possible to determine a physical reason for the symptoms; i.e., there is no objective evidence of tissue damage. A consequence of the symptoms is an inability (or reluctance) to perform activities of daily living, including work activities. In essence, LBT is a symptomatic complaint (usually) without evidence of specific tissue or structural damage, which may be associated with some degree of dysfunction; it is well recognized that there is poor correlation between the degree of symptoms and the extent of purported physical stress prior to onset.

Work-related LBT has to be viewed against this high background level of reporting, in the general population, of a symptom which has an undetermined pathology, a propensity for recurrence, and a variable tendency to progress to disability. It is the link between these features (damage, reporting/recurrence, and disability) which is the focus of the following discussion.

## 20.3  Spinal Damage

The structure presumed at greatest risk for damage from heavy work (including WBV) is the intervertebral disc, and this has been the major subject of investigation. While it has been indicated that heavy work is associated with lumbar disc degeneration (Riihimaki et al., 1990), much may depend on the measurement methods, imaging techniques, and the population studied. For instance, a review on this subject (Troup and Edwards, 1985) noted that reports of severe disc degeneration occurring in miners are not entirely consistent; numerous studies support the link, but at least one found that it was osteophytes rather than discal changes that were related to duration of mining work (Caplan et al., 1966). This association has been noticed in another mining population (author's unpublished data). Interestingly, Olympic weight lifters do not seem to show an excess of radiological changes (Hult, 1954). Assuming a link between work and disc degeneration, it might reasonably be assumed that disc degeneration will have an adverse effect on spine function (e.g., flexibility), and thus exposure to physical stressors (*via* occupation or leisure) will result in diminished flexibility. However, it would seem that lumbar flexibility, while being somewhat related to disc height, is not substantially influenced by either degeneration or exposure to physical stressors (Burton et al., 1996a). Recent investigation involving multivariable statistical techniques has confirmed that disc degeneration — as measured from magnetic resonance images (MRI) — is influenced only modestly by work history; the greatest proportion of the explained variances in degeneration score can be accounted for by genetic influences, though age did have some expected influence (Battié et al., 1995).

Disc herniations (as opposed to generalized degenerative changes) are not apparently associated with heavy work (Kelsey, 1975), but sitting work, particularly involving driving, may carry an increased risk (Kelsey et al., 1984). However, it should be borne in mind that the presence of herniations is poorly correlated with symptoms (Boos et al., 1995). Disc herniations were found in 76% of an asymptomatic control group (matched for age, sex, and work-related factors) compared with 96% in a symptomatic group; the presence of symptoms was related to neural compromise and psychosocial aspects of work, but not to the exposure to physical stressors (Boos et al., 1995).

Certainly we would expect to find, based on *in vitro* studies, signs of damage to the discs and vertebral bodies from exposure to mechanical overload (Brinckmann et al., 1988), but this has yet to be confirmed.

Perhaps the level of physical stressors commonly encountered in workers is simply below the threshold needed to create the predicted damage, or it may be that the measurement techniques currently available are insufficiently precise to identify it. Alternatively, it may be that the normal biological repair processes obscure the evidence (Brinckmann, 1985). A new method for quantifying overload damage from radiographs has become available (Brinckmann et al., 1994), by which it is possible to compare radiographs from cohorts exposed to heavy work or WBV with those exposed to light or sedentary work. Overload damage has only been detected in cohorts exposed to extremely heavy work, or to undamped vehicular vibration; cohorts performing strenuous work within current ergonomic guidelines did not show overload damage (Brinckmann et al, 1998). Also, certain strenuous sports activities have been shown to induce disc disruption (e.g., fast bowlers in cricket [Elliott et al., 1993]).

Severe trauma leading to vertebral body fractures, such as can happen in parachute landing (Murray-Leslie et al., 1977) or ejection from military aircraft (Laurell and Nachemson, 1963), may present few if any symptoms. Even when symptoms do ensue, recovery is relatively swift with no lasting disability or reduced effectiveness (Laurell and Nachemson, 1963).

Structures other than vertebrae and discs (i.e., muscles and ligaments) may, of course, be damaged (Adams and Dolan, 1995), but for the most part we do not have objective means for detecting such damage. Experience in areas other than the low back indicate that muscle and ligament injuries (in the absence of complete rupture) have a healing time of the order of 4 to 8 weeks. Perhaps 85% of those taking time off work with back trouble will have returned to work during that time (Clinical Standards Advisory Group. 1994), but that leaves a sizeable number for whom an obvious physiological explanation for the persisting disability is lacking.

Current knowledge is insufficient, in the vast majority of cases of LBT, for accurate identification of the structure or structures that are "injured," or the extent to which they might be injured. Some measure of strain to the soft tissues supporting the spine (possibly involving neuromuscular control mechanisms) are the most likely candidates. New modeling techniques are suggesting that spine stability is closely related to muscle activity. Deficient intrinsic spine muscles or a lack of motor control seem to reduce the mechanical stability of the spine which could increase the risk of straining muscles or ligaments (Cholewicki and McGill, 1996). The question remains whether back strains are most often the result of some specific element of work or the result of other circumstances in life. The evidence shows that the back can certainly be injured in various ways, but the injury model can be seen to be lacking explanatory power where prolonged disability is concerned.

Irrespective of whether damage to spinal structures can be identified or quantified, there is no doubt that workers do get painful backs and some will believe it is their work which is to blame. In a general survey of workers, 61% stated their back disorder was caused by their work, and 39% felt it was made worse by their work; workers report back disorders twice as frequently as any other disorder (Clinical Standards Advisory Group. 1994). A study, of sicklisted blue collar workers found that 60% of patients believed that work demands had caused their back trouble, but neither an assessment of workload (e.g., lifting, bending) nor calculated compression loads predicted the rate of return to work or sick leave during follow-up (Lindstrom et al., 1994). Low back pain attributed to work has been reported to be more common in female nurses than female teachers, but non-occupationally attributed back pain had the same prevalence in both groups (Cust et al., 1972). The finding that worker ratings of job-required physical abilities are correlated with back injury rates is interesting (Skovron et al., 1991) but, because both parameters are subject to possible reporting bias, the relationship may, in part, represent subjective attribution. Confounding of physical workload with psychosocial workload has been found among various branches of agricultural work with differing prevalence rates, rendering clear identification of risk factors difficult (Hildebrandt, 1995).

## 20.4   Injury, Recurrence, and Work Loss

The biomechanical approach to identification of occupational risk factors for LBT has shifted from static models of spinal loading to dynamic models which use accelerations and velocity to estimate forces acting on the spine (Halpern, 1992), while the more sophisticated approaches are advocating methods which incorporate muscle synergy and cocontraction patterns (Cholewicki et al., 1995). This sort of modeling fits better with the epidemiology. Some data suggest that the risk of LBT is associated with dynamic lifting (Bigos et al., 1986), and this parameter has been studied in detail by Marras and colleagues (Marras et al., 1993). This latter study was the first to link epidemiological findings with quantitative biomechanical findings in a large working population. A multiple logistic regression model revealed that a combination of five trunk motion and workplace factors distinguished between high and low risk; the factors were lifting frequency, load moment, trunk lateral velocity, trunk twisting velocity, and trunk sagittal angle. The outcome variable (risk of LBT) was derived from medical and injury records, which admittedly suffer from confounding effects of reporting bias and inadequate information about previous trouble. While a causative link was not proved, the association between biomechanical factors and risk was clear. When looking at worker-rated job demands (including ratings of dynamic components) and back injury rates, it is apparent that high-risk jobs can have quite diverse demands, i.e., not all jobs with high injury rates require the same physical abilities (Halpern, 1992). A study of 1800 nurses in Belgium and The Netherlands, using a quasi-objective rating system for task demands, showed a significantly lower prevalence of back trouble (and other musculoskeletal complaints) in the Dutch nurses despite the fact that their average workload was substantially greater than their Belgian counterparts. Overall, symptoms and work loss in the previous 12 months were not related to workload, nor was attribution that work was causative. It transpired that the Dutch nurses differed strikingly from the Belgian nurses on a range of psychosocial variables; they reported less depressive symptoms and were significantly more positive about pain, work, and activity (Burton et al. 1997).

It is known that a previous history of back pain is highly predictive of future episodes (Troup et al., 1987; Burton et al., 1989; Bigos et al., 1991), so it might be more illuminating to investigate the relationship between occupational physical stressors and first-time back injury rather than with prevalence rates (i.e., study of workers without any back trouble history). Because of methodological difficulties, reports of such studies are rare. Videman and colleagues (Videman et al., 1984) studied nursing aides and trained nurses. They found that young nursing aides (with high work loads but low likelihood of previous LBT history) had a higher prevalence than the trained nurses who did less manual handling, suggesting that skill and training may be important. But it was also noted that the aides tended to have children at an earlier age, thus domestic spinal loading was possibly a confounding influence. While the annual incidence rate of LBT in nurses has been found to be higher than in teachers, the annual prevalence and point prevalence rates were the same (Leighton and Reilly, 1995), though first-onset at a younger age in nursing aides compared with teachers has been reported (Cust et al., 1972). It has been shown that experienced industrial workers do show a reduced risk for LBT compared with inexperienced workers, but this is not necessarily because they experience reduced spinal loads, rather, their smoother motions may be related to muscular coordination aiding spinal stability (Granata et al., 1996).

A large general population study (Croft et al., 1995) has found that new episodes of LBT are more likely for those who are psychologically distressed; a relationship that held true even for first onsets. In the large prospective study at the Boeing plant (Bigos et al., 1991) it was found that reported first injuries were not related specifically to job demands, rather to psychosocial factors (i.e., low job satisfaction and an elevated hysteria scale on the Minnesota Multiphasic Personality Inventory); the workers at Boeing, though, worked in an environment where job tasks were not particularly stressful for the back. Some clarification of the issue could emerge from study of first-onsets in workers exposed to the same substantial physical stressors over long periods; police officers in Northern Ireland have proved to be a useful group for study (Burton et al., 1996c). They compulsorily wear body armor (weighing >8 kg) for up to 12 hours per day; they do so irrespective of rank, and they return to the same work on recovery from back trouble. This police force was compared with an English police force which did not wear armor. It was found that the physical stress of wearing body armor reduced the survival time to first-onset, and there was a slight increase in the hazard over time. It was also found that spending longer than two hours per day in vehicles comprised a separate risk for first-onset of LBT, but this hazard did not increase; it is interesting to note that the effect of exposure to armor and vehicles was not additive. Surprisingly, the proportion of officers with persistent (chronic) back complaints did not depend on the length of exposure (to the same stressors) since first-onset, rather chronicity was associated with psychosocial factors (distress and blaming work) (Burton et al., 1996c).

Physical stress on the spine may actually be beneficial. It has been reported that a physical fitness program improved cardiovascular performance, strength, and flexibility, and was associated with reduced workers' compensation costs for back injuries (Cady et al., 1979), though this finding has not been confirmed by others (Battié et al., 1989). Porter has suggested that heavy manual work (in miners) may benefit the spine by developing ligamentous and annular strength, thus restraining encroachment of a disc protrusion into the spinal vertebral canal (Porter, 1987). A later cadaveric study gave some support to this notion; the compressive strength of the spines from young men who had been physically active tended to be greater than those who had been less active (Porter et al., 1989).

There may, of course, be factors other than task-induced stress that could be important in respect of causation. In a large prospective study of employees in various industries, Troup and colleagues (Troup et al., 1981) found that the current attack of back pain arose with no evidence of injury in over 50%. For those with an "injury," two thirds were truly accidental, i.e., they occurred in association with some identifiable event. Falls were actually as common a cause as handling injuries, and there was no indication that those currently injured by handling were more likely to have had a previous handling injury. Falls were also associated with longer periods of sick leave and a greater propensity for recurrence. Overall, recurrence was found to be common; a figure of 44% in the first year dropped to 31% in the second year, suggesting recurrence rates may be more a feature of the natural history of the disorder than a reflection of continued exposure to the work environment (Troup et al., 1981). Support for this notion has come from other studies, involving multivariable analyses, which have found little relationship between recurrence and work demands (Troup et al., 1987; Burton et al., 1989; Bigos et al., 1991). As mentioned above, the best predictor of future trouble seems to be a previous history, with perception of work demands being more important than objective measurement (Troup et al., 1987), and dissatisfaction with work being a significant factor; the term re-injury may be a misnomer (Bigos et al., 1991). More recently, an enhanced instrument, the Psychosocial Aspects of Work questionnaire, has been described which separately evaluates job satisfaction, social support, and mental stress (Symonds et al., 1996). It has been possible to show that workers with current LBT have a lower score for job satisfaction and social support but, surprisingly, absenteeism (Symonds et al., 1996) and work heaviness (Burton et al., 1997) were not related to these parameters.

While absenteeism may not be related to attitudes about work itself, other attitudes and beliefs do seem to be relevant. Psychosocial factors such as negative beliefs about the inevitable consequences of LBT (measured by the Back Beliefs Questionnaire [Symonds et al., 1996]), inadequate pain control strategies (Symonds et al., 1996), fear–avoidance beliefs (Waddell et al., 1993) and belief that work was causative (Burton et al., 1996c) have all been found to relate to absenteeism. The relationship between attribution of cause, job satisfaction, and pain perception is complex and, in part, related to job level; issues related to attribution are important factors for compliance with intervention strategies (Linton and Warg, 1993). Accordingly, it has been found that a simple educational intervention program (comprising workplace broadcasting of a pamphlet stressing the benign nature of LBT, the importance of activity and desirability of early work return) is capable of creating a positive shift in beliefs with a concomitant reduction in extended absence (Symonds et al., 1995).

The question of when and how to return workers with LBT to their job has attracted considerable attention. It has been popularly held that too early a return to the same task would risk recurrence of symptoms (or do further damage). Indeed back injured workers may perceive their back problem as lifelong trouble, and believe that their back injury has made them more vulnerable to reinjury and disability (Tarasuk and Eakin, 1994). There is accumulating evidence that this belief is not only false, but detrimental. The use of work restrictions does not necessarily correlate with reduced symptoms following return to work. A 3-year follow-up of reported occupational musculoskeletal injuries (including LBT) found that those whose workloads had been reduced did not report fewer problems (Kemmlert et al., 1993). A 1-year follow-up report concerning the use of restrictions for nonspecific back pain found that restrictions were associated with a reduced probability of returning to the original work; they did not reduce work absence, and they did not significantly reduce recurrences (Skovron et al., 1997). In fact, a successful rehabilitation program for patients with subchronic back pain has advocated early return to unrestricted duties as part of a combined graded activity/behavioral therapy approach (Lindstrom et al., 1992). Prospective study of even the most severe chronic disabling spinal disorder workers' compensation patients who completed a functional restoration program and returned to work, found that they were at relatively low risk for recurrence (Garcy et al., 1996). When looking at predictors for chronicity, using multivariable analyses in a mixed population of workers, occupation was not found to be significant, rather chronicity was related to age and to the duration of the first spell (Burton et al., 1989). The study of working police officers revealed that change of duties after developing LBT was rare, and persistence of symptoms was found to be unrelated to the length of exposure (to the same physical stressors at work) following first-onset (Burton et al., 1996c).

Even when there is known damage to lumbar structures, activity restriction seems not to be necessary. A prospective study of 50 patients operated for lumbar disc herniation, who were urged to return to full activity as soon as possible, showed the following features: mean return to work time was 1.7 weeks with 25% returning within 1 to 2 days; 97% returned to their previous job and had returned to full work by 8 weeks; no patient changed jobs because of symptoms; only 6% had a recurrence (follow-up averaged 3.8 years) (Carragee et al., 1996). Clinical studies in workers' compensation back pain patients have found that delayed functional recovery was associated with psychosocial factors more than with perceived task demand (Hadler et al., 1995), and that longer spells off work were associated with a poor outcome (Lancourt and Kettelhut, 1992). In compensation cases, settlement of the claim, interestingly, does not result in a reduction in morbidity (Greenough and Fraser, 1989). Generally it would seem that workers over 50 years old return to work faster than younger ones; neither gender nor length of employment influenced recovery rates, but a lower hourly wage was associated with a longer return to work period (Reid et al., 1996).

The reluctance to confront normal physical challenges seen in back-disabled workers has been termed activity intolerance, which is variously linked to individual response to pain, the belief that a specific injury must be the cause of the pain, and behavioral roles such as suffering. This has led to the proposal that treatment and benefits should not, in general, be (pain) complaint-contingent but time-contingent such that there is a clear incentive to return to work within an allotted time (IASP, 1995).

The question obviously arises as to the origin of the various relevant psychosocial traits — do they develop before or after onset of symptoms? This issue will be addressed in the following chapters, but there is clinical evidence that psychological profiles predictive of chronicity are present early in the course of the back pain experience (within the first three weeks) (Burton et al., 1995) and will even precede it in a few people (Mannion et al., 1996). The profiles identified clinically (coping strategies and individual beliefs, as opposed to job satisfaction) are in the same domain as those found to be related to work loss (Symonds et al., 1996).

## 20.5 Effectiveness of Ergonomic Intervention

Most ergonomic interventions will focus on strategies to reduce spinal loading and have been discussed elsewhere in this book. What reports there are to support the belief that ergonomic intervention will reduce the impact of occupational low back pain have been considered largely anecdotal (Smedley and Coggon, 1994), and there are significant risks of ergonomics being confounded with rationalization leading to an "ergonomic pitfall" (Winkel and Westgaard, 1996). The only intervention which has been formally evaluated is worker training in manual handling techniques; while lifting techniques can be improved, the effect on injury rates has not been clearly demonstrated (Smedley and Coggon, 1994). Rigorous controlled trials of ergonomic intervention programs, with morbidity as the outcome, remain to be conducted. Meanwhile, there is evidence that personnel issues may be more important than physical ergonomics. A large industrial study has found that ergonomic solutions failed to reduce lost workdays, and that reliance on case monitoring and wellness orientation actually increased work loss. However, diligent safety programs and an emphasis on systematic return-to-work programs did reduce lost work-days (Hunt and Habeck, 1993). Similarly, a comparison of a personnel program (designed to increase communication between claimants and their employer, their doctor and the Workers' Compensation Board) with a back program (designed to reduce back injuries through intensive feed-back training) found that it was the personnel program which was by far the most effective (Wood, 1987).

## 20.6   Concluding Remarks

There is now sufficient evidence to seriously challenge the power of the injury/damage model to explain the current phenomenon of occupational back trouble. Although epidemiological studies can link the incidence (first recalled onset) of back pain with certain physical stressors (notably spinal loading and vehicle use [Burton et al., 1996c]), the fact remains that these injuries should, by virtue of known physiological processes, resolve within the natural healing time of 4 to 8 weeks. Doubtless many such injuries do resolve naturally, but the problem (for science as well as society) is that a considerable number recur or progress to chronic disability. There is strong evidence that recurrence and disability are not generally related to physical stressors, rather they may be mediated by psychosocial phenomena. Biomechanics may help to explain initial injury mechanisms, but currently offers little in the way of explanation for persisting trouble.

It can be accepted that much work is physically demanding, and may (frequently) lead to some discomfort and pain. These transient symptoms may be a normal consequence of life, but if the worker believes that the job is to blame there is the potential for psychosocial factors to intervene. A proportion of back injured workers having inappropriate beliefs about the nature of their problem, and its relation-ship to work, will develop fear–avoidance behaviors because of inadequate pain coping strategies. They then begin to function in a disadvantageous way and drift into chronic disability. Once a worker has developed back trouble, it would seem that therapeutic programs combining physical challenges to the back with operant conditioning are more successful than traditional approaches involving rest. Just as bed rest is known to be detrimental to recovery from an episode of back trouble (Malmivaara et al., 1995), it may be that the same principle applies to prolonged work absence. Certainly, undue rest can be expected to result in the deficient motor control mechanisms which heighten biomechanical suscep-tibility to injury or reinjury (Cholewicki and McGill, 1996).

## 20.7   Summary

On balance, there is evidence to support the notion that biomechanics-based ergonomic improvements to the workplace may have some potential to limit first-time back injury; therefore they should be deployed where practicable. The possible role of ergonomics for reducing recurrence rates seems at best equivocal, but there is no convincing evidence that continuance of work is detrimental in respect of disability. Because it is likely that much back pain is only work-related inasmuch as people of working age get painful backs, it is becoming clear that reducing spinal loads or awkward postures is likely to have only a small impact on the overall pattern of back pain. Non-biomechanical approaches (organi-zational and psychosocial) seemingly are more effective in maintaining workability. Primary prevention may be an unrealistic goal, but reducing the costly burden of chronic disability may well be possible; strategies which involve psychosocial advice to overcome activity intolerance (IASP, 1995), reduce fear avoidance, and promote activity (along with therapeutic motor-control training) may well transpire to be effective.

## References

Adams, M.A. and Dolan, P. 1995. Recent advances in lumbar spinal mechanics and their clinical signif-icance. *Clin Biomech* 10:3-19.

Andersson, G.B.J., Svensson, H., and Oden, A. 1983. The intensity of work recovery in low back pain. *Spine* 8:880-884.

Balagué, F., Skovron, M.L., Nordin, M., Dutoit, G., and Waldburger, M. 1995. Low back pain in school-children: a study of familial and psychological factors. *Spine* 20:1265-1270.

Battié, M.C., Bigos, S.J., and Fisher, L.D. 1989. A prospective study of the role of cardiovascular risk factors and fitness in industrial back pain complaints. *Spine* 14:141-147.

Battié, M.C., Videman, T., Gibbons, L., Fisher, L., Manninen, H., and Gill, K. 1995. Determinants of lumbar disc degeneration: a study relating lifetime exposures and MRI findings in identical twins. *Spine* 20:2601-2612.

Bendix, A.F., Bendix, T., Ostenfeld, S., Bush, E., and Andersen, A. 1995. Active treatment programs for patients with chronic low back pain: a prospective, randomized, observer-blinded study. *European Spine Journal* 4:148-152.

Bigos, S.J., Spengler, D.M., Martin, N.A., Zeh, J., Fisher, L., Nachemson, A., and Wang, M.H. 1986. Back injuries in industry: a retrospective study. II. Injury factors. *Spine* 11:246-251.

Bigos, S.J., Battié, M.C., Spengler, D.M., Fisher, L.D., Fordyce, W.E., Hansson, T., Nachemson, A.L., and Wortley, M.D. 1991. A prospective study of work perceptions and psychosocial factors affecting the report of back injury. *Spine* 16:1-6.

Bigos, S.J. and Battié, M.C. 1990. Industrial low back pain. Risk factors, in Weinstein, J.N. and Wiesel, S.W. (Eds.) *The Lumbar Spine*, pp. 846-859. Philadelphia: Saunders.

Boos, N., Reider, V., Schade, K., Spratt, N., Semmer, M., and Aebi, M. 1995. The diagnostic accuracy of magnetic resonance imaging, work perception, and psychosocial factors in identifying symptomatic disc herniations. *Spine* 20:2613-2625.

Brinckmann, P. 1985. Pathology of the vertebral column. *Ergonomics* 28:77-80.

Brinckmann, P., Biggemann, M., and Hilweg, D. 1988. Fatigue fracture of human lumbar vertebrae. *Clin Biomech* 3 (Suppl. 1):s1-s23.

Brinckmann, P., Frobin, W., Biggemann, M., Hilweg, D., Seidel, S., Burton, K., Tillotson, M., Sandover, J., Atha, J., and Quinell, R. 1994. Quantification of overload injuries to the thoracolumbar spine in persons exposed to heavy physical exertions or vibration at the workplace: Part 1 — The shape of vertebrae and intervertebral discs — study of a young, healthy population and a middleaged control group. *Clin Biomech* 9, (Suppl. 1):s5-s83.

Brinckmann, P., Frobin, W., Biggemann, M., Tillotson, K.M., Burton, A.K. 1998. Quantification of overload injuries to thoracolumbar vertebrae and discs in persons exposed to heavy physical exertions or vibration at the workplace. Part II. Occurrence and magnitude of overload injuries in exposed cohorts *Clin Biomech* 13, (Suppl. 2): s1-s42.

Burton, A.K. 1987. *Patterns of lumbar sagittal mobility and their predictive value in the natural history of back and sciatic pain*, Doctoral Thesis: Huddersfield Polytechnic/CNAA.

Burton, A.K., Tillotson, K.M., and Troup, J.D.G. 1989. Prediction of low-back trouble frequency in a working population. *Spine* 14:939-946.

Burton, A.K., Tillotson, K.M., Main, C.J., and Hollis, S. 1995. Psychosocial predictors of outcome in acute and subchronic low back trouble. *Spine* 20:722-728.

Burton, A.K., Battié, M.C., Gibbons, L., Videman, T., and Tillotson, K.M. 1996a. Lumbar disc degeneration and sagittal flexibility. *J Spinal Disord* 9:418-424.

Burton, A.K., Clarke, R.D., McClune, T.D., and Tillotson, K.M. 1996b. The natural history of low-back pain in adolescents. *Spine* 21:2323-2328.

Burton, A.K., Tillotson, K.M., Symonds, T.L., Burke, C., and Mathewson, T. 1996c. Occupational risk factors for the first-onset of low back trouble: a study of serving police officers. *Spine* 21: 2612-2620

Burton, A.K., Symonds, T.L., Zinzen, E., Tillotson, K.M., Caboor, D., Van Roy, P., and Clarys, J.P. 1997. Is ergonomics intervention alone sufficient to limit musculoskeletal problems in nurses? *Occup Med* 47:25-32

Burton, A.K. and Sandover, J. 1987. Back pain in Grand Prix drivers: a "found" experiment. *Appl Erg* 18:3-8.

Cady, L.D., Bischoff, D.P., O'Connell, E.R., Thomas, P.C., and Allan, J.H. 1979. Strength and fitness and subsequent back injuries in firefighters. *J Occup Med* 21:269-272.

Caplan, P.S., Freedman, L.M.J., and Connelly, T.P. 1966. Degenerative joint disease of the lumbar spine in coal miners — a clinical and X-ray study. *Arthr Rheumatism* 9:693-701.

Carragee, E.J., Helms, E., and O'Sullivan, G.S. 1996. Are postoperative activity restrictions necessary after posterior lumbar discectomy? A prospective study of outcomes in 50 consecutive cases. *Spine* 21:1893-1897.

Cholewicki, J. and McGill, S.M. 1996. Mechanical stability of the *in vivo* lumbar spine: implications for injury and chronic low back pain. *Clin Biomech* 11:1-15.

Cholewicki, J., McGill, S.M., and Norman, R.W. 1995. Comparison of muscle forces and joint load from an optimization and EMG assisted lumbar spine model: towards development of a hybrid approach. *J Biomechanics* 28:321-332.

Clinical Standards Advisory Group. 1994. *Epidemiology review: the epidemiology and cost of back pain*, London: HMSO.

Croft, P.R., Papageorgiou, A.C., Ferry, S., Thomas, E., Jayson, M.I.V., and Silman, A.J. 1995. Psychologic distress and low back pain: evidence from a prospective study in the general population. *Spine* 20:2731-2737.

Cust, G., Pearson, J.C.G., and Mair, A. 1972. The prevalence of low back pain in nurses. *International Nursing Review* 19:169-179.

Elliott, B.C., Davis, J.W., Khangure, M.S., Hardcastle, P., and Foster, D. 1993. Disc degeneration and the young fast bowler in cricket. *Clin Biomech* 8:227-234.

Garcy, P., Mayer, T., and Gatchel, R.J. 1996. Recurrent or new injury outcomes after return to work in chronic disabling spinal disorders: tertiary preventative efficacy of functional restoration treatment. *Spine* 21:952-959.

Granata, K.P., Marras, W.S., and Kirking, B. (1996) Influence of experience on lifting kinematics and spinal loading, in Anonymous *20th Annual Meeting*, Georgia Tech, Atlanta, USA: American Society of Biomechanics.

Greenough, C.G. and Fraser, R.D. 1989. The effects of compensation on recovery from low-back injury. *Spine* 14:947-955.

Hadler, N.M., Carey, T.S., and Garrett, J. 1995. The influence of indemnification by works' compensation insurance on recovery from acute backache. *Spine* 20:2710-2715.

Halpern, M. (1992) Prevention of low back pain: basic ergonomics in the workplace and clinic, in Nordin, M. and Vischer, T.L. (Eds.) *Common Low Back Pain: Prevention of Chronicity*, pp. 705-730. London: Baillier Tindall.

Hemmingway, H., Shipley, M.J., Stansfield, S., and Marmot, M. 1994. *Society for Back Pain Research*, Leeds: November 11.

Hildebrandt, V.H. 1995. Musculoskeletal symptoms and workload in 12 branches of Dutch agriculture. *Ergonomics* 38:2576-2587.

Holmstrom, E.B., Lindell, J., and Mortiza, U. 1992. Low back and neck/shoulder pain in construction workers: Occupational workload and psychosocial risk factors. Part 2: Relationship to neck and shoulder pain. *Spine* 17:672-677.

Hult, L. 1954. Cervical, dorsal and lumbar spinal syndromes. *Acta Orthop Scand* Suppl. 17:1-102.

Hunt, A. and Habeck, R. 1993. *The Michigan Disability Prevention Study*, Kalamazoo, Mich. W.E. Upjohn Institute for Employment Research.

IASP. 1995. *Back Pain in the Workplace: Management of Disability in Nonspecific Conditions*, Seattle: IASP Press.

Kelsey, J.L. 1975. An epidemiological study of the relationship between occupations and acute herniated lumbar intervertebral discs. *Int J Epidemiol* 3:197-205.

Kelsey, J.L., Githens, P.B., and O'Connor, T. 1984. Acute prolapsed lumbar intervertebral disc. An epidemiologic study with special reference to driving automobiles and cigarette smoking. *Spine* 9:608-613.

Kemmlert, K., Orelium-Dallner, M., Kilbom, Å. and Gamberale, F. 1993. A three-year follow-up of 195 reported occupational over-exertion injuries. *Scand J Rehabil Med* 25:16-24.

Kohles, S., Barnes, D., Gatchel, R.J., and Mayer, T.G. 1990. Improved physical performance outcomes after functional restoration treatment in patients with chronic low-back pain: Early vs. recent training results. *Spine* 15:1321-1324.

Lancourt, J. and Kettelhut, M. 1992. Predicting return to work for lower back pain patients receiving workers compensation. *Spine* 17:629-640.

Laurell, L. and Nachemson, A. 1963. Some factors influencing spinal injuries in seat ejected pilots. *Aerospace Med* 7:726-729.

Leighton, D.J. and Reilly, T. 1995. Epidemiological aspects of back pain: the incidence and prevalence of back pain in nurses compared to the general population. *Occup Med* 45:263-267.

Lindstrom, I., Ohlund, C., Eek, C., Wallin, L., Peterson, L., and Nachemson, A. 1992. Mobility strength and fitness after a graded activity program for patients with subacute low back pain: A randomized prospective clinical study with a behavioral therapy approach. *Spine* 17:641-652.

Lindstrom, I., Ohlund, C., and Nachemson, A. 1994. Validity of patient reporting and predictive value of industrial physical work demands. *Spine* 19:888-893.

Linton, S.J. and Warg, L. 1993. Attributions (beliefs) and job satisfaction associated with back pain in an industrial setting. *Percept Motor Skills* 76:51-62.

Malmivaara, A., Hakkinen, U., Heinrichs, M., Koskenniemi, L., Kuosma, E., Lappi, S., Paloheimo, R., Servo, C., Vaaranen, V., and Hernberg, S. 1995. The treatment of acute low back pain — bed rest, exercises, or ordinary activity? *New Eng J of Med* 332:351-355.

Mannion, A.F., Dolan, P., and Adams, M.A. 1996. Psychological questionnaires: do "abnormal" scores precede or follow first-time low back pain? *Spine* 21: 2603-2611.

Marras, W.S., Lavender, S.A., Leurgens, S.E., Rajulu, S.L., Allread, W.G., Farthallah, F.A., and Ferguson, S.A. 1993. The role of dynamic three-dimensional trunk motion in occupationally-related low back disorders: The effects of workplace factors trunk position and trunk motion characteristics on risk of injury. *Spine* 18:617-628.

Murray-Leslie, C.F., Lintott, D.J., and Wright, V. 1977. The spine in sport and veteran military parachutists. *Ann Rheum Dis* 36:332-342.

Nachemson, A.L. (1996) Low back pain in the industrial world, in Aspden, R.M. (Ed.) *Lumbar Spine Disorders*, pp. 1-5. Chesterfield: Arthritis & Rheumatism Council for Research.

Papageorgiou, A.C., Croft, P.R., Ferry, S., Jayson, M.I.V., and Silman, A.J. 1995. Estimating the prevalence of low back pain in the general population: Evidence from the South Manchester Back Pain Survey. *Spine* 20:1889-1894.

Porter, R.W. 1987. Does hard work prevent disc protrusion? *Clin Biomech* 2:196-198.

Porter, R.W., Adams, M.A., and Hutton, W.C. 1989. Physical activity and the strength of the lumbar spine. *Spine* 14:201-203.

Reid, S., Haugh, L.D., Hazard, R.G., and Tripathi, M. 1996. *Recovery Rates of Workers After Occupational Low Back Injury*, Burlington, VT. June 25-29: Int Soc Study Lumbar Spine.

Riihimaki, H. 1989. Radiographically detectable lumbar degenerative changes as risk indicators of back pain. A cross-sectional epidemiological study of concrete reinforcement workers and house painters. *Scand J Work Environ Health* 15:280-285.

Riihimaki, H., Tola, S., Videman, T., and Hanninen, K. 1989. Low-back pain and occupation. A cross-sectional questionnaire study of men in machine operating, dynamic physical work, and sedentary work. *Spine* 14:204-209.

Riihimaki, H., Mattison, T., Zitting, A., Wickstrom, G., Hanninen, K., and Waris, P. 1990. Radiologically detectable changes of the lumbar spine among concrete reinforcement workers and house painters. *Spine* 15:114-123.

Riihimaki, H., Viikari-Juntura, E., Moneta, G., Kuha, J., Videman, T., and Tola, S. 1994. Incidence of sciatic pain among men in machine operating dynamic physical work and sedentary work: A three-year follow up. *Spine* 19:138-142.

Skovron, M.L., Nordin, M., Halpern, N., and Cohen, H. 1991. *Do Worker Ratings of Job-Required Physical Abilities Correlate with Back Injury Rates?* Heidelberg, May 12-16: ISSLS.

Skovron, M.L., Hiebert, R., Nordin, M., Brisson, P.M., and Crane, M. 1997. Work restrictions and outcome of non-specific low back pain. *Spine,* in press.

Smedley, J. and Coggon, D. 1994. Will the manual handling regulations reduce the incidence of back disorders? *Occup Med* 44:63-65.

Spengler, D.M., Bigos, S.J., Martin, N.A., Zeh, J., Fisher, L., and Nachemson, A. 1986. Back injuries in industry: a retrospective study. I. Overview and cost analysis. *Spine* 11:241-245.

Suadicani, P., Hansen, K., Fenger, A.M., and Gyntelberg, F. 1994. Low back pain in steelplant workers. *Occup Med* 44:217-221.

Svensson, H. and Andersson, B.J. 1982. Low back pain in forty to forty-seven year old men. I. Frequency of occurrence and impact on medical services. *Scand J Rehabil Med* 14:47-53.

Symonds, T.L., Burton, A.K., Tillotson, K.M., and Main, C.J. 1995. Absence resulting from low back trouble can be reduced by psychosocial intervention at the workplace. *Spine* 20:2738-2745.

Symonds, T.L., Burton, A.K., Tillotson, K.M., and Main, C.J. 1996. Do attitudes and beliefs influence work loss due to low back trouble? *Occup Med* 46:25-32.

Tarasuk, V. and Eakin, J.M. 1994. Back problems are for life: perceived vulnerability and its implications fro chronic disability. *J Occup Rehabil* 4:55-64.

Troup, J.D.G., Martin, J.W., and Lloyd, D.C.E.F. 1981. Back pain in industry — a prospective survey. *Spine* 6:61-69.

Troup, J.D.G., Foreman, T.K., Baxter, C.E., and Brown, D. 1987. The perception of back pain and the role of psychophysical tests of lifting capacity. *Spine* 12:645-657.

Troup, J.D.G. and Edwards, F.C. 1985. *Manual Handling and Lifting: an Information and Literature Review with Special Reference to the Back,* London: Health & Safety Executive, HMSO.

Troup, J.D.G. and Videman, T. 1989. Inactivity and the aetiopathogenesis of musculoskeletal disorders. *Clin Biomech* 4:174-178.

Videman, T., Numinen, T., Tola, S., Kuorinka, I., Vanharranta, H., and Troup, J.D.G. 1984. Low back pain in nurses and some loading factors of work. *Spine* 9:400-404.

Waddell, G. (1992) Biopsychcosocial analysis of low back pain. In: Nordin, M. and Vischer, T.L. (Eds.) *Common Low Back Pain: Prevention of Chronicity,* pp. 523-557. London: Baillere Tindall.

Waddell, G., Newton, M., Henderson, I., Somerville, D., and Main, C.J. 1993. A fear avoidance belief questionnaire (FABQ) and the role of fear-avoidance beliefs in chronic low back pain and disability. *Pain* 52:157-168.

Wells, N. 1985. *Back Pain. Publication No. 78,* London: Office of Health Economics.

Wikstrom, B., Kjellberg, A., and Landstrom, U. 1994. Health effects of long-term occupational exposure to whole-body vibration: a review. *Int J Industrial Ergonomics* 14:273-292.

Winkel, J. and Westgaard, R.H. 1996. Editorial: a model for solving work related musculoskeletal problems in a profitable way. *Appl Erg* 27:71-78.

Wood, D.J. 1987. Design and evaluation of a back injury prevention program within a geriatric hospital. *Spine* 12:77-82.

# 21

# Fall-Related Occupational Injuries

**Stephen N. Robinovitch**
*San Francisco General Hospital*

## 21.1  Introduction

In the United States, falls account for approximately one third of all injury-related hospitalizations and one tenth of all injury-related deaths, making them the leading cause of accidental injury and second leading cause of traumatic death (Rice, 1989; Public Health Service, 1993). The annual cost associated with such injuries has been estimated at a staggering $37.3 billion, over one third of which involves direct medical expenses (Rice, 1989). Therefore, in terms of both frequency and cost, falls closely rival motor vehicle accidents (MVA) as a leading cause of injury. However, aside perhaps for the problem of fall-related hip fractures in the elderly (which represent approximately one quarter of all fall-related injury costs [Praemar, 1992]), they have received considerably less attention from injury prevention experts.

To design effective strategies for preventing fall-related injuries, we must consider two sets of risk factors: those which lead to an increased incidence of falls, and those which increase risk for injury in the event of a fall. Research tools for identifying such risk factors generally fall under the categories of epidemiology and biomechanics. The former (discussed in Section 21.3 of this chapter) provides important information on the circumstances surrounding fall injuries, thereby identifying host and environmental targets for intervention. The latter (discussed in Section 21.4) complements epidemiological data by identifying, through the analysis of experimental or mathematical models of the injury-causing environment, how injury risk is influenced by factors such as motor control, Newtonian dynamics, and mechanical properties of biological tissues. Prevention strategies (discussed in Section 21.5) with the highest likelihood for success are often those derived from, and therefore compatible with, a combined epidemiological and biomechanical perspective.

## 21.2  General Considerations

Three general groups of individuals are at high risk for fall-related injuries: the elderly over age 65, children between ages 6 to 10, and individuals whose occupation requires them to work at heights (Figure 21.1). For each of these, injury results from the transfer of gravitational potential energy to tissue strain energy (and the corresponding generation of mechanical stresses and strains) that exceeds the tissue's energy-absorbing

**FIGURE 21.1** Overview of factors influencing injury risk during typical falls in the elderly individual, child, and construction worker.

capacity. However, among the three groups, important differences exist in risk factors for fall-related injury. In particular, injury risk in the elderly is generally dominated by *host* factors such as balance, strength, reaction time, vision, and bone strength (Tinetti, 1994), while injury risk in the workplace is generally dominated by *environmental* factors such as poor lighting, slippery surfaces, workplace clutter, or lack of adequate fall prevention equipment (McVittie, 1995). Consequently, ergonomic or human factors-based strategies are likely to be highly effective in preventing fall-related injuries in the workplace, while this has, as yet, been untrue for the elderly (Fleming, 1993; Parker, 1996; Tinetti, 1993; Waller, 1978).

**TABLE 21.1** Annual Fall-Related Occupational Injuries and Deaths in the United States, by Occupation

| Industry | Fatal Injuries[a] | | Nonfatal Injuries[b] | |
|---|---|---|---|---|
| | Occurrences | Percent of All Injuries | Occurrences (Thousands) | Percent of All Injuries |
| Agriculture/Forestry | 63 | 7.9 | 6.9 | 16.7 |
| Mining | 10 | 6.4 | 3.9 | 18.9 |
| Construction | 335 | 32.0 | 43.0 | 19.7 |
| Manufacturing | 62 | 8.8 | 62.7 | 10.8 |
| Transportation/Public Utilities | 26 | 3.0 | 41.8 | 17.3 |
| Wholesale Trade | 14 | 5.5 | 24.2 | 14.6 |
| Retail Trade | 26 | 3.9 | 80.5 | 20.4 |
| Services | 71 | 9.6 | 99.5 | 19.3 |

[a] Source: Bureau of Labor Statistics, U.S. Department of Labor, Census of Fatal Occupational Injuries, 1995.

[b] Source: Bureau of Labor Statistics U.S. Department of Labor, Survey of Occupational Illnesses and Injuries, 1994. Data reflect a sum of injuries due to falls to a lower level and falls to the same level.

## 21.3 Epidemiology

### Fatal Injuries

*Incidence.* In 1995, falls accounted for 10% of fatal work-related injuries in the United States (653 of 6210 total deaths), ranking them alongside MVA, violent assault, and being struck by an object as a leading cause of accidental work-related death (Bureau of Labor Statistics, 1996a). As shown in Table 21.1, approximately one half of these deaths occurred in the construction industry, where falls accounted for 32% of work-related fatalities, more than any other cause (Bureau of Labor Statistics, 1996a; National Safety Council, 1994; McVittie, 1995; Ore, 1996). Construction-related occupations with the highest risk for fall-related fatality include roofers, structural metal workers, carpenters, and painters. In each of these, falls account for about one half or more of total fatalities (Bureau of Labor Statistics, 1996a; Sorock, 1993; Suruda, 1992).

*Circumstances.* Not surprisingly, over 95% of fall fatalities in the workplace involve falls from elevation (Bureau of Labor Statistics, 1996c), and accordingly, working at elevation is the single greatest risk factor for a fatal fall in the workplace. A review of 63 fatal falls in the Quebec construction industry between 1989 and 1992 (McVittie, 1995) found that the most common working surfaces from which falls occurred were skeletal structures such as building frameworks and roof trusses (25%), unfinished floors (24%), scaffolds (24%), roofs (18%), and ladders (6%). OSHA reported slightly different results in the analysis of 1148 fall-related deaths in the U.S. construction industry between 1985 and 1989 (Occupational Safety and Health Administration, 1992), citing roofs as the most common working surface (26%), followed by scaffolds (21%), skeletal structures (14%), unfinished floors (10%), and ladders (6%).

A second important class of risk factors for fall fatalities is lack of proper protective equipment. Indeed, several investigators have observed that the vast majority of fall-related fatalities involve the absence or improper use of fall restraint equipment, and noncompliance with recommended equipment and job site standards. For example, in McVittie's study of fall-related injuries in the Ontario construction industry (McVittie, 1995), no obvious fall risk existed in 84% of deaths, which apparently could have been prevented with proper use of safety equipment. In such cases, safety harnesses were often available but not being used. In another 13% of cases, falls occurred through poorly guarded floor openings. Only one case resulted from failure of a fall restraint system, and this involved the worker fastening his safety harness to a hoisting line rather than a lifeline. Similarly, in Suruda's study of fatal falls to construction painters (Suruda, 1992), over one half involved falls from scaffolds, and the vast majority of these (88%) involved the painter falling off the scaffold, as opposed to the scaffold collapsing. In 74% of cases, OSHA issued safety violations, commonly citing the lack of scaffold guardrails, nets, lifelines, and safety belts. Indeed, the single most common cause of death was failure to connect a safety belt to a lifeline.

**TABLE 21.2**　Incidence Rates (Injuries per 10,000 Worker-Years) for Fall-Related Nonfatal Injury by Nature of Injury and Body Part Affected, 1994

| Injury | Fall to Lower Level | Fall on Same Level |
|---|---|---|
| Type | | |
| Fractures | 2.7 | 4.7 |
| Dislocations | 0.4 | 0.6 |
| Sprains and Strains | 4.6 | 11.7 |
| Lacerations | 0.3 | 0.8 |
| Bruises | 2.2 | 6.6 |
| Intracranial Injury | 0.1 | 0.3 |
| Back Pain | 0.3 | 1.0 |
| Soreness, Non-Back | 0.5 | 1.6 |
| Location | | |
| Head | 0.4 | 1.4 |
| Neck | 0.1 | 0.3 |
| Trunk | 4.3 | 11.3 |
| Pelvis | 0.4 | 1.1 |
| Upper Extremity | 1.6 | 4.6 |
| Lower Extremity | 4.4 | 9.4 |

*Source:* Bureau of Labor Statistics, U.S. Department of Labor, Survey of Occupational Illnesses and Injuries, 1994.

## Nonfatal Injuries

*Incidence.* According to the Bureau of Labor Statistics, falls were the cause of 18% of all work-related injuries in 1994, trailing only overexertion and being struck by an object as a leading cause of injury (Bureau of Labor Statistics, 1996b). Between 1992 and 1994, the overall frequency for disabling fall-related injuries was fairly constant at 49 per 10,000 person-years. Sectors with the greatest number of fall-related injuries were the service and trade industries (Table 21.1). However, incidence rates were highest for the construction, transportation, agriculture, and mining industries, with respective risks of 96, 73, 64, and 62 injuries per 10,000 worker-years.

In contrast to fall fatalities, over two thirds of nonfatal fall injuries were due to falls onto the same level (Bureau of Labor Statistics, 1996b). However, we might generally expect that, the higher the fall height, the greater the injury severity, and indeed, lost workdays were considerably higher for injuries caused by falls to a lower level (10 days vs. 6 days for falls on the same level). As shown in Table 21.2, the most frequent types of injury were strains and sprains, followed by fractures and contusions, and the body part most affected was the trunk (including back and shoulders), followed by the lower extremity, upper extremity, and head. While a useful reflection of overall injury trends, Table 21.2 fails to show that falls represent the leading cause of work-related spinal cord injury (Rosenberg, 1993) and brain injury (Heyer, 1994), surely among the most devastating of injuries. Not surprisingly, workers in construction and agriculture are at greatest risk for these injuries.

*Circumstances.* When compared to fall fatalities, risk factors for nonfatal falls in the workplace are considerably more diverse, as reflected by the small variation among industries in injury incidence (Table 21.1). McVittie (McVittie, 1995) found that risk factors for falls in the construction industry generally fell under two categories: in-transit activities and poor housekeeping. Risk-factors associated with the former included uneven terrain and the use of steps and ladders, while those associated with the latter included the presence of waste materials, debris, clutter, poor lighting, and slippery surfaces.

## 21.4   Biomechanics

In considering the biomechanics of falls and fall-related injuries, it is useful to consider falls as having three stages (Hayes, 1993): (1) an *initiation* stage, involving a slip, trip, or loss of balance, and potential attempts to regain upright posture; (2) a *descent* stage, where movements may be attempted in preparation for landing; and (3) a *contact* stage, where impact occurs between the body parts and the ground, resulting in the generation of reaction forces and dissipation of the body's kinetic energy.

### Fall Initiation

Numerous studies have been conducted on the control of balance, involving application of destabilizing perturbations to the feet or trunk (see reviews by Alexander, 1994; Prieto, 1993; Winter, 1990). Most have focused on balance correction through sway, or standing balance maneuvers (Alexander, 1992; Era, 1995; Kuo, 1993; Maki, 1987; Romick-Allen, 1988; Stelmach, 1989), which involves activation of lower extremity muscles in either a proximal-to-distal sequence (hip strategy), or distal-to-proximal sequence (ankle strategy) (Manchester, 1989; Woollacott, 1993). However, several investigators have observed that sway-based recovery is feasible only for small perturbations to posture (i.e., small displacements of the body's center-of-gravity from the base-of-support formed by the feet), such as those occurring during normal walking and standing. Consequently, in the event of a slip or trip, stepping emerges as the primary means for balance recovery (Maki, 1996; Nashner, 1980; Thelen, 1997; Wolfson, 1986).

To assess the factors which influence balance recovery by stepping after a slip, we recently measured body movements as subjects (aged 22 to 35 years) stood barefoot on a large gymnasium mat and attempted to prevent falls after the mattress was made to unexpectedly translate (Hsiao, 1998). Throughout the testing session, we randomly varied both the direction of the perturbation (by having the subject stand forward, backward, or sideways to the perturbation direction), and the strength of the perturbation (randomized between 4 acceleration levels). To evoke "natural" responses, no practice trials were allowed, and subjects were only instructed that (1) in the event of platform movement, they should "try to prevent themselves from falling," and (2) prior to platform movement, they should maintain their gaze directed forward and at eye level. In each trial, a three-dimensional motion measurement system acquired the positions of 20 skin surface markers located throughout the extremities and trunk.

As expected, stepping was the predominant balance recovery technique; only a single trial involved stabilization of posture through sway (Figure 21.2). However, the effectiveness of stepping in preventing a fall was highly directional-dependent, with subjects being more than twice as likely to fall when the perturbation was directed posteriorly (i.e., the mattress translated in the anterior direction), as opposed to anteriorly or laterally. Furthermore, falls due to posterior perturbations commonly involved trunk and/or pelvis impact, while this was rare in falls due to anterior or lateral perturbations (in these trials, impact was almost always restricted to the hands and/or knees). Finally, females were twice as likely to fall as males, and in all but one fall trial, a failed attempt at balance recovery by stepping was observed (Figure 21.3).

Implicit in the observation that stepping represents the primary means for balance recovery after a slip or trip is the suggestion that fall risk is inherently greater on scaffolds and ladders, since the small surface area of these environments nearly eliminates one's ability to recover balance by stepping. This highlights the importance of educating workers to correctly perceive this difference in fall risk, and the importance of utilizing fall restraint devices when working in such environments.

### Fall Descent Kinematics

Even in a fall from standing height, sufficient energy exists to fracture the wrist (Chiu, 1996; Myers, 1993) or hip (Courtney, 1994; Robinovitch, 1991). However, only a small percentage of such falls actually result in serious injury (Melton, 1988). This suggests that highly effective movement strategies exist for preventing

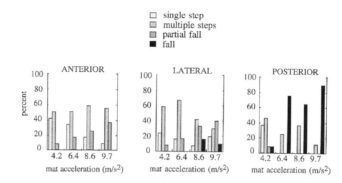

**FIGURE 21.2** Balance recovery and falling responses during simulated slipping experiments. Seventy-seven percent of trials involved balance recovery through one or more steps. During anterior and lateral perturbations, loss-of-balance was more likely to lead to a "partial fall" (defined as contact to one or both knees and/or wrists) than a "fall" (defined as contact to the pelvis and trunk), while the reverse was true for posterior perturbations. (From Hsiao ET, Kearns M, Robinovitch SN. 1998. Analysis of movement strategies during unexpected falls. *J Biomechanics*, 31:1-9. With permission.)

**FIGURE 21.3** Stick-figure (oblique view) image of a typical fall from a posterior perturbation. Note the failed attempt to recover balance by stepping at approximately t = 0.45 s, the initial upward and then downward movement of the upper extremity, the small degree of trunk rotation during descent, and the near-simultaneous impact to the wrist and pelvis at approximately t = 0.75 s. (From Hsiao ET, Kearns M, Robinovitch SN. 1998. Analysis of movement strategies during unexpected falls. *J Biomechanics*, 31: 1-9. With permission.)

injury during a fall, and that injury risk may indeed be governed by impairment of such responses due to host or environmental factors.

To assess the nature of protective responses during falls, we analyzed body movements during the 28 falls occurring during the slipping experiments described in the previous section (Hsiao, 1997). Our results confirmed our suspicion that, rather than being random and unpredictable, body segment kinematics during falls are organized into specific movement sequences which facilitate safe landing. These could be classified into two major types. The first was a complex yet highly repeatable sequence of upper extremity movements which allowed subjects to impact their wrist at nearly the same instant as the pelvis (average time interval between contacts = 38 ms; Figure 21.4). This involved an immediate upward movement of the upper extremity (perhaps reflecting a startle response), followed by a rapid downward movement, and finally a second upward acceleration just prior to impact (Figure 21.3). This last deceleration substantially

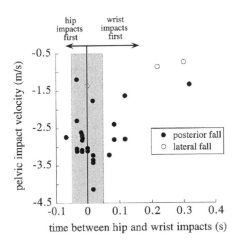

**FIGURE 21.4** Vertical velocity of the pelvis at the moment of impact as a function of the time interval between wrist and pelvis contacts. Positive x-axis values reflect the wrist impacting before the pelvis, and negative values on the y-axis reflect downward movement. In approximately 70% of trials, the time difference between hip and wrist impacts was less than 50 ms (gray band), suggesting a sharing of impact energy between the two body regions. (From Hsiao ET, Kearns M, Robinovitch SN. 1998. Analysis of movement strategies during unexpected falls. *J Biomechanics*, 31: 1-9. With permission.)

decreased the vertical velocity of the wrist at contact velocity, which averaged 2.6 m/s, or 66% of its peak downward velocity during descent. Impact to both wrists was observed in 26 of the 28 falls, with an average interval of 70 ms between contacts. Average vertical pelvis velocity at impact was 2.5 m/s, or 83% of its peak downward velocity during descent. Head impact was observed in only five falls, three of which involved the same subject.

These results suggest (as does common experience) that the upper extremity plays a major role in allowing safe landing during a fall. For example, it is reasonable to assume that without the occurrence of upper extremity impact, all cases of failed balance recovery would have resulted in impact to the head and trunk. Therefore, braking the fall with the outstretched hand allowed for complete avoidance of upper body impact in anterior trials, and reduced the incidence of upper body impact by over fourfold in lateral trials. Second, over two thirds of falls involved wrist contact within 50 ms of pelvis contact (Figure 21.4), and previous studies have shown that approximately 50 ms is required to reach peak force during a fall on the hip (Robinovitch, 1991) or wrist (Chiu, 1996). This suggests that impact configurations are chosen to allow *sharing* of impact energy between these body parts, and a subsequent reduction in localized impact forces and injury risk.

The second class of protective response observed in these trials was marked trunk rotation in falls due to lateral (but not anterior or posterior) perturbations (Figure 21.5). This commenced late in the descent stage of the fall, and resulted in impact with the body anteriorly facing the contact surface. This, in turn, allowed subjects to contact the impact surface with both right and left upper extremities, and avoid impact to the lateral aspect of the hip (thereby decreasing one's risk for hip fracture [Greenspan, 1994; Nevitt, 1993]). In posterior perturbations, subjects apparently sensed the infeasibility of a 180-degree trunk rotation to anteriorly face the impact surface, and instead focused on safe landing on the buttocks and wrists.

## Impact Mechanics

During a fall, the risk for injury to a specific anatomical structure depends on the ratio of the force applied to the structure divided by the force required to cause failure. To estimate failure forces of anatomical structures, cadaveric materials may be loaded to failure in a mechanical testing system which simulates the fall impact environment. Such studies have measured failure loads for the distal radius,

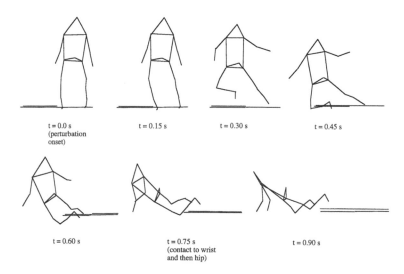

**FIGURE 21.5**　Stick figure image of a typical side fall. Note that trunk rotation about an inferior–superior axis allowed for impact to the anterior–lateral aspect of the pelvis rather than the hip, at nearly the same instant that contact occurred to the wrist. (From Hsiao ET, Kearns M, Robinovitch SN. 1998. Analysis of movement strategies during unexpected falls. *J Biomechanics,* 31: 1-9. With permission.)

which average $1800 \pm 700$ N (Myers, 1993; Spadaro, 1994), and the hip, which average $4200 \pm 1600$ (S.D.) N (Courtney, 1994; Weber, 1992).

In order to estimate fall impact forces, consider the mechanics of a simple mass–spring representation of the body, as shown in Figure 21.6. In configuration (a), the system is stationary at height $h$ (measured in units of m), possessing zero kinetic energy and a gravitational potential energy of $mgh$, where $m$ is the mass in kg, and $g$ is the gravitational constant, 9.81 m/s². Configuration (b) represents the beginning of the impact phase. The system has descended a height $h$, and the spring just contacts the ground but has not yet undergone compression. Therefore, the system has zero potential energy and a kinetic energy of $mv^2/2$, where $v$ is the downward velocity in m/s. By equating total energy between configurations (a) and (b), it is readily observed that $v = \sqrt{2\,gh}$ (recall that free-fall velocity is dependent only on descent height, and not system mass). Finally, in configuration (c), the spring has compressed maximally (by an amount $x$, measured in m), and is about to rebound upward. Therefore, the system has zero downward velocity and kinetic energy, and (neglecting the difference in gravitational potential energy between configurations (b) and (c)) a potential energy of $kx^2/2 = F^2/(2k)$. Equating this to the total energy in configurations (a) or (b), the peak impact force on the system is:

$$F = v\sqrt{mk} = \sqrt{2kmgh}. \qquad (21.1)$$

Equation (21.1) illustrates that the main requirement in determining impact force is to derive reasonable estimates for the system's effective mass $m$ and stiffness $k$. In previous studies, we estimated such values based on the measured response of human subjects during safe, simulated falls on the hip (Robinovitch, 1991) and outstretched hand (Chiu, 1996). Average values for $m$ and $k$ during falls onto the hip were 35 kg and 71 kN/m, respectively, while those for falls on the outstretched hand were 29 kg and 8.7 kN/m. Figure 21.7 shows resulting estimates of peak contact force when floor stiffness is infinitely stiff, and primary contact occurs to either the upper extremity (filled circles) or hip (filled squares). Under such conditions, hip contact force is over threefold greater than upper extremity contact force, due to the substantially higher effective stiffness of the body during a fall on the hip. Also shown in Figure 21.7 are peak contact forces for falls onto a rather soft surface, possessing a stiffness of 10 kN/m.

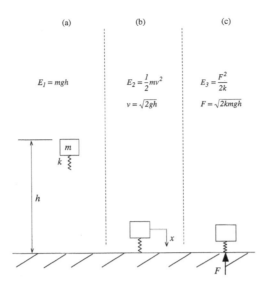

**FIGURE 21.6**  Idealized spring–mass model of impact during a fall. Simple energy considerations imply that contact force scales with the square-root of descent height, effective mass, and effective stiffness.

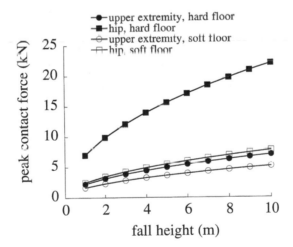

**FIGURE 21.7**  Effect of fall height and ground stiffness on estimated contact forces during falls. Contact force is based on $F = \sqrt{2k_{eq}mgh}$ (see Figure 21.6), where $k_{eq} = kk_f/(k + k_f)$ is the effective series stiffness (in kN/m) of the body spring $k$ and ground spring $k_f$, $m$ is the effective mass of the body (in m), $g = 9.81$ m/s², and $h$ is descent height (in m). Estimated values for $m$ and $k$ are 35 kg and 71 kN/m for impact to the hip (Robinovitch, 1991) and 29 kg and 8.7 kN/m for impact to the outstretched hand (Chiu, 1996). Note that, during impact to an infinitely stiff surface ($k_f = \infty$; filled symbols), the much stiffer hip region experiences contact forces threefold higher than those generated during a fall on the upper extremity. However, during impact to a soft surface ($k_f = 10$ kN/m; open symbols), the series stiffness governing hip impact is greatly reduced, and peak hip impact forces exceed upper extremity forces by only 1.5-fold.

These were calculated by replacing $k$ in Equation (21.1) with the equivalent series stiffness $kk_f/(k + k_f)$. Note that, under this scenario, hip contact force is reduced 65%, while upper extremity force is reduced by only 27%. Consequently, a soft impact surface substantially reduces one's risk for hip fracture during a fall (or injury to other "stiff" body parts, such as the head or back), but offers considerably less protection against upper extremity injury.

## 21.5 Prevention

The single most effective means for lowering the incidence of serious fall-related injuries in the workplace is to ensure appropriate use of personal fall protection equipment. This follows from the following facts discussed in Section 21.3: (1) that the vast majority of serious fall injuries and fatalities involve descents from high elevation; (2) that in the majority of cases, fall protection equipment was not being used at the time of the injury; and (3) that few fall-related deaths result from restraint system failure.

A major challenge for ergonomics researchers is to therefore engineer systems which ensure usage of fall protection equipment in high-risk environments. This requires the conversion of fall protection systems from an active form of protection, which relies on workers' choice in utilizing it, to an automatic (or passive) form of protection, in which user choice, and thus risk for noncompliance, is eliminated (Committee on Trauma Research, 1985). To illustrate the point here, consider the role of seatbelts in preventing injury during motor vehicle accidents. While it is well-established that seatbelts reduce injury risk, the existence of a seatbelt in itself provides no actual protection, unless the occupant chooses to use it (active protection). In contrast, an automobile designed so the transmission operates only when seatbelts are engaged (passive protection) is likely to be inherently safer, since it removes user choice as a variable. An analogous system for work at high elevation is, for example, one which prevents workers from ascending to elevated workspaces without the use of an effective fall protection system.

Simultaneous with efforts to develop such systems, improvements should be undertaken in job training and work site inspection for fall hazards. Equipment should be regularly tested for structural integrity, and inspected for evidence of wear or damage. Risk factors for standing height falls should also be minimized, with special attention to proper lighting, and the elimination of clutter and slippery surfaces. Workers and supervisors should also be alert to intrinsic risk factors for falls, such as alcohol or medication use, and impairments in vision, balance, strength, and joint range-of-motion.

Education should focus on the principles of load-transfer and load-sharing among restraint system components, and the need to ensure that the load-bearing capacity of each component (e.g., anchoring sites, belts, lifeline attachment fixtures, ropes) meet appropriate factors of safety. Workers should also be educated regarding the importance of short lifelines, since the loads imposed on a tethering system increase with descent height. In using scaffolding, attention should be paid to manufacturer recommendations on the use of counterweights, safety brakes, tie-off of support devices, and floor load capacity. Different anchor points should be used to support scaffolding and harness or body belt lifelines. Workers should also be trained on proper use of ladders, including reaching practices, use of three-point contact, avoidance of unstable surfaces, and limitations on ladder inclination and ascent height.

Education on fall mechanics may also promote safe work practices, and eliminate misconceptions regarding risk for falls and fall-related injuries. For example, workers should be instructed on the relationship between injury tolerance limits and fall descent height, and be made aware that postural stability is inherently lower in the constrained environment of a scaffold or ladder, since reduced surface area impairs one's ability to recover balance by stepping.

## References

Alexander NB, Shepard N, Gu MJ, Schultz A. 1992. Postural control in young and elderly adults when stance is perturbed: kinematics. *J Gerontol Med Sci* 47: M79-87.

Alexander NB. 1994. Postural control in older adults. *J Am Geriatr Soc* 42: 93-108.

Bureau of Labor Statistics. 1996a. *Census of Fatal Occupational Injuries, 1995.* Washington, D.C., U.S. Department of Labor.

Bureau of Labor Statistics. 1996b. *Characteristics of Injuries and Illnesses Resulting in Absences from Work, 1994.* Washington, D.C., U.S. Department of Labor.

Bureau of Labor Statistics. 1996c. *Issues in Labor Statistics: Issue 96-1: Construction Falls.* Washington, D.C., U.S. Department of Labor.

Chiu J, Robinovitch SN. 1996. Transient impact response of the body during a fall on the outstretched hand. BED-Volume 33, *1996 Advances in Bioengineering,* Proceedings from the ASME International Mechanical Engineering Congress and Exposition. New York: American Society of Mechanical Engineers. p. 269-270.

Committee on Trauma Research, Commission on Life Sciences, National Research Council and Institute of Medicine. 1985. *Injury in America: A Continuing Public Health Problem.* Washington, D.C.: National Academic Press.

Courtney AC, Wachtel EF, Myers BR, Hayes WC. 1994. Effects of loading rate on strength of the proximal femur. *Calcif Tissue Int* 55: 53-58.

Era P, Heikkinen E. 1995. Postural sway during standing and unexpected disturbance of balance in random samples of men of different ages. *J Gerontol* 40:287-295.

Fleming E, Pendergast DR. 1993. Physical condition, activity pattern, and environment as factors in falls by adult care facility residents. *Arch Phys Med Rehab* 74: 627-630.

Greenspan SL, Myers ER, Maitland LA, Resnick NM, Hayes WC 1994. Fall severity and bone mineral density as risk factors for hip fracture in ambulatory elderly. *JAMA* 271:128-133.

Hayes WC, Myers ER, Morris JN, Gerhart TN, Yett HS, Lipsitz LA. 1993. Impact near the hip dominates fracture risk in elderly nursing home residents who fall. *Calcif Tissue Int* 52: 192-198.

Heyer NJ, Franklin GM. 1994. Work-related traumatic brain injury in Washington State, 1988-90. *Am J Public Health* 84: 1106-1109.

Hsiao ET, Robinovitch SN. 1998. Common protective movements govern unexpected falls from standing height. *J Biomechanics,* 31: 1-9.

Kuo AD, Zajac FE. 1993. A biomechanical analysis of muscle strength as a limiting factor in standing posture. *J Biomechanics* 26S:137-150.

Maki BE, Holliday PJ, Fernie GR 1987. A posture control model and balance test for the prediction of relative postural stability. *IEEE Transactions on Biomedical Engineering* 34:797-810.

Maki BE, McIlroy WE, Perry SD. 1996. Influence of lateral destabilization on compensatory stepping responses. *J Biomechanics* 29:343-353.

Manchester D, Woollacott M, Zederbauer-Hylton N, Marin O. 1989. Visual, vestibular and somatosensory contributions to balance control in the older adult. *J Gerontol Med Sci* 44:M118-M127.

McVittie DJ. 1995. Fatalities and serious injuries. *Occup Med: State of the Art Rev* 10:285-293.

Melton LJ, Chao EYS, Lane J. 1988. Biomechanical aspects of fractures, in Riggs BL, Melton LJ. (Eds.), *Osteoporosis: Etiology, Diagnosis, and Management,* pp. 111-131, New York, NY, Raven Press.

Myers ER, Hecker AT, Rooks DS, Hipp JA, Hayes WC. 1993. Geometric variables from DXA of the Radius predict forearm fracture load *in vitro. Calcif Tissue Int* 52:199-204.

Nashner LM. 1980. Balance adjustments of humans perturbed while walking. *J Neurophysiol* 44:650-664.

National Safety Council. 1994. *Accident Facts,* 1994 Edition. Itasca, IL: National Safety Council.

Nevitt MC, Cummings SR. 1993. Type of fall and risk of hip and wrist fractures: the study of osteoporotic fractures. *J Am Geriatr Soc* 41:1226-1234.

Occupational Safety and Health Administration. *Construction Accidents: Workers Compensation Database 1985-88.* U.S. Department of Labor, Washington, D.C.

Ore T, Stout NA. 1996. Traumatic occupational fatalities in the U.S. and Australian construction industries. *Am J Indust Med* 30:202-206.

Parker MJ, Twemlow TR, Pryor GA. 1996. Environmental hazards and hip fractures. *Age and Ageing* 25:322-325.

Praemar A, Furner S, Rice DP. 1992. Costs of musculoskeletal conditions, in *Musculoskeletal Conditions in the United States,* pp. 143-170, Park Ridge, IL, American Academy of Orthopaedic Surgeons.

Prieto TE, Myklebust JB, Myklebust BM. 1993. Characterization and modeling of postural steadiness in the elderly: A review. *IEEE Trans Rehab Eng* 1:26-34.

Rice DP, MacKenzie EJ, and associates. 1989. *Cost of Injury in the United States: A Report to Congress.* Institute for Health and Aging, University of California San Francisco, and Injury Prevention Center, Johns Hopkins University.

Robinovitch SN, Hayes WC, McMahon TA. 1991. Prediction of femoral impact forces in falls on the hip. *J Biomech Eng* 113:366-374.

Romick-Allen R, Schultz AB. 1988. Biomechanics of reactions to impending falls. *J Biomechanics* 21:591-600.

Rosenberg NL, Gerhart K, Whiteneck G. 1993. Occupational spinal cord injury: demographic and etiologic differences from non-occupational injuries. *Neurology* 43 (July): 1385-1388.

Sorock GS, Smith EO, Goldoft M. 1993. Fatal occupational injuries in the New Jersey construction industry, 1983 to 1989. *J Occupational Med* 35:916-921.

Spadaro JA, Werner FW, Brenner RA, Fortino MD, Fay LA, Edwards WT. 1994. Cortical and trabecular bone contribute strength to the osteopenic distal radius. *J Orthop Res* 12:211-218.

Stelmach GE, Teasdale N, Di Fabio RP, Phillips J. 1989. Age-related declines in postural control mechanisms. *Intl J Aging Human Development* 29:205-223.

Suruda AJ. 1992. Work-related deaths in construction painting. *Scand J Work Environ Health* 18:30-33.

Thelen DG, Wojcik LA, Schultz AB, Ashton-Miller JA, Alexander NB. 1997. Age differences in using a rapid step to regain balance during a forward fall. *J Gerontol Med Sci* 52A:M8-M13.

Tinetti ME, Baker DI, Garrett PA, Gottschalk M, Koch ML, Horwitz RI. 1993. Yale FICSIT: risk factor abatement strategy for fall prevention. *J Am Geriatr Soc* 41:315-320.

Tinetti ME. 1994. Prevention of falls and fall injuries in elderly persons: A research agenda. *Prev Med* 23:756-762.

U.S. Department of Health and Human Services, Public Health Service. 1993. Healthy People 2000: National Health Promotion and Disease Prevention Objectives, Washington, D.C.

Waller JA. 1978. Falls among the elderly-human and environmental factors. *Accid Anal Prev.* 10:21-33.

Weber TG, Yang KH, Woo R, Fitzgerald RHJr. 1992. Proximal femur strength: Correlations of the rate of loading and bone mineral density. *ASME*, BED 22:111-114.

Winter DA, Patla AE, Frank JS. 1990. Assessment of balance control in humans. *Medical Prog Technol* 16:31-51.

Wolfson LI, Whipple R, Amerman P, Kleinberg A. 1986. Stressing the postural response: A quantitative method for testing balance. *J Am Geriatr Soc* 34:845-850.

Woollacott MH. 1993. Age-related changes in posture and movement. *J Gerontol* 48S:56-60.

# 22

# Low Back Pain (LBP) Glossary: A Reference for Engineers and Ergonomists

Simon M. Hsiang
*Liberty Mutual Research Center for Safety and Health*

Raymond M. McGorry
*Liberty Mutual Research Center for Safety and Health*

## 22.1 Foreword

Low back pain (LBP) is a pervasive problem in many societies and low back pain disability (LBPD) is the most expensive and pervasive health and safety problem arising in industry (Spengler et al., 1986a; Webster and Snook, 1990). However, there is no clear definition, well-accepted pathology, diagnostic scheme or treatment protocol for the problem. With occurrences both in and out of the workplace and in both the presence and absence of biomechanical stressors arising from work, the problem is regarded as belonging to several academic domains. Consequently, ergonomists, sociologists, psychologists, and engineers are attempting to determine the relevant approaches from their individual disciplines. The large volume of LBP research papers produced by widely divergent disciplines is indicative of the scope, and the urgent need to control the problem.

In attempting to deal with the immense and sometimes controversial literature, many ergonomists and engineers interested or involved in the studies of LBP are often exposed to documents replete with medical vocabularies and unfamiliar theories.

For the benefit of these professionals, in their review of the literature, this glossary of LBP terminology and research has been compiled. The reader should be aware of the complexity and the subjective nature of the large volume of LBP literature. This document is not meant to substitute for a medical dictionary, or for review of the current literature, but rather as a desk reference for those requiring assistance with terms or hypotheses encountered during an article review.

## 22.2 Introduction

The terms compiled in this glossary can be classified into six highly interrelated categories. They are: *anatomy* of the vertebral column, the *mechanism* of disease process, *diagnosis*, *diagnostic tools*, *treatment* plans based on diagnosis, and terms specific to *low back pain* (see Figures 22.1 and 22.2).

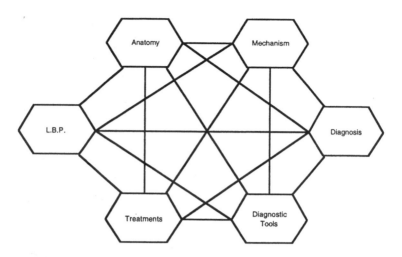

**FIGURE 22.1**    Schematic of the six categories of terms of low back pain.

*Anatomy* deals with terms that are descriptive of and specific to the structure and physical properties of the vertebrae, the intervertebral disc, the spinal cord and its nerve roots, and the muscles and ligaments that attach to it.

*Mechanism* describes the normal and abnormal function (physiology) of the anatomy, providing insight into the effect of disease process on the mechanics of the spinal column.

*Diagnostic* tools have been developed to provide insight into the anatomy and function to aid the medical practitioner in developing an appropriate diagnosis and subsequent treatment plan.

*Diagnosis* is a medical judgment based on the qualitative and quantitative findings produced by the diagnostic tools. Development of the appropriate diagnosis provides insight into the disease process and as a result provides goals and direction to the treatment plan.

*Treatment* covers a spectrum of techniques and procedures, performed by a wide array of medical professionals, based on any number of philosophies.

*Low back pain* is a category that encompasses terms related to the cause, nature, and description of the pain.

Two cases are provided to demonstrate the interrelationships that exist between the terms included in this glossary. In these cases the terms are linked to illustrate two different pathways from activity to the mechanism and finally the diagnosis and medical management of LBP. Consider the case where lifting and activities of daily living (ADL) create compressive and torsional forces acting on the spine. These forces are borne by both the anterior (disc) and posterior (apophysis) portion of the vertebral column. Case 1, as shown in Figure 22.3, uses the apophyseal (facet) joint as an example to trace the low back pain pathway for the posterior aspect of the spine. Case 2, as shown in Figure 22.4, uses the disc to trace the same pathway for the anterior spine. Note that both examples are for illustrative purposes only. They are *not* all-inclusive and are *not* intended as suggested limitations to the use of diagnostic tool or treatment protocols.

## 22.3    Terms and Definitions

**Acupuncture** — based on Chinese medicine techniques. Involves insertion of thin needles of various length through the skin at sites believed (1) to relieve symptoms, (2) to induce surgical anesthesia, and (3) to promote healing. Sometimes used to relieve low back pain.

**Analgesic** — (1) insensitive to pain, (2) relieving pain, and (3) an agent whose intended action is to alleviate pain (see pharmacological agents for low back pain).

**Anaphylaxis (Anaphylactic)** — a manifestation of sudden allergic reaction (see antigen).

**Ankylosis** — immobility or abnormal fixation of a joint. May be congenital, hereditary, or resulting from disease, injury, aging process (Sampson and Davis, 1988), or induced artificially by surgical fixation or fusion. Ankylosis reduces spinal segment range of motion, and often is painful. It can sometimes be asymptomatic.

**Ankylosing spondylitis** — inflammation of the spine that leads to stiffening or fusion of the vertebral joints.

**Annulus** — a ring or ring-like structure.

**Annulus fibrosus** — The fibrous outer ring-like portion of the intervertebral disc (Bogduk, 1983, 1991; Brinckmann, 1986; Hickey and Hukins, 1983).

Anatomy
- Can be classified into the inner portion and the outer portion. The inner portion of the annulus is mainly type II collagen (fibrocartilaginous). These collagen fibers are attached to the vertebral endplate, forming an internal capsule surrounding the nucleus pulposus. The outer portion is largely type I collagen (fibroelastic) and spans the rims of the vertebral bodies.
- The annulus fibrosis is comprised of 16 to 20 layers called lamellae. Collagen fibers in a single lamella of the annulus fibrosus have the same orientation. They are typically oriented at an angle of 65° to the long axis of the vertebral column. The orientation of the fibers alternates from 65° to –65° in successive lamellae.

Mechanism
- Both apophyseal joints and annulus fibrosus can resist shear force. The apophyseal joints have facets oriented to block forward translation by bony contact. The annulus fibrosus acts as a substantial ligament capable of resisting shear due to its high density of collagen.
- The orientation of collagen fibers may be biomechanically essential for the annulus to withstand: (1) the tensile force due to trunk flexion; (2) the radial load due to axial loading; (3) the torsional stress due to trunk rotation; (4) the shear stress due to anterior or posterior translation of the vertebrae; (5) or the combination of these forces.

Failure
- The annulus fibrosis can fail under conditions of low magnitude cyclic loading when exposed to a combined pattern of flexion and rotation (Gordon et al., 1991).
- The annulus may creep in time under compression with buckling and bulging of the annulus fibrosus and narrowing of the disc space. Creeping may become symptomatic as a result of foraminal or canal stenosis. In some cases of annular disruption, CT scan and discography provocation have demonstrated good correlation (Vanharanta et al., 1988b) (see creep and stenosis).
- The annulus fibrosus may creep due to the shear force when the mechanism of apophyseal joints is disabled by spondylolysis. Creeping may cause the upper vertebrae in the segment to slide progressively forward, as the condition progresses to spondylolisthesis (see spondylolysis and spondylolisthesis).

**Anterior longitudinal ligament** — the broad, strong band of longitudinal fibers that extends along the anterior surface of the vertebral bodies. It extends from the cervical spine to the sacrum (see ligament, and spinal ligament).

**Antigen** — any substance that the body regards as foreign and potentially dangerous and against which it initiates a specific immune response.

**Apophyseal joint of the spine** — commonly used as a synonym for the zygapophyseal or facet joint between adjacent vertebrae. The function of the apophyseal joint is to allow limited movement between vertebrae and to protect the disc from excessive shear forces, flexion, and axial rotation (Adams and Hutton, 1983b; Oxland et al., 1992). Excessive loading of this joint stretches the joint capsule and can itself be a cause of low back pain (Yang and King, 1984a). These joints contain rudimentary menisci which have been suspected of causing pain in entrapment disorders and may respond to treatment with manipulation (Bogduk and Engle, 1984).

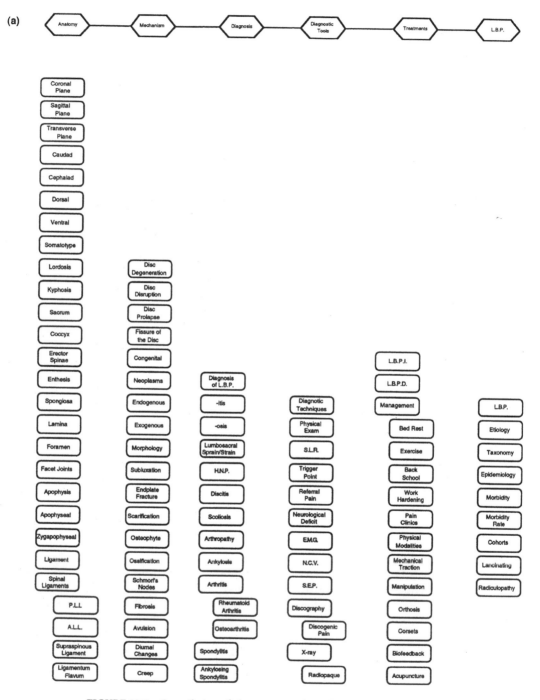

**FIGURE 22.2**    Compilation of glossary terms in each of the six categories.

**Apophysis** — projection or swelling from a bone.

**Arachnoid** — resembling a spider's web in consistency.

**Arachnoid membrane** — the middle of the three membranes (meninges) covering the spinal cord and brain (see meninges).

**Arachnoiditis** — inflammation of the arachnoid membrane; can progress to fibrosis or scarring that binds the nerve roots of the cauda equina (see cauda equina, sciatica).

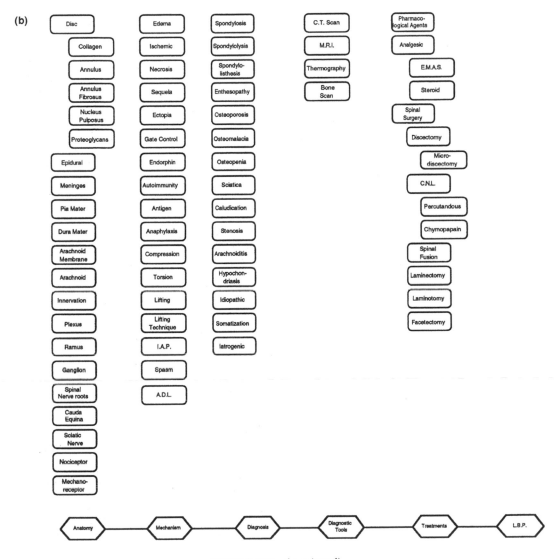

**FIGURE 22.2**   (continued)

**Arthritis** — inflammation of a joint (see osteoarthritis, rheumatoid arthritis).

**Arthropathy** — disease affecting a joint.

**Autoimmunity** — a condition characterized by an immune response directed against the body's own tissues.

**Avulsion** — the tearing or forcible separation of part of a structure.

**Back school** — a patient education program usually administered by health care professionals with the goal of achieving the greatest functional level for the patient. This usually includes education about the anatomy of the spine, mechanisms of back pain, proper body mechanics for lifting and in activities of daily living (ADLs), and safe exercise techniques (Saal and Saal, 1989).

**Bed rest for low back pain** — restriction of movement and activity during the acute phase of an LBP episode. Current medical management suggests bed rest for only a brief period (two days or less) (Deyo et al., 1986; Nachemson, 1985).

**Biofeedback** — the process of providing audio, visual, or tactile feedback of some physiological measure (e.g., skin temperature, heart rate, EMG, blood pressure) to the patient. The goal is to teach the patient self modulation or voluntary control of the measures, often with a resultant reduction or cessation of the associated symptoms (Asfour et al., 1989).

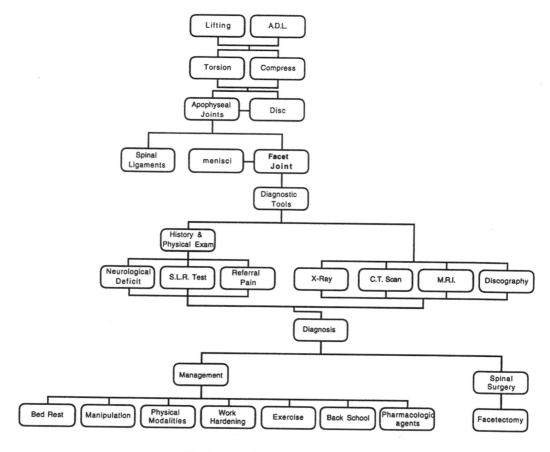

**FIGURE 22.3**   Schematic of Case 1: Facet joint.

**Bone scan** — an imaging technique during which a radioactively tagged metabolite is injected into the body and becomes incorporated into the skeletal system of the individual. Regions of increased metabolic activity and increased blood supply (e.g., inflammation, healing fractures, cancerous growths) can be visualized using nuclear medicine imaging techniques.

**Cancellous** — lattice like; applied to the bony tissue laid down by osteoblasts during development of bone and in the consolidation stage of fracture repair.

**Caudal** — toward the tail, opposite to cephalad.

**Cauda equina** — Latin for "mare's tail." The terminus of the spinal cord in the lower lumbar spine forming an array of nerve roots and occupying the vertebral canal below the cord.

**Cephalad** — toward the head, opposite to caudal.

**Chemonucleolysis (CNL)** — dissolution of the nucleus pulposus of an intervertebral disc by injection of a proteolytic agent (see chymopapain). Can be effective in the treatment of herniation of a disc (Farfan, 1985). Some of the complications or side effects of this procedure can be severe allergic reactions, migration of the enzyme to surrounding tissues, tissue destruction, postoperative LBP and spasm, and a slowing of nerve conduction (Wehling et al., 1989; Bromley et al., 1984; Fager, 1984; Agre et al., 1984).

**Chymopapain** — an enzyme that dissolves soft tissue, especially the intervertebral disc nucleus. The mechanism of the injection of chymopapain is the hydrolysis of proteoglycans resulting in a decrease in intradiscal pressure, and subsequent dissolution of disc components. A temporary change in the mechanical properties of the disc may result from the injection (Spencer et al., 1985; Nordby, 1983) (see chemonucleolysis).

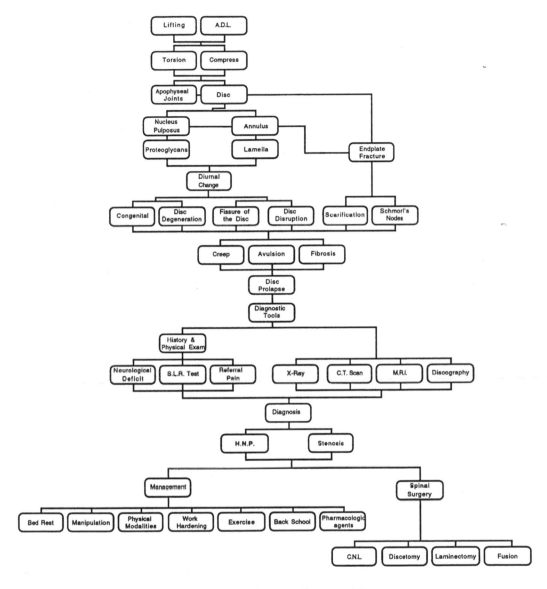

**FIGURE 22.4** Schematic of Case 2: HNP.

**Claudication (neurogenic)** — pain due to neurologic or vascular compromise.

- Neurogenic claudication — pain of a neurologic origin occurring in the back, buttocks, and legs. It increases with walking or other weight-bearing activities (Dong and Porter, 1989) (see straight leg raise test).
- Vascular claudication — leg pain secondary to a circulatory insufficiency. No pain in a limb when at rest. The pain starts after walking has begun. Intensification of the pain will not occur without activity. Pain stops after a period of rest.

**Coccyx** — commonly known as the tail bone; the last four bones of the spine, generally fused.

**Cohort** — a group of individuals who share a common exposure or experience.

**Collagen** — a group of fibrous proteins that are major constituents of connective tissues (e.g., skin, tendon, bone, cartilage).

**Compression loading of the disc** — the load applied along the long axis of the vertebral column (Marras et al., 1984). It is the force vector normal to the body of the disc. Several mutually dependent factors can contribute to compressive loading of the intervertebral disc (Barnes et al., 1989; Dillard et al., 1991; Dolan et al., 1988; Marras and Reilly, 1988; Schultz, 1986): (1) weight of the body; (2) forces due to external loading; (3) forces secondary to muscular contraction (Bogduk et al., 1992; Cartas et al., 1992; Goel et al., 1993; Ladin et al., 1989; Poulsen, 1981; Schultz and Andersson, 1981; Schultz et al., 1981); (4) restorative forces of the spinal ligaments; (5) forces secondary to intra-abdominal pressure (Davis, 1981; Gracovetsky et al., 1981; Örtengren et al., 1981; Bogduk and Macintosh, 1984). It may be that the $L_3/L_4$ lumbar disc is exposed to greater compressive loading than the $L_5/S_1$ disc. This is due to the greater mechanical advantage provided by the more pronounced $L_5$ spinous process which serves as the attachment for ligaments and extensor muscles (Gracovetsky et al., 1977; Gracovetsky and Farfan, 1986).

Capacity

- The forces required to produce failure purely by compression alone are quite large, on the order of 10 kN. Such forces are not incurred under normal circumstances (Hutton and Adams, 1982).
- There are studies suggesting that the maximal compressive force on the $L_5/S_1$ disc could be used to predict incidence rates of LBP (Chaffin and Park, 1973; Herrin et al., 1986). However, LBP due to annulus rupture from acute strain on the tissue appears improbable, and the ultimate compressive strength is not a reliable measure since it varies over a wide range.
- There is some evidence that an individual's level of physical activity may positively influence the compressive strength of the lumbar disc and vertebral body (Porter et al., 1989).

Mechanism (Lin et al., 1978)

- The annulus can passively transmit weight (compressive loads) between consecutive vertebral bodies because the sheer mass of its fibers renders them effective space fillers.
- Weight bearing between two vertebrae thus requires the cooperative action of both the nucleus pulposus and the annulus fibrosus.

Failure

- Sudden large magnitude compressive forces tend to cause anterior disc prolapse, while smaller magnitude cyclic compressive loading tends to result in bulging of the lamella of the posterior annulus (Adams and Hutton, 1985; Lavender et al., 1993; McNally et al., 1993).
- When a spinal joint is under compressive loading or flexed beyond its normal range, failure may result due to microfracture of a vertebral endplate (Adams and Hutton, 1982) and/or excessive strain on the underlying vertebra (Brinckmann et al., 1983).
- Interdependency has been documented between the surface area of the endplate and the density of the vertebral body (Brinckmann et al., 1989). Likewise, an interdependence has been reported between the vertebral bone density and the mechanical properties of the disc (Keller et al., 1989).
- Repeated microfractures and scarring of the bone underlying the vertebral endplates may induce a spot deterioration of the disc (Horst and Brinckmann, 1981). Uneven pressure gradients due to the spot deterioration propagate unrecoverable disc bulge. Deterioration is due to fluid loss and resultant change in viscoelastic properties. The mechanical failure of the endplate is highly dependent on the rigidity of the endplate and the underlying trabecular bone (Holmes et al., 1993). The consequences are decreased disc height and destabilization of the motion segments.
- The deterioration may continue with herniation of nuclear material into the vertebral canal, where it irritates or compresses nerve roots.
- Under compressive loading the facet joints are the foremost structure to yield at the limit of torsion. When compression of the disc is combined with torsion, the probability of failure increases.

**Computer assisted tomography (CT or CAT) scan** — an imaging technique involving a series of sequential X-ray exposures (slices) taken at discrete space intervals. The resultant tissue density information can be manipulated and stored by the computer. These data can be digitally enhanced to provide clinically useful information (Dvorak et al., 1987). The integration of a series of these exposures can be used to create three-dimensional images. CT images can provide evidence of soft tissue encroachment, such as disc herniation (Ninomiya et al., 1992); however, MRI has become the "gold standard" for identifying disc herniation except in some cases where CT with myelography is needed. MRI provides greater resolution of soft tissues and has become more widely used in the imaging and diagnosis of spinal pathology where soft tissue involvement is suspected (see MRI).

**Congenital** — referring to a condition present at the time of birth.

**Coronal plane** — relating to the plane that bisects the body into anterior and posterior sections.

**Corsets** — an orthopedic device that encircles and supports a part, as worn in certain spinal injuries or deformities (Lantz and Schultz, 1986a, 1986b) (see orthosis).

**Creep** — increasing strain (elongation) due to constant loading under the elastic limit of the material over a long period of time

**Creep of the intervertebral disc** — Under experimental conditions, the disc with its nucleus removed initially exhibits the same **compressive stiffness as** an intact disc (Adams and Hutton, 1983a). The problem with an isolated annulus fibrosus is that, in time, it will **creep; its lamellae** buckle inward and outward, and the disc space is decreased (Adams et al., 1987; Panjabi et al., 1984; Shirazi-Adl et al., 1984). Prolonged sitting or standing postures, and exposure to whole body vibration may cause creep detectable as a temporary decrease in stature (Brinckmann and Grootenboer, 1991).

**Diagnosis of Low Back Pain** — Diagnostic Terminology for LBP based on Fardon et al., (1993):

| Nonspecific pain syndromes | { | Deconditioning syndrome<br>Facet syndrome<br>Idiopathic lumbar pain<br>Low back pain<br>Lumbago<br>Lumbar pain syndrome<br>Lumbar syndrome<br>Mechanical derangement<br>Mechanical low back pain<br>Sciatica | Disc specific disorders | { | Degenerative disc disease<br>Degenerative disc syndrome<br>Disc disease<br>Disc disruption<br>Disc herniation<br>Disc protrusion<br>Herniated nucleus pulposus<br>Internal disc derangement<br>Internal disc disorder |
|---|---|---|---|---|---|

| Degenerative Arthritic | { | Arthritis<br>Degenerative arthritis<br>Degenerative joint disease<br>Degenerative spine disease<br>Facet arthritis<br>Facet arthrosis<br>Facetitis<br>Facet spondylosis<br>Hypertrophicarthritis<br>Spondylosis | Muscle / tendon disorder | { | Acute and chronic pain<br>Fibromyositis<br>Sprain<br>Sprain/strain<br>Strain<br>Tendinitis |
|---|---|---|---|---|---|

Stability $\begin{cases} \text{Degenerative instability} \\ \text{Lumbar (spinal) instability} \\ \text{Segmental instability} \end{cases}$     Stenosis $\begin{cases} \text{Central stenosis} \\ \text{Stenosis} \end{cases}$

Miscellaneous $\begin{cases} \text{Failed low back surgery} \\ \text{Fracture nerve root compression} \\ \text{Osteoporosis} \\ \text{Post laminectomy syndrome} \\ \text{Radiculopathy} \\ \text{Scoliosis} \\ \text{Spondylolisthesis} \\ \text{Spondylolysis} \end{cases}$

**Diagnostic techniques/tools for low back pain** (Wiesel et al., 1984; Nachemson, 1985; Hadler, 1984; Haig, 1992) —

* History, physical exam, and self report;
* Imaging techniques — X-rays (Panjabi et al., 1992a; Scavone et al., 1981), CT scan, MRI (Bernhardt et al., 1992), Thermograph (Mills et al., 1986);
* Enhanced imaging — discography, myelography, bone scan;
* Electrophysiologic testing — EMG (electromyography), NCV (nerve conduction velocity), and SEP (somatosensory evoked potentials) (Tonzola and Ackil, 1981; Haldeman, 1984; Eisen and Hoirch, 1983).

**Disc** — The intervertebral disc is intimately connected to two adjacent vertebral bodies. Composed of soft tissue, the disc serves to transmit loads down the vertebral column, as well as provide an energy absorptive capacity. Due to its flexibility and its interposition between the vertebrae, it allows for greater range of motion of the spine.

Gross anatomy (Inoue, 1981; Markolf and Morris, 1974)
* The disc consists of three parts: (1) a semifluid nucleus pulposus, consisting of water bound by a matrix of proteoglycans and collagen; (2) an annulus fibrosus, consisting of some 16 to 20 lamellae of collagen fibers surrounding the nucleus pulposus; and (3) vertebral endplates, covering the top and bottom surfaces of the disc and connecting it to the vertebral bodies.

Neuroanatomy (Bogduk et al., 1981; Goren et al., 1990; Panjabi et al., 1988b; Yoshizawa et al., 1980)
* Nerve fibers (including nociceptors) exist within the outer third of the annulus. The middle third of the annulus fibrosus may or may not be innervated. The inner third is not innervated (Bogduk, 1983, 1991).
* The sources of nerve fibers in the annulus fibrosus are the sinuvertebral nerves posteriorly, from the ventral rami posterolaterally, and from the gray rami communicants anteriorly and laterally (Bogduk et al., 1992).

**Disc degeneration** — a term describing the physical deterioration resulting from an injury to the disc (Thompson et al., 1990) or from the gradual wear and tear associated with the aging process (Hickey et al., 1986) and gender (Hsu et al., 1990; Kim et al., 1991; Miller et al., 1988; Pedrini-Mille et al., 1990; Tanaka et al., 1993; Vanharanta et al., 1989).

Mechanical property alterations
- The structural deterioration results in a change in viscoelastic properties (from resilient, with a high water concentration, to nonresilient, with a low water concentration) of the disc (Horst and Brinckmann, 1981; Panagiotacopulos et al., 1987; Hukins et al., 1990; McNally et al., 1992).
- The nucleus pulposus loses fluid and becomes less elastic (Adams and Hutton, 1983a).
- The normally firm and compact annulus fibrosis becomes fragmented and can surrender its structural integrity (Hickey and Hukins, 1982).
- The bone of the vertebral bodies adjacent to a degenerated disc exhibits a decrease in mechanical properties under compressive loading (Hansson et al., 1987). Relatively greater degeneration in the lumbar discs of cadavers occurs at lower spinal levels (Panjabi, 1988).

Diagnosis
- Measurement of disc height from radiographic studies has been found to be a poor method of detecting early, potentially symptomatic, degenerative changes (Vanharanta et al., 1988a).
- A relationship has been described between disc degeneration and osteoarthritis of the facet joint. It is believed that arthritis follows disc degeneration because of changes in the loading of the facet (Butler et al., 1990; Noren et al., 1991).

**Disc disruption** — degradation of the nucleus pulposus and inner annulus fibrosus without change in contour or size of the disc. There is a resultant degradation of the mechanical properties of the disc (Bogduk, 1983; Bogduk, 1991).

Mechanism of disc disruption
- Nuclear degradation may progress to involve erosion of the annulus fibrosus and fissures in the radial directions.
- The annulus fibrosus may be left to bear weight alone, because the bracing effect of the nucleus on the annulus fibrosus is reduced.
- The nerve endings in the annulus fibrosus can be sensitized by substances produced during the inflammatory response at thresholds lower than would be anticipated if the mechanical process operated alone (McCarron et al., 1987).

Physical evidence
- Very little physical evidence of disc disruption can be found directly. Since the outer perimeter (boundary) of the disc remains intact and normal in contour, no element of disc bulge or herniation is observed. Discography is required for definitive diagnosis (Sachs et al., 1987).

Symptoms (without frank herniation of the nucleus)
- Pain may be due to the chemical nociception and would be aggravated by any movement that mechanically stressed the affected disc.
- Inflammatory cells penetrate the annulus fibrosus of disrupted discs, whereupon chemical mediators may trigger nociceptive nerve endings.
- Nerve endings in the annulus fibrosus may be exposed to enzymes and breakdown products involved in the degradative process of the disc.

Diagnosis
- No abnormal neurologic signs will be present because the disc disruption does not involve nerve root irritation or compression.
- CT scans and myelography are usually normal because the outer perimeter of the disc is intact.

**Discectomy** — surgical removal or excision of all or part of the disc (Spengler, 1982a; Eismont and Currier, 1989). There is a change in disc contour and size following discectomy (Brinckmann and Horst, 1985). Disc height, radial bulging, and intradiscal pressure all decrease following discectomy (Brinckmann and Grootenboer, 1991). The inflammatory focus and the source of mechanical pain is often eliminated with excision of the problematic disc (see micro-discectomy). Total disc excision may destabilize the motion segment, and surgical fusion may be required (Goel et al., 1985, 1986; Frymoyer, 1981).

**Disc prolapse or disc herniation** — extrusion of the nucleus pulposus of an intervertebral disc. The extruded material may be mixed elements of the annulus fibrosus and nucleus pulposus (see HNP).

**Discitis** — inflammatory disease of the disc.

**Discogenic pain** — pain caused by derangement of an intervertebral disc.

**Discography** — an imaging technique for direct visualization of disc structure and pathology. This procedure involves X-ray photography of an intervertebral disc after direct injection of an absorbable radiopaque contrast medium (Sachs et al., 1987).

Pain provocation
- Discography pain provocation is usually conducted on the suspected disc and at least one adjacent disc (Weinstein et al., 1988a; Calhoun et al., 1988; Friedman and Goldner, 1955). During the procedure the patient is asked to distinguish between the painful and pain-free discs (Vanharanta et al., 1987).
- Patients undergoing discography sometimes suffer reproduction of their back pain.

Application
- Some clinicians and researchers question if discography provides information that cannot be obtained by other methods (Mooney et al., 1988; Nachemson, 1989). Others question the reliability and utility of the technique (Nachemson, 1989; Holt, 1968).
- Test specificity is reported to be markedly improved by combining analysis of the discographic image with information acquired during provocation (Walsh et al., 1990).
- A combined analysis using computed tomography (CT scan) and discography is reported to provide better documentation of the site and size of a prolapsed disc (Ninomiya et al., 1992; Anntti-Poika et al., 1990).
- Discography, like any invasive procedure, is not without its risks. One serious complication is discitis secondary to infection (Fraser et al., 1987).

**Diurnal changes (in the disc)** — It is hypothesized that the daily cyclic change in stature is due to expungement and reabsorbtion of fluid by the disc (Adams et al., 1986). This is thought to be related to the change of compressive loading of the disc associated with standing and sleep positioning (Tyrrell et al., 1985; Krag et al., 1990). Research indicates that the greatest rate of shrinkage or swelling occurs during the first hour of standing or lying, respectively. However, disc height measurements can vary greatly (Andersson et al., 1981; Adams and Hutton, 1983a; Adams et al., 1987; Adams et al., 1990; Adams and Dolan, 1991).

**Dorsal** — (1) related to the back; (2) posterior; a position more toward the back surface.

**Dura mater** — the tough outermost of the three membranes (meninges) covering the spine and brain (see epidural).

**Ectopic** — misplaced, occurring in an unnatural location, especially if congenital.

**Edema (oedema)** — excessive accumulation of fluid in the intercellular tissue spaces.

**EMAS** — endorphin mediated analgesia system (see endorphin).

**Electromyography (EMG)** — the measurement of the electrochemical potential produced during: (1) rest; (2) voluntary contraction; and (3) electrical stimulation of the skeletal muscle. Although EMG is used to estimate muscular exertion force, the limits of its accuracy and precision have not yet been clearly defined. EMG can also be used as an indirect measure of the function of the motor nerves which control the tested muscle (Schultz et al., 1982b; Khatri et al., 1984).

**Endogenous** — originating or produced within the organism.

**Endorphin** — a neurotransmitter produced in the central nervous system that acts as an analgesic (see EMAS).

**Endplate fracture or disruption** — refers to fissuring or loss of continuity of the cartilage interface between the disc and adjacent vertebral bodies. An endplate fracture may heal and cause no further problems. It is possible for an endplate fracture to set into motion a series of sequelae that manifest as pain and a variety of end stages. These symptoms may be demonstrated by pain provocation during discography (see discography) (Hsu et al., 1988).

**Enthesis** — (1) the junction of tendon and bone; (2) the use of a man-made substance as a substitute for lost or removed tissue.

**Enthesopathy** — any rheumatic disease resulting in inflammation of enthesis.

**Epidemiology of low back pain** — Epidemiology is the science concerned with the study of the factors influencing the frequency and distribution of disease or injury in human populations. Epidemiologic studies of low back pain have been limited due to the difficulty in defining exposures and outcomes. Three domains of risk factors for the epidemiology of low back pain have been established (Allan and Waddell, 1989; Bigos et al., 1986a, 1986b; Damkot et al., 1984; Frymoyer and Gordon, 1989; Nachemson et al., 1991; Sandover, 1988; Skovron, 1992).

Task factors (Pearcy et al., 1985; Troup et al., 1981)
- Vibration (Althoff et al., 1991; Kelsey, 1975; Kelsey et al., 1984b; Weinstein et al., 1988b; Wilder et al., 1982).
- Lifting (Kelsey et al., 1984b; Mirka and Marras, 1993; Mundt et al., 1993; Svensson and Andersson, 1983).
- Monotonous work (Magnusson et al., 1990; Svensson and Andersson, 1983).
- Work load (Sairanen et al., 1981; Stubbs, 1981; Videman et al., 1984).

Environmental factors
- Job satisfaction (Bigos et al., 1991a).
- Prolonged sitting/Sedentary jobs (Kelsey, 1975; Svensson and Andersson, 1983; Videman et al., 1990; Williams et al., 1991).

Individual factors (Bigos et al., 1991c)
- Cigarette smoking (Battié et al., 1989b; Kelsey et al., 1984a).
- Physical characteristics.
- Cardiovascular conditioning (Battié et al., 1989b).
- Age and gender (Videman et al., 1990; Videman et al., 1984).
- Pregnancy and menstruation (Ostgaard et al., 1993; Svensson et al., 1990; Videman et al., 1984).
- Strength (Battié et al., 1989b; Troup, 1991b).
- Psychological characteristics (Bigos et al., 1991b; Ostgaard et al., 1991).
- Psychosocial status (Troup, 1991a).

**Epidural (extradural)** — on or over the dura mater; often used in reference to the space outside the dura, a common site for administration of anesthetics (see dura mater).

**Erector spinae** — large rope-like muscle mass made up of many small, segmental muscles that run parallel to the spine. They act in concert to extend the trunk, creating an extension moment and compressive forces within the vertebral column (Schultz et al., 1982a; Macintosh and Bogduk, 1987, 1991; McGill, 1991). The extensor moment created by the erector spinae muscle force is a function of trunk position since the muscle moment arm varies with flexion angle (Cartas et al., 1992; Goel et al., 1993; Macinstosh et al., 1993; Tan et al., 1993; Triano and Schultz, 1987).

**Etiology of low back pain** — Etiology is the study of the factors and causes of disease. The potential causes of LBP can be categorized into mechanical, chemical, biological, infectious, genetic, degenerative, and psychosocial causes (Spengler, 1982b, 1983; Spengler and Freeman, 1979).

**Exercise treatment for low back pain** — usually prescribed to improve flexibility of the spine and strengthening of the trunk musculature. This may involve flexion or extension exercises, or both (Saal and Saal, 1989; Elnaggar et al., 1991; Nachemson, 1985).

**Exogenous** — originating or produced externally.

**Facet joints of the spine** — the paired synovial (moveable) joints between adjacent vertebrae, located on the apophyseal protuberances on the posterolateral aspect of the vertebral body. Its integrity and proper function depend on support from surrounding structures such as the joint capsule, ligaments, the intervertebral disc, and the muscles that cross the joint (Carrera, 1980a, 1980b; Davis and Carragee, 1993; Panjabi et al., 1992b, 1993; Van Schaik et al., 1985; Shirazi-Adl et al., 1987; Yang and King, 1984b).

**Facetectomy** — surgical excision of the articular facet of the spine (Wiltse, 1977). This procedure can be performed for correction of severe scoliosis. A major concern following this type of surgery is the stability of the spinal segment (Abumi et al., 1989b).

**Fibrosis** — formation of fibrous tissue, as in repair of or replacement of parenchymotous (organic, non-framework, non-supporting) material; pl. fibrosus.

**Fissure of the disc** — a channel, cleft, or groove through the annulus which forms secondary to injury. Degradation of the nuclear matrix destroys its cohesiveness. This may allow the nucleus to extrude through the fissure. Formation of a fissure is a prerequisite phenomenon for HNP to occur (see HNP).

**Foramen** — an opening or passage (Panjabi et al., 1983).

**Ganglion** — an aggregation of nerve cells.

**Gate control theory** — a theory on the modulation of pain perception by the central nervous system first proposed by Melzak and Wall (Melzack and Wall, 1965; Wall and Melzack, 1994). The main hypothesis of this theory is that sensory neurons trigger central control systems which inhibit or facilitate the input of nociceptors and their transmission pathways.

**Herniated nucleus pulposus (HNP)** — commonly known as a "slipped disc" or "herniated disc," or more technically as a prolapsed disc. Extrusion of material from the nucleus pulposus mixed with elements of the annulus fibrosus (Haig et al., 1993).

Failure mechanism
- Antecedent injury may damage the nucleus and produce a passageway through the annulus fibrosus. Heavy load lifting or long-term exposure to excessive vibration may be contributing factors to development of HNP (Wilder et al., 1982; McCarron et al., 1987).
- HNP may start when a radial fissure completely erodes the annulus. Spinal nerve roots may be compressed mechanically by the prolapsed material (Spencer et al., 1984; Adams and Hutton, 1985). The prolapsed material may elicit an inflammatory response that secondarily caused edema of the nerve roots. The blood vessels which supply the nerve may be compressed either mechanically or as a result of the edema, thereby rendering the nerve root ischemic.
- Recent research supports the idea that the annulus has a limited capacity for healing. Surgically created defects in the lumbar discs of dogs seem to show that large lesions can heal by filling with a solid plug of fibrous tissue and prevent leakage of nuclear fluid. Small wounds (comparable to a radial fissure), tend to heal more slowly, and continue to leak (Hampton et al., 1989). Other researchers have reported on the spontaneous resolution of extruded disc material over time without surgery (Saal and Saal, 1989).
- Disc herniations are categorized based on the degree of displacement of the nuclear material. The four levels ranked in order of increasing severity are (Spengler, 1982a):
  1. Intraspongy herniation (herniation through the endplate into the underlying cancellous, or spongy, bone of the vertebral body),
  2. Protrusion (displacement beyond a line drawn between the margin of two adjacent vertebral bodies),
  3. Extrusion (displacement into the foramina or canal),
  4. Sequestration (separation of the extruded material).

Symptoms
- HNP is present in a small percentage of incidence of low back pain, and conversely not all instances of HNP produce low back pain and neurologic symptoms. It is hypothesized that chemical nociception starts when a radial fissure reaches the middle and outer thirds of the annulus fibrosus and encounters nerve endings (Mooney, 1987). It may be that a combination of pressure and chemical irritation is necessary for HNP to produce pain.
- HNP patients often present with lancinating pain in the lower limb, severely limited straight-leg raising, and objective neurologic signs of numbness or weakness in the distribution of the affected nerve root.

- Stenosis of the disc space occurs when there is prolapse into the spinal canal or near the nerve roots. This can cause pressure and can become symptomatic if it compromises a spinal nerve, the spinal cord, or its roots. Pressure can be applied to the nerve root by a disc protrusion without direct compression of the nerve against the posterior elements of the foramen (Spencer et al., 1983).

Diagnosis

- Clinical examination can reliably identify HNP. Additional investigations, such as electromyography and radiology, are not essential for the purpose of diagnosis. Computed tomographic (CT) scans, magnetic resonance imaging (MRI), and myelography are of value in determining the exact site and extent of the prolapse (Morris et al., 1986).

Treatment

- The objectives of treatment are: (1) relieving pain; (2) increasing mobility; and (3) minimizing the impairment and disability associated with LBP.
- For a small percentage of patients with HNP, surgical intervention may be necessary. The surgical intervention can include chemonucleolysis, percutaneous discectomy, microdiscectomy, open laminotomy, and laminectomy.

**Hypochondriasis** — a mental disorder characterized by morbid anxiety about health, body function, and sensations.

**Iatrogenic** — describing a disease or dysfunction that has resulted from treatment. Iatrogenic discitis is an intensely painful condition, resulting from a disc infected by bacteria introduced by needles used for discography.

**Idiopathic** — of unknown origin.

**Innervation** — the nerve supply for a body part.

**Intra-abdominal pressure (IAP)** — pressure that develops within the abdominal cavity secondary to forces produced by the diaphragm, chest wall, and abdominal musculature.

Mechanism

- A great deal of controversy exists whether IAP directly or indirectly produces a reactive force counter to the compressive forces developed in the vertebral column during lifting (Gracovetsky et al., 1995). Some research supports a contrary opinion that IAP under some conditions and postures may actually increase compressive loading (Nachemson et al., 1986).
- IAP can be generated by other means as well, such as the Valsalva maneuver. This occurs when the breath is held while "bearing down" as if one were to move one's bowels.
- The use of back belts during lifting activities is often recommended as an ergonomic intervention. Wearing a back belt while lifting constrains expansion of the abdomen. In theory this should result in an increase in IAP and a subsequent decrease in the compressive load upon the lumbar disc. The efficacy of a back belt is quite controversial at present, and the exact mechanism of IAP development is not clear.

**Ischemia** — insufficient blood flow or inadequate blood supply.

**-itis** — suffix meaning "inflammation of."

**Kyphosis** — a convex or posterior curvature of the spine; the normal curvature of the thoracic spine.

**Lamina** — a thin bony plate; part of the dorsal region of a vertebra.

**Laminectomy** — surgical excision of the posterior arch (lamina) of the vertebrae. This procedure is performed to relieve spinal or nerve root pressure secondary to a space occupying lesion (e.g., herniated disc). Laminectomy is a futile intervention for disc disruption, because the lesion lies within an otherwise intact disc (Wiltse, 1977).

**Laminotomy** — division of lamina of a vertebra without removal or excision of the arch. The rationale for this procedure is the same as for laminectomy or for exploration of the intervertebral space.

**Lancinating pain** — sharp stabbing or cutting pain.

**Lifting (manual lifting) and low back pain** — The current theory of the relationship between LBP and lifting is based on three assumptions (Buseck et al., 1988; Jäger and Luttmann, 1989):

- Manual lifting poses a risk of LBP to many workers (Bigos et al., 1986a). The risks vary due to postural characteristics, repetition rates, and the load of the specific task (Nordin et al., 1984).
- LBP is more likely to occur when workers lift loads that exceed their physical capacities (Battié et al., 1989a; Poulsen, 1981; Herrin et al., 1986).
- The physical capacities of workers vary substantially (Kishino et al., 1985).

The implications of these three assumptions are:

- LBP is a potential outcome of various activities. Lifting may be one of them.
- Manual lifting can be evaluated as a risk factor for LBD. The measurement of the exposure includes type (e.g., lifting or other activities) and virulence (e.g., lifting over the physical capacity).
- Each individual has his or her own physical capacity, and such capacity can be approximated by statistics (percentile), such as,

    The maximal permissible compressive forces on the spine (Chaffin and Park, 1973),
    The maximal permissible energy expenditure,
    The maximal acceptable weight of load (Troup et al., 1987; Snook and Ciriello, 1991).

- The deviation of physical capacities among individuals can be identified and measured.

**Lifting technique and low back pain** — several variables have been reported to be important in the evaluation of the spinal loading of various lifting techniques (Hart et al., 1987).

- Initial position of the load (Marras et al., 1993).
- Initial posture (stoop vs. squat) (Omino and Hayashi, 1992).
- Foot position (straddle vs. parallel) (Anderson and Chaffin, 1986).
- Velocity and acceleration of lift (Marras et al., 1993).
- Muscular recruitment and fatigue (Trafimow et al., 1992; Lavender et al., 1992; Schultz et al., 1982b).
- Compound movements (e.g., flexion with torsion) (Kelsey et al., 1984b; Markolf, 1972; Marras et al., 1993; Omino and Hayashi, 1992).
- Lumbodorsal curvature (lordotic vs. kyphotic) (Panjabi and White, 1980) — Controversy exists as to the role of lumbar lordosis in lifting. One theory is that preservation of the lordosis in conjunction with intra-abdominal pressure (IAP) strengthens the spine and is critical to prevention of injury during lifting (Aspden, 1989). To the contrary, another theory states that decreased lordosis may increase the tension in the posterior spinal ligaments thus decreasing the muscular forces required and subsequently decreasing the resultant compressive forces on the spine (Gracovetsky et al., 1989, 1990). The risk of injury may be influenced more by the degree of flexion than the choice of lifting posture (stoop vs. squat) (Potvin et al., 1991). The mechanical advantage provided the trunk muscles is also dependent on the assumed posture (Schultz et al., 1984).

**Ligament** — a band of strong, fibrous connective tissue that attaches bone to bone and serves to provide stability, transmit loads, and restrict motion (Yamamoto et al., 1990; Panjabi and White, 1980). The posterior ligaments of the vertebral column act passively to assist in lifting (Gracovetsky et al., 1981). The spinal ligaments have different biochemical composition and fiber density and orientation based on their specific function within the spinal column (Hukins et al., 1990) (see spinal ligament).

**Ligamentum flavum** — connects the laminae of adjacent vertebrae (see ligament, spinal ligament).

**Lordosis** — the inward curvature of the spine; the normal curvature of the cervical and lumbar spine (Gracovetsky and Farfan, 1986, Gracovetsky et al., 1987).

- Different postures result in different loading patterns of the passive tissues, such as the ligaments and fascia. In some circumstances these structures can contribute significantly to the development of axial torque (McGill, 1992).
- Posture can directly influence the amount of curvature. The height of the heel has been shown to change the curvature of the lumbar spine (Bendix et al., 1984).
- The degree of lumbar lordosis has not been directly correlated to low back pain (Hansson et al., 1985).

**Low back pain (LBP)** — a symptom of any one of several pathologies of the lumbosacral spine, with varying terminology, etiology, evaluation and medical management techniques (Cavanaugh and Weinstein, 1994; Ghosh, 1988; Hadler, 1987; Loeser, 1980; Troup, 1981; Waddell et al., 1980; White and Gordon, 1982).

**Low back pain research** — There are many different hypotheses as to the origin of LBP. In general the potential causes of LBP may be categorized as mechanical, chemical, biological, infectious, genetic, and degenerative (Boden et al., 1991; Carr et al., 1985; Ghosh, 1988). Ergonomists and engineers have focused on the mechanical basis of LBP since the variables in that domain are more readily defined and manipulated. Good scientific method dictates that research must build upon the knowledge at each of the levels below it in the critical path (Pope, 1990; Weinstein, 1988a; Zohn, 1988):

Level I: The pain — Is the pain of spinal origin, and is it real? Usually based on medical history, self reporting, standardized pain questionnaires, and physical exam.

Level II: The anatomy — Can the pain signal be transmitted? Is the presence of pain receptors consistent with the present understanding of the anatomy?

Level III: The autopsy and mechanical testing — What are the internal and external forces needed to cause sufficient injury to stimulate the pain receptors? The techniques used in cadaver lumbar motion segment studies have not been standardized, making prediction based on this body of research difficult (Keller et al., 1990).

Level IV: Biomechanical modeling — Can the forces required to cause damage as determined from mechanical testing, be generated in a person during work activities?

**Low back pain impairment (LBPI)** — refers to the physical limitations in movement, strength and function of an individual due to LBP (Waddell et al., 1984; Mayer et al., 1988; Deyo, 1988).

**Low back pain disability (LBPD)** — refers to the loss of capacity or time for an individual to function at work or in activities of daily living. It is not synonymous with LBPI, since impaired individuals may be able to function normally, and conversely an individual with minimal measurable impairment may be significantly disabled with respect to work and/or activity of daily living. This dichotomy is secondary to psychosocial factors, socioeconomic factors and specific job requirements (Andersson et al., 1983; Bigos et al., 1992; Deyo and Diehl, 1983; Barnes et al., 1989; Polatin et al., 1989; Rowe, 1971; Spengler et al., 1986a, 1986b, 1986c; Waddell et al., 1984; Waddell and Main, 1984).

**Lumbosacral sprain/strain** — one of the more common diagnoses associated with an episode of low back pain often related to a memorable injury of moderate severity. An imprecise diagnosis, it is often used prior to ruling out other, more specific diagnoses.

**Manipulation/Mobilization/Chiropractic practices for low back pain** — manual techniques involving movement of spinal segments for the purpose of adjusting spinal alignment, improving ROM, or decreasing pain (Cassidy and Kirkaldy-Willis, 1988; Hadler et al., 1987; Haldeman, 1980; Khalil et al., 1992).

**Mechanical traction** — the application of a tensile load along the long axis of the spine with the goal of promoting relaxation of the back musculature and decompressing the spine.

**Mechanoreceptor** — a specialized nerve ending that is stimulated by mechanical pressure, pressure changes, oscillation or movement.

**Medical management of low back pain** — the goals of intervention are the reduction of symptoms and the restoration of functional levels. The outcome of treatment for LBP is very variable, with a portion of the population typically becoming chronic (Gatchel et al., 1992; Wiesel et al., 1984; Spitzer et al., 1987; Deyo, 1983).

- Activity modification: (see bed rest, back school, and pain clinics).
- Oral medication (see pharmacological agents).
- Physical medicine: (see acupuncture, biofeedback, corsets/orthotics, exercise, mobilization/manipulation/chiropractic, mechanical traction, physical modalities, work hardening).

**Meninges** — the three membranes that cover the spinal cord and brain. They are the dura mater, pia mater, and arachnoid membrane.

**Meniscus** — a dense fibrocartilage pad present in some synovial joints (e.g., knee, facet) which acts to protect the cartilage surfaces of the joint and to distribute the load between the articular surfaces. A meniscus is typically attached at one end to the joint capsule and/or ligaments and thus receives some degree of vascular supply and innervation.

**Micro-discectomy** — the use of microsurgical techniques and a small incision to perform a partial removal of a herniated disc. The smaller incision and decreased disruption to surrounding tissues allows for a more rapid recovery phase.

**Morbidity** — state of being diseased.

**Morbidity rate** — the number of instances of a disease over a specified period of time per unit of population. It is usually expressed as $x$ cases per year per 1,000, 10,000, or 100,000 of population.

**Morphology** — the science of forms and structures of organisms.

**Myelography** — a radiographic imaging technique for visualization of the spinal cord, involving the injection of a radiopaque dye into the area around the meninges of the spinal cord. It is particularly useful in detecting impingement on the spinal canal and nerve roots.

**MRI (magnetic resonance imaging)** — a noninvasive imaging technique using electromagnetic radiation particularly useful for visualization of soft (non-radio-opaque) tissues. The MRI provides greater resolution of intrinsic disc abnormality than a CT scan due to its ability to detect water content in the disc (Bernhardt et al., 1992; Hickey et al., 1986; Panagiotacopulos et al. 1987a 1987b) (see CT scan).

**NCV (nerve conduction velocity)** — a direct measure of the conductivity of motor nerves in the region being tested. Often performed in conjunction with EMG testing.

**Necrosis** — death of one or more cells.

**Neoplasm** — new growth of cells or tissues, may be benign or malignant (cancerous).

**Neurological deficit** — a defect or dysfunction of the central or peripheral nervous system, such as diminished deep tendon reflexes or decreased sensation.

**Nociceptor** — a specialized peripheral nerve ending which senses painful stimuli.

**Nucleus pulposus** — the central, more viscous portion of the disc.

Anatomy
- A semifluid mass of proteoglycans in a collagen fiber matrix.
- The nucleus is intrinsically cohesive and resists herniation, even under experimental conditions when a posterior channel is cut into the annulus fibrosus.

Immunological basis of spinal pain syndromes
- The nucleus pulposis lacks a blood supply and is never exposed to the circulation. In a disc that suffers an endplate fracture, proteoglycans may be exposed to the body's immune system for the first time, triggering an inflammatory response (Pennington et al., 1988). The gelatinous matrix of the nucleus has not been studied according to contemporary standards of immunology, but available evidence suggests that the matrix does have antigenic properties.
- Prolapsed nuclear material elicits an inflammatory response if it enters the epidural space or the vertebra spongiosa in the case of traumatic Schmorl's nodes. Patients exhibit changes in lymphocyte migration and antibody profiles consistent with an antigenic response.

Mechanism
- The main function of the nucleus pulposis is to transmit loads from one vertebra to another.
- Functioning as a volume of fluid within an enclosed space, the nucleus is incompressible. The attempted expansion of the nucleus pulposus effectively braces the annulus fibrosus from within, preventing it from buckling.
- When exposed to a compression load, the nucleus deforms by attempting to spread in a radial direction, but this displacement is arrested by the annulus fibrosus. Tension develops in the annulus fibrosus to resist radial expansion of the nucleus pulposus.

**Orthosis** — orthopedic device for assisting the function of part of the body without replacing it (see corset).

**-osis** — suffix meaning "disease of."

**Ossification** — the formation of bone.

**Osteoarthritis** — commonly described as a "wear and tear" disease of the joints. Chronic, degenerative disease particularly of weight-bearing joints characterized by erosion of the cartilage and bony overgrowth.

**Osteomalacia** — a softening of the bone related to demineralization of the matrix, often associated with a deficiency of vitamin D.

**Osteopenia** — loss of bone mass.

**Osteophyte** — a bony or osseous outgrowth or excrescence (tumor).

**Osteoporosis** — a decrease in bone mass, often related to prolonged bed rest, post-menopausal endocrine changes, and prolonged periods of weightlessness among astronauts. It is often treated with hormone therapy and dietary supplements in post-menopausal women. Weight-bearing activities are recommended for cases resulting from prolonged bed rest, as bone remodels in response to the stresses imposed on it.

**Pain clinics and low back pain** — specialized facilities utilizing a multidisciplinary approach to the treatment of pain. Emphasis is placed on improving functional levels, using activity and behavior modification, and pain management techniques rather than treatment of symptoms (Aronoff, 1982, 1983).

**Percutaneous** — through the skin.

**Pharmacological agents for low back pain** — the medical intervention for low back pain may involve the use of pharmacological agents. These drugs, prescribed for several reasons, are classified into three major categories:

- Analgesic — a category of drugs whose main action is reduction of pain. They can be narcotic (morphine based, and by prescription only) or non-narcotic. They may be prescription (e.g., Darvon™ — propoxyphene hydrochloride) or over the counter (e.g., Tylenol™ — acetaminophen, aspirin).
- Anti-inflammatory — a category of drugs whose main action is reduction of the inflammatory response with a secondary effect of pain reduction. These drugs are further divided into two subcategories based on their pharmacology:

  Steroidal — these are typically prescription corticosteroids or their derivatives which are analogs of the hormones produced by the adrenal cortex (i.e., prednisone). Their action is to directly modulate the inflammatory response via the endocrine system. There use is usually limited to severe episodes due to the significant number and severity of side effects.

  Nonsteroidal anti-inflammatory drug (NSAID) — a category of prescription such as Feldene™ — piroxicam, Voltaren™ — diclofene, Naprosyn™ — naproxen and nonprescription such as Motrin™, Advil™ — ibuprofen. The action of this class of drugs is to inhibit the inflammatory response. The side effects of NSAIDs tend to be less severe than steroids.

- Muscle relaxants — a category of drugs whose action is to decrease muscle spasm, thereby reducing associated muscle pain (e.g., Flexeril™, Soma™, Robaxin™).

**Physical exam and history of low back pain** (Hadler, 1984; Haig, 1992; McCombe et al., 1989; Roland and Morris, 1983a 1983b) —

- Patient self report
    Medical History (Bigos et al., 1992).
    Pain diagram — patient drawing of pain type and distribution.
    Pain questionnaire — developed to assess the nature of the pain as well as its psychological and psychosocial impact (Herron and Pheasant, 1982; Waddell et al., 1980).
    Visual analog scale — subjective patient report of pain level using a calibrated scale.
- Referral pain/trigger point — palpation (probing by touch) of the muscles in or around the location of pain, for local tenderness.
- Range of motion (ROM) — a measure of the freedom of movement of a joint (Keeley et al., 1986; Pope et al., 1985). A relationship between decreased lumbar ROM and a previous history of LBP is reported to exist, but no significant predictive value for future LBP has been demonstrated (Battié et al., 1990).
- Straight leg raise test (see straight leg raise test, SLR).
- Neurologic/reflex evaluation — assessment of sensory and motor function, including deep tendon reflexes (i.e., knee jerk), muscle strength. Provides an indication of the function of the nerves controlling the tested area.

**Physical modalities** — the application of physical agents with the goal of decreasing pain, decreasing inflammation, increasing local circulation to promote healing, and/or increasing flexibility. These include: electrical stimulation, heat or ice applications, and ultrasound (Kane and Taub, 1975).

**Pia mater** — the innermost of the three meninges (membranes) covering the spinal cord and brain.

**Plexus** — network of nerves or blood vessels.

**Posterior longitudinal ligament** — the ligament extending along the posterior surface of the vertebral body (within the spinal canal) from the cervical spine to the sacrum (see ligament, spinal ligament).

**Proteoglycans** — substance found in the matrix of connective tissues and in synovial fluid, vitreous humor of the eye, and mucous secretions.

**Radiculopathy** — any disease of the motor or sensory root of a spinal nerve. Distribution of symptoms and findings is characteristic for each spinal segment. There are mono- and polyradiculopathies (Saal and Saal, 1989).

**Radiopaque** — impenetrable by X-ray.

**Ramus (rami** pl.) — branch.

**Rheumatoid arthritis** — chronic systemic disease characterized by inflammation of the joint linings, bony changes and eventually deformity usually involving multiple joints.

**Sacrum** — the five fused vertebrae between the lumbar vertebrae and the coccyx. They comprise the posterior aspect of the pelvis.

**Sagittal plane** — the plane through the body which bisects it into left and right parts.

**Scarification** — making a number of superficial incisions in the skin or other tissue.

**Schmorl's nodes** — disc protrusions through the endplate and into the vertebral body (Hansson and Ross, 1983). They are reported to be healed fractures of the vertebral endplates (Farfan et al., 1972).

**Sciatic nerve** — longest nerve of the body formed from lower lumbar and sacral nerve roots and running deep in the posterior aspect of the thigh. It is responsible for sensation of the posterolateral thigh, calf, and parts of the foot, as well as motor control of some of the muscles of the foot and calf.

**Sciatica** — common terminology for the pain and sensory changes along the distribution of the sciatic nerve, usually resulting from compression or trauma to the nerve or the nerve roots that form it (Heliovaara et al., 1986; Allan and Waddell, 1989).

**Scoliosis** — abnormal curvature of the spine. This can be an accentuation or decrease of the normal regional curvatures, or rotation or curvature of the spine in an abnormal plane or direction.

**Sequela (sequelae** pl.) — disorder due to a preceding disease or accident.

**Shear loading of the disc** — the component of the resultant force acting on the disc which is perpendicular to the long axis of the spinal segment. It is the complement of the compressive force. All the factors contributing to disc compression may also contribute to shear loading. The magnitude of shear loading depends on the orientation of the disc and body posture. Because there is little dynamic information about trunk muscles and their orientation to the disc, for biomechanical modeling straight line representation of the muscle force perpendicular to the disc is assumed. As a result, shear loading may be underestimated.

**Somatotype** — the classification of persons by body shape or type (morphology). The categories are: ectomorph (thin, wiry), mesomorph (well proportioned), endomorph (soft, rotund).

**Somatosensory evoked potentials (SEP)** — an electrodiagnostic procedure used to assess sensory neurons in peripheral and spinal cord pathways. This technique is sometimes used to evaluate stenosis.

**Somatization** — conversion of anxiety into physical symptoms.

**Spasm** — involuntary muscular contraction.

**Spinal fusion** — the surgical fixation of two or more adjacent vertebrae to provide structural stability or maintain normal disc space (Abumi et al., 1989a). Fusion is indicated in the presence of clinical instability of one or more lumbar motion segments (Farfan and Gracovetsky, 1984). Fusion provides no significant advantage over laminectomy alone in the case of the simple herniated disc (White et al., 1987). The mechanics of spinal fusion are complex, and the procedure can produce adverse effects, such as higher stresses at the adjacent vertebrae (Goel et al., 1988). Likewise, clinical research has not yet demonstrated what degree of stability is necessary for optimal healing and progression to solid fusion (Panjabi, 1988; Panjabi et al., 1988a).

- Anterior interbody fusion — a transabdominal spinal fusion interferes little with the integrity of the lumbar spine as compared to the posterior approach. There are some advantages of this procedure: back muscles are not disabled, the posterior elements remain intact, and the contents of the vertebral canal are not exposed. This latter advantage avoids the potential complication of arachnoiditis and preserves the sinuvertebral nerves, avoiding the risk of neuroma formation (tumor of the nerve fiber).
- Posterior interbody fusion — spinal fusion surgery involving access to the vertebrae through the back musculature and ligaments.

**Spinal ligaments** — see ligament, anterior longitudinal ligament, posterior longitudinal ligament, ligamentum flavum, and supraspinous ligament (Panjabi et al., 1982; Ratcliffe, 1980).

**Supraspinous ligament** — the strong, fibrous cord joining the spinous processes from the seventh cervical vertebrae to the sacrum (see ligament).

**Spinal nerve roots** — the nerve trunks that exit and enter the spinal cord at the intervertebral space (Bogduk, 1983).

**Spinal surgery** — surgical intervention for the treatment of severe pain, spinal instability, and/or neurologic involvement (Pope and Panjabi, 1985; Posner et al., 1982). Surgery is usually the treatment of last resort, when conservative therapies have failed.

**Spondylitis** — inflammation of one or more vertebrae.

**Spondylolisthesis** — a condition of forward slippage of a lumbar vertebrae which may or may not be associated with instability (Grobler et al., 1993a, 1993b). It can be categorized as follows (Pearcy and Shepherd, 1985):

- Isthmic spondylolisthesis — the slippage is due to loss of the constraint by the neural arch, secondary to a spondylolysis, most commonly occurring in the last lumbar vertebrae (Grobler et al., 1993a).
- Degenerative/pseudo spondylolisthesis — forward slippage, typically at $L_4$ and $L_5$ levels due to a degenerative process, without any defect in the neural arch. This may also be due to a developmental disposition related to the orientation of the facet joint (Grobler et al., 1993b).
- Retrospondylolisthesis — misalignment of one vertebra relative to the adjacent vertebra characterized by backward displacement.
- Displastic, traumatic, and pathologic spondylolisthesis are less common diagnoses.

**Spondylolysis** — a defect in the neural arch, typically at $L_5$ and commonly believed to be due to repetitive low-grade trauma over prolonged periods. Short duration exposures may be of little clinical significance (Semon and Spengler, 1981; Crock, 1981).

**Spondylosis** — ankylotic state or fusion of the vertebrae. Disease of the spine often associated with disc degeneration. There is a spectrum of changes effecting one or multiple levels, initially signified radiologically by loss of disc height, osteophytes arising from the margins of the vertebral bodies, and osteoarthritic changes in the apophyseal joints (Edwards and LaRocca, 1985).

**Spongiosa** — spongy. The sustantia spongiosa ossium is the thin bony matrix that makes up the center of the intervertebral body.

**Stenosis, spinal** — narrowing of the spinal canal, nerve root canals, or tunnels of the intervertebral foramina. The narrowing not only reduces the anteroposterior and lateral diameters, but also alters the cross-sectional configuration of the spinal canal (Schonstrom et al., 1985; Spengler 1983). Decreased cross-sectional area causes an associated increase in local pressure, with a resultant effect on local nerve function (Schonstrom and Hansson, 1988; Grabias, 1980; Crock, 1981).

- When the spinal canal and lateral nerve root canal are narrowed, a relatively small disc herniation can produce clinically significant symptoms, such as sciatica, which would not have occurred if the canal had been of adequate dimensions. A nerve root may be compromised in a stenotic canal even during normal movement patterns (Panjabi et al., 1984).
- Stenosis may result from occupational factors or trauma, or from non-occupational factors such as malnutrition or genetic predisposition (Clark et al., 1985).
- There are gender related differences in the dimensions of the canal (Vanharanta et al., 1985).

**Steroid** — a lipid-based organic compound. Steroidal hormones are synthesized by the sex glands (e.g., estrogen, testosterone) and the adrenal cortex (corticosteroids, e.g., cortisone). There are a number of synthetic analogs which also have physiologic activity. Corticosteroids are potent anti-inflammatory agents, and in some cases may be given orally or by direct injection to the target site.

**Straight leg raise (SLR) test** — a clinical evaluation technique where the thigh with the knee extended straight is gradually bent (flexed) at the hip. Onset or aggravation of low back pain or sciatica during the maneuver is considered a positive test result and may be indicative of lumbar disc disease (Troup, 1981).

**Subluxation** — partial dislocation of a joint; bone ends are misaligned but still in contact.

**Taxonomy** — the science dealing with the identification, naming, and classification of organisms.

**Thermography** — an imaging technique currently being investigated as an indirect measure of autonomic nervous system function. It is a measure of skin surface temperature using the infrared spectrum.

**Torsional loading of the disc** — torque resulting from rotation about or near the long axis of the spinal column (Bogduk, 1991; Farfan, 1984). Torsional loading occurs frequently in common activities such as the rotation of the pelvis during walking, and rotation of the trunk when throwing a ball (Farfan et al., 1970; Farfan and Gracovetsky, 1984; Gracovetsky and Farfan, 1986; Marras and Mirka, 1992; Miller et al., 1988).

Mechanism (Panjabi and White, 1980)
- The rotational range of motion of a lumbar segment is limited to less than three degrees by the blocking of the apophyseal joint (Adams and Hutton, 1981). Both the disc and the apophyseal joint provide resistance to rotation. (Gunzburg et al., 1991a).
- When the apophyseal joint fails, the range of rotational movement increases. This occurs at the expense of increased axial torque and lateral shear forces on the disc itself (Adams and Hutton, 1981). Fracture or dislocation of this joint is a rare occurrence (Kramer and Levine, 1990).
- An isolated disc can tolerate only three degrees of axial rotation before the collagen fibers of the annulus suffer injury. For a given direction of rotation, only half of the collagen fibers in the annulus fibrosus are appropriately oriented to resist the rotation.

Failure

- Torsion injury to the annulus fibrosus is increased if axial rotation occurs in combination with flexion. Flexion prestresses the posterior annulus fibrosus. Consequently, far less axial rotation is required to strain collagen fibers beyond their physical limit.
- In flexion, the articular processes of the apophyseal joints are subluxated, and the apophyseal joints afford less resistance to axial rotation. As a result, during flexion–rotation, the annulus fibrosus is maximally stressed while least protected by the posterior elements (Miller et al., 1983). Trunk rotation ROM is decreased while the trunk is flexed (Gunzburg et al., 1991a, 1991b).
- Torsion is more recurrent and more detrimental than a compressive force. After injury, if the joint never heals fully, it is more likely to fail with repeated torsion. When torsion of the disc is combined with compression, the probability of failure increases.
- Torsional injury may include a rupture of ligamentous tissue, the annulus, and associated damage to the facet.
- Under continued rotation beyond three degrees, disc failure occurs with avulsion of the laminae of the annulus from the endplate. The nucleus and endplate remain intact. Gradually, the annulus develops radial fissures and the lumbar segment may become unstable. Furthermore, the torsional deformity can cause the neural elements in the intervertebral canal to displace to one side, stretching the nerve roots.

Diagnosis/Imaging

- Neurologic examination, CT scans, myelography, and MRI are normal because the lesion is restricted to the annulus fibrosus and does not involve nerve root compression.
- During discography, contrast medium and local anesthetic are injected into the painful site. If pain is relieved by the local anesthetic, and if the contrast medium is confined to the annulus fibrosus, the source of pain is assumed to be the annulus fibrosus.

Symptoms

- The pain will be aggravated by any movements that stress the annulus fibrosus, particularly flexion and rotation in the same direction that produced the lesion. Healing may require several months of protection from reinjury coupled with gradual and gentle reintroduction and restoration of function.

**Trabecula** — any of the bands of tissue that pass from the outer part of an organ to its interior, dividing it into separate chambers.

**Trabecular bone** — spongy bone.

**Transverse plane** — the cross-sectional plane through the body perpendicular to the long axis.

**Trigger point** — small, focal region of a muscle which is tender to touch. This sensitivity is assumed to be related or in reaction to regional pain.

**Ventral** — pertaining to the anterior or frontal surface of the body.

**Work hardening** — a therapeutic approach to functional training for return to work. This involves reduction of the patient's work tasks to component parts, progressive training in these components, followed by recombination of the work task and progression to maximum functional work level.

**X-ray (radiography)** — an imaging technique in which a body part is placed between an X-ray source and a photographic film plate. The X-rays are absorbed by the body part in direct proportion to the density of the tissue in its path. The radiograph (a photographic negative) is therefore a "map" of tissue density. Bone and mineralized tissue are well defined, while less dense substances such as body fluids and muscle are poorly imaged. X-ray can be used to estimate the dimension of vertebrae with adequate accuracy (Malmivaara et al., 1986).

**Zygapophyseal** — pertaining to an articular process of a vertebra.

# References

Abumi, K., Panjabi, M. M. and Duranceau, J., 1989, Biomechanical evaluation of spinal fixation devices. Part III Stability provided by six spinal fixation devices and interbody bone graft, *Spine*, **14**, 1249-1255.

Abumi, K., Panjabi, M. M., Kramer, K., Duranceau, J., Oxland, T. and Crisco, J. J., 1990, Biomechanical evaluation of lumbar spinal stability after graded facetectomies, *Spine*, **14**, 1142-1147.

Adams, M. A. and Dolan P., 1991, A technique for quantifying the bending moment acting on the lumbar spine in vivo, *Journal of Biomechanics*, **24**, 117-126.

Adams, M. A. and Hutton, W. C., 1981, The relevance of torsion to the mechanical derangement of the lumbar spine, *Spine*, **6**, 241-248.

Adams, M. A. and Hutton, W. C., 1981, Prolapsed intervertebral disc: A hyperflexion injury, *Spine*, **7**, 184-191.

Adams, M. A. and Hutton, W. C., 1983a, The effect of posture on the fluid content of lumbar intervertebral discs, *Spine*, **8**, 665-671.

Adams, M. A. and Hutton, W. C., 1983b, The mechanical function of the lumbar apophyseal joints, *Spine*, **8**, 327-330.

Adams, M. A. and Hutton, W. C., 1985, Gradual disc prolapse, *Spine*, **10**, 524.

Adams, M. A., Dolan, P. and Hutton, W. C., 1987, Diurnal variations in the stresses on the lumbar spine, *Spine*, **12**, 130-137.

Adams, M. A., Dolan, P. and Hutton, W. C., 1988, The lumbar spine in backward bending, *Spine*, **13**, 1019-1026.

Adams, M. A., Dolan, P., Hutton, W. C. and Porter, R. W., 1990, Diurnal changes in spinal mechanics and their clinical significance, *The Journal of Bone and Joint Surgery* (Br), **72**, 266-270.

Agre, K., Wilson, R. R., Brim, M. and McDermott, D. J., 1984, Chymodiactin postmarketing surveillance demographic and adverse experience data in 29,075 patients, *Spine*, **9**, 479-485.

Allan, D. B. and Waddell, G., 1989, Understanding and management of low back pain, *Acta Orthop Scand*, **60**, 1-23.

Althoff, I., Brinckmann, P., Frobin, W., Sandover, J. and Burton, K., 1992, An improved method of stature measurement for quantitative determination of spinal loading: Application to sitting postures and whole body vibration, *Spine*, **17**, 682-683.

Anderson, C. K. and Chaffin, D. B., 1986, A biomechanical evaluation of five lifting techniques, *Applied Ergonomics*, **17**, 2-8.

Andersson, G. B. J., Schultz, A., Nathan, A. and Irstam, L., 1981, Roentgenographic measurement of lumbar intervertebral disc height, *Spine*, **6**, 154-158.

Andersson, G. B. J., Svensson, H. and Oden, A., 1983, The intensity of work recovery in low back pain, *Spine*, **8**, 880-884.

Antti-Poika, I., Soini. J., Talroth, K., Yrjonen, T. and Konttinen, Y., 1990, Clinical relevance of discography combined with CT scanning, *The Journal of Bone and Joint Surgery (Br)*, **72-B**, 480-485.

Aronoff, G. M., 1982, Pain clinic #2 pain units provide an effective alternative technique in the management of chronic pain, *Orthopaedic Review*, **11**, 95-100.

Aronoff, G. M., 1983, The role of the pain center in the treatment for intractable suffering and disability resulting from chronic pain, *Seminars in Neurology*, **3**, 377-381.

Asfour, S. S., Khalil, T. M., Waly, S. M., Goldberg, M. L., Rosomoff, R. S. and Rosomoff, H. L., 1990, Biofeedback in back muscle strengthening, *Spine*, **15**, 510-513.

Aspden, R. M., 1989, The spine as an arch: A new mathematical model, *Spine*, **14**, 266-274.

Barnes, D., Smith, D., Gatchel, R. J. and Mayer, T. G., 1989, Psychosocioeconomic predictors of treatment success/failure in chronic low-back pain patients, *Spine*, **14**, 427-430.

Battié, M. C., Bigos, S. J., Fisher, L. D., Hansson, T. H., Jones, M. E. and Wortley, M. D., 1989, Isometric lifting strength as a predictor of industrial back pain reports, *Spine*, 14, 851.

Battié, M. C., Bigos, S. J., Fisher, L. D., Hansson, T. H., Nachemson, A. L., Spengler, D. M., Wortley, M. D. and Zeh, J., 1989, A prospective study of the role of cardiovascular risk factors and fitness in industrial back pain complaints, *Spine*, **14**, 141-147.

Battié, M. C., Bigos, S. J., Fisher, L. D., Spengler, D. M., Hansson, T. H., Nachemson, A. L. and Wortley, M. D., 1990, The role of spinal flexibility in back pain complaints within industry: A prospective study, *Spine*, **15**, 768-773.

Bendix, T., Sorensen, S. S. and Klausen, K., 1984, Lumbar curve, trunk muscles, and line of gravity with different heel heights, *Spine*, **9**, 223-227.

Bernhardt, M., Gurganious, L.R., Bloom, D. L. and White, A. A. III, 1993, Magnetic resonance imaging analysis of percutaneous discectomy: A preliminary report, *Spine*, **18**, 211-217.

Bigos, S. J., Battié, M. C. and Fisher, L. D., 1991a, Methodology for evaluating predictive factors for the report of back injury, *Spine*, **16**, 669-670.

Bigos, S. J., Battié, M. C., Fisher, L. D., Hansson, T. H., Spengler, D.M. and Nachemson, A. L., 1992. A prospective evaluation of preemployment screening methods for acute industrial back pain, *Spine*, **17**, 922-926.

Bigos, S. J., Battié, M. C., Spengler, D. M., Fisher, L. D., Fordyce, W. E., Hansson, T. H., Nachemson, A. L. and Worley, M. D., 1991b, A prospective study of work perceptions and psychosocial factors affecting the report of back injury, *Spine*, **16**, 1-6.

Bigos, S. J., Battié, M. C., Spengler, D. M., Fisher, L. D., Fordyce, W. E., Hansson, T., Nachemson, A. L. and Zeh, J., 1992, A longitudinal, prospective study of industrial back injury reporting, *Clin Orthopaed*, **279**, 21-34.

Bigos, S. J., Spengler, D. M., Martin, N. A., Zeh, J., Fisher, L., and Nachemson A., 1986b, Back injuries in industry: a retrospective study: III. employee-related factors, *Spine*, **11**, 252-256.

Bigos, S. J., Spengler, D. M., Martin, N. A., Zeh, J., Fisher, L., Nachemson, A. and Wang, M. H., 1986a, Back injuries in industry: a retrospective study: II injury factors, *Spine*, **11**, 246-251.

Boden, S. D., Wiesel, S. W., Laws, E. R. and Rothman, R. H., 1991, *The Aging Spine*, (Philadelphia: W. B. Saunders Co.), 21-22.

Bogduk, N. and Engle, R., 1984, The menisci of the lumbar zygapophyseal joints: A review of their anatomy and clinical significance, *Spine*, **9**, 454-460.

Bogduk, N. and Macintosh, J. E., 1984, The applied anatomy of the thoracolumbar fascia, *Spine*, **9**, 164-170.

Bogduk, N., 1983, The innervation of the lumbar spine, *Spine*, **8**, 286-293.

Bogduk, N., 1991, The lumbar disc and low back pain, *Neurosurgery Clinics of North America*, **2**, 791-803.

Bogduk, N., Macintosh, J. E. and Pearcy, M. J., 1992, A universal model of the lumbar back muscles in the upright position, *Spine*, **17**, 897-913.

Bogduk, N., Tynan, W. and Wilson, A. S., 1981, Innervation of lumbar intervertebral disks, *Anat. Soc. G.B. & I*, 39-56.

Brinckmann, P. and Grootenboer, H., 1991, Change of disc height, radial disc bulge, and intradiscal pressure from discectomy: An *in vitro* investigation on human lumbar discs, *Spine*, **16**, 641-646.

Brinckmann, P. and Horst, M., 1985, The influence of vertebral body fracture, intradiscal injection, and partial discectomy of the radial bulge and height of human lumbar discs, *Spine*, **10**, 138-145.

Brinckmann, P., 1986, Injury of the annulus fibrosus and disc protrusions: an *in vitro* investigation on human lumbar discs, *Spine*, **11**, 149-153.

Brinckmann, P., Biggemann, M. and Hilweg, D., 1989, Prediction of the compressive strength of human lumbar vertebrae, *Spine*, **14**, 606-610.

Brinckmann, P., Frobin, W., Hierholzer, E. and Horst, M., 1983, Deformation of the vertebral end-plate under axial loading of the spine, *Spine*, **8**, 851-856.

Bromley, J. W., Varma, A. O., Santoro, A. J., Cohen, P., Jacobs, R. and Berger, L., 1984, Double-blind evaluation of collagenase injections for herniated lumbar discs, *Spine*, **9**, 486-488.

Buseck, M., Schipplein, O. D., Andersson, G. B. J. and Andriacchi, T. P., 1988, Influence of dynamic factors and external loads on the moment at the lumbar spine in lifting, *Spine*, **13**, 918-921.

Butler, D., Trafimow, J. H., Andersson, G. B. J., McNeill, T. W. and Huckman, M. S., 1990, Discs degenerate before facets, *Spine*, **15**, 111-113.

Calhoun, E., McCall, I. W., Williams, L. and Pullicino, V. N., 1988, Provocation discography as a guide to planning operations on the spine, *The Journal of Bone and Joint Surgery (Br)*, **70-B**, 267-271.

Carr, D., Gilbertson, L., Frymoyer, J., Krag, M. and Pope, M., 1985, Lumbar paraspinal compartment syndrome: A case report with physiologic and anatomic studies, *Spine,* **10**, 816-820.

Carrera, G. F., 1980a, Lumbar facet joint injection in low back pain and sciatica, *Radiology*, **137**, 661-664.

Carrera, G. F., 1980b, Lumbar facet joint injection in low back pain and sciatica, *Radiology*, **137**, 665-667.

Cartas, O., Nordin, M., Frankel, V. H., Malgady, R. and Sheikhzadeh, A., 1993, Quantification of trunk muscle performance in standing, semistanding, and sitting postures in healthy men, *Spine*, **18**, 603-609.

Cassidy, J. and Kirkaldy-Willis, W., 1988, Manipulation in *Managing Low Back Pain*, Ed. William H. Kirkaldy-Willis, New York: Churchill Livingstone. 287-291.

Cavanaugh, J. M. and Weinstein, J. N., 1994, Low back pain: epidemiology, anatomy and neurophysiology, Ed. Patrick D. Wall and Ronald Melzack. *Textbook of Pain*, 3rd edition, Churchill Livingstone, New York, NY. 441-454.

Chaffin, D. B. and Park, K. S., 1973, A longitudinal study of low-back pain as associated with occupational weight lifting factors, *American Industrial Hygiene Association Journal*, **34**, 513-525.

Clark, G. A., Panjabi, M. M. and Wetzel, F. T., 1985, Can infant malnutrition cause adult vertebral stenosis, *Spine,* **10**, 165-170.

Crock, H. V., 1981, Normal and pathological anatomy of the lumbar spinal nerve root canals, *The Journal of Bone and Joint Surgery*, **63-B**, 487-490.

Damkot, D. K., Pope, M. H., Lord, J. and Frymoyer, J. W., 1984, The relationship between work history, work environment and low-back pain in men, *Spine*, **9**, 395-399.

Davis, A. A. and Carragee, E. J., 1993, Bilateral facet dislocation at the lumbosacral joint: A report of a case and review of literature, *Spine*, **18**, 2540-2544.

Davis, P. R., 1981, The use of intra-abdominal pressure in evaluating stresses on the lumbar spine, *Spine*, **6**, 90-92.

Deyo, R. A. and Diehl, A. K., 1983, Measuring physical and psychosocial function in patients with low-back pain, *Spine*, **8**, 635-642.

Deyo, R. A., 1983, Conservative therapy for low back pain, *JAMA*, **250**, 1057-1062.

Deyo, R. A., 1986, Diehl AK; Rosenthal M. How many days of bed rest for acute low back pain? A randomized clinical trial. *NEJOM*, **315**, 1064-1070.

Deyo, R. A., 1988, Measuring the functional status of patients with low back pain, *Arch Phys Med Rehabil*, **69**, 1044-1053.

Dillard, J., Trafimow, J., Andersson, G. B. J. and Cronin, K., 1991, Motion of the lumbar spine: Reliability of two measurement techniques, *Spine*, **16**, 321.

Dolan, P., Adams, M. A. and Hutton, W. C., 1988, Commonly adopted postures and their effect on the lumbar spine, *Spine*, **13**, 197-201.

Dong, G. X. and Porter RW., 1989, Walking and cycling tests in neurogenic and intermittent claudication, *Spine*, **14**, 965-969.

Dvorak, J., Panjabi, M., Gerber, M. and Wichmann, W., 1987, CT-functional diagnostics of the rotatory instability of upper cervical spine: 1. An experimental study on cadavers, *Spine*, **12**, 197-205.

Edwards, W. C. and LaRocca, S. H., 1985, The developmental segmental sagittal diameter in combined cervical and lumbar spondylosis, *Spine*, **10**, 42-49.

Eisen, A. and Hoirch, M., 1983, The electrodiagnostic evaluation of spinal root lesions, *Spine*, **8**, 98-106.

Eismont, F. J. and Currier, B., 1989, Surgical management of lumbar intervertebral-disc disease, *The Journal of Bone and Joint Surgery*, **71-A**, 1266-1271.

Elnaggar, I. M., Nordin, M., Sheikhzadeh, A., Parnianpour, M. and Kahanovitz, N., 1991, Effects of spinal flexion and extension exercises on low-back pain and spinal mobility in chronic mechanical low-back pain patients, *Spine*, **16**, 967-972.

Fager, C. A., 1984, The age-old back problem new fad, same fallacies, *Spine*, **9**, 326-328.

Fardon, D., Pinkerton, S., Balderstron, R., Garfin, S., Nasca, R. and Salib, R., 1993, Terms used for diagnosis by English speaking spine surgeon, *Spine*, **18**, 274-277.

Farfan, H. F. and Gracovetsky, S., 1984, The nature of instability, *Spine*, **9**, 714-719.

Farfan, H. F., 1984, The torsional injury of the lumbar spine, *Spine*, **9**, 53.

Farfan, H. F., 1985, The use of mechanical etiology to determine the efficacy of active intervention in single joint lumbar intervertebral joint problems: Surgery and chemonucleolysis compared: A prospective study, *Spine*, **10**, 350-358.

Farfan, H. F., Cossette, J.W., Robertson, G. H., Wells, R. V. and Kraus, H., 1970, The effects of torsion on the lumbar intervertebral joints: the role of torsion in the production of disc degeneration, *The Journal of Bone and Joint Surgery*, **52-A**, 468-497.

Farfan, H. F., Huberdeau, R. M. and Dubow, H. I., 1972, Lumbar intervertebral disc degeneration the influence of geometrical features on the pattern of disc degeneration a post mortem study, *The Journal of Bone and Joint Surgery*, **54-A**, 492-510.

Fraser, R. D., Osti, O. L. and Vernon-Roberts, B., 1987, Discitis after discography, *The Journal of Bone and Joint Surgery*, **69-B**, 26-35.

Friedman, J. and Goldner, M. Z., 1955, Discography in evaluation of lumbar disk lesions, *Radiology*, **65**, 653-661.

Frymoyer, J. W. and Gordon, S. L., 1989, Research perspectives in low-back pain, Report of a 1988 workshop, *Spine*, **14**, 1384-1390.

Frymoyer, J. W., 1981, The role of spine fusion, *Spine*, **6**, 284-307.

Gatchel, R. J., Mayer, T. G., Hazard, R. G., Rainville, J. and Mooney, V., 1992, Functional restoration: pitfalls in evaluating efficacy, *Spine*, **17**, 988-995.

Ghosh, P., 1988, *The Biology of the Intervertebral Disc, I & II*, Boca Raton, FL: CRC Press, Inc.

Goel, V. K., Goyal, S., Clark, C., Nishiyama, K. and Nye, T., 1985, Kinematics of the whole lumbar spine: Effect of discectomy, *Spine*, **10**, 543-554.

Goel, V. K., Kim. Y. E., Lim, T-H. and Weinstein, J. N., 1988, An analytical investigation of the mechanics of spinal instrumentation, *Spine*, **13**, 1003-1010.

Goel, V. K., Kong, W., Han, J. S., Weinstein, J. N. and Gilbertson, L. G., 1993, A combine finite element and optimization investigation of lumbar spine mechanics with and without muscles, *Spine*, **18**, 1531-1541.

Goel, V. K., Nishiyama, K., Weinstein, J. N. and Liu, Y. K., 1986, Mechanical properties of lumbar spinal motion segments as affected by partial disc removal, *Spine*, **11**, 1008-1012.

Gordon, S. J., Yang, K. H., Mayer, P. J., Mace, A. H. Jr., Kish, V. L. and Radin, E. L., 1991, Mechanism of disc rupture: A preliminary report, *Spine*, **16**, 450-456.

Goren, G. J., Baljet, B. and Drukker, J., 1990, Nerves and nerve plexuses of the human vertebral column, *The American Journal of Anatomy*, **188**, 282-296.

Grabias, S., 1980, Current concepts review the treatment of spinal stenosis, *The Journal of Bone and Joint Surgery*, **62-A**, 308-313.

Gracovetsky, S. and Farfan, H., 1986, The optimum spine, *Spine*, **11**, 543-573.

Gracovetsky, S., Farfah, H. and Helleur, C., 1985, The abdominal mechanism, *Spine*, **10**, 317-324.

Gracovetsky, S., Farfan, H. F. and Lamy, C., 1977, A mathematical model of the lumbar spine using an optimization system to control muscles and ligaments, *Orthopaed Clin North Am*, **8**, 135-153.

Gracovetsky, S., Farfan, H. F. and Lamy, C., 1981, The mechanism of the lumbar spine, *Spine*, **6**, 249-262.

Gracovetsky, S., Kary, M., Levy, S., Said, R. B., Pitchen, I. and Helie, J., 1990, Analysis of spinal and muscular activity during flexion/extension and free lifts, *Spine*, **15**, 1333-1339.

Gracovetsky, S., Kary, M., Pitchen, I., Levy, S. and Said, R. B., 1989, The importance of pelvic tilt in reducing compressive stress in the spine during flexion-extension exercises, *Spine*, **14**, 412-416.

Gracovetsky, S., Zeman V., Carbone, A. 1987, Relationship between lordosis and the position of the center of reaction of the spinal disc. *J Biomed Eng*, **9**, 237-248.

Grobler, L., Robertson, P. A., Novotny, J. E. and Ahern, J. W., 1993a, Decompression for degenerative spondylolisthesis and spinal stenosis at $L_{4-5}$: The effects of facet joint morphology, *Spine*, **18**, 1475-1482.

Grobler, L., Robertson, P. A., Novotny, J. E. and Pope, M. H., 1993b, Etiology of spondylolisthesis: Assessment of the role played by lumbar facet morphology, *Spine*, **18**, 80-91.

Gunzburg, R., Gunzburg, J., Wagner, J. and Fraser, R. D., 1991a, Radiologic interpretation of lumbar vertebral rotation, *Spine*, **16**, 660-664.

Gunzburg, R., Hutton, W. and Fraser, R., 1991b, Axial rotation of the lumbar spine and the effect of flexion: an in vitro and in vivo biomechanical study, *Spine*, **16**, 22-28.

Hadler, N. M., 1984, *Diagnosis and Treatment of Backache. Medical Management of the Regional Musculoskeletal Diseases*, New York, NY: Grune & Stratton Inc., 3-52.

Hadler, N. M., 1987, Regional musculoskeletal diseases of the low back. Cumulative trauma versus single incident, *Clinical Orthopaedics and Related Research*, **221**, 33-41.

Hadler, N. M., Curtis, P., Gillings, D. B. and Stinnett, S., 1987, A benefit of spinal manipulation as adjunctive therapy for acute low-back pain A stratified controlled trial, *Spine*, **12**, 703-706.

Haig, A. J., 1992, Diagnoses and treatment options in occupational low-back pain, *Occupational Medicine*, **7**, 641-653.

Haig, A. J., Weismann, G., Haugh, L. D., Pope, M. and Grobler, L. J., 1993, Prospective evidence for change in paraspinal muscle activity after herniated nucleus pulposus, *Spine*, **18**, 926-930.

Haldeman, S., 1980, *Spinal Manipulative Therapy in the Management of Low Back Pain, Low Back Pain*, 2nd ed. Philadelphia, PA: J.B. Lippincott Company, 245-275.

Haldeman, S., 1984, The electrodiagnostic evaluation of nerve root function, *Spine*, **9**, 42-48.

Hampton, D., Laros, G., McCarron, R. and Franks, D., 1989, Healing potential of the annulus fibrosus, *Spine*, **14**, 398.

Hansson, T. and Ross, B., 1983, The amount of bone mineral and Schmorl's nodes in lumbar vertebrae, *Spine*, **8**, 266.

Hansson, T. H., Keller, T. S., and Panjabi, M. M., 1987, A study of the compressive properties of lumbar vertebral trabeculae: effects of tissue characteristics, *Spine*, **12**, 56-62.

Hansson, T. H., Bigos, S., Beecher, P. and Wortley, M., 1985, The lumbar lordosis in acute and chronic low-back pain, *Spine*, **10**, 154.

Hart, D. L., Stobbe, T. J. and Jaraiedi, M., 1987, Effect of lumbar posture on lifting, *Spine*, **12**, 138-145.

Heliovaara, M., Vanharanta, H., Korpi, J. and Troup, J. D. G., 1986 Herniated lumbar disc syndrome and vertebral canals, *Spine*, **11**, 433.

Herrin, G. D., Jaraiedi, M. and Anderson, C. K., 1986, Prediction of overexertion injuries using biomechanical and psychophysical models, *Am. Ind. Hyg. Assoc. J*, **47**, 322-330.

Herron, L. D. and Pheasant, H. C., 1982, Changes in MMPI profiles after low-back surgery, *Spine*, **7**, 591-597.

Hickey, D. S. and Hukins, D. W. L., 1982, Aging changes in the macromolecular organization of the intervertebral disc: An X-ray diffraction and electron microscopic study, *Spine*, **7**, 234-242.

Hickey, D. S. and Hukins, W. L., 1980, Structure of fetal annulus, *Anat. Soc. G.B. & I*, 81-89.

Hickey, D. S., Aspden, R. M., Hukins, D. W. L., Jenkins, J. P. R. and Isherwood, I., 1986, Analysis of magnetic resonance images from normal and degenerate lumbar intervertebral discs, *Spine*, **11**, 702-708.

Holmes, A. D., Hukins, D. W. L., and Freemont, A. J., 1993, End-plate displacement during compression of lumbar vertebra-disc-vertebra segments and the mechanism of failure, *Spine*, **18**, 128-135.

Holt, E. P., 1968, The question of lumbar discography, *The Journal of Bone and Joint Surgery*, **50-A**, 720-726.

Horst, M. and Brinckmann, P., 1981, Measurement of the distribution of axial stress on the end-plate of the vertebral body, *Spine*, **6**, 217-232.

Hsu, K. Y., Zucherman, J. F., Derby, R., White, A. H., Goldthwaite, N. and Wynne, G., 1988, Painful lumbar end-plate disruptions: A significant finding, *Spine*, **13**, 76-78.

Hsu, K. Y, Zucherman, J., Shea, W., Kaiser, J., White, A., Schofferman, J. and Amelon, C., 1990, High lumbar disc degeneration: Incidence and etiology, *Spine*, **15**, 679-682.

Hukins, D. W. L., Kirby, M. C., Sikoryn, T. A., Aspden, R. M. and Cox, A. J., 1990, Comparison of structure, mechanical properties, and functions of lumbar spinal ligaments, *Spine*, **15**, 787-795.

Hutton, W. C. and Adams, M. A., 1982, Can the lumbar spine be crushed in heavy lifting? *Spine*, **7**, 586-590.

Inoue, H., 1981, Three dimensional architecture of lumbar intervertebral discs, *Spine*, **6**, 139-146.

Jager, J. and Luttmann, A., 1989, Biomechanical analysis and assessment of lumbar stress during load lifting using a dynamic 19-segment human model, *Ergonomics*, **32**, 93-112.

Kane, K. and Taub, A., 1975, A history of local electrical analgesia, *Pain*, **1**, 125-138.

Keeley, J., Mayer, T. G., Cox, R., Gatchel, R. J., Smith, J. and Mooney, V., 1986, Quantification of lumbar function: part 5: reliability of range-of-motion measures in the sagittal plane and an *in vivo* torso rotation measurement technique, *Spine*, **11**, 31-35.

Keller, T. S., Hansson, T. H., Abram, A. C., Spengler, D. M. and Panjabi, M. M., 1989, Regional variations in the compressive properties of lumbar vertebral trabeculae. Effects of disc degeneration, *Spine*, **14**, 1012-1019.

Keller, T. S., Holm, S. H., Hansson, T. H. and Spengler, D. M., 1990, 1990 Volvo award in experimental studies: The dependence of intervertebral disc mechanical properties on physiologic conditions, *Spine*, **15**, 751-761.

Kelsey, J. L., 1975, An epidemiological study of the relationship between occupations and acute herniated lumbar intervertebral discs, *International Journal of Epidemiology*, **4**, 197-205.

Kelsey, J. L., Githens, P. B., O'Conner, T., Weil, U., Calogero, J. A., Holford, T. R., White, A. A. III., Walter, S. D., Ostfield, A. M. and Southwick, W. O., 1984, Acute prolapsed lumbar intervertebral disc: An epidemiologic study with special reference to driving automobiles and cigarette smoking, *Spine*, **9**, 608-613.

Kelsey, J. L., Githens, P. B., White, A. A., Holford, T. R., Walter, S. D., O'Connor, T., Ostfeld, A. M., Weil, U., Southwick, W. O. and Calogero, J. A., 1984, An epidemiologic study of lifting and twisting on the job and risk for acute prolapsed lumbar intervertebral disc, *Journal of Orthopaedic Research*, **2**, 61-66.

Khalil, T. M., Asfour, S. S., Martinez, L. M., Waly, S. M., Rosomoff, R. S. and Rosomoff, H. L., 1992, Stretching in the rehabilitation of low-back pain patients, *Spine*, **17**, 311-317.

Khatri, B. O., Baruah, J. and McQuillen, M. P., 1984, Correlation of electromyography with computed tomography in evaluation of lower back pain, *Arch Neurol*, **41**, 594-597.

Kim, Y. E., Goel, V. K., Weinstein, J. N. and Lim, T-H., 1991, Effect of disc degeneration at one level on the adjacent level in axial mode, *Spine*, **16**, 331-335.

Kishino, N. D., Mayer, T. G., Gatchel, R. J., Parrish, M. M., Anderson, C., Gustin, L. and Mooney, V., 1985, Quantification of lumbar function: Part 4: Isometric and isokinetic lifting simulation in normal subjects and low-back dysfunction patients, *Spine*, **10**, 922-927.

Krag, M. H., Cohen, M. C., Haugh, L. D. and Pope, M. H., 1990, Body height change during upright and recumbent posture, *Spine*, **15**, 202-207.

Kramer, K. M. and Levine, A. M., 1989, Unilateral facet dislocation of the lumbosacral junction. A case report and review of the literature, *The Journal of Bone and Joint Surgery*, **71-A**,1258-1261.

Ladin, Z., Murthy, K. R. and De Luca, C. J., 1989, 1989 Volvo award in biomechanics: Mechanical recruitment of low-back muscles: Theoretical predictions and experimental validation, *Spine*, **14**, 927-938.

Lantz, S. A. and Schultz, A. B., 1986a, Lumbar spine orthosis wearing I. Restriction of gross body motions, *Spine*, **11**, 834-837.

Lantz, S. A. and Schultz, A. B., 1986b, Lumbar spine orthosis wearing II. Effect on trunk muscle myoelectric activity, *Spine*, **11**, 838-842.

Lavender, S. A., Marras, W. S., Miller, R. A., 1993, The development of response strategies in preparation for sudden loading to the torso, *Spine*, **18**, 2097-2105.

Lavender, S. A., Tsuang, Y. H., Andersson, G. B. J., Hafezi, A. and Shin, C. C., 1992, Trunk muscle cocontraction: the effects of moment direction and moment magnitude, *Journal of Orthopaedic Research*, **10**, 691-700.

Lin, H. S., Liu, Y. K. and Adams, KH., 1978, Mechanical response of the lumbar intervertebral joint under physiological (complex) loading, *The Journal of Bone and Joint Surgery*, **60-A**, 41-55.

Macintosh, J. E. and Bogduk, N., 1991, The attachments of the lumbar erector spinae, *Spine*, **16**, 783-792.

Macintosh, J. E. and Bogduk, N.,1987, 1987 Volvo award in clinical sciences: The morphology of the lumbar erector spinae, *Spine*, **12**, 658-660.

Macintosh, J. E., Bogduk, N. and Pearcy, M. J., 1993, The effects of flexion on the geometry and actions of the lumbar erector spinae, *Spine*, **18**, 884-893.

Magnusson, M., Granqvist, M., Jonson, R., Lindell, V., Lundberg, U., Wallin, L. and Hansson, T., 1990, The loads on the lumbar spine during work at an assembly line: The risks for fatigue injuries of vertebral bodies, *Spine*, **15**, 774-779.

Malmivaara, A., Videman, T., Kuosma, E. and Troup, J. D. G., 1986, Radiographic vs. direct measurements of the spinal canal of the thoracolumbar junctional region (T10-L1) of the spine, *Spine*, **11**, 574-578.

Markolf, K. L. and Morris, J. M., 1974, The structural components of the intervertebral disc: A study of their contributions to the ability of the disc to withstand compressive forces. *The Journal of Bone and Joint Surgery*, **56-A**, 675-687.

Markolf, K. L., 1972, Deformation of the thoracolumbar intervertebral joints in response to external loads, *The Journal of Bone and Joint Surgery*, **54-A**, 511-533.

Marras, W. S. and Mirka, G. A., 1992, A comprehensive evaluation of trunk response to asymmetric trunk motion, *Spine*, **17**, 318-326.

Marras, W. S. and Reilly, C. H., 1988, Networks of internal trunk-loading activities under controlled trunk-motion conditions, *Spine*, **13**, 661-667.

Marras, W. S., King, A. I. and Joynt, R. L., 1984, Measurements of loads on the lumbar spine under isometric and isokinetic conditions, *Spine*, **9**, 176-187.

Marras, W. S., Lavender, S. A., Leurgans, S. E., Rajulu, S. L., Allread, W. G., Fathallah, F. A. and Ferguson, S. A., 1993, The role of dynamic three-dimensional trunk motion in occupationally-related low back disorders: The effects of workplace factors, trunk position, and trunk motion characteristics on risk of injury, *Spine*, **18**, 617-628.

Mayer, T. G., Barnes, D., Kishino, N. D., Nichols, G., Gatchel, R. J., Mayer, H. and Mooney, V, 1988, Progressive isoinertial lifting evaluation: I. A standardized protocol and normative database, *Spine*, **13**, 993-997.

McCarron, R. F., Wimpee, M. W., Hudkins, P. G. and Laros, G. S., 1987, The inflammatory effect of nucleus pulposus: A possible element in the pathogenesis of low-back pain, *Spine*, **12**, 760.

McCombe, P. F., Fairbank, J. C., Cockersole, B. C. and Pynsent, P. B., 1989, Reproducibility of physical signs in low-back pain, *Spine*, **14**, 908-918.

McGill, S. M., 1991, Kinetic potential of the lumbar trunk musculature about three orthogonal orthopaedic axes in extreme postures, *Spine*, **16**, 809-815.

McGill, S. M., 1992, The influence of lordosis on axial trunk torque and trunk muscle myoelectric activity, *Spine*, **17**, 1187-1193.

McNally, D. S. and Adams, M, A., 1992, Internal intervertebral disc mechanics as revealed by stress profilometry, *Spine*, **17**, 66-73.

McNally, D. S., Adams, M. A. and Goodship, A. E., 1993, Can intervertebral disc prolapse be predicted by disc mechanics? *Spine*, **18**, 1525-1530.

Melzack, R. and Wall, P. D., 1965, Pain mechanisms: a new theory, *Science*, **150**, 971-979.

Miller, J. A. A., Haderspeck, K. A. and Schultz, A. B., 1983, Posterior element loads in lumbar motion segments, *Spine*, **8**, 331-337.

Miller, J. A. A., Schmatz, C. and Schultz, A. B., 1988, Lumbar disc degeneration: Correlation with age, sex, and spine level in 600 autopsy specimens, *Spine*, **13**, 173-178.

Mills, G. H., Davies, G. K., Getty, C. J. M. and Conway, J., 1986, The evaluation of liquid crystal thermography in the investigation of nerve root compression due to lumbosacral lateral spinal stenosis, *Spine*, **11**, 427-432.

Mirka, G. A. and Marras, W. S., 1993, A stochastic model of trunk muscle coactivation during trunk bending, *Spine*, **18**, 1396-1409.

Mooney, V., 1987, Presidential address, International Society for the Study of the Lumbar Spine, Dallas, 1986 — Where is the pain coming from? *Spine*, **12**, 754-759.

Mooney, V., Haldeman, S., Nasca, R. J., White, A. H., Nix, J. E., Wiltse, L. L., Selby, D. K., Kostuik, J. P., Krag, M. H., Ray, C. D., Simmons, J. W., Drabing, J., Yong-Hing, K., Russell, G. S., Cauthen, J. and Saal, J. A., 1988, Position statement on discography, *Spine*, **13**,1343.

Morris, E. W., DiPaolo, M., Vallance, R. and Waddell, G., 1986, Diagnosis and decision making in lumbar disc prolapse and nerve entrapment, *Spine*, **11**, 436-439.

Mundt, D. J., Kelsey, J. L., Golden, A. L., Pastides, H., Berg, A. T., Sklar, J., Hosea, T. and Panjabi, M. M., 1993, Northeast Collaborative Group on Low Back Back Pain. An epidemiologic study of non-occupational lifting as a risk factor for herniated lumbar intervertebral disc, *Spine*, **18**, 595-608.

Nachemson, A. L., 1985, Advances in low-back pain, *Clinical Orthopaedics and Related Research*, **200**, 266-278.

Nachemson, A. L., 1989, Editorial comment lumbar discography — where are we today? *Spine*, **14**, 556-557.

Nachemson, A. L., Andersson, G. B. J. and Schultz, A. B., 1986, Valsalva maneuver biomechanics; Effects on lumbar trunk loads of elevated intrabdominal pressures, *Spine*, **11**, 476-479.

Nachemson, A. L., Troup, J. D. G., Videman, T., Bigos, S. J., Battié, M. C., Fisher, L. D., Cats-Baril, W. L., Frymoyer, J. W., Pope, M. H., Nelson, R. M., Spangfort, E. and Waddell, G., 1991, Symposium: research methods in occupational low-back pain, *Spine*, **16**, 667-686.

Ninomiya, M. and Muro, T., 1992, Pathoanatomy of lumbar disc herniation as demonstrated by computed tomography/discography, *Spine*, **17**, 1316-1322.

Nordby, E. J., 1983, Chymopapain in intradiscal therapy, *The Journal of Bone and Joint Surgery*, **65-A**, 1350-1353.

Nordin, M., Ortengren, R., and Andersson, G.B.J., 1984, Measurements of trunk movements during work, *Spine*, **9-5**, 465-469.

Noren, R., Trafimow, J., Andersson, G. B. J. and Huckman, M. S., 1991, The role of facet joint tropism and facet angle in disc degeneration, *Spine*, **16**, 530-532.

Omino, K. and Hayashi, Y., 1992, Preparation of dynamic posture and occurrence of low back pain, *Ergonomics*, **35**, 693-707.

Örtengren, R., Andersson, G. B. J. and Nachemson, A. L., 1981, Studies of relationships between lumbar disc pressure, myoelectric back muscle activity, and intra-abdominal (intragastric) pressure, *Spine*, **6**, 98-103.

Ostgaard, H. C., Andersson, G. B. J. and Karlsson, K., 1991, Prevalence of back pain in pregnancy, *Spine*, **16**, 549.

Ostgaard, H. C., Andersson, G. B. J., Schultz, A. B. and Miller, J. A. A., 1993, Influence of some biomechanical factors on low-back pain in pregnancy, *Spine*, **18**, 61-65.

Oxland, T. R., Crisco, J. J. III., Panjabi, M. M. and Yamamoto, I., 1992, The effect of injury on rotational coupling at the lumbosacral joint: A biomechanical investigation, *Spine*, **17**, 74-80.

Panagiotacopulos, N. D., Pope, M. H., Bloch, R. and Krag, M. H., 1987, Water content in human intervertebral discs: Part II. Viscoelastic behavior, *Spine*, **12**, 918-924.

Panagiotacopulos, N. D., Pope, M. H., Krag, M. H. and Bloch, R., 1987, Water content in human intervertebral discs: Part I. Measurement by magnetic resonance imaging, *Spine*, **12**, 912-917.

Panjabi, M. M. and White, A. A., 1980, Basic biomechanics of the spine, *Neurosurgery*, **7**, 76-93.

Panjabi, M. M., 1988, Biomechanical evaluation of spinal fixation devices: I. A conceptual framework, *Spine*, **13**, 1129-1134.

Panjabi, M. M., Abumi, K., Duranceau, J. and Crisco, J. J., 1988a, Biomechanical evaluation of spinal fixation devices: II Stability provided by eight internal fixation devices, *Spine*, **13**, 1135-1140.

Panjabi, M. M., Geol, V., Oxland, T., Takata, K., Duranceau, J., Krag, M. and Price, M., 1992b, Human lumbar vertebrae: Quantitative three-dimensional anatomy, *Spine*, **17**, 299-306.

Panjabi, M. M., Krag, M. H. and Chung, T. Q., 1984, Effects of disc injury in mechanical behavior of the human spine, *Spine*, **9**, 707-713.

Panjabi, M. M., Oxland, T., Takata, K., Goel, V., Duranceau, J. and Krag, M., 1993, Articular facets of the human spine: Quantitative three-dimensional anatomy, *Spine*, **18**, 1298-1310.

Panjabi, M. M., Takata, K. and Goel, V. K., 1983, Kinematics of lumbar intervertebral foramen, *Spine*, **8**, 348-352.

Panjabi, M. M., Brown, M., Lindahl, S., Irstam, L. and Hermens, M., 1988b, Intrinsic disc pressure as a measure of integrity of the lumbar spine, *Spine*, **13**, 913-917.

Panjabi, M. M., Chang, D. and Dvorak, J., 1992a, An analysis of errors in kinematic parameters associated with *in vivo* functional radiographs, *Spine*, **17**, 200-205.

Panjabi, M. M., Goel, V. and Takata, K., 1982, Physiologic strains in the lumbar spinal ligaments: An *in vitro* biomechanical study, *Spine*, **7**, 192-203.

Pearcy, M. and Shepherd, J., 1985, Is there instability in spondylolisthesis, *Spine*, **10**, 175-177.

Pearcy, M., Portek, I. and Shepherd, J., 1985, The effect of low-back pain on lumbar spinal movements measured by three-dimensional X-ray analysis, *Spine*, **10**, 150-153.

Pedrini-Mille, A., Weinstein, J. N., Found, E. M., Chung, C. B. and Goel, V. K., 1990, Stimulation of dorsal root ganglia and degradation of rabbit annulus fibrosus, *Spine*, **15**, 1252-1256.

Pennington, J. B., McCarron, R. F. and Laros, G. S., 1988, Identification of IgG in the canine intervertebral disc, *Spine*, **13**, 909-912.

Polatin, P. B., Gatchel, R. J., Barnes, D., Mayer, H., Arens, C. and Mayer, T. G., 1989, A psychosociomedical prediction model of response to treatment by chronically disabled workers with low-back pain, *Spine*, **14**, 956-961.

Pope, M. H. and Panjabi, M., 1985, Biomechanical definitions of spinal instability, *Spine*, **10**, 255-256.

Pope, M. H., 1990, Bioengineering — The bond between basic scientists, clinicians, and engineers: The 1989 presidential address, *Spine*, **15**, 214-217.

Pope, M. H., Bevins, T., Wilder, D. G. and Frymoyer, J. W., 1985, The relationship between anthropometric, postural, muscular, and mobility characteristics of males ages 18-55, *Spine*, **10**, 644-648.

Pope, M. H., Svenson, M., Andersson, G. B. J., Broman, H. and Zetterberg, C., 1987, The role of prerotation of the trunk in axial twisting efforts, *Spine*, **12**, 1041-1045.

Porter, R. W., Adams, M. A. and Hutton, W. C., 1989, Physical activity and the strength of the lumbar spine, *Spine*, **14**, 201-203.

Posner, I., White, A. A. III., Edwards, W. T. and Hayes, W. C., 1982, A biomechanical analysis of the clinical stability of the lumbar and lumbosacral spine, *Spine*, **7**, 374-389.

Potvin, J. R., McGill, S. M. and Norman, R. W., 1991, Trunk muscle and lumbar ligament contributions to dynamic lifts with varying degrees of trunk flexion, *Spine*, **16**, 1099-1107.

Poulsen, E., 1981, Back muscles strength and weight limits in lifting burdens, *Spine*, **6**, 73-75.

Ratcliffe, J. F., 1980, The arterial anatomy of the adult human lumbar vertebral body: a microarteriographic study. *J. Anat.*, 131, 57-59.

Roland, M. and Morris, R., 1983a, A study of the natural history of low-back pain: part II: development of guidelines for trials of treatment in primary care, *Spine*, **8:2**, 145-150.

Roland, M. and Morris, R., 1983b, A study of the natural history of low-back pain: part I: development of a reliable and sensitive measure of disability in low-back pain, *Spine*, **8:2**, 141-144.

Rowe, M, L., 1971, Low back disability in industry: update position, *JOM*, **13**, 476-478.

Saal, J. and Saal, J., 1989, Nonoperative treatment of herniated lumbar intervertebral disc with radiculopathy: an outcome study, *Spine*, **14**, 431-437.

Sachs, B. L., Vanharanta, H., Spivey, M. A., Guyer, R. D., Videman, T., Rashbaum, R. F., Johnson, R. G., Hochshuler, S. H. and Mooney V., 1987, Dallas discogram description: A new classification of CT/Discography in low-back disorders, *Spine*, **12**, 287-294.

Sairanen, E., Brushaber, L. and Kaskinen, M., 1981, Felling work, low-back pain and osteoarthritis, *Scand J Work Environ Health*, **7**, 18-30.

Sampson, H. W. and Davis, J. S., 1988, Histopathology of the intervertebral disc of progressive ankylosis mice, *Spine*, **13**, 650-654.

Sandover, J., 1988, Behaviour of the spine under shock and vibration: a review, *Clin Biomech*, **3**, 249-256.

Scavone, J. G., Latshaw, R. F. and Rohrer, G. V., 1981, Use of lumbar spine films, Statistical evaluation at a university teaching hospital, *JAMA*, **246**, 1105-1108.

Schonstrom, N. and Hansson, T., 1988, Pressure changes following constriction of the cauda equina: An experimental study in situ, *Spine*, **13**, 385-388.

Schonstrom, N. S. R., Bolender, N-F. and Spengler, D. M., 1985, The pathomorphology of spinal stenosis as seen on CT scans of the lumbar spine, *Spine*, **10**, 806-811.

Schultz, A. B. and Andersson, G. B. J., 1981, Analysis of loads on the lumbar spine, *Spine*, **6**, 76-82.

Schultz, A. B., 1986, Loads on the human lumbar spine, *Mechanical Engineering*, 36-41.

Schultz, A. B., Sorensen, S-E. and Andersson, G. B. J., 1984, Measurements of spine morphology in children, ages 10-16, *Spine*, **9**, 70-73.

Schultz, A. B., Andersson, G. B. J., Ortengren, R., Bjork, R. and Nordin, M., 1982a, Analysis and quantitative myoelectric measurements of loads on the lumbar spine when holding weights in standing postures, *Spine*, **7**, 390-397.

Schultz, A. B., Andersson, G., Ortengren, R., Haderspeck, K. and Nachemson, A., 1982b, Loads on the lumbar spine — Validation of a biomechanical analysis by measurements of intradiscal pressures and myoelectric signals, *The Journal of Bone and Joint Surgery*, **64-A**, 713-720.

Schultz, A. B., Haderspeck, K. and Takashima, S., 1981, Correction of scoliosis by muscle stimulation: biomechanical analyses, *Spine*, **6**, 468-476.

Semon, R. L. and Spengler, D., 1981, Significance of lumbar spondylolysis in college football players, *Spine*, **6**, 172.

Shirazi-Adl, S. A. and Drouin, G., 1987, Load-bearing role of facets in a lumbar segment under sagittal plane loadings, *Journal of Biomechanics*, **20**, 601-613.

Shirazi-Adl, S. A., Shrivastava, S. C. and Ahmed, A. M., 1984, 1983 Volvo award in biomechanics: Stress analysis of the lumbar disc-body unit in compression: A three-dimensional nonlinear finite element study, *Spine*, **9**, 120-134.

Skovron, M. L., 1992, Epidemiology of low back pain, *Baillierss Clinical Rheumatology*, **6**, 559-573.

Snook, S. H. and Ciriello, V. M., 1991, The design of manual handling tasks: revised tables of maximum acceptable weights and forces, *Ergonomics*, **34**, 1197-1213.

Spencer, D. L., Irwin, G. S. and Miller, J. A. A., 1983, Anatomy and significance of fixation of the lumbosacral nerve roots in sciatica, *Spine*, **8**, 672-679.

Spencer, D. L., Miller, J. A. A. and Bertolini, J. E., 1984, The effect of intervertebral disc space narrowing on the contact force between the nerve root and a simulated disc protrusion, *Spine*, **9**, 422-426.

Spencer, D. L., Miller, J. A. A., and Schultz, A. B., 1985, The effects of chemonucleolysis on the mechanical properties of the canine lumbar disc, *Spine*, **10**, 555-561.

Spengler, D. M. and Freeman, C. W., 1979, Patient selection for lumbar discectomy: An objective approach *Spine*, **4**, 129-134.

Spengler, D. M., 1982a, Lumbar discectomy: Results with limited disc excision and selective foraminotomy, *Spine*, **7**, 604-607.

Spengler, D. M., 1982b, *Low Back Pain, Assessment and Management*, Grune & Stratton, New York, N.Y.

Spengler, D. M., 1983, The clinical spectrum of lumbar spinal stenosis, *Orthopaedic Surgery*, **2**, 2-6.

Spengler, D. M., Bigos, S. J., Martin, N. A., Zeh, J., Fisher, L. and Nachemson, A., 1986a, Back injuries in industry: a retrospective study I. Overview and cost analysis, *Spine*, **11**, 241-245.

Spengler, D. M., Bigos, S. J., Martin, N. A., Zeh, J., Fisher, L. and Nachemson, A., 1986c, Back injuries in industry: a retrospective study III. Employee-related factors, *Spine*, **11**, 252-256.

Spengler, D. M., Bigos, S. J., Martin, N. A., Zeh, J., Fisher, L., Nachemson, A. and Wang, M. H., 1986b, Back injuries in industry: a retrospective study II. Injury factors, *Spine*, **11**, 246-251.

Spitzer, W. O., Leblanc, F. E., Dupuis, M., Abenhaim, L., Belanger, A. Y., Bloch, R., Bombarider, C., Cruess, R. L., Drouin, G., Duval-Hesler, N., Laflamme, J., Lamoureux, G., Nachemson, A., Page, J. J., Rossignol, M., Salmi, L. R., Salois-Arsenault, S., Suissa, S. and Wood-Dauphinnee, S., 1987, Scientific approach to the assessment and management of activity-related spinal disorders: A monograph for clinicians: Report of the Quebec task force on spinal disorders, *Spine*, **12**, 1S.

Stubbs, D. A., 1981, Trunk stresses in construction and other industrial workers, *Spine*, **6**, 83-89.

Svensson, H. O., Andersson, G. B. J., Hagstad, A. and Jansson, P-O., 1990, The relationship of low-back pain to pregnancy and gynecologic factors, *Spine*, **15**, 371-375.

Svensson, H. O. and Andersson, G. B. J., 1983, Low-back pain in 40 to 47 year old men: Work history and work environment factors, *Spine*, **8**, 272-276.

Tan, J. C., Parnianpour, M., Nordin, M., Hofer, H. and Willems, B., 1993, Isometric maximal and submaximal trunk extension and different flexed position in standing: Triaxial torque output and EMG, *Spine*, **18**, 2480-2490.

Tanaka, M., Nakahara, S. and Inoue, H., 1993, A pathologic study of discs in the elderly. Separation between the cartilaginous endplate and the vertebral body, *Spine*, **18**, 1456-1462.

Thompson, J. P., Pearce, R. H., Schechter, M. T., Adams, M. E., Tsang, I. K. Y. and Bishop, P. B., 1990, Preliminary evaluation of a scheme for grading the gross morphology of the human intervertebral disc, *Spine*, **15**, 411-415.

Tonzola, R. F. and Ackil, A. A., 1981, Usefulness of electrophysiological studies in the diagnosis of lumbosacral root disease, *Annals of Neurology*, **9**, 305-308.

Trafimow, J. H., Schipplein, O. D., Novak, G. J. and Andersson, G. B., 1992, The effects of quadriceps fatigue on the technique of lifting, *Spine*, **18**, 364-367.

Triano, J. J. and Schultz, A. B., 1987, Correlation of objective measure of trunk motion and muscle function and low-back disability ratings, *Spine*, **12**, 561-565.

Troup, J. D. G., 1981, Briefly noted: Straight-leg raising (SLR) and the qualifying tests for increased root tension: Their predictive value after back and sciatic pain, *Spine*, **6**, 526-527.

Troup, J. D. G., 1991a, Briefly noted: Definitions of occupational low-back pain in Great Britain, *Spine*, **16**, 667-668.

Troup, J. D. G., 1991b, Measurement of strength and endurance: The psychophysical lift test, *Spine*, **16**, 679.

Troup, J. D. G., Foreman, T. K., Baxter, C. E. and Brown, D., 1987, 1987 Volvo award in clinical sciences: the perception of back pain and the role of psychophysical tests of lifting capacity, *Spine*, **12**, 645-657.

Troup, J. D. G., Martin, J. W. and Lloyd, D. C. E. F., 1981, Back pain in industry: A prospective survey, *Spine*, **6**, 61-69.

Tyrrell, A. R., Reilly, T. and Troup, J. D. G., 1985, Circadian variation in stature and the effects of spinal loading, *Spine*, **10**, 161-164.

Van Schaik, J. P. J., Verbiest, H. and Van Schaik, F. D. J., 1985, The orientation of laminae and facet joints in the lower lumbar spine, *Spine*, **10**, 59-63.

Vanharanta, H., Guyer, R. D., Ohnmeiss, D. D., Stith, W. J., Sachs, B. L., Aprill, C., Spivey, M., Rashbaum, R. F., Hochschuler, S. H., Videman, T., Selby, D. K., Terry, A. and Mooney, V., 1988, Disc deterioration in low-back syndromes: A prospective, multi-center CT/discography study, *Spine*, **13**, 1349-1361.

Vanharanta, H., Korpi, J., Heliovaara, M. and Troup, J. D. G., 1985, Radiographic measurements of lumbar spinal cord size and their relation to back mobility, *Spine*, **10**, 461-466.

Vanharanta, H., Sachs, B. L., Ohnmeiss, D. D., Aprill, C., Spivey, M., Guyer, R. D., Rashbaum, R. F., Hochschuler, S. H., Terry, A., Selby, D., Stith, W. J. and Mooney, V., 1989, Pain provocation and disc deterioration by age: A CT/Discography study in a low-back pain population, *Spine*, **14**, 420-423.

Vanharanta, H., Sachs, B. L., Spivey, M. A., Guyer, R. D., Hochschuler, S. H., Rashbaum, R. F., Johnson, R. G., Ohnmeiss, D. and Mooney, V., 1987, The relationship of pain provocation to lumbar disc deterioration as seen by CT/discography, *Spine*, **12**, 295-298.

Vanharanta, H., Sachs, B. L., Spivey, M., Hochschuler, S. H., Guyer, R. D., Rashbaum, R. F., Ohnmeiss, D. D. and Mooney, V., 1988b, A comparison of CT/discography, pain response and radiographic disc height, *Spine*, **13**, 321-324.

Videman, T., Nurimen, M. and Troup, J. D. G., 1990, 1990 Volvo award in clinical sciences: Lumbar spinal pathology in cadaveric material in relation to history of back pain, occupation, and physical loading, *Spine*, **15**, 728-740.

Videman, T., Nuriminen, T., Tola, S., Kuorinka, I., Vanharanta, H. and Troup, J. D. G., 1984, Low-back pain in nurses and some loading factors of work, *Spine*, **9**, 400-464.

Waddell, G. and Main, C. J., 1984, Assessment of severity in low-back disorders, *Spine*, **9**, 204-208.

Waddell, G., Main, C. J., Morris, E. W., Di Paola, M. and Gray, I. C. M., 1984, Chronic low-back pain, psychologic distress, and illness behavior, *Spine*, **9**, 209-213.

Waddell, G., McCulloch, J. A., Kummel, E. and Venner, R. M., 1980, Nonorganic physical signs in low-back pain, *Spine*, **5** (2), 117-125.

Wall, P. D. and Melzack, R., 1994, *Text Book of Pain*, Churchill Livingston, Edinburgh, U.K.

Walsh, T. R., Weinstein, J. N., Spratt, K. F., Lehmann, T. R., Aprill, C. and Sayre, H., 1990, Lumbar discography in normal subjects, *The Journal of Bone and Joint Surgery*, **72-A**, 1081-1088.

Webster, B. and Snook, S., 1990, The cost of compensable low back pain, *JOM*, **32** (1), 13-15.

Wehling, P., Pak, M. A., Cleveland, S. J. and Schulitz, K. P., 1989, The influence on spinal cord evoked potentials of chymopapain applied to the rat lumbar spine canal, *Spine*, **14**, 65-67.

Weinstein, J. N., 1988, The perception of pain, *Managing Low Back Pain*, New York: Livingston Churchill, 83-90.

Weinstein, J. N., Claverie, W. and Gibson, S., 1988a, The pain of discography, *Spine*, **13**, 1344-1348.

Weinstein, J. N., Pope, M., Schmidt, R. and Seroussi, R., 1988b, Neuropharmacologic effects of vibration on the dorsal root ganglion: An animal model, *Spine*, **13**, 521-525.

White, A. A. III. and Gordon, S. L., 1982, Synopsis: Workshop on idiopathic low-back pain, *Spine*, **7**, 141-147.

White, A. H., Von Rogov, P., Zucherman, J. and Heiden, D., 1987, Lumbar laminectomy for herniated disc: A prospective controlled comparison with internal fixation fusion, *Spine*, **12**, 305-307.

Wiesel, S. W., Feffer, H. L. and Rothman, R. H., 1984, Industrial low-back pain: A prospective evaluation of a standardized diagnostic and treatment protocol, *Spine*, **9**, 199-203.

Wilder, D. G., Woodworth, B. B., Frymoyer, J. W. and Pope, M. H., 1982, Vibration and the human spine, *Spine*, **7**, 243-254.

Williams, M. M., Hawley, J. A., McKenzie, R. A. and Van Wijmen, P. M., 1991, A comparison of the effects of two sitting postures on back and referred pain, *Spine*, **16**, 1185-1191.

Wiltse, L. L., 1977, Surgery for intervertebral disk disease of the lumbar spine, *Clinical Orthopaedics and Related Research*, **129**, 22-45.

Yamamoto, I., Panjabi, M. M., Oxland, T. R. and Crisco, J. J., 1990, The role of the iliolumbar ligament in the lumbosacral junction, *Spine*, **15**, 1138-1141.

Yang, K. H. and King, A. I., 1984a, Mechanism of facet load transmission as a hypothesis for low-back pain, *Spine*, **9** (6), 557-565.

Yoshizawa, H., O'Brien, J. P., Smith, W. T. and Trumper, M., 1980, The neuropathology of intervertebral discs removed for low-back pain, *J. Pathology*, **132**, 95-104.

Zohn, D. A., 1988, *Musculoskeletal Pain: Diagnosis and Physical Treatment*, Boston: Little Brown, 185-247.

# Part II

## Administrative Controls

# Section I

## Ergonomics Surveillance

# 23

# Fundamentals of Surveillance for Work-Related Musculoskeletal Disorders

**Vern Putz-Anderson**
*U.S. Department of Health and
Human Services*

**Katharyn A. Grant**
*U.S. Department of Health and
Human Services*

## 23.1  Introduction

### Scope

This chapter will address occupational **surveillance**\* as it relates to the practice of industrial ergonomics and the prevention of work-related musculoskeletal disorders. Surveillance is a "continuous analysis, interpretation, and feedback of systematically collected data, generally using methods distinguished by their practicality, uniformity, and frequently their rapidity, rather than by their accuracy or completeness" (Last, 1995). Occupational surveillance provides the data needed to identify, control, and prevent work-related injuries and illnesses. Epidemiologists also use surveillance data to study "the distribution and determinants of health-related states or events in defined populations." Methods for conducting occupational surveillance programs are well documented, and reviews of surveillance concepts and methods have been presented elsewhere (Baker et al., 1989; Baker and Matte, 1994; Greife et al., 1995). This chapter will limit the discussion of surveillance to those issues that affect the work of the industrial ergonomist and the prevention of musculoskeletal disorders.

---

\*This chapter contains a glossary of the terms in bold type.

## Purpose

The purpose of occupational surveillance is to track patterns of health and disease in groups of workers and to identify risk factors that influence these trends. Ultimately, this information should be used to direct the implementation of measures to prevent and/or control work-related disorders. Surveillance can answer questions such as, "Is there a problem?" or "Is there *still* a problem?" Establishing surveillance procedures is generally the first step in establishing an industrial ergonomics program. Surveillance serves to stimulate and focus prevention efforts and provides a method to assess the impact of corrective action.

## 23.2   Collecting, Analyzing, Intervening

Effective surveillance systems include the following components: (1) data collection, (2) analysis of the data, i.e., a mechanism to evaluate the meaning of the health/injury or hazard data, and (3) some action or response to ensure that surveillance activities are translated into preventive action. The response action may be directed toward individuals (e.g., providing medical treatment for symptomatic workers), or toward groups of workers (e.g., eliminating hazardous workplace conditions).

Several factors govern the strategy for collecting, analyzing, and acting upon surveillance data. These factors include the scope and urgency of the perceived problem, the resources available to the investigators, and the types of information systems already set in place. Many surveillance programs begin with efforts to document the number of work-related musculoskeletal disorders that have occurred in recent history (i.e., establish a baseline). Data for this effort can be obtained from a review of **existing records**. If existing records are incomplete or are unreliable as sources of surveillance information, or if additional information about current (rather than historical) conditions is needed, **questionnaires** or **worker surveys** may be used to determine how many workers are experiencing symptoms that could be caused by work activities. Additionally, because prevention of work-related musculoskeletal disorders depends on the identification and elimination of hazardous working conditions, **hazard surveys** or simple checklists can be used to identify jobs where risk factors for musculoskeletal disorders may be present.

For example, during an annual review of the OSHA 200 logs, an occupational nurse notices there has been a sudden increase in the number of workers reporting severe back strains on a new job. The surveillance plan would likely call for an immediate inspection of the worksite, along with interviews of workers and the supervisors to assist in identifying hazards. Once hazard control techniques had been identified and implemented, worker and hazard surveys would be administered at periodic intervals to determine the long-term effectiveness of the intervention.

Additional information about each of these data sources (existing records, worker surveys, and hazard surveys) will be provided later in this chapter.

## Establishing Definitions and Criteria

Before data can be collected and analyzed, it must be defined. An obvious (and unfortunately common) problem in the surveillance of work-related musculoskeletal disorders is the lack of consistency or standardization in the way these disorders are defined. The resulting confusion can lead to inaccuracies or inconsistencies in the data that make comparisons between locations and over time difficult if not impossible.

Defining what is meant by terms like "musculoskeletal disorder" or "ergonomic hazard" can be especially difficult. "Musculoskeletal disorders" encompass a broad spectrum of illnesses; in a recent review, NIOSH (1993) listed more than 160 different musculoskeletal conditions that can be caused or aggravated by various work activities. Diagnosis of these conditions is often complicated since many of these disorders have nonspecific symptom patterns, long latency periods and complex etiologies (Schierhout and Myers, 1996). For surveillance purposes, work-related musculoskeletal disorders have been defined from self-reported symptoms, clinical signs, specialized medical tests, impairment, or disability.

**TABLE 23.1** Examples of Case Definitions Used in Studies of Work-Related Musculoskeletal Disorders

| Disorder | Defined as: | Determined by: | Applied to: | Reference |
|---|---|---|---|---|
| Carpal tunnel syndrome (CTS) | Pain or numbness in the hands or wrists, pain in the hand or wrists that awakens at night, or tingling in the hands and fingers. | Questionnaire | Supermarket cashiers | Margolis and Kraus, 1987 |
| Cumulative trauma disorders (CTDs) | Inflammation or irritation of joints, tendons, or muscles (excluding strains, sprains, or dislocations), resulting from overexertion or nonimpact repetitive motion, occurring over a protracted or unknown period of time. | Review of workers' compensation records | Industrial workers in Ohio | Tanaka et al., 1988 |
| Work-related carpal tunnel syndrome | Combination of 1. symptoms affecting the median nerve distribution of the hand (e.g., pain, paresthesia, numbness), 2. objective physical or electro-diagnostic findings suggestive of CTS, and 3. history of employment in a job involving frequent or repetitive hand movements, forceful hand exertions, awkward hand postures, use of vibrating tools, or prolonged pressure over the wrist or palm of the hand. | Clinical history and physical examination | Patients referred for neurophysio-logic testing | Katz et al., 1991 |
| Shoulder tendinitis | Pronounced palpable pain of the muscle attachment or pronounced pain reaction to isometric contraction in any of the rotator cuff or biceps muscles. | Clinical examination | Construction workers | Stenlund et al., 1993 |
| Work-related musculoskeletal disorders | Symptoms of pain, aching, stiffness, numbness, tingling, or burning: 1. Not due to accident or sudden trauma; 2. Developed since working in current job; 3. Occurring within past year; 4. Lasting more than one week, or occurring at least once a month; 5. Reported as moderate or worse on a 5-point intensity scale. | Questionnaire | Newspaper workers | Bernard et al., 1994 |

Examples of **case definitions** used in various studies are found in Table 23.1. Two important considerations in the selection of a case definition are its **sensitivity** and **specificity**. The sensitivity of a case definition is the likelihood that the definition will identify diseased individuals. The specificity of a case definition is the probability that it will not label healthy individuals as diseased. In general, case definitions based solely on symptoms (e.g., pain, tingling in a particular body part) tend to be more sensitive but less specific than definitions that rely on physical findings or diagnostic tests; however, for surveillance purposes, it may be preferable to use definitions that are more inclusive or sensitive (Katz et al., 1991). In the face of uncertainty, the use of sensitive definitions will encourage investigators to examine suspect jobs and begin interventions at the earliest indication of a problem.

## Implementing Standardized Reporting Procedures

Providing information in an accessible and usable format is critical to the success of a surveillance effort. Unfortunately, the data needed for surveillance are usually not centralized in a single location. Performing a comprehensive analysis of costs associated with absenteeism, turnover, injury morbidity, compensation, lost productivity, and poor quality can require a search of numerous record systems, many of which employ their own coding procedures. Although relational database management systems can greatly simplify this process, their success depends on the establishment of common linkages between record systems. One approach employed by some plant medical departments and occupational health clinics is to use the coding system provided in the International Classification of Diseases (ICD). Such a coding system should lead to greater consistency in collecting, analyzing, and reporting health data.

## Calculating Rates and Percentages

Surveillance data are often reported as rates or percentages. Expressing the occurrence of injuries, illnesses, symptoms, complaints, etc. as a rate or percentage allows investigators to compare the occurrence across jobs/departments/plants that employ different numbers of workers and across periods of fluctuating employment. When the surveillance goal is to identify high-risk jobs that need attention, rates are most useful when they are computed for individual departments or even on similar jobs within a department.

The **incidence rate** is the rate of work-related musculoskeletal disorders appearing for the first time during a specific period (usually a year). The value commonly used in occupational health is the number of illnesses per 100 full-time workers per year and is calculated as follows:

$$\frac{\text{\# new cases during the past 12 months} \times 200,000 \text{ hours}}{\text{\# work hours during the past 12 months}}$$

The information needed to compute the denominators for the rates can usually be obtained from personnel or payroll records. The assumption is that each employee works 2000 hours per year (8 hours a day, 5 days a week, 50 weeks a year). If the number of hours worked in the past 12 months is not known, it can be estimated by multiplying the number of full-time equivalent workers employed in the job, department, or plant by 2000 hours.

The **prevalence rate** is the percentage of all cases of disease/symptoms/complaints at a specific instance in time, regardless of when they first appeared. It is calculated as follows:

$$\frac{\text{total \# cases at a given point in time}}{\text{\# workers at the same point in time}}$$

The **severity index** uses the number of lost or restricted workdays due to illness as a surrogate for the seriousness of the disorder. The severity index is calculated as follows:

$$\frac{\text{total \# lost or restricted workdays in the past 12 months} \times 200,000 \text{ hours}}{\text{\# work hours during the past 12 months}}$$

The magnitude of the severity index can be influenced by medical treatment practices, the health benefits available to employees, and the opportunity for transfer to jobs that are less stressful. The severity index can be skewed by unusually long illnesses suffered by a few employees.

## Interpreting Incidence, Prevalence, and Severity

Although closely related, incidence and prevalence rates convey somewhat different information. Prevalence depends on the incidence and the duration of the disease from its onset to its resolution. For example, if the incidence of back pain is low, but recovery is slow, the prevalence will be high relative to the incidence. Conversely, if the incidence of musculoskeletal disorders among workers is high, the prevalence may be low relative to the incidence if workers recover quickly, or if they leave the workforce because of their condition (Hennekens and Buring, 1987).

From a surveillance perspective, incidence and prevalence rates serve as valuable *prevention tools* for guiding interventions to mitigate hazardous workplace conditions. Incidence and prevalence rates, however, are also being used as *management tools* to gauge the performance of supervisors and health and safety staff who are held responsible for workplace injuries and illnesses. Such measures, when used as an indicator of performance or compliance, can inadvertently penalize employers or safety and health personnel who have recently introduced an ergonomics program. Typically, the incidence and prevalence rates will initially increase in organizations in which ergonomics programs are implemented. This is often due to an increase in training-related awareness by the employees of the work-relatedness of their conditions and represents better reporting rather than a true increase in the disorders. Moreover, the goal of an ergonomics program is to encourage employees to report problems early to allow immediate treatment and intervention, and thus reduce lost time and the risk of permanent injury. This goal may be undermined when incidence and prevalence rates are used as the sole measure to evaluate the effectiveness of an ergonomics program. For this purpose, the severity index may provide a more appropriate yardstick, since it reflects failures in early detection and prevention that represent real costs to the company. These failures are evident in the records of workers whose injuries have gone undetected, unrecognized, or unreported until the injuries reach a level of severity where restricted duty or lost time result.

## Establishing Priorities for Intervention

Ultimately, surveillance should direct the allocation of resources toward groups at highest risk for musculoskeletal injury. A common question that arises in discussions of surveillance is "When does surveillance data become compelling?" i.e., when is there enough information to warrant intervention? Some suggest that even a single reported musculoskeletal disorder should be sufficient cause to trigger more focused evaluations of workplace conditions (ANSI, 1996). This recommendation is based on the recognition that formal surveillance activities usually detect only a small proportion of the musculoskeletal problems and that one reported case may lead to several times as many unreported cases.

If multiple problem areas are identified and resources are constrained, it is often necessary to rank order jobs for further analysis and intervention. Jobs can be ranked by incidence, prevalence, or hazard severity; many survey instruments are designed to provide guidance for directing intervention efforts. For example, a checklist developed for the automobile industry uses a series of questions and a three-level scale to rate the postural stress associated with a specific job (Keyserling et al., 1992; 1993). Jobs that receive one or more "stars" (indicating significant exposure to postural stress) are considered to have priority for additional investigation. Giving precedence to jobs that employ many people, or jobs where major changes are already planned, can also be a sensible and cost-effective approach.

## 23.3   Data Sources

Data sources for conducting occupational surveillance can be conveniently grouped into three classifications: (1) existing records, (2) worker surveys, and (3) hazard surveys.

## Existing Records

Record-based surveillance involves reviewing and analyzing existing records or data systems for evidence of work-related musculoskeletal disorders. Because of its availability, these data can provide an initial gauge of the status of workers' health without a substantial investment of time or labor. Potential information sources include Occupational Safety and Health Administration (OSHA) 200 logs, on- and off-site medical records, workers' compensation records, and insurance claim data. Other records that can provide helpful information include absentee records, job transfer applications, employee grievances, or job satisfaction surveys. Although this information has been described as "passive surveillance data" (Fine et al., 1984), this description is not meant to minimize the importance of reviewing available records. Record reviewing is far from a passive endeavor, requiring great diligence in interpreting and coding the information in a consistent manner. Rather, the term serves to differentiate record reviewing from the "active" process of generating data of interest using targeted surveys.

*OSHA Records.* OSHA requires most employers to maintain records of work-related injuries and illnesses. Since 1978, the standard form for keeping these records has been the *Log and Summary of Occupational Injuries and Illnesses* (OSHA No. 200). In February 1996, OSHA proposed changes in record keeping requirements, including replacement of the OSHA 200 form with a new recording form, known as OSHA No. 300. These changes had not been adopted at the time of this writing. Whether or not new requirements are enacted, OSHA-required records should be maintained at each facility or establishment where work activities are performed.

Under OSHA record keeping guidelines, musculoskeletal disorders must be recorded if (1) they were caused or exacerbated by work activities, and (2a) there is at least one physical finding (e.g., positive Tinel's or Phalen's test, swelling, loss of motion), or (2b) there is at least one subjective symptom (e.g., pain, numbness, burning) that resulted in medical treatment, lost or restricted workdays, or transfer/rotation to another job (OSHA, 1990). Determination of these conditions may be made by a physician, nurse, or other health care provider. Musculoskeletal conditions are generally recorded on the existing OSHA 200 form as an occupational illness under column "7f" ("disorders associated with repeated trauma"). These disorders are caused, aggravated, or precipitated by repeated motion, vibration, or pressure.

A review of OSHA logs will generally yield a count of the number of musculoskeletal disorders recorded within a given time frame. Although almost all worksites keep an OSHA log, its utility for musculoskeletal disorder surveillance can be limited. The form does not require a detailed description of the worker's job or the disorder, making it difficult to determine the exact cause and nature of the injury. Also, because most musculoskeletal problems tend to develop over time (i.e., do not result from a specific event), workers or medical personnel may not realize that these disorders are work-related, and hence, recordable.

The OSHA logs do provide a convenient basis for making internal comparisons within a company. The rates of injuries and illnesses can be compared over time and at different sites to assist in determining trends in injuries and illnesses. An industry can also use the data to compare itself against a national benchmark, such as the experience of other companies in the same Standard Industrial Classification (SIC).

For example, Company XYZ, a nursing home provider, employed 250 workers in 1991. This same year, workers reported 20 injuries and illnesses that resulted in a total of 100 lost workdays. Using payroll records, the company determined that employees had worked 450,000 hours during this period. They calculated an injury severity rate of 42.1 lost workdays per 100 full-time employees. Although the severity rate is high, this rate is less than the national severity rate of 61.2 for companies in SIC 805, individual and family services.

To find the national rate for the company's SIC code, refer to the Department of Labor's *Occupational Injuries and Illnesses in the United States by Industry*, available from the U.S. Government Printing Office each spring.

*On-and Off-Site Medical Records.* Some companies maintain an onsite health clinic; others contract with external medical providers to provide care to workers injured at the job site. In either case, reports from these services, whether they are first-aid reports, dispensary logs, or employee medical records may provide useful information about potential work-related musculoskeletal disorders. At a minimum, these records may supplement information contained in the OSHA log.

For example, a review of the dispensary records at a poultry processing plant revealed no OSHA recordable injuries among workers on the trimming line in the previous year. However, the first aid medical reports showed that several employees who worked on that line on a daily basis requested anti-inflammatory medications and ice packs.

The utility of medical records for surveillance purposes can vary — they may or may not describe the reason for the visit or the condition underlying the prescribed treatment. Also, unless the information is routinely summarized, reviewing many medical records can be highly inefficient. Because medical records can contain sensitive information about employees, their contents must be treated confidentially. Routine access should be limited to health care personnel and public health agencies. Others should not have access without consent of the affected individuals.

*Workers' Compensation Records.* In many states, companies are required to obtain workers' compensation insurance to cover the medical and indemnity costs of employees who sustain injuries arising out of, and in the course and scope of their work. Where they exist, workers' compensation records can provide valuable information about the direct costs (medical and indemnity) associated with work-related musculoskeletal disorders. These costs include (1) payments made to outside hospitals, clinics, physicians, and other licensed medical personnel for the diagnosis, treatment (including surgery), and rehabilitation of the injury, (2) payments made to the injured worker as compensation for lost work time, and (3) payments made as a lump sum settlement for permanent disability. The actual cost of musculoskeletal disorders is often higher than those covered by workers' compensation insurance. For example, the cost of medical treatments rendered directly by the employer or charged to the employee's health plan is not included. As a result, the total financial burden of these disorders to a company's balance sheet often goes unrecognized and unaddressed.

Workers' compensation claims are filed under specific rules and regulations that vary from state to state. Each workers' compensation law specifies what, how, and when work-related injuries and illnesses must be reported. To be covered, the worker must first choose to report the injury. Once the worker files a claim, there are issues of eligibility that govern whether the worker will receive benefits. Surveillance data based on the number of claims filed will differ from those based on the number of claims paid. As a result, workers' compensation data will often underestimate the true rates of work-related injuries and illnesses (Schwartz, 1987).

Misclassification and coding errors are also serious problems in using workers' compensation data. In an examination of workers' compensation data in Washington State, Franklin et al. (1991) found that only 72% of claims for carpal tunnel surgery were given an ICD code compatible with carpal tunnel syndrome.

Despite these limitations, workers' compensation data provide some significant advantages as a source of surveillance information. First, all records in the data set relate to conditions of suspected occupational origin. Also, workers' compensation records usually describe the circumstances of the disorder in a way that provides an understanding of the cause of the condition (Baker and Matte, 1994). If case identification leads to improvement of workplace conditions, the cost of compensation should be greatly reduced in future years.

*Payroll Records.* Payroll records are useful from two standpoints. First, payroll records can be used to determine the number of hours worked by employees on a particular job. This value is often used as a crude measure of employees' exposures to job hazards and is required for the incidence/prevalence/severity rate calculations described previously in this chapter. Second, payroll records can be used to identify jobs or departments where absenteeism or turnover is high. Although high turnover can result from several causes, physical stress is a common reason for leaving a job. If less stressful jobs are available, workers may choose to quit or bid out of their present, more stressful jobs. In a study by Lavender and Marras (1994), high rates of job turnover were identified as a useful indicator of jobs that posed a risk of overexertion injury leading to low back disorders. Although turnover can be affected by factors beyond the physical hazards posed by the job (e.g., the availability of higher paying positions, psychosocial factors), the results suggest that the sensitivity of surveillance programs based on existing records can be improved by using turnover rates to supplement injury rate data.

*Summary.* The quality and utility of existing record systems for surveillance purposes can vary. Musculoskeletal disorders often go unreported, and depending on the data source, striking differences in the incidence of work-related musculoskeletal disorders can be found (CDC, 1989). Even when records are complete, a lag will often exist between the appearance of a hazard and the onset of injury; therefore, records may not give an accurate picture of the current situation. Finally, linking injury/disease data with exposure to a hazard can be especially challenging. Job titles are often poor indicators of exposure to risk factors for musculoskeletal disorders. Additional data needed to link disorders to specific tasks or job processes may not exist.

## Worker Surveys

Most work-related musculoskeletal disorders produce symptoms of pain or discomfort. Likewise, workers can provide valuable information about job attributes that cause fatigue or pain. Therefore, one of the most direct and effective methods for collecting surveillance data is to administer questionnaires or other surveys to workers. These techniques are sometimes called "active surveillance" measures because the investigator plays an active role in soliciting and collecting information specifically for surveillance purposes. (In traditional public health surveillance, e.g., for infectious diseases such as tuberculosis, "active surveillance" means going out to hospitals and clinics to review patient records to detect cases. To avoid confusion, we do not use the term "active surveillance" here.)

Worker surveys can take several forms. They can be lengthy or quite short; they can be oral (i.e., administered by an interviewer) or written; and they may rely heavily on pictures or charts. Examples of worker surveys include the "body part discomfort scale" (Corlett and Bishop, 1976), the *Nordic Musculoskeletal Questionnaire* (Kuorinka et al., 1987), the *NIOSH Health Hazard Upper Extremity Questionnaire* (Hales et al., 1992), and the computerized "discomfort assessment system" (Saldana et al., 1994). Common features of surveys include the following:

- Use of body part diagrams, where workers can indicate the location of pain or other symptoms;
- Questions about the onset and duration of the symptoms;
- Questions about the nature of job activities;
- Use of numerical rating scales to indicate the severity of pain, fatigue, or discomfort.

Worker surveys are sometimes underrated as useful sources of surveillance data. A main reason stems from concerns that workers either may under- or over-report their symptoms. Inaccurate reporting may be more of an indication of a poorly worded survey, a lack of understanding, fear of job loss or recrimination, or simply a stoic attitude about discomfort. In situations in which worker surveys yield evidence of over-reporting of an occupational problem, investigators should not dismiss the findings until they are certain they understand the reason for their findings. In such cases organizational problems may be intertwined with safety and health problems (Schierhout and Myers, 1996).

To avoid concerns about reporting biases in worker surveys, some investigators have combined questionnaires with physical examinations of workers to identify musculoskeletal conditions (Bernard et al., 1994; Hales et al., 1994). These studies show that while symptoms of discomfort or pain may not always reflect an underlying pathology, the vast majority of employees with moderate to severe musculoskeletal symptoms have at least one positive physical finding on a concurrent physical examination (Baron et al., 1992).

Worker surveys should be conducted (1) when there is evidence from any data source of increased musculoskeletal injury or illness in the facility, (2) when new jobs or tasks are begun, or (3) when employees change jobs. The latter two surveys provide useful baseline data for determining the impact of the change on employee health. For example, one method to evaluate the effectiveness of a redesigned workstation would be to compare the pattern of shoulder, neck, and back discomfort recorded before the intervention with the pattern of discomfort for those same body segments after the improvements have been installed.

The advantages of using worker surveys as a source of information include the following: (1) The investigator can exert more control over the data collected. Once the investigator has decided what questions need to be answered, he or she can select the survey items that will elicit the most complete information that is needed. Because the investigator has control over the survey, there is also less chance that the information will change or become biased between administrations of the survey. (2) Surveys are easy to administer in the field — workers can complete the surveys at their convenience, and responses can be kept anonymous. These features can encourage high participation rates and candid responses, although the opportunities for follow-up surveillance are more limited. (3) Worker surveys can be readministered periodically to allow for early recognition of musculoskeletal disorders.

Worker surveys also have some inherent limitations. First, the time and resources required to develop and administer surveys may make them more costly than surveillance activities that rely on existing records. Second, the survey information may be unreliable if there is a lack of trust between management and workers. Third, the design of the questionnaire can have a dramatic effect on the quality of the responses and the number of workers who complete the survey. Key design issues include the length of the questionnaire, the phrasing of the questions, and method of administration. In a diverse or multilanguage workforce, translating the questionnaires into several languages may be necessary. Finally, responses can be influenced by the workers' expectations about how much effort is required to perform a certain job. For example, a young warehouse worker may rate his job as only moderately strenuous, whereas an older worker using the same scale in a busy office environment may perceive his work to be very stressful in terms of work demands.

## Hazard (Risk Factor) Surveys

Ideally, actions to prevent work-related musculoskeletal disorders should proceed before injuries and/or symptoms develop. In the last decade, a significant amount of research has been undertaken to improve our understanding of the risk factors that lead to the development of work-related musculoskeletal disorders (Gerr et al., 1991; Rempel et al., 1992). The process of examining jobs for these risk factors is known as hazard surveillance. Even without clear medical evidence that musculoskeletal problems exist, hazard surveillance activities can provide the data needed to begin an effective primary prevention program.

Because surveillance for ergonomic hazards is a relatively new notion, few industries maintain records that explicitly identify hazards for each job. However, many companies maintain job descriptions that identify the skills or abilities required to perform jobs in their facilities. In recent years, many industries have updated their job descriptions to include statements defining "Essential Job Functions" to comply with the Americans with Disabilities Act (ADA). According to the ADA, the descriptions should identify essential functions, or fundamental job duties, and the physical and mental abilities needed to perform these functions. If functional job descriptions are available, they can provide useful information for identifying potentially stressful jobs or jobs requiring unique skills or special endurance.

For example, a job description in a large assembly plant may list the ability to perform manual lifts, assemble parts, and use an impact wrench as requirements. Based on this description, an investigator could infer that the job might require high force exertions, awkward postures, repetitive hand movements, and exposure to segmental hand vibration.

Whether or not job descriptions exist, hazard surveillance efforts depend heavily on walk-through surveys. Investigators need to observe job activities, speak with workers and supervisors to obtain job information not apparent from observation, and use checklists to score job features against a listing of risk factors. The walk-through survey is differentiated from more formalized job analysis efforts by the amount of detailed information collected. The purpose of the walk-through is to identify risk factors that might otherwise go unnoticed and provide additional basis for prioritizing jobs for further evaluation.

Examples of checklists that might be used for hazard surveillance are provided in Tables 23.2, 23.3, and 23.4. Although most of the checklists are designed for use by non-experts, some minimal level of training is usually needed to use checklists properly. Hazard checklists can vary in length and in scope. At one extreme, there are "generic" checklists that are widely applicable, i.e., for use on nearly all jobs in all industries (OSHA, 1995). These checklists can be contrasted with more focused checklists, designed to evaluate a specific job or industry, e.g., carpenters or nurses' aides (Bhattacharya, 1992; Engkvist et al., 1995). While some checklists are intended to serve primarily as mnemonics (i.e., to remind users to evaluate a particular job characteristic), other checklists incorporate a scoring system for indicating the risk associated with a particular job or process. Attempts to validate various checklists as scoring instruments have yielded mixed results (Lifshitz and Armstrong, 1986; Keyserling et al., 1992; 1993; Shierhout et al., 1994; Kemmlert, 1995).

Hazard surveys should be administered (1) whenever a job, task, or process is changed substantially, (2) when new jobs are introduced, and (3) periodically (especially after new cases of musculoskeletal disorders are reported) to detect whether trends exist across jobs that use similar equipment, tools, or processes (ANSI, 1996). Hazard surveys can also be incorporated into regular safety and health inspections at the facility, expanding the scope of these inspections to include identification of musculoskeletal disorder risk factors.

**TABLE 23.2**    Ergonomic Hazard Identification Checklist*

[Answer Questions Based on the Primary Job Activities of Workers in Facility]

Never — Worker is never exposed to the condition

Sometimes — Worker is exposed to the condition less than 3 times daily.

Usually — Worker is exposed to the condition 3 times or more daily.

| | Never | Some < 3 | Usually > 3 | If USUALLY, list jobs to which answer applies here |
|---|---|---|---|---|
| 1. Do workers perform tasks that are externally paced? | | | | |
| 2. Are workers required to exert force with their hands (e.g., gripping, pulling, pinching)? | | | | |
| 3. Do workers use hand tools or handle parts or objects? | | | | |
| 4. Do workers stand continuously for periods of more than 20 mins? | | | | |
| 5. Do workers sit for periods of more than 30 mins without a chance to stand or move around freely? | | | | |
| 6. Do workers use keyboards, mice, joysticks, etc. for continuous periods of more than 30 mins? | | | | |
| 7. Do workers kneel for more than 5 min (one or both knees)? | | | | |
| 8. Do workers perform activities with hands raised above shoulder height? | | | | |
| 9. Do workers perform activities while bending or twisting at the waist? | | | | |
| 10. Are workers exposed to vibrations? | | | | |
| 11. Do workers lift or lower objects between floor and waist height, or above shoulders? | | | | |
| 12. Do workers lift or lower objects more than once/min. for continuous periods of more than 15 minutes? | | | | |
| 13. Do workers lift, lower, or carry objects weighing >12 lbs that are not held close to body? | | | | |
| 14. Do workers lift, lower or carry objects weighing more than 50 lbs.? | | | | |

*Developed by Grant, Habes, Fernandez, and Putz-Anderson, NIOSH, Cincinnati, Ohio, 1994.
Any response of: *"Usually"* to Questions 1-14 = Potential Risk Present: Follow-up

## 23.4   Conclusions

Surveillance is essential to the prevention/control of musculoskeletal disorders in the workplace. Unfortunately, surveillance efforts will ultimately be wasteful and ineffective if there are major breakdowns in any of the three major components: data, analysis, or intervention. In general, no single data source provides enough information to direct a program for preventing work-related musculoskeletal disorders. Therefore, effective surveillance programs make use of multiple data sources to identify problem areas and determine intervention priorities. Even in the absence of health data, hazard surveys conducted in workplaces where there are significant or well-defined hazards can provide the data needed to mount an effective primary prevention program for work-related musculoskeletal disorders.

    Once established, surveillance should become an ongoing process. As corrective actions are taken, surveillance data can provide the information needed to show the beneficial effects of these efforts. By integrating surveillance efforts with existing quality assurance and cost containment programs, their utilization and success will be maximized.

**TABLE 23.3** Manual Material Handling Checklist for Lifting, Carrying, Pushing, or Pulling*

| Risk Factors | | Yes | No |
|---|---|---|---|
| 1. General | | | |
| 1.1 | Does the load handled exceed 50 lbs.? | [ ] | [ ] |
| 1.2 | Is the object difficult to bring close to the body because of its size, bulk, or shape? | [ ] | [ ] |
| 1.3 | Is the load hard to handle because it lacks handles or cutouts for handles, or does it have slippery surfaces or sharp edges? | [ ] | [ ] |
| 1.4 | Is the footing unsafe? For example, are the floors slippery, inclined, or uneven? | [ ] | [ ] |
| 1.5 | Does the task require fast movement, such as throwing, swinging, or rapid walking? | [ ] | [ ] |
| 1.6 | Does the task require stressful body postures, such as stooping to the floor, twisting, reaching overhead, or excessive lateral bending? | [ ] | [ ] |
| 1.7 | Is most of the load handled by only one hand, arm, or shoulder? | [ ] | [ ] |
| 1.8 | Does the task require working in environmental hazards, such as extreme temperatures, noise, vibration, lighting, or airborne contaminants? | [ ] | [ ] |
| 1.9 | Does the task require working in a confined area? | [ ] | [ ] |
| 2. Specific | | | |
| 2.1 | Does lifting frequency exceed 5 lifts per minute? | [ ] | [ ] |
| 2.2 | Does the vertical lifting distance exceed 3 feet? | [ ] | [ ] |
| 2.3 | Do carries last longer than 1 minute? | [ ] | [ ] |
| 2.4 | Do tasks which require large sustained pushing or pulling forces exceed 30 seconds duration? | [ ] | [ ] |
| 2.5 | Do extended reach static holding tasks exceed 1 minute? | [ ] | [ ] |

Comment: "Yes" responses are indicative of conditions that pose a risk of developing low back pain. The larger the percentage of "yes" responses, the greater the possible risk.

* Developed by Thomas Waters, Ph.D., CPE, NIOSH, Cincinnati, Ohio, 1994.

**TABLE 23.4** Physical Exertion Questionnaire

The purpose of this questionnaire is to assess the amount of exertion you use in your job. Exertion is defined by (1) force, (2) repetitive motion, and (3) whole-body activity. The following pages contain three lists of activities, one for force, one for repetition, and one for whole-body activity. Use the activities in these lists to estimate the amount of exertion you use in your job.

To complete the questionnaire, read the activities on each page. Then compare the amount of force, whole-body activity, or repetitive motion (according to the particular list) the activities require, with the amount of force, whole-body activity, or repetition your work activities require. If you don't spend much work time doing **exactly** these activities, you may spend time doing activities that require **about the same amount** of exertion as the activities in the lists. Read the activities and think about an average workday. Then, in the boxes beside the activities, write in the number of hours and minutes during a typical workday, on the average, you spend doing activities that require about the same amount of exertion as those at each intensity level. **The total amount of time on each page should add up to about the length of your average workday.** Some of the intensity levels do not have activities to describe them. These levels stand for an amount of exertion that is halfway between the activities above and below. Write in the amount of time on those lines also. Think carefully about the amount of time you spend in low intensity activities, which may be more difficult to remember than high intensity activities.

As an example of completing the questionnaire, how much time during an average workday do you spend in activities that require about the same amount of whole-body activity as running up stairs? If you usually spend about 4 hours and 30 minutes during an average day, write this in the box beside the activity. Then estimate the amount of time you spend doing activities at the other intensity levels and write the time in the boxes. Please do not leave any of the boxes empty. Put a 0 in the box if you don't spend time doing activities at that intensity level.

**Remember:** The activities are only **examples** of amounts of exertion. The question is "Do you use the same amount of exertion as required by the different activities and not whether you perform the activities listed." The activities are only a guide for you to help you estimate your level of exertion, relative to the listed activity.

### Whole-Body Activity Scale

On this page, you will indicate how much whole-body activity your job requires. Whole-body activity is any activity that increases heart rate and breathing, and involves movement mostly of the legs, such as pedaling a bicycle, or movement of the entire body. The activities below require whole-body activity; those near the top of the list require more whole-body activity than those near the bottom.

TABLE 23.4 (continued)    Physical Exertion Questionnaire

Compare your work activities to the activities below in terms of how much whole-body activity your job requires. Then decide how much time on the average, you spend doing activities that require about the same amount of whole-body activity as those at each intensity level. *To help you estimate the intensity of your whole-body activity, think about how your body feels when you perform the activities in the scale, and compare this to the way your body feels when you perform your work activities.*

Remember: The activities below are only examples of amounts of whole-body activity for you to compare with your work activities. We are NOT interested in whether you perform the specific activities listed in the scale. Also, think only about the amount of whole-body activity used to perform these activities, and not about the amount of force or repetitive motion the activities require.

| Duration | Level | Equivalent Amount of Whole-Body Activity |
|---|---|---|
| | 6 | Running up stairs |
| | 5 | Between Level 6 and Level 4 |
| | 4 | Climbing a vertical ladder |
| | 3 | Carrying boxes or packages from your house/apt to your car |
| | 2 | Making a bed (straightening the sheets and blankets) |
| | 1 | Rolling over from back to stomach while lying in bed |
| | 0 | No whole-body activity |

**The total number of hours and minutes in the table should equal about the length of your average workday.**

### Repetitive Motion Scale

In this table, we are interested in how much repetitive motion your job requires. Repetitive motion is movement that requires use of the same muscles and body parts to perform the same movements or sequence of movements over and over.

The activities below require repetitive motion. Those near the top of the list require more repetitive motion, meaning that the movements or sequence of movements take only a short time to complete before being repeated. Activities near the bottom of the list require less repetitive motion, and it takes longer before the same movement is repeated.

Please read the activities below and think about your **average** workday. Then write in the number of hours and minutes during a typical workday, **on the average**, you spend doing work activities that require about the same amount of repetitive motion as those at each intensity level. Please remember: The activities below are only examples of repetitive motion to be compared with your work activities. We are NOT interested in whether you perform the specific activities listed in the scale. Also, think only about the amount of repetitive motion used to perform these activities, and not about the amount of whole-body activity or force the activities require.

| Duration | Level | Equivalent Amount of Repetitive Motion |
|---|---|---|
| | 6 | Stirring water with a spoon<br>Manually grating cheddar cheese using a cheese grater |
| | 5 | Between Level 6 and Level 4 |
| | 4 | Hammering nails into soft wood<br>Cleaning windows/mirrors with a cloth or paper towel |
| | 3 | Between Level 4 and Level 2 |
| | 2 | Taking notes using a pen/pencil and paper<br>Removing groceries from a paper grocery bag<br>Answering telephones and writing messages |
| | 1 | Between Level 2 and Level 0 |
| | 0 | No repetitive motion |

**The total number of hours and minutes in the table should equal about the length of your average workday.**

TABLE 23.4 (continued)    Physical Exertion Questionnaire

### Force Scale

On this page, you will indicate how much muscle force your job requires. In the boxes below enter the number of hours and minutes during a typical workday, on the average, you spend doing activities that require about the same amount of force as those at each intensity level.

The activities near the top of the list require more force, and activities near the bottom of the list require less force.

*To help you estimate the amount of force you use in your job, think about how your body feels when you perform the activities in the scale, and compare this to the way your body feels when you perform your work activities. Remember:* The activities are only examples of amounts of force. We are NOT interested in whether you perform the specific activities listed in the scale. Also, think only about the amount of force used to perform these activities, and not about the amount of whole-body activity or repetitive motion the activities require.

| Duration | Level | Equivalent Amount of Force |
|---|---|---|
| | | Pushing a refrigerator on a flat, smooth surface, such as tile, linoleum, or a wood floor (the refrigerator is not on wheels, a cart, or a dolly) |
| | | Between Level 6 and Level 4 |
| | | Scooping hard, frozen ice cream out of a container<br>Opening a new jar (a jar that has never been opened) of pickles or jelly |
| | | Lifting a 12-pack of beer or pop using one hand<br>Unscrewing a bottle cap on a new bottle or container or pop<br>(a bottle or container of pop that has never been opened) |
| | | Crushing an aluminum soda can or beer can with one hand (crushing it side-to-side, not top-to-bottom)<br>Lifting a full gallon of milk<br>Opening a car door with one hand |
| | 1 | Pulling the end of a cord out of an electrical wall outlet<br>Lifting a telephone receiver<br>Turning a doorknob |
| | 0 | Lifting a quarter<br>Bending a piece of typing paper or notebook paper<br>Lifting a cottonball |

**The total number of hours and minutes in this table should equal about the length of our average workday.**

Cole, L.L. (1996). *Construction and Validation of a Musculoskeletal Risk Questionnaire.* Unpublished doctoral dissertation, University of Cincinnati, Cincinnati, Ohio.

## Defining Terms

**Case definition:** Set of decision-making criteria, intended to assist health care providers in identifying work-related injuries and illnesses so that further investigation and preventive activities can be initiated.

**Existing records:** Records created for other purposes that may be useful in surveillance efforts. These records can include OSHA 200 logs, medical or health care records, workers' compensation claims, payroll records, sickness and accident reports, etc.

**Hazard surveys:** Assessments of jobs, workplaces, or processes for the purpose of identifying risk factors (e.g., biomechanical stress, vibration) that can lead to development of musculoskeletal disorders. Can be used to identify intervention targets before injuries or diseases have occurred.

**Incidence rate:** The rate of work-related musculoskeletal disorders appearing for the first time during a specific period.

**Prevalence rate:** The percentage of all cases of disease/symptoms/complaints at a given point in time.

**Sensitivity:** The probability that a case definition or screening procedure will identify individuals with the condition (disease) of interest.

**Severity index:** The rate of lost or restricted workdays due to musculoskeletal illness occurring within a specific period.

**Specificity:** The probability that a case definition or screening procedure will not label a healthy individual as diseased.

**Surveillance:** The continuous analysis, interpretation, and feedback of systematically collected data, essential to the planning, implementation and evaluation of occupational safety and health programs. Surveillance methods are often distinguished by their practicality, uniformity, and rapidity, rather than by their accuracy or completeness.

**Worker surveys:** Questionnaires and interview procedures developed to solicit information about signs, symptoms, and risk factors for musculoskeletal disorders from workers.

# References

ANSI, 1996. *ANSI Z-365 Control of Work-Related Cumulative Trauma Disorders, Part 1: Upper Extremities* (Working Draft, January 1, 1996). National Safety Council, Itasca, Illinois.

Baker, E.L., Honchar, P.A., and Fine, L.J. 1989. Surveillance in occupational illness and injury: concepts and contents. *Am. J. Public Health* 79(suppl):9-11.

Baker, E.L. and Matte, T.P. 1994. Surveillance for occupational hazards and disease, in *Textbook of Clinical Occupational and Environmental Medicine*, Eds. L. Rosenstock and M.R. Cullen, p. 61-67. W.B. Saunders Company, Philadelphia.

Baron, S., Hales, T., and Fine, L. 1992. Evaluation of a questionnaire to assess the prevalence of work-related musculoskeletal disorders. *Arbete Och Halsa* 17:39-41.

Bhattacharya, A. 1992. *Walkthrough Ergonomics Checklist for Carpentry Tasks*. Greater Cincinnati Occupational Health Center, Cincinnati, Ohio.

Bernard, B., Sauter, S., Fine, L., Petersen, M., and Hales, T. 1994. Job task and psychosocial risk factors for work-related musculoskeletal disorders among newspaper reporters. *Scand. J. Work Environ. Health* 20:417-426.

Centers for Disease Control, 1989. Occupational disease surveillance — carpal tunnel syndrome. *MMWR* 38:485-489.

Corlett, E.N. and Bishop, R.P. 1976. A technique for assessing postural discomfort. *Ergonomics* 19(2):175-182.

Engkvist, I.L., Hagberg, M., Wigaeus-Hjelm, E., Menckel, E., and Ekenvall, L. 1995. Interview protocols and ergonomics checklist for analyzing overexertion back accidents among nursing personnel. *Appl. Ergonom.* 26(3):213-220.

Fine, L.J., Silverstein, B.A., Armstrong, T.J., Anderson, C.A., and Sugano, D.S. 1986. Detection of cumulative trauma disorders of the upper extremity in the workplace. *J. Occup. Med.* 28:674-678.

Franklin, G.M., Haug, J., Heyer, N., Checkoway, H., and Peck, N. 1991. Occupational carpal tunnel syndrome in Washington State, 1984-1988. *Am. J. Public Health* 81:741-746.

Gerr, F., Letz, R., and Landrigan, P.J. 1991. Upper-extremity musculoskeletal disorders of occupational origin. *Annu. Rev. Publ. Health* 12:543-566.

Greife, A., Halperin, W., Groce, D., O'Brien, D., Pedersen, D., Myers, J.R., and Jenkins, L. 1995. Hazard surveillance: its role in primary prevention of occupational disease and injury. *Appl. Occup. Environ. Hyg.* 10(9):737-742.

Hales, T., Sauter, S., Peterson, M., Fine, L., Putz-Anderson, V., Schleifer, L., Ochs, T., and Bernard, B. 1994. Musculoskeletal disorders among visual display terminal users in a telecommunications company. *Ergonomics* 37(10):1603-1621.

Hennekens, C.H. and Buring, J.E. 1987. *Epidemiology in Medicine*. Little, Brown and Company, Boston.

Katz, J.N., Larson, M.G., Fossel, A.H., and Liang, M.H. 1991. Validation of a surveillance case definition of carpal tunnel syndrome. *Am. J. Public Health* 81:189-193.

Kemmlert, K. 1995. A method assigned for the identification of ergonomic hazards — PLIBEL. *Appl. Ergonom.* 26(3):199-212.

Keyserling, W.M., Brouwer, M., and Silverstein, B.A. 1992. A checklist for evaluating risk factors resulting from awkward postures of the legs, trunk and neck. *International J. of Industrial Ergonomics* 9:282-301.

Keyserling, W.M., Stetson, D.S., Silverstein, B.A., and Brouwer, M.L. 1993. A checklist for evaluating ergonomic risk factors associated with upper extremity cumulative trauma disorders. *Ergonomics* 36(7):807-831.

Kuorinka, I., Jonsson, B., Kilbom, Å., Vinterberg, H., Biering-Sorensen, F., Andersson, G., and Jorgensen, K. 1987. Standardized Nordic questionnaire for the analysis of musculoskeletal symptoms. *Appl. Ergonom.* 18(3):233-237.

Last, J.M. (Ed.) 1995. *A Dictionary of Epidemiology,* 3rd ed. Oxford University Press, New York.

Lifshitz, Y. and Armstrong, T.J. 1986. A design checklist for control and prediction of cumulative trauma disorders in intensive manual jobs. *Proceedings of the Human Factors Society, 30th Annual Meeting.* Human Factors Society, Santa Monica, pp. 837-841.

Margolis, W. and Kraus, J.F. 1987. The prevalence of carpal tunnel syndrome symptoms in female supermarket cashiers. *J. Occup. Med.* 29(12):953-956.

NIOSH, 1993. Comments from the National Institute for Occupational Safety and Health on the Occupational Safety and Health Administration Proposed Rule on Ergonomic Safety and Health Management. 29 CFR Part 1910, Docket No. S-777.

OSHA, 1990. *Ergonomics Program Management Guidelines for Meatpacking Plants.* U.S. Department of Labor, Occupational Safety and Health Administration, Washington, D.C.

OSHA, 1995. OSHA Draft Proposed Ergonomic Protection Standard: Summaries, Explanations, Regulatory Text, Appendices A and B, March 20, 1995. *Occupational Safety and Health Reporter* 24(42):S3-S248.

Rempel, D.M., Harrison, R.J., and Barnhart, S. 1992. Work-related cumulative trauma disorders of the upper extremity. *JAMA* 267(6):838-842.

Saldana, N., Herrin, G.D., Armstrong, T.J., and Franzblau, A. 1994. A computerized method for assessment of musculoskeletal discomfort in the workforce: a tool for surveillance. *Ergonomics* 37(6):1097-1112.

Schierhout, G.H., Myers, J.E., and Bridger, R.S. 1993. Musculoskeletal pain and workplace ergonomic stressors in manufacturing industry in South Africa. *Int J. Ind. Ergon.* 12:3-11.

Schierhout, G.H. and Myers, J.E. 1996 Is self-reported pain an appropriate outcome measure in ergonomic-epidemiologic studies of work-related musculoskeletal disorders? *Am. J. Ind. Med.* 30:93-98.

Schwartz, E. 1987. Use of workers' compensation claims for surveillance of work-related illness — New Hampshire, January 1986-March 1987. *MMWR* 36(43):713-720.

Stenlund, B., Goldie, I., Hagberg, M., and Hogstedt, C. 1993. Shoulder tendinitis and its relation to heavy manual work and exposure to vibration. *Scand J. Work Environ. Health* 19:43-49.

Tanaka, S., Seligman, P., Halperin, W., Thun, M., Timbrook, C.L., and Wasil, J.J. 1988. Use of workers' compensation claims data for surveillance of cumulative trauma disorders. *J. Occup. Med.* 30(6):488-492.

# 24

# Injury Surveillance Database Systems

Carol Stuart-Buttle
*Stuart-Buttle Ergonomics*

## 24.1   Introduction

A database is a collection of organized, related data typically in electronic form that can be accessed and manipulated by computer software. When the interface between the data and user is developed beyond just managing the data, but with a particular function in mind, a database system evolves. Features of the interface may include data dictionaries, data security, statistical and graphical capabilities, and report writing.

Injury surveillance database systems focus on the collection of data that are pertinent to injuries and illnesses, the management of that data, and data interpretation. The goals of interpretation may include determining trends, promoting prompt treatment and directing prevention strategies. The general use of the term injury refers to all types of injuries and illnesses. Conducting epidemiological studies or other research that attempts to find causal relationships of injuries may be practical with large organizations but not for most of industry; therefore research use of databases are not discussed in this chapter. Injury surveillance databases and the relationship of surveillance to ergonomics are addressed from the practicing industrial safety and health professional's perspective.

Surveillance of the workplace for injuries and potential injuries is an important aspect of the ergonomics process as injuries are an indicator of poor design. Often, the effectiveness of ergonomic processes are measured by the rates of injury and lost workdays. Most employers are required to record work-related injuries and illnesses in accordance with the Occupational Safety and Health (OSH) Act (OSHA, 1970). As personal computers (PCs) have become commonplace in industry, the most basic information pertaining to the OSH Act can be easily computerized; in other words, a company database can be established. Considerable time and effort can be saved in the update and analysis of the data with the right software interface.

The rate of change and growth in the computer software market is so great that for many industries the task of exploring, evaluating, and choosing a database system for the company is formidable. This chapter does not attempt to predict the future of information systems, nor does it try to summarize the existing safety and health software since such information rapidly changes. However, some safety and health software directories are provided. The main points and discussion of this chapter focus on the fundamental components of injury surveillance as they relate to ergonomics and tips on how to assess commercial injury surveillance database systems. A section touches on the development of an injury surveillance database only to provide an opportunity to understand injury surveillance in the context of system development.

## 24.2  Injury Surveillance

### History

"Surveillance is the ongoing and systematic collection, analysis, and interpretation of data related to health." (Baker, 1989a). Surveillance historically began in the public health setting, as a method of watching out for certain serious diseases or illnesses such as syphilis and smallpox, so that isolation could be instituted at the first signs of symptoms. Since 1950 the term "surveillance" was formally applied to the systematic collection of relevant disease data with the purpose to improve the control of specific diseases. During the 1960s, surveillance in the U.S. expanded such that it applied not only to communicable diseases but also to noninfectious diseases including environmental and occupational hazards (Langmuir, 1976).

During the 1960s, some states developed reporting requirements for selected occupational diseases or used data from the workers' compensation system for surveillance. Not until the federal OSH Act of 1970 was there a standardized scheme to monitor injuries and illnesses (Wegman, 1985). It was not surprising that the chemical industry was a pioneering group in developing occupational surveillance systems, since there was much potential for disease. Hazard surveillance, as a means of predicting work-related health problems in the chemical industry, was developed to complement the basic surveillance of recording injuries and illnesses required by the OSH Act. Many companies then started hazard profiles that included information such as industry demographics (employment size in the geographic area), use of chemicals in the workplace, levels of the chemicals to which workers were exposed, and data on the dose–response relationship for each chemical (Froines, 1986).

The hazard surveillance model has been applied to ergonomics-related issues in the workplace. One problem in applying such a model to ergonomics is that suspected risk factors for musculoskeletal disorders are treated discretely without taking into account the interactions of the risk factors. The effects of the interactions and the dose–response relationships are still being discovered.

"Monitoring" is another term that refers to the methods to anticipate a disease or medical disorder before it occurs (Yodaiken, 1986). Monitoring complements medical surveillance in a way similar to that of hazard surveillance. The difference lies in semantics and the use of the word "hazard" which implies that the dose–response is known. As stated above, it is difficult to determine when one or more risk factors are hazardous for musculoskeletal disorders.

Monitoring may be considered as a more general term, one that refers to looking at the workplace in order to anticipate problems and prevent their occurrence. An illustration of the practical difference between medical surveillance and monitoring is the distinction between a symptom survey and discomfort survey (Stuart-Buttle, 1994). A symptom survey identifies the employees with physical problems through the collection of detailed information about symptoms. On the other hand, a discomfort survey attempts to find out the discomfort that employees are experiencing while performing their jobs. Discomfort indicates a physical stress at performing the job, not necessarily an active medical problem. The implication of a discomfort survey is that discomfort could be a precursor to injury, hence a discomfort survey is a monitoring method for determining where to focus prevention efforts.

Another common distinction of the term surveillance is whether it is "passive" or "active" (BNA, 1993a). Passive surveillance is the analysis of data from existing databases, such as the Occupational Safety and Health Administration (OSHA) 200 logs that record work-related injuries and illnesses, or workers' compensation records. Active surveillance is collecting data to "actively" seek information, such as using surveys to determine the physical conditions that individuals have not reported, or defining aspects of the jobs that may be risk factors for injury. The distinction between passive and active surveillance is almost the same as that between surveillance and monitoring, or medical surveillance and hazard surveillance.

## Goals and Objectives

Injury surveillance should entail both active and passive surveillance. Existing injury data as well as additional information collected about the workers and their jobs must be analyzed. The goals of a surveillance system include "the prevention of occupational disease *(injuries)* through control of the causative agents" (Sundin, 1986). Surveillance helps to direct the prevention programs to control or eliminate preventable disorders (Baker, 1989b).

The objectives to fulfill the goals of surveillance are to locate and monitor groups of workers who are exposed to risk, to determine the risk factors and take corrective action, or reduce exposure to those factors. Discovering the relationships between exposure and the injury is a research domain which may be undertaken by some corporations with substantial databases or as state or university projects. The interaction and dosage of risk factors for musculoskeletal disorders are not yet known, but industry can take action based on the best knowledge to date and refine the decisions as research progresses. Gerard Scannell, president of the National Safety Council asks, "How do you direct your safety and health program if you don't know what is happening in your workplace? The proper analysis of recorded injuries and illnesses will drive the programs" (Smith, 1995). In addition, surveillance is the primary method of assessing the efficacy of prevention measures that are instituted to control the identified problems (Hanrahan, 1989). Therefore, a company benefits from collecting the data that are useful in reducing the risk of injuries and improving job performance.

## 24.3  Sources of Primary Injury Surveillance Data

### Injury Data

There are several primary sources of injury data that companies either keep or to which they have access without having to survey the workforce.

- Injury logs that are mandated by OSHA
- Workers' compensation records
- Medical records
- Accident records

## OSHA Recordkeeping

Since the OSH Act of 1970 companies have had to keep a record of their injuries. In 1981 OSHA began to look at the log data as indicators for an inspection. Industry responded to this new focus in one of two ways. They either improved the safety and health in the facility or became more selective in what was recorded on the logs. OSHA began to discover significant under-reporting on companies' logs, and stronger enforcement action was instituted in 1986 (Tyson, 1991). Total penalties have become so substantial that companies now pay much more attention to keeping accurate records (BNA, 1993b, 1994). However, some of the difficulties of accurate recordkeeping have come to light and influence injury surveillance.

The increase in cumulative trauma disorders in the workplace highlights the difficulty of keeping the records up to date. As a medical condition progresses, the diagnosis sometimes changes, but the OSHA records may not reflect the new diagnosis. For example, if an injury originally recorded as a strained wrist eventually led to carpal tunnel syndrome, the diagnosis is typically not changed in the logs. This may be a common oversight if the records are kept by nonmedical personnel or if there is poor communication with the health care professional (HCP). Computerized databases can help a company, even if there is no medical department, as the system could be designed to prompt for confirmation of diagnosis while a case is open.

Despite guidelines on recordkeeping (BLS, 1986), there remains considerable confusion among employers about when and how to record occupational injuries and illnesses. The ambiguity of some of the requirements has been a point of litigation, and one company issued a digest of official interpretations of the recordkeeping guidelines primarily based on legal decisions in enforcement cases or in the courts (Duvall, 1996). OSHA proposed revised recordkeeping requirements and a new guidebook on how to interpret the requirements (OSHA, 1996). The proposal suggested significant improvements that included new versions of the OSHA 200 and 101 forms and simpler recording requirements.

In addition to recording injuries, OSHA and other government and state agencies mandate that several types of records be kept by the person or people responsible for safety and health. Records pertaining to hearing conservation, tracking of and training in the use of personal protective equipment (PPE), and compliance with confined space are just a few examples of the requirements. These records may be easily tracked and integrated into the safety and health database system.

## Workers' Compensation Data

Small industries commonly rely on workers' compensation data supplied by the company's insurance carrier as an injury surveillance system. The benefit of this practice is the convenience of the data, especially if the insurance carriers summarize the information. There are several downsides to using only these data. Not all injuries become workers' compensation cases, and therefore the injury profile is only partial. By the time a claim is processed and appears on a summary, a considerable length of time has passed during which preventative action could have been initiated. However, the claims on the workers' compensation insurance originate in the company, so that the company can have immediate access to the information if good records are kept. But, if the workers' compensation data can be tracked effectively by the company, then so can the accident data that might not become workers' compensation. The company would have richer data on which to base prevention and improvements.

The cost of injuries is one type of information that workers' compensation data do provide that is difficult to obtain any other way. Cost data are very useful for calculating cost benefit equations for improvements. Since workers' compensation databases are often large, they can be used effectively for major epidemiological studies or to complement large-scale surveillance (Seligman, 1986).

## Medical Records

Medical records can be the most reliable source of injury data since there is no interpretation required for recording as there is for the OSHA logs and workers' compensation filing. As discussed later in this chapter, a difficulty that can arise is access to the medical records. Very few workplaces have onsite medical services, and medical records can reflect both work- and nonwork-related injuries which can confuse data collection for injury surveillance.

First aid treatment is often recorded by a company but may not be entered in an employee's medical record. Although by definition the incidents are minor, there may be indication of a design problem. For example, it is not uncommon to find frequent treatment for grazed knuckles. If this is happening to the same employee or to a group of employees under the same circumstances, it could be a clearance or access issue.

## Accident Records

Accident reports are filled out by many companies and are the earliest records of an injury within a company's safety and health process. Not all injuries from accidents are recorded on the OSHA forms or claimed on workers' compensation, so that accident reports potentially provide the most comprehensive record of injuries, although maybe without medical detail. If good root cause analyses are conducted and recorded on the accident reports there may be valuable information indicating job redesign.

# Job Descriptions

Job information is useful from many standpoints. The causes of injuries need to be identified in order to prevent them. Although the causes for musculoskeletal injuries are not well defined, identifying risk factors of the job can be helpful. For example, risk factors for cumulative trauma disorders include extreme postures, high forces, high repetitions, inadequate recovery times and excessive duration. Some of the reasons to consider building a database of job descriptions are to:

- Have a list of risk factors present in each job
- Track employees' exposure
- Aid in the initial placement of employees
- Determine a job rotation system to create a balance of the job demands
- Provide information to the HCPs to help them with decisions regarding the work-relatedness of conditions
- Provide the context for functional capacity evaluations or to tailor workhardening or conditioning programs
- Guide appropriate placement of employees returning to work and to determine temporary alternative jobs or jobs that could be modified
- Help determine jobs that may be modified for accommodating a person under the Americans with Disabilities Act (ADA) (EEOC, 1990).

The downside of developing a job description database is the expense of establishing it and keeping it up to date. Such costs will vary considerably according to the type and size of industry. Some industries change their processes frequently, perhaps seasonally. To keep job descriptions current in a rapidly changing environment may not be practical.

Computerized injury surveillance database systems could provide a tremendous benefit for a small company that is communicating with offsite services. If an electronic network is established, the records can be efficiently updated. Information about the jobs could be on-line for the medical service to access, including formats such as computer-captured video.

# Worker Tracking

Keeping up with job descriptions may be a challenge for some industries, as is tracking the worker. Many of the companies that developed computerized databases early on stated that tracking the workers was the most complex and difficult task (Sugano, 1982; Hagstrom, 1982; Smith, 1982; Kuritz, 1982). Effective tracking required frequent updating of employee assignments including the length of time at the job in the specific area. In addition, each job in every area required a profile so that there was a history of exposure. The job profiles or descriptions also needed to be kept up to date by periodic job analyses or audits.

Employee tracking remains worthwhile, except that the experience of the companies that were in the forefront of developing surveillance databases, found that not only was employee tracking difficult but also very costly to implement. Therefore, each company must assess the importance of worker tracking, that is, to determine the type and extent of the hazards of the industry and to develop surveillance that is the most reasonable, practical, and cost effective. The trade-offs to consider with the four basic qualities of an employee tracking system follow (Sugano, 1982):

- *Uniformity within the exposure group.* Are the exposures the same across many jobs, and if so, may some of the descriptions be combined?
- *Accuracy of tracking the worker.* What are the means for communicating when an employee moves area or job? How does job rotation affect exposure? What differences are there between plants?
- *Adequacy of job and location description.* How often does the job change, and how frequently should the database be updated? Is environmental monitoring conducted regularly? How difficult is it to keep the physical, environmental, and material descriptions up to date?
- *Cost effectiveness.* How much money and manpower is needed to develop and maintain the database? Is too much data being collected and too often?

At present, worker tracking for exposure to risk factors for musculoskeletal injuries remains an area of research. There is no model by which to analyze the data and to provide an answer to the question "how much exposure to each risk factor or combination of factors is too much." Risk models that do exist are based on task analyses rather than monitoring the worker throughout a shift.

## 24.4　Benefits of an Injury Surveillance Database System

A complete and standardized reporting system for occupational injuries does not exist in the U.S. (Brewer, 1990). Almost all employers with more than 11 employees are mandated by OSHA to keep a current record of injuries and illnesses and details of how the injury occurred. The government forms may be substituted by ones that are equivalent. Exposure information, such as the demands of the job, a history of the different jobs the employee has held, or the job characteristics such as the state of the environment or the presence of chemicals, is typically not included. However, even without the enhanced job information, the two mandated OSHA forms contain considerable amounts of data that are not easily analyzed by hand. Computerization of only the OSHA forms can be helpful. What is important to remember is that the quality of the data depends on the information that is entered. The OSHA forms, filled thoroughly and thoughtfully, provide a better company database. A basic computerized injury surveillance database system can have the following benefits:

- Standardized and accurate recordkeeping, improving the quality of health records, and compliance with OSHA
- Efficient handling of data, reducing paper work and data re-entry
- Easy and effective reporting system
- Quick statistical analyses for the evaluation of potential problems and the measurement of effectiveness of interventions
- Enhanced tracking ability, particularly of employees' job exposures, and such tracking also facilitates epidemiological studies
- Quick verification in disability and compensation cases
- Potential for integration with other data, e.g., those used by the safety or industrial hygiene professional
- Ease of cost tracking for cost benefit analyses
- Efficient corporate response to litigation, community complaints, and public relations requirements.

There are many commercial software packages available for computerizing the OSHA forms combined with basic surveillance functions. The sophistication of the available programs vary considerably, from large database systems with an injury surveillance module combined with advanced statistical analyses, to simple OSHA form recording and tracking. A few software directories that are regularly updated are listed in For Further Information.

## Cost Benefit Analysis of Database Investment

A cost benefit analysis for investing in an injury surveillance database system is not easy to conduct (Wrench, 1990). Some of the benefits are listed above but assigning a dollar value can be difficult. At the planning stage the time saved can only be anticipated; however, software vendors can sometimes provide a percentage figure of time saved by their systems. Wrench (1990) listed some cost aspects to consider in the cost benefit equation.

- *Direct costs*
  - Initial costs that include the stages of development, such as needs and requirements analysis, systems analysis and design, programming and software development, hardware, software, installation, training, and system implementation.
  - Operating costs that include personnel to use and maintain the system, backup systems, training, and supplies. In addition there are costs related to updating the system as software changes, or when regulations alter, which affect decision algorithms, data entry, and form outputs of a database.
- *Indirect costs*
  - Space requirements may increase with purchase; air-conditioning may be needed for larger systems; security measures and additional insurance may be necessary.

## Integration of Database Systems

The data collection mandated by OSHA may be useful, but, most industries would benefit from a broader collection of data and greater integration with other company functions (Finucane, 1982; Committee Report, 1992; Holzner, 1993; Dieterly, 1995). Early on, during the development of surveillance within the chemical industry it was realized that exposure measures were needed to interpret the medical findings, thus integrating medical surveillance programs with industrial hygiene surveillance (Parkinson, 1986). A survey conducted by OSHA also found medical surveillance programs were most effective when they were integrated into a comprehensive and systematic approach for identifying and addressing workplace hazards (BNA, 1991, 1993c). A comprehensive approach requires the involvement of different expertise within a company and hence involves more than one discipline.

Injury surveillance can be incorporated into comprehensive safety and health programs. The Occupational Medical Practice Committee (OMPC) of the American College of Occupational and Environmental Medicine (ACOEM) provided guidelines for the scope of occupational and environmental health programs. The committee identified 14 essential components of a program (Committee Report, 1992).

1. Health and evaluation of employees.
2. Diagnosis and treatment of occupational and environmental injuries or illnesses, including rehabilitation.
3. Emergency treatment of nonoccupational injury or illness.
4. Education of employees in jobs where potential occupational hazards exist, including job-specific instruction, instruction on methods of prevention, and recognition of possible adverse health effects.
5. Implementation of programs for the use of indicated personal protective devices — ear protection (plugs, muffs), safety spectacles, respirators, etc.
6. Evaluation, inspection, and abatement of workplace hazards.
7. Toxicological assessments, including advice on chemical substances that have not had adequate toxicological testing.
8. Biostatistics and epidemiology assessments.

9. Maintenance of occupational medical records.
10. Immunization against possible occupational infections.
11. Medical interpretation and participation in development of governmental health and safety regulations.
12. Periodic evaluation of the occupational or environmental health program.
13. Disaster preparedness planning for the workplace and the community.
14. Assistance in rehabilitation of alcohol and drug-dependent employees or those with emotional disorders.

Many of these items can be addressed by other professionals such as those in industrial hygiene, safety, or ergonomics. The OMPC stressed that the practitioner in occupational medicine should understand how to enlist and collaborate with the skills of colleagues from the many other professions involved with industry. The above list is a useful one for a company to consider. Some elements will be more important than others, depending on the industry. If many services are contracted offsite, the overall program may not be managed by an HCP but perhaps by another professional, for example in safety or industrial hygiene.

## 24.5    Ergonomics and Injury Surveillance

Injury surveillance is of interest to several professions in industry such as safety, industrial hygiene, medicine, and risk management, as well as ergonomics. The different disciplines may have a part in contributing to the database, or one department may be responsible for it overall, while others access it. Different professions may develop their own interface or module with the same database so as to manipulate and interpret the data as they wish. If a person wears the hats of several job functions, for example all aspects of safety and health, he or she they might work with a number of discrete modules, such as safety, industrial hygiene, and ergonomics. However, a small company may choose to integrate the modules to a basic safety and health one. The scale of module for each discipline depends on the level of detail desired or degree of expertise of the person using the system. If expertise is outsourced, then an in-house module in that field may not be needed.

Many injuries are related to poor design of the workplace or jobs, which is the domain of ergonomics. In particular, musculoskeletal injuries such as low back disorders and cumulative trauma disorders have become informally termed *ergonomics-related disorders*. This is a limited view as other injuries can occur due to poor design; for example, accidents can arise because of inadequate accessibility or because of an error provoked by poor layout and labeling of controls and displays. However, there is considerable focus on musculoskeletal injuries as they are one of the 10 leading work-related health problems affecting workers in the U.S. and such injuries are the leading cause of disability of people in their working years. The cost, based on lost earnings and workers' compensation, exceeds that of any single health disorder (NIOSH, 1986). The most common source of injury and illness for lost workday cases in manufacturing was reported to be the "worker motion or position." Overexertion was the primary event causing injury and illness across all the divisions of industry (Synergist, 1995). Injury surveillance, therefore, is an important method to determine indicators of poor job design and the need for ergonomics intervention. The data are also important as an effectiveness measure of interventions in an ergonomics process.

### Ergonomics Process

An ergonomics process in a company commonly has components of medical management which includes injury surveillance; worksite analysis which is active surveillance for risk factors; prevention and control which also entails measures of effectiveness of the controls; and training and education (OSHA, 1990). Good injury surveillance promotes prompt treatment and directs the prevention programs. The greater the understanding of the causes of the injuries, the more proactive can be the prevention. If prevention measures are not taken, then the system consists only of treatment and remains reactive to the occurrences of the workplace. Surveillance systems that are set up from only a medical standpoint can be weak on the prevention aspect, even if there are good data. The "who, what, when, and where" (Garrett, 1982) of injuries can often be captured on the accident reports and on the OSHA forms or equivalent if they are thoroughly completed. Typically, the "how and why" of an injury are not well reported and are dependent upon the quality of the accident investigation and whether it includes an ergonomics perspective. Some caution is needed in relying upon a medical focus for improving the workplace. The workplace system may suffer if workstation changes are in the form of accommodation or excessive tailoring on a case by case basis, as a response to medical injuries. A balance must be kept between appropriate design of workstations and the overall function of the system, including organizational design aspects. Involvement of an interdisciplinary team helps to ensure the work system is addressed. Disciplines such as professionals in safety, engineering, management, health care, industrial hygiene, and ergonomics, as well as the operators often participate in the teams.

Responsibility has to be assigned to an individual or a group within the company to apply the results of the surveillance data by looking at the workplace, identifying the problems and root causes, and developing potential solutions. The process does not stop there, as interventions have to be implemented and then followed to ensure effectiveness. It is more likely that prevention measures will not be taken when the medical facilities are offsite, which is the case for the majority of workplaces. It is the company's responsibility to bridge the communication gap between the company and offsite services. Assigning a company contact person often helps in promoting communication.

## Ergonomics Module of a Database System

An ergonomics module that interfaces with an injury surveillance database system has its own list of needs to fulfill the ergonomics function. The main tasks are:

- Collecting data, apart from injury data, that indicate ergonomics intervention or help with the interpretation of injury surveillance data
- Interpretation of the data, for example, prioritization methods and graphical trends or statistical analysis of surveys
- Tracking of information such as, project status, timelines, workplace changes, effectiveness, and training

To accomplish data collection and interpretation in the context of task analyses, an ergonomics module may also interface with technical software such as the National Institute of Occupational Safety and Health (NIOSH) Lifting Index. Job profiles may be generated from job analyses or analyses results added to profiles that may have been developed by another department, such as human resources. Data that can be gathered apart from injuries include:

- Turnover rates
- Absenteeism rates
- Productivity rates and quotas
- Quality-related figures
- Workers' compensation costs
- Job profiles or descriptions
- Employees' jobs record or work history
- Discomfort survey results
- Ergonomics audit results
- Employee suggestions or reports of problems
- Ergonomic analyses results

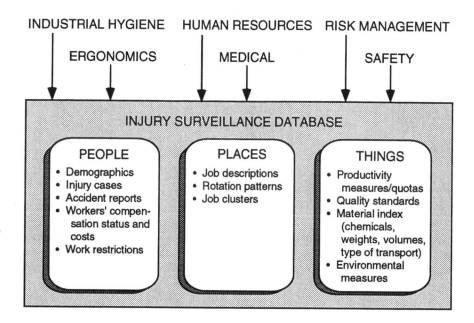

**FIGURE 24.1** A fundamental model of an injury surveillance database system and possible user groups. Listings under People, Places, and Things are examples of database information of common interest to the users. (Adapted from Hillman, G. 1982. ECHOES: IBM's environmental, chemical and occupational evaluation system, in *Medical Information Systems Roundtable*. p. 827-835. *JOM* 24(10) Supplement.)

A simple way to categorize most of this information is by using three groups: People, Places and Things as illustrated in Figure 24.1. An ergonomics module would add ergonomics software tools of choice, statistical analysis programs, analyses and survey results, and tracking systems. Specific customization will be required to interface the desired software of an ergonomics module to an injury surveillance database, whether that database is company developed or a commercial package. The degree of software tailoring and the extent of the company's capabilities and resources will determine how much would be in an ergonomics module. It may be practical to keep some components independent such as the ergonomics tools.

## 24.6 Early Databases

The earliest computerized surveillance databases were purchased or developed during the 1970s primarily by the chemicals and petroleum industry. Computer-based information systems became a necessity to manage the data growth more effectively. Manual data systems in large corporations were hand-written and paper-based, lacked consistency and standardization and could not adequately capture the data on important monitoring aspects such as job history and workplace exposures (Finucane, 1982). The models that were conceived were comprehensive and sophisticated, and many currently continue their development (Holzner, 1993). The focus of the systems were occupational health and, at a minimum, incorporated (Finucane, 1982):

- Detailed worker and job histories and demographic data
- An inventory of potential exposures and associated adverse health effects related to specific workplace location
- Worksite exposure data
- Employee medical information collected throughout the worker's career

These extensive models depended on effective interaction between the departments and the readily available retrieval and analysis of the information. Wolkonsky (1982) described the following information as necessary for a health data system at Amoco:

- *Employee* — job history, location codes, sickness, and disability
- *Claims* — workers' compensation payments, death certificates, internal cost reports
- *Medical* — personal and family history, physical examination, laboratory results, immunizations
- *Industrial Hygiene* — personal samples, area samples, potential exposures by location, exposure levels
- *Toxicology* — animal research, materials safety data sheets
- *Safety* — accident data, physical conditions, supervisor data, DOT data

These early medical surveillance database systems remain but are continually up-graded and developed as computer hardware and software change, regulations alter, and knowledge about exposures and risk assessment models progress through research.

## Challenges that Were Met

Descriptions of the early computerized systems highlighted many difficulties in establishing comprehensive database systems. Of the large and sophisticated models being developed by the major corporations in the country, none were the same. Those who attempted to purchase another's system ended up making significant modifications and customizations to fit the needs of the company (Joiner, 1982). However, despite the individuality of the systems there were some common factors that made them successful. The common factors of the systems were: flexibility, interaction, user friendliness, modular design, valid databases, economy, innovation, key staff, commitment, and phased approach.

One of the first difficult decisions for the developers was whether to develop a system in-house or buy a commercial system and modify it to the meet the company's needs. Building the system gradually by modules was a common approach, but the modules of the database systems varied considerably between companies. On a basic level, the modules can be summarized as People, Places, and Things (Hillman, 1982). Deciding how much information about each area was needed and was cost effective, and what hardware to accomplish the goal varied according to company philosophy and existing computer experience or investments.

Early developers reported some modules as challenging to design. In particular, tracking workers was the most complex and difficult task (Sugano, 1982; Hagstrom, 1982; Smith, 1982; Kuritz, 1982). Centralization of information, such as personnel data, was another issue for a few companies. Despite the large corporate size, not all systems, such as the personnel system, were consistent or centralized across facilities (Hagstrom, 1982). Privacy and confidentiality of medical information were of general concern and were resolved in different ways. Some companies encrypted information in case it was routed to the wrong terminal, while others limited access and used passwords. Many of the issues and difficulties that were met by those who developed the early medical surveillance systems are pertinent today. Some of these issues are further addressed in the context of developing or choosing an injury surveillance database system.

## 24.7  Defining the Database Model

For a company to define the model of an injury surveillance database system to be developed or purchased requires more than a clear understanding of the goals and objectives of the system. Development of a system must be a team effort (Wrench, 1990). The computer software professionals need to work closely with the safety and health professionals.

An assessment of the company itself is necessary to ensure that the goals of the surveillance system integrate with the company's needs, vision, and resources. There are common phases of development for a computerized health surveillance program. Designing the system and defining the components are just one stage (Joiner, 1982). If commercial software packages are being deliberated, take particular care to consider them in context of the overall system because the result can be a different program for each departmental function. If later the decision is to combine the programs, it may be technically too difficult to do so and efficiency is lost in the long run. The following program development phases are also pertinent for defining programs to purchase:

- Feasibility study
- Planning and identifying potential users
- Designing the system and defining components (system development and modification)
- Implementing
- Operating
- Maintaining
- Upgrading the system

## Designing the System and Defining Components

The step of designing the system and defining the components requires particular comment because it is important for determining the scope of an injury surveillance system. Deciding whether injury surveillance should be part of a larger system or if it should be a small discrete system may not be an easy task for a company. This step of program development is especially useful also for the decision process necessary prior to purchasing a program.

1. Define the present injury surveillance system in the company, however manual that may be, and critique for aspects that you wish to maintain and determine areas of improvements.
2. Define the basic components of the injury surveillance database system that you would like to have. Keep in mind the primary goal of an injury surveillance database system; that is, determining the who, what, where, when of employee exposure (Garrett, 1982) and using that information to guide prevention. As a database, this may be organized conceptually as three subsets: People, Places, Things (Hillman, 1982) (Figure 24.1).
3. Determine the functions or modules of the groups that could utilize the database and compare to the needs of the company; e.g., an industrial hygiene function is possible but may not be needed as a full module by the company.
4. Compare the present surveillance system in the company with the defined database model. Prioritize the development of the modules.
5. Assess the status of computerization in the company. Allocate which modules warrant computerization now vs. in the future. Not all aspects of a system need to be computerized. Some aspects may not be cost effective to develop, depending on the size of the company, the number of people using the information, and the criticality of the information to risk exposure for the workers in the industry. However, allow for growth in both software and hardware.

## Company Size

The size of a company influences the models and decisions about injury surveillance database systems. In the past, database systems utilized mainframes and analyzed volumes of data, putting them on a scale and at a cost prohibitive to smaller industries. With the advent of powerful personal computers (PC) and easy networking, computerization is available to most companies. Despite the availability, many companies have not devoted resources to developing injury surveillance database systems despite the time-saving advantages. One possible reason is that the system is not developed in a manual form in those companies. Basic recordkeeping that entails filling out the OSHA forms does not constitute a simple injury surveillance system, unless there is some analysis of the data and preventive action is taken. Another possible reason that resources have not been devoted to surveillance is that a company experiences so few recordable injuries that attention to injury surveillance is deemed unwarranted. A third reason may be that improving the efficiency of surveillance is not a top priority relative to other company improvements. It is important to remember, however, that computerization only recently has been so available and relatively affordable.

### Large Companies

Large corporations were the first to develop injury surveillance database systems. This is not surprising for several reasons. There was a large volume of data due to the size of the company. The companies had a number of potential exposures from chemicals and the environment, and they had the resources for epidemiological studies.

### Small and Medium Companies

Establishment sizes of 50 to 249 employees incur the highest injury incidence rates (per 100 full-time workers). Often, less than 100 employees is considered a small company and up to 1000 employees is medium-sized. The Small Business Administration (SBA), however, defines small as typically less than or equal to 500 employees (Stuart-Buttle, 1996). Regardless of definition, companies of about 50 to 500 employees in size should be particularly vigilant of their injury statistics. Despite national statistics, there will be some small companies with such low injury prevalence that a formal injury surveillance system is not necessary (other than mandated recordkeeping).

Some of the characteristics of a small industry indicate that conducting injury surveillance may be difficult (Stuart-Buttle, 1996). The main common characteristics are:

- Less formality
- Responsibility for several positions
- Greater responsiveness
- Less specific knowledge
- More management involvement
- Less data-oriented approach

However, it is precisely some of these characteristics, such as the responsibility for several positions, that would make an investment in an injury surveillance database system such an aid. The system could provide the benefits listed earlier and enhance an individual's and the company's efficiency.

## Onsite or Offsite Medical and Surveillance Services

Only 6.3% of U.S. workplaces provide onsite medical surveillance that includes treatment and recording of occupational injuries (BNA, 1995). These facilities are large corporations and medium-sized companies with a sufficient number of employees or injuries or both for an onsite medical department to be cost-effective. The remainder of the workplaces use offsite treatment services that may or may not track data for the company. Usually a company employee, typically a person without medical knowledge, fulfills the basic clerical duty of recording the OSHA logs and filling out workers' compensation forms.

If the OSHA logs and accident records are the minimum data for surveillance tracking, the quality of the information is very important if it is to be useful. The questionable quality of offsite surveillance has been raised by NIOSH. Murthy of NIOSH suggested that physicians away from the worksite were less likely to probe a patient about work history and often had very little occupational medicine training to assess work-related conditions effectively (BNA, 1995). Although the company is dependent upon the HCP for accuracy and detail of diagnosis, the designated contact person for the company can exert some control by communicating the requirements of the company to the medical service (Stuart-Buttle, 1996). Close communication between the medical service and company is needed for all injury cases to keep the information accurate and up to date.

Brewer et al. (1990) described a clinic-based occupational injury surveillance system that served many industries. The system connected several clinics that treated employees for even minor injuries. All injuries, regardless of whether they were compensable or recordable on the OSHA log, were recorded. Statistical analyses were routinely conducted and results provided to the companies. The companies learned what type of injuries were occurring, the frequency, part of the body, severity, and how the injury was caused. Of all the cases analyzed within the system, 78.4% of the patients received care for minor traumas which were classified as cuts or lacerations, sprains, strains, or contusions. Such analyses showed that collecting data of first aid treatment is beneficial and that a company could reduce costs by preventing the occurrence of minor trauma. The study by Brewer et al. also demonstrated the feasibility of clinic-based occupational injury surveillance as a means of assisting employers with the control of work-related conditions.

Medical screening is often used to complement environmental control measures (Baker, 1990). The purpose of screening is to detect disorders early and implement appropriate intervention. Screening also serves as a monitor of the effectiveness of prevention measures already instituted. Many medical screening functions, such as auditory testing and pulmonary function screening, can be conducted and supported by an offsite contract service. Tracking and time tabling the routine screening of employees is easily performed by a database.

## Types of Database Models

An injury surveillance database system can be developed anywhere along the continuum of simple to sophisticated. It may be a system that is customized, commercially purchased, or contracted as an offsite service. Whichever type of system is chosen, it is most likely to be modular. The most basic injury surveillance database model fulfills the recordkeeping function and provides some descriptive statistics of the data. Many PC packets provide the interface to enter the information required for the OSHA forms, analyze the data, and produce summary graphs by department, for the month, quarter, or year and sometimes across years. Printouts of the filled forms are also commonly available.

Beyond a recordkeeping level, the fundamental model can be represented by the three interacting modules of Hillman (1982): Places, People, Things (Figure 24.1). This still could be kept simple: Places includes a basic coding of the areas or jobs, possibly with a description; People includes demographics (gender, age, date of hire, job), and health data, both evaluations and incidents; and Things includes information files of chemicals or materials that are in the company. A small company may have very little information in the Places and Things files. A large corporation could expand this model with considerable databases of information under all three modules.

With this concept, potentially there may be many users of the database, depending what data is entered in the first place. Human resources, safety, medical, industrial hygiene, ergonomics, and risk management are some of the possible parties interested in an injury surveillance database system. Each group interfaces with the system from its own perspective, using its own programs to extract appropriate data, perform statistical functions and risk assessments as required, and present data in the format desired and appropriate for the user's job function. Injury surveillance is dependent upon someone being responsible for looking at the data of the injury reports that has been entered into the system. The responsibility may be almost any one of the users indicated, as the job function depends on the company size and structure. Each interface of a user group is a module that could be developed as the system evolves according to the company's needs.

## Confidentiality

A database that is shared between many users raises concern about confidentiality and the need to limit access to certain information. Systems with distinct modules for each user group typically make each module a limited access to a defined group of people, usually by use of a password. A network can be designed to monitor passwords and terminal identification. Medical data are particularly sensitive information. Joyner and Pack (1982) described the Shell Oil Company's system that encrypts the data and stores it centrally. Decryption programs were installed in the medical department to decode output to the terminal, therefore precluding disclosure of confidential information in the event the output inadvertently arrived at another terminal. Another technique that has been adopted to ensure confidentiality is the scrambling of social security numbers (Hillman, 1982; Wolkonsky, 1982).

## Data Entry

Many of the early large and sophisticated systems were invested with efficient methods of data entry. For example, the Amoco occupational health system recorded a patient's medical history and examination on mark-sense forms that were computer input sheets that were entered by optical scanner. This eliminated any possible error from transcription (Wolkonsky, 1982). Questionnaires or surveys can be efficiently collected in a similar way. An overall benefit of a database, however, is the reduced re-entry of data. Personnel data, in particular, is typically entered many times if there is no network to a personnel database.

## Computerization

There are many commercial software programs which can suit a variety of companies. (See For Further Information for software directories.) However, some industries choose to develop their own injury surveillance database system. The level of computerization that already exists in the company influences the choice of a customized program. Often there are many programs that already exist in a company, for example, production and quality control systems, payroll, accounting, and maintenance department systems. With the increase of computers in the workplace many companies have invested in management information system (MIS) departments to develop and maintain the hardware and software. Older systems were originally written for mainframes and minicomputers, and most newer systems are written for PCs. If an additional system is intended to be added to a network, it is sometimes easier to develop it rather than purchase a program that may be difficult to make compatible and to support in the future.

The compatibility of commercial software to an existing system is influenced by the age of the system and to what extent the software language has been updated. Therefore, companies with relatively new computer systems are more likely to be able to take advantage of the large commercial market, tailor the packages with fewer technical problems, and support the systems with in-house personnel. Considering the available choices of sophisticated programs, purchase of a system can save considerable design and development costs. The choice of system also depends upon the level of computer skill of the users or resources of the company. The Centers for Disease Control and Prevention (CDC) developed "Epi Info," a series of public-domain PC software programs in IBM-compatible format (DOS) for word processing, database, and statistics work in public health. (See For Further Information.) The programs are designed to be easily modified to meet company needs but still require some computer knowledge, especially since many computer users are only familiar with the Microsoft Windows operating system. In addition, before modifying the program, the user must clearly define the functions desired of Epi Info which requires knowledge of injury surveillance. Despite being in the public domain and having excellent technical support, the programs may not be compatible with some companies.

Not every company has extensive computerization or a network. Some may have computerization but choose to have an independent safety and health system. As stated above, the majority of workplaces do not have onsite medical services or extensive exposure issues, so the company's needs are primarily to track and analyze first aid logs, OSHA logs, and accident-related records. Job descriptions would be beneficial for injury analysis, but, as previously discussed, they can be difficult to maintain for some industries. Many commercial software programs are available that provide a recordkeeping function with OSHA log entry and analysis. (See For Further Information.) There are several safety or industrial hygiene functions that could be efficient to computerize depending on the company size, and for which there is available software, for example, maintaining Material Safety Data Sheets (MSDS), tracking PPE, and tracking employee training (Johnson, 1996).

An expanding area of computers is internet capability. The internet provides a source of information for data such as MSDSs, as well as updates on regulations. As more professional associations have representation on the world wide web, the internet may serve to help resolve technical questions.

## 24.8   Statistical Treatment of Data

A computer can perform sophisticated data analyses only if specifically instructed to do so. A good understanding of statistical methods is required to develop a data analysis program, and if necessary, a statistician should be consulted. There are commercial statistical packages that may be interfaced with a database system, and some of the packages guide a user in interpreting data. There is potential for a vast array of analyses of the data in an injury surveillance database, depending on the extent and sophistication of the overall system. For example, if there is a focus on industrial hygiene issues, numerous calculations may be required to interpret data from workplace monitoring. The safety and health professional needs some knowledge of statistics to interpret data beyond the OSHA logs. The commercial software market is rapidly responding to demand by producing programs specifically for industrial hygienists, such as those that assist with industrial hygiene formulae (Cameron, 1996). There are other programs on the market designed to meet the technical needs of the ergonomist and safety professional.

The majority of workplaces are small businesses in which those responsible for safety and health often do not have a background in statistics. In addition, they may not know what to look for in the basic data collected from OSHA logs and accident records. Some fundamental information is, what volume and type of injuries are occurring, to which parts of the body, while performing which jobs or tasks. Most of these questions can be answered by descriptive statistics. The commercial market offers many programs that focus on OSHA recordkeeping and provide basic data analysis with graphic output. After careful assessment of these programs there is likely to be a software package that meets the company's needs. However, even more useful information can be obtained if data collection and analysis extend beyond the minimum requirements for compliance. The following are a few points that will help to make the data more useful. The points may also help when assessing software programs.

- Convert the number of injuries into an incidence rate as described in OSHA's recordkeeping guideline (BLS, 1986). This provides a figure that has a common denominator with other companies and national statistics. The national injury incidence rate calculated by the Bureau of Labor Statistics (BLS) or the rate for the company's Standard Industrial Classification (SIC) code can serve as benchmarks for the company. Incidence rate can be calculated for groups of types of injuries and also for departments. Direct comparison between departments can be made when the numbers are expressed as incidence rates.

- The number of lost workdays is usually considered an indication of severity of the injury. An incident rate can also be calculated for the lost workdays. At times a given incidence may be so severe that it accounts for most of the recorded lost workdays. In such situations the incidence rate should be calculated with and without that particular case. The proposed recordkeeping revision does not require the number of restricted days to be recorded, however, a company may find it useful to record them (OSHA, 1996). When injuries begin to be detected early and undergo conservative treatment, then lost workdays are less likely. However, an increase in restricted workdays is more probable. Recording the restricted workdays provides some measure of progress, especially if the number of restricted workdays are then improved.

- Record the side of the body that was injured. This is not often recorded yet it can be useful to know. If the information is related to a record of the employee's dominant hand a useful picture of the stresses and injuries at the job begin to be built, which helps with job analyses and prevention strategies.

- Look at the incidence rate by age, gender, length of time on the job, and whether the employees are full-time or part-time. All of these analyses will help to determine if a certain group is susceptible. Those who are very new on the job are typically suffering from lack of conditioning. If this is found to be the case, strategies can be implemented to gradually introduce the worker to the job. For example, orientation sessions could be scheduled throughout the first week rather than in the first couple of days.

- Look at the injury rate by supervisor or line leader if these are different groupings from the department or area. This can be very useful in determining if a certain team has a problem and suggests there may be some administrative issues to investigate. Similarly, data should be compared between shifts. Often any differences are related to management.

- Incorporate workers' compensation costs into the database and determine the mean for type of injury and cause and for each area or job. This helps with cost benefit analyses and prioritizing areas to be improved. Sometimes the incidence rate in a plant has increased, but the workers' compensation costs decreased. This may indicate that more people are reporting before the condition is severe and more expensive to treat.

- Track the work status of all those injured, whether they are off work on workers' compensation or on work restrictions. Develop a list on a regular basis, such as once a week, to ensure that each case is kept up to date and progressed toward full return to work, preferably at the previous job. It is not unusual for an employee to "fall between the cracks" and be on workers' compensation for a period of time without any functional change or decision about the case.

- Identify the causes and mechanisms of injuries based on the accident record, and in particular, relate the findings to the areas of high incidence rates.

- Look at the data to see if there are patterns according to time of day, days of the week, the month, quarter, or year. Compare across the years as well.

- If possible, collect first aid data as well as the data mandated by OSHA. Treating minor cuts and bruises takes time and there can also be an associated loss of productivity. If this information is collected and analyzed similar to the logs, there may be some useful emerging patterns that can help with identifying the causes and lead to prevention measures. Examples of common problems are blunt tools that cause the user to slip, or inadequate clearance around equipment that provokes bruising.

More sophisticated analyses can be conducted on data from OSHA forms. However, some understanding of statistics is needed to make use of such features in a software program, even if the interface is user friendly. A larger company may benefit from a data analysis that indicates when the occurrence of injuries is "out of control" relative to a determined baseline. Boyd (1988) presented one approach based on injury costs per exposure hour. Goldberg, (1993) describes an injury surveillance database with some useful, advanced analysis features. A description of some of these features follows.

- Actual exposure hours were collected from payroll to calculate the incidence rate. This allowed for a better picture of incidence rate related to overtime hours, especially since overtime was inconsistent and potentially a reason for increased injuries.

- Regression analysis helped determine the factors leading to lost and restricted workdays. The software interface displayed a list of independent variables and interactions from which to choose. For example, the number of restricted workdays could be looked at by gender, age, and length of employment. Regression analysis could also be used to predict the number of expected lost or restricted workdays for an injured employee, and this helped management plan for the situation.

- Time series modeling provided insight as to whether there were certain periods of higher incidence rates. Accidents are sometimes based on cyclical seasonal weather change, employment relationships, or perhaps rapid expansion.

- A forecasting feature smoothed the data to remove known cyclical fluctuations and provided indication of the data trend, whether the incidence rate was increasing or decreasing.

- Inclusion of cost and billing information was used to generate cost/benefit analyses.

### Report Generation

Except for mandated government forms, reports are primarily for company use, (e.g., OSHA forms). The reports are not generally required for scientific presentations but rather analysis and communication. Two levels of reports are needed: one for the professional to explore and interpret the data, and another that summarizes the findings for presentation to management. The system needs to be flexible enough such that a summary report can be focused on specific findings.

The maxim "pictures speak a thousand words" is true when it comes to concise presentation. Good graphic presentation straight from the software program saves considerable time, especially when the alternative is transporting the data to another program. However, it is possible to set up standard links to a program that produces charts and graphs, so that graphs can be routinely produced.

## 24.9   Choosing a Commercial Injury Surveillance Database

One size does not fit all when it comes to setting up an injury surveillance database system (Wrench, 1990). If a commercial program is purchased, it will require professional tailoring to fit the needs of the company. For example, along with social security numbers, company identification numbers are commonly used for personnel tracking, yet every company has a different numbering system which requires modification of that data field. Likewise, the job coding system is different in every company.

Involve the employees as much as possible in decisions about a database. The employees know the information needed for their jobs and can contribute significantly to the needs assessment and system design and to evaluating existing programs. In addition, the success of a system is dependent upon their acceptance of it.

The following are some aspects to consider when choosing a commercial software program (Menzel, 1994).
*Evaluating needs and resources*

- Identify data management goals and objectives.
- Define requirements. (This may be accomplished by mapping out the existing manual system.) Include anticipated growth and potential new functions.
- Define company constraints, such as type of hardware system, technical equipment, and the characteristics of other company databases.
- Assess company resources (personnel, budget, hardware, and software).
- Develop a list of criteria, based on the needs and resources evaluation, by which to assess the software programs. There may not be a program that satisfies each criterion so developing a spreadsheet can be useful for deciding the trade-offs for a final selection. A scoring system could also be used to determine several final contenders.

*Finding the software*

- Use directories that may be available through professional safety and health associations (see For Further Information).
- Research articles on software in safety and health magazines.
- Respond to software advertisements in occupational safety and health magazines.
- Ask colleagues about the software they use and their experience with it.
- Request demonstration disks and onsite demonstrations.

*Evaluating the software*

- Assess the basic features of the programs against the criteria. This will narrow the selection for more critical view.
- Evaluate for qualities such as:
  - ease of use
  - provision of high quality, onsite training

- variety and quality of reports
- ability to transfer data from other programs
- potential degree of customization
- presence of security systems to safeguard confidentiality and data integrity
- reasonable initial and continuing cost
- Select a few final contenders based on the extent to which criteria are satisfied.
- With program specifications attached to the proposal, request price proposals with details of features, service, and renewal fees.
- Request names of current users that can be contacted.
- Assess the responsiveness and quality of the support available.
- Review the length of time each company has been in operation.
- Assess their financial stability and long-term viability. (This is not easy but the number of clients being supported may give an indication.)

Selecting the vendor

- Negotiate the price
- Negotiate the extent of customization and support
- Negotiate for additional features such as training

Evaluating the effectiveness

- Ensure that the program is thoroughly reviewed before the first anniversary (when maintenance fees usually begin).
- If necessary, negotiate for modifications before paying maintenance fees.

When implementing computerization of a system ensure that parallel systems are run for at least a month after full installation, even if the computerized system is off-the-shelf (Bonnett, 1982). The accuracy of the system must be checked before abandoning the old methods. If there are software modifications or connections made to an existing network, it is better to assure that the system is running smoothly before relying upon it. New hardware should be tried also before being relied upon.

## 24.10 Summary

The cost benefits of injury surveillance database systems are becoming more apparent since there has been greater focus on accurate recordkeeping, and as computerization has become more affordable. There are several benefits including standardized and accurate recordkeeping, time saving and efficient data entry and analysis, enhanced tracking abilities, and effective guidance to prevention strategies.

Considerable and careful thought should be given by a company to the design of the injury surveillance database system. The design is dependent upon the company's needs, size, constraints, potential growth, and resources. An assessment of the company's needs prevents a quick purchase of a commercial system that might not have the growth potential, adequate support and customization services to make the program a successful addition to the company.

## References

Baker, E. L., Honchar, P. A., and Fine, L. J. 1989(a). I. Surveillance in occupational illness and injury: concepts and content, in *Surveillance in Occupational Safety and Health*, Ed. E. L. Baker, p. 9-11. *AJPH* 79 Supplement.

Baker, E. L. 1989(b). IV. Sentinel Event Notification System for Occupational Risks (SENSOR): The concept, in *Surveillance in Occupational Safety and Health*, Ed. E. L. Baker, p. 18-20. *AJPH* 79 Supplement.

Baker, E. L. 1990. Role of medical screening in the prevention of occupational disease. *JOM.* 32(9):787-788.

Brewer, R. D., Oleske, D. M., Hahn, J., and Leibold, M. 1990. A model for occupational injury surveillance by occupational health centers. *JOM.* 32(8):698-702.

Bonnett, J. C. and Pell, S. 1982. Du Pont's health surveillance systems, in *Medical Information Systems Roundtable.* p. 819-823. JOM 24(10) Supplement.

Boyd, A. H. and Herrin, G. D. 1988. Monitoring industrial injuries: A case study. *JOM.* 30(1):43-48.

Bureau of Labor Statistics (BLS). 1986. *Recordkeeping Guidelines for Occupational Injuries and Illnesses.* OMB. No. 1220-0029, U.S. Government Printing Office, Washington, D.C.

Bureau of National Affairs (BNA). 1991. Preliminary results from OSHA survey said to show benefit of "integrated" plan. *Occupational Safety and Health Reporter.* 5-22-91:1712-1713.

Bureau of National Affairs (BNA). 1993a. All worksites need passive surveillance, job checklist for risk factors, ANSI group told. *Occupational Safety and Health Reporter.* 1-20-93:1444-1445.

Bureau of National Affairs (BNA). 1993b. OSHA's egregious penalty policy for recordkeeping violations upheld. *Occupational Safety and Health Reporter.* 2-10-93:1601-1602.

Bureau of National Affairs (BNA). 1993c. Comprehensive risk management programs effective at finding illness, draft report says. *Occupational Safety and Health Reporter.* 3-31-93:1899-1900.

Bureau of National Affairs (BNA). 1994. Recordkeeping violations affirmed against general dynamics as non-serious. *Occupational Safety and Health Reporter.* 3-16-94:1353-1354.

Bureau of National Affairs (BNA). 1995. NIOSH staffer says onsite surveillance plays important role in treating workers. *Occupational Safety and Health Reporter.* 11-8-95:817.

Cameron, M. 1996. IH calculator. *The Synergist.* April:17-18.

Committee Report. 1992. Scope of occupational and environmental health programs and practice. *JOM.* 34(4):436-440.

Dieterly, D. L. 1995. Industrial injury cost analysis by occupation in an electric utility. *Human Factors.* 37(3):591-595.

Duvall, M. N. 1996. Digest of official interpretations of the bureau of labor statistics recordkeeping guidelines for occupational injuries and illnesses (second revised edition). *Occupational Safety and Health Reporter.* 1-31-96:1144-1171.

EEOC, 1990. *Americans with Disabilities Act of 1990.* Public Law 101-336 101st Congress, July, 26. U.S. Government Printing Office, Washington, D.C.

Finucane, R. D. and McDonagh, T. J. 1982. Foreword, in *Medical Information Systems Roundtable.* p. 781-782. *JOM* 24(10) Supplement.

Froines, J. R., Dellenbaugh, C. A., and Wegman, D. H. 1986. Occupational health surveillance: A means to identify work-related risks. *AJPH.* 76(9):1089-1096.

Garrett, R. W. 1982. Environmental tracking at Eli Lilly and Company, in *Medical Information Systems Roundtable.* p. 836-839. *JOM* 24(10) Supplement.

Goldberg, J. H., Leader, B. K., and Stuart-Buttle, C. 1993. Medical logging and injury surveillance database system. *Int J of Indus Ergonomics.* 11:107-123.

Hagstrom, R. M., Dougherty, W. E., English, N. B., Lochhead, T. J., and Schriver, R. C. 1982. SmithKline Environmental Health Surveillance System, in *Medical Information Systems Roundtable.* p. 799-803. JOM 24(10) Supplement.

Hanrahan, L. P. and Moll, M. B. 1989. VIII. Injury surveillance, in *Surveillance in Occupational Safety and Health*, Ed. E. L. Baker, p. 38-45. *AJPH* 79 Supplement.

Hillman, G. 1982. ECHOES: IBM's environmental, chemical and occupational evaluation system, in *Medical Information Systems Roundtable.* p. 827-835. *JOM* 24(10) Supplement.

Holzner, C. L., Hirsh, R. B., and Perper, J. B. 1993. Managing workplace exposure information. *Am Ind Hyg Assoc J.* 54(1):15-21.

Johnson, L. F. 1996. A world of great choices. *Occupational Health and Safety.* October: 70 and 107.

Joiner, R. L. 1982. Occupational health and environmental information systems: basic considerations, in *Medical Information Systems Roundtable.* p. 863-866. *JOM* 24(10) Supplement.

Joyner, R. E. and Pack, P. H. 1982. The Shell Oil Company's computerized health surveillance system, in *Medical Information Systems Roundtable*. p. 812-814. *JOM* 24(10) Supplement.

Kuritz, S. J. 1982. The Ford Motor Company environmental health surveillance system, in *Medical Information Systems Roundtable*. p. 844-847. *JOM* 24(10) Supplement.

Langmuir, A. D. 1976. William Farr: founder of modern concepts of surveillance. *Int. J of Epidemiology*. 5(1):13-18.

Menzel, N. N. 1994. Occupational health software: Selecting the right program. *AAOHN Journal*. 42(2):76-81.

National Institute for Occupational Safety and Health (NIOSH). 1986. *Proposed National Strategy for the Prevention of Musculoskeletal Injuries*. DHHS (NIOSH) No. 89-129. Cincinnati, OH.

Occupational Safety and Health Administration (OSHA). 1970. *Occupational Safety and Health Act*. Public Law 91-596 91st Congress S. 2193 Dec. 29. U.S. Government Printing Office, Washington, D.C.

Occupational Safety and Health Administration (OSHA). 1996. OSHA notice of proposed rulemaking to revise occupational injury and illness recording and reporting requirements. (61 FR 4030, Feb. 2, 1996) *Occupational Safety and Health Reporter*. 2-7-96:1219-1270.

Occupational Safety and Health Administration (OSHA). 1990. *Ergonomics Program Management Guidelines for Meatpacking Plants*. OSHA 3123. Department of Labor/OSHA, Washington, D.C.

Parkinson, D. K. and Grennan, M. J. Jr. 1986. Establishment of medical surveillance in industry: Problems and procedures. *JOM*. 28(8):773-777.

Saldaña, N. 1990. DAS: A graphical computer system for the collection of musculoskeletal discomfort data. *Proceedings of the Human Factors Society 34th Annual Meeting*. p. 1097.

Seligman, P. J., Halperin, W. A. E., Mullan, R. J., and Frazier, T. M. 1986. Occupational lead poisoning in Ohio: Surveillance using workers' compensation data. *AJPH*. 76(11):1299-1302.

Smith, R. B. 1995. Recordkeeping rule aims for accuracy, wiser use of injury and illness data. *Occupational Health and Safety*. January:37-68.

Smith, F. R., Gutierrez, R. R., and McDonagh, T. J. 1982. Exxon's health information system, in *Medical Information Systems Roundtable*. p. 824-826. JOM 24(10) Supplement.

Stuart-Buttle, C. 1994. A discomfort survey in a poultry-processing plant. *Applied Ergonomics*. 25(1):47-52.

Stuart-Buttle, C. 1999. How to set up ergonomic processes: A small industry perspective, in *The Occupational Ergonomics Handbook*, Eds. W. Karwowski and W. S. Marras, CRC Press, Boca Raton, FL.

Sugano, D. S. 1982. Worker tracking — a complex but essential element in health surveillance systems, in *Medical Information Systems Roundtable*. p. 783-784. *JOM* 24(10) Supplement.

Sundin, D. S., Pedersen, D. H., and Frazier, T. M. 1986. Occupational hazard and health surveillance. Editorial. *AJPH*. 76(9):1083-1084.

The Synergist, 1995. Graphically speaking. *The Synergist*. October:7.

Tyson, P. R. 1991. Record-high OSHA penalties. *Safety and Health*. March:17-20.

Wegman, D. H. and Froines, J. R. 1985. Surveillance needs for occupational health. Editorial. *AJPH*. 75(11):1259-1261.

Wolkonsky, P. 1982. Computerized recordkeeping in an occupational health system: The Amoco system, in *Medical Information Systems Roundtable*. p. 791-793. *JOM* 24(10) Supplement.

Wrench, C. P. 1990. *Data Management for Occupational Health and Safety: A User's Guide to Integrating Software*, Van Nostrand Reinhold, New York.

Yodaiken, R. E. 1986. Surveillance, monitoring, and regulatory concerns. *JOM*. 28(8):569-571.

## For Further Information

ACOEM. 1996. *Directory of Occupational Health and Safety Software*. Version 9.0. American College of Occupational and Environmental Medicine, Arlington Heights, IL. Telephone: 847 228-6850.

Brauer, R. (Ed). 1996. *Directory of Safety Related Computer Resources*. American Society of Safety Engineers, Chicago, IL. Telephone: 847 699-2929.

Canadian Centre for Occupational Health and Safety (CCOHS). 1997. Directory of Health and Safety Software (Diskette). AIHA Press, Fairfax, VA. Telephone: 703 849-8888.

Centers for Disease Control and Prevention (CDC). 1996. *Epi Info, Version 6.04a.* Brixton Books, 740 Marigny Street, New Orleans, LA 70117. (504) 944-1074. Internet: http://www.cdc.gov

# 25

# OSHA Recordkeeping

Stephen J. Morrissey
*Oregon OSHA Consultative Services*

## 25.1 Introduction

The history of occupational safety has been recorded as progress primarily initiated by injury, loss, and disaster. Despite many well-meaning efforts at controlling the human and social–economic losses in the workplace, it was not until the passage of the Occupational Safety and Health Act of 1970 that a comprehensive, integrated safety and health program at the national level was present in the United States. The OSHA Act, while safety and health oriented, has clear social and economic goals in its basic philosophy as reflected in the enabling legislation which in part states that the purpose of the Act was "… to assure so far as possible every working man and woman in the Nation safe and healthful working conditions and to preserve our human resources."

Over the years the regulatory aspects of OSHA have become more complicated and widespread with changing industries and job demands, public pressures for changes in safety law, and as the work environments have become better understood. This increased regulatory presence has not been without industry complaint. However, since OSHA has been in existence there have been clear drops in the numbers of workers injured, killed, and suffering adverse health effects, all of which translate into lower costs of doing business (Gray and Scholz, 1993; U.S. Department of Labor, 1995, 1997).

Fundamental to the effective control of job-related injuries and illnesses has been the development of a comprehensive safety and health recordkeeping standard. The Occupational Safety and Health Act of 1970 gave OSHA the authority to develop such a program, and in 1971 the occupational injury and illness recordkeeping regulations, 29 CFR Part 1904, were published. These regulations set comprehensive and mandatory requirements for reporting job-related injuries and illnesses. The basic goals of OSHA recordkeeping are to establish the risks present in the work environment; to direct employer attention to tasks and areas that need attention or which have unacceptable injury–illness levels; and to help guide research and enforcement activities to better serve industry and the employee.

However, despite the breadth and detail of these recordkeeping goals, the overall recordkeeping requirements have been kept limited to reduce the burden on employers and to help safety and health professionals identify and then respond to high-risk industries and workplaces. Over the years, OSHA has set forth and enforced the recordkeeping regulations. The Bureau of Labor Statistics (BLS) has prepared survey and recordkeeping forms and information and conducted the Annual Survey of Occupational Injuries and Illnesses. Information from this survey is used by OSHA to identify industries with high injury rates. In 1990, the duties performed by the BLS were transferred to OSHA.

This chapter will discuss the basics of occupational safety and health recordkeeping as mandated by OSHA. It will not discuss the parallel systems used by workers' compensation systems, as these systems differ widely by state and many major changes in compensation systems are being undertaken or planned. There are also major changes proposed in the actual mechanics of OSHA recordkeeping and in definitions of critical factors that are used to record data. At the time of writing, only preliminary proposals have been presented for these changes, and these proposed changes in recordkeeping will be identified and discussed later in this paper. To obtain the most current information, the interested reader is encouraged to contact his or her local or regional OSHA office or access the OSHA world wide web site at: http://www.osha.gov or http://www.osha-slc.gov.

## 25.2    Who Must Keep Records

Compliance with OSHA recordkeeping is required of private sector employers in the 50 States, the District of Columbia, Puerto Rico, the Virgin Islands, American Samoa, Guam, and the Trust Territories of the Pacific Islands. In general, employers who employed at least 11 or more employees at any time in the previous calendar year must comply with OSHA recordkeeping regulations. If employers have more than one establishment or place of business, it is the combined total number of employees at all places of work that is considered in determining the number of employees in the previous calendar year. If this number is 11 or more at any time in the previous calendar year, then the company must comply with OSHA recordkeeping requirements, and records must be kept at *each* individual place of employment. Employers with fewer than 11 employees at any one time in the previous calendar year are generally exempt from recordkeeping, as are the self-employed, partners with no employees, those that employ domestic help in their private residences, and employers engaged in religious activities, services, or rites. However, on occasion, the Bureau of Labor Statistics (BLS) will require companies with fewer than 11 employees or in other generally noncovered occupations to keep records in order to develop more accurate hazard data for their Annual Survey of Occupational Injuries and Illnesses.

Employers who are required to keep records must do so at each of their establishments, regardless of the number of employees at the site, but they *do not have to keep records for employees of other firms, independent contractors, or companies temporarily present on the worksite.* If an injury or illness occurs to the employees of the other firms on the worksite, these injuries or illnesses must be recorded on the records of the company for which the injured employee works, or that of the company which has the most immediate control and supervision of work activities. Thus, if the employer is using temporary or leased workers, and has immediate control of their day-to-day work activities, then the employer must record any injuries or illness on the employer's own records. Employers must also present copies of the OSHA 200 and 101 upon request of OSHA inspectors and may, under the direction of other codes, have to provide these data to employees or their legal representatives.

## 25.3    Which Employers Do Not Have to Keep Records

Despite the apparent simplicity of the recordkeeping rule of 11 or more employees, not all industries or occupations are required to follow OSHA recordkeeping requirements. Most manufacturing, agricultural, construction, and service industries are required to keep records if they have the required number of employees. However, OSHA recordkeeping is generally not required for employers in what are considered low-risk occupations. These low-risk industries or occupations include the retail trade, finance, insurance, real estate, and various service industries in the Standard Industrial Codes (SIC) 52-89. However, there are some employers within this group who must follow recordkeeping regulations. These covered industries are building materials and garden supplies, SIC 52; general merchandise and food stores, SIC 53 and 54; hotels and other lodging places, SIC 70; repair services, SIC 75 and 76; amusement and recreation services, SIC 79; and health services, SIC 80. Also exempt from recordkeeping requirements are those industries which have an average lost workdays cases injury rate over a three-year recording period that is at or below 75% of the U.S. private sector injury rate. Even though these industries are exempt from general federal recordkeeping requirements, they may be subject to state recordkeeping requirements, must comply with all other safety and health requirements set by OSHA, and as noted above, may be required by the BLS to participate in special studies to establish injury and illness rates.

Employers who do not have to keep formal OSHA records either because of the number of employees or by virtue of their SIC code still must meet reporting and recordkeeping requirements in the event of the death of an employee or if multiple employees are hospitalized as a result of a work-related incident or exposure. Basically these requirements are to notify the local OSHA office within eight hours in the event of an employee death or within eight hours in the event of an incident that results in the hospitalization of three or more individuals.

Employers who are subject to other federal government safety and health regulations must also comply with basic OSHA recordkeeping requirements. However, these employers can use modified reporting forms as long as they are equivalent to the required OSHA 200 and 101 forms discussed later in this paper.

Churches or other religious organizations or their affiliated operations do not have to comply with recordkeeping requirements for those individuals who perform religious rites or activities. However, churches or other religious organizations must comply with OSHA recordkeeping requirements for those individuals not engaged in performance of religious rites or activities. In some cases, volunteers who are not clergy or direct participants in religious services, may be covered by recordkeeping requirements.

Charitable and non-profit organizations, if they meet the number of employees criteria, must comply with recordkeeping requirements.

More detailed information on specific employers, industries, and SIC codes that are required to keep records, and to what extent, is available in the two primary documents on recordkeeping (U.S. Department of Labor, 1986 a,b).

It must be noted again that this discussion applies only to OSHA mandated recordkeeping. Recordkeeping requirements mandated by insurance or workers' compensation laws are often separate and different from those of OSHA, and employers should consult with their own state to determine applicable laws and codes.

## 25.4 What is An Employer? An Employee?

A critical aspect of recordkeeping is the determination of who is an employer, an employee, or what is a place of employment. In Section 3(5) of the Act, an employer is defined as "…a person or persons engaged in business that directly or indirectly affects commerce and has one or more employees and who is not the United States or any State or political subdivision of a State."

An employee is any individual who receives compensation of any form from the employer for services. This also applies to part-time, temporary workers or leased workers or employees. Volunteers are exempt from recordkeeping if they serve of their own free will and do not receive any compensation for their efforts. Temporary workers are a particular issue these days as more and more employers use them to fill staffing needs. Despite common views, work-related injuries and illness experienced by temporary employees are reported on the records of the employer who has the most immediate supervision, direction, or control of them. This is almost always the employer who is using the temporary worker. Eventual liability for any citable violations of safety and health codes may, however, be distributed between both the immediate employer and the temporary worker supplier.

There are many more highly specific questions that can be raised that relate to the role and definitions of an employer and employee, and many of these are discussed in the recordkeeping reference, *Recordkeeping Guidelines for Occupational Injuries and Illnesses* (U.S. Department of Labor, 1986b). These documents are available from the U.S. Government Printing Office, local and regional OSHA offices, and from OSHA's world wide web sites (http://www.osha.gov; http://www.osha-slc.gov).

A place of employment is any location at which services or industrial operations are performed or where employees report to work, work at, or where they are paid. Records have to be kept for each individual workplace.

## 25.5  What Records Have to be Kept, Should be Kept?

While the regulatory coverage under OSHA is extensive and sometimes complicated, the recordkeeping is relatively simple, with only two forms required. These two forms are:

*OSHA 200: Log and Summary of Occupational Injuries and Illnesses*: The OSHA 200 form is widely available and will not be reproduced here because of its large size (11 inches by 17 inches). The OSHA 200 is used to record and classify all recordable injuries and illness and contains three basic sections. Section I records information on the date of the injury, who was injured, how, and where (Columns A–F). Section II records the nature and severity of the injury/illness, how much time was lost, and whether the injury resulted in a fatality (Columns 1–6). The third section (Columns 7–13) contains information on occupational illnesses experienced by the employee. Section 7 has seven subsections (a to g) that further define the nature of the occupational illness.

Data for any injury or illness must be filled in within six days after the incident is discovered and, in the case of a fatality or multiple hospitalizations (three or more workers), the local OSHA office must be notified within eight hours. At the end of each calendar year, the data on this form are totaled at the bottom, signed, and dated. The summary sections describing overall lost workdays (Columns 1–6) must be posted in a conspicuous location at each work location for the entire month of February in the following calendar year. The completed forms must be kept at each work location for the next five calendar years, even if an establishment changes owners. If, in years subsequent to the one in which an injury or illness is recorded, there need to be changes in a recorded entry based on new information, then the original entry must be lined out (not erased) and a corrected entry made. If errors in recording are discovered, or if unreported cases are identified, these need to be entered.

Conversely, if it is later found that an entry is a nonreportable entry, it should be lined out but not erased. While only recordable injuries have to be entered on the OSHA 200, many companies enter all injuries and illnesses on the form to help identification, understanding, and control of hazards that may be present in the workplace.

*OSHA 101, Supplemental Record of Occupational Injuries and Illnesses*: For every recordable event entered on the OSHA 200 log, it is necessary to record additional information on the OSHA 101 form or its equivalent. This form gives additional information that describes how the accident or illness occurred, the work processes involved, substances or exposures present, and more detail on the worker and the exact nature of the injuries or illnesses experienced. This form is also commonly used by workers' compensation carriers and various state organizations for filing and evaluating claims. In general, entries on the OSHA 101 must be completed within six working days of the date the employer received information on the event. A major problem with the OSHA 101 and 200 forms is that the data from these forms are not always properly interrelated, or even related to information describing the actual accident. This lack of correspondence can complicate or slow down understanding and control or elimination of hazards.

## 25.6  Posting Requirements

Employers who are required to keep records must keep OSHA 101 and 200 forms for each physical location where work is performed. If an employer has multiple locations, then separate sets of records must be kept for each distinct location. Further, if within a physical work location distinctly different or separate activities are performed, separate records must be kept for each area. As noted earlier, the OSHA 200 must be summarized and posted for the entire month of February in a place that is used by all employees.

Failure to keep proper records can lead to substantial fines from OSHA. Over the past few years, some recordkeeping penalties have approached a million dollars, so the consequences can be substantial from a regulatory aspect. The more real penalties that occur from improper recordkeeping are the loss of control of the economics of production. Studies have found that companies that keep good records have good safety and health programs, lower overall production costs, and higher profitability (Gray and Scholz, 1993). The changes in OSHA regulations proposed *at the time of the writing of this chapter* all support a reduction in penalties or enforcement schedules for companies that have a well-developed safety and health program. OSHA considers careful recordkeeping and tracking of progress made in addressing accidents, injuries, and illness as the cornerstone of any such program.

## 25.7   Essential Definitions and Concepts for Recordkeeping

*Recordable Injury/Illness:* By OSHA definition, a recordable event is any "... work-related death and illness, and those work-related injuries which result in: Loss of consciousness, restriction of work or motion, transfer to another job, or require medical treatment beyond first aid." (U.S. Department of Labor, 1986a). It is important to understand that a case being recordable does not necessarily mean that the employer is at fault, has violated any OSHA standards, that the worker is at fault, or that any injuries are of a compensable nature. Recording an event only means that the consequences to the worker meet the legal definitions for being recorded. Once a case is determined to be recordable, it must be recorded in the correct columns on the OSHA 200 and 101 forms or their equivalent.

Other tests to define the recordability of an injury or illness are that an injury or illness is recordable if it results in *any* of the following: (1) the inability of the worker to perform all parts of the normal work assignments with the same ability as before the incident; (2) transfer of the worker, even temporarily to another job; (3) physical damage to the structures of the body, such as fractures, cuts requiring stitches, burns and bruises requiring repeated visits to a health care provider, and continuing infections; (4) loss of consciousness; or (5) an illness or injury that requires treatment by licensed medical personnel or physicians.

*Work Related:* An injury or illness is considered to be work-related if it resulted from a single or cumulative exposure or event in the work environment. The work environment is the immediate workplace and any and all locations where employees are engaged in work-related activities, or have to be present at, as part of their normal work activities. When workers normally have to travel between different worksites as part of work or to perform a work duty, the transportation may also be considered as a work activity, and injuries that occur during this transportation may be recordable. With many types of injuries and illnesses, determination of work-relatedness can be complicated, and the interested reader is referred to either of the documents on recordkeeping (U.S. Department of Labor, 1986 a,b) for more detailed information. Determination of work-relatedness is usually easier for injuries than for illnesses, particularly for those illnesses whose symptoms only appear after long exposures or long latency periods after a critical exposure.

*First Aid:* With respect to OSHA, first aid is "...any one-time treatment, and any follow up visit for the purpose of observation of minor scratches, cuts, burns, splinters, and so forth, which do not ordinarily require medical care. Such one-time treatment and follow up visits are considered as first aid, even though provided by a physician or registered professional personnel."

The critical distinction between medical treatment (a recordable case) and first aid treatment (a nonrecordable case) depends not only on what treatment is provided but also on the severity of the injury involved. A brief listing of different types of injuries and whether they classify as being first aid or as medical treatment is presented as Table 25.1 and further discussion of the recordability of injuries and illnesses is given in the two basic documents on recordkeeping (U.S. Department of Labor, 1986 a,b).

**TABLE 25.1**    Determining an Injury's Recordability

| Type of Injury | Type of Treatment Needed and Subsequent Recordability First Aid Only, Not Recordable | Requires Medical Treatment Recordable |
|---|---|---|
| **Cuts, lacerations, abrasions, splinters, punctures** | Bandaging on any visit to health care provider (HCP) Application of antiseptic on first visit to HCP Application or ointments on any visit to prevent drying or cracking of skin Removal of foreign bodies by tweezers or other simple techniques Removal of non-embedded foreign bodies in eye by irrigation | Stiches Butterfly sutures Medical treatment of infections or application of antiseptic on second or subsequent visits to HCP Removal of foreign bodies requiring the skilled services of physician |
| **Fractures** | X-ray is taken as a precaution and found to be negative. | X-ray shows fracture. Cast or other professional method of immobilization of limb is applied. |
| **Strains, sprains, and dislocations** | Use of an elastic bandage on strain on the first visit to HCP Use of hot or cold compresses on strain on the first visit to HCP | Application of casts or other professional means of immobilizing injured part, including rigid splints Use of hot or cold compresses for treatment of strains, sprains, or dislocation on second or subsequent visits to HCP Use of diathermy and whirlpool treatment Hot wax treatments |
| **Thermal or chemical burns.** Any burn is recordable if the worker cannot perform any of his or her normal duties even if medical treatment is not required. | Treatment by HCP for first degree burn | Treatment of all second- and third-degree burns. |
| **Bruises, contusions.** Any bruise or contusion is recordable if the worker's range of motion is affected in any manner which prevents him or her from performing any of the regularly assigned duties, regardless of whether medical treatment is rendered. | Soaking or application of cold compresses to a bruise on the first visit to HCP | Treatment of a bruise by draining of collected blood Soaking or application of cold compresses on second or subsequent visits to HCP |
| **Miscellaneous procedures** | Tetanus shots Observation of injury on subsequent visits | Any hospitalization, even if only for observation will usually result in lost workdays and is thus recordable Use of prescription drugs |

*Injury:* Is any injury or disorder such as a cut, fracture, sprain/strain, etc., that results from a work accident or from a *single, instantaneous exposure* to the work environment. This includes exposures to all work processes, insects, animals, and chemical or toxic agents.

*Illness:* Is any abnormal condition or disorder other than an injury that is a consequence of exposure to environmental factors related to work and the work environment. This includes cumulative, acute and chronic illness or diseases resulting from any exposure to agents in the work environment by inhalation, absorption, ingestion, direct contact, or direct exposure. There are seven different categories of occupational illness which are recorded on the OSHA 200 form. These illness categories are: Occupational skin diseases or disorders; dust disease of the lungs; respiratory conditions due to toxic agents; poisoning (systematic effects of toxic materials); disorders due to physical agents other than toxic materials; disorders associated with repeated trauma (noise-induced hearing loss, vibration-related injuries such as Raynaud's phenomenon, and other conditions arising from repetitive motion such as bursitis, synovitis, tenosynovitis); and, all other occupational illnesses.

Criteria to identify and evaluate the work-relatedness of injuries or illnesses related to repeated trauma (cumulative trauma), such as tendinitis, synovitis, bursitis, back injuries, and carpal tunnel syndrome, are more fully developed in the *Ergonomics Program Management Guidelines for Meatpacking Plants* (U.S. Department of Labor, 1993). This document, which is available on OSHA's world wide web servers or from OSHA centers, describes the stages in developing an ergonomics program to reduce the losses associated with ergonomics injuries.

*Restricted Work or Motion:* A work or motion restriction occurs when, as a consequence of a work-related injury or illness, the employee cannot perform any part of, or all of, the normal job assignments during any part of, or all of, a workday or workshift. The common practice of having someone move to another job for a while to rest up thus constitutes a work restriction and requires the accident or illness to be recorded. The key point to understand is that if, as a consequence of a work-related accident, the employee cannot do the same job or work assignments at the same rate as before the accident, then the accident is recordable.

## 25.8    Recording Data on OSHA Forms

Once an injury or illness is determined to be recordable it must be evaluated to determine its type and outcomes. There are three types of recordable cases: Fatalities; accidents resulting in lost workdays; and cases that are recordable but do not result in lost workdays. Each case must be entered into the correct columns on the OSHA 200 form.

For fatalities, the entry is obvious, but determination of what is to be entered for lost workdays can be more complicated. Lost workdays occur either because the worker could not be at work due to the injury or illness, was unable to perform the normal job duties for a period of time, or the worker was assigned to another job until he or she recovered (light duty, modified duty, restricted work). In all cases, the total number of days to be recorded is the total number of days that worker was not present due to the injury or illness not including the initial day of injury/illness, or any days that the worker would not have normally worked (holidays, weekends, special days off, vacations) as a result of the injury or illness. If the worker's injury extends into another year, the amount of time that will be lost in the next year must be estimated and entered for the year in which the injury occurred.

Once the total number of days missed is correctly entered, it becomes important to properly and fully detail the remainder of the OSHA 200 and 101 to facilitate tracking of problems in the workplace as well as to meet recordkeeping requirements. While the various columns on the OSHA 200 form are self-evident, a common error is to incorrectly or too briefly describe the events that led to the accident. As much as possible, the physical location where an accident occurred (loading dock station 3; not "receiving"), actual job title of the individual involved in the accident (assistant press operator, machine #3; not machine operator), and a clear description of the injury (laceration to top of left hand; not cut hand) should be included on the 200 form to facilitate tracking of accidents, injuries, illnesses, and losses.

*Data Developed From OSHA 200 Form:* There are several statistics that are developed from the OSHA 200 form to describe the relative and absolute injury and illness experience for a particular facility, company, or industry. These data are used by OSHA to direct inspections and research and by insurance providers in rate setting.

All OSHA statistics are based on an Incidence Rate statistic (Equation 25.1) that is the number of injuries, illness, or lost workdays normalized to reflect a worksite with 100 full-time employees. As all employers who must comply with OSHA must use the same basic criteria for recordkeeping, this allows direct comparison of individual company loss experience with respect to the respective industry or region, across time, or even within a company. The statistic that is most commonly used by OSHA to direct inspections, define research needs and also by insurers to set and adjust rates is the Lost Workdays Case Incident Rate (LWDCIR). The LWDCIR is calculated by entering the appropriate date into the following equation:

$$\text{Incidence Rate } = \frac{\text{N} \times 200{,}000}{\text{HW} \times 2{,}000} \tag{25.1}$$

where

N         = total number of recordable injuries, illnesses, or lost workdays (sum of Columns 2 and 9)

HW        = total number of hours worked by all full-time and part-time employees during the previous calendar year as derived from payroll data

*200,000* = the normalizing factor to correct data to a 100 full-time employee workforce that works 40 hours a week, 50 weeks a year

## 25.9    Proposed Changes in OSHA Recordkeeping

Over the past 10 years, numerous criticisms have been leveled against OSHA's recordkeeping requirements by industry, government, and academic policy and study groups. Some of the most common criticisms included that the current system was too complicated, particularly with respect to definition of injury and illness classifications; that the current system encouraged under-reporting or misreporting of injuries and illnesses; tracking of changes in the system was hard to follow; difficulty of the forms to facilitate good accident investigation; and, the difficulty of establishing "work-relatedness" of injuries and illnesses.

In February of 1996, OSHA published its proposed modifications to recordkeeping, "29 CFR Parts 1904 and 1952: Occupational Injury and Illness Recording and Reporting Requirements; Proposed Rule" for comment in the February 2, 1996, *Federal Register.* This proposal, which will be debated and discussed for the next year or so, involves major changes in several aspects of recordkeeping. Important definitions and criteria that are being proposed for change are briefly described in the following paragraphs. In addition to changed definitions, the recording forms would be changed to the OSHA 300 (replacing the OSHA 200) and the OSHA 301 (replacing the OSHA 101). These proposed forms (from the *Federal Register* of February 2, 1996) are shown as Figures 25.1 and 25.2. Bearing in mind that this legislation is being discussed and changed *at the time of publication of this book,* the interested reader should consult the *Federal Register* or a local OSHA office for more information.

*OSHA 300 and 301 Forms:* The new OSHA 300 form (Figure 25.1) is designed to fit on standard 8½ inch by 11 inch paper and is significantly streamlined in definitions and instructions for use as well as providing a column for the employer's use. Some of these improvements are described in the following sections and are designed to make recordkeeping easier and more consistent. The use of computerized equivalents is specifically allowed in the proposed regulations to increase speed and accuracy of recordkeeping as well as to facilitate better database manipulation and study. Similarly, the OSHA 301 (Figure 25.2), has been redesigned to facilitate ease and consistency of data entry and linking to other reports.

## 25.10    Specific Changes in Recordkeeping Requirements

All employers must maintain an OSHA 300 log or its equivalent at each establishment. Employers with multiple establishments may maintain a consolidated log for establishments with no more than 20 employees. Recordable injuries or illness must be recorded within seven calendar days of receiving information that a recordable injury or illness has occurred. The employer will also maintain an Injury and Illness Incidence Record, OSHA 301 or its equivalent, and it must be updated within seven calendar days of receiving information that a recordable injury or illness has occurred.

Before the end of January of each calendar year, the employer will post a year-end summary of all occupational injuries and illness for each establishment for the previous year. This form will remain posted for the *entire year* and be updated as new information is determined. This form *must be signed by a responsible company official indicating that they have examined the summary log and that the year-end summary is true, accurate, and complete.* Records must be kept for three years following the calendar year they represent rather than the current five-year period. During this retention period, the OSHA 300 and 301 must be updated if new information is discovered.

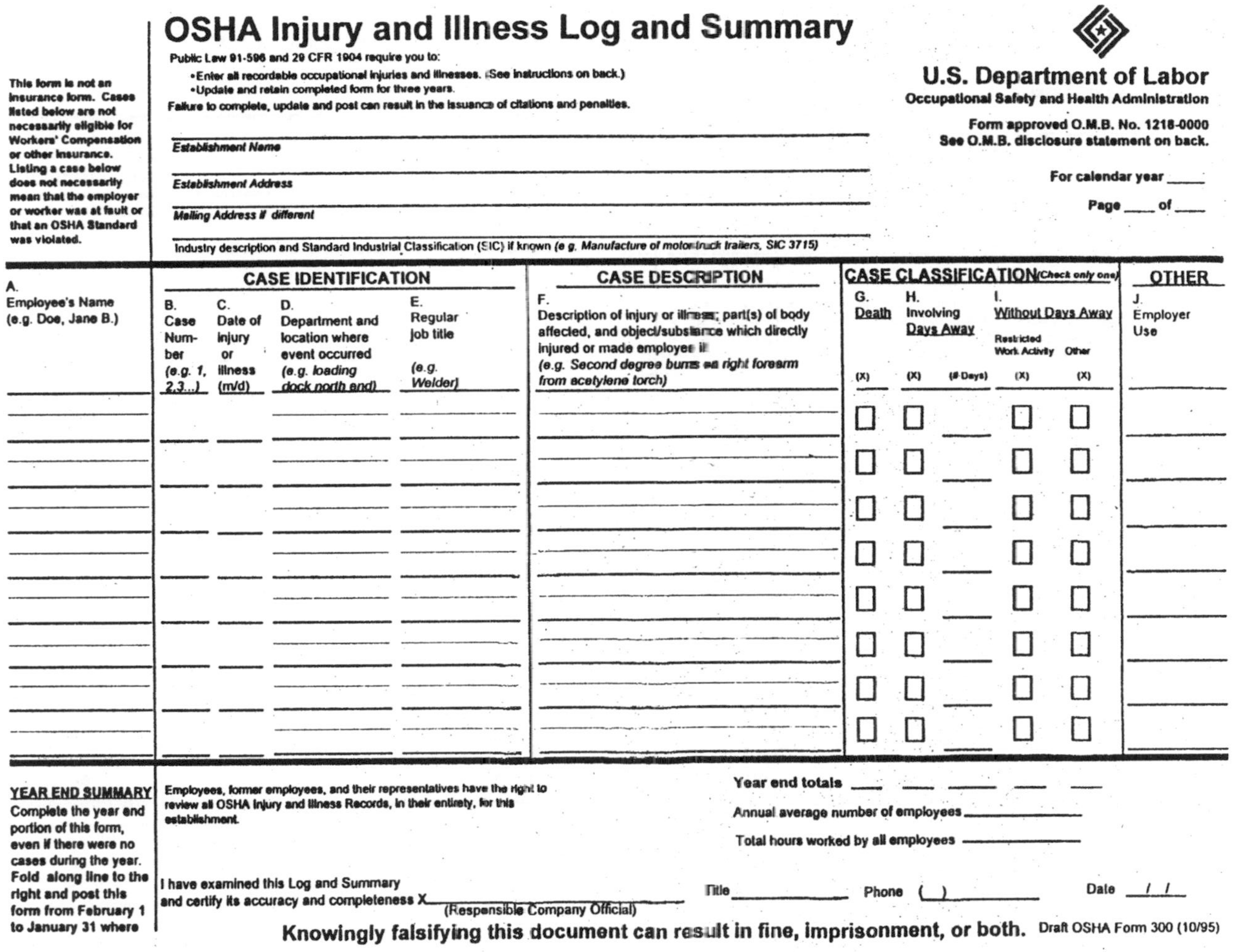

FIGURE 25.1 Proposed OSHA 300 form that will replace OSHA 200. (From *Federal Register*, Friday, February 2, 1996; 29 CFR Parts 1904 and 1952: Occupational Injury and Illness Recording and Reporting Requirements; Proposed Rule; pages 4030–4067.)

# OSHA Injury and Illness Incident Record

Public Law 91-596 and 29 CFR 1904 require you to update and retain completed form for three years.

Failure to complete this form can result in the issuance of citations and penalties.

Employees, former employees, and their representatives have the right to review all OSHA Injury and Illness Records, in their entirety, for this establishment.

This form is not an insurance form. Cases listed below are not necessarily eligible for Workers' Compensation or other insurance. Listing a case below does not necessarily mean that the employer or worker was at fault or that an OSHA Standard was violated.

**U.S. Department of Labor**
Occupational Safety and Health Administration

Form approved O.M.B. No. 1218-0000
See O.M.B. disclosure statement on back.

_____ Case number from OSHA Form 30

## Employee

1. **Last name**      First name      MI

2. Male ☐   Female ☐    3. Date of birth   /   /

4. Home address

5. Date hired   /   /

## Health Care Provider

6. Name of health care provider

7. If treatment off-site, facility name and address

8. Hospitalized overnight as in-patient?
(If emergency room only, mark "no")    yes ☐   no ☐

## Employer Use (Optional)

| **Completed by** | |
|---|---|
| Name | Title |
| Phone (  ) | Date |

## Illness or Injury

9. Specific injury or illness
(e.g. Second degree burn or Toxic hepatitis )

10. Body part(s) affected (e.g. Lower right forearm)

11. **Date of Injury or Illness:**   /   /    12. If employee died, date of death   /   /

13. If the case involved days away from work or restricted work,
activity, enter the date the employee returned to work at full capacity:   /   /

14. Time of event:      15. Time employee began work:
(Specify a.m. or p.m.)      (Specify a.m. or p.m.)

16. All equipment, materials, or chemicals employee was using when the event occurred.
(e.g. Acetylene cutting torch, metal plate)

17. Specify activity the employee was engaged in when the event occurred
(e.g. Cutting metal plate for flooring) Indicate if activity was part of normal job duties.

18. How injury or illness occurred. Describe the sequence of events and include
any objects or substances that directly injured or made the employee ill.
(e.g. Worker stepped back to inspect work and slipped on some scrap metal.
As she fell, worker brushed against the hot metal)

Draft OSHA Form 301 (10/96)

FIGURE 25.2 Proposed OSHA 301 form that will replace OSHA 101. (From *Federal Register*, Friday, February 2, 1996; 29 CFR Parts 1904 and 1952: Occupational Injury and Illness Recording and Reporting Requirements; Proposed Rule; pages 4030–4067.)

If an establishment changes hands, the new owners are responsible for recording only those injuries and illnesses that occur after the purchase date. They still have to keep the previous owners' records for the required period of time.

*Access to Records*: The employer must, upon request by a government representative, provide copies of OSHA Forms 300 and 301 or their legal equivalents, year-end summaries for their own employees, and injury and illness records for "subcontractor employees." When the request for records is in person, the information must be provided in hard copy within four hours. If the request is in writing, the information must be made available within 21 days unless requested otherwise.

Upon request, employees, former employees, and/or their representatives may see the OSHA 300 and 301 forms or their legal equivalents. These forms must be made available for viewing by the close of business on the next scheduled workday. An individual or his or her representative may also request to see the OSHA 301 form for his or her own injury or illness.

*Days Away from Work (Lost Workdays)*: is defined as the total number of days the employee could have worked but did not because of the injury/illness. These days do not include the day the worker was injured or days the employee would not have normally worked because of holidays or weekends. The maximum number of days away from work will be entered as 180 or "180+" in the days lost column. For injuries that last into the next calendar year, the number of days lost should be estimated and entered into the days lost column for the year in which the injury/illness occurred. The 180 or 180+ days lost limits still apply.

*Employee*: an employee is any full-time, part-time, temporary, leased, limited service, "independent contractor" subject to day-to-day supervision, and those corporate officials who receive compensation of any form. Day-to-day supervision is defined as specifying how all aspects of output, production, and work processes are done. This definition now formally includes temporary or leased workers, independent contractors, or migrant laborers.

*First Aid*: refers to any treatment in the following comprehensive list that is provided by any heath care provider for a work-related injury or illness:

1. Visit(s) to health care provider limited to observation.
2. Diagnostic procedures including use of prescription medicines solely for diagnostic purposes.
3. Use of nonprescription medications, including antiseptics.
4. Simple administration of oxygen.
5. Administration of tetanus or diphtheria shot(s) or booster(s).
6. Cleansing, flushing, or soaking wounds on skin surface.
7. Use of wound coverings such as bandages, gauze pads, etc.
8. Use of any hot/cold therapy for local relief, except for musculoskeletal disorders. (Musculoskeletal disorders are covered in a special section, Appendix B, in the Proposed Rules.)
9. Use of any totally non-rigid, non-immobilizing means of support, such as elastic bandages.
10. Drilling of a nail to relieve pressure for subungual hematoma.
11. Use of eye patches.
12. Removal of foreign bodies not embedded in the eye, if only irrigation or removal with a cotton swab is needed.
13. Removal of splinters or foreign material from areas other than the eyes by irrigation, tweezers, cotton swabs, or other simple means.

Any procedure other than those listed above is considered as being medical treatment, and thus defines the injury or illness as recordable.

*Injury or Illness*: is any sign, symptom, or laboratory abnormality which indicates an adverse change in an employee's anatomical, biochemical, physiological, functional, or psychological condition.

*Work-Related*: is still a very complicated question, but under the proposed rules, an injury or illness is work-related if an event or exposure in the *work environment* either caused or contributed to the resulting condition or aggravated a pre-existing condition. The events or exposures must occur at the employer's establishment or as a consequence of a work activity performed as a condition of employment. More detailed descriptions of the criteria for work-relatedness are given in Appendix A of the Proposed Rules.

*Restricted Work Activity*: means that the employee is not capable of performing his or her normal work activities or the activity he or she was performing at the time of the injury or illness, at full capacity, for a full workshift.

*Recordable Injury or Illness*: a recordable injury or illness is one that meets all of the following four criteria:

1. An injury or illness, as defined above, exists.
2. The injury or illness is work related.
3. The injury or illness is new. A new injury or illness does not result from the recurrence of a pre-existing condition if no new or additional workplace incident or exposure or occurs.
4. The injury or illness meets one or more of the following conditions:
   a. results in death or loss of consciousness,
   b. results in day(s) away from work, restricted work activity or job transfer,
   c. requires medical treatment beyond first aid,
   d. is a recordable condition as listed in Appendix B of the proposed rules.

Only if an injury or illness meets *all* of these tests is it recordable.

## 25.11   Summary and Conclusions

The recordkeeping requirements developed by OSHA set the basic criteria for controlling economic and human loses in the workplace. While recordkeeping may seem to be a nonproductive, time-consuming task, it is the law, and when properly used, will result in a more profitable organization.

## References

Gray, W. and Scholz, J., Does regulatory enforcement work? *Law and Society Review*, July, 1993.

U.S. Department of Labor, Bureau of Labor Statistics (BLS), 1997. Workplace Injuries and Illnesses in 1995. Statistical Report from the Bureau of Labor Statistics, Safety and Health Statistics, March 12, 1997.

U.S. Department of Labor, Bureau of Labor Statistics. *A Brief Guide to Recordkeeping Requirements for Occupational Injuries and Illnesses.* June, 1986a. U.S. Government Printing Office.

U.S. Department of Labor, Bureau of Labor Statistics. *Recordkeeping Guidelines for Occupational Injuries and Illnesses.* September, 1986b. U.S. Government Printing Office.

U.S. Department of Labor, Occupational Safety and Health Administration, 1995. The Cumulative Impact of Current Congressional Reform on American Working Men and Women. Special Report prepared for presentation by Robert B. Reich, Secretary of Labor, August, 1995.

U.S. Department of Labor, Occupational Safety and Health Administration. *Ergonomics Program Management Guidelines for Meatpacking Plants (OSHA 3123).* 1993 (reprinted). U.S. Government Printing Office.

# 26

# Body Discomfort Assessment Tools

Leon M. Straker
*Curtin University of Technology*

## 26.1   Overview

This chapter reviews various body discomfort assessment tools that can be used in industry for the purpose of preventing work-related musculoskeletal disorders. Prior to describing the various tools available the theoretical background to discomfort assessment is reviewed to assist the ergonomics practitioner with decisions about when to use discomfort as a measure and which tool to use.

The aim of ergonomics is to provide a match between the person and the environment. In order to evaluate the success of any ergonomics input or intervention, tools are required which provide information on the degree of match or mismatch.

One aspect of the match between people and their environment that has become increasingly important is the physical aspect. An important reason for this is that it has been claimed that the high rates of work-related musculoskeletal disorders (such as work-related back problems and work-related neck and upper limb disorders) are at least partially attributable to physical mismatches.

Unfortunately, the etiology of these work-related musculoskeletal disorders is complex and controversial. This lack of etiology clarity makes an assessment of risk difficult. However, it is widely believed that discomfort is a useful risk indicator.

Conceptually, discomfort is an attractive risk indicator as it uses the body's own feedback system to detect possible problems.

Possible sources of discomfort resulting from musculoskeletal stress include: tension in muscles, nerves, blood vessels, ligaments, and joint capsules; compression of the same body tissues; local chemical changes associated with muscle fatigue; local chemical changes related to restricted blood flow and partial ischemia; disruption of nerve conduction resulting from pressure; and secondary inflammation.

As mechanical stress on tissue and local chemical changes are both thought to be sources of tissue damage and pathology, the potential utility of discomfort as a risk indicator is clear.

However, discomfort may also be influenced by psychological and social factors. This is seen by some as a major disadvantage, but perhaps it should not be seen as a disadvantage as the level of disability a certain pathology creates is also related to psychological and social factors.

Discomfort is thought to be especially useful for assessing situations where the impact of physical mismatch may be greatest on small muscles and where static muscle activity is required. This is beneficial, because small muscle problems are not detected well with other common risk assessment tools, such as biomechanical modeling, and gross physiological indicators, such as heart rate and body temperature.

The following section reviews some concepts fundamental to the use of discomfort as an indicator of risk of musculoskeletal disorder.

## 26.2   Fundamental Concepts

### The Discomfort Phenomenon/Construct

There has been little ergonomics research and discussion reported about the phenomenon of discomfort. Therefore, many of the fundamental concepts of discomfort measurement discussed here are drawn from the extensive research on the measurement of pain.

One of the important issues from pain measurement research is the specificity of terms. Pain researchers have found that different people interpret terms differently. This has important implications for discomfort measurement in ergonomics as the ergonomics literature has not been very specific about the use of the term discomfort. For example, some authors have considered pain and discomfort as synonymous. However, the error of this assumption is well illustrated in a study by Bates et al. (1989).

The study by Bates et al. was mainly aimed at investigating the effect of arm posture on worker performance at a computer task. They therefore had their 32 subjects perform a computer-based choice reaction time task for 8 minutes while working in both a "good" arm posture (no shoulder flexion) and a "poor" arm posture (45° shoulder flexion). However, they were also interested in the possible interactions of discomfort (from poor posture) and performance. The ergonomics literature available at the time of their study used discomfort and pain interchangeably; however, their understanding of the pain literature suggested this was not valid. To investigate the interchangeability assumption, their subjects were asked to rate both their shoulder discomfort and their shoulder pain, with a balanced order of recording.

Figure 26.1 shows the relationship between the discomfort and pain intensities recorded by Bates et al. It is clear that discomfort intensity tended to increase before pain did, suggesting discomfort is more sensitive at lower noxious stimuli levels.

However, Figure 26.1 also shows that the relationship between discomfort and pain was a nonlinear relationship. This suggests that different subjects interpreted the two terms, pain and discomfort, differently. The implication of this is that scales which use multiple nouns may not be valid.

Figure 26.2 shows the visual analog scale intensity ratings of discomfort and pain at the end of working in the 45° shoulder posture. The significant difference (38mm, $t_{31} = -6.04$, p = .0001) confirms that discomfort is a more sensitive scale than pain, although they may both be accessing the same feelings. This study also confirmed the utility of discomfort in gathering early information about possible physical ergonomics mismatches.

Another example of possibly invalid discomfort measurement scales is where some authors have considered discomfort and comfort to be on the same continuum. However, there is little evidence to suggest that this is true and some ergonomists have suggested that perhaps comfort is a separate entity which is more (or less?) than just the absence of discomfort. Branton (1969) illustrates this with the analogy that the absence of pain does not necessitate the presence of pleasure.

A final point about the importance of terminology in scales is that the interpretation of terms is likely to not only vary between individuals, but also between cultural groups. Thus, multiple term scales which may be valid in one culture may not be valid in another culture. With many modern societies being comprised of individuals from varied ethnic and cultural backgrounds, the clear implication for ergonomics is that

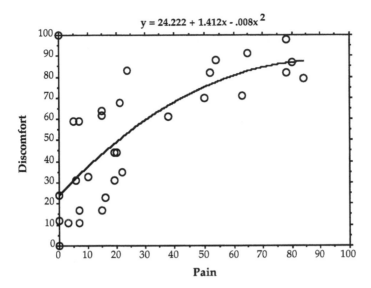

FIGURE 26.1 Relationship between discomfort and pain.

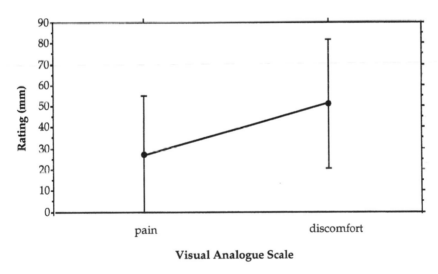

**Visual Analogue Scale**

**FIGURE 26.2** Pain and discomfort ratings (mean and standard deviation) after working in 45° shoulder flexion for 8 minutes.

widely applicable discomfort assessment tools should restrict the number of terms (nouns) used to ensure more consistent use of the tool across cultural groups.

## The Description of Discomfort

To adequately describe discomfort four aspects need to be covered — intensity, quality, location, and temporal pattern. For example, sitting on a hard chair for several hours may result in discomfort which could be described as a numb, cold feeling in the areas extending approximately 15 cm out from each ischial tuberosity which is of low to moderate intensity and which began after about 15 minutes of sitting and increased to the end of the first hour then remained at a constant level until arising from the chair, when the discomfort subsided to minimal intensity after 5 minutes. The following sections describe how the different attributes of discomfort can be measured.

## Intensity

Measurement of the intensity of discomfort has usually been attempted by asking the worker to rate the intensity on a scale commonly termed a subjective scale. Although there is a large number of possible subjective scales, they can be grouped into: verbal rating scales; visual analog scales; numeric rating scales; and graphic rating scales. There have also been a number of attempts to try to use a more "objective" measure of intensity. Thus, discomfort intensity is inferred from changes in behavior (using a behavior rating scale), or changes in correlated biomechanical and physiological entities. Examples of possibly suitable correlates are: estimates of static muscle tension (using biomechanical modeling); and estimates of muscle fatigue (using amplitude and frequency shifts in power spectrums of muscle electrical activity). The various types of scales and their relative advantages and disadvantages are described below.

### Biomechanical and Physiological Correlates

If discomfort is thought to arise from mechanical loads around joints, then it is reasonable to estimate those loads using position data and biomechanical modeling. Some studies have demonstrated a good correlation between joint load and discomfort ratings (Boussenna, Corlett, and Pheasant 1982). Similarly, if discomfort is thought to be due to sustained or high-magnitude muscle activity, then electromyography can provide an objective measure. Other physiological correlates which could be used are heart rate, blood pressure, respiratory rate, skin conductance, sweat rate, and skin temperature.

The advantage of these measures is their lack of reliance on worker reports.

However, it should always be remembered that it is not discomfort which is being measured. Rather, a correlation is assumed between the measure taken and discomfort. Another disadvantage is the potentially culture-specific nature of any correlation. For example, people in some cultures understand comfort to equate to dynamic balance (and resultant moderate muscle activity) and not the lack of muscle activity commonly accepted as "comfort" by people in Western cultures.

### Behavior Rating Scales

Some ergonomists have suggested measuring discomfort intensity by using observation of behavior thought to be indicative of discomfort reaching a certain intensity, such as fidgeting. For example, Branton (1969) suggested sitting was for a purpose and that discomfort can be seen as an interference which, when it reaches a sufficient intensity, results in changes to sitting posture. Thus, an increased number of postural changes would be considered to indicate an increase in discomfort intensity.

Shackel et al. (1969) considered the use of time-sampling of posture changes and duration as an objective measure of discomfort. However, the labor intensive nature of such observation was not thought to be feasible for their purposes. However, with newer technologies of electrogoniometry and digital motion analysis this may now be more feasible.

Corlett and Bishop (1976) recorded machine use and machine idle times for two weeks before and after ergonomic intervention. The increased overall percentage of work vs. idle time and the increased length of work periods were interpreted as suggesting a decrease in physical stress. That is, a reduction of discomfort allowed the workers to work productively for longer periods which increased overall production. It was also used to argue the cost-benefit of the changes as the cost of the machine changes were paid for in a few days of higher productivity.

One advantage of behavioral scale assessment is that it is independent of a worker's capacity and willingness to verbalize feelings. It also provides task interference information and thus can be more readily used in productivity-based justifications for ergonomic interventions. A major disadvantage is the assumption that the postural movements are due to discomfort reaching a certain intensity. For example, frequent movements could be the result of a good work habit of not maintaining static postures, and be a desirable characteristic to be encouraged in any ergonomic intervention.

### Verbal Rating Scales

There are two types of verbal rating scale: one in which a single noun is used to describe the construct (in this case "discomfort") and multiple adjectives are used to indicate changes in intensity, and another in which different nouns are used.

| no discomfort | minimal discomfort | moderate discomfort | severe discomfort | maximal discomfort |
|---|---|---|---|---|

**FIGURE 26.3**  Single noun verbal rating scale for discomfort intensity.

| relaxed | comfortable | neutral | uncomfortable | painful |
|---|---|---|---|---|

**FIGURE 26.4**  Multiple noun verbal rating scale for discomfort intensity.

no discomfort  extreme discomfort

**FIGURE 26.5**  Visual analog discomfort scale.

Figure 26.3 shows a single noun, multiple adjective verbal rating scale for discomfort which uses "no, minimal, moderate, severe, and maximal" to indicate increasing intensities of discomfort. Commonly five or seven categories are used.

Figure 26.4 shows an example of a multiple noun scale where the terms "relaxed, comfortable, neutral, uncomfortable, and painful" are used to indicate increasing intensities of discomfort.

For both types of verbal rating scale a rating of discomfort intensity is collected from the worker either by their circling a descriptor or their verbally reporting a descriptor. Analysis of these data is by frequency distributions and rank order nonparametric statistics.

The advantages of verbal rating scales are that they are relatively straightforward and easy for workers to understand.

One disadvantage of verbal rating scales is the relatively small number of points on the scale, resulting in only gross changes in the intensity of discomfort being detected. Another disadvantage is the assumption that feelings like discomfort can be verbalized (which has led some ergonomists to trial cross modality matching and suggest behavioral scales). Multiple noun scales have the added disadvantage of introducing error from different interpretations of the different nouns. For example, one worker may consider "numb" to equate to a higher intensity of discomfort than "stiff," while another worker may interpret them in the opposite order. Such errors will hinder the evaluation of ergonomic interventions because an improvement post intervention may not be detected due to the different uses of the scale by workers.

## Visual Analog Scales

Visual analog scales consist of a line, usually 100 mm in length, with a label at each end (often termed "anchors"). Figure 26.5 shows a Visual Analog Discomfort Scale which uses the anchors "no discomfort" and "extreme discomfort" to indicate the ends of the continuum. Another common anchor for the high intensity end is "discomfort as bad as it could be." Both vertical and horizontal lines can be used.

To indicate the level of discomfort, a worker places a mark on the line to indicate the intensity. The intensity rated is then measured as the distance from the left-hand end of the line to the mark placed by the worker. When the measuring is recorded in mm, the scale effectively has 101 levels of discomfort. Robust parametric statistics, such as analysis of variance, are often used for analysis, although the data are strictly not interval data and may be skewed (for example, if most workers record very low or no discomfort).

The advantages of visual analog discomfort scales include their ease of administration, sensitivity, and amenability to statistical analysis.

Pain measurement research has suggested that a possible disadvantage of visual analog scales is that some workers may have difficulty conceptualizing how to indicate perceptions of discomfort intensity

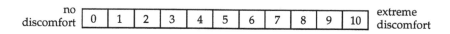

**FIGURE 26.6**   Visual numeric rating scale.

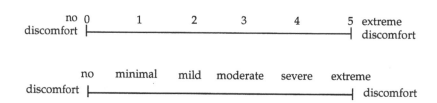

**FIGURE 26.7**   Graphic rating scales.

on a line. Groups thought to have greater difficulty are older people and those without formal education (those who may be less familiar with abstracting concepts).

### Numeric Rating Scales

Numeric rating scales are similar to visual analog scales except they have a discrete number of categories and can be either visual or verbal. Common examples use 0 to 10 in one-unit intervals to give an 11-unit scale or 0 to 100 in one-unit intervals to give a 101 unit scale. The scale has anchors similar to analog scales. Figure 26.6 shows a visual numeric rating scale with 11 levels of intensity.

Workers rate their discomfort intensity either by marking a number or verbally reporting a number. Data collected are less parametric than visual analog scale ratings so nonparametric statistical analysis should be used.

Advantages of numeric rating scales include that they are simple to administer and the verbal scale can be used during a manual task without interference with posture.

Disadvantages include that the 0 to 10 point scale has limited sensitivity, and workers often tend to clump ratings on the 1 to 100 scale around deciles.

### Graphic Rating Scales

Graphic rating scales are a mixture of a visual analog scale and either a numeric or verbal rating scale. The scale thus consists of a vertical or horizontal line with anchors (as for a visual analog scale) with the addition of either numbers or adjectives along the line. Figure 26.7 shows examples of both types.

Workers place a mark on the line to represent their rating of discomfort intensity. Nonparametric statistical tests should be used.

An advantage of graphic rating scales is that the extra labels may assist a worker having difficulty using a visual analog scale.

However, it has been demonstrated that there is a problem with clustering of results around the labels along the line.

## Quality

The quality of discomfort can probably only be assessed by allowing different nouns to be used by the worker. Different qualities of discomfort may include: tingling; burning; searing; numbness; coldness; stiffness; heat; cramping; prickling; stabbing; and gnawing. Although quality of pain is widely used in health assessment, quality of discomfort has not been regularly used by ergonomists. Perhaps this is because the implications of the different possible qualities are unclear, whereas the implications of intensity, location, and temporal pattern are usually clear.

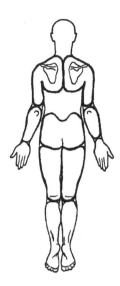

FIGURE 26.8    Body map for reporting discomfort location.

## Location

The location of discomfort is commonly collected either through the use of a body map or by specific reference to a body part. Thus, where only the discomfort of a specific body part is of interest, it is usually made clear to the worker that the other information collected about discomfort (intensity, quality, and temporal pattern) is just related to that body part. However, when discomfort at a number or sites, or at any site, is of interest, the worker is usually asked to indicate where each and every discomfort is felt. Workers can indicate the body parts either by shading on a body map or by reporting the name or number of a body part. Figure 26.8 shows an example of a body map which separates the body into 13 parts.

## Temporal Pattern

The temporal pattern of discomfort is often measured by collecting information about discomfort at different times. Depending on the reason for investigating discomfort, the time between collections may vary from a number of minutes (presumably for quite severe or intense tasks), to a number of hours (if a daily fluctuation is of interest), to a number of days or longer. Multiple recording of aspects of discomfort can be achieved by either using separate data collection sheets for each time (thus keeping the worker blind to previous recordings), or by recording on the same data collection sheet (enabling the worker to compare with previous recordings).

Another issue related to assessing the temporal pattern of discomfort is the importance of the period between worker experience and data collection. Branton (1969) suggested that because post-experience discomfort reporting relies on kinesthetic memory, discomfort information should be collected while the worker is experiencing the discomfort. Pain research has clearly demonstrated the importance of immediacy for best validity.

## Summary of Important Fundamentals of Discomfort Measurement

Based on the discussion above, the following fundamentals of discomfort measurement can be distilled:

- Discomfort measurement is likely to be useful in the assessment of information about physical matches and mismatches.
- Consistent use of the sole noun "discomfort" will assist the validity of assessment.

- Discomfort is a subjective experience and can therefore only be measured by worker report.
- Intensity, location and temporal pattern are important attributes of discomfort.
- A Visual Analog Discomfort Scale is probably the most widely applicable discomfort intensity scale.

The following section provides examples of how discomfort tools have been used in ergonomics research and practice.

## 26.3  Application — Examples of Discomfort Tools Used in Ergonomics

Some examples of the use of discomfort tools in ergonomics are described below. A list of other reports of discomfort tool use is provided in the Further Reading section.

### Whole Body Discomfort Scales

Shackel et al. (1969) used a multiple noun graphic rating scale, which they called a "General Comfort Scale," to measure the "discomfort" caused by different chair designs. The rating scale had 11 items arranged at 10 mm intervals on a vertical line as shown in Figure 26.9. Chair raters were asked to place a mark on the vertical line to express their rating. Scoring was achieved by rounding the mark to the nearest 5 mm, giving a 0 to 20 scale. Using nonparametric statistics, Shackel et al. were able to rank chairs for "discomfort" and compare user ratings with expert opinion.

Although the scale appeared to work reasonably well for the purposes of the study, the earlier discussion presented in this chapter suggests that a single noun scale would have been more valid, and may have allowed better differentiation of the chairs.

Corlett and Bishop (1976) evaluated ergonomic changes to pedestal spot welder machines by measuring "discomfort" before and after machine modifications using an "Overall Comfort Scale." Three operators were asked to rate the intensity of their overall comfort using a seven-point graphic rating scale with end labels "extremely comfortable" and "extremely uncomfortable." Coding of intensity was achieved by dividing each scale interval into 2 to give 14 measurement units. Prior to modification, one machine resulted in mean operator ratings of around 12, which was reduced to around 7.5 after modification. From the earlier discussion, use of a continuum from extreme comfort to extremely uncomfortable may not be valid because comfort may be a separate state.

Bhatnager et al. (1985) used a slightly different approach to assessing whole body discomfort. They used Drury and Coury's (1982) "Body Part Discomfort Frequency" (total number of body parts with non-zero discomfort) and "Body Part Discomfort Severity" (mean severity of body parts with non-zero discomfort) to gain an indicator of overall body discomfort. Their fascinating study showed a negative relationship between discomfort and productivity at an inspection task. There has been some discussion

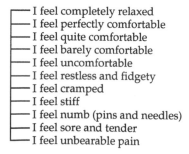

**FIGURE 26.9**  General Comfort Scale. (From Shackel, B., Chidsey, K. D., and Shipley, P. (1969). The assessment of chair comfort. *Ergonomics*, 12 (2), 269-306. With permission.)

in the ergonomics literature about whether it is the number of body parts experiencing discomfort, or the average severity of discomfort in various body parts, or the peak severity of discomfort in one body part which determines overall ratings of discomfort. Until this issue is resolved, it is probably useful to use a tool which collects the specific information required so that no unproven assumptions are needed.

Visser and Straker (1994) collected general body discomfort intensity ratings using a "General Body Visual Analog Discomfort Scale." The written instructions to workers were to "Place a dash (/) at a point along the line that best corresponds to your *present* feeling of *overall* discomfort." The scale was the same as shown in Figure 26.5. The tool was easy to use as was demonstrated by the remarkably high compliance rate for the study. (Other details of this study are described below.)

## Body Part Discomfort Location and Intensity Scales

Besides collecting whole body discomfort, Shackel et al. (1969) also collected body part discomfort data. Chair raters were shown a body map divided into 15 numbered body parts. Alongside the map were 5 boxes, the first of which was marked "3 most comfortable" and the last marked "3 most uncomfortable." Chair raters were asked to identify the 3 most comfortable body parts and note these in the first box and then cross them off on the body map. They were then to identify the next 3 most comfortable and record the numbers for those body parts in the next box and cross them off the body map. This was continued until all body parts were noted. Chair raters were also allowed to rate the 3 most comfortable then the 3 most uncomfortable at each successive round of rating, if that was found easier. Shackel et al. tried to rank chairs by various body part discomforts, but as the buttocks experienced the most discomfort for each chair the differences between chairs were not able to be detected. The authors also reported the technique to be extremely laborious. The difficulty of use, lengthy procedure, and lack of sensitivity were disadvantages of this assessment tool.

Corlett and Bishop (1976) also collected body part discomfort data. Their method was to ask operators to identify the body part with the most discomfort, then the body part with the next most discomfort, and so on until all body parts had been ranked. The data was used to compare the locations of discomfort experienced when using 2 different machines. A machine which required considerable right foot force showed asymmetry in discomfort and greater lower limb discomfort compared with a machine which required a much smaller foot force. The data were therefore useful in assisting to identify the features of the machinery which contributed to a physical mismatch with the worker. The assessment tool was easier to use and quicker than that of Shackel et al., though it still lacked some sensitivity.

Straker et al. (1997b) collected ratings of body part discomfort from manual handlers performing a range of tasks. The data collection form used a body part map for location information and multiple visual analog scales for intensity information. Figure 26.10 shows the form used.

Besides statistical analysis of the information, Straker et al. used graphic representation to facilitate comparison of the location and intensity of discomfort between different tasks. For example, the body diagram on the left side of Figure 26.11 shows the group average intensity for each of 13 body parts during a task requiring box carrying at shoulder height (high carry). The areas of highest discomfort are the shoulders and wrists/hands. This can be compared to the body on the right side in Figure 26.11 which summarizes the group discomfort data for a task requiring box carrying in a stooped posture (low carry). The change of highest discomfort areas to low back, pelvis, and thighs in the low carry task is consistent with biomechanical modeling of areas of greatest stress. This assessment tool was easy and quick to administer and provided a high level of sensitivity.

## Temporal Pattern

Visser and Straker (1994) used a "General Body Visual Analog Discomfort Scale" to collect ratings from 56 dental workers at 6 different times over a working day (on arrival, morning break, prior to lunch break, upon returning from lunch break, mid afternoon, and on completion of work). Figure 26.12 shows some of the results of Visser and Straker's study and clearly indicates both the trend for increasing discomfort as the workday progressed and the ameliorative effect of the lunch break.

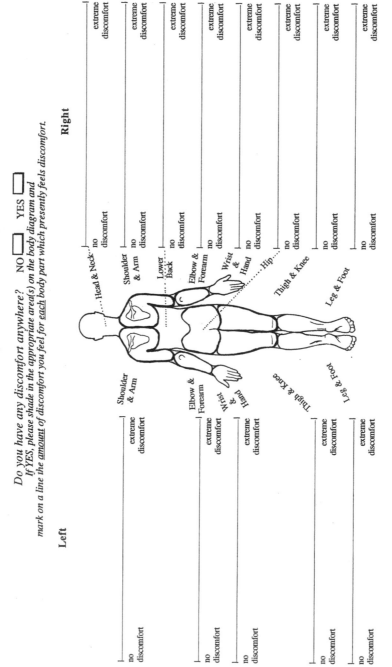

**FIGURE 26.10**   Body part discomfort location and intensity form.

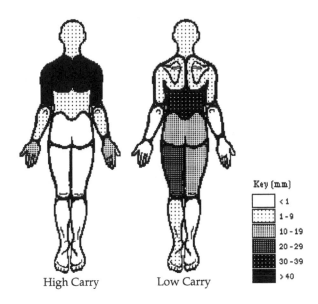

High Carry    Low Carry

Key (mm)
□ < 1
▨ 1 - 9
▤ 10 - 19
▦ 20 - 29
▩ 30 - 39
■ > 40

**FIGURE 26.11**   Body part discomfort for high and low carry tasks.

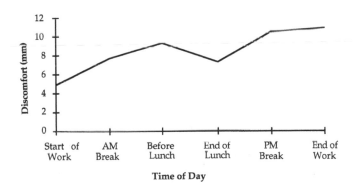

**FIGURE 26.12**   Mean general body discomfort of dental professionals for 6 different times of day.

Visser and Straker also compared the differences in the development of whole body discomfort when dental workers where provided with a specially designed chair which incorporated an arm rest. Figure 26.13 clearly shows the benefit of the introduction of an arm rest in reducing discomfort in the afternoon.

Straker et al. (1997a) conducted a study to further investigate the relationship between arm posture and discomfort and computer operator performance, as begun by Bates et al. They were interested in the changes in discomfort over a 20-minute task. A verbal "Numeric Rating Scale-101" was used which allowed the operators to verbally report the intensity of the discomfort they were experiencing without needing to change postures. Straker et al.'s scale was an adaptation of the scale recommended by Jensen et al. (1986) for pain. Operators were verbally instructed to "Indicate to me the number between '0' and '100' that best describes your right shoulder discomfort. A '0' would mean no discomfort and a '100' would mean discomfort as bad as it could be at your right shoulder." These ratings were collected every 5 minutes from 21 female operators working in either no shoulder flexion or 30° shoulder flexion. Figure 26.14 shows how the data collected was able to clearly show a more rapid rise in discomfort in the poorer posture.

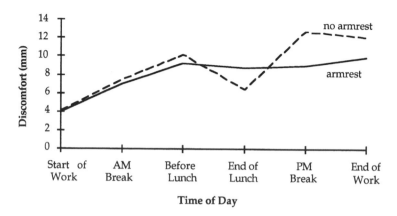

**FIGURE 26.13**   Comparison of mean general body discomfort of dental professionals across the day with and without an armrest.

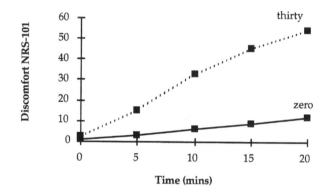

**FIGURE 26.14**   Changes in shoulder discomfort intensity while working in 0° and 30° shoulder flexion.

These examples have shown that a wide range of discomfort tools have been used in ergonomics research and practice to provide information about the physical match between workers and their environment.

To assist ergonomists in the decision of which discomfort assessment tool to use, a number of criteria are presented.

## Criteria for Selection of an Appropriate Tool

A discomfort assessment tool should have high utility, validity, and sensitivity. Utility can be considered in two phases, data collection and data analysis. High utility in data collection requires the tool to be easy for workers to use correctly, quick to administer, and with minimal interference with workers' performance of their tasks. One aspect of widespread ease of use is that the tool should require minimal language skills. Thus, a tool used in an English-speaking country should require little English competence so that the tool can be used across workgroups without difficulty and without jeopardizing validity. Ease of use will improve the quality of collected data by minimizing errors.

High utility in data analysis requires the data to be readily amenable to statistical analysis and graphical representation. Numerical data with either nonparametric characteristics or close to parametric characteristics will facilitate easy statistical analysis. Currently, parametric analysis of complex statistical models tends to be easier, giving an advantage to a tool which collects parametric-like data. Graphical representation of data is important for the communication of ergonomics findings to managers and workers.

Validity is critical, as no validity means the data are of no value. As previously mentioned, the only valid measure of a person's experience is the report of that person. Although there is little discomfort-specific research on tool validity, pain measurement research has produced considerable evidence on the validity of certain tools, such as the visual analog scale for intensity assessment.

However, because discomfort is commonly used in ergonomics to imply a problem in the physical match between worker and work, the strong correlation of discomfort to biomechanical and physiological risk indicators such as joint torque and electromyograph power spectral shifts provides corroboratory evidence of the validity of discomfort.

Reliability is an important corequisite of validity. Van der Grinten (1991) provides evidence of reasonable reliability of a discomfort assessment tool, and this was the only study of discomfort tool reliability found.

Finally, a discomfort assessment tool needs to have sensitivity appropriate to the workers' capacities for discrimination and the assessment purpose. It is unlikely that workers can reliably differentiate 1000 levels of discomfort intensity, and even 100 levels may be ambitious. However, 10 levels are often not sensitive enough for comparisons between work situations unless radical ergonomic interventions have occurred.

## 26.4   Conclusion and Recommendations

To help ergonomists choose and use an appropriate assessment tool, this chapter has provided a concise review of the fundamental concepts, provided examples of how discomfort assessment tools have been used, and provided criteria to select an appropriate assessment tool.

Discomfort is a valuable variable for ergonomists to use to assess the physical match between workers and their work. Several decades of practical experience by ergonomists and research in the area of pain, has resulted in the development of easy-to-use, valid, and sensitive discomfort assessment tools.

This author recommends the use of a Visual Analog Discomfort Scale or Verbal Numerical Rating Scale for assessment of intensity; body map or specific instructions for assessment of discomfort location; and repeated measurements for the assessment of the temporal pattern of discomfort.

### Acknowledgments

The author would like to thank past students, Grace Szeto, Michael Bates, Marshall Stockden, Mark Petrich, Jodie Visser, and Jean Mangharam for stimulating the conceptualization of this chapter.

### References/Further Reading

Studies Cited in Chapter

Bates, M., Petrich, M., and Stockden, M. (1989). *Posture, pathology, pain and performance.* Bachelor of Applied Science Research Report, Curtin University of Technology, Perth Australia.

Bhatnager, V., Drury, C. G., and Schiro, S. G. (1985). Posture, postural discomfort and performance. *Human Factors*, 27 (2), 189-199.

Boussenna, M., Corlett, E. N., and Pheasant, S. T. (1982). The relation between discomfort and postural loading at the joints. *Ergonomics*, 25 (4), 315-322.

Branton, P. (1969). Behaviour, body mechanics and discomfort. *Ergonomics*, 12, 316-327.

Corlett, E. N. and Bishop, R. P. (1976). A technique for assessing postural discomfort. *Ergonomics*, 19 (2), 175-182.

Drury, C. G. and Coury, B. G. (1982). A methodology for chair evaluation. *Applied Ergonomics*, 13 (3), 195-202.

Jensen, M. P., Karoly, P., and Braver, S. (1986). The measurement of clinical pain intensity: a comparison of six methods. *Pain*, 27, 117-126.

Shackel, B., Chidsey, K. D., and Shipley, P. (1969). The assessment of chair comfort. *Ergonomics*, 12 (2), 269-306.

Straker, L. M., Pollock, C. M., and Mangharam, J. (1997a). The effect of shoulder posture on performance, discomfort and muscle fatigue while working on a Visual Display Unit. *International Journal of Industrial Ergonomics*, 20, 1-10.

Straker, L. M., Stevenson, M. G., and Twomey, L. T. (1997b). A comparison of single and combination manual handling tasks risk assessment: 2. discomfort, rating of perceived exertion and heart rate measures. *Ergonomics*, 40 (6), 656-669.

van der Grinten, M. P. (1991). Test-retest reliability of a practical method for measuring body part discomfort, in Y. Quennec and R. Daniellou (Eds.), *Designing for Everyone. Proceedings of the 11th Congress of the International Ergonomics Association*, (pp. 54-56). Paris: Taylor & Francis.

Visser, J. L. and Straker, L. M. (1994). An investigation of discomfort experienced by dental therapists and assistants at work. *Australian Dental Journal*, 39 (1), 39-44.

## Further Reading

Cameron, J. A. (1996). Assessing work-related body-part discomfort: Current strategies and a behaviorally oriented assessment tool. *International Journal of Industrial Ergonomics*, 18, 389-398.

Chapman, C. R., Casey, K. L., Dubner, R., Foley, K. M., Gracely, R. H., and Reading, A. E. (1985). Pain measurement: An overview. *Pain*, 22, 1-31.

Hagberg, M., and Sundelin, G. (1986). Discomfort and load on the upper trapezius muscle when operating a wordprocessor. *Ergonomics*, 29 (12), 1637-1645.

Life, M. A. and Pheasant, S. T. (1984). An integrated approach to the study of posture in keyboard operation. *Applied Ergonomics*, 15 (2), 83-90.

Marley, R.J. and Kumar, N. (1994). An improved musculoskeletal discomfort assessment tool, in F. Aghazadeh (Ed.). *Advances in Industrial Ergonomics and Safety VI* (pp. 45-52). London: Taylor & Francis.

Scott, J., and Huskisson, E. C. (1976). Graphic representation of pain. *Pain*, 2, 175-184.

Van der Grinten, M. P. and Smitt, P. (1992). Development of a practical method for measuring body part discomfort, in F. Adhazadeh (Ed.), *Advances in Industrial Ergonomics and Safety IV* (pp. 311-318). London: Taylor & Francis.

# Section II

## Medical Management Prevention

# 27

# Medical Management of Work-Related Musculoskeletal Disorders

**Thomas Hales**
*National Institute for Occupational
Safety and Health (NIOSH)*

**Patricia Bertsche**
*The Ohio State University*

## 27.1   Introduction

The Bureau of Labor Statistics (BLS) reports that in 1994 nearly two thirds of the workplace illnesses were disorders associated with repeated trauma (one category of musculoskeletal disorders) (BLS, 1995). These figures do not include low back disorders associated with overexertion, which accounted for 380,000 lost time cases in 1993. The number of repeated trauma cases reported in 1994 was 332,000, a 10% increase from the 1993 figure. In fact, since 1982, the number of reported disorders associated with repeated trauma has been increasing each year (BLS, 1995). Not surprisingly, many health care providers (HCPs) find evaluating and treating these employees consumes an increasing proportion of their time and energy.

To prevent or reduce symptoms, signs, impairment, or disability associated with work-related mus-culoskeletal disorders (WRMSDs), employers, in collaboration with HCPs, should develop a medical management program which is outlined in Figure 27.1. This chapter provides assistance to employers setting up a medical management program and to HCPs managing these cases in two ways — first, by outlining the general principles and listing the components of a program needed to adequately evaluate

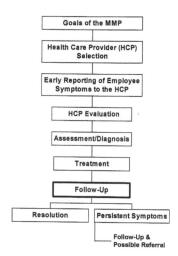

**FIGURE 27.1**    Overview of a medical management program (MMP).

and treat affected employees; second, by providing HCPs with practical guidance and forms to collect the appropriate information. These forms can then be incorporated into the employee's medical record.

## 27.2    Terminology

Before addressing the various components of a medical management program, the term musculoskeletal disorder must be defined. MSDs are disorders of the muscles, tendons, peripheral nerves, or vascular system not directly resulting from an acute or instantaneous event (e.g., slips or falls). These disorders are considered to be work-related when the work environment and the performance of work contribute significantly, but as one of a number of factors, to the causation of a multifactorial disease (WHO, 1985). Physical risk factors that cause or aggravate MSDs and that may be present at the workplace include, but are not limited to: repetitive, forceful, or prolonged exertions; frequent or heavy lifting; pushing, pulling, or carrying of heavy objects; fixed or awkward work postures; contact stress; localized or whole-body vibration; cold temperatures; and poor lighting leading to awkward postures. These workplace risk factors can be intensified by work organization characteristics, such as inadequate work–rest cycles, excessive work pace and/or duration, unaccustomed work, lack of task variability, machine-paced work, and piece rate.

## 27.3    Selection of a Health Care Provider (HCP)

An HCP is a practitioner operating within the scope of his or her license, registration, certification, or legally authorized practice. The evaluation and treatment of employees with WRMSDs should be performed by an HCP with experience and/or training in managing these disorders. Many HCPs are capable of providing these services, including physicians, occupational health nurses, physical therapists, occupational therapists, and hand therapists. Employers and employees may be more familiar with the services of physicians, therefore Table 27.1 provides information regarding some of the other HCPs who might be directly providing the care or coordinating the care of employees with WRMSDs. Considerations for the employer to use in selecting an HCP include:

- Specialized training and experience in ergonomics and the treatment of work-related musculoskeletal disorders
- Current working knowledge of the worksite and the specific industry
- Willingness to periodically tour the worksite

**TABLE 27.1** Non-Physician Health Care Providers Who Might Be Involved in the Medical Management of Work-Related Musculoskeletal Disorders (Not intended to be all-inclusive)

| Profession | Scope of Practice | Training/Experience | Services They Provide |
|---|---|---|---|
| Occupational Health Nurse (OHN) | An OHN is a Registered Nurse (RN), independent licensure with scope defined by individual state boards of nursing; certification is voluntary (COHN); Advanced practice nurses (nurse practitioners) treat independently or provide medical treatment with protocol depending on requirements of state licensing board. RNs refer to physicians and other health care providers when treatment beyond their scope of practice is required. | Basic education includes complete assessment (history and physical examination) of all body systems; OHNs have academic and/or continuing education in assessment of the musculoskeletal and nervous systems and diagnosis, treatment, and rehabilitation of work-related musculoskeletal disorders. | Assessment, treatment of common work-related musculoskeletal disorders, particularly in early stages (under protocol when required by state statute), referral to other appropriate health care providers as needed, and rehabilitation including case management; Preventive services include trend analysis, education and training, and involvement in the job improvement process including job analysis. |
| Occupational Therapist (OT) | 49 states, the District of Columbia, Guam, and Puerto Rico have laws regulating the profession; The American Occupational Therapy Certification Board's national certification exam is a basic requirement in the states/jurisdictions that license or certify OTs. Generally, an OT may independently provide services, however, in certain states, occupational therapy laws/regulations require physician referral for services for specific medical conditions. | OTs have either a bachelor's or master's degree and pursue continuing education and extensive on-the-job training to specialize in work-related musculoskeletal disabilities; OTs have a comprehensive background in the biological and behavioral sciences; knowledge and application of the components of human performance including psychosocial, neurological, cognitive, perceptual, and motor function. | OTs use standardized tests, observational skills, activities and tasks designed to evaluate specific work-related skills, functional abilities, physical abilities, and behaviors. Examples of assessments include: functional capacity evaluation, physical capacity testing, examination of essential functions of a job. Other services include work hardening and involvement in the job improvement process such as job analysis and workstation and tool modification. |
| Physical Therapist (PT) | PTs licensed in all states, the District of Columbia, Puerto Rico, and the U.S. Virgin Islands; Direct physician oversight is not required. Of the 53 jurisdictions, 44 jurisdictions permit physical therapy evaluation without physician referral. | PTs have either a bachelor's or graduate degree and pursue continuing education to specialize in prevention and rehabilitation of work-related musculoskeletal disorders. PTs' basic education includes courses in anthropometrics, biomechanics, ergonomic interventions, kinesiology, movement and posture analysis, the components of human psychophysical performance, orthotic prescription, fabrication, and application of supportive devices. | PTs evaluate a variety of conditions such as abnormalities of body alignment and movement patterns; impaired motor function and learning; impaired sensation; limitations of joint motion; muscle weakness; and pain. PTs perform tests and measures such as batteries of work performance; assessment of work hardening or conditioning; determination of dynamic capabilities and limitations during specific work activities. Involvement in the job improvement process including analysis of jobs or activities, and workstation or tool modifications. |
| Hand Therapist (HT) | A Hand Therapist is either an OT or PT who voluntarily becomes certified by the Hand Therapy Certification Commission. Certified HTs specialize in upper extremity rehabilitation. | HTs have specialized training and experience in assessment and rehabilitation of work-related musculoskeletal disorders. | Services include diagnostic work up of quantitative sensory testing to determine peripheral neuropathy, grip strength, and motor testing to determine the localization of muscular tenderness areas of inflammation; physical or functional capacity evaluations. HTs apply treatments such as thermotherapy, ultrasound, and electric stimulation; re-education home exercise programs, splintage, pain management, soft tissue mobilization and myofascial release. HTs are skilled in work task analysis and therefore are well suited for involvement in the job improvement process. |

- Willingness to communicate with the employer and employees (Louis, 1987; Haig et al., 1990)
- Experience in the case management of work-related musculoskeletal disorders
- Willingness to consider conservative therapy prior to surgery
- History of successful treatment of work-related musculoskeletal disorders

## 27.4  Early Reporting of Symptoms and Access to Health Care Providers

The case management process begins with an employee informing his or her employer of the presence of musculoskeletal symptoms or signs. Generally, the earlier that symptoms are identified, an evaluation completed, and treatment initiated, the likelihood of a significant disorder developing is reduced. Early treatment of many MSDs has been shown to reduce their severity, duration of treatment, and ultimate disability (Haig et al., 1990; Wood, 1987; Wiesel et al., 1984; Mayer et al., 1987). There can be various workplace situations influencing an employee's decision to report symptoms. These situations can result in employees over-reporting, or under-reporting, symptoms. In either case, to prevent severe disorders from occurring, employees must not be subject to reprisals or discrimination based on reporting symptoms to their supervisors.

Supervisors and foremen are not trained to evaluate and assess MSDs. To prevent supervisors or other plant personnel from performing triage, employees reporting persistent musculoskeletal symptoms (e.g., symptoms lasting seven days from onset, or symptoms that interfere with the employee's ability to perform the job) should have the opportunity for a prompt HCP evaluation. If an HCP is available at the workplace, this initial assessment should be offered when the employee reports symptoms or at least within two days. If the HCP is offsite, the employer should make available an assessment to the employee promptly, but no later than a week after the signs or symptoms are reported. This is not meant to imply that employers should wait seven days from onset of all employee's symptoms before referring the employee to an HCP. There are foreseeable circumstances where immediate evaluation by an HCP would be warranted. For example, an employee who reports to the supervisor that he/she is experiencing severe low back pain with numbness and tingling radiating down his/her leg, an inability to sleep due to the pain, and obvious difficulty walking should immediately be referred to the HCP.

## 27.5  Health Care Providers Familiarity with Employee's Job

HCPs who evaluate employees, determine an employee's functional capabilities, and prepare opinions regarding work-relatedness and work-readiness, must be familiar with employee jobs and job tasks. Being familiar with employee jobs not only assists HCPs in making informed case management decisions, but also demonstrates to employers and employees the importance HCPs place on making informed decisions, assists with the identification of workplace hazards that cause or aggravate MSDs, assists with the identification of alternate duty jobs, and can help establish the proper diagnosis for the employee's condition.

Critical to this process is open lines of communication with the employer, employee, and the HCP. The employer should appoint a contact person who is familiar with plant jobs and workplace risk factors to communicate and coordinate with the HCP. In addition, HCPs should perform a plant walk-through. Once familiar with plant operations and job tasks, the HCP can periodically revisit the facility to remain knowledgeable about working conditions. Other approaches to become familiar with jobs and job tasks include review of job analysis reports, job surveys or risk factor checklists, detailed job descriptions, job safety analyses, photographs and/or videotapes accompanied by narrative or written descriptions, and interviewing the employee.

# 27.6  Evaluation of the Employee

The HCP evaluation of the symptomatic employee should contain a relevant occupational and health history, a physical examination, laboratory tests appropriate to the reported signs or symptoms, and conclude with an initial assessment/diagnosis. If the HCP providing the initial evaluation does not have the training or experience to make a preliminary assessment or diagnosis, the employee should be referred to an HCP with such training and experience. The content of the evaluation is outlined below with a recording form available (see Form 1).

1. Characterize the symptoms and history
   - Onset (date; circumstance; abrupt vs. gradual, etc.)
   - Duration and frequency
   - Quality (pain; tingling; numbness; swelling; tenderness, etc.)
   - Intensity (mild; moderate; severe; other rating scales)
   - Location
   - Radiation
   - Exacerbating and/or relieving factors or activities (both on-the-job and off-the-job)
   - Prior treatments
2. Relevant considerations:
   - Demographics (e.g., age; gender; hand dominance)
   - Past medical history (e.g., prior injuries or disorders related to the affected body part)
   - Recreational activities, hobbies, household activities
   - Occupational history with emphasis on the (a) job the employee was performing when the symptoms were first noticed, (b) prior job if the employee recently changed jobs, (c) amount of time spent on that job, and (d) whether the employee was working any other "moonlighting" or part-time jobs.
3. Characterize the job:
   Becoming familiar with an employee's job is a critical component of the HCP evaluation and treatment process. In addition to collecting the information from the plant contact person and plant walk-through (described above), employees should be interviewed regarding their work activities. The employee should be asked to describe their required job tasks with respect to known workplace risk factors for MSDs and the duration of exposure such as hours per day, days per week and shift work. Workplace risk factors for MSDs include repetitive, forceful, or prolonged exertions; frequent or heavy lifting or lifting in awkward postures (e.g., twisting, trunk flexion, or lateral bending); pushing, pulling, or carrying of heavy objects; fixed or awkward work postures; contact stress; localized or whole-body vibration; cold temperatures; and others. The employee should also be asked if there has been any recent changes in their job, such as longer hours, increased pace, new tasks or equipment, or new work methods which may have caused or contributed to the current illness.
4. Physical examination:
   The physical examination should be targeted to the presenting symptoms and history. Components of the exam include inspection (redness, swelling, deformities, atrophy, etc.), range of motion, palpation, sensory and motor function (including functional assessment), and appropriate maneuvers (e.g., Finkelstein's). It is important to note that clinical examinations may not identify the specific structure affected, nor find classic signs of inflammation (e.g., redness, warmth, swelling). This should not be surprising since the role of inflammation in the pathophysiology of these disorders is unclear (Nirschl, 1990). For further information on the content of an appropriate exam, or the technique to perform the exam, please consult the following references: AHCPR, 1994; ASSH, 1990; Hoppenfield, 1976; Tubiana et al., 1984.

5. Assessment and diagnosis:

For each employee referred for an assessment, the HCP should make a specific diagnosis consistent with the current International Classification of Diseases, or the HCP should summarize the findings of his or her assessment. Terms such as repetitive motion disorders (RMDs), repetitive strain injury (RSI), overuse syndrome, cumulative trauma disorders (CTDs), and work-related musculoskeletal disorder (WRMSD) are not ICD diagnoses and, although useful as general terms, should not be used as medical diagnoses. Given the difficulty in establishing the specific structure affected, many diagnoses should describe the anatomic location of the symptoms without a specific structure diagnosis (e.g., unspecified neck symptoms or disorders should be listed as ICD-9 723.9; unspecified disorders of the soft tissues should be listed as ICD-9 729.9). When a specific anatomical structure can be ascertained, most of these conditions involve the muscles or tendons (unspecified disorders of muscle, ligament, and fascia should be listed at ICD-9 728.9; unspecified disorders of synovium, tendon, and bursa should be listed as ICD-8 727.9). Table 27.2 provides a listing of ICD-9 codes.

The HCP should assist in determining whether occupational risk factors are suspected to have caused, contributed to, or exacerbated the condition. Factors helpful in making this determination are:

• Is the medical condition known to be associated with work?
• Does the job involve risk factors (based on job surveys or job analysis information) associated with the presenting symptoms?
• Is the employee's degree of exposure consistent with those reported in the literature?
• Are there other relevant considerations (e.g., unaccustomed work, overtime, etc.)?

## 27.7   Treatment of the Employee

Before initiating treatment, the HCP should document the specific treatment goals (e.g., symptom resolution or restoring of functional capacity), expected duration of treatment, dates for follow-up evaluations, and time frames for achieving the treatment goals. Resting the symptomatic area, and treatment of soft tissue and tendon disorders are the mainstays of conservative treatment. Despite the wide application of some therapeutic modalities, many are untested in controlled clinical trials.

### Resting the Symptomatic Area

Reducing or eliminating employee exposure to musculoskeletal risk factors through engineering and administrative controls in the workplace is the most effective way to rest the symptomatic area while allowing employees to remain productive members of the workforce (Upfal, 1994). Until effective controls are installed, employee exposure to workplace risk factors can be reduced through restricted duty and/or temporary job transfer. The specific amount of work reduction for employees on restricted duty must be individualized; however, the following principles apply: the degree of restriction should be proportional to the condition severity and to the frequency and duration of exposure to relevant risk factors involved in the original job. HCPs are responsible for determining the physical capabilities and work restrictions of the affected worker. The employer is responsible for finding a job consistent with these temporary restrictions. The employer's contact person (who is knowledgeable about the employee's job requirements and their associated risk factors) is critically important to this process. The contact person should communicate and collaborate with the HCP so that appropriate job placement of the employee occurs during the recovery period. Written return-to-work plans ensure that the HCP, the employee, and the employer all understand the steps recommended to promote recovery, and ensure that the employer understands what his or her responsibility is for returning the employee to work. A form is included to collect and distribute this written plan (Form 2). The HCP is also responsible for employee follow-up to document a reduction in symptoms during the recovery period.

**TABLE 27.2** Specific ICD-9 Diagnoses Referred to as Musculoskeletal Disorders by ICD-9 Numbers

| | |
|---|---|
| **Tendon, synovium, and bursa disorders** | **727** |
| Trigger finger (acquired) | 727.03 |
| Radial styloid tenosynovitis (deQuervain's) | 727.04 |
| Other tenosynovitis of hand and wrist | 727.05 |
| Specific bursitides often of occupational origin | 727.2 |
| Unspecified disorder of synovium, tendon, and bursa | 727.9 |
| **Peripheral enthesopathies** | **726** |
| Rotator cuff syndrome, supraspinatus syndrome | 726.10 |
| Bicipital tenosynovitis | 726.12 |
| Medial epicondylitis | 726.31 |
| Lateral epicondylitis (tennis elbow) | 726.32 |
| Unspecified enthesopathy | 726.9 |
| **Disorders of muscle, ligament, and fascia** | **728** |
| Game-Keepers thumb | 728.8 |
| Muscle spasm | 728.85 |
| Unspecified disorder of muscle, ligament, and fascia | 728.9 |
| **Other disorders of soft tissues** | **729** |
| Myalgia, myositis, fibromyositis | 729.1 |
| Swelling of limb | 729.81 |
| Cramp | 729.82 |
| Unspecified disorders of soft tissue | 729.9 |
| **Osteoarthritis** | **715** |
| **Mononeuritis of upper limb** | **354** |
| Carpal tunnel syndrome (median nerve entrapment) | 354.0 |
| Cubital tunnel syndrome | 354.2 |
| Tardy ulnar nerve palsy | 354.2 |
| Lesions of the radial nerve | 354.3 |
| Unspecified mononeuritis of upper limb | 354.9 |
| **Peripheral vascular disease** | **443** |
| Raynaud's syndrome | 443.0 |
| Hand-Arm Vibration Syndrome | 443.0 |
| Vibration White Finger | 443.0 |
| **Arterial embolism and thrombosis** | **444** |
| Hypothenar hammer syndrome | 444.2 |
| Ulnar artery thrombosis | 444.21 |
| **Nerve root and plexus disorders** | **353** |
| Brachial plexus lesions | 353.0 |
| Cervical rib syndrome | 353.0 |
| Costoclavicular syndrome | 353.0 |
| Scalenus anticus syndrome | 353.0 |
| Thoracic outlet syndrome | 353.0 |
| Unspecified nerve root and plexus disorder | 353.9 |
| **Spondylosis** (inflammation of the vertebrae) | **721** |
| Cervical without myelopathy | 721.0 |
| Cervical with myelopathy | 721.1 |
| Thoracic without myelopathy | 721.2 |
| Lumbarsacral without myelopathy | 721.3 |
| Thoracic or lumbar with myelopathy | 721.4 |
| **Intervertebral disc disorders** | **722** |
| Displacement of cervical disc | 722.0 |
| Displacement of thoracic or lumbar disc | 722.1 |
| Degeneration of the cervical disc | 722.4 |
| Degeneration of the thoracic or lumbar disc | 722.5 |
| Intervertebral disc disorder with myelopathy | 722.17 |
| **Disorders of the cervical region** | **723** |
| Cervicalgia (pain in neck) | 723.1 |
| Cervicobrachial syndrome (diffuse) | 723.3 |
| Unspecified neck symptoms or disorders | 723.9 |
| **Unspecified Disorders of the Back** | **724** |
| Low back pain | 724.2 |

**FORM 1 — Occupational and Health History Recording Form for Musculoskeletal Disorders**

Name: _____ Dept: _____ Job Title: _____

Age: ____ yrs   Gender: ____ F ___ M      Length of time at the plant: _____ mo/yrs

Dominate Hand: ____ R ____ L ___ Both      Length of time on-the-job: _____ mo/yrs

**Symptom Characterization:**

Onset: Date: _____ Abrupt vs. Gradual: _____

Quality: (let employee describe, check all that apply)

___ pain      ___ tenderness ___ weakness ___ soreness ___ numbness

___ tingling ___ burning ___ swelling ___ cramping ___ throbbing

Duration: _____ Frequency: _____

Intensity: (mild, moderate, or severe) _____

Location: (R = right, L = left) ___ neck      ___ upper arm ___ lower arm ___ back ___ upper leg ___ foot

(Check all that apply)      ___ shoulder ___ elbow      ___ hand/wrist ___ hip ___ lower leg

Radiation: (R = right, L = left) ___ neck      ___ upper arm ___ lower arm ___ back ___ upper leg ___ foot

(Check all that apply)      ___ shoulder ___ elbow      ___ hand/wrist ___ hip ___ lower leg

Exacerbating or relieving activities (both on-the-job and off-the-job):

Exacerbating: 1) _____ 2) _____ 3) _____

Relieving:    1) _____ 2) _____ 3) _____

**Past Medical History** (prior injuries or disorders):

1) _____ 3) _____

2) _____ 4) _____

**Recreational Activities, Hobbies, Household Activities:**

1) _____ 3) _____

2) _____ 4) _____

**Occupational History:**

1) _____ 3) _____

2) _____ 4) _____

**Characterize the Job:**

Forceful, repetitive or sustained **exertions** can be estimated from production standards, employee ratings of efforts required to complete job tasks, descriptions of work objects and tools, weights of work objects and tools, and length of the workday. Extreme, repetitive or sustained **postures** can be estimated from a description of work methods and equipment. Employees can demonstrate the posture required for each step of the job task, or simulate the workstation in the examining room. Insufficient rest, pauses, or **recovery time** and be estimated from a description of rest breaks, production standards, work flow, and work organization factors. Extreme levels, repeated or long exposure to **vibration** can be estimated from a description of hand tools, or equipment. **Cold temperatures**, repeated or long exposure to cold can be based on temperature measurements, estimated from a description of the work environment, and the duration of time spent in cold areas.

| Physical Stress | Property | | |
| --- | --- | --- | --- |
| | Magnitude | Repetition Rate | Duration |
| Force | | | |
| Joint Angle | | | |
| Recovery | | | |
| Vibration | | | |
| Temperature | | | |

**FORM 2 — Musculoskeletal Disorder Management Plan**
***Forward Only Work Related Medical Information to the Employer***

Date of Assessment: _____

Name: _____ Date of Birth: _____
Employer: _____ Contact Person: _____ Phone: _____ FAX: _____

Diagnosis/Assessment:_____
_____

Treatment Plan: (e.g., medications/dosage, splints, physical or occupational therapy including frequency and duration of treatment, etc.)
_____
_____

Next Appointment: _____
Other Scheduled Appointments:
_____

**WORK STATUS**
Is the Employee able to perform his/her regular work?

____ Yes, Full duty
____ No, Remove from Work Environment until _____
____ No, Modified or Alternate Work until _____

(Complete Activity Checklist below for Job Modifications)

**Name:** _____

**Description of Restricted Work Activity**

| Activity | Duration | Frequency |
|---|---|---|
| a. Sitting | ____ Hrs. Per Day | ____ Hrs. at a Time |
| b. Standing | ____ Hrs. Per Day | ____ Hrs. at a Time |
| c. Walking | ____ Hrs. Per Day | ____ Hrs. at a Time |
| d. Lift/Carry:____ lbs. | ____ Hrs. Per Day | ____ Times Per Hr. |
| e. Climbing Stairs | ____ Hrs. Per Day | ____ Times Per Hr. |
| f. Climbing Ladders | ____ Hrs. Per Day | ____ Times Per Hr. |
| g. Kneeling | ____ Hrs. Per Day | ____ Times Per Hr. |
| h. Bending at Waist | ____ Hrs. Per Day | ____ Times Per Hr. |
| I. Squatting | ____ Hrs. Per Day | ____ Times Per Hr. |
| j. Twisting | ____ Hrs. Per Day | ____ Times Per Hr. |
| k. Pull/Push: ____ lbs. | ____ Hrs. Per Day | ____ Times Per Hr. |
| l. Reach Above Shoulder | ____ Hrs. Per Day | ____ Times Per Hr. ____ L ____ R |
| m. Extended Reaching | ____ Hrs. Per Day | ____ Times Per Hr. ____ L ____ R |
| n. Neck bend/twisting | ____ Hrs. Per Day | ____ Times Per Hr. |
| o. Elbow/Forearm Twist | ____ Hrs. Per Day | ____ Times Per Hr. ____ L ____ R |
| p. Hand/Wrist Bending | ____ Hrs. Per Day | ____ Times Per Hr. ____ L ____ R |
| q. Pinch Gripping | ____ Hrs. Per Day | ____ Times Per Hr. ____ L ____ R |
| r. Forceful Grasping | ____ Hrs. Per Day | ____ Times Per Hr. ____ L ____ R |
| s. Continuous Keyboard Use | ____ Hrs. Per Day | ____ Times Per Hr. |
| t. Vibrating Tool/Equip Use | ____ Hrs. Per Day | ____ Times Per Hr. ____ L ____ R |
| u. Ankle/Foot Bend/Twist | ____ Hrs. Per Day | ____ Times Per Hr. ____ L ____ R |
| v. Cold Temperature | ____ Hrs. Per Day | |

Other Restricted Job Tasks (including frequency and duration): _____
_____

Other Specific Job Recommendations: _____
_____

Health Care Provider Name: _____
Address: _____ City/State/Zip: _____
Phone: ( ___ )_____ FAX: ( ___ )_____
Copy of Form Given to Employee: ____ Yes        ____ No
Health Care Provider Signature: _____        Date: _____

Complete removal from the work environment should be avoided unless the employer is unable to accommodate the prescribed work restrictions. Research has documented that the longer the employee is off work, the less likely he/she will return to work (Vallfors, 1985). In these cases, the employer's contact person and the employee should be in day-to-day contact, and the employee can be encouraged to participate in a fitness program that does not involve the injured anatomical area.

Wrist immobilization devices, such as wrist splints or supports, can help rest the symptomatic area in some cases. These devices are especially effective off the job, particularly during sleep. They should be dispensed to individuals with MSDs only by HCPs with the training and experience in the positive and potentially negative aspects of these devices. Wrist splints, typically worn by patients with possible carpal tunnel syndrome, should not be worn at work unless the HCP determines that the employee's job tasks do not require wrist deviation or bending. Struggling against a splint can exacerbate the medical condition due to the increased force needed to overcome the splint. Splinting may also cause other joint areas (elbows or shoulders) to become symptomatic as work technique is altered. Recommended periods of immobilization vary from several weeks to months depending on the nature and severity of the disorder. Immobilization should be prescribed judiciously and monitored carefully to prevent iatrogenic complications (e.g., disuse muscle atrophy).

The *prophylactic* use of immobilization devices worn on or attached to the wrist or back is not recommended. Research indicates wrist splints have not been found to prevent distal upper extremity musculoskeletal disorders (Rempel, 1994). Likewise, there is no rigorous scientific evidence that back belts or back supports *prevent* injury, and their use is not recommended for prevention of low back problems (NIOSH, 1994; Mitchell et al., 1994). Where the employee is allowed to use a device that is worn on or attached to the wrist or back, the employer, in conjunction with a HCP, should inform each employee of the risks and potential health effects associated with their use in the workplace, and train each employee in the appropriate use of these devices. (McGill, 1993)

The HCP should advise affected employees about the potential risk of continuing non-modified work, or spending significant amounts of time on hobbies, recreational activities, and other personal habits that may adversely affect their condition (e.g., requires the use of the injured body part). However, as mentioned above, the employee should engage in a fitness program designed for exercise and aerobic conditioning that does not involve the injured anatomical area.

## Thermal (more frequently cold) Therapy

Such treatment is generally considered useful in the acute phase of some MSDs. Cold therapy may be contraindicated for other conditions (e.g., neurovascular).

## Oral Medications

Aspirin or other nonsteroidal anti-inflammatory agents (NSAIA) are useful in reducing the severity of symptoms either through their analgesic or anti-inflammatory properties. Their gastrointestinal and renal side effects, however, make their prophylactic use among asymptomatic employees inappropriate, and may limit their usefulness among employees with chronic symptoms. In short, NSAIAs should not be used prophylactically.

It must be noted that the effectiveness of Vitamin B-6 for treatment of musculoskeletal disorders has not been established (Amadio, 1985; Stransky et al., 1989; Spooner et al., 1993). Additionally, at this time there is no scientifically valid research that establishes the effectiveness of Vitamin B-6 for *preventing* the occurrence of musculoskeletal disorders.

## Stretching and Strengthening

A valuable adjunct in individual cases, this approach should be under the guidance of an appropriately trained HCP (e.g., physiatrists, physical and occupational therapists). Exercises that involve stressful motions or an extreme range of motions, or that reduce rest periods may be harmful.

## Hot Wax

At this time there is no scientific evidence regarding the effectiveness of hot wax treatments as a preventative measure or as a therapeutic modality.

## Steroid Injections

For some disorders resistant to conservative treatment, local injection of a corticosteroid by an experienced physician may be indicated. The addition of a local anesthetic agent to the injection can provide valuable diagnostic information.

## Surgery

With an effective ergonomics and medical management program, surgery for work-related MSDs should be needed rarely. Surgical intervention should be used for objective medical conditions and should have proven effectiveness. While the indications for prompt or emergency surgical intervention may still be present (e.g., ulnar artery thrombosis), surgery should be reserved for severe cases (e.g., very high levels of pain resulting in significant functional limitations) not responding to an adequate trial of conservative therapy.

# 27.8    Follow-up and Return to Work

## Follow-up

Many, if not most, WRMSDs improve with conservative measures. HCPs should follow up the symptomatic employee to document improvement, or to reevaluate employees who have not improved. The time frame for this follow-up depends on the symptom type, duration, and severity. A clinical exam or telephone contact with the employee should be made once a week, followed by a complete reevaluation within ten days from the last examination if the employee's symptoms are not improving. Where HCPs are available at the workplace, monitoring the symptomatic employee should occur every 3 to 5 working days depending on the clinical severity of the disorder (Wiesel et al., 1984; Wiesel et al., 1994).

In reassessing employees who have not improved, the following should be considered:

- Is the diagnosis correct?
- Are the treatment goals appropriate?
- Have the MSD risk factors on and off the job been addressed?
- Is referral appropriate?

If the job's relevant risk factors have been eliminated but the employee's symptoms persist, it is important for the HCP to realize that employee reactions to pain and functional limitations may prolong the recovery period. Strategies to help the employee cope with the pain and stress associated with these disorders should be incorporated into the employee's treatment plan. The time frames for considering referral depends on the primary HCP's training and expertise, in addition to the type, duration, and severity of the condition. In general, severe symptoms with objective physical examination findings interfering with an employee's ability to perform his/her job should be referred to an appropriate HCP specialist sooner than milder symptoms without objective findings.

## Return to Work

If an employee's treatment plan required time away from work, the next step is to return the employee to work in a manner that will minimize the chance for re-injury. Employees returning to the same job without a modification of the work environment are at risk for a recurrence. Key to the return to work process is open communication among the employee, the HCPs, and management. This will allow:

(1) prompt treatment, (2) an expedient return to work consistent with the employee's health status and job requirements, and (3) regular follow-up to manage symptoms and modify work restrictions as appropriate. The principles guiding the return to work determination include the type of MSD condition, the severity of the MSD condition, and the MSD risk factors present on the job.

Employees with MSDs who have difficulty remaining at work or returning to work in the expected timeframes are candidates for rehabilitation therapy. Rehabilitation refers to the process in which an injured worker follows a specific program that promotes healing and helps him or her return to work. During the rehabilitation process, psychosocial factors (factors present both on the job, and off the job, that can compromise an individual's ability to cope with symptoms, physical disorders, and functional limitations) should be addressed.

## 27.9    Screening

Currently there is no scientific evidence that validates the use of preassignment medical examinations, job simulation tests, or other screening tests as a valid predictor of which employees are likely to develop MSDs (Frymoyer, 1992; Werner et al., 1994; Cohen et al., 1994). Literature findings are mixed on the use of preplacement strength testing as a valid predictor of back injury.

## 27.10    Conclusion

The financial and human costs of work-related musculoskeletal disorders to our society are staggering. This chapter on the medical management of these disorders should help employers and HCPs wishing to prevent or reduce the severity of these disorders, resulting in a healthier, more productive workplace.

### Acknowledgments/Disclaimer

We would like to thank the individuals and professional associations who contributed to the draft ANSI Z-365 standard, and the draft OSHA Ergonomic Protection Standard. Ms. Bertsche's contribution to this chapter occurred while a visiting scientist at the National Institute for Occupational Safety and Health (NIOSH), from the U.S. Department of Labor Occupational Safety and Health Administration (OSHA). This chapter represents the views of the authors and does not constitute official policy of NIOSH.

### References

AHCPR (1994). Acute Low Back Problems in Adults: Assessment and Treatment. U.S. Dept of Health and Human Services, Public Health Service, Agency for Health Care Policy and Research. Rockville, MD. Publication No. 95-0643.

Amadio PC (1985). Pyridoxine as an adjunct in the treatment of carpal tunnel syndrome. *J Hand Surg;* 10A:237-241.

ANSI (1995). ANSI Z-365 Control of Work-Related Cumulative Trauma Disorders Part 1: Upper Extremities. American National Standards Institute. Chicago, IL: Working draft 4/17/95.

ASSH (1990). *The Hand: Examination and Diagnosis,* 3rd ed. American Society for Surgery of the Hand. New York, NY; Churchill Livingstone.

BLS (1995). Workplace injuries and illnesses in 1994. U.S. Department of Labor, Bureau of Labor Statistics. Washington, D.C.

Bongers PM, De Winter CR, Kompier MA, Hildebranndt VH (1993). Psychosocial factors at work and musculoskeletal disease. *Scand J Work Environ Health;* 19:297-312.

Cohen JE, Goel V, Frank JW, Gibson ES (1994). Predicting risk of back injuries, absenteeism, and chronic disability. *J Occup Med;* 36(10):1093-1099.

Day DE (1987). Prevention and return to work aspects of cumulative trauma disorders in the workplace. *Seminars in Occup Med;* 2(1):57-63.

Frymoyer JW (1992). Can low back pain disability be prevented? *Bailliere's Clinical Rheumatology*; 6(3):595-607.

Haig A, Linton P, McIntosh M, Moneta L, Mead P (1990). Aggressive early medical management by a specialist in physical medicine and rehabilitation: effect on lost time due to injuries in hospital employees. *J Occup Med*; 32(3):241-244.

Hoppenfield S (1976). *Physical Examination of the Spine and Extremities*. Norwalk, CT: Appleton Century Crofts.

Kiefhaber TR, Stern PH (1992). Upper extremity tendinitis and overuse syndromes in the athlete. *Clinics in Sports Med*; 11(1):39-55.

Louis DS (1987). Cumulative trauma disorders. *J Hand Surg*; 12A(5):823-825.

Mayer TG, Gatchel RJ, Hayer H et al. (1987). A prospective two-year study of functional restoration in industrial low back injury: an objective assessment procedure. *JAMA*; 258:1763-1767.

McGill SM (1993). Abdominal belts in industry: A position paper on their assets, liabilities, and use. *Am Indust Hygiene Assoc*; 54(12):752-554.

Mitchell LV, Lawler FH, Bowen D, Mote W, Ajundi P, Purswell J (1994). Effectiveness and cost-effectiveness of employer issued back belts in areas of high risk for back injury. *J Occup Med*; 36(1):90-94.

NIOSH (1994). Workplace use of back belts. Review and recommendations. U.S. Department of Health and Human Services, Public Health Service, Centers for Disease Control and Prevention, National Institute for Occupational Safety and Health. Cincinnati, OH; Publication No. 94-122.

Nirschl RP (1990). Patterns of failed healing in tendon injury, in Leadbetter WB, Buckwalter JA, Gordon SL (eds). Sports-Induced Inflammation. Park Ridge, IL: American Orthopaedic Society, pp 577-585.

OSHA (1995). Draft Ergonomics Protection Standard. U.S. Department of Labor, Occupational Safety and Health Administration. Washington, D.C. March, 1995

Ranney D (1993). Work-related chronic injuries of the forearm and hand: Their specific diagnosis and management. *Ergonomics*; 36(8):871-880.

Rempel D, Manejlovic R, Levinsohn DG, Bloom T, Gordon L (1994). The effect of wearing a flexible wrist splint on carpal tunnel pressure during repetitive hand activity. *J Hand Surg*; 19(1):106-110.

Spooner GR, Desai HB, Angel JF, Reeder BA, Donat JR (1993). Using pyridoxine to treat carpal tunnel syndrome. Randomized control trial. *Canadian Family Practice*; 39:2122-2127.

Stransky M, Rabin A, Leva NS, Lazaro RP (1989). Treatment of carpal tunnel syndrome with vitamin B-6: A double blind study. *Southern Med J*; 82(7):841-842.

Tubiana R, Thomine JM, Mackin E (1984). *Examination of the Hand and Upper Limb*. Philadelphia, PA: W.B. Saunders Co.

Upfal M (1994). Understanding medical management for musculoskeletal injuries. *Occup Hazards*; Sept:43-47.

Vallfors B (1985). Subacute and chronic low back pain; Clinical symptoms, absenteeism, and work environment. *Scand J Rehab Med*; Suppl 11:1-99.

Werner RA, Franzblau A, Johnston E (1994). Quantitative vibrometry and electrophysiological assessment in screening for carpal tunnel syndrome among industrial workers. *Arch Phys Med Rehabil*; 75:1228-1232.

WHO (1985) Identification and control of work-related diseases. World Health Organization Geneva: WHO Technical report; 174:7-11.

Wiesel SW, Feffer HL, Rothman RH (1984). Industrial low back pain, a prospective evaluation of a standardized diagnostic and treatment protocol. *Spine*; 9:199-203.

Wiesel SW, Boden SD, Feffer HL (1994). A quality based protocol for management of musculoskeletal injuries. *Clinical Orthopaedics and Related Research*; 301:164-176.

Wood DJ (1987). Design and evaluation of a back injury prevention program within a geriatric hospital. *Spine*; 12:77-82.

# 28

# Ergonomic Programs in Post-Injury Management

Tarek M. Khalil
*University of Miami*

Elsayed Abdel-Moty
*University of Miami*

Renee Steele-Rosomoff
*University of Miami*

Hubert L. Rosomoff
*University of Miami*

0-8493-1800-9/03/$0.00+$1.50
© 2003 by CRC Press LLC

# 28.1   Introduction

The management of work-related injuries and illnesses continues to be a challenge to ergonomists and health and safety professionals. This chapter outlines a selected number of ergonomic contributions in rehabilitation and disability prevention. The rationale, concepts, and methods developed for patient evaluation, work conditioning and job simulation, workplace design, worksite analysis, neuromuscular conditioning, and other topics are presented. Many of these intervention techniques were developed and tested by the ergonomics team at the University of Miami Comprehensive Pain and Rehabilitation Center (CPRC), and have proven valuable to post-injury management and return to work. It is difficult to provide detailed descriptions of the various ergonomic interventions in post-injury management in a few pages. Therefore, the authors have included an expanded list of references that may help the reader get more detailed information about the approaches used.

# 28.2   Workplace Injuries

Injuries or pain happen either due to specific accidents (trauma or a single stress of high magnitude) or cumulative events (a series of repeated stressors or micro traumas). If the resulting effects of an injury extend beyond three months, an acute injury is classified as a chronic condition (Pain Commission, 1987).

Injuries in the workplace often occur due to inadequate design of work environments, tools, and tasks. Workplace inadequacy also means that environments, tools, and tasks are designed without consideration for the people who must use them. Injuries can also develop due to a variety of causes including unsafe acts or human errors. These can also be dealt with through ergonomic interventions. According to the Bureau of Labor Statistics, there were approximately 6,252,000 occupational injuries and 514,000 occupational illnesses in 1994 (Safety Management, 1996). These figures translate into a total rate of 8.4 occupational injuries or illnesses per 100 full-time workers. There are certainly many more injuries and ergonomic disorders in the nonoccupational environment.

# 28.3   Ergonomics Disorders

Poor ergonomic design of tasks, equipment, or workplaces have resulted in a newly recognized group of injuries recently called ergonomic disorders. These are disorders of the musculoskeletal and nervous systems occurring in upper or lower extremities. They may be caused by repetitive motions, forceful exertions, vibration, sustained or awkward postures, and/or mechanical compression. Among ergonomic disorders are: low back pain, soft tissue injuries, eye fatigue and strain, sensitivity to light, blurred vision, changes in color perception, numbness of fingertips, headache, neck strain, skin problems, stress, and cumulative trauma disorders (also known as repetitive motion injuries, overuse disorders, or motion strain.)

Musculoskeletal injuries are the most common workplace injuries with back pain leading the charts in terms of incidence, cost, and suffering. Backaches can strike almost anyone. Although the exact incidence of low back pain is unknown, it is obviously high. It has been estimated that 70 to 80% of the population will develop some form of back pain during their life time. Annual estimates of the new cases of low back pain (LBP) have ranged from 10 to 15% of the United States population. LBP accounts for 25% of all disabling injuries (Rosomoff and Rosomoff, 1991). It has been reported to consume about 80% of total health care expenditures associated with musculoskeletal injuries.

Impairments of the back are the most frequent cause of activity limitation in persons under the age of 45, and they rank as the third most common cause of disability after heart and arthritic conditions in people age 45 years and over. The National Institute for Occupational Safety and Health (NIOSH) cited that back injuries account for about 20% of all injuries and illnesses in the workplace and that these injuries cost the nation an estimated $20 to 50 billion per year (NIOSH, 1991).

Cumulative trauma disorders (CTD) have been identified as the "industrial disease of the information age" and the "work-related disease of the 1980s and 1990s." CTDs are the most frequently reported on-the-job ailment. Millions of Americans have been afflicted by this family of disorders. It has been estimated that by the year 2000, over half the workforce could be vulnerable.

The latest federal statistics put the number of work-related ergonomic CTD injury cases at 514,000, up from just 30,000 in the mid-1980s (Safety Management, 1996). Carpal tunnel syndrome (CTS) is a frequently reported work illness. CTS is a special case of CTDs. It refers to injury to the median nerve of the wrist. Associated symptoms include numbness, tingling, burning sensations in the fingers, pain and wasting of muscles at the base of the thumb, dry or shiny palms, and clumsiness. CTS develops when the effective cross section of the carpal tunnel in the wrist is decreased or there is an increase in the volume and pressure within the tunnel due to inflammation or edema. While there is an increasing number of CTS cases which have been associated with workplace factors, some believe that the majority of CTS cases are not occupationally related and that certainly not all wrist problems are necessarily CTS.

The staggering statistics make rehabilitation and effective management of low back pain, cumulative trauma disorders, and all types of musculoskeletal or soft tissue injury a challenge to health care professionals and ergonomists.

## 28.4  Causes of Workplace Injuries/Illnesses

Workplace injuries can be caused by a host of factors. Musculoskeletal work injuries afflict people who perform light tasks (sedentary) and those involved in heavy labor. The magnitude of the stressor is an important factor, however a common characteristic among injury profiles is that job tasks require excessive bending, twisting, reaching, and deviation from neutral postures.

Most ergonomic disorders have been associated with injuries to the soft tissues in the human body. Soft tissue injury (for example: tension, sprain, spasms, weakness, strain, contracture, and inflammation) is the most common cause of pain (Quebec Study, 1987).

Soft tissue injury (STI) can result from activities such as lifting heavy objects, slipping and falling, bending and twisting, driving, accidents, and the manner by which work activities are carried out (also known as body mechanics). Soft tissue injuries can also develop gradually due to unknown causes and due to disuse and muscle weakness (Abdel-Moty et al., 1992a). In one study, we have shown that 20% of a sample of 200 patients with LBP were not able to attribute their painful condition to a specific incident, indicating that "pain developed gradually" (Abdel-Moty et al., 1990a).

STI can also develop gradually as a result of the "wear and tear" when activities are done with forceful muscular exertion, awkward postures, fast actions, sustained activity, and/or with frequent repetitive motion (case of CTDs) (Abdel-Moty et al., 1991b). Several human activities and job tasks are, therefore, more likely to result in STIs. Examples are: checkout scanning in supermarket, keypunching, working at an assembly line, manufacturing, meat processing, butchers, hairdressers, sewing and knitting, packing, stapling, tight grasping at the steering wheel, polishing and buffing, surface grinding, painters, dentists, interpreters, postal workers, milking cows, use of a spray gun, and sports activities.

Work-related injuries can also develop due to medical conditions and dysfunction of body structures and systems (e.g., infections, congenital disorders, neurological factors) (Khalil et al., 1993a).

In particular, computers have been blamed for causing or contributing to workplace injuries. Computers are becoming a household item in millions of offices, business, and homes for the purpose of increasing productivity. However, the effects on health have been negative. Prolonged keypunching, constant switching of the eyes from the document to the computer screen, improper combination of computer tasks and other activities, and inadequate visual attributes of the operator are but some of the conditions that have resulted in disabling physical and psychological conditions of the workplace user.

Numerous guidelines have been recommended to deal with workplace and ergonomic injuries. These will be discussed throughout this chapter.

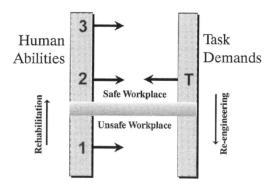

**FIGURE 28.1**   The premise of the ergonomic approach to injury/reinjury prevention is matching human abilities to task demands. When task demands exceed human abilities (T>1), human safety is at risk. There is a need, then, to improve human abilities through conditioning, exercise, and rehabilitation. When human abilities meet or exceed task demands (T = 2 or T<3), human safety is improved. Task demands can be decreased through ergonomic, engineering, and management controls.

## 28.5   Prevention of Workplace Injuries — The Ergonomic Approach

The thrust of the ergonomic approach to injury prevention is matching human abilities/limitation to the demands of the task (Figure 28.1). In doing so, ergonomists utilize knowledge of engineering, health, behavioral, physical, and management sciences in an integrated fashion for analyzing, designing, testing, and evaluating systems to improve human performance, reduce risk, improve productivity, and reduce workers' compensation costs.

Ergonomic contributions to injury prevention can be identified under three stages:

1. Primary prevention stage — where the prevention of the onset of injury becomes a priority and a desired goal for the avoidance of subsequent disability and lost workdays.
2. Secondary prevention stage — where the goal is the prevention of disability, quick functional restoration, and an expeditious return to work and productive life style of the injured worker.
3. Tertiary prevention stage — where prevention of re-injury becomes essential in order to ensure no recurrence of the injury-disability cycle.

Ergonomics contributions at each level of injury prevention are discussed below with emphasis on secondary and tertiary prevention approaches, herewith referred to as post-injury management.

## 28.6   Primary Prevention of Workplace Injuries

By far, this is considered to be the most effective approach to curing the problem of work injuries and illnesses. However, despite the increasing efforts to make workplaces safer, injuries still happen at an alarming rate. In response to the problem, several government, academic, commercial, and industrial entities have devoted tremendous resources toward injury prevention. The underlying approach has been early identification and recognition of the problem. This is important especially if done at a stage that allows prevention of injury through ergonomic work reorganization and work redesign (Kroemer, 1994).

To enhance efforts associated with primary prevention of injury, several bills, regulations, and standards have been issued by local, state, and federal agencies. In 1990, OSHA introduced its "OSHA Program Management Guidelines" which emphasize the need for:

1. Routine worksite analysis to recognize, identify, and correct ergonomic hazards including review and analysis of injury and illness records.
2. Hazard prevention and control through design measures to prevent or control hazards; engineering controls for workplace design; work practice controls such as work rest schedule; use of personal protective equipment; and administrative controls.
3. Medical management; injury/illness recordkeeping; early recognition and reporting; systematic evaluation and referral; treatment and return to work; systematic monitoring; and adequate staffing and facilities.
4. Training and education (both general and job specific) of the supervisors, managers, and engineers and maintenance personnel.

Also, NIOSH declared that the most effective way to prevent back injury is to implement an ergonomic program that focuses on redesigning the work environment and work tasks to reduce the hazards. One such effort was the introduction of the "lifting equation" and guidelines for manual material handling. Repetitive material handling has been considered to increase vulnerability to injuries/claims involving the back. NIOSH, therefore, developed its *Work Practice Guide* which has undergone two revisions since 1981. The guide provided direction as to the amount of weight to be handled under various conditions. Obviously, this biomechanical approach has its practical limitations. Employee capabilities, skills, past injuries, stress, and similar factors should be considered together with this mathematical determination of lifting limits. Additionally, guidelines alone are not an effective substitute for employee training in safety, a well-engineered workplace, and a "fit" worker.

The American National Standards Institute (ANSI) is another agency which advocates human factors in workplace design for health and safety. ANSI published technical standards for implementation of human factors engineering principles and practices in the design of visual display terminals (VDT), associated furniture, and the office environment in which they are placed (ANSI, 1988).

## 28.7 Post-Injury Management — Prevention of Disability

The primary objective of ergonomic contributions in this stage is "to design effective intervention strategies for the restoration of functional abilities and immediate return to work and productive life style." Another objective is deterrence of further aggravation of an existing condition or injury.

This philosophy was adopted by the University of Miami Comprehensive Pain and Rehabilitation Center as early as 1982 when ergonomics was first introduced in a multidisciplinary pain management team. This marked the beginning of a new era for ergonomics research and application in health care systems and pain treatment in particular. The basis for introducing ergonomics in post-injury management is demonstrated in Figure 28.2. In this model, injury management according to the traditional medical approach alone may be effective in treating the symptoms of pain, but it may fail to address and correct the cause of the injury. Through the use of a "rehabilitation through technology" approach, ergonomists assist in effective treatment as well as returning the patient to a modified work environment with improved safety and better ergonomic design. Therefore, the potential for reinjury is greatly reduced or eliminated. Since its introduction in pain management, ergonomics has gained wide acceptance, and its contributions have proved vital in addressing issues of primary importance. Working with other health care professionals who are usually involved in musculoskeletal and low back pain rehabilitation, ergonomists have played equally important roles in the successful rehabilitation and management of injuries.

### The Role of the Medical Community in Post-Injury Management

Several medical treatment approaches have been advocated by different care providers to deal with musculoskeletal injuries; especially low back pain problems. Successful rehabilitation and management of musculoskeletal injuries should integrate the different medical treatment disciplines in order to accomplish such goals as: restoration of function; pain reduction; and, consequently, increasing productivity

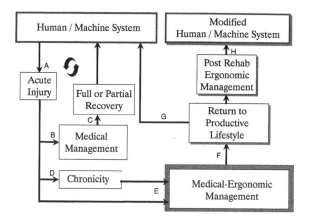

**FIGURE 28.2**    (A) Injury occurs as a result of a faulty human-machine system. (B&C) Medical management attempts to resolve acute injuries in order to return the injured worker to the *same* job. Since no measures are taken to change the cause of injury, the worker is at a risk of reinjury. (D) Acute injuries may develop into chronic conditions if left untreated or when mismanaged. (E&F) The medical–ergonomic management of both acute and chronic injuries or illnesses will assist the injured individual to return to a productive life style. (G) The inclusion of ergonomics in the post-rehabilitation stage will ensure the individual's return to an ergonomically correct human machine system, thus minimizing the potential of reinjury.

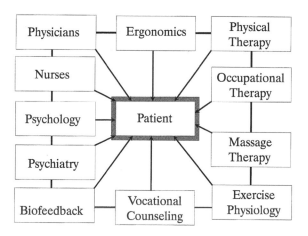

**FIGURE 28.3**    Comprehensive pain and rehabilitation programs employ a multidisciplinary approach. In this model, ergonomists work jointly with other health care providers to address patient issues and needs for full restoration.

and reducing rising disability costs. The health care professions that are usually involved in dealing with an important problem such as low back pain rehabilitation include: the physician(s), physical therapists, occupational therapists, nurses, vocational counselors, psychologists, and psychiatrists. Another profession which has shown to be equally important in a successful low back pain rehabilitation effort is ergonomics (Figure 28.3) (Khalil et al., 1993a).

## The Role of the Clinical Ergonomist in Injury Management

The application of ergonomics principles and methods to the injured population is far from simple. A major difficulty often encountered in applying ergonomic methods in a clinical setting is that these methods are developed for the "healthy" population, representing military personnel, industrial workers,

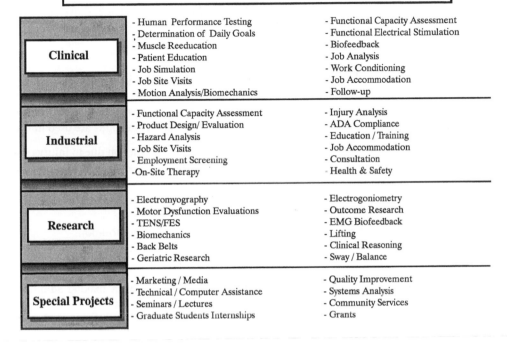

**FIGURE 28.4** Listing of ergonomic contributions to post-injury management in the various areas.

and volunteers. Therefore, it becomes important to bridge the gap between traditional ergonomic research and its application in health care systems. This requires a thorough understanding of the type of injury, the specifics of the population at hand, and of medical, social, and legal issues (Khalil et al., 1991b; Khalil et al., 1992b). For example, in ergonomic research, the evaluation of muscle strength has been based on the use of maximum voluntary exertion protocols where subjects are asked to increase their efforts to a maximum level. Applying this protocol to patients with back injury may be detrimental to their health and their safety, may be psychologically unacceptable to the patient, or may result in irreversible damage leading to medical as well as legal complications. In this case, it becomes necessary to develop methods of strength testing which are more appropriate and specific to the type of injury such as the acceptable maximum effort procedures developed by Khalil et al. (1987). Conversely, and as their awareness of the role of ergonomics is increased, the medical staff will begin to redefine their roles and contributions in order to make use of available ergonomic approaches not previously known to them.

## 28.8   Ergonomic Interventions in Post-Injury Management

Ergonomic contributions to injury management can be classified under the following categories in the order of their application (Abdel-Moty et al., 1990b; Khalil et al., 1993b) (Figure 28.4):

1. Physical, functional, and work capacity assessments.
2. Determination of patients' daily treatment goals.
3. Electrically assisted methods to improve neuromuscular functioning.
4. Work conditioning and job simulation.
5. Workplace design and analysis including job site visits.
6. Biomechanical analysis of human motion (gait, sway, kinetics, and kinematics).
7. Patient education (safety, body mechanics, ergonomics).

8. Applied clinical research to develop innovative solutions to clinical problems (electronic goniometry, stretching, traction, outcome).

The methods used in each area have been successfully implemented at the University of Miami Comprehensive Pain and Rehabilitation Center (CPRC) at South Shore Hospital in Miami Beach, Florida and are discussed in more detail below.

# 28.9　Evaluation of Physical, Functional, and Work Capacities

Objective evaluation of human performance is a step in an ergonomic program aimed at matching people to their environments, machines, and tasks. In a rehabilitation setting, evaluations are used in order to:

1. Assess the current level of function and the degree of the loss of function due to injury or illness. This helps the patient understand the degree of their functional loss and the clinician to establish a baseline for evaluation of treatment efficacy
2. Evaluate and monitor patients' progress during rehabilitation to ensure that treatment goals are met and that the patient is part of the therapeutic plan of care
3. Direct treatment to address problem areas and focus on functional restoration
4. Evaluate treatment outcome following rehabilitation
5. Examine the patient's ability to re-enter a productive life style

Also, objective assessment of functional capacities can be a useful tool in making administrative, clinical, and legal decisions regarding work readiness.

## Components of the Human Performance Profile

The patient's performance profile may consist of one or more of the following measures (Abdel-Moty et al., 1991c; Abdel-Moty et al., 1992b; Fishbain et al., 1994, 1995):

a. Measures of *physical* capacities such as isometric and dynamic strength, flexibility, mobility, posture, sway and balance, psychomotor abilities, and gait.
b. Measures of *physiological* capacities such as muscular endurance and cardiovascular endurance.
c. Measures of *functional* capacities such as tolerance to sitting, standing, walking, climbing, lifting, carrying, pushing, pulling, driving, stooping, crouching, and squatting.
d. Measures of work capacities specific to the patient's ability to perform job demands.

Measurements of the human performance profile are reported relative to basic scores upon initiation of treatment, progress from beginning to final scores, and in comparison to "norms." Behaviors during evaluation are observed and recorded with respect to cooperation, consistency, effort, motivation, patient comments, and pain behavior. At the CPRC, this profile is established for each patient at three intervals: upon admission to the rehabilitation program (inpatient or outpatient); two weeks into the program; and at the end of the 4-week treatment program. The profile can also be compared to the physical demands dictated by the job thus assisting in job placement for the prevention of further injury.

The issues surrounding patient evaluation are numerous and can be found in Abdel-Moty et al. (1996). Among these are:

1. Operational definitions with respect to the types of evaluations (physical capacity evaluation vs. functional capacity assessment, work-related assessments)
2. Patient's issues (pain, perceptions, behaviors, motivation, effort, secondary gains, use of medications, prior exposure to similar testing, contraindications, use of assistive devices)
3. Relationship to the Americans with Disabilities Act (ADA) — the ADA requires that employment/post-hire evaluations using functional capacity assessment (FCA) to be job specific and task oriented (Abdel-Moty et al., 1993a)
4. Administrative issues of insurance, authorization, referrals, and scheduling

5. Evaluator's issues (bias, objectivity, training, experience, qualifications)
6. Methodological issues (patient safety, testing equipment, protocols, instructions, sequence of testing)
7. Statistical issues (validity, reliability, sensitivity)
8. Other issues (testing environment, presence of third-party, such as attorneys, during testing, reporting of results)

By accurately establishing performance parameters of the patients, it is possible to chart their course of treatment through the use of the concept of daily goals.

## 28.10 Scientific Determination of Patients' Daily Treatment Goals

Restoration of functional abilities of patients requires knowledge about patients' baseline level of performance as well as target goals. Baseline values represent the levels of functional abilities at which patients enter the rehabilitation process. Typically, baseline values are determined for categories of muscle strength, ranges of motion, tolerances, mobility, and soft-tissue factors, as well as a host of behavioral and vocational variables. Treatment goals are usually determined based on vocational and/or avocational objectives as well as on expected levels of functional abilities representing norms for healthy subjects (e.g., the expected angle for lumbar flexion is 90 degrees).

The objective of rehabilitation is to treat, condition, and build the patients' functional abilities in order for them to achieve the desired levels of performance constituting their treatment goals.

In most rehabilitation settings, the progression from the initial (admission) levels to the final goals (i.e., daily achievement during treatment) is determined through subjective clinical judgment of the patient's current status. Additionally, in some instances patients' progression may be seen as tied to the length of hospitalization or rehabilitation.

In today's atmosphere of health care management awareness, cost effectiveness, and health reforms, optimization of treatment approaches and maximization of service delivery becomes a priority. Using a daily goal model can help maximize treatment plans and determine optimal pathways patients should follow during rehabilitation. A prediction modeling approach provides the treating team, as well as the patient, with scientific tools for determining, realistically and objectively, patients' daily performance achievement during treatment. It also enables effective utilization of resources, and hence cost effectiveness, for both the patient and the health care delivery system.

At the CPRC, ergonomists developed a novel approach to maximizing patients' progress during rehabilitation through the use of computerized modeling and optimization approaches. Named the computerized daily goals (CDG) model, it was developed to determine optimal pathways a patient should follow during chronic low back pain rehabilitation. The model utilizes statistical projection methods which take into consideration the patient's initial performance level and the desired goals (Abdel-Moty et al. 1993b) (Figure 28.5).

This nonlinear model derives its coefficients from retrospective data collected on over 1000 patients with chronic pain who were admitted for treatment at the CPRC, successfully completed the 4-week rehabilitation program, and returned to a productive life style. All patients underwent evaluations of functional abilities at three points during treatment: upon admission, after 2 weeks, and at the end of the 4-week treatment program. Patients were evaluated and their performance was analyzed in the following categories:

1. Tolerances (sitting, standing, walking, squatting, kneeling, stair climbing).
2. Strength (lifting, carrying, reaching, pulling, pushing).
3. Flexibility (trunk, cervical).

The purpose of this type of classification was to determine quantitative trajectories of progress in the various categories as a result of the intervention protocols. Following necessary testing and validation, a model was constructed and implemented on a personal microcomputer. Currently, once initial levels of

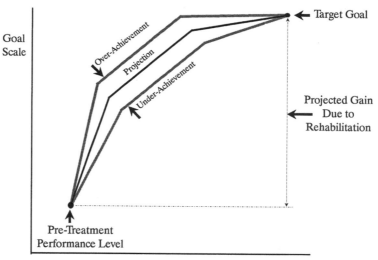

**FIGURE 28.5** The scientific determination of patients' daily projected goals and clinical pathways requires accurate determination of a patient's initial performance levels as well as knowledge of a desired. Projections are then made over the length of the patient's treatment. The "range of projection" allows for flexibility in order to accommodate individual patients.

performance have been measured and treatment goals have been determined, the daily goals program is used to provide a personalized print-out of the expected daily performance in the different categories. Final goals are reached just prior to patients' discharge from the program. The daily goals program predicts daily increments and corresponding dosage of patients' therapeutic activities. Optimal progression from initial tolerances to final goals is then printed and used by the patient and the medical staff to determine activities needed to achieve desired daily performance in the various categories throughout treatment.

In order to demonstrate its usefulness, a sample of 200 patients who completed the treatment program was analyzed. About 82% of patients who were studied followed the projected levels, with another 9% exceeding projections and showing improvements at rates higher than expected. In about 3% of the cases examined, the "printed" goals were lowered due to unexpected changes in the patients' medical condition. In another 5% of the cases the projected goals were modified (mostly decreased) due to changes in return-to-work plans.

## 28.11  Electrically Assisted Methods (EMG, Biofeedback, Muscle Reeducation, and Functional Electric Stimulation)

Soft tissue injury and the resulting pain often produce compensatory changes in posture and in the manner by which patients perform daily activities. The failure to treat chronic pain patients effectively is, in part, due to an inability to identify and pinpoint the source of these altered motor patterns affecting functional abilities. In order to assist the patient and the medical team with treatment goals, several technology-based methods and intervention protocols were developed by clinical ergonomists. Among these are protocols for biofeedback, neuromuscular reeducation, and functional electrical stimulation.

One method was developed to identify and effectively treat motor dysfunction in patients with chronic pain conditions. This innovative multidisciplinary evaluation/treatment approach is based on the use of on-line computerized electromyographic (EMG) methods to study multiple muscles involved in a chain of precise activities specific to recruiting these muscles. The EMG signals of the various muscles are examined for: baseline activity, symmetry, magnitude, frequency contents, synchrony, timing, and patterns. Patients' behaviors are also observed. EMG findings are then compared to relevant clinical findings. The team of medical and ergonomics professionals discusses the significance of the findings, composes interpretations, and recommends treatment plans or addresses outcomes as applicable. Patient-specific, as well as condition-specific, multidisciplinary approaches are then generated to deal with the problems during daily treatment. The overall objective is to decrease pain and improve function. This is accomplished through using EMG and other electrically assisted methods to increase sensory perception (of muscles and joints); increase neuromuscular recruitment; increase strength and endurance; and re-establish synchrony, symmetry, pattern, and synergy of muscle activity to increase functional capacities. These protocols are incorporated into patients' daily treatment and clinical pathways. These treatment approaches fall under the categories of functional electric stimulation, biofeedback, and neuromuscular reeducation.

## Functional Electric Stimulation

Functional electric stimulation (FES) is used when minimal muscle recruitment or reduced voluntary control is detected upon medical or ergonomic testing. FES is a good example of an approach for muscular conditioning and strengthening through which muscles can be strengthened "passively" without placing excessive demands on the patient especially in the presence of pain. FES is the process of applying an external electrical stimulus to a muscle or muscle groups in order to induce muscle contraction. FES found many applications in rehabilitation, physical therapy, sports medicine, and recently in physical fitness. Its usefulness ranged from increasing local blood flow to strengthening weak and healthy muscles and enabling paraplegics to walk. We have also used FES successfully to treat conversion-disorders-type paralysis, for electric testing of motor responses, and as a motor dysfunction treatment method. The strength of FES arises from its ability to induce maximal muscular contraction without any voluntary effort on the part of the treated individual. This later aspect is of value for patients with chronic pain whose pain is often aggravated through regular exercise or who are unable to initiate the voluntary effort necessary for muscular conditioning due to disuse.

The use of FES as a muscle strengthening alternative proved useful and shows great potential in the rehabilitation and physical restoration of muscles weakened by disuse or pain. The current technology has allowed for the development of electrical stimulation units that are safe for use with minimal adverse effects. Studies on FES indicate that this passive intervention strategy is quite effective in numerous cases (Khalil et al., 1991a; Abdel-Moty et al., 1994). The use of FES may be less effective in some cases due to such reasons as fear of electricity, impaired sensory perception, skin conditions, lesions, muscles of the chest wall, and cervical paraspinals.

It should be emphasized here that FES is not a substitute for regular exercise. FES is used to "jump start" the neuromuscular system function. Once the patient has gained sufficient power to initiate voluntary movement comfortably, he or she is to engage in active exercises of strengthening, flexibility, and endurance training. In this regard recent advertisements which market muscle stimulators so "you never have to do sit-ups again" and "the cellulite breakthroughs" must be evaluated objectively and scientifically, especially with respect to patients' perception of such products.

## Electromyography and Biofeedback

Another electrically assisted tool for improving functional abilities of injured individuals is the use of electromyography (EMG). Recently there has been a renewed interest in the use of surface electromyographic (EMG) recording for the assessment and evaluation of patients with chronic pain. In general, a carefully processed EMG signal can be a useful tool in the quantitative measurement of muscular performance. In chronic low back pain rehabilitation, we found EMG to be valuable in the study of muscle dynamics, for purposes of muscle reeducation and biofeedback, and in the evaluation of patients' response to specific treatments (Khalil et al., 1991a). In earlier reports, it has been shown that the initial EMG levels of pain patients upon admission to the rehabilitation program were low and increased following muscle reeducation and physical restoration (Khalil et al., 1993a). The significant increase in EMG was associated with significant increase in strength and reduction of pain.

Biofeedback (biological feedback) is the process of using specialized instruments to give people information about their biological systems (temperature, heart rate, muscle activity, etc.). It is a set of treatment/training techniques used to increase awareness and voluntary control of biological conditions and relate them to human physical and emotional well being. Biofeedback (BF) is useful in the relief of stress, tension, headaches, muscular dysfunction, and for improvement of muscle strength (Khalil et al., 1987; Asfour et al., 1990).

Using EMG biofeedback, patients perform therapeutic maneuvers while affected muscles are monitored. Muscles are then compared bilaterally, and the information is used to facilitate the patient's awareness of muscular performance. Patients are also evaluated without the use of feedback to determine carry-over and awareness level. Ergonomists contribute in this area by making the technology and measurement methods available to clinicians. The developed methods can then be used to improve the patients' ability to coordinate muscle activity, re-establish proper reciprocal inhibition, re-establish functional synergy including appropriate force couples and sequential contractions, decrease the need for inappropriate muscular or postural compensations, and achieve recruitment levels and baseline activity.

## Work Conditioning and Job Simulation

The concepts of job simulation and work conditioning are important to effective return-to-work strategies. The objective here is to improve the physical and functional abilities of post-injury individuals while practicing job tasks. It is important to raise the individual's tolerances to adequate levels in order to maximize his or her potential for return to work. This is accomplished through exercise, training, work conditioning, and job simulation. In clinical settings, the ergonomists assist the health care delivery team in developing realistic job simulations within the rehabilitation establishment to permit patients to perform critical job tasks, preferably under medical supervision. An individual or a patient is taught to perform his or her job task properly, which assures them that he or she is capable of carrying them out. This also allows the treating physician to certify that the patient has been physically rehabilitated to handle task demands.

During this activity, the ergonomist evaluates motions, time, and forces required to perform the various tasks and relates these demands to the functional abilities of the individual. For this purpose, it is not possible, nor necessary, to replicate jobs. It is also not necessary to simulate a full workday unless a work tolerance evaluation is requested.

Work conditioning programs are highly structured, task focused, goal oriented, individualized, and multidisciplinary/interdisciplinary in nature. The following are factors which affect the design of a work conditioning program:

1. Intensity (the level of exertion at which a patient is expected to perform). The amount of weight to be carried, distance to be covered, and other physical demands are a function of, not only the patient's status during treatment, but also his or her projected goal. The ergonomist seeks input from the rehabilitation team regarding the patient's medical status in order to determine work intensity in job simulation.
2. Environment (actual or simulated). In many instances the rehabilitation team may decide that it is necessary to observe the patient performing work tasks on the job. This gives an insight into work behaviors, ability to manage stress on the job, and helps develop realistic treatment plans and ergonomic recommendations.
3. Personnel involved (medical, ergonomic). In most cases, the ergonomist designs and conducts work conditioning activities with input from the multidisciplinary team. It is not unusual to find a team of the ergonomist, the occupational therapist, physical therapist, the vocational counselor, the biofeedback specialist, and a psychologist working simultaneously with a patient during job simulation. Each professional brings to this process a unique input and perspective and assists in

immediate problem solving. The degree of involvement of each discipline depends on the activities required.

4. Approach (progressive nature of work simulation sessions). A patient who is deconditioned and who has a physically demanding job to return to (e.g., heavy work), will need a different approach to work conditioning than a patient who will return to a sedentary job. For example, at the CPRC, the objective of rehabilitation and ergonomic interventions is to raise the patient's abilities to maximum attainable physical and functional abilities. The inclusion of work demands in these abilities will modify the approach to development of the plan of care for the patient. The plan of care also includes essential goals (or prerequisites) to be achieved before work conditioning activities begin. These are:
   a. Increase patient's awareness of posture and body mechanics.
   b. Increase flexibility and mobility.
   c. Increase strength and endurance.
   d. Improve stress management skills.
   e. Improve pain control, safety, and preventive medicine techniques.
   f. Modify behaviors toward work, employment, and return to work.
   g. Involve patient in work conditioning activities early in the treatment program.

The ergonomist provides direct supervision during the work conditioning sessions as well as encouragement, support, and reinforcement to the patient while monitoring behaviors.

## Workplace Design

This type of ergonomic intervention aims at assessing the relationship between human characteristics (e.g., posture, body mechanics) and musculoskeletal stresses with emphasis on work issues. This is a rather complex process due to the large number of human and environmental factors involved (Khalil et al., 1990). The ergonomic premise here is that awkward postures and poor body mechanics can result in a multitude of health problems. With patients who already have an injury or illness, the aches and pains are magnified. The task of the ergonomist is to assist the patient or employer to design/modify the workstation to ensure:

1. Proper engineering design from an ergonomic point of view.
2. Proper posture of the workplace user.
3. Good body mechanics when performing job tasks.

These issues are also interrelated. For example, poor workplace design can result in poor work habits by not allowing good posture and proper body mechanics. Additionally, proper design does not guarantee stress-free environment if good posture and proper body mechanics are not practiced.

The process of workplace design within a clinical setting consists of the following components:

1. Preliminary data collection of patients' information, job-related information, and a self description of a typical workday.
2. Initial evaluation. This is a session of simulated activities during which essential job tasks are conducted, recorded, and analyzed. Emphasis is on posture, body mechanics, and muscular work during repetitive activities. This evaluation aims at identifying tasks as potentially stressful. During this session, no feedback or suggestions for modifications or adjustments are given. Observations and analysis of video graphic data are used to isolate the critical risk factors with respect to the worker, the tools, the furniture, and other workplace parameters.
3. Intervention phase. Once risk factors are identified, the ergonomist begins to develop the interventions specific to reducing/resolving the risk factors. Once more, the interventions usually address the engineering design of the workplace as well as the posture, body mechanics, and patients' ability to follow through (learning). Engineering measures aim at adjusting, modifying, changing, and/or replacing current heights, layout, equipment, tools, and design characteristics.

It is important at this stage to explain to the patient the rationale behind the recommendations in order to increase his or her awareness and ability to generalize to situations outside work. It is also not necessary to always make recommendations for new equipment or assistive technologies (e.g., cushions, ergonomic chairs, etc.). In most cases, workplaces can be reasonably accommodated to meet ergonomic needs. An essential ergonomic rationale is that people may not have the "ideal ergonomically correct" setup wherever they go. It is, therefore, the responsibility of the ergonomist to teach patients methods of improving safety and comfort without the need for expensive adjustments.

A tool we have used to aid in the process of analyzing and recommending workplace adjustments and modifications is SWAD (sitting workplace analysis and design) (Abdel-Moty and Khalil, 1991). SWAD is a computer program resembling artificial intelligence. The inputs to the program are workplace users' demographics, 16 anthropometric dimensions, workplace dimensions, work tools, and the priority and frequency of use of each work element. SWAD then combines this information with a knowledge base of ergonomic principles and guidelines and a set of inference procedures. The output of SWAD is the recommended workplace dimensions, heights, reaches, foot rest, chair parameters, VDT parameters, and optimal placement of all work tools. SWAD packages anthropometric data and ergonomic principles into a software available to the medical community and to workplace users. It enables customization of workstation adjustment without trial and error.

Also, in this intervention phase, patients' posture and body mechanics are corrected and the effect on reduced muscle activity is demonstrated through the use of EMG. Patients are shown how to correct movements, minimize awkward postures, and refine their body mechanics. In most instances, patients can now perform the same job tasks with minimum discomfort.

4. Follow-up evaluation. In order to document the effectiveness of the interventions, a follow-up evaluation is performed which consists of the patient performing the same simulated activities performed during the initial evaluation. Also, during this session, no feedback is given. Data are then analyzed and results are compared to baseline information. In all cases, significant resolution in the risk factors are observed, muscle activity is decreased reflecting improved economy of motions and less stress, and patients are able to perform essential work functions with no increase in pain or discomfort.

## Job Site Analysis

The analysis of worksites can be extremely helpful and should follow a systematic method. The process starts with a data collection phase during which health records are examined in order to identify priority problem areas/tasks. Data collection also includes photographing work areas, recording postures and motions, measuring forces and vibration data, and documenting repetitiveness and frequencies of job cycles.

The data collected are then analyzed and examined to determine the risk factors specific to motion economy, postures, and body mechanics. Quantitative data are also analyzed to determine if, for example, forces and vibration data are within exposure ranges.

Recommendations can then be offered to improve working conditions and/or reduce repetitiveness, vibration, forces, etc. For example, in order to reduce vibration the ergonomists may recommend the use of dampen technology, reduce contact surface area, reduce exposure to driving force, isolate vibration or use material that absorbs vibration. Another common risk factor that usually requires ergonomic intervention is task repetitiveness. Common recommendations to reduce the exposure to high-frequency activities can be: reducing the number of cycles, augmenting human activities with machines through automation and mechanization (e.g., use of scanners), enlarging jobs (e.g., providing help to workers), and when possible alternating use of extremities and job rotation.

Recommendations often include workers' training in proper posture and body mechanics for performing job tasks. Workers are either taught to maintain the neutral postures of their body joints or are provided with technological aides to minimize awkward deviations of joints.

In order to reduce muscle force, the ergonomists may recommend using machines to replace human effort, improving quality of tools to reduce force, changing tool design to allow use of stronger muscles (e.g., the thumb is stronger than any single finger), modifying design to use multiple muscle groups (e.g., group fingers are stronger than the thumb), distributing force over a larger area (e.g., using trigger strips), reducing force by having a slip prevention design in handheld tools, improving tool balance, use of proper gloves (e.g., not too tight at the wrist), and utilizing spring action in tools such as scissors and clippers.

A useful tool in data collection is the use of ergonomic checklists. These are specially designed forms that facilitate problem identification and documentation. While it may be possible to prepare a generic checklist, job environments are not alike, thus necessitating customized checklists. An ergonomic checklist should be designed to focus on the human component of the workplace rather than the workplace itself. For example, it should be asked "Does work surface height permit satisfactory postures of the arms?" rather than "What is work surface height?"

## Biomechanical Analysis, Motion Analysis

Biomechanics is the discipline dedicated to the study of the living body as a structure which can function properly only within the confines of both the laws of Newtonian mechanics and the biological laws of life. Occupational biomechanics deals with understanding the complex mechanisms of human interaction with the industrial environment. A number of biomechanical models have been developed which allow the prediction of tissue load indices. Some of these mathematical models deal with evaluating the muscle force and joint reactions for different static postures or during motion. Other models of the lumbosacral region are used to study the combined stresses and strains on the local ligaments, muscles, and disc tissue (Khalil and Ramadan, 1987; Andersson and Chaffin, 1986; Adams et al., 1994; Hindle et al., 1990; McGill and Norman, 1987).

Biomechanical models use different criteria for the evaluation of the muscle and joint forces necessary for maintaining the human body's equilibrium during various activities. Some researchers use such models for the design of manual handling tasks and for the development of a physical job evaluation methodology. Biomechanical models are usually validated mathematically, experimentally, or through the use of EMG. The use of biomechanical models in post-injury management is helpful for patient education and for worksite analysis.

## Patient Education

Another task of the clinical ergonomist is to develop patient education programs and materials based on ergonomics concepts. Patient education is an essential component in the comprehensive rehabilitation process. This is especially important with chronic pain patients who have been treated and evaluated in many facilities and have failed various rehabilitation efforts. In addition to the education on pain, myofascial syndrome, health, diet, relaxation, dealing with flare-ups, activities of daily living, and stress management issues, patients are educated in workplace/home safety. Patient education takes many forms: lecture type, group discussions, and hands-on training.

## Applied Clinical Research: Innovations in Problem Solving

In the area of applied clinical research, ergonomists evaluate the effectiveness of treatment regimens upon the restoration of functional abilities and upon the reduction of pain, thus providing objective rationalization and justification of treatment. Research is also performed to develop and evaluate devices useful in diagnosis and treatment. Quantitative methods based on recognized approaches are used to assist medical professionals in identifying the usefulness of tools and equipment often prescribed for use in pain management (Zaki et al., 1990).

Another type of research has been necessitated by today's health care environment. This is clinical outcome research. This involves the development of a computerized program evaluation system to manage and analyze data regarding patient treatment, satisfaction, and the effectiveness of rehabilitation (Zaki et al., 1992). Data collection instruments and questionnaires are developed to assess patients' self-report of pain and functional status upon admission to the program, at the conclusion of treatment, and at regular intervals following discharge (3 months, 6 months, and 1 year). Data are used, not only to evaluate program effectiveness, but also to measure and improve the quality of care. Ergonomists provide the research methods, computer expertise, and statistical approach to manage this type of information.

## 28.12    Ergonomics and the ADA

The Americans with Disabilities Act (ADA, 1992) is a civil rights protection act that provides comprehensive protection to individuals with disabilities. The ADA addresses disability issues in relation to employment, public accommodation, transportation, and telecommunication. The application of ergonomics methods to individuals with disabilities necessitates complete knowledge of the ADA and its provisions. A review of the ADA is beyond the scope of this chapter; however, a selected number of relevant issues are discussed here.

According to the ADA, an individual with a disability is (ADA, 1992):

1. A person who has a physical or mental impairment that substantially limits one or more major life activities
2. Person who has a record of physical or mental impairment that substantially limits one or more major life activities
3. A person who is regarded as having such an impairment

The ADA prohibits discrimination in all employment practices, including job application procedures, hiring, firing, advancement, compensation, and training. It applies to recruitment, advertising, tenure, layoff, leave, fringe benefits, and all other employment-related activities. An employer may not make a pre-employment inquiry on an application form or in an interview as to whether, or to what extent, an individual is disabled. The employer may ask a job applicant whether he or she can perform particular job functions. If the applicant has a disability known to the employer, the employer may ask how he or she can perform job functions that the employer considers difficult or impossible to perform because of the disability, and whether an accommodation would be needed. A job offer may be conditioned on the results of a medical examination, provided that the examination is required for all entering employees in the same job category regardless of disability, and that information obtained is handled according to confidentiality requirements specified in the ADA. After an employee enters duty, all medical examinations and inquiries must be *job related* and necessary for the conduct of the employer's business. The employment provision of the ADA will, therefore, govern the ergonomist's approach to assisting an employer in designing job applications, job descriptions, and employment offers. It also affects the approach to functional capacity assessment since evaluations have to be job specific. This, in turn, means that the use of computerized testing equipment for the evaluation of "generic" functional abilities may no longer be valid. Evaluation setting should simulate the work situation. Testing protocols will need to be changed and modified to meet compliance. The ergonomist can also assist the employer in developing practical "post-offer" screening tools in order to test the applicant's ability to perform essential job functions (with or without reasonable accommodations).

The ADA expressly permits employers to establish qualification standards that will exclude individuals who pose a direct threat — i.e., a significant risk — to the health and safety of others, if that risk cannot be lowered to an acceptable level by reasonable accommodation. However, an employer may not simply assume that a threat exists; the employer must establish through objective, medically supportable methods that there is genuine risk that substantial harm could occur in the workplace. By requiring employers to make individualized judgments based on reliable medical evidence rather than on generalizations, ignorance, fear, patronizing attitudes, or stereotypes, the ADA recognizes the need to balance the interests of people with disabilities against the legitimate interests of employers in maintaining a safe workplace. The ergonomist can assist in this area by establishing workplace safety criteria in relation to human factors while taking medical information into account.

Another core issue in the ADA is reasonable accommodation. Reasonable accommodation is any modification or adjustment to a job or the work environment which should be made to enable a qualified applicant or employee with a disability to perform essential job functions. This is another area of primary relevance to ergonomics. Examples of reasonable accommodation include making existing facilities used by employees readily accessible to and usable by an individual with a disability; restructuring a job; modifying work schedules; acquiring or modifying equipment; providing qualified readers or interpreters; or appropriately modifying examinations, training, or other programs. Reasonable accommodation also may include reassigning a current employee to a vacant position for which the individual is qualified, if the person becomes disabled and is unable to do the original job. An employer, however, is not required to make an accommodation if it would impose an "undue hardship" on the operation of the employer's business.

## 28.13 Post-Injury Management of the Aging Population

Another population which requires special ergonomic attention is the injured elderly. One of the most pressing needs in the 1990s is to develop effective ways to serve an increasingly aging population.

In the coming decades the greatest growth is projected for the population aged 55 and over. In fact between the years 1990 and 2000 this group will increase by 11.5%, a gain of over 6 million persons. Within this group the greatest growth will be among persons 75 and over, an increase of 26.2% (U.S. Bureau of the Census, 1988). A question of major policy and fiscal importance is how well will the large population of elderly be able to live and function independently? It is well established that age-related changes in functional capacities affect the ability of older individuals to successfully complete daily living activities.

Many elderly persons reside in nursing homes and need some type of assistance to perform basic activities of daily living. Many others have difficulty functioning well at home and experience a high rate of accidents. The most frequent types of accidents are falls associated with stairs and steps, floors and floor coverings, bathtubs and showers, and ladders (Consumer Product Safety Commission, 1985). Many of the accidents could be prevented with the implementation of appropriate ergonomic intervention strategies. A wide variety of remedial techniques is already available for hazard reduction. The role of the ergonomist is to provide the information and products regarding older individuals to caregivers and medical professionals. Currently there are more than 15,000 assistive products on the market and a number of home safety catalogs available; however, most elderly home care providers and health care specialists are unaware of their existence. The ergonomist assists with the training and education. Not only must older people be aware of assistive tools and safety strategies, they must also understand how to use them. Elderly people must also be aware of safety issues in the living environment and must learn how to deal with them. Certainly, technological aids are not effective without appropriate training. Furthermore, this training needs to be directed at the health care specialist, designers of homes for the elderly, as well as potential users.

Understanding the human factors-related changes in performing daily activities for the aged population is one of the major aims of the ergonomist (Khalil et al., 1992b). Application of this knowledge to the design of tasks, equipment, and the home environment in which the elderly live contributes to the enhancement of their quality of life.

The ergonomic approach in this case will consist of:

1. Studying and understanding accident and hazardous environments of the elderly population (e.g., stairs and steps, bathtubs and showers, use of tools and appliances)
2. Identifying appropriate hazard intervention strategies

3. Developing and evaluating education and safety training material to ensure successful implementation of the hazard interventions
4. Developing techniques to educate and train relevant individuals caring for the elderly

## 28.14    Ergonomic Interventions Post-Rehabilitation (Prevention of Reinjury)

The continuous spectrum of including ergonomics in injury prevention goes beyond the clinical setting into post-rehabilitation. The thrust of the ergonomic approach continues to follow up on the worker once on the job or at home. Strategies used in the post-rehabilitation stage may include:

1. Immediate matching of the worker to the work.
2. Job modification based on the recommendations of the clinical ergonomist.
3. Ongoing screening of worker's capabilities.
4. Development of effective health maintenance programs including onsite stretching and home exercise programs.
5. Continuity of care through follow-up questionnaires at predetermined intervals. (This is a requirement of the Commission on the Accreditation of Rehabilitation Facilities, CARF.)
6. Potential involvement with the employer to extend the ergonomic safety interventions to other workers.

## 28.15    Who Should Do Ergonomics in Post-Injury Management Programs?

This is by no means a rhetorical question. It represents the reality of practicing ergonomics in clinical settings. The dilemma facing the ergonomics professions recently has been that anyone can say he or she is an ergonomist after attending brief training seminars. The field has become attractive to unqualified individuals who claim association to the profession regardless of education or experience. Recently, efforts have been made to address this issue. However, more needs to be done to combat the problem of unqualified individuals presenting themselves as ergonomists.

## 28.16    Who Pays for Ergonomics Service in a Clinical Setting?

Rehabilitation settings and hospital facilities follow a stringent system of reimbursement through different types of insurance (e.g., self-insurance, third-party insurance). In most states, workers' compensation provisions allow for reimbursement of necessary and justifiable ergonomic services. There must be a physician referral for the service, and authorization must be obtained in advance. Charges are then posted according to specific codes describing the procedure(s) performed, the length of the session, work performed, patients' interactions, and outcome data. In doing so, the ergonomist follows the hospital/rehabilitation facility's policies and procedures regarding documentation, reporting, and billing systems.

## 28.17    Summary

Ergonomics is the missing piece of the puzzle in post-injury management programs. Since 1982, the University of Miami Comprehensive Pain and Rehabilitation Center realized the importance of including ergonomics rationale, expertise, personnel, and resources in the day-to-day struggle to return injured individuals to a productive life style.

The clinical ergonomists have demonstrated that it is possible to return the disabled individual to full function with a properly designed multidisciplinary therapeutic rehabilitation program. The contribution of ergonomics to post-injury management has proved to be quite valuable. At the CPRC, ergonomists interact with patients as well as other members of the health care team to restore function to the patient. The involvement of ergonomists in activities of patient evaluation, job simulations, body mechanics teachings, patient education, and rehabilitation research should be an integral part of pain management settings.

The capacity to utilize ergonomics in a clinical setting extends beyond treatment to optimization of the treatment and prognosis of treatment outcome. It also extends to the post-rehabilitation stage where re-injury can be avoided without the need for expensive technology to solve workplace problems.

It remains the responsibility of the ergonomics profession, its organizations, and governing bodies to protect the field from less qualified individuals who encroach on the practice of ergonomics in general, and in clinical settings in particular.

In order to be involved in post-injury management, ergonomists will need to become familiar with health care trends, reforms, and language. They will require training in and exposure to a host of rehabilitation and medical issues ranging from hospital policies to documentation and health insurance. Above all, they will need to adapt to the clinical setting and develop interactive ties with other health care providers. The reward of helping patients overcome their disability and restore their identity not only justifies but compels the inclusion of the science and practice of ergonomics in rehabilitation and all phases of post-injury management.

# For Further Information

The book *Ergonomics in Back Pain: A Guide to Prevention and Rehabilitation* (Van Nostrand Reinhold, NY, 1993) provides detailed descriptions of the methods and approaches presented in this chapter.

# Questions

For specific questions, call Elsayed Abdel-Moty, Ph.D., at (305) 532-7246, or write to the Ergonomics Division, University of Miami Comprehensive Pain and Rehabilitation Center, 600 Alton Road, Miami Beach, Florida 33139 USA.

# References

Abdel-Moty, E. and Khalil, T.M., 1991. Computer-aided design and analysis of the sitting workplace for the disabled. *International Disability Studies*. 13:121-124.

Abdel-Moty, E., Khalil, T., Goldberg, M., Rosomoff, R. and Rosomoff, H., 1990a. Posture and pain: health effects and ergonomics interventions, in *Advances in Industrial Ergonomics and Safety II*, Ed. B. Das, p. 117-124, Taylor & Francis, London.

Abdel-Moty, E., Khalil, T.M., Rosomoff, R.S. and Rosomoff, H.L., 1990b. Ergonomics considerations and interventions, in *Painful Cervical Trauma: Diagnosis and Rehabilitative Treatment of Neuromusculoskeletal Injuries*, Eds. C. D. Tollison and J.R. Satterthwaite, p. 214-229, Williams & Wilkins, Baltimore.

Abdel-Moty, E., Khalil, T.M., Asfour, S.S., Sadek, S., Rosomoff, R.S. and Rosomoff, H.L., 1991a. Workers compensation and non-worker's compensation chronic pain patients responsiveness to rehabilitation, in *Advances in Industrial Ergonomics & Safety III*, Eds. W. Karwowski and J.W. Yates, p. 467-474, Taylor & Francis, London.

Abdel-Moty, E., Khalil, T., Diaz, E., Sadek, S., Rosomoff, R. and Rosomoff, H., 1991b. Ergonomic job analysis for patients with chronic low back pain during rehabilitation, in *Designing for Everyone*, Eds. Y. Queinnec and F. Daniellou, p. 1638-1640, Taylor & Francis, London.

Abdel-Moty, E., Khalil, T.M., Fishbain, D., Rosomoff, R.S. and Rosomoff, H.L., 1991c. Functional capacity assessment of low back pain patients, in *Advances in Industrial Ergonomics & Safety III*, Eds. W. Karwowski and J.W. Yates, p. 475-482, Taylor & Francis, London.

Abdel-Moty, E., Diaz, E., Khalil, T.M., Abou Elseoud, M., Steele-Rosomoff, R. and Rosomoff, H.L., 1992a. Ergonomic job analysis for patients with cervical trauma during rehabilitation, in *Advances in Industrial Ergonomics and Safety IV*, Ed. S. Kumar, p. 1195-1200, Taylor & Francis, London.

Abdel-Moty, E., Khalil, T.M., Sadek, S., Dilsen, E.K., Fishbain, D., Steele-Rosomoff, R. and Rosomoff, H.L., 1992b. Functional capacity assessment: a test battery and its use in rehabilitation, in *Advances in Industrial Ergonomics and Safety IV*, Ed. S. Kumar, p. 1171-1178, Taylor & Francis, London.

Abdel-Moty, E., Fishbain, D., Khalil, T., Sadek, S., Cutler, R., Steele Rosomoff, R. and Rosomoff, H., 1993a. Functional capacity and residual functional capacity and their utility in measuring work capacity. *The Clinical Journal of Pain*. 9:168-173.

Abdel-Moty, E., Khalil, T., Steele-Rosomoff, R. and Rosomoff, H., 1993b. Maximizing progress during low back pain rehabilitation, in *Advances in Industrial Ergonomics and Safety V*, Eds. R. Nielsen and K. Jorgensen, p. 331-335, Taylor & Francis, London.

Abdel-Moty, E., Fishbain, D., Goldberg, M., Cutler, B., Zaki, A., Khalil, T., Peppard, T., Steele Rosomoff, R. and Rosomoff, H., 1994. Functional electrical stimulation treatment of postradiculopathy associated muscle weakness. *Arch Phys Med Rehabil*. 75: 680-686.

Abdel-Moty, E., Compton, R., Steele-Rosomoff, R., Rosomoff, H.L. and Khalil, T.M., 1996. Process analysis of functional capacity assessment. *J Back and Musculoskeletal Rehabil*, 9:223-236.

Adams, M.A., McNally, D.S. and Dolan, P., 1994. Posture and the compressive strength of the lumbar spine. *Clin Biomech*, 9:5-14.

Americans with Disabilities Act, 1992. *Technical Assistance Manual on the Employment Provision (Title) of the ADA*. U.S. Equal Employment Opportunities Commission.

Andersson, C.K. and Chaffin, D.B., 1986. A biomechanical evaluation of five lifting techniques. *Appl Ergonom*, 17:2-8.

Asfour, S.S., Khalil, T.M., Waly, Goldberg, M.L., Rosomoff, R.S. and Rosomoff, H.L., 1990. Biofeedback in back muscle strengthening. *Spine*, 15(6):510-513.

ANSI, 1988. *American National Standard for Human Factors Engineering of VDT Workstations*. Human Factors Society, Inc., Santa Monica.

Consumer Product Safety Commission January 1985. *Safety for Older Consumers: Home Safety Checklist*. Washington, D.C.

Department of Commerce, Bureau of the Census, 1988. *Statistical Abstracts of the United States*. Washington, D.C.: U.S. Government Printing Office.

Department of Labor, Bureau of Labor Statistics, 1982. *Back Injuries Associated with Lifting*. Bulletin No. 2144, Washington, D.C.

Federal Register, 1992. *Ergonomics, Safety and Health Management. Advanced Notice of Proposed Rulemaking*. Vol 57, No. 149, August 3.

Fishbain, D.A., Abdel-Moty, E., Cutler, R., Khalil, T., Sadek, S., Steele Rosomoff, R. and Rosomoff, H., 1994. Measuring residual functional capacity in chronic low back pain patients based on the Dictionary of Occupational Titles. *Spine*, 19(8):872-880.

Fishbain, D.A., Khalil, T.M., Abdel-Moty, E., Cutler, R., Sadek, S., Steele Rosomoff, R. and Rosomoff, H.L., 1995. Physician limitations when assessing work capacity: A review. *J Back and Musculoskeletal Rehabil*, 5:107-113.

Hindle, R.J., Pearcy, M.J., Cross, A.T. and Miller, D.H.T., 1990. Three-dimensional kinematics of the human back. *Clin Biomech*, 5:218-228.

Khalil T.M. and Ramadan M.Z., 1987. Biomechanical evaluation of lifting tasks: A microcomputer-based model. *Comput Indust Eng*, 14:1.

Khalil, T.M., Goldberg, M.L., Asfour, S.S., Abdel-Moty, E., Steele, R. and Rosomoff, H.L., 1987. Acceptable maximum effort (AME): A psychophysical measure of strength in back pain patients. *Spine*, 12(4):372-376.

Khalil, T.M., Abdel-Moty, E. and Asfour, T.M., 1990. Ergonomics in the management of occupational injuries, in *Industrial Ergonomics: Case Studies*. Eds. B.M. Pulat and D.C. Alexander, p. 41-53, Industrial Engineering and Management Press, Norcross, GA.

Khalil, T.M., Abdel-Moty, E., Diaz, E., Rosomoff, R.S. and Rosomoff, H.L., 1991a. Electromyographic symmetry pattern in patients with chronic low back pain and comparison to controls, in *Advances in Industrial Ergonomics & Safety III*, Eds. W. Karwowski and J.W. Yates, p. 483-490, Taylor & Francis, London.

Khalil, T.M., Asfour, S.S. and Moty, E.A., 1991b. Clinical ergonomics: ergonomic practice in health care setting, in *Designing for Everyone*, Eds. Y. Queinnec and F. Daniellou, p. 314-316, Taylor & Francis, London.

Khalil, T.M., Abdel-Moty, E., Sadek, S., Dilsen, E.K., Steele-Rosomoff, R. and Rosomoff, H.L., 1992a. Postural sway and balance in healthy subjects and in patients with chronic pain, in *Advances in Industrial Ergonomics and Safety*, Ed. S. Kumar, p. 925-932, Taylor & Francis, London.

Khalil, T.M., Abdel-Moty, E., Zaki, A. M., Dilsen, E.K., DeVito, C., Steele-Rosomoff, R. and Rosomoff, H.L., 1992b. Reducing the potential for fall accidents among the elderly through physical restoration, in *Advances in Industrial Ergonomics and Safety IV*, Ed. S. Kumar, p. 1127-1134, Taylor & Francis, London.

Khalil, T.M., Abdel-Moty, E., Zaki, A. M., Velez, B., Dilsen, E.K., Diaz, E., Steele-Rosomoff, R. and Rosomoff, H.L., 1992c. Effect of secondary gain issues on performance and response to rehabilitation of workers compensation chronic low back pain patients, in *Advances in Industrial Ergonomics and Safety IV*, Ed. S. Kumar, p. 1187-1194, Taylor & Francis, London.

Khalil, T.M., Abdel-Moty, E., Rosomoff, R.S. and Rosomoff, H.L., 1993a. *Ergonomics in Back Pain: A Guide to Prevention and Rehabilitation*. Van Nostrand Reinhold, New York.

Khalil, T.M., Abdel-Moty, E., Steele-Rosomoff, R. and Rosomoff, H., 1993b. The role of ergonomics in the prevention and treatment of myofascial pain, in *Diagnosis and Comprehensive Treatment of Myofascial Pain: Handbook of Trigger Point Management*. Ed. E.S. Rachlin, p. 487-523, Mosby Year Book, St. Louis.

Khalil, T.M., Abdel-Moty, E., Diaz, E., Steele-Rosomoff, R. and Rosomoff, H., 1994. Efficacy of physical restoration in the elderly. *Experimental Aging Research*, 20(3):189-199.

Kroemer, K., Kroemer, H. and Kroemer-Elbert, K., 1994. *Ergonomics: How to Design for Ease and Efficiency*. Prentice Hall, Inc., Englewood Cliffs, N.J.

McGill, S.M. and Norman, R.W., 1987. Effects of anatomically detailed erector spinae model on L4/L5 disc compression and shear. *J Biomech*, 20(6):591-600.

NIOSH, 1981. *A Work Practice Guide for Manual Lifting, Technical Report #81-122*. National Institute for Occupational Safety and Health, U.S. Department of Health and Human Services, Cincinnati, Ohio.

NIOSH, 1991. *Work Practices Guide for Manual Lifting. NIOSH Technical Report Draft*, National Institute for Occupational Safety and Health, U.S. Department of Health and Human Services, Cincinnati, Ohio.

OSHA, 1990. *Ergonomics Program Management Guidelines*. U.S. Department of Labor, Occupational Safety and Health Administration.

Pain Commission, 1987. *Pain and Disability: Clinical, Behavioral, and Public Policy Perspectives*. Institute of Medicine, Committee on Pain Disability and Chronic Illness Behavior. Eds. M. Osterweis, A. Kleinman and D. Mechan. National Academy Press, Washington, D.C.

Quebec Study, 1987. Scientific approach to the assessment and management of activity-related spinal disorders. A monograph for clinicians. Report of Quebec Task Force on Spinal Disorders. *Spine*, 12-7S.

Rosomoff H.L. and Rosomoff R.S., 1991, Comprehensive multidisciplinary pain center approach to the treatment of low back pain. *Neurosurgery Clinics of North America*, 2(4):877-890.

*Safety Management*, 1996. Bureau of Business Practice, Waterford, CT, pp. 8.

SSA, 1986. *Social Security Administration: Report of the Commission on Evaluation of Pain*. Department of Health and Human Services, Washington, D.C.

Zaki A.M., Goldberg M.L., Khalil, T.M. et al., 1990, Comparison between the one and two inclinometer techniques for measuring range of motion, in *Advances in Industrial Ergonomics and Safety II*, Ed. B. Das, p. 135-142, Taylor & Francis, New York.

Zaki, A.M., Khalil, T.M., Abdel-Moty, E., Steele-Rosomoff, R. and Rosomoff, H.L., 1992. Profile of chronic pain patients and their rehabilitation outcome, in *Advances in Industrial Ergonomics and Safety IV*, Ed. S. Kumar, p. 1179-1186, Taylor & Francis, New York.

# 29

# Physical Ability Testing for Employment Decision Purposes

Charles K. Anderson
*Advanced Ergonomics, Inc.*

## 29.1   Introduction

Much of ergonomic activity is focused on designing or altering the demands of the job so that there is better match with the capabilities of the workforce. Sometimes, this approach reaches a point where further change in the job is either cost-prohibitive or technically infeasible at the moment. An alternative approach is to consider matching workers to the job demands on the basis of their physical abilities. For instance, if the job requires individuals lift cases weighing 60 pounds and there is no way to reduce the case weight, one approach would be to assess job candidates' ability to lift 60 pounds as part of the process of determining whether the individual is able to perform the job.

A number of studies have documented the effectiveness of physical ability testing as a means of identifying individuals who will be able to safely perform a given job (Anderson et al., 1980; Anderson et al., 1987; Anderson, 1989a,b; Anderson, 1992a,b; Arnold et al., 1982; Ayoub 1983; Cady et al., 1979; Chaffin et al., 1977; Herrin et al., 1982; Jackson et al., 1984; Jackson et al., 1991a,b; Jackson et al., 1992; Keyserling 1979; Keyserling et al., 1980a,b; Laughery et al., 1984; Laughery et al., 1988; Liles et al., 1984; Reilly et al., 1979). The experience of the author is that injury rates for new hires typically fall 20 to 40% with the implementation of testing (Anderson, 1996).

There are numerous issues that need to be considered when implementing a physical ability test battery to assure that it will be effective. One of the first concerns is to assure that the test battery will truly assess what is intended to be assessed — the individual's ability to perform the job. This means that a thorough job analysis must be performed, tests carefully chosen, and efforts taken to validate that the battery truly predicts job performance. Without this foundation, the employer may find that the battery provides no useful information while being a source of additional hiring cost and time delay.

In many countries, including the United States, there is a legal mandate that test batteries be valid predictors of job performance, particularly if it can be anticipated that protected groups, such as females and older individuals, will be less likely to pass the battery. In the United States, lack of compliance with the various pieces of legislation addressing employment testing can result in a company having to pay back wages to all individuals denied employment on the basis of the test, being required to offer those individuals employment, and perhaps in being liable for punitive damages of up to $300,000. Hence, there are moral, financial, and legal considerations when implementing a physical ability test battery.

The courts prefer that the emphasis be on the ability to perform the job rather than the risk of an injury or illness that may occur sometime in the future. The basis for this preference is the observation that employers have sometimes decided not to hire individuals with particular health conditions, such as a "bad back," because of the employer's perception that the individual is at increased risk of injury. If the basis for establishing the battery is risk of injury, the technical assistance manual for Title I of the Americans with Disabilities Act, which deals with employment discrimination, emphasizes that injury must be expected to occur in the near future, be severe, and have a significantly higher likelihood in the population to be denied employment than the risk in the general population (EEOC, 1992).

All of these considerations can be met with a carefully constructed implementation and ongoing management practice. The implementation consists of the job analysis, test battery design, and validation. Ongoing practices refer to the steps taken to assure that all individuals are given equal consideration and the manner in which other information, such as prior experience on a similar job, is integrated in the decision-making process.

## 29.2   Implementation

### Job Analysis

One of the most critical aspects of the job analysis is to identify the essential functions of the job. These are the functions that define the purpose of the job and are typically the elements that have to be performed in order for an employee to be considered satisfactory. In most cases, the essential functions are the ones that are performed frequently, though there may be situations where a function that is performed infrequently is still considered essential. For example, the essential function of a warehouse selector is to lift cases from storage racks to a pallet that will be shipped to a client. A second essential function is to drive a motorized vehicle that transports the pallet throughout the warehouse. These tasks comprise the bulk of the selector's activity over the course of the shift. As a counter-example, one of the essential functions of a firefighter is to respond to emergencies. The amount of time actually spent in emergency response may be small, but it is essential that the firefighter be able to perform these duties. An example of a nonessential function for the warehouse selector would be the task of cleaning up broken cases since selectors rarely need to perform that task and it is an essential function of the janitor.

When developing a physical ability test battery, the degree to which it can be anticipated that there will be a substantial portion of individuals who will be unable to perform the task is also an important consideration. If the task can be performed by virtually everyone, then there is little economic justification for testing applicants since virtually everyone would pass the test.

The cost/benefit analysis can be quantified by calculating the cost of not being able to perform the task and the probability of an individual not being able to perform the task, and then balancing these against the cost of testing all applicants for the ability to perform the task. For instance, consider a situation where 100 employees are hired per year. If there is a 1% probability that an individual could not perform the task, and the inability to perform the task results in a $2,000 loss associated with hiring and training the individual, then there is an expected cost of $2,000 per year for not using the test (100 new-hires × 1% chance of not being able to perform the job × $2,000 cost). If the cost of the test is $30 per applicant and 101 applicants must be tested to find 100 who pass (99% pass rate), the expected cost of testing would be $3,030 per year. Hence, in this example, it is less costly to not test and accept the risk that 1% of all the workers will not be able to perform the job (expected cost of $2,000 per year) rather than test all candidates ($3,030 per year). In contrast, if it can be anticipated that only 80% of new-hires truly have the ability to perform the job, then the cost of not testing rises to $40,000 per year (100 new-hires × 20% chance of not being able to perform × $2,000 cost). The cost of testing would now be $3,750 ($30 per test × 125 applicants tested to find 100 who pass). Under this scenario there is better than a 10-to-1 benefit/cost ratio for implementation of testing ($40,000 without testing vs. $3,750 with testing).

A third consideration during the job analysis is whether the essential functions that are potentially difficult for a substantial portion of the population can be modified through some form of ergonomic intervention (i.e., reasonable accommodation). For instance, wholesale grocery delivery drivers have to lift cases weighing up to 60 pounds on a routine basis. In addition, to access the cases, they must raise the rear door on their trailer. When brand-new trailers arrive from the manufacturer, virtually no force is required to open the door. However, the door track can become bent from forklifts running into the track while loading pallets onto the trailer. This can lead to a situation where a force of up to 200 pounds has to be exerted to open the door. Obviously, an essential function of the driver's job is to open the trailer door, but is it reasonable to design a test that assesses the driver's ability to exert 200 pounds to open a jammed door? An alternative is to require preventative maintenance rather than require the driver to exert 200 pounds. This would reduce the demand for this essential function back into the region where it can be anticipated that virtually all candidates will be able to perform the task. Hence, in this example, the critical task for the purposes of test battery design for these drivers would be the ability to lift 60-pound cases.

## Test Battery Design

It has already been mentioned that tests should be included only for those essential functions that will be difficult for a substantial portion (5 to 10% or more) of the applicant population. There are three additional considerations in designing the test battery.

### Job-Relatedness

First, it is critical that the tests bear a high degree of job-relatedness to the essential functions of the job that are used as the basis for the test battery design. For instance, when testing the ability of an individual to lift a 60-pound case, the most job-related test would be to have the individual literally demonstrate the ability to lift a 60-pound case by having him or her move the case through the region (for example, floor to table level) that is found on the job. In contrast, having the applicant perform a series of exercises such as push-ups, chin-ups, and sit-ups would not be a job-related test of the ability to lift a 60-pound case.

### Basis for Establishing the Cutoff Score

The second consideration in test selection is the ease with which the cutoff score can be determined. The cutoff score is the level of test performance required in order to be considered as having demonstrated the ability to perform the job. Ideally, there is a clear relationship between the cutoff score and the performance of the job. For instance, if the test for the ability to lift a 60-pound case consists of lifting a case from floor level to table level, the most job-related cutoff for weight lifted would be 60 pounds.

In contrast, an alternative method is to use normative data. For example, some companies strap the applicant into a fixture that isolates his or her movement to flexion and extension about the low back. The test consists of measuring the moment, or torque, that the applicant can then generate with these muscles. It is difficult to directly measure the moment about the low back required for a given individual to lift a 60-pound case, since the moment will be a function of the posture used, how rapidly the individual lifts the case, the anthropometry of the individual, and a host of other factors. The only alternative is to measure the moment created by a number of incumbents and develop a distribution of their scores. Conceivably, one could then choose the lowest score of the incumbents and require that the applicant at least demonstrate the ability to demonstrate that level of strength. A problem with this approach is that there is no guarantee that the scores demonstrated by the incumbents have a bearing on what the job requires. For instance, if all of the incumbents are regular participants in heavy weight training, their strength may be a reflection of the weight training rather than the job requirement.

A second example illustrating the problem of using cutoff scores based on normative data comes from test batteries used for evaluating police officer candidates. Some batteries for police officers require that applicants perform a run over a short distance to simulate the essential function of chasing suspects fleeing on foot. The score for the test is the time taken to cover the distance. The question becomes "How fast must an applicant complete the distance in order to be able to perform the job?" One approach that has been used is to determine whether the applicant's time is above the median for individuals of their sex and age. The justification has been that one would want their police officers to be at least as fit as an average person of their sex and age. Unfortunately, this does not speak to whether the officer can perform the essential function of the job.

Most jobs lend themselves to determining the cutoff scores directly from the job requirements, as described in the first example given in this section — handling 60-pound cases. Likewise, there typically are testing formats available that will allow the test administrator to assess the ability at issue in a manner that can be directly compared to the job requirement. Hence, the problems that arise with utilizing the "normative data" approach can generally be avoided once the issue is recognized.

### Accuracy, Reliability, and Safety

The third consideration is that the tests have a high degree of accuracy, reliability and safety. An accurate test is one that precisely measures the attribute it purports to measure. A reliable test is one that yields the same results when it is repeated over time and by different test administrators. Reliability is enhanced when objective test measures are used rather than subjective assessments or opinions of the test administrator.

Safety is an important consideration, particularly if maximal strength tests are being used. Typically, steps can be taken to reduce the risk of injury during the test. For instance, an alternative to having the participant demonstrate the maximum amount of weight they can lift is to ask them to only demonstrate the ability to lift the weight required by the job. A second safety modification would be to have the participant first demonstrate the lift with an empty case, and then add weight to the box in fixed increments so that there is the chance to detect inability to lift the required weight without actually having the individual attempt the lift with the full weight required on their first effort.

## Validation

### Statistical Validation Studies

The strongest form of validation is a prospective statistical analysis. The typical form of this type of analysis involves administering the test to a group of individuals who are about to be hired into the position in question, although the test scores are not used as part of the hiring decision. The performance of these individuals is then tracked over the course of their employment. Measures of performance might be productivity, retention, injury rate, or supervisor evaluation. Care must be taken when selecting a measure that it reflect the important aspects of job performance and be reliable. Performance for individuals who pass the test would then be compared to performance for individuals who fail. As an example, Figure 29.1 shows the worker compensation injury rate for new-hires who failed a physical test battery compared to the injury rate for their peers who passed the battery designed for grocery warehouse selectors (Anderson, 1989a). The injury rate for those who failed the test was about double the rate for those who passed.

The effectiveness of the battery can be projected by comparing the performance of the entire group of new-hires, which reflects the performance anticipated without implementation of the battery, to the performance of those individuals who passed the battery. This latter group reflects the performance that would be expected with implementation of the battery. Table 29.1 summarizes the results calculated in this manner for three prospective statistical studies performed by the author (Anderson, 1989b; Anderson, 1992a,b).

A less strong form of statistical validation is to compare performance for new-hires who began employment in time periods prior to and after implementation of the battery. The primary shortfall of this approach is that there is no control for factors that did not stay constant between the two time periods, such as work load, job structure, or management policies. The benefit of this type of analysis is that there is at least some indication of the effectiveness of implementing the battery when it is technically or economically infeasible to perform a prospective validation study. As noted earlier, the author has found that worker compensation incidence rates for new-hires typically fall 20 to 40% with implementation of a well-designed test battery (Anderson, 1996).

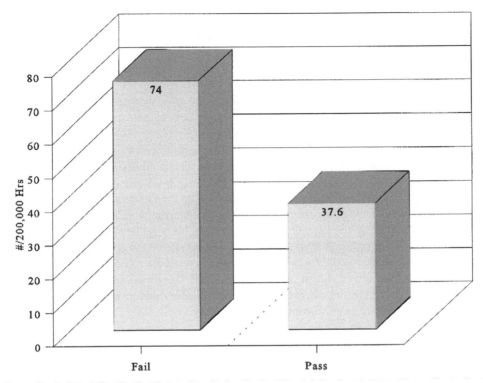

**FIGURE 29.1** Worker compensation incidence rate by pass/fail status.

**TABLE 29.1** Summary of Expected Impact of Test Battery Implementation

|  | Grocery Distribution | Soft Drink Distribution | Retail Distribution |
|---|---|---|---|
| Decrease in Incidence Rate | 18% | 26% | 17% |
| Decrease in Lost Time Due to Injuries | n/a | 11% | 27% |
| Decrease in Workers' Compensation Cost | n/a | 55% | n/a |
| Increase in Retention | 10% | 15% | 7% |
| Increase in Productivity | n/a | n/a | 7% |

n/a — data not available for all locations in the study

### Content Validation Studies

The alternative to a statistical validation is a content validation. A content validation study consists of demonstrating that the content of the job is reflected in the content of the test battery, and that the cutoffs on the tests reflect the job demands. It is valuable to be able to also make reference to any statistical validation studies in similar industries that further support the content validity of the battery at issue.

In the United States, the Equal Employment Opportunity Commission states that a content validation is acceptable when a statistical validation is technically infeasible (EEOC, 1978). For example, this could be the case when there is a small number of new-hires on which to base the study, or the job demands are not homogeneous.

## 29.3   Ongoing Practices

It is critical to not only carefully construct the test battery, but also the management policies that surround its use. Four aspects are particularly crucial to assure that the manner in which the test battery is used is fair to all applicants.

## Equal Treatment

The fundamental consideration is that all applicants must be treated equally. For instance, all applicants for the job for which the test battery has been designed must be given the same opportunity to test. It is not considered appropriate to only give the battery to females, individuals over the age of 40, or those considered disabled under the premise that these are individuals who are most likely to be unable to perform the job. Furthermore, they should all be given the same information prior to the test. This information might include instructions to wear loose-fitting clothes, avoid a heavy meal prior to the test, and have adequate rest the night prior to their appointment.

Another dimension of equal treatment is that the same scoring process must be used for all individuals. It is generally recommended that the same cutoff be used for all individuals without exception. Problems arise when managers attempt to factor in their own judgment by allowing some individuals who fall below the cutoff to be employed, particularly when the decision as to who shall benefit is arbitrary.

Finally, all individuals who fail should be given the same opportunity to be reconsidered. This opportunity could be structured on the basis of their performance, but must be equally applied. For instance, all individuals who score within 95% of the cutoff might be offered the chance to retest immediately. Those who score between 80 and 95% might be allowed to retest after four weeks, whereas those who score less than 80% have to wait at least eight weeks to retest. There should be a valid basis for the time periods, such as the percent of improvement in performance that could be expected at particular points in time with a well-designed physical conditioning program.

## Prior Experience

There is only one exception to the policy of rigid interpretation of the cutoff score which occurs when an individual who fails the test has demonstrated the ability to perform the job during prior experience in similar employment. It is important to remember that the best indicator of an individual's ability to perform a job is actual demonstration of that performance. Lacking direct evidence of ability to perform the job, one might consider performance on a test battery designed to provide information about that ability. The test battery is not necessarily a perfect indicator, so it is possible that an individual who can perform the job might fail the battery. Clearly, the battery should be designed so that this possibility is minimized. This is the reason for selecting tests that bear the closest resemblance to the way the job is actually performed, carefully selecting the cutoff score and taking all possible steps to assure that the test scores are valid predictors of job performance.

In the circumstance where there is additional information available about an individual's ability to perform the job, such as prior experience, this information can be used to override a failing test result. It is critical, though, that all individuals in similar circumstances be treated the same way. This can be promoted by establishing clear definitions of what would be considered adequate previous experience that can be consistently applied in each case that arises. For instance, if an individual applying for a warehouse selector position has previous experience as a selector yet fails the battery, the manager might ascertain if the previous experience involved handling cases of similar weight at a similar pace for a similar shift length using similar equipment (e.g., manual lifting vs. picking full pallets with a forklift). If all aspects of the previous job and the job at issue are similar, then the prior experience could be used as a legitimate basis for hiring the individual in spite of the failing test results.

## 29.4 Summary

Physical ability testing can be an effective way to improve the match between job demands and worker ability, thereby enhancing performance and reducing the risk of injury. Thorough job analysis is required to guarantee that the test battery selected is a valid predictor of the ability to perform the essential functions of the job. Furthermore, it is important to assure that the management policies related to the utilization of the battery guarantee equal application of its use and interpretation. This includes to whom the battery is administered, the information that applicants are given prior to the test, the way their results are interpreted, and the opportunity for retesting that is extended to those who fail the test. When the test battery is designed with these factors in mind, it can meet the legal requirements and be an effective management tool. Results from a wide range of applications indicate that there typically is a 20 to 40% reduction in worker compensation injuries associated with implementation of a well-designed test battery.

# References

Anderson, C.K. Strength and endurance testing for pre-employment placement. K. Kroemer, Ed. *Manual Material Handling: Understanding and Preventing Back Trauma*, American Industrial Hygiene Association, Akron, 1989a.

Anderson, C.K. Impact of physical ability screening for grocery warehousing. Technical Report, Advanced Ergonomics, Inc., 1989b.

Anderson, C.K. Impact of physical ability testing on injuries and retention for the Coca-Cola Bottlers Association Technical Report, Advanced Ergonomics, Inc., 1992a.

Anderson, C.K. Impact of physical ability testing on workers' compensation injury rate and severity: target distribution centers. Technical Report, Advanced Ergonomics, Inc., 1992b.

Anderson, C.K. The advanced ergonomics physical ability testing program: a seven year review. Technical Report, Advanced Ergonomics, Inc., 1996.

Anderson, C.K.; Catterall, M.J. The impact of physical ability testing on incidence rate, severity rate and productivity. S.S. Asfour, Ed. *Trends in Ergonomics/Human Factors IV*, Elsevier Science Publishers, North Holland, 1987.

Anderson, C.K.; Herrin, G.D. Validation study of pre-employment strength testing at Dayton Tire and Rubber Company. Technical Report, University of Michigan Center for Ergonomics, 1980.

Arnold, J.D.; Rauschenberger, J.M.; Soubel, W.G.; Guion, R.M. Validation and utility of a strength test for selecting steel workers. *Journal of Applied Psychology*, 67(5): 588-604, 1982.

Ayoub, M.A. Design of a pre-employment screening program. Kvalseth, T.O., Ed. *Ergonomics of Workstation Design*, Butterworths & Co. London, 1983.

Cady, L.D.; Bischoff, D.P.; O'Connell, E.R.; Thomas, P.C.; Allan, J.H. Strength and fitness and subsequent back injuries in fire fighters. *Journal of Occupational Medicine*, 21(4): 269-272, 1979.

Chaffin, D.B.; Herrin, G.D.; Keyserling, W.M.; Foulke, J.A. *Pre-Employment Strength Testing in Selecting Workers for Materials Handling Jobs*. Cincinnati, OH, NIOSH Physiology and Ergonomics Branch. Contract No. CDC 99-74-62, 1977.

Equal Employment Opportunity Commission (EEOC). *Title I of the Americans with Disabilities Act: EEOCs Technical Assistance Manual*. 1992.

Equal Employment Opportunity Commission (EEOC). *Uniform Guidelines on Employee Selection Procedures*. Title 29, CFR, Part 1607, 1978.

Herrin, G.D.; Kochkin, S.; Scott, V. Development of an employee strength assessment program for United Airlines. Technical Report, The University of Michigan Center for Ergonomics, 1982.

Jackson, A.S.; Osburn, H.G.; Laughery, K.R. Validity of isometric strength tests for predicting performance in physically demanding tasks. *Proceedings of the Human Factors Society 28th Annual Meeting*, 452-454, 1984.

Jackson, A.S.; Osburn, H.G.; Laughery, K.R. Validity of isometric strength tests for predicting endurance work tasks of coal miners. *Proceedings of the Human Factors Society 35th Annual Meeting*, 753-767, 1991a.

Jackson, A.S.; Osburn, H.G.; Laughery, K.R.; Vaubel, K.P. Strength demands of chemical plant work tasks. *Proceedings of the Human Factors Society 35th Annual Meeting*, 758-762, 1991b.

Jackson, A.S.; Osburn, H.G.; Laughery, K.R.; Vaubel, K.P. Validity of isometric strength tests for predicting the capacity to crack, open, and close industrial valves. *Proceedings of the Human Factors Society 36th Annual Meeting*, 688-691, 1992.

Keyserling, W.M. *Isometric Strength Testing in Selecting Workers for Strenuous Jobs.* Ph.D. Dissertation, University of Michigan, Center for Ergonomics, 1979.

Keyserling, W.M.; Herrin, G.D.; Chaffin, D.B. Isometric strength testing as a means of controlling medical incidents on strenuous jobs. *Journal of Occupational Medicine*, 22(5): 332-336, 1980a.

Keyserling, W.M.; Herrin, G.D.; Chaffin, D.B.; Armstrong, T.A.; Foss, M.L. Establishing an industrial strength testing program. *American Industrial Hygiene Association Journal*, 41: 730-736, 1980b.

Laughery, K.R.; Jackson, A.S. Pre-employment physical test development for roustabout jobs on offshore platforms. Technical Report, Kerr McGee Corporation, 1984.

Laughery, K.R.; Jackson, A.S.; Fontenelle, G.A. Isometric strength tests: Predicting performance in physically demanding transport tasks. *Proceedings of the Human Factors Society 32nd Annual Meeting*, pp. 695-699, 1988.

Liles, D.H.; Deivanayagam, S.; Ayoub, M.M.; Mahajin, P. "A job severity index for the evaluation and control of lifting injury. *Human Factors*, 26(6):683-693, 1984.

Reilly, R.R.; Zedeck, S.; Tenopyr, M.L. Validity and fairness of physical ability tests for predicting performance in craft jobs. *Journal of Applied Psychology*, 64(3): 262-274, 1979.

# 30

# Preplacement Strength Screening

Valerie J. Rice
U.S. Army Research Institute of
Environmental Medicine

According to federal employment laws, it is illegal to disqualify an employee for reasons of race, color, religion, gender, national origin, or disability status. Accordingly, employers are using preplacement screening tests to determine an individual's ability to safely perform job tasks. This chapter reviews the pertinent literature related to preplacement screening and derives guidelines for developing job-specific screening. Both management and workers benefit from the increased productivity and safety resulting from a well-constructed preplacement screening process.

## 30.1 Introduction

When hiring new personnel, management has a vested interest in having workers succeed in their jobs. Workers who will have an acceptable level of productivity and will perform the job safely are needed to run an efficient operation. In physically demanding jobs, there is a concern that smaller, weaker individuals may be at greater risk for injury and may be less productive. Job redesign, worker selection, and training are used in an attempt to ensure safe execution of job requirements. Of these approaches, it is preferred to design the equipment, job, or task so the majority of workers can perform the task safely and efficiently. In this way, the risks of injury are controlled by reducing the probability of occurrence. If the task or tools cannot be redesigned, then selection procedures can be used to identify personnel who meet the physical requirements of the job. For example, the tasks of fire fighters, police, rescue workers, and soldiers cannot always be anticipated or redesigned in order to permit persons with lower levels of physical fitness and strength to complete them. Instead, worker selection procedures are used to select personnel for these physically demanding jobs.

Pre-employment and preplacement screening refer to the process of administering a test or set of tests to job applicants in order to discern whether they can safely perform the job in question. Three assumptions must be accepted before a strength-based screening program is introduced. First, it is assumed the job and job tasks have been evaluated to determine whether ergonomic redesign was appropriate and any necessary redesign is complete. Second, the individuals designing the preplacement strength testing program must believe that they can accurately test a person's physical performance in some manner and use those test results to predict successful job performance. Third, it is assumed that an inverse relationship exists between injury risk and an individual's ability to meet the physical requirements of a job, and that this relationship can be demonstrated objectively.

The ergonomist must also be acutely aware of the legal issues involved in preplacement strength screening. The majority of screening programs have the appearance of inherent bias for age, gender, and disability classification. Therefore, legislation has been enacted in an attempt to eliminate these biases and provide guidance to base personnel selection criteria on sound business and ethical considerations.

## 30.2 Literature Review

A number of researchers have examined methods of determining and/or predicting a person's ability to perform a job or task. Static strength tests (Chaffin et al., 1978; McMahan, 1988) dynamic strength tests (Kroemer, 1983; Menon and Freivalds, 1985; Pytel and Kamon, 1981), work simulations (Arnold et al., 1982), calisthenics (Sharkey, 1981), aerobic capacity (Sharp et al., 1980), body composition and anthropometrics (Misner et al., 1987), and other physical performance measures (Reilly et al., 1979) have been suggested as evidence of ability to perform physically demanding work. The research is limited, however, as many of the predictions have centered on simple lifting capacity, rather than multistep complex tasks such as those involved in most jobs.

For example, Pytel and Kamon (1981) examined the relationship between several dynamic strength measures (lift, back and arm strength) and a one repetition maximum (1-RM) lift from the floor to a 113 cm height with a psychophysical assessment of a maximal acceptable lift to the same height at a rate of 6 lifts/min. A prediction equation was developed using dynamic lift strength and gender, accounting for 94% of the variance. The authors concluded the study demonstrated the usefulness of a simple, portable "isokinetic" dynamic strength measuring device; however, it is unknown whether the same device can be used to predict actual job performance. Teves et al. (1985) examined the use of body composition and various strength measures to predict a one repetition maximum lift to 132 cm for military soldiers. This is similar to lifting a box onto the bed of a 2½ ton truck. They found that lean body mass and a lift to a height of 183 cm, using an incremental dynamic lift device, were the best predictors ($R^2 = 0.47$). They predicted the ability to complete one lift, and no task analysis was mentioned; therefore, it is unknown whether this was an essential task for the military. Similar difficulties are seen in other studies predicting individual (Kumar, 1995; Mital and Ayoub, 1980) and team lifting capacities (Fox, 1982; Karwowski, 1988; Rice et al., 1995).

At face value, perhaps the best measure of job performance is to have the applicant perform the job. However, having each applicant work at the job, even for a short period, is time consuming, costly, and may put the applicant at undue risk of injury. Instead, criterion tasks, in the form of work simulations or task components, are used to predict job performance. It is assumed that testing applicants on an aspect of job performance that is critical to job success is a valid approach, as it seems to satisfy content requirements (Wigdor et al., 1991). It also appears to ensure the constructs of job performance are represented and as long as characteristic samples of *each* critical element of a job are included, the validity "should be axiomatic" (Wigdor et al., 1991, p. 59). In addition to using work simulations to predict work performance, many screenings use work simulations as the performance to be predicted. In other words, instead of being on the predictor side of the equation, the work simulation is on the criterion side of the equation as the standard against which the preplacement screening measures will be tested. In these cases, various constructs such as strength or aerobic capacity are used to predict performance on simulated tasks.

The use of isometric strength tests to predict performance is appealing because they are relatively easy to administer, standardized procedures have been established, test–retest reliability has been established (Hershenson, 1979; Keyserling et al., 1980a), and they have been shown to be safe in a series of industrial studies (Chaffin, 1974; Hershenson, 1979; Kamon and Goldfuss, 1978). They have also been found to be predictive of performance on dynamic tasks (Aghazadeh and Ayoub, 1985; Ayoub et al., 1980; Kamon et al., 1982; Kroemer, 1982; Kroemer, 1983; McDaniel et al., 1983). However, accurate predictions of dynamic tasks from isometric strength tests are not always found, as the differences in mechanical and physiological processes that occur during static and dynamic strength testing make prediction difficult (Kroemer, 1970). In addition, strength in one working posture is not necessarily correlated with strength in another posture (Hershenson, 1979), and maximal isometric lifting strength is not necessarily a predictor of psychophysically acceptable levels of lifting (Garg et al., 1980).

## Predicting Job Performance from Preplacement Screening

*Simulated job performance.* A review of over 21 research studies designed to predict work performance revealed the use of static (isometric), dynamic (isotonic), and combined strength tests (task simulations) to predict job performance, with varying degrees of success (see Table 30.1 at end of chapter). Predictions accounted for anywhere from 12 to 90% of the variance. The most successful predictions, those that accounted for 60% of the variance or more demonstrated several consistencies (Arnold et al., 1982; Celentano et al., 1984; Davis et al., 1982; Jackson et al., 1984; Rice and Sharp, 1994; Robertson, 1982; Robertson and Trent, 1985; Schonfeld et al., 1990; Sharkey, 1981; Stalhammar et al., 1992; Wilmore and Davis, 1979). All performed a well-constructed job analysis, with a breakdown of job tasks into component parts, using multiple assessment techniques: interviews with ratings of key tasks for frequency, duration, and difficulty; direct observation, videotaping, measurements of masses moved and forces exerted; and identification of pace and frequency. As a result, all of the highly successful evaluations examined multiple constructs determined from the job analysis.

Sharkey (1981) (Table 30.1) found that unless the full complement of pertinent measures (with a combination of constructs) was used, screening did not guarantee positive results. The inclusion of multiple factors increases the predictive ability, due to correctly identifying essential components of the task. For example, Rice and Sharp (1994) evaluated seven strength measures, three Army Physical Fitness Test scores, and three physical descriptors for their ability to predict performance on two stretcher-carry tasks (Table 30.1). The best predictors for repeated short distance stretcher-carrying were 2-mile run time (aerobic factor) and peak handgrip (strength factor) ($R^2$ range from 0.74 to 0.78). For a continuous carry, a sustained mean grip strength was the best predictor of performance ($R^2$ range from 0.74 to 0.78). The results illustrate the necessity for tailoring preplacement tests to accurately reflect job demands, as different predictors were selected for each type of stretcher-carrying task.

It also appears that use of consolidated or summed scores either as the dependent or independent variable may strengthen the predictive ability of a model. This procedure has been used with an isometric test battery administered to coal miners (Laughery et al., 1986), roughneck and roustabout positions in oil drilling and production operations (Jackson, 1984), and operator positions in refining and chemical manufacturing (Jackson, 1986 as cited in Jackson, 1994). In their study of oil drilling operations, Jackson et al. (1984) used three strength tests and one arm endurance test to predict performance in several work simulations (Table 30.1). A composite score for the strength battery was used, resulting in correlations between the isometric strength tests and work sample tests ranging from 0.67 to 0.93, with equal validity for men and women. In another study by the same authors (Jackson et al., 1991) (Table 30.1), a composite of the isometric strength tests was found to be a valid predictor of endurance work tasks, again with the measure being valid for both men and women (accounting for up to 87% of the variance, Table 30.1). Other studies have also used this technique with positive results (Laughery et al., 1988), with one study concluding that task performance was dependent on strength rather than gender, and the higher the sum of an individual's isometric strength, the greater the probability of success on work simulations (Jackson et al., 1991). The commonalities in the studies conducted by these authors are their meticulous attention to task analysis, inclusion on all pertinent performance constructs, and the use of a composite score to predict performance on job simulations. However, since the use of a composite score can artificially raise the $R^2$ value, it is debatable whether its use adds or detracts from a prediction. Arnold et al. (1982) used a composite score on the other side of the equation (Table 30.1). General physical abilities thought to be required for work simulations of moving/carrying, lifting, wheelbarrowing, and shoveling were used as a test battery to predict a composite score derived from several work simulations. Each of the simulations was evaluated by observers and objective measures were used for as many tasks as possible (strength, number of objects moved during a timed session). The best single predictor was the strength measure attained with the arm dynamometer, accounting for 67% to 72% of the variance of a composite score.

*On-the-job performance.* Table 30.1 also contains studies that examine the ability to predict performance and productivity while in training or on the job (Anderson, 1988; Anderson and Catterall, 1989; Arnold et al., 1982; Bernauer and Bonanno, 1975; Doolittle et al., 1988; Hogan, 1985; Reilly et al., 1979). Certainly attrition is one aspect of productivity, since retraining is required following most job changes. In one study of transfers from jobs with heavy demands vs. jobs with light demands, it was found that men left the heavy demand jobs at a rate almost double to those with light demands (34% vs. 57%), and women's transfers from heavy jobs were more than triple those from light jobs (23% vs. 75%, Kowal, 1983). Robertson (1992) discussed the idea that personnel whose strength capacity marginally meets the strength requirements of a task may fatigue at a significantly faster rate. For jobs requiring repeated heavy strength exertions, it can be assumed individuals with lower strength will be put at a performance disadvantage and increased risk of injury.

However, results of other studies do not support the relationship between worker strength and productivity. The job demands and loads on the lumbosacral spine of National Guard mechanics were compared to static strength physical capacity testing of the same workers (arm lift, back lift, and floor lift) (Pedersen et al., 1989). Results indicated up to a 38% job mismatch between requirements and capabilities; however, it was not clear how the workers were able to perform the job with such a mismatch. They were currently working (and presumably performing successfully) in those jobs, although they reported a high incidence of low back pain. It is possible they completed their work through development of alternate methods for task completion or through informal selection (letting the stronger individuals perform the heavier jobs), but the usefulness of the study results are questionable without this information. General physical fitness has not been found to be predictive of successful job performance. Wilmore and Davis (1979) found no difference in general physical fitness levels between California Highway Patrol officers who received high vs. low supervisory ratings. Nor did they find differences in the physical fitness of police officers compared with other researcher's results for the general public.

Researchers have also examined the ability to predict success during training. Bernauer and Bonanno (1975) investigated applicants for pole climbing jobs using body fat, arm strength, trunk strength, cardiovascular fitness, response time, and body balance. The mean scores for the step test and balance were found to differ for successful and unsuccessful applicants. Hogan (1985) also evaluated successful training completion (Table 30.1), using a test battery of 23 measures and five widely recognized physical fitness batteries. Performance results were compared with receiving either a pass or fail, and with nine performance rating scales (completed by three independent instructors). Multiple regression revealed that cardiovascular fitness, muscular endurance, and flexibility were highly related to program completion and overall potential ratings (Table 30.1). Results from the five physical fitness batteries yielded multiple R's ranging from 0.38 to 0.60 for prediction of final outcome (pass/fail). Although a comparison with actual job performance might be preferable, these results indicate that models can be developed for predicting training attrition.

Few studies have gone the extra distance to try and predict actual job performance. Instead many researchers assume that being able to complete a task simulation of the most demanding portion of the job is indicative of an individual's ability to perform the job. This may be accurate, but becomes questionable if a one-time administration of a task is used as the criterion, when repeated performances are required on the job. Those research evaluations that have included this last step to ensure that their measure is predictive of successful job performance appear more persuasive for court challenges of preplacement testing.

Anderson and Catterall (1989) studied job productivity and turnover of initial applicants (Table 30.1). They conducted a prospective validation involving grocery warehousemen and demonstrated that use of a screening battery decreased turnover and increased productivity. It was also demonstrated that the more stringent the cutoff criteria, the greater the productivity. The test battery consisted of two isometric strength tests and a step test, based on a thorough job analysis. Criteria were established based on the heaviest cases to be moved (60 to 100 lbs) and an estimated aerobic capacity three times the average metabolic requirement of the job (15 kcal/min).

*Considerations in predicting performance.* One difficulty in reviewing the literature is that researchers vary in the methods they use and how they report their data. For example, several authors did not report the r values or used correlations without any regression analysis. Although a correlation or prediction may be significant ($p < 0.05$), if it only accounts for 30% of the variance, the model may be questionable in court challenges of its ability to foretell job performance.

Another challenge is where to place the cutoff criterion. It would appear that determining the cutoff criterion by the exact job requirements would be the ideal. For example, if the job requires a 75-lb. lift, then have the job applicant perform a 75-lb. lift.

The more challenging position comes with tasks requiring repetitive heavy physical demands, as it is unlikely that preplacement testing could incorporate four lifts per minute for a full day due to time, financial constraints, and safety concerns. Instead, if some examiners observe an applicant straining, out of breath, or demonstrating a high heart rate and slow recovery, they report that information to the physician for consideration in the final medical report. Should an applicant be denied a particular position based on this analysis and challenge the medical and preplacement screening, the problem will be in demonstrating that although the person could perform the basic task, these observations would indeed indicate an inability to do the job or increased risk to themselves or others.

A better procedure would be to establish a rigorous statistical model that is predictive of performance. For example, Anderson and Catterall (1989) demonstrated the more stringent the pass cutoff score, the more likely personnel were to stay for 8 weeks and the more productive they were.

Ayoub (1982) introduced a method to determine cut scores for exercise intensity relative to the individual. He suggested that cutoff scores for heart rate be based on a percentage of their heart rate maximum. He suggested that cutoff scores for heart rate as "those falling below 75% are acceptable, those between 75% and 80% are marginal, and those above 80% are unacceptable." For strength, he suggests a cutoff of 50% of a person's maximum strength; "applicants falling below 50% are acceptable, those between 50% and 60% are marginal, and those above 60% are unacceptable."

Dr. Ayoub noted the rationale for pre-employment screening, and these cutoff scores are explainable, integrative (consider all aspects of the physical and physiological properties of human work capacity), and can be used for jobs characterized by brief exertions as well as sustained performance. However, his recommendations are conservative and appear more focused on sustained activity rather than brief periods of high activity.

Caution should be used if assessing personnel who are currently performing the job to establish cutoff criteria. This could lead to misleading conclusions, as these persons have developed the skills, strength, and knowledge regarding task requirements. Use of a cutoff score that allows for improvement through a training program could be one solution.

The military services accept the fact that physical strength will be necessary and include physical training as part of a soldier's entry level instruction, job requirements, and performance assessment. Civilian counterparts may be less likely to invest the time and funding in such an effort, instead recommending that persons who are unable to qualify for a given position may be able to train on their own and reapply for a position.

TABLE 30.1  Prediction of Performance

| Study | Subjects | Initial Assessment and Tests | | | Relationships/Predictions | | |
| | | Job Analysis | Work Simulation | Tests | Work Simulation | On-the-Job Performance | Results |
| --- | --- | --- | --- | --- | --- | --- | --- |
| Anderson, 1988 | 613 | not described | | strength-arm pull, aerobic step test | | productivity | Chi²: p < 0.05 for drivers, p < 0.02 for selectors |
| Anderson and Catterall, 1989 | ♂ 518 | critical tasks: onsite analysis, measurements, posture analysis | | pull-up, low bar pull, step test | | probability of staying on job 8 wks, productivity levels | Chi²: p < 0.05. Lower pass rates more likely to quit by 8 wks, higher pass scores had higher avg. productivity at week 8 |
| Arnold et al., 1982 | ♂ 168 ♀ 81 | Interview, task quantification | | arm, back, and leg strength; balance; leg lifts; push-ups; squat thrusts; pull-ups; flexibility; and a step test | composite score: move/carry, lift, wheelbarrow, and shovel | observer ratings | regression: arm dynamom. → 67 – 72% variance composite score |
| Bernauer and Bonanno, 1975 | 241 pole climber applicant | not mentioned | | 40 item test battery | | successful completion of climbing school | difference in mean scores between success/no-success applicants w/step test & balance (p < 0.05) |
| | 300 applicant ♂ ♀ | | | body fat, body weight, grip, sit-up, recovery HR, reaction time, beam walking | | | regression: body weight & lifting → 76% var.body weight, lifting & gender → 92% variance |
| Celentano et al., 1984 | ♂ 23 ♀ 18 | interviews, task quantification | | static & dynamic strength, aerobic endurance, anthropometry | breech-block trade task | | Laboratory strength/power → 90% variance Non-laboratory → 54% variance |
| Davis et al., 1982 | 100 fire-fighters (♂?) | not mentioned | | anthropometrics, grip strength, sit-ups, chin-ups, standing long jump, push-ups, hamstring flexibility, physiological measures | ladder extension & retraction, carry, hose pull, rescue, forcible entry | | Laboratory fatigue resist. → 80% variance Non-laboratory → 60% variance |

| Study | Sample | Job analysis | Tests | Criterion | Results |
|---|---|---|---|---|---|
| Doolittle et al., 1988 | 48 line-workers (♂ ?) | interviews, observation, movement classification, forces & masses, pace & frequency, energy cost via heart rate | modified biceps curl, lateral rise, row, lat pull, chin-ups, submaximal aerobic step test | performance ratings on productivity, work w/others, supervision, safety, physical ability, technical | correlation: composite score → overall perform rating 35% variance |
| Hogan, 1985 | ♂ 46 | not mentioned | 23 fitness measures and 5 physical fitness batteries | successful completion of diving training, supervisor rating scales | regression: cardiovascular fitness, muscular endurance, and flexibility → program completion 40% variance & overall potential 41% variance |
| Jackson et al., 1984 | ♂ 25 ♀ 25 | conducted, process not described | isometric strength:grip strength, arm and back lift, arm endurance (arm $VO_2$ max) | roof bolting, bag lift & carry, block carry, shoveling | regression values not reported. mult. correlation: composite stren. → bag carry 41% variance, → block carry 69% variance, → shoveling 56% variance composite stren. + arm endur → shovel 64% variance |
| | ♂ 25 ♀ 25 | conducted, process not described | isometric strength:grip strength, arm and back lift, arm endurance (arm $VO_2$ max) | pipe transport | composite stren. → pipe transport ♂ 77-87% var. ♀ 46-64% var. |
| Jackson et al., 1991 | ♂ 118 ♀ 66 | conducted, process not described | isometric grip, arm lift, and torso lift summed | tool lift, scaffold building, railroad tank car cap replacement, one and two arm tool hold. Composite strength score predictive of work simulations | ANOVA: $p < 0.01$ isometric str. higher for those passing work sample test. regression: equations given, probability of passing work sample given for each strength category |
| Jackson et al., 1991b (U. Houston/Rice) | ♂ 28 ♀ 28 | conducted, process not described | isometric grip, arm lift, torso lift, and arm endurance (arm $VO_2$ max), body weight (fat, lean) | shoveling, bag lift & carry | regression: $R^2$ not reported. mult. correlation: composite strength score → shovel 56% variance, comp str + arm endurance → shovel 56% variance |

**TABLE 30.1 (continued)**    Prediction of Performance

| Study | Subjects | Initial Assessment and Tests | | | Relationships/Predictions | | |
|---|---|---|---|---|---|---|---|
| | | Job Analysis | Work Simulation | Tests | Work Simulation | On-the-Job Performance | Results |
| Laughery et al., 1988 | ♂ 25 ♀ 25 | interviews, observations, physical measurements | | isometric strength tests: grip arm lift, back lift, arm press. | transporting a 15.9 kg box up and down a stair, also a 22.7 kg box | | mult. correlation: composite score with work power (kg/m/min) 22.7kg/15.9kg both genders → 41/41% var. ♂ = 18/12% var ♀ = 49/42% var |
| Oseen et al., 1992 | ♀ 45 | | | static and dynamic strength, anthropometrics, body fat, and body image | casualty evacuation; jerry can lift, carry, and empty; box lift, a trench dig, weighted load march | | correlation(r = ?; p < 0.05) |
| Reilly et al., 1979 | ♂ 83 ♀ 45 | interviews, task quantification | | static (grip, arm) and dynamic (trunk, arm) strength, reaction time, balance, aerobic step-test, flexibility, body density | overall total performance (OTP): pole testing, pole climbing (2), ladder placing (2), pole climb with drop wire removal | 6 months job completion (retention), supervisor job ratings (r = 0.36) | regression: dynamic arm strength → OTP 12% variance; dynamic stren. → retention 13%; dynamic arm strength, reaction time, step test → OTP 17% variance |
| | ♂ 132 ♀ 78 | interviews, task quantification | | static (grip, arm) and dynamic (trunk, arm) strength, reaction time, balance, aerobic step-test, flexibility, body density | time to complete training task (TTC), supervisor rating, training completion, tasks w/o errors, accident in 1st 6 months | 6 months job completion (retention) | regression: body density, balance, static stren. (PAT)→ TTC 20% var.; → train. dropout 14% var, →tasks w/o errors 13%, → OTP 10%, → retention 11%, → accidents 2% var |
| Rice and Sharp, 1994 | ♂ 12 ♀ 11 | process not described | | 7 strength tests, 3 fitness scores (sit-up, push-up, run), 3 physical descriptors | repeated stretcher carry — mass casualty, continuous stretcher carry — remote site | | regression: grip + run → repeated carry 74% var for harness & 78% var for hand, grip → continuous hand carry 74% var, bench press → cont. harness carry 33% variance |
| Robertson, 1992 | ♂ 274 ♀ 259 | criterion tasks, survey, interview, observation, measurement (force, grip points, distance, direction) | | anthropometrics, static, dynamic, and power strength measures | carrying, lifting, and pulling | | regression: separate gender groupings validity coefficient = 0.30's — 0.50's, combined gender 0.60's — 0.80's |

| Study | N | Job analysis | Tests/measures | Criterion tasks | Results |
|---|---|---|---|---|---|
| Robertson and Trent, 1985 | ♂ 274 ♀ 259 | survey, observation, measurement | anthropometrics, static, dynamic, and power strength measures | carrying, lifting, pushing and pulling | predictive validity for separate gender groupings ranged from 0.3's to 0.50s, combined ranged 0.60's — 0.80's. Greatest single predictor = arm pull |
| Schonfeld et al., 1990 | ♂ 20 | | treadmill VO$^2$ max, isokinetic knee flexion and extension, power arm and leg tests, grip strength, push-ups, sit-ups, sit-and reach, body composition | stairclimb, chopping simulation, and victim drag | regression: most powerful predictor = treadmill time. Various test combinations → simulations 2–91% var. |
| Sharkey, 1981 | 121 | process not described | body fat, chin-ups, sit-ups in 60 sec, push-ups in 60 sec, and a 50 lb pack test | hiking, building a fireline, deploying & moving hoses, shoveling & throwing sand, carrying | Combined tests → 71% variance; Muscular fitness + experience → 59% variance |
| Stalhammar et al., 1992 | ♂ 103 | not mentioned | anthropometrics, skinfold body fat, hand grip (static, dynamic, endurance), max. voluntary contraction of trunk flexor/extensor, range of motion | - rating of acceptable lifting load for an 8 h day; - rating of acceptable lifting load for simulated postal parcel sorting | correlation & regression - left lateral bend, dynamic grip endurance, and back stren/wt ratio → 63% variance - dynamic grip endurance, back stren/wt ratio, and trunk flexor stren/wt ratio → 44% variance |
| Wilmore and Davis, 1979 | ♂ 217 ♀ 13 | critical job task analysis | laboratory: strength (bench press, leg press, two arm curl and grip), flexibility, body comp., treadmill max-aerobic. field: strength (bench press, grip), flexibility, power (vertical jump), run time, predicted body composition | barrier surmount; dummy drag from car | correlation & regression: laboratory test → barrier surmount & dummy drag (46 & 44% var). Field test → barrier surmount & dummy drag (38 & 32%). |

→ = predicted

## Using Preplacement Screening to Predict Injury Rates

There is some evidence for an association between physically demanding jobs and injury rates, which could lead to the assumption that workers who are better able to meet the physical demands are less prone to injury. Physical fitness was found to be predictive of back injuries for fire fighters; however, the injuries sustained by the fit group were more severe than those of the "unfit" group (Cady et al., 1979). Load transportation in narrow spaces was found to be associated with musculoskeletal (especially spinal) injury (Stalhammar et al., 1992), and increased incidence and severity of low back pain has been found in people who work in jobs requiring lifting and moving of heavy loads (Wickstrom, 1978; Kosiak et al., 1968; Rowe, 1969). Although most individuals with back problems cannot associate the onset of their symptoms with an accident or unusual activity, those who can cite a specific event, cite lifting (Cady et al., 1979; Snook et al., 1978). In a retrospective review of back injuries, it was demonstrated that the most common event associated with back injury claims was manual materials handling, and improper lifting was most frequently listed as the cause (Bigos et al., 1986).

A previous history of back pain (Biering-Sorensen, 1984; Bigos et al., 1986) and/or a change of occupation to heavier work (Nordgren et al., 1980) have been suggested as potential predictors of consequent back pain. Persons experiencing back pain within the last 6 months have been found to have a higher incidence of subsequent back problems (Battié et al., 1989), which is consistent with the idea that recurrences (of back injury) are more likely in the 2 to 3 years following an injury. However, Venning et al. (1987) found work-related factors such as the job category and exposure levels have greater contributions to back injury than did a previous history of back pain.

Table 30.2 contains a synopsis of studies relating preplacement screenings with injuries. Strength has been identified as a potential risk factor for back injuries by a number of researchers (Biering-Sorensen, 1984; Cady et al., 1979; Chaffin et al., 1978; Doolittle and Kaiala, 1986; Keyserling et al., 1980a; Nachemson and Lindh, 1969). In fact, it has been over 20 years since researchers identified an association between an increased chance of injury with strength capabilities of workers (Chaffin, 1974; Chaffin, 1971; Rowe, 1971). However, Hagberg et al. (1995) reviewed 25 articles dealing with preplacement or pre-employment screening and concluded that "at present there is no scientific evidence to support the use of pre-employment or preplacement screening tests to predict the development of work-related musculoskeletal disorders, nor are they justified from an ethical and, in some countries, legal point of view" (p. 336).

Several articles provide direct support for this viewpoint. Battié et al. (1989) examined isometric lifting strength as a predictor of industrial back pain. During a 4-year follow-up, those with *greater* isometric strength were found to be at greater risk for injury than were weaker workers. However, after controlling for age, no significant difference was seen. These researchers concluded that isometric lifting strength testing was ineffective in identifying individuals at risk for industrial back problems. Anecdotally, in military operations, it has been suggested that natural selection occurs with stronger workers executing the heavier work more often. Dueker et al. (1994) measured isokinetic trunk flexion and extension, as well as repetitive isotonic lifting (Table 30.2), and monitored the workers for 6 years. No difference was found between any isokinetic scores of injured vs. non-injured workers. Mostardi et al. (1992) used an isokinetic (Cybex) lifting device to measure lifting capacity of nurses. After following the workers for 2 years, the authors concluded that isokinetic lifting strength was not related to occupational back injury. Several elements were missing from these studies. The design of the strength measures was not based on a precise task analysis (at least none was reported), and no other constructs (aerobic capacity, movement quality) were included.

Several studies have demonstrated the positive effect of an individual's strength capacity meeting or exceeding job requirements (Chaffin, 1974; Chaffin et al., 1978). In Chaffin's seminal work, he examined subjects' maximal acceptable isometric lifting capacity and compared it with job requirements. The hypothesis was that those with higher job strength ratios (job lift requirements ÷ average acceptable lift) would be more likely to experience low back pain. The hypothesis was upheld, as a sharp increase in the mean low back pain incidence rates was found for those jobs populated by persons without strength equal to or exceeding that required by the job. The medical incidence rate of workers with insufficient strength to meet job demands was nearly three times the rate of workers who were matched to their jobs (Chaffin, 1974). Three subsequent prospective studies used isometric tests that were similar to job requirements and compared those results with the job requirements (Chaffin et al., 1978; Keyserling et al., 1980a; Keyserling et al., 1980b). Results revealed greater injury rates when the job demands approached or exceeded the person's abilities. The authors concluded that workers whose strength abilities were less than job requirements suffered higher incidence rates in general than workers whose strength matched or exceeded job demands. However, excess strength was not found to be of protective value. Anderson (1988) found incidence rates were 34% greater for persons who did not meet job strength requirements (below criteria) compared with those who met or exceeded the requirements. He also noted that workers below criteria had more lost duty time and higher worker compensation costs (although the data were not statistically analyzed), and workers above criteria were more productive. The authors suggested the use of a screening battery could result in an increase in productivity and a reduction of back injuries, over-exertion injuries, incurred compensation costs, and lost time. Although the results demonstrated that workers who met the criteria were less likely to report back or over-exertion injuries, this was not true for all five jobs studied, even though they were considered physically demanding.

Other methods of decreasing injuries in the workplace have been suggested in early reports by Snook et al. (1978) and later by Doolittle and Daniel (1989). Their suggestions focused on limiting the requirements of the task. For Snook et al. (1978) the limitations were based on his research using maximal acceptable strength tasks, while Doolittle and Daniels' limitations used the NIOSH lifting formula and previous research by Ayoub et al. (1984). Snook et al. (1978) determined that a worker is three times more susceptible to low back injury if exposed to manual handling tasks that less than 75% of the population find acceptable and, therefore, that 67% of the injuries could be avoided through ergonomic design. Troup et al. (1987) also found that lifting capability, expressed as a rating of acceptable load, was related to the development of future low back pain of bakery workers. The physically demanding tasks were compared with the 1981 NIOSH lifting guide, and the authors' recommended preassignment screening to select workers who are physically capable of performing the job in order to prevent injuries. They further recommended that workers not exceed 75% of maximum capability based on Ayoub's recommendations (Ayoub, 1982). Rather than a pure cutoff score, they suggested ranking candidates into four performance levels: below minimum (applicant taxed at more than 40% maximum aerobic capacity and more than 95% strength capacity), marginal (40% aerobic capacity and 85% strength), acceptable (33% to 40% aerobic and 75% strength), and better qualified (<33% maximum aerobic capacity and <65% strength). Similar procedures were used for screening fire fighters (Doolittle and Kaiyala, 1987) and for identifying physically demanding jobs for line workers (Doolittle et al., 1988); however, neither article included research to demonstrate that their method actually reduced injuries of workers.

One of the more convincing reports was published in a recent study by Karwowski et al. (1994) (Table 30.2). They examined the relationship between the NIOSH 1991 Lifting Index, estimated compressive and shear forces on $L_5/S_1$, and back injury incidence rates. Strong, positive correlations were found between the maximum compressive forces and the maximum lifting index for the job ($r = 0.88$, $p = 0.05$), with similar findings between the average compressive forces and the average lifting index. Strong correlations were also found between the average recommended weight limit (RWL) and the maximum incidence rate of low back injury ($r = .97$, $p = 0.004$), and between the Lifting Index (ratio between the actual load lifted and the RWL) and incidence rate for 1992 ($r = .94$, $p = 0.1$) and 1993 ($r = .97$, $p = 0.007$). Although no prediction equations were used, Reimer et al. (1994) found that use of a worker fitness evaluation, based on job-related lifting and trunk function, reduced injury rates over a 4-year follow-up period.

It appears that there is a relationship between preplacement strength measures and injuries, but the associations were not as strong or as consistent with those found between preplacement testing and performance. Once again, it appears that basing the criterion tasks on a well-designed task analysis is essential. Although not as obvious, it also appears beneficial for other constructs such as aerobic capacity to be included in the assessment. Certainly, additional rigorous evaluation of this assumption is warranted.

**TABLE 30.2** Prediction of Injury

| Study | Subjects | Initial Assessment and Tests | | | Relationships/Predictions | | |
|---|---|---|---|---|---|---|---|
| | | Job Analysis | Work Simulation | Tests | Work Simulation | Work-Related Injury | Results |
| Anderson, 1988 | 613 total | not described | | strength-arm pull, aerobic step test | | contact, back, and over exertion injury reports | Chi²: $p < 0.05$ for 1 job for back, over-exertion injuries, 2 jobs for all injuries reported |
| Battié et al., 1989 | ♂ 1726 ♀ 452 | not mentioned | | Isometric torso, arm, and leg lift strength | | reported back problems (pain) | correlation: back injury not related to strength |
| Chaffin et al., 1978 | 900 jobs ♂ 446 ♀ 105 | biomechanical evaluation, hand placement during heavy strength demands | hand placement during lifting (job position test) | torso, arm, leg, and job position lifting tests | | contact, back, musculoskeletal injuries | As strength requirement approaches/exceeds strength, incident and severity rates increase 3:1 Statistical sig. not reported |
| Chaffin and Park, 1973 | 411 total ♂ & ♀ 103 jobs | most stressful lifting tasks (weight, frequency) | | age, weight, stature, previous back injuries isometric lift | lifting strength rating (max load required ÷ strength of large/strong man) | low back incidence rates | higher strength requirements, very low and very high frequency rates, previous low back pain, and low lift ratio related to higher incidence rates. statistical sig. not reported |
| Doolittle et al., 1988 | 48 line-workers (♂ ?) | interviews, observation, movement classification, forces & masses, pace & frequency, energy cost via heart rate | | modified biceps curl, lateral rise, row, lat pull, chin-ups, submaximal aerobic step test | | lost duty days | multiple correlation: composite score → lost days (21% of variance) |
| Griffin and Troup, 1984 | ♂ 234 ♀ 116 | | | anthropometry, psychophysical lifting capacity, weight estimation, handling accuracy, isometric lifting strength, trunk strength | | | Diff. in mean scores between those with and those without back pain. Both genders: acceptable lift, accep lift/body wt. Chi² for situps for men only |

| Study | Sample | Job analysis method | Tests/measures | Outcome measure | Results |
|---|---|---|---|---|---|
| Karwowski et al., 1994 | not reported | | | low back injury incidence rate | multiple correlation: (86% of variance) |
| Keyserling et al., 1980 | ♂ 54 ♀ 27 | biomechanical job analysis, observation by subject experts, measurements | NIOSH lifting index, compressive and shear forces on lumbosacral joint task based requirements for strength: arm, back, push, pull | control group did not have to pass cutoff score (actual job force requirements), experimental group did | Chi²: marginal significance (p < 0.1) with all medical visits, not for musculo-skeletal visits |
| Keyserling et al., 1980 | ♂ 309 ♀ 35 | biomechanical job analysis, observation, measurements | 9 strength tests: lift, push, pull, arm, back | all medical visits, musculoskeletal visits | Chi²: weaker group had between 1.25 and 2.71 times the incidence rate for all visits. Between 1.6 and 3.1 for musculoskeletal visits (p < 0.05) |
| Reilly et al., 1979 | ♂ 83 ♀ 45 ♂ 132 ♀ 78 | interviews, task quantification | six tests of pole testing, pole climbing, and ladder placing | 6 months job completion (r = 0.36), supervisor job ratings (r = 0.36) | task perform (r = 0.45), time-to-complete a training task (r = 0.45), training completion (r = 0.38) regression: number of accidents (r = −0.15) |
| Reimer et al., 1994 | ♂ 33 control ♀ 1001 workers | onsite job analysis (observation, measurement) | trunk flex/ext, body wt, isokinetic lifting, range of motion | pre- and post-implementation low back injury rates - total injury rates | yr 1 = 32% decreased injury rate yr 2 = 41% decr injury rate yr 1 = 18% yr 2 = 21% yr 3 = 28% yr 4 (projected) = 37% |

## 30.3    Establishing a Preplacement Screening Program

The goal of a preplacement program is to match the job requirements with a worker's capacity in order to ensure productivity and safety. As stated by Keyserling et al. (1980b), preplacement testing should be safe, reliable, quantitative, practical (time, equipment, financial expenses, supervision/manpower–ease of administration, availability), related to job requirements, and predictive of job performance and/or risk of future illness or injury. Preplacement screening may result in fewer older individuals, women, and persons with disabilities being hired for certain jobs. As long as the assessment batteries are carefully developed, validated, and alternate methods of completing a task are considered so that it is evident the preplacement screening is based on the job requirements, this should not be an issue of undue concern.

The steps in developing a preplacement screening program are in outline form in Appendix A (preceding the Reference list for this chapter). During the job analysis, it is important to consider the skill levels and degree of competency required in each job. In order to do this, ergonomists must be directly involved in the job analysis. They should interview the subject matter experts (i.e., the workers, supervisors, administrators, and safety personnel; consult training manuals), observe, and conduct careful assessments of the job to include frequency, duration, and difficulty, measurements of masses moved and forces exerted; and identification of pace, frequency, and postural constraints. The ergonomist should also be aware of confounding factors such as noise and calibration of equipment. It is important to reiterate that each preplacement test will most likely differ for distinct jobs, all of which may be physically demanding. No studies have shown that the same preplacement/prescreening evaluations have been predictive of job performance or injury rates in multiple physically demanding jobs. Instead, the more successful reports have come from authors who tailored their assessments to the jobs in question (Jackson, 1994).

From the job analysis, an accurate description of the task requirements (essential job functions) with "limiters" should be identified. Every essential function does not need to become part of the preplacement evaluation. For example, if one of the essential functions requires the worker to lift a 40-lb. box, carry it 15 ft, and place it on a 4-ft. shelf and a second essential function requires the worker to lift a 30-lb. box to 2 ft, it can be assumed that an applicant who can complete the former task, can also complete the latter task. The first task (lift and carry) is a "limiter" Once a task analysis is complete, the description of the tasks should be presented to the company subject matter experts and revised as necessary. The preplacement job simulations and test battery should then be derived from the task descriptions of the essential functions, and the model of evaluation and criterion measures should be identified.

Kroemer et al. (1988) described various models to identify limiting factors, such as a linear anatomical model of the long bones, joints, and muscles to describe volume, mass, and muscle strength; physiological models using oxygen consumption, energy expenditure, and circulatory demands; orthopedic models using a prediction of musculoskeletal injuries; biomechanical models which may use anatomical, anthropometric, and orthopedic measures; and psychophysical models which evaluate synergistic mental and physical functions. One of these models, or a combination of them, must be selected for use in developing a preplacement screening evaluation.

The type of model selected will determine the test measures that will be used. For example, a medical examination may be used when employing an orthopedic model, and a lumbar motion monitor may be used to assess biomechanical limitations, such as the load-bearing capacities of the spine (Marras et al., 1992; Marras et al., 1995). Hogan (1991) conducted a review of literature and a series of evaluations on the physical requirements of tasks as reflected in job analysis. As a result of these evaluations, Hogan suggested that the structure of physical performance has three major components: strength, endurance, and movement quality. Hogan concluded that all occupational task performances fit within these three constructs. Although examination of specific subcomponents may be necessary, Hogan (1991) suggests using this taxonomy to organize worker selection procedures, and hence the test measures, for physically demanding jobs. Each model and test measure has its own strengths and weaknesses and its own proponents and adversaries.

Generally, it may be easier to develop a job simulation and consequently identify subelements that can then be used in a test battery. Many researchers prefer the use of a test battery, rather than task simulations in their preplacement screening. Test batteries are repeatable, require less skill from the participants (and often from the evaluators), can be used in a variety of locations, and require less equipment compared with task simulations. Of course, task simulations have greater face validity. Once the job simulation and test batteries have been developed, they should be compared using statistical analysis to determine whether one is representative of the other. If they are closely related, then only one need be used to predict on-the-job performance. Although the ADA and EEOC requirements do not include validation studies *per se*, it behooves those conducting the preplacement screening to be aware of the predictive ability of their simulations and test batteries. If possible, it would be wise to conduct a validation study in which either the job simulation, the test battery, or both are statistically examined for their relationship (association and predictive ability) with on-the-job performance. Thus, the ergonomist can establish that the preplacement testing does what it purports to do, and potentially, the results will be more defensible.

Once a job simulation or test battery is selected and implemented, it is desirable to conduct a follow-up analysis on the effectiveness of the preplacement screening through the evaluation of both productivity and injury data. Although the investigators can select any amount of time for a follow-up period, it is recommended that it be a minimum of 18 months to attempt to eliminate a placebo effect.

As mentioned previously, the preplacement tests should be reliable (giving the same evaluation at different times should yield the same result), and the safety of the applicants should be of utmost importance. Isometric strength tests have exhibited test–retest coefficients-of-variation ranging between 5% and 15% (Hershenson, 1979, Keyserling et al., 1980b), and have been shown to be safe in a series of industrial studies (Hershenson, 1979; Chaffin, 1974; Kamon et al., 1978). For task simulations, test–retest reliability evaluations may become part of the validation process. Simulations must also be fair. If they are too complex, learning and skill development can contaminate the assessment capabilities.

It will not always be possible to include each of the steps in Appendix A, due to funding or contract restraints of the hiring company. It is imperative that the researcher/ergonomist developing the preplacement screening evaluation be comfortable with the level and stringency of the particular program. It is possible that both the hiring company and the involved contractor could be questioned, should an applicant decide to challenge the validity of the preplacement process. The "developer" of the preplacement screening evaluation is not relieved of the duty to meet test validation requirements, even if the test user (hiring company) did not request the test be validated or asked for a lesser standard.

An inordinate amount of testing is undesirable due to time and funding costs, and the use of validation testing should assist in honing the preplacement evaluation. A direct relationship between the preplacement evaluation and task demands should be unequivocal. However, the requirement for business necessity may be more lenient if there is a high degree of risk to the job or if all or substantially all, women would be unable to perform the job safely and efficiently because of the strenuous manual labor required. For this reason, a validation of injury potential should be just as rigorous as the validation of the prediction of work performance.

## 30.4 Summary

According to federal employment laws, it is illegal to disqualify an employee for reasons of race, color, religion, gender, national origin, or disability status. Accordingly, employers are using preplacement screening tests to determine an individual's ability to safely and effectively perform a job. To date, there have been successful challenges to preplacement screening which failed to test important job actions (Jackson, 1994). Using a well-defined, scientific approach, it is possible to design a preplacement screening battery that is directly related to (and predictive of) job performance. It is also possible to use the preplacement screening to prospectively determine its impact on injuries, workers' compensation, lost workdays, and turnover. Following the evaluation and completion of ergonomic design alterations, a preplacement screening program should be the next step in ensuring a match between work demands and worker abilities for physically demanding positions. Both management and workers will benefit from the resulting increases in productivity and safety.

## Appendix A

I. Job Analysis
  A. Job descriptions
  B. Observation/Videotaping
  C. Interview/Survey (subject matter experts, management/workers)
    1. Difficulty
    2. Frequency
    3. Importance
    4. Other means to accomplish tasks
  D. Direct measurement
    1. Heights
    2. Weights
    3. Distances
    4. Frequency
    5. Pace
  E. Develop task list/description of essential functions

II. Presentation and Alteration
  A. Present to company
    1. Officials
    2. Supervisors
    3. Workers
  B. Modify task list, as needed
  C. Suggest uses of task list
    1. Preplacement screening
    2. Return to work of injured employees
    3. Graded introduction to task demands for new workers
    4. Americans with Disabilities Act compliant job advertisements

III. Preplacement Test Development
  A. Develop simulation (criterion tasks with standards of performance) and may include "task limiters"
    1. Lifting, carrying, pushing/pulling, digging, or climbing,
      a. Single
      b. Repetitive
      c. Location of beginning/ending object placement
      d. Object description (handles, size, mass, etc.)
      e. Posture requirements
      f. Distance
    2. Aerobic demands
    3. Other
  B. Identify task elements from task simulation
    1. Range of motion
    2. Strength
    3. Endurance
    4. Coordination
    5. Positioning/posture
  C. Development of preplacement test battery from task elements
  D. Development of necessary instrumentation
IV. Test new (avoid "cumulative" injury confounds) and current workers

    A. On job task simulations

    B. On preplacement test battery

V. Validation of task simulation, test battery, and on-the-job performance

    A. Compare scores on task simulation and test battery

        1. Multiple correlations

        2. Regression equations

        3. Other statistical methods as needed

    B. Validation of task simulation and test battery with on-the-job work performance

        1. Follow workers for specified period of time and record on-the-job data

            a. Descriptive data (age, weight, job, etc.)

            b. Work performance (productivity, supervisor ratings)

            c. Lost duty time

            d. Turnover/retention

        2. Record potential confounding factors, such as pertinent changes at company

            a. New safety programs

            b. Reorganization

            c. Other

        3. Compare scores on task simulations and test battery with on-the-job work performance

            a. Multiple correlations

            b. Regression equations

            c. Other methods as needed

VI. Validation studies of test battery and task simulation with injury data

    A. Follow-up study for specified period of time (suggested *minimum* of 18 months)

        1. Tabulate data of new and current workers

            a. Descriptive data (age, weight, job, etc.)

            b. Injury rates (categorized by type of injury/illness)

            c. Lost duty time due to injuries

            d. Turnover/retention

        2. Record potential confounding factors, such as pertinent changes at company

            a. New safety programs

            b. Reorganization

            c. Other

    B. Compare test battery and task simulation with injury data

        1. Multiple correlation

        2. Regression

        3. Other methods as needed

VII. Report findings to company management

VIII. Revise preplacement strength test battery and/or simulation in accordance with findings

IX. Repeat steps IV-X, as necessary

# References

Aghazadeh, F. and Ayoub, M. M. (1985). A comparison of dynamic and static strength models for prediction of lifting capacity. *Ergonomics*, 28, 1409-1417.

Anderson, C. K. (1988). *Strength and endurance testing for pre-employment placement*. (Available from C.K. Anderson, Back Systems, Inc., 5520 LBJ Freeway #200, Dallas, TX 75240.)

Anderson, C. K. and Catterall, M. J. (1989). A prospective validation of pre-employment physical ability tests for grocery warehousing, in A. Mital (Ed.), *Advances in Industrial Ergonomics and Safety* (pp. 57-63). New York: Taylor & Francis.

Arnold, J. D., Rauschenberger, J. M., Soubel, W. G., and Guion, R. M. (1982). Validation and utility of a strength test for selecting steelworkers. *Journal of Applied Psychology*, 67(5), 588-604.

Ayoub, M. A. (1982). Control of manual lifting hazards: III. Preemployment screening. *Journal of Occupational Medicine*, 24(10), 751-761.

Ayoub, M. M., Mital, A., Asfour, S. S., and Bethea, J. J. (1980). Review, evaluation, and comparison of models for predicting lifting capacity. *Human Factors*, 22, 257-269.

Battié, M. C., Bigos, S. J., Fisher, L. D., Hansson, T. H., Jones, M. E., and Wortley, M. D. (1989). Isometric lifting strength as a predictor of industrial back pain reports. *Spine*, 14(8), 851-856.

Bernauer, E. M. and Bonanno, J. (1975). Development of physical profiles for specific jobs. *Journal of Occupational Medicine*, 17(1), 27-33.

Biering-Sorensen, F. (1984). Physical measurements as risk indicators for low back trouble over a one-year period. *Spine*, 9, 106-119.

Bigos, S. J., Spengler, D. M., and Martin, N. A. (1986). Back injuries in industry: A retrospective study II Injury factors. *Spine*, 11(3), 246-251.

Cady, L. D., Bischoff, D. P., O'Connell, E. R., Thomas, P. C., and Allan, J. H. (1979). Strength and fitness and subsequent back injuries in fire fighters. *Journal of Occupational Medicine*, 21, 269-272.

Celentano, E. J., Nottrodt, J. W., and Saunders, P. L. (1984). The relationship between size, strength, and task demands. *Ergonomics*, 27(5), 481-488.

Chaffin, D. B. (1971). Human strength capability and low back pain. *Journal of Occupational Medicine*, 16, 248-254.

Chaffin, D. B. (1974). Human strength capability and low back pain. *Journal of Occupational Medicine*, 16(4), 248-254.

Chaffin, D. B., Herrin, G. D., and Keyserling, W. M. (1978). Preemployment strength testing: An updated position. *Journal of Occupational Medicine*, 20(6), 403-408.

Davis, P. O., Dotson, C. O., and Santa Maria, D. L. (1982). Relationship between simulated fire fighting tasks and physical performance measures. *Medicine and Science in Sports and Exercise*, 14(1), 65-71.

Doolittle, T. L. and Kaiyala, K. (1986). Strength and musculo-skeletal injuries of fire fighters. *Proceedings of the 19th Annual Conference of the Human Factors Association of Canada* (pp. 49-52). Vancouver: British Columbia: Human Factors Association of Canada.

Doolittle, T. L. and Kaiyala, K. (1987). A generic performance test for screening fire fighters, in S. Kumar (Ed.), *Trends in Ergonomics and Human Factors IV*. New York: Elsevier.

Doolittle, T. L. and Daniel, B. (1989). Physical demands of bakery workers. *Proceedings of the Human Factors Society 33rd Annual Meeting* (pp. 682-686). Santa Monica, CA: Human Factors Society.

Doolittle, T. L., Spurlin, O., Kaiyala, K., and Sovern, D. (1988). Physical demands of lineworkers. *Proceedings of the Human Factors Society 32nd Annual Meeting* (pp. 632-636). Santa Monica, CA: Human Factors Society.

Dueker, J. A., Ritchie, S. M., Knox, T. J., and Rose, S. J. (1994). Isokinetic trunk testing and employment. *Journal of Occupational Medicine*, 36(1), 42-48.

Fox, R. R. (1982). *A psychophysical study of bimanual lifting*. Masters thesis in industrial engineering, Texas Tech University, Lubuck, TX.

Garg, A., Mital, A., and Aurelius, J. R. (1980). A comparison of isometric strength and dynamic lifting capability. *Ergonomics*, 23, 13-27.

Hagberg, M., Silverstein, B., Wells, R., Smith, M., Hendrick, H. W., Carayon, P., and Perusse, M. (1995). *Work Related Musculoskeletal Disorders (WMSDs): A Reference Book for Prevention*. London: Taylor & Francis.

Hershenson, J. D. (1979). Cumulative injury: A national problem. *Journal of Occupational Medicine*, 21(10), 674-676.

Hogan, J. (1985). Tests for success in diver training. *Journal of Applied Psychology*, 70(1), 219-224.

Hogan, J. (1991). Structure of physical performance in occupational tasks. *Journal of Applied Psychology*, 76(4), 495-507.

Jackson, A. S., Osburn, H. G., and Laughery, K. R. (1984). Validity of isometric strength tests for predicting performance in physically demanding tasks. *Proceedings of the Human Factors Society 28th Annual Meeting* (pp. 452-454). Santa Monica, CA: Human Factors Society.

Jackson, A. S., Osburn, H. G., Laughery, K. R., and Vaubel, K. P. (1991). Strength demands of chemical plant work tasks. *Proceedings of the 35th Annual Human Factors Society Meeting* (pp. 758-762). Santa Monica, CA: Human Factors Society.

Jackson, A. S. (1994). Preemployment physical evaluation, in J. O. Holloszy (Ed.), *Exercise and Sport Sciences Reviews* (pp. 53-90). Philadelphia: Williams and Wilkins.

Kamon, E. and Goldfuss, A. J. (1978). In-plant evaluation of the muscle strength of workers. *American Industrial Hygiene Association Journal*, 39(10), 801-807.

Kamon, E., Kiser, D., and Pytel, J. (1982). Dynamic and static lifting capacity and muscular strength of steelmill workers. *American Industrial Hygiene Association Journal*, 43, 853-857.

Karwowski, W. (1988). Maximum load lifting capacity of males and females in teamwork. *Proceedings of the Human Factors Society 32nd Annual Meeting* (pp. 680-682). Santa Monica, CA: Human Factors Society.

Karwowski, W., Caldwell, M., and Gaddie, P. (1994). Relationships between the NIOSH (1991) lifting index, compressive and shear forces on the lumbosacral joint, and low back injury incidence rate based on industrial field study. *Human Factors and Ergonomics Society 38th Annual Meeting* (pp. 654-657). Santa Monica, CA: Human Factors and Ergonomics Society.

Keyserling, W. M., Herrin, G. D., and Chaffin, D. B. (1980a). Isometric strength testing as a means of controlling medical incidents on strenuous jobs. *Journal of Occupational Medicine*, 22, 332-336.

Keyserling, W. M., Herrin, G. D., Chaffin, D. B., Armstrong, T. J., and Foss, M. L. (1980b). Establishing an industrial strength testing program. *American Industrial Hygiene Association Journal*, 41, 730-736.

Kosiak, M., Aurelius, J. R., and Harfiel, W. F. (1968). The low back problem: An evaluation. *Journal of Occupational Medicine*, 10, 588-593.

Kowal, D.M. (1983). Validation and utility of a work capacity test battery for selection and classification of military personnel. Washington, D.C.: DCP Office of Assistant Secretary of Defense (MRA & L).

Kroemer, K. H. E. (1970). Human strength: Terminology, measurement and interpretation of data. *Human Factors*, 12, 279-313.

Kroemer, K. H. E. (1982). *Development of LIFTTEST, A dynamic technique to assess the individual capability to lift material*. (Final Report, NIOSH Contract 210-79-0041). Blacksburg, VA: Ergonomics Laboratory, Industrial Engineering and Operations Research Department, Virginia Polytechnical Institute and State University.

Kroemer, K. H. E. (1983). An isoinertial technique to assess individual lifting capability. *Human Factors*, 25, 493-506.

Kroemer, K. H. E., Snook, S. H., Meadows, S. K., and Deutsch, S. (1988b). *Ergonomic Models of Anthropometry, Human Biomechanics, and Operator-Equipment Interfaces*. Washington, D.C.: National Academy Press.

Kumar, S. (1995). Development of predictive equations for lifting strengths. *Applied Ergonomics*, 26(5), 327-341.

Laughery, K. R., Hayes, T. L., Jackson, A. S., Osburn, H. G., and Hogan, J. C. (1986). Physical abilities and performance tests for coal miner jobs. *Proceedings of Human Factors 30th Annual Meeting* (pp. 377-391). Santa Monica, CA: Human Factors Society.

Laughery, K. R., Jackson, A. S., and Fontenelle, G. A. (1988). Isometric strength tests: Predicting performance in physically demanding transport tasks, in *Human Factors and Ergonomics Society 32nd Annual Meeting* (pp. 695-699). Santa Monica, CA: Human Factors Society.

Marras, W. S., Fathallah, F. A., Miller, R. J., Davis, S. W., and Mirka, G. A. (1992). Accuracy of a three-dimensional lumbar motion monitor for recording dynamic trunk motion characteristics. *International Journal of Industrial Ergonomics*, 9, 75-97.

Marras, W. S., Lavender, S. A., Leurgans, S. E., Rajulu, S. L., Allread, W. G., Fathallah, F. A., and Ferguston, S. A. (1995). Biomechanical risk factors and trunk motion. *Ergonomics*, 38(2), 377-410.

McDaniel, J. W., Kendis, R. J., and Madole, S. W. (1983). *Weight Lift Capabilities of Air Force Basic Trainees*. (TR No. 83-0001). Wright Patterson Air Force Base: Air Force Aerospace Medical Research Laboratory.

McMahan, P. B. (1988). Strength testing may be an effective placement tool for the railroad industry, in F. Aghazadeh (Ed.), *Trends in Ergonomics/Human Factors V* (pp. 787-794). North-Holland: Elsevier.

Menon, K. and Freivalds, A. (1985). Repeatability of dynamic strength tests. *Proceedings of the Human Factors Society 29th Annual Meeting* (pp. 517-520). Santa Monica, CA: Human Factors Society.

Misner, J. E., Plowman, S. A., and Boileau, R. A. (1987). Performance difference between males and females on simulated fire-fighting tasks. *Journal of Occupational Medicine*, 29, 801-805.

Mital, A. and Ayoub, M. M. (1980). Modeling of isometric strengths and lifting capacity. *Human Factors*, 22, 285-290.

Mostardi, R., Noe, D., Kovacik, M., and Porterfield, J. (1992). Isokinetic lifting strength and occupational injury. *Spine*, 17, 189-193.

Nachemson, A. L. and Lindh, M. (1969). Measurement and abdominal and back muscle strength with and without low back pain. *Scandinavian Journal of Rehabilitation Medicine*, 1, 60-65.

Nordgren, B., Schele, R., and Linroth, K. (1980). Evaluation and prediction of back pain during military service. *Scandinavian Journal of Rehabilitation Medicine*, 12, 1-8.

Pedersen, D. M., Clark, J. A., Johns, R. E., White, G. L., and Hoffman, S. (1989). Quantitative muscle strength testing: a comparison of job strength requirements and actual worker strength among military technicians. *Military Medicine*, 154, 14-18.

Pytel, T. L. and Kamon, E. (1981). Dynamic strength test as a predictor for maximal and acceptable lifting. *Ergonomics*, 24, 663-672.

Reilly, R. R., Zedeck, S., and Tenopyr, M. L. (1979). Validity and fairness of physical ability tests for predicting performance in craft jobs. *Journal of Applied Psychology*, 64(3), 262-274.

Reimer, D. S., Halbrook, B. D., Dreyfuss, P. H., and Tibiletti, C. (1994). A novel approach to preemployment worker fitness evaluations in a material-handling industry. *Spine*, 19(18), 2026-2032.

Rice, V. J. and Sharp, M. A. (1994). Prediction of performance on two stretcher-carry tasks. *Work: A Journal of Prevention Assessment and Rehabilitation*, 4(3), 201-210.

Rice, V. J., Sharp, M. A., Nindl, B., and Bills, R. (1995). Prediction of two-person team lifting capability. *Proceedings of the Human Factors and Ergonomics Society 39th Annual Meeting* (pp. 645-649). Santa Monica, CA: Human Factors and Ergonomics Society.

Robertson, D. W. (1982). *Development of an Occupational Strength Test Battery (STB)*. (AD-A114 247). San Diego: CA, Navy Personnel Research and Development Center.

Robertson, D. W. and Trent, T. (1985). *Documentation of Muscularly Demanding Job Tasks and Validation of an Occupational Strength Test Battery (STB)*. (MPL TN 86-1). San Diego: CA, Navy Personnel Research and Development Center.

Robertson, D. W. (1992). Development of job performance standards for muscularly demanding military tasks, in Kumar S. (Ed.), *Advances in Industrial Ergonomics and Safety IV* (pp. 1299-1304). Washington, D.C.: Taylor & Francis.

Rowe, L. M. (1969). Low back pain in industry. *Journal of Occupational Medicine*, 11, 161-169.

Rowe, M. L. (1971). Low back disability in industry: Updated position. *Journal of Occupational Medicine*, 13, 476-478.

Schonfeld, B. R., Doerr, D. F., and Convertino, V. A. (1990). An occupational performance test validation program for fire-fighters at the Kennedy Space Center. *Journal of Occupational Medicine*, 32(7), 638-643.

Sharkey, B. J. (1981). Fitness for wildland fire fighters. *The Physician and Sports Medicine*, 9(4), 93-102.

Sharp, D. S., Wright, J. E., Vogel, J. A., Patton, J. F., Daniels, W. L., Knapik, J., and Kowal, D. M. (1980). *Screening for Physical Capacity in the U.S. Army: An Analysis of Measures Predictive of Strength.* (T8/80). Natick, MA: Exercise Physiology Division, U.S. Army Research Institute of Environmental Medicine.

Snook, S. H., Campanelli, R. A., and Hart, J. W. (1978). A study of three preventive approaches to low back injury. *Journal of Occupational Medicine*, 20, 478-481.

Stalhammar, H. R., Leskinen, T. P. J., and Nurmi, P. A. (1992). Psychophysical tests, isometric and dynamic muscle force measurements as determinants of aircraft loaders' functional capacity, in S. Kumar (Ed.), *Advances in Industrial Ergonomics and Safety IV* (pp. 683-691). Washington, D.C.: Taylor & Francis.

Teves, M. A., Wright, J. E., and Vogel, J. A. (1985). *Performance on Selected Candidate Screening Test Procedures Before and After Army Basic and Advanced Individual Training.* (T13/85). Natick, MA: Exercise Physiology Division, U.S. Army Research Institute of Environmental Medicine.

Troup, J., Foreman, T., Baxter, C., and Brown, D. (1987). The perception of back pain and the role of psychophysical tests of lifting capacity. *Spine*, 7, 645-657.

Venning, P. J., Walter, S. D., and Stitt, L. W. (1987). Personal and job-related factors as determinants of incidence of back injuries among nursing personnel. *Journal of Occupational Medicine*, 29(10), 820-825.

Wickstrom, G. (1978). Effect of work on degenerative back disease: A review. *Scandinavian Journal of Work Environment and Health*, 4(Suppl. 1), 1-12.

Wigdor, A. K. and Green, B. F. (1991). *Performance Assessment for the Workplace.* Washington, D.C.: National Academy Press.

Wilmore, J. H. and Davis, J. A. (1979). Validation of a physical abilities field test for the selection of state traffic officers. *Journal of Occupational Medicine*, 21(1), 33-40.

# 31

# Assessment of Worker Functional Capacities

Glenda L. Key
*KEY Method*

## 31.1   Introduction

There are nearly 500,000 U.S. workers each year who, as a result of injury, are unable to resume their jobs for long periods of time. Low back pain will afflict roughly 80% of the population at some time in their lives (Wheeler and Hanley, 1995). Workers' compensation costs have reached $70 billion per year, tripling since 1980. The direct cost of the situation is staggering and in each case the degree of loss is unpredictable.

The knowledge of what a worker's functional capabilities are, is one of the most valuable pieces of information that a professional can have for reducing workers' compensation costs and prevention of injury in the workplace today. There are two occasions in the employment of a worker when it is critical to know just what the physical work capability of that worker is. One occurs during the decision to hire, and the other is upon return to work of an injured employee. The focus of this chapter will be on these two occasions as highlighted in Figure 31.1, the Worker Care Spectrum representing assessment of worker functional capabilities.

## 31.2   Worker Assessments

### Definitions

#### Functional Capacity Assessment

The functional capacity assessment (FCA) is a return-to-work testing process that determines an individual's physical functional work-related capabilities through measuring, recording, and analyzing data gathered during a standardized physical testing procedure. There are many different formats of functional

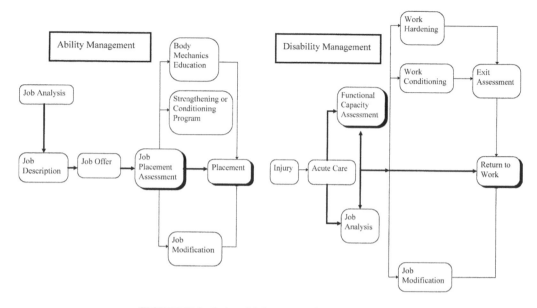

**FIGURE 31.1** Industrial therapy worker care spectrum.

capacity assessments and almost as many names for them. The term used may be physical capacity assessment, functional capacity evaluation, work capacity evaluation, or work assessment. For the purposes of this chapter, the term functional capacity assessment is used.

The activities of an FCA include materials handling functions of lifting, carrying, pushing, and pulling. It also includes tolerances of posture such as standing, sitting, kneeling, crawling, and activities of walking and stair climbing (Key, 1995a).

Simulation of the movements and postures during the weighted activities and standardizing for consistency of instruction, weight loading protocols, and determining termination points are primary issues that need to be considered in a functional capacity assessment. An FCA allows for the testing of the individual to confirm meeting or not meeting the minimum physical requirements of the job.

As quickly as possible, the workers should be returned to their jobs and workplaces — even if they are unable to be fully productive. Athletes do not wait until they are fully recovered before they start back in practice or training. They work their bodies to the point where they still maintain safety and will not reinjure themselves. Their capabilities are built up gradually until they are back in their "game." It does not mean that they play without pain. It does not mean that they can do everything as well as or in the same manner as before. This approach applies to the injured worker as it does to the athlete.

## Job Placement Assessment

The job placement assessment (JPA) provides data prior to hiring that assists managers in making decisions resulting in significantly decreased incident rates in the future. When armed with the information of the requirements of the job as provided through a job analysis and job description, and the physical abilities of the individual as provided through a job placement assessment, it is relatively easy to determine whether persons should be hired into a specific job or not. If they meet the requirements of the job, the answer is "Yes." If they do *not* meet the requirements of the job, the answer is "No."

"The job placement assessment is a series of specific, objective, and standardized protocols followed in a consistent progression to allow for objective, accurate, and repeatable results" (Frey, 1995). These results identify the prospective employee's work capabilities in the areas of lifting, carrying, pushing, pulling, and other job-specific activities.

The continuous and significant increase cost of health care and workers' compensation has added to the already inherent need to have healthy, productive employees. This process begins with hiring individuals who are able to physically perform the requirements of the job. How can employers tell if their next hire will be their best employee or become their worst workers' compensation and litigation nightmare?

The employee has the most to gain and lose if decisions are made based on anything other than accurate data. At the time of hire, the prospective employee has the right to be judged through a nondiscriminatory process. Generally, his or her personal need is to be employed and be a productive part of society. The expectation is that the hiring process will be fair and that the job activity will not be a precursor of an injury. Once injured, whether on the job or not, the worker relies on the medical system and those individuals carrying out company policies to establish the return-to-work process. The desire to be productive, the fear of reinjury, the unfamiliar role of patient, and the acceptability of not working can all be contributors to the complex course of action. It is of substantial importance that the match of the worker to the work be accurate to prevent any reinjury yet place the worker in the highest level of productivity possible — for economic and social reasons.

At these two points in the Worker Care Spectrum, the employee is especially vulnerable to the decisions of others. Employees often report that their experience in a worker assessment, JPA or FCA, is the first time that decisions are being made based on their actual participation, with themselves and their activities in control of the end points. To have their capabilities objectively tested in a participatory manner is important to them.

## Components

Recommendations of an individual's physical work ability are made based on assessing components in the following three categories.

1. Weighted capabilities
2. Tolerance and endurance parameters
3. Validity of participation determinants

### Weighted Capabilities

These tests include analyzing an individual's ability to perform specific work-related materials handling functions such as lifting at three standard heights, carrying, pushing, and pulling. A detailed list of components is found in Figure 31.2. Cardiac responses, kinesiological changes and repetition of posture and posture changes are documented and included in the formulas of decision of the individual's capabilities. This further confirms cardiovascular and body mechanics safety of the employee or prospective employee at that level of physical activity. Important questions are answered through these thorough testing techniques. For example, can the employee lift the 48 lbs. that are required for the job? How can one know unless the worker is tested?

### Tolerances and Endurance

Testing includes and recommendations are given in the categories of standing tolerance, sitting tolerance, and workday tolerance. Kneeling, crawling, and activities such as walking and stair climbing are also covered. A detailed list of components is found in Figure 31.2. For example, can your employee return to an eight hour workday or is he or she limited to a lesser amount? And if so, what is the safe limitation of a workday?

### Validity of Participation Determinants

It is important to objectively, scientifically, and statistically determine and demonstrate when individuals are not being forthright in the representation of their capabilities (Grossman, 1985; Osterweis et al., 1987). It is known that some individuals will demonstrate less than their full capability. They may be attempting to prolong their time off work or may be attempting to falsely maximize their case closure settlement.

---

## Weighted Capabilities

| | |
|---|---|
| Lifting desk to chair 30" to 18" | Lifting chair to floor 18" to 0" |
| Carrying | Lifting above shoulder 30" to 60" |
| Pulling | Pushing |

---

## Tolerances and Endurance

| | |
|---|---|
| Balancing | Bending |
| Cervical mobility | Circuit board tolerance |
| Climbing | Grip strength |
| Crouching | Crawling |
| Fine manipulation | Fastener board tolerance |
| Keyboard tolerance | Firm grasping |
| Simple grasping | Kneeling |
| Repetitive foot motion | Reaching |
| Sitting | Simple grasping |
| Standing | Squatting |
| Tool station work tolerance | Stooping |
| Walking | Work day tolerances |

---

**FIGURE 31.2**

The percentage of people participating in this kind of exaggeration and falsification is less than the public generally perceives it to be. Figure 31.3 presents the percentages in each category of participation determinant based on a sample of 43,000 injured clients (Worker Data Bank, 1994–1997).

It is important to the valid participators to have their honest attempts proven to be full effort demonstrations. The injustice is not only when the conscious and deceptive low participator is not revealed, but also when the honest, full effort, low capability level participator is misjudged.

Validity of participation determinants have been developed by one vendor of FCA equipment and protocols (Gilliand et al., 1986; Personnel Decisions, Inc., 1994). The individual's level of participation is determined through the use of algorithms and decision science objectively identifying the degree of an individual's consistency or inconsistency. The formula includes data from a database made up of over 100,000 cases of this standardized, protocolized, FCA (Worker Data Bank, 1994–1997). These data have provided valuable correlations, one of which is between the amount someone can lift at specific heights and how much they can push and pull. Scientific literature and the same database have provided statistics on certain vital signs, such as heart rate, which respond predictively as the individual approaches and reaches exertion level (Astrand and Rhyming, 1954; Gilliand et al., 1986; Worker Data Bank, 1994–1997).

The delineations of performance levels as introduced by KEY Method in Minneapolis, Minnesota are (Gilliand et al., 1986; Key, 1984; Worker Data Bank, 1994–1997):

1. Valid participation. This determinant means that the individual participated with full effort and the results and recommendations reflect safe capabilities.
2. Invalid participation. This as a determinant means that the participant consciously and intentionally provided less than full effort. The numbers produced by this individual reflect less than full capability. The results can be used in addressing the issue directly with the participant or, if necessary, in litigation.
3. Conditionally valid participation. An individual with this determinant has demonstrated less than full capability. The results reflect their *perception* of capability even though they can physically do more. The numbers and capabilities are, therefore, safe, but may also reflect an unconscious psychological barrier.
4. Conditionally invalid participation. The individual with this determinant has demonstrated work levels that are beyond what would be considered full, safe levels for extended work periods. When left to their own end point determinations of work activity, they exceed the safe levels.

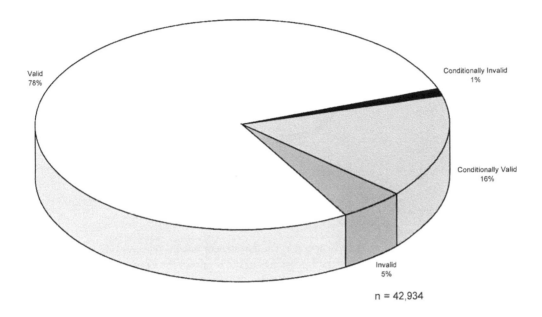

**FIGURE 31.3** Functional capacity assessment validity determinants.

**TABLE 31.1**   Consistency of KEY Method Assessment Results

|  | Validity by Percent in Each United States Territory | | | |
| --- | --- | --- | --- | --- |
|  | West | Central | North East | South East |
| Valid | 78 | 78 | 78.5 | 76 |
| Invalid | 5 | 3 | 5 | 5 |
| Conditionally Valid | 15 | 17 | 15 | 17 |
| Conditionally Invalid | 1 | 2 | 1.5 | 1 |

Freidman two-way analysis of variance by ranks. Region by participation (%)
$k = 4, n = 4; X^2 = 7.5 \ X_r^2 + .9$

   The importance of standardization, statistical analysis of data, and carrying the level of data in a database to produce determinants of participation is paramount.

## Standardization

When looking for an assessment, standardization and validity of participation determinants are of consummate importance. They are the primary elements of defense against the occurrence of litigation and the primary elements of defense should the case eventually go to court. One of the standardized systems maintains a complete data bank of all assessments performed (Worker Data Bank, 1994–1997).

   To be assured of such level of standardization, studies should be provided to those using the assessment that demonstrate reliability and consistency. The consistency of the KEY Method assessment results was analyzed and is represented in Table 31.1. Analysis using 2-way analysis by ranks found "no differences" in providers' results across the United States. Statistically, any differences were found to be *extremely not significant* (Aitken, 1996; Portney and Watkins, 1993, Worker Data Bank, 1994–1997).

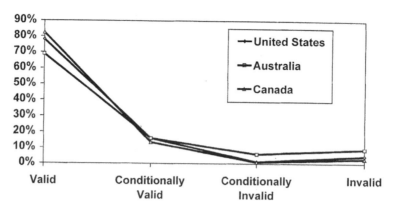

**FIGURE 31.4**   Descriptive summary validity.

Another analysis demonstrating consistency in the assessment results, representing standardization, is the graph in Figure 31.4. This confirms the consistency of the results of the assessment when administered in three different countries: the United States, Canada, and Australia.

## Worker Assessment Principles

Principles to look for in selecting a functional capacity assessment include:

1. The assessment must contain standards for identifying validity of participation of the individual being tested
2. The methodology administered must be consistent from tester to tester and test to test
3. Standardized equipment must be used and the same procedure followed with each assessment
4. The administrator of the assessment must be thoroughly trained and objective
5. The processing of the information gathered during testing must be standardized

The psychology of the individual needs to recognized in the assessment as well as the kinesiology of the activities (Schmidt et al., 1989). There is a growing body of literature supporting the theory that a statistical relationship does exist between low back pain and an individual's psychological factors as tested by the Minnesota Multiphasic Personality Inventory (MMPI) (Block et al., 1996). The MMPI has also been able to predict the occurrence of job-related low back pain or when poor response to surgery would be the outcome (Bigos et al., 1991; Blair et al., 1994; Block et al., 1996; Keller and Butcher, 1991; Schmidt et al., 1989).

## Assessment Reports

A report should provide details of the results of the assessment and then compare those results with the physical demands of the job. Through this matching exercise, the decision maker can make an informed, unbiased, and defensible decision relating to the return to work of a previously injured employee.

It is also helpful to receive a visual display of the major highlights of the results, as in Figure 31.5. Graphic displays of these major areas can be most helpful in assisting the decision-making process for rehire or case management. The individual's assessment results are compared with the same profile of individual in the data bank. They also may be compared with the job requirements. Another option is to compare norms from the injured database with similar profiles of individuals in the uninjured database.

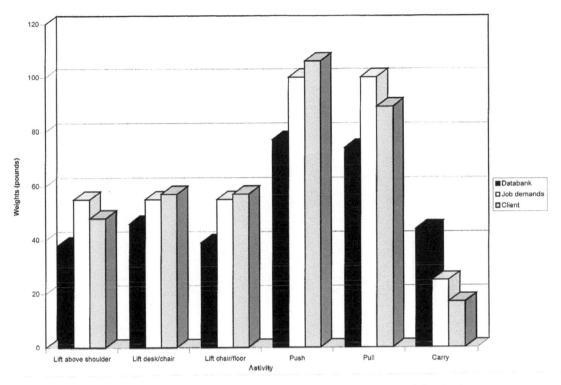

**FIGURE 31.5** Injured worker FCA results compared to job requirements and data bank norms.

## 31.3 The Provider of Worker Assessments

The provider of these assessments can be instrumental in increasing return-to-work percentages, reducing reinjury rates, decreasing lost workdays, and yielding other short and long-term cost reductions.

Physical and occupational therapists are the primary providers of worker assessments, job placement assessments, and return-to-work functional capacity assessments. The motive for providing quality results in functional assessments lies primarily in the accuracy of the results and the defensibility in the courts. The therapist needs to provide the most objective, unbiased data. These data are then used as an assist in the hiring and returning-to-work decisions. As a result, therapists need to be able to support their method with objectivity, standardized equipment, and consistent protocols and procedures.

## 31.4 Outcomes

Outcome surveys or studies should be available for review for one to trust the results and the legitimacy of the recommendations of a worker assessment (Dobrzykowski, 1995; Key, 1995b). The predictive ability of an FCA needs to be demonstrated based on a track record of return-to-work without reinjury.

The primary outcomes that one should be looking for include:

1. Decreased reinjury rates
2. Decrease lapse of time from date of injury to date of return to work
3. Decrease incidents and cost of litigation

The State of Colorado studied the impact of FCAs on shortening the amount of time required for vocational evaluation of injured workers. H. D. Waite found that the inclusion of the KEY Method FCA

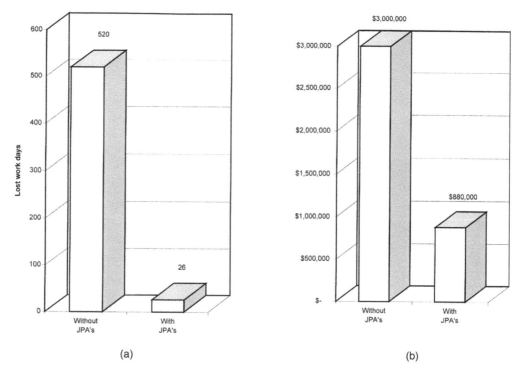

**FIGURE 31.6**   (a) Lost work days. (b) Workers' compensation costs.

resulted in a median of 18 fewer days of rehabilitation. This would save the State of Colorado over $200,000 annually (Waite, 1987).

## JPA Outcomes

Having the data and using them to safely match workers' capability levels to the job demands, or being in a position to know the modifications necessary to facilitate a match results in fewer injuries and subsequent lower health care costs, lower workers' compensation complications, preserved productivity, worker morale, and job satisfaction. These issues along with that of the assessments' predictive capabilities regarding future injuries are the primary issues in outcome reports. Examples assist in demonstrating the effects (Karwowski and Salvendy, 1998).

### Outcome — Case A

In 1988 a paper manufacturer in Minnesota instituted job placement assessments to help stem the costs related to workers' compensation claims and lost workdays. In analysis two years later, the experiences of 70 employees hired before the use of JPAs were compared to 70 hired after the initiation of administering JPAs. Use of the JPAs lowered both lost workdays and workers' compensation costs (see Figure 31.6) (Frey, 1995).

### Outcome — Case B

A state transportation authority discovered that job placement assessments enable them to predict if an employee is at risk of injury. Of the 36 employees injured from 1985 to 1992, 75% had been categorized as "at risk" by JPAs performed when they were hired. Fourteen providers across the state administered the same system of JPA. Personnel Decisions, Inc. reports, "While the analysis is based on a relatively few cases, the results are statistically significant. The chi square for this cross-tabulation is 15.4, which is significant at p = .00045" (Personnel Decisions, Inc., 1994).

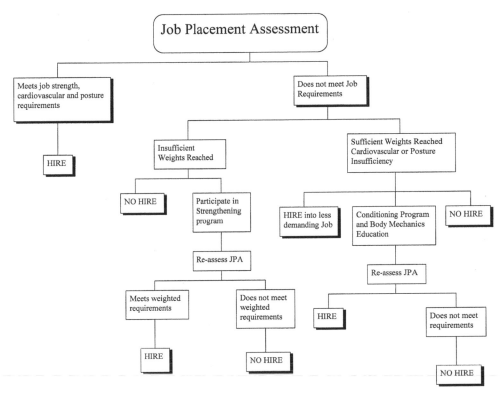

**FIGURE 31.7**

### Outcome — Case C

A major trucking firm used the JPA for two separate hiring locations over a continuous eighteen-month period. Since implementing the administration of the JPA on each candidate and hiring based on the results, there were no new injuries (Worker Data Bank, 1994–1997).

## 31.5    After the Assessment Results Are In

The results of job placement assessments and functional capacity assessments offer options for pathways to follow, depending on the data assembled. The decision tree layout of Figures 31.7 and 31.8 demonstrate optional pathways available.

## 31.6    Diversification Options

As change occurs in industry so will industry's approach to minimizing work-related injuries. To meet these changing needs medical providers are diversifying the services they now offer to industry and the mode of delivery of those services. Today's marketplace offers some examples of successful diversification.

### Mobile Assessments

For those who offer functional capacity assessments or job placement assessments, a beneficial adaptation has been to perform these assessments at the job site rather than in a clinic. This is done by utilizing a mobile testing unit that uses the same standardized test procedures as those which are practiced in the clinical setting. By having the JPAs performed onsite, the individual responsible for the "hire/no-hire" decision has immediate access to the assessment results. Because of this, no administrative or production

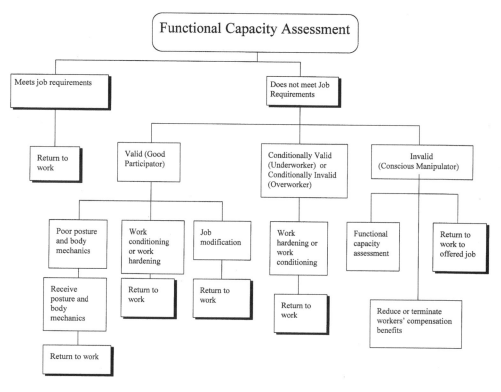

**FIGURE 31.8**

time is lost waiting for reports. They can be printed up immediately upon completion of the JPA. With today's tight economic issues and tight worker market it is important that the hire be made quickly. Onsite JPAs provide a system by which many employees can test through and begin work the same day.

Having the return-to-work assessment, the FCAs, performed onsite promotes the bond between the injured worker and the workplace. It assists in the process of employee acceptance, co-worker to co-worker. Management and nonmanagement observe that the individual is indeed participating in the return-to-work program. Bringing the worker back to the site for testing also allows the employee to reunite with other colleagues and co-workers.

Additionally, therapists are able to develop a closer working relationship with management and thereby better understand and meet the company's needs.

## Mobile Occupational Health Clinic

Taking this idea a step further, some have adapted large vans or buses to house a mobile occupational health clinic. This includes facilities for examinations and physicals, functional capacity assessments, job placement assessments, job analysis, ongoing treatment, and educational materials, as well as facilities for drug, hearing, and vision screenings. This mobility presents industry with a broad offering for delivery of services. One such mobile unit travels up to 200 miles to deliver such services to hundreds of industries in a rural region.

## Onsite Services

Onsite industrial therapist services are becoming a very cost-effective way for industries to more thoroughly meet the needs of their company. When the provider of services shows up for work every day at the plant site, there is a deepened knowledge level of the problems and a quicker pathway to the solutions. Services onsite may be delivered by contracting with a local clinic or therapist. It may be to the company's advantage to have the individual or individuals be employees of the company.

Onsite services can vary from full-time positions to a few hours per week depending on the size of the company and the workers' risk for injury. This approach requires that the company designate space for the medical provider to assess and treat workers. The most apparent benefits of onsite therapy services are the immediacy of services and the decreased worker down time. Musculoskeletal injuries (i.e., repetitive motion injuries, back injuries, sprain/strain injuries) receive treatment faster and therefore allow the patient to return to work faster.

Prompt treatment is the key to returning the worker to work as quickly as possible. In addition to helping the worker overcome injuries, prompt treatment can also impact the bottom line by cutting the costs of finding a replacement worker and reducing workers' compensation premiums. An intangible benefit of onsite services is the therapist's ability to effectively communicate the workers' situations to administration. This independent opinion can alleviate tensions which often develop between injured workers and their supervisors. It is these tensions that can lead to exaggerated symptoms and litigation.

The services that can be provided to industry and the worker through this approach are limited only by the willingness of the parties involved. Most services offered in a medical clinic can be offered at the plant site. The number of employees and the incident rate usually dictate the choice of services brought in-house.

The entire Worker Care Spectrum as represented in Figure 31.1 can be provided along with drug, hearing and vision screening, medical physicals, ergonomics consulting, fitness and wellness programs, inoculations, and traditional general medical treatment. The company also experiences economic benefits through the immediacy of evaluation and treatment of resultant earlier return-to-work, less time off work, and maintaining the bond between the worker and the worksite. Diversification can take many forms. These are but a few examples which are being offered to industry in today's market and demonstrate insight of how challenges will be met in the future.

## 31.7 Conclusion

Financial reward for losses as result of work activities is evidenced as far back as Greek warriors and pirates. More that 2000 years ago, the families of warriors who lost their lives in battle were compensated. Pirates were also provided a scheduled award for loss of a limb, but only after a successful attack and after the captain had taken his share of the recovered booty. Today's workers' compensation system is much more complicated. In 1986, Liberty Mutual reported $6807 as the mean cost per case for *low back pain*. In 1989, this cost had risen to $8321. This 1989 amount is more than twice the amount for the average workers' compensation claim which was $4075 (Webster and Snook, 1990). Liberty estimates the cost of time lost by injured workers to be $18.7 billion. Liberty also reports the cost of disability for corporations to be 6 to 12 percent of their corporate payroll. In 1993, 8.5 million people had disabling injuries (Liberty Mutual, 1998). It has been shown that the majority of costs incurred when an employee is injured comes from nonmedical costs (Quebec Task Force, 1987). A 1987 study revealed that only 14% of the total cost incurred was direct medical. The remaining 86% was attributed to indirect costs including the costs surrounding the replacement of an employee. Considering that industry directly pays for lost time, it is incumbent that industry know what an individual's capability is upon hiring and upon return-to-work from an injury.

Armed with accurate data, the human resource department can be sure that potential employees who do not have the capacity for the work will not be hired. Also, when an injured worker returns to work, it will be as soon as they are able to do so without risk of reinjury.

The employer needs increasingly more objective determination of job requirements and of worker capabilities to defensibly support their decisions of hire and of return-to-work once injured. By focusing on what the worker *can* do, job placement assessments and functional capacity assessments can trigger a more accurate and proactive response from the employer. This includes proper placement upon hiring and easier placement and accommodation, if necessary, once injured (Key, 1995a).

# References

Aitken MJ, Creighton University School of Pharmacy and Allied Health, Department of Occupational Therapy, Omaha, Nebraska June, 1996.

Astrand PO, Ryhming I, A nomogram for calculation of aerobic capacity from pulse rate during submaximal work, *J of Applied Physiology* 7:218-221, 1954.

Bigos SJ, Battié MC, Spengler DM, et al. A prospective study of work perceptions and psychosocial factors affecting the report of back injury. *Spine* 16:1-6, 1991.

Blair JA, Blair RS, Rueckert P, Pre-injury emotional trauma and chronic back pain — an unexpected finding. *Spine,* 19: 10, May 15, 1994.

Block AR, Vanharanta H, et al. Discogenic pain report, influence of psychological factors, *Spine* 21(3):334-338, February 1996.

Dobrzykowski E, Data collection and use in industrial therapy, in *Industrial Therapy,* Key GL (Ed.) St. Louis, Mosby, 1995, p 42-60.

Frey DH, Job placement assessments and pre-employment screening, in *Industrial Therapy,* Key, GL (Ed.) St. Louis, Mosby, 1995, pp 110-122.

Gilliand RG, Sevy BA, Ahlgren A, *A Study of Statistical Relationships Among Physical Ability Measures of Injured Workers Undergoing KEY Functional Assessments,* Minneapolis, Minnesota 1986.

Grossman P, Respiration, stress, and cardiovascular function, *Psychophysiology* 20(33):284-300, 1983.

Karwowski E, Salvendy G, Ed. *Ergonomics and Manufacturing: Raising Productivity Through Work Place Improvement,* Society of Manufacturing Engineers, 1998.

Keller LS, Butcher JN, *Assessment of Chronic Pain Patients with the MMPI-2.* Minneapolis, MN: University of Minnesota Press, 1991.

Key GL, Functional capacity assessment, in *Industrial Therapy,* Key GL (Ed.) St. Louis, Mosby, 1995a.

Key GL, The impact and outcomes of industrial therapy, in *Industrial Therapy,* Key GL (Ed.) St. Louis, Mosby, 1995b, p 220-254.

Key GL, *Key Functional Assessment Policy and Procedures Manual and Training and Resource Manual,* Minneapolis, MN, 1984.

Liberty Mutual, For Your Employees: Workers' Compensation, available at: www.libertymutual.com/business/workcomp.html, accessed 10-6-98.

Osterweis M, Kleinman A, Mechanic D, *Pain and Disability, Clinical, Behavioral, and Public Policy Perspectives,* Washington, DC, National Academy Press, 1987.

Personnel Decisions Inc. *Key Functional Assessment Pre-employment Screening Battery as a Predictor of Job Related Injuries,* Minneapolis, MN, January, 1994.

Portney LG, Watkins MP, *Foundations of Clinical Research,* Norwalk, Conn., Appleton & Lange, 1993.

Quebec Task Force Study: Scientific approach to the assessment and management of activity-related spine disorders, *Spine,* 12:7S, 1987.

Schmidt AJM, Gierlings EH, Madelon LP, Environmental and interoceptive influences on chronic low back pain behavior. *Pain* 38:137-43, 1989.

Waite HD, *Use of a New Physical Capacities Assessment Method to Assist in Vocational Rehabilitation of Injured Workers,* University of Colorado, Thesis, 1987.

Webster BS, Snook, SH, The cost of compensable low back pain, *J Occup Med,* 32:13-15, 1990.

Wheeler AH, Hanley EN, Spine update: nonoperative treatment for low back pain — rest to restoration, *Spine,* 20(3): 375-378, February 1, 1995.

*Worker Data Bank,* KEY Method, Minneapolis, Minnesota. 1994, 1995, 1996, 1997.

# 32

# Ergonomics and Rehabilitation

Ewa Nowak
*Institute of Industrial Design*

## 32.1 Introduction

Based on a criterion of body "efficiency," the human population can be divided into three main groups:

- People of the highest (extreme) efficiency
- People of an average efficiency (within normal range)
- People of efficiency below normal range — these are the *disabled*

Efficiency undergoes changes during ontogenesis. It is determined genetically, but can also change (within the limits of so-called reaction standard) depending on the environmental, social, material, cultural, and other conditions. Body efficiency can be improved through physical exercises, better nutrition, and a more hygienic way of life. In the case of the disabled, *rehabilitation* is in charge of this question.

*Disability*, according to International Classification posed by WHO (1980) is defined as follows: "A restriction or lack of ability (resulting from impairment) to perform an activity in the manner or within the range considered normal for a human being."

Rehabilitation (after Kumar, 1989) is defined as "a science of systematic multidimensional study of disordered human neuropsychosocial and/or musculoskeletal function(s) and its/their remediation by physicochemical and/or psychosocial means." Rehabilitation can be treated as a process that, by making use of the latest medical, technical, and social sciences achievements, restores, within the optimum limits, the efficiency of *impaired* or *handicapped* organs. The aim, and at the same time the result, of rehabilitation is enabling the man to come back to normal life in society. Two steps can be distinguished in this process. At the same time they constitute two basic aims accepted by the World Health Organization in the World Program of Action (1984) (Figure 32.1).

The first step is rehabilitation. As a result of rehabilitation, the disabled person regains the optimum level of mental, physical, and social functioning. Of course, the process, in addition to being enabling, also strives to reduce pain and suffering. The second step is equalization of opportunities for disabled people. This involves a process through which physical environment, housing and transportation, social and health services, educational and work opportunities, cultural and social life, including sports and recreational facilities, are made accessible to all (Kumar, 1992).

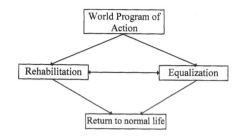

**FIGURE 32.1**    Activities of the World Program of Action.

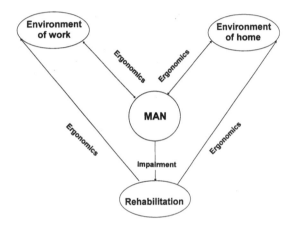

**FIGURE 32.2**    Correlation of ergonomics and rehabilitation in home and work environments.

Particularly helpful in this second step is ergonomics. While rehabilitation improves the efficiency of the body, ergonomics, through adjusting and optimizing external environmental factors to the capabilities of the person, enhances work comfort and increases its efficiency and effectiveness. Investigations of Canadian scientists (Kumar, 1992) as well as my own observations confirm the fact that rehabilitation and ergonomics applied together considerably intensify and advance both the progress of the rehabilitation process and the return to normal life.

Exercises, as well as medical treatment, supported by properly constructed workstands and objects of everyday use, intensify significantly the effect of rehabilitation. Under the influence of cumulated factors an impaired organ or function attain their initial capabilities to a higher degree. Ergonomics is of extreme importance in attaining "equal chances." This is of assistance in the process of integrating the disabled into society. It helps to achieve the goal of the World Health Organization, i.e., full participation of the disabled in social life and development. Figure 32.2 illustrates interrelation between ergonomics and rehabilitation in work and home environments.

It seems that ergonomics and its possibilities are not sufficiently used for the needs of the disabled. Several methods and measuring techniques applied in ergonomics can be used, e.g., in rehabilitation. Some of these should be completed, partially changed, and adapted. It is necessary to develop closer cooperation between designers and specialists from the fields of ergonomics and rehabilitation.

Even at the present moment there appears a new direction of ergonomics aimed at the needs of disabled people. This is *rehabilitation ergonomics* (Kumar, 1992; Nowak, 1993). This can be roughly defined as an interdisciplinary field of science that aims at adjusting tools, machines, equipment, and technologies as well as material work and life environments including objects of daily use and rehabilitation equipment to the psychophysical needs of the disabled. Rehabilitation ergonomics takes part both in the rehabilitation process and in equalizing chances. This chapter presents ergonomic possibilities and their achievements in these spheres.

## 32.2   Ergonomics and Equalization of Opportunities

### Somatic Characteristics of the Disabled as a Determinant of Spatial Structures

Defining the possibilities and necessities of the disabled is a necessary condition for designing the material work and life environment for this population. Data that characterize somatic structure constitute basic information. The influence of disabilities on shaping the body structure was studied by numerous anthropologists (Floyd et al., 1966; Pheasant, 1986; Boussena and Davis, 1987; Goswami et al., 1987, 1994; Laubach et al., 1981; Molenbroek, 1987; Nowak, 1988, 1989, 1994; Samsonowska-Kreczmer, 1988; Jarosz, 1996; Mięcsowicz, 1990; Łuczak et al., 1993; Das and Kozey, 1994; Lebiedowska et al., 1994).

The largest disproportions between the healthy and the disabled population can be found in a group of people with motor dysfunction. This is understandable since the dysfunction results from the past or currently developing diseases that lead to joint disturbances of the osseous, ligament and joint, muscular, and nervous systems. These disturbances lead to deformities and somatic changes of particular parts of the body, and this affects the final shape and dimensions of the body and its motorics.

Other factors that restrain the development and growth of the body including restriction of motion activities, neglected nursing, improper or lack of rehabilitation, as well as stresses connected with pain, frequent stays in hospitals and rehabilitation centers etc., accompany the pathological process. Descriptions of investigations of people with lower extremity dysfunction are usually found in the literature, and they usually are wheelchair users. It should be realized, however, that this group embraces people with various degrees of motor efficiency limitations. This depends not only on a type and stage of a disease but also on the time of its appearance. Therefore, researchers dealing with this problem face great difficulties in selecting subjects and in describing results scientifically. This may be the cause of the small number of studies undertaken in this field. This particularly concerns studies, the results of which are to provide data for designing.

It turns out, that height measurements are measured in relation to various reference bases. This fact creates a difficulty not only for comparison purposes, but also creates a difficulty for designers who would like to utilize investigation results. A synthesis of existing data was done for design purposes.

Table 32.1 comprises height measurements measured in relation to the seat plane (Bs), and Table 32.2 to the floor level (B). *Anthropometric* data differ significantly regardless of the fact that these concern various populations. The influence of diseases resulting in the necessity of using the wheelchair affects shaping the body. Pheasant (1986) indicates that the body proportions of wheelchair users resemble those of elderly people above age 65. This was confirmed by the results of investigations carried out by Molenbroek (1987).

Not only anthropometric measurements of the disabled are important for the needs of design but also differences between the disabled and healthy population. The majority of authors indicate that the body structure of disabled men and women differs significantly from the able-bodied population (Pheasant, 1986; Samsonowska-Kreczmer, 1988; Nowak, 1988, 1989; Jarosz, 1996; Das and Kozey 1994). This problem was illustrated by the example of data on the Polish population (Jarosz, 1996). Using the threshold values of the 5th percentile, a linear anthropometric model of disabled women in the sitting position was drawn. This model was compared to the model of the 5th percentile of the able-bodied (Figure 32.3).

It appears that the above measurements show smaller values for the disabled than for healthy people. The differences are significant and amount to 110 mm for the seated stature (Bs-V), for eye level (Bs-en) to 113 mm, for shoulder height (Bs-a) to 126 mm, and for elbow height (Bs-r) to 57 mm. For arm reach measurements the differences amount for arm reach forward (Bs-phIII) to 204 mm, and to 90 mm for arm overhead reach (Bsd-phIII). Similar results were obtained by Nowak (1988, 1989) by comparing the population of disabled young people aged 15 to 18 with the lower extremity dysfunction, with young of the same age representing the Polish population (Nowak, 1988).

Floyd (1966) and Bouisset and Moynot (1985) indicate that the smaller seated stature (Bs-v) measurements of the disabled can result from the deformities of the osseous system as well as from the fact that,

**TABLE 32.1**   Structural Anthropometric Data for Males and Females with Respect to the Seat

| | | Author | | | | | | | | | | | |
|---|---|---|---|---|---|---|---|---|---|---|---|---|---|
| | | Boussena and Davis | | Das and Kozey | | Goswami et al. | | Jarosz | | Molenbroek | | Nowak | |
| Dimension per/mm/ | | Men | Women | Men | Women | Men | | Men | Women | Men | Women | Men | Women |
| Seated Stature | 5 | 824 | 794 | 734 | 647 | — | | 769 | 668 | 761 | 702 | 744 | 708 |
| | 95 | 962 | 912 | 963 | 857 | | | 960 | 894 | 919 | 858 | 972 | 890 |
| Eye Height | 5 | — | — | 496 | 546 | — | | 667 | 570 | 643 | 585 | 630 | 592 |
| | 95 | | | 717 | 744 | | | 857 | 789 | 810 | 763 | 857 | 783 |
| Shoulder | 5 | — | — | 468 | 423 | 330 | | 495 | 433 | 520 | 479 | 474 | 461 |
| Height | 95 | | | 676 | 597 | 564 | | 682 | 619 | 649 | 601 | 647 | 592 |
| Elbow Height | 5 | 177 | 176 | 108 | 105 | 136 | | 144 | 133 | 168 | 156 | 158 | 139 |
| | 95 | 269 | 266 | 312 | 257 | 212 | | 297 | 281 | 289 | 270 | 289 | 309 |
| Knee Height | 5 | 483 | 450 | — | — | — | | 468 | 407 | — | — | 453 | 442 |
| | 95 | 586 | 539 | | | | | 605 | 530 | | | 572 | 532 |
| Popliteal | 5 | 381 | 364 | — | — | 343 | | 383 | 315 | 401 | 361 | 386 | 371 |
| Height | 95 | 473 | 453 | | | 465 | | 513 | 454 | 503 | 460 | 502 | 462 |
| Trunk Depth | 5 | — | — | 198 | 143 | — | | 180 | 191 | 211 | 219 | 165 | 182 |
| | 95 | | | 281 | 182 | | | 340 | 315 | 344 | 368 | 270 | 286 |
| Popliteal Depth | 5 | 421 | 418 | — | — | 356 | | 461 | 418 | 401 | 405 | 435 | 424 |
| | 95 | 522 | 516 | | | 447 | | 636 | 571 | 525 | 524 | 555 | 545 |
| Shoulder | 5 | 383 | 368 | 354 | 291 | — | | 353 | 310 | — | — | 337 | 316 |
| Breadth | 95 | 482 | 434 | 439 | 355 | | | 425 | 394 | | | 439 | 410 |
| Overhead | 5 | — | — | 1072 | 947 | — | | 1028 | 882 | 828 | 733 | 1022 | 963 |
| Reach | 95 | | | 1415 | 1090 | | | 1324 | 1192 | 1214 | 1113 | 1320 | 1195 |
| Reach Forward | 5 | 568* | 552ˣ | — | — | — | | 653 | 558 | — | — | 668 | 617 |
| | 95 | 677* | 630ˣ | | | | | 840 | 713 | | | 861 | 768 |

* measured from the acromiale point
ˣ measured from the acromiale point

**TABLE 32.2**   Reported Anthropometric Data of Disabled People with Respect to the Floor

| | | Floyd et al | | Jarosz | |
|---|---|---|---|---|---|
| Dimensions per/mm/ | | Men | Women | Men | Women |
| Floor to vertex | 5 | 1260 | 1180 | 1299 | 1198 |
| | 95 | 1410 | 1355 | 1490 | 1424 |
| Floor to eye | 5 | 1150 | 1080 | 1197 | 1100 |
| | 95 | 1290 | 1235 | 1387 | 1319 |
| Floor to shoulder | 5 | 965 | 910 | 1025 | 963 |
| | 95 | 1100 | 1065 | 1212 | 1149 |
| Floor to elbow | 5 | 625 | 610 | 674 | 663 |
| | 95 | 745 | 730 | 827 | 811 |
| Floor to top of height | 5 | 620 | 565 | 607 | 598 |
| | 95 | 680 | 635 | 672 | 667 |
| Floor to top of foot | 5 | 120 | 165 | — | — |
| | 95 | 180 | 215 | | |
| Floor to vertical grip reach | 5 | 1550 | 1460 | 1558 | 1412 |
| | 95 | 1785 | 1680 | 1854 | 1722 |

as a result of back muscle paralysis, difficulties with maintaining the straight position of the body appear. This problem seems to be very serious for paraplegics. A similar interpretation can be accepted for reach measurements. According to Pheasant (1986) the same analogies of changes in body proportions occur in wheelchair users and in elderly people. In old age, similarly to people with motor organ dysfunction, deficient muscular tonicity of the chest and belly appears. This results in pectoral kyphosis increasing.

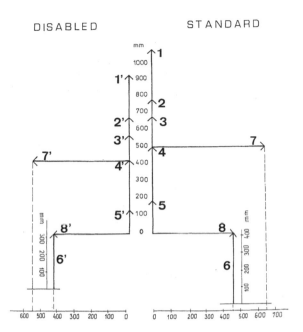

**FIGURE 32.3** Anthropometric model 5th percentile of disabled and standard women (in the sitting position), where 1,1′ — Arm overhead reach; 2,2′ — Stature; 3,3′ — Eye height; 4,4′ — Shoulder height; 5,5′ — Elbow height; 6,6′ — Popliteal height; 7,7′ — Arm reach forward; 8,8′ — Popliteal depth. (From Dr. Emilia Jarosz. With permission.)

At the same time, the processes of intervertebral cartilage flattening and of the back shortening occur. This leads to the C shape of the spine (Nowak, 1980, 1988). Samsonowska-Kreczmer (1988) indicated that the stooping back and sunken chest occur in young people with motor organ dysfunction. Another reason for diminishing the seated stature measurements is that the buttock and thigh muscles atrophy, resulting from immobility of the back and the lower extremities. Studies conducted by Jarosz (1993) indicate that for 55% men and 65% women thigh thickness measurement is below the lower limit of the healthy population standard. Similar results concerning adults were obtained by Goswami et al. (1987), and concerning the young by Nowak (1988, 1989) and Miesowicz (1990). In comparison with the stature and reach characteristics, shoulder breadth characteristics (a-a) appear different. Most scientists confirm the fact that the value of this characteristic is higher for wheelchair users than for the healthy population (Goswami et al., 1987; Boussena and Davies, 1987; Nowak, 1989; Jarosz, 1996). Wężyk (1989) and Miesowicz (1990) point out that children with cerebral palsy have larger shoulder breadth values in comparison to those of healthy children. Without exploring the reason for this phenomenon, this is, without doubt, important information for clothes designers. This question will be discussed in a later section.

Essential characteristics exerting an influence on workspace shaping are functional characteristics of the upper extremities, i.e., reaches. Values observed in these characteristics (Tables 32.1 and 32.2) are significantly lower in persons with lower extremity dysfunction, although their upper extremities are qualified as "efficient." Lower reach values result not only from lower values of the arm and forearm length, but also from limitations in shoulder and elbow joints. In connection with the above, the disabled have difficulty in performing the movements of abduction and extension (Nowak, 1988). Grabowska et al. (1986) found that in the case of persons suffering from rheumatoid arthritis, workspace of the upper extremity is 7 to 10% lower if a shoulder joint is constrained and 25 to 33% lower when there is a constraint of the elbow joint movements. It is obvious that disorders of these two joints increase the limitation of the upper extremity movements, and thus the efficient workspace is significantly reduced. This is confirmed by investigations conducted by Nowak (1988, 1989) and Jarosz (1996). It should be pointed out that particular values of reach of the disabled refer to the straight position of the body. Thus, they can be increased through trunk movements forward and lateral. Floyd's investigations (1966) proved

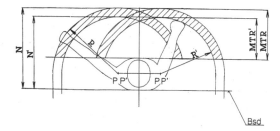

**FIGURE 32.4** Maximum transverse reach for the all-Polish population (MTR) and the disabled population (MTR′), where R,R′ — reach radius; P,P′ — hypothetical axis of rotation; N,N′ — reach forward.

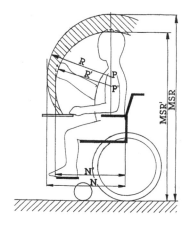

**FIGURE 32.5** Maximum sagittal reach for the all-Polish population (MSR) and the disabled population (MSR′), where R,R′ — reach radius; P,P′ — hypothetical axis of rotation; N,N′ — reach forward.

that the difference between reach forward of disabled women in the straight position of the body and the same reach in the position of maximum bend forward amounted to 142 mm, and in case of the lateral reach with the trunk lateral bend — 135 mm. Reaches forward of the disabled are limited by the foot-rest of the wheelchair. The difference between the point of reach of disabled women investigated and the beginning of the wheelchair amounted to 242 mm. That is, it exceeded the difference between reaches in the straight position and reaches in the bend forward stated by Floyd. That means that the wheelchair user is not able to reach any devices in the frontal position if there is not enough space for the wheelchair under the working plane.

Nowak (1989), based on the Das and Grady's method (1983), developed a simple method of defining workspace for arms. This space was determined for disabled young people with dysfunction of the lower extremities, using the wheelchair. Figures 32.4 and 32.5 show the graphic way to determine this space in the sagittal plane (maximum sagittal reach — MSR) and in the transverse plane (maximum transverse reach — MTR). These figures show the difference for both reaches between disabled and healthy young people. Differences in maximum reach measurements were significant and amounted for the 5th percentile to 300 mm. Results of the study were used for ergonomic analysis of schools and for designing school workshops, laboratories, and rehabilitation centers. This method can be recommended for workspace design for the disabled. It is simple and easy to use. Using only five anthropometric characteristics, one can obtain the graphic representation of workspace of any population or an individual person.

While designing the functional space for the paraplegic, one should bear in mind that they have difficulty with the stability of position, and they can easily fall out of the wheelchair if they bend forward too much (Przybylski, 1979). Upper extremities moving forward disturb the balance of the body, which is more difficult to redress because of the paralysis of the muscles of the spine and pelvis, stabilizing the body in the sitting position (Bouisset and Moynot, 1985).

Molenbroek's investigations (1987) provide data determining the values of reach of the elderly. Limitation of the movement range of particular joints and decrease in the values of reach increase as age increases. Molenbroek (1987) proved that differences between the values of the arm reach overhead for 50-year-old and 95-year-old Dutchmen amount (for the value of the 5th percentile) to approximately 300 mm. These are the directions for designing living interiors meant for the elderly. It is important for determining the ability to reach the top shelf of a wardrobe, the highest buttons in a lift, or a window handle. These data prove that while designing for people aged 50 to 110, one should place all kinds of switches at the height not exceeding 1474 mm. Above this height, the elderly have difficulty operating them.

According to Pheasant (1986), a wheelchair user (whose upper extremities are unimpaired) can reach a zone from around 600 to 1500 mm in a sideways approach, but considerably less "head on." It may well be that the location of fittings within this limited zone will prove entirely acceptable for the ambulant users of the building, but in case of working surface heights no such easy compromise is possible.

Another task is determining appropriate steps and handrails adjusted to the dimensions of the disabled, who can move without aid. Petzal (1993) recommends installing a handrail at the height of 900 mm above the floor. Based on the investigations of different step variants, he suggests the height of a step within 150 to 200 mm and the width from 250 to 300 mm.

The above author, on the basis of anthropometric data, research tests, and observations, determined placement of handrails in a bus adjusted to the needs of ambulant disabled people, as well as the size of seats and the distance from one to another.

As the review of investigations has proved, the body structure of disabled men and women differs considerably from that of the healthy population. Ergonomics, providing data concerning the body structure of disabled people, makes it possible to adjust designs of products and spatial structures to the abilities and predispositions of this group of users.

## Hand Characteristics as the Basis for Designing Objects of Everyday Use

Statistics of many countries report that a significant proportion of the population with disabilities requires help from family, friends, volunteers, or paid care-givers. Almost one half (45.6%) of all disabled people require assistance with heavy household work, and 22.4% with daily housework (Kumar, 1995). In terms of the cost, though the figures were not given, it is surmised that it is significant. Many able-bodied people have to spend their time to assist. The cost of this time along with the cost of the paid health care worker can add up to a significant magnitude. It may be useful to point out that the larger cost for these people is not medical but maintenance, attendant care, nursing home, and home care expenses on top of the loss of productivity due to inability to work. Environment modification and assistive devices may offer disabled people a chance to decrease these costs and be productive members of society.

Therefore, the problem of making daily life more efficient for the disabled is of extreme importance from the point of view of economics. The possibility of designing objects of every day use with the help of ergonomics can play a significant part in this field. One of the basic aims of ergonomics is to define the needs and capabilities of the user. These capabilities, in comparison with the able bodied, are usually limited and result from dysfunction of a particular organ. The dysfunction of performing manual functions is felt most acute during everyday activities. The hand is a unique organ, adjusted to performing manipulation activities. Efficiency of performing these activities depends on the grasping capabilities of the hand. The grasping function is one of the most important functions allowing a person to perform daily activities. Dysfunction in grasping makes it impossible for a person to perform the majority of activities, including such fundamental ones as eating and drinking. Figure 32.6 shows the types of hand grips used (Nowak, 1993). The possibility of performing particular grips depends on the type and degree of hand dysfunction. Thus, it is important to know handgrip capabilities of the disabled user, and these capabilities should be taken into consideration by a designer while designing an object. Stabilization of an object in the disabled hand should be secured by the grip, that at the same time allows an object to be manipulated. Normally this is attained by adjusting the shape and size of an object or its handle to

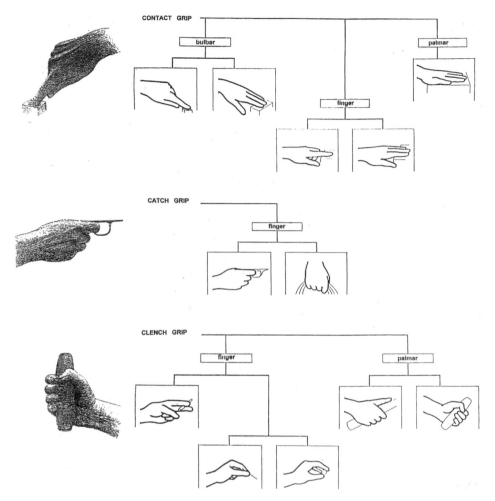

**FIGURE 32.6** Hand grips classification. (From Nowak, E. 1993. Hand grip classification, in the catalogue *Human Dimension — Ergonomics and Design*, Institute of Industrial Design, Warsaw: 30 [in Polish]. With permission.)

the capabilities of the deformed hand. Investigations that define the contact area of the hand and the object or its handle, depending on its shape and size, are quite popular in this field. Figure 32.7 (Frejlich, 1993) shows pictures of the contact area of the hand and the handle with the shape of a circle cross section that depends on diameter measurements of the handle. The larger cross section, the larger the contact area. Frejlich proved (1983) that handles with square, rectangular, and triangle cross sections attain smaller contact areas in comparison to handles with circle cross sections. This is important information for designers, since the shape and size of handles are of great significance for an object to be held, stabilized, and manipulated. This was confirmed by Benktzon's investigations (1993). The most important for her was to define the appropriate contact between the hand and a handle. Benktzon, with the cooperation of numerous specialists from the Ergonomi Design Gruppen (Benktzon, 1993), analyzed objects of everyday use and offered new solutions for these objects. These can be used both by the elderly and the disabled. Several of these objects were presented by the Ergonomi Design Gruppen at the exhibition "Design and Ergonomics," organized in Warsaw in 1993. Figure 32.8 presents a new solution for a walking stick. The design was developed basing on a keen analysis of the manner of gripping by the disabled with hand dysfunction. The contact area between the hand and the new anatomical handle has increased considerably, thus providing much better support. In practice, seemingly small differences can be of vital importance for function, safety, and comfort.

The new design (Fenix) is characterized by:

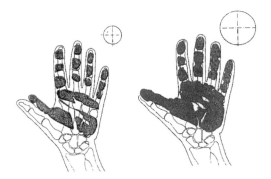

**FIGURE 32.7** Pictures of contact area of the hand with the handle of a circular cross section. (From Frejlich, C. 1993. Analysis of the active contact surface of the hand with a handle depending on its cross-section shape and size, in the catalogue *Human Dimension — Ergonomics and Design,* Institute of Industrial Design, Warsaw: 100 [in Polish]. With permission.)

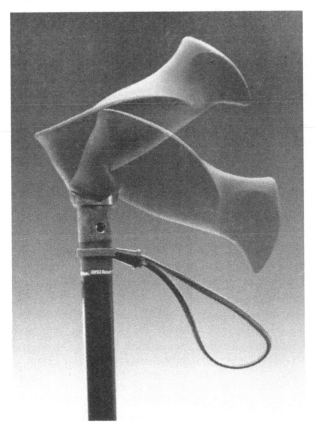

**FIGURE 32.8** The walking stick handles; made by RFSU, Sweden, designed by Ergonomi Design Gruppen. (From *Catalogue of the exhibition "Human Dimension — Design and Ergonomics."* 1993. Institute of Industrial Design, Warsaw. With permission.)

- Larger load-bearing areas for the hand, without the usual consequence of the thumb being pushed out of place
- The stick being positioned at the back of the handle which allows the fingers to be flexed together
- Good friction and soft grip surface to enhance comfort
- The angle of the handle being adjustable to each individual by means of a ball-and-socket joint

**FIGURE 32.9**   Cutlery for the disabled using one hand; made by RFSU, Sweden, designed by Ergonomi Design Gruppen. (From *Catalogue of the exhibition "Human Dimension — Design and Ergonomics."* 1993. Institute of Industrial Design, Warsaw. With permission.)

Figure 32.9 presents a design developed by the same group. This is a set of cutlery for people using only one hand. The fork has small prongs, and the spoon has a serrated edge that allows cutting of food. There is also a range of cutlery designed to accommodate many "special needs," along with a goblet and a curved rimmed plate. All these designs demonstrate the goal of integrating style and "special needs" design; people should be not stigmatized because their needs can be classed as "special."

## Elements of Ergonomics in Clothing Design

Clothes for the disabled constitute a social problem of particular importance. A person suffering from disability or permanent illness should get such clothing that makes his or her life easier, is functional, meets his or her emotional needs, and, at the same time, is adjusted to the limitations caused by a disability. People with motion dysfunction have the most problems with clothes. Clothing designed for this group of people should assure appropriate physiohygienic properties, heat comfort, as well as ease of manipulation. Concepts of model-constructions solutions should take into account not only the usefulness of products, but also the economic effect. Designs must be simple in construction and easy to manufacture. The form of clothing is to integrate the group of the disabled with society, to conceal a disability, and to give the feeling of satisfaction at possessing clothes appropriate to needs and expectations.

Basing on ergonomic criteria, we can determine two main types of clothing:

- Clothing for totally inefficient people, who are attended by other persons
- Clothing for people able to self-service

In both these cases ergonomic investigation are helpful.

Determination of the somatic characteristic of the disabled is the starting point for clothing design. Figure 32.10 (a,b) shows the measurements that determine the necessary dimensions of a working suit. As was proved earlier, disability and the motion limitations connected with it result in changes in the shape, dimensions, and proportions of the body.

Investigations of 15- to 18-year-olds (Nowak, 1988, 1989) with motor dysfunction of the lower extremities proved that the body dimensions of this group differ considerably from those of the healthy population of the same age. Their body weight and height are lower. This was confirmed by other scientists' investigations (Malinowski, 1975; Pheasant, 1986; Samsonowska-Kreczmer, 1990; Jarosz,

**(a)**

**(b)**

**FIGURE 32.10**　Examples of anthropometric characteristics defining clothes dimensions. (a) At maximum backward trunk bending; (b) at maximum lateral trunk bending. (From Batogowska, A. 1993. Selected maximum measurements for the needs of clothing design. In the catalog *Human Dimension — Ergonomics and Design*, Institute of Industrial Design, Warsaw: 64 (in Polish). With permission.)

1993). On the other hand, it was proved that this group of people has considerably higher values of shoulder breadth characteristic. Walking with the help of crutches and walking sticks, or driving a wheelchair is performed thanks to the work of the upper extremities and the trunk muscles. It is also supposed that intensive physical training of these muscles during rehabilitation can stimulate the growth of the clavicles in length. This results in the considerable increase in the shoulder breadth characteristic values. In addition to the body dimensions, determination of motor abilities is of vital importance. Figure 32.11 shows the measurement of the upper extremity movement range (Nowak, 1978). Basing on ergonomic investigations, assumptions for clothing design are determined. For example, clothing for the wheelchair user must fulfill (among others) the following requirements: (1) appropriate clearance allowing the hands to move easily up and forward and appropriately larger shoulder breadth. This is ensured by the special construction of sleeve and armpit cut; (2) increased transverse dimensions related to the enlarged, muscular chest; (3) widened sleeve finish related to the considerably enlarged biceps and triceps; (4) adjustment of the bottom part of clothing length and cut (in case of a blouse, shirt, and vest) to the sitting position of its user. This is attained through the removal of excess fabric in front. Figures 32.12, 32.13, and 32.14 show example solutions of the upper and lower parts of clothing for a wheelchair user. The projects were designed in the clothing department of the Institute of Industrial Design in Warsaw (Dąbrowska-Kiełek; Szymańska-Petrykowska, 1991).

Interesting ergonomic investigations for the needs of clothing design for the disabled were conducted by Sperling and Karlsson (1989). They determined demands on clothing fasteners for long-term care patients:

**Functional Demands**

The clothing fastener should:

- Be located in the patient's optimum grip area (Figure 32.3)
- Be easy to understand and identify, visually as well as tactally
- Be possible to handle with one hand
- Be easy to grip and hold
- Not demand more grip strength or precision handling than the patient is able to produce
- Stand body movements without opening
- Have dimensions suitable for handling as well as clothing construction

**FIGURE 32.11**   Measurement of upper extremity movements range performed in working clothes.

**FIGURE 32.12**   Exemplary solution of the upper part of clothes for men using a wheelchair; designed by Jolanta Szymańska-Petrykowska, made by the Institute of Industrial Design. (From Dąbrowska-Kiełek, J., Szymańska-Petrykowska, J. 1991. Clothes and shoes. *Designer's Vademecum — Problems of the Disabled*, Vol. 5 (in Polish). With permission.)

## Comfort and Safety Demands

The closure should:

- Not scratch or rub
- Not cause such a pressure against the skin that might lead to pressure-sores

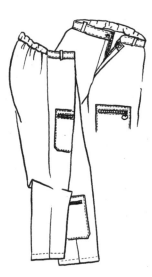

**FIGURE 32.13** Design of trousers for men and women using a wheelchair; designed by Jolanta Szymańska-Petrykowska, made by the Institute of Industrial Design. (From Dąbrowska-Kiełek, J., Szymańska-Petrykowska, J. 1991. Clothes and shoes. *Designer's Vademecum — Problems of the Disabled*, Vol. 5 (in Polish). With permission.)

**FIGURE 32.14** Design of a skirt for women using a wheelchair; designed by Jolanta Szymańska-Petrykowska, made by the Institute of Industrial Design. (From Dąbrowska-Kiełek, J., Szymańska-Petrykowska, J. 1991. Clothes and shoes. *Designer's Vademecum — Problems of the Disabled*, Vol. 5 (in Polish). With permission.)

## Sewing and Maintenance Demands

The closure should:

- Be easy to fasten
- Be lasting and easy to replace
- Stand washing (chemicals, high temperature)
- Be inexpensive

The aim of this study was to facilitate a more independent daily life for long-term care patients. For this group of people it is of great importance to be able to dress and undress without assistance. The design and position of clothing fasteners often lead to a restriction of the functional capacity of the patient. Three different types of fasteners were evaluated:

- Oval asymmetric button for a horizontal buttonhole (25 × 19 mm)
- Oval symmetric button for a vertical buttonhole (28 × 19 mm)
- A finger strap, which was designed to be handled with one hand and nonprehensile movements (36 × 14 mm)

In a subsequent study, an adapted oval button and a "finger strap" alternative to hook-and-eye were designed and evaluated, together with a standard button. The oval button in combination with a vertical buttonhole improved the function for most patients, and the front position for the fasteners was superior to a diagonal or lateral position on the chest. The finger strap was of advantage to patients with hemiplegia and joint complaints but was difficult for many of the subjects to understand, being a technical innovation.

Many investigators and designers of clothing for the elderly and the disabled (Benktzon, 1980; Dąbrowska-Kiełek; Szymańska-Petrykowska, 1991; Nowak, 1996) underline the fact that in the process of clothing design they are governed by the idea that clothing, in addition to its functional characteristics, should give its user physical comfort. The disabled want to live and work among healthy people. Therefore, they have to accept the clothes they wear. Clothing should not deform the shape of the body, but just the opposite, it should cover anatomical defects.

## 32.3    Ergonomic Methods for the Needs of Rehabilitation

Rehabilitation is a complex and complicated process. It consists of restoring, to a maximum possible degree, motor efficiency of the human body. Rehabilitation employs many means and research techniques used in various fields of science, such as medicine, sociology, psychology, mechanics, and biology. It would seem that rehabilitation processes could also benefit from drawing more freely on achievements of anthropometry. An objective of this subsection is to present methods of dynamic anthropometry used in ergonomics, and to put forward proposals for suitable modification of those methods to satisfy the requirements of rehabilitation processes.

Anthropometry can be particularly useful in diagnosing and assessing the motor efficiency of the human body. Rheumatic diseases and mechanical injuries result in pathological changes in joints, ligaments, tendons, muscles, and lead to considerably restricted movement ranges. Thus, the assessment of motor efficiency of the affected joints is essential for monitoring the rehabilitation processes.

The simplest way of assessing motor efficiency is by comparing a restricted motion range of an affected joint with its initial range of motion, i.e., before injury or disease. Unfortunately, after injury or disease has already prevailed, it is impossible to determine what the initial motion range was in the healthy patient. Also, it may hardly be expected that a rehabilitated patient had his initial motion ranges measured just before the injury.

Thus, motor efficiency of a rehabilitated patient can only be assessed based upon the data of the healthy population. A method based upon this kind of data and allowing the quantitative assessment of rehabilitation progress was developed by Nowak (1992). The results of investigations conducted for the needs of ergonomics were used. The investigations comprised ranges of motion of arm, leg, hand, foot, and head. The following maximal movements were measured:

- Flexion and extension of the arm and leg
- Flexion and extension, adduction and abduction, supination, and pronation of the extended hand
- Flexion and extension, supination, and pronation of the grasping hand
- Flexion and extension, adduction, and abduction of the foot
- Flexion and extension, bending to the right and left, right and left turning of the head

Motion ranges for these movements were measured by means of a set of measuring devices. The investigations embraced 355 men and 215 women aged 18 to 65. The results of the investigations were used to calculate standards, i.e., biological bases of reference for particular motion ranges. Standards were developed for three age categories and adequate subordinate classes of movements.

These are:

| Age Groups | Motion Classes |
|---|---|
| 1. 18 up to 30 years | 1. Wide range of motion (W) |
| 2. 31 up to 40 years | 2. Average range of motion (A) |
| 3. 41 up to 65 years | 3. Small range of motion (S) |

The following formulae were used to calculate the ranges of motion in three classes (Batogowska, 1977):

$$W \quad — \text{Wide range of motion} \quad W = (\beta_1, \beta_2)$$
$$A \quad — \text{Average range of motion} \quad A = (\beta_2, \beta_3) \qquad (32.1)$$
$$S \quad — \text{Small range of motion} \quad S = (\beta_3, \beta_4)$$

Values of $\beta$ for i = 0, 1, 2, 3, 4 were found from equations:

$$\begin{aligned} \beta_1 &= P_{95} \\ \beta_2 &= x + 1/2\ s \\ \beta_3 &= x - 1/2\ s \\ \beta_4 &= P_5 \end{aligned} \qquad (32.2)$$

where
$P_5$ = value of the 5th percentile
$P_{95}$ = value of the 95th percentile
$x$ = average value
$s$ = standard deviation

Angular values of the hand grasping characteristic are shown in Table 32.3. It includes the values attained by a healthy population of adults aged 18 to 65. The values of movement ranges are grouped in three motion classes, i.e., from the maximum to the minimum value — classes W, A, and S. In each class, minimum and maximum extreme values are given, calculated according to formulae given earlier. Three age categories are subordinated to the classes of movement shown. The youngest persons, aged 18 to 30, belong to class W, characterized by the maximum values. Class A includes adults aged 31 to 40, and class S includes the eldest subjects, aged 41 to 65, whose movement range in particular joints has the smallest values.

**TABLE 32.3** Movement Ranges of the Grasping Hand for Three Classes of Motion: Wide (W), Average, (A), and Small (S) for Polish Population 18–65 Years

| | | W 18–30 years | | A 31–40 years | | S 41–65 years | |
|---|---|---|---|---|---|---|---|
| | | Men | Women | Men | Women | Men | Women |
| Flexion | min. | 96° | 99° | 75° | 75° | 50° | 48° |
| | max. | 119° | 126° | 95° | 98° | 74° | 74° |
| Extension | min. | 65° | 62° | 46° | 43° | 31° | 28° |
| | max. | 88° | 85° | 64° | 61° | 45° | 42° |
| Pronation | min. | 100° | 108° | 81° | 87° | 64° | 64° |
| | max. | 122° | 129° | 99° | 107° | 80° | 86° |
| Supination | min. | 94° | 100° | 69° | 82° | 59° | 65° |
| | max. | 112° | 118° | 93° | 99° | 68° | 81° |

The values of movement ranges for particular classes shown in the table can be practically applied in the rehabilitation process. If the age of a patient is known, one can read from the table the value of a range of movement in a given joint that can be attained by the joint, i.e., the range of movement that it should have had prior to the injury. The maximum value is particularly significant in our case. It shows the angular value that should be aimed at in the rehabilitation process. The maximum value of range movement provides a basis for the evaluation of rehabilitation progress in a selected age group and is expressed by $\phi$.

After assessing the value $\phi$, the value of the range of movement of a subject investigated before the rehabilitation process should be defined. This state is denoted by the symbol $\phi'$ and is indicated by measurement, using a suitable protractor.

The final step is to define the "decrease in the movement range." Knowing the values $\phi$ and $\phi'$ one can calculate the absolute decrease in the movement range (Ar), and a percentage decrease in the movement range (Pr) using the following formula:

$$Ar = (\phi - \phi') \tag{32.3}$$

$$Pr = \phi - \phi'/\phi \times 100 \tag{32.4}$$

The value of a percentage decrease in movement range (Pr) is inversely proportional to the ability of movement (Ab) which can be defined by the equation:

$$Ab = 100 - Pr\% \tag{32.5}$$

The above values Ar, Pr, and Ab provide a basis for a direct evaluation of the progress of the rehabilitation process. Based upon the above, the motion ability can be assessed in many various stages of rehabilitation. This assessment may be employed in:

- Diagnosing a disease
- Evaluating progress and monitoring the rehabilitation process
- Determining a permanent decrease in movement ranges

The assessment may be either continuous or administered at random, depending upon the requirements.

The application of the method described above is presented in a graph (Figure 32.15) which shows the rehabilitation process of a wrist joint. A considerable movement restriction was caused by injury of the hand in the wrist joint and surgery that followed the injury. Results of systematically taken measurements of hand flexion (a) and extension (b) shown in Figure 32.15, illustrate a process of gradual improvement. Percentage of angular decrease in the range of hand flexion was 14.6% and hand extension 30.8%. Thus, the post-injury efficiency of the wrist joint can be determined as 85.4% during flexion and 69.2% during extension.

Motion efficiency of individual movements can be assessed using different scales, depending upon requirements, disease and general mobility. The scale is also dependent upon the individual preferences of the assessor. Scales of three, five, or ten grades may be used in assessing the motion efficiency. A five-grade scale may be considered optimal after Zeyland-Malawka et al. (1968):

- Very good: decrease up to 20% — efficiency from 80%
- Good: decrease up to 40% — efficiency from 60%
- Poor: decrease up to 60% — efficiency from 40%
- Very poor: decrease more than 60% — efficiency less than 40%

The above proposal does not preclude applications of other assessment scales, which may be found convenient.

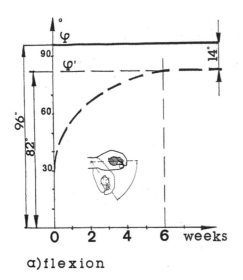
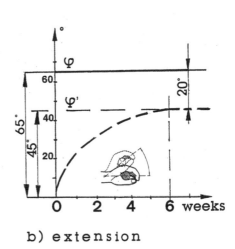

a) flexion                                    b) extension

FIGURE 32.15    An example of the wrist joint rehabilitation at the movements of hand flexion (a) and extension (b).

The measurement of force is, besides motion ranges, one of the parameters determining the efficiency of the person. Ergonomics studies the amount of force exerted by the hand or the foot. The results of investigations are used in designing hand- or foot-operated control devices. Some of the results can be used in rehabilitation for the assessment of the efficiency of an impaired organ or organ with dysfunction resulting from a disease. The assessment is based, as in the case of motion ranges, on the standards of force range, developed with the use of data concerning the healthy population. These standards make biological bases of reference to be strived for by a person increasing the efficiency of an impaired organ. Investigations conducted by many scientists confirm the fact that there is a correlation between the exerted force and circumferential measurements as well as body weight (Laubach, 1966; Hunsicker, 1970; Mali-nowski, 1975; Jarosz, 1993; Łuczak et al., 1993; Batogowska, 1993).

Persons with motor dysfunction, as it was proved in an earlier section, have lower values of somatic characteristics — the same concerns children and young people. Rehabilitation makes it possible to increase the efficiency of particular motor organs and activate respective groups of muscles. Measurement of force can be an excellent indicator of the rehabilitation progress. Following this line of thought, the measuring stand to test the hand efficiency was developed in the Institute of Industrial Design in Warsaw (Nowak, 1995). This efficiency is determined, among others, by measurement of force exerted by the hand grasping a cylindrical grip. Figure 32.16 shows the schematic diagram of the measuring stand. The principle of measurement is based upon the hydraulic system. The measuring system cooperates with a computer unit. The appropriate program makes it possible to calculate and visualize tested parameters. The subject can watch the effect of measurement on the monitor. This can increase motivation to achieve better results when the stand is used as a rehabilitation device, creating natural biofeedback. In the case of children, rehabilitation progress can be stimulated by games. There is a special computer program developed for this purpose.

The increase of motor efficiency is an important factor in the rehabilitation of children with cerebral palsy. Myszkowski and  Stężała (1989) put forward a proposal of an interesting testing stand. It consists of the set of control devices — manipulators using the handgrip function to perform repeated movements. The set of manipulators and the respective psychomotor computer tests allow the investigation and rehabilitation of children regardless of the degree of motor dysfunction of the hands.

The examples reported in this subsection prove that methods used in ergonomics are useful in the rehabilitation process. These methods, as well as the results of investigations, can be used for assessing an impaired or injured organ. Measuring stands can serve as rehabilitation devices.

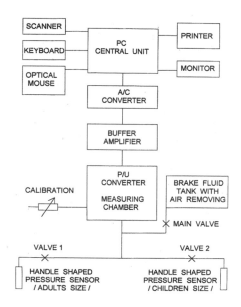

**FIGURE 32.16** Schematic diagram of measuring stand for estimation of hand capabilities.

## Defining Terms

**Anthropometry:** (anthropos — a man, metreo — I measure). A set of techniques used in anthropological investigations, it embraces methods of physical measurements of the man.

**Rehabilitation:** A science of systematic multidimensional study of disordered human neuropsychosocial and/or musculoskeletal function(s) and its (their) remediation by physicochemical and/or psychosocial means.

**Impairment:** Any loss or abnormality of psychological and anatomical structure or function (WHO, 1980).

**Disability:** A restriction or lack of ability (resulting from impairment) to perform an activity in the manner or within the range considered normal for a human being (WHO, 1980).

**Handicap:** A disadvantage for a given individual, resulting from an impairment or disability that limits or prevents the fulfillment of role that is normal (depending on age, sex, and social and cultural factors) for that individual.

**Rehabilitation ergonomics:** An interdisciplinary field of science aiming at adjusting material work and life environment, including articles of everyday use and rehabilitation equipment, to the psychophysical requirements of the disabled. Rehabilitation ergonomics assists the rehabilitation process, accelerating integration of the disabled into society.

## References

Batogowska, A. 1993. Selected maximum measurements for the needs of clothing design. In the catalog *Human Dimension — Ergonomics and Design*, Institute of Industrial Design, Warsaw: 64 (in Polish).

Benktzon, M. 1993. Designing for our future selves: the Swedish experience. *Applied Ergonomics*. 24(1): 19-27.

Bouisset, S., Moynot, C. 1985. Are paraplegics handicapped in the execution of a manual task? *Ergonomics*. 28(7): 299-308.

Boussena, M., Davies, B.T. 1987. Engineering anthropometry of employment rehabilitation centre clients. *Applied Ergonomics*. 18(3): 223-228.

*Catalogue of the exhibition "Human Dimension — Design and Ergonomics."* 1993. Institute of Industrial Design, Warsaw.

Das, B., Grady, M. 1983. Industrial workplace layout design. An application of engineering anthropometry. *Ergonomics*. 26(5): 433-447.

Das, B., Kozey, J. 1994. Structural anthropometry for wheelchair mobile adults. *12th Triennial Congress of IEA, Toronto. Vol. 3 (Rehabilitation Ergonomics)*: 63-65.

Dąbrowska-Kiełek, J., Szymańska-Petrykowska, J. 1991. Clothes and shoes. *Designer's Vademecum — Problems of the Disabled*, Vol. 5 (in Polish).

Floyd, W.F. 1966. A study of the space requirements of wheelchair users. *Paraplegia*. 1(4): 24-37.

Frejlich, C. 1993. Analysis of the active contact surface of the hand with a handle depending on its cross-section shape and size, in the catalogue *Human Dimension — Ergonomics and Design*, Institute of Industrial Design, Warsaw: 100 (in Polish).

Goldsmith, S. 1967. *Designing for the Disabled*. London: RIBA Publications Ltd.

Goswami, A., Ganguli, S., Chatterjee, B.B. 1987. Anthropometric characteristics of disabled and normal Indian men. *Ergonomics*. 30(5): 817-823.

Grabowska, Z., Salwa, J., Seyfried, A. 1968. Studies on the work zone limitation in patients with rheumatoid arthritis. *Informative Newsletter, Research Department of ZSI*, No 5: 7-11.

Holden, J.M., Fernie, G. 1989. Specification for a mass producible static lounge chair for the elderly. *Applied Ergonomics*. 20(1): 187-199.

Hunsicker, P., Greey, G. 1970. Studies in human strength. *Research Quarterly*, No 2: 109-122.

Jarosz, E. 1996. Determination of the workspace of wheelchair users. *International Journal of Industrial Ergonomics*, 17: 123-133.

Kumar, S. 1989. Rehabilitation and ergonomics: Complimentary disciplines. *Canadian Journal of Rehabilitation*, 3: 99-111.

Kumar, S. 1992. Rehabilitation: An ergonomic dimension. *International Journal of Industrial Ergonomics*. Elsevier. 9(2):97-108.

Kumar, S. 1995. Rehabilitation Ergonomics: Rationale, Means and Justification. *IEA World Conference, Rio de Janeiro*: 84-89.

Laubach, L.L. 1981. Anthropometry of aged male wheel-dependent patients. *Annals of Human Biology*, 8(1): 25-29.

Lebiedowska, M.K., Graff, K., Syczewska, M., Kalinowska, M., Polisiakiewicz, A., Lebiedowski, M.J. 1994. Mathematical models as a method of child growth process assessment. *12th Triennial Congress of IEA, Toronto, Vol. 3 (Rehabilitation Ergonomics)*: 52-54

Łuczak, E., Mięsowicz, I., Szczygieł, A. 1993. Somatic characteristics of children and the young with celebral palsy. AWF Cracov, *Scientific Annals*, Vol. XXVI: 121-142.

Malinowski, A. 1975. Differentiation of muscular force of the right and left hand in adult men and women according to the age. Poznań, Academy of Physical Training. *Monographs*, No 8.

Mięsowicz, I. 1990. Somatic development of handicapped children. *Annals of Special Pedagogics*, Vol. 1: 157-170.

Molenbroek, J.F.M. 1987. Anthropometry of elderly people in the Netherlands: research and applications. *Applied Ergonomics*, 18(3): 187-199.

Myszkowski, R., Stężała, D. 1989. Scientific system for measuring the level of psychomotoric efficiency of children with celebral palsy. *Technical Tasks of Medicine* XX(3): 166-176.

Nowak, E. 1976. Determination of the upper extremities workspace for the needs of workstands design. Institute of Industrial Design, *Works and Materials*, Vol. 30. Warsaw.

Nowak, E. 1978. Determination of the spatial reach area of the arms for workspace design purposes. *Ergonomics*, No 7: 493-507.

Nowak, E. 1988. Physical development of children and the young aged 4-18. Data for design purposes. Institute of Industrial Design, *Works and Materials*, Vol. 75. Warsaw.

Nowak, E. 1988. Method of workspace determination. *Institute of Industrial Design News-Design*, Vol.1: 3-8.

Nowak, E. 1989. Workspace for disabled people. *Ergonomics*, No 9: 1077-1088.

Nowak, E. 1992. Practical application of anthropometric research in rehabilitation. *International Journal of Industrial Ergonomics*, No 9: 109-115.

Nowak, E. 1993. Ergonomics and design for the needs of the disabled. *Materials of the Conference "Equalizing opportunities of the disabled."* Warsaw (in Polish).

Nowak, E. 1993. Hand grip classification, in the catalogue *Human Dimension — Ergonomics and Design*, Institute of Industrial Design, Warsaw: 30 (in Polish).

Nowak, E. 1994. Anthropometric measurements of the young for the needs of clothing design. *12th Triennial Congress of IEA, Toronto, Vol. 3 (Rehabilitation Ergonomics)*: 58-59.

Nowak, E. 1995. Workstand for measuring hand efficiency. *Materials of the Conference of PTF Physiotherapeutists of the Hand*, Poznań, September 21-23.

Nowak, E. 1996. The role of anthropometry in design of work and life environments of the disabled population. *International Journal of Industrial Ergonomics*, 17: 113-121.

Petzall, J. 1993. Ambulant disabled persons using buses: experiments with entrances and seats. *Applied Ergonomics*, No 24: 313-326.

Pheasant, S. 1986. *Bodyspace. Anthropometry, Ergonomics and Design.* Taylor & Francis. London and Philadelphia.

Przybylski, B. 1979. *Flats for the Disabled.* Publishing House of CZSBM, Warsaw.

Samsonowska-Kreczmer, M. 1988. Measurements of disabled young people. Data for clothing design. *Institute of Industrial Design News — Design*, No 4: 5-8.

Samsonowska-Kreczmer, M. 1990. *Anthropometric Measurements of Children with Cerebral Palsy.* Institute of Industrial Design, Warsaw.

Sperling, L., Karlsson, M. 1989. Clothing fasteners for long-term-care patients. Evaluation of standard closures and prototypes on test garments. *Applied Ergonomics*, 20(2): 97-104.

Wężyk, E. 1989. *Physical Development of Children and the Young with Cerebral Palsy.* Wrocław: Department of Anthropology of Wrocław University. (Master's thesis).

World Program of Action for the Benefit of the Disabled. *Problems of Occupational Rehabilitation* CNB CZSI, No 4.

Thoren, M. 1994. Clothing made to fit the disabled users. *12th Triennial Congress of IEA, Toronto, Vol. 3 (Rehabilitation Ergonomics)*: 187-189.

Zeyland-Malawka, E., Gładkowska, E., Henicz, T., Domańska, B. 1968. Method of investigation and assessment of motion range and usefulness of the hand. *Physical Culture*, No 6: 246-256.

# 33

# Update on the Use of Back Belts in Industry: More Data — Same Conclusion

Stuart M. McGill

*University of Waterloo*

The use of abdominal belts in industrial settings continues to be the topic of lively debate. The premier question still remains "Should abdominal belts be prescribed to workers in industry to perform manual materials handling tasks?" In a paper published a few years ago (McGill, 1993), I reviewed the available scientific literature pertaining to the use of back belts in industry with the objective of formulating a policy for belt prescription. The intent was not to take a position, either unconditionally for or against belt usage, but rather weigh the potential assets and liabilities for the development of belt prescription guidelines. The purpose of this chapter is to briefly review the literature previously reported, together with the most recent scientific data to see if our position on belt prescription has changed.

Abdominal belts and lumbar supports continue to be sold to industry in the absence of a regulatory requirement to conduct controlled clinical trials similar to that required of drugs and other medical devices. Many claims have been made as to how abdominal belts could reduce injury. For example, some have suggested that the belts remind people to lift properly. Some have suggested that belts may possibly support shear loading on the spine that results from the effect of gravity acting on the handheld load and mass of the upper body when the trunk is flexed. Compressive loading of the lumbar spine has been suggested to be reduced through the hydraulic action of increased intra-abdominal pressure associated with belt wearing. Belts have been suspected of acting as a splint, reducing the range of motion, and thereby decreasing the risk of injury. Still other hypotheses as to how belts may affect workers include (1) providing warmth to the lumbar region, (2) enhancing proprioception via pressure to increase the perception of stability, and (3) reducing muscular fatigue. These issues, together with others, will be addressed in this chapter.

A recent publication from NIOSH (1994) entitled *Workplace Use of Back Belts* contained critical reviews of a substantial number of scientific reports evaluating back belts and concluded that back belts do not prevent injuries among uninjured workers nor do they consider back belts to be personal protective equipment. While this is generally consistent with our position stated in 1993, my personal position for belt prescription is somewhat more moderate.

The following sections have subdivided the scientific studies into clinical trials and those that examined biomechanical, psychophysical, and physiological changes from belt wearing. Finally, based on the evidence, guidelines are recommended for the prescription and usage of belts in industry.

## 33.1 Field Trials

Many clinical trials that were reported in the literature were fraught with methodological problems and suffered from the absence of a matched control group, no post-trial follow-up, limited trial duration, and insufficient sample size. While the extreme difficulty in executing a clinical trial is acknowledged, only a few trials will be reviewed in this chapter.

The first trial reviewed here was reported by Walsh and Schwartz (1990), in which 81 male warehouse workers were divided into three groups: a control group (n = 27); a group that received a half-hour training session on lifting mechanics (n = 27); and a group that received the one-hour training session and wore low back orthoses while at work for the subsequent six months (n = 27). Instead of using more common types of abdominal belts, this research group used orthoses with hard plates that were heat molded to the low back region of each individual. Given the concern that belt wearing was hypothesized to cause the abdominals to weaken, the abdominal flexion strength of the workers was measured both before and after the clinical trial. The control group and the training-only group showed no changes in abdominal flexor strength nor any change in lost time from work. The third group, which received both training and wore the belts, showed no changes in abdominal flexor strength or accident rate, but did show a decrease in lost time. However, it appears that the increased benefit was only to those workers who had a previous low back injury.

In a larger clinical trial reported by Reddell (1992) and colleagues, 642 baggage handlers who worked for a major airline were divided into 4 treatment groups: a control group (n = 248); a group that received only a belt (n = 57); a group that received only a one-hour back education session (n = 122); and a group that received both a belt and a one-hour education session (n = 57). The trial lasted eight months and the belt used was a fabric weight lifting belt, 15 cm wide posteriorly and approximately 10 cm wide anteriorly. There were no significant differences between treatment groups for total lumbar injury incident rate, lost workdays, or workers' compensation rates. While the lack of compliance by a significant number of subjects in the experimental group was cause for consideration, those who began wearing belts but discontinued their use had a higher lost-day case injury incident rate. In fact, 58% of workers belonging to the belt-wearing groups discontinued wearing belts before the end of the eight-month trial. Further, there was an increase in the number and severity of lumbar injuries following the trial of belt wearing.

The clinical trial reported by Mitchell and colleagues (1994) was a retrospective study administered to 1,316 workers that performed lifting activities in the military. While this study relied on self-reported physical exposure and injury data over six years prior to the study, the authors did note that the costs of a back injury that occurred while wearing a belt were substantially higher than if injured otherwise.

A most recent study, reported by Kraus and colleagues (1996), surveilled the low back injury rates of nearly 36,000 employees of the Home Depot Stores in California from 1989 to 1994. As reported, the company implemented a mandatory back belt use policy. Although the authors claim that belt wearing reduced the incidence of low back injury, analysis of the data and methodology suggests a much more cautious interpretation may be warranted. The concern is based on two issues: the lack of a robust effect and co-interventions (in addition to belts). The data suggest that the beneficial effect was limited to men with 1 to 3 years of employment (but not for longer or shorter lengths of employment) and for women employed 1 to 2 years — this is a very narrow band of affected employees. However, of greatest concern is the lack of scientific control over co-interventions to ferret out the true belt wearing effect — there was no comparable non-belt wearing group, which is critical given that the belt wearing policy was not the sole intervention at Home Depot. For example, over the period of the study, the company increased the use of pallets and forklifts (changing physical demands), installed mats for cashiers, implemented post-accident drug testing, and enhanced worker training. In fact, the company made a conscious attempt to enhance safety in the corporate culture. This was a large study and the authors deserve credit for the massive data reduction and logistics. However, despite the title and claims that back belts reduce low back injury, this uncontrolled study cannot answer the question about the effectiveness of belts.

In summary, difficulties in executing a clinical trial are acknowledged: the placebo effect is a concern, as it is difficult to present a true double-blind paradigm to workers since those who receive belts certainly know so; and there are logistical constraints on duration, diversity in occupations, and sample size. However, the data reported in the better-executed clinical trials cannot support the notion of universal prescription of belts to all workers involved in manual handling of materials to reduce the risk of low back injury. There is weak evidence to suggest that those already injured may benefit from belts (or molded orthoses) with a reduced risk of injury recurrence. However, there does not appear to be support for uninjured workers wearing belts to reduce the risk of injury, and in fact, there appears to be an increased risk of injury during the period following a trial of belt wearing. Finally, there appears to be some evidence to suggest that cost per back injury may be higher if the worker was wearing a belt than if injured otherwise.

## 33.2  Biomechanical Studies

Biomechanical studies have examined changes in low back kinematics, posture, and issues of specific tissue loading. Two studies in particular have suggested that wearing an abdominal belt can increase the margin of safety during repetitive lifting: Lander et al. (1992) and Harman et al. (1989). Both of these papers reported ground reaction force and measured intra-abdominal pressure while subjects performed repeated lifting of barbells. Both reports observed an increase in intra-abdominal pressure when abdominal belts were worn. These researchers assumed that intra-abdominal pressure is a good indicator of spinal forces, which is highly contentious. Nonetheless, they assumed the higher recordings of intra-abdominal pressure indicated an increase in low back support which, in their view, justified wearing belts. Spinal loads were not directly measured or calculated in these studies.

Several studies have questioned the hypothesized link between elevated intra-abdominal pressure and reduction in low back load. For example, using an analytical model and data collected from three subjects lifting various magnitudes of loads, McGill and Norman (1987) noted that a build-up of intra-abdominal pressure required additional activation of the musculature in the abdominal wall, resulting in a net increase in low back compressive load and not a net reduction of load as had been previously thought. In addition, Nachemson et al. (1986) published some experimental results that directly measured intra-discal pressure during the performance of valsalva maneuvers documenting that an increase in intra-abdominal pressure increased, not decreased, the low back compressive load. Therefore, it would seem erroneous to conclude that an increase in intra-abdominal pressure due to belt wearing reduces compressive load on the spine. In fact, it may have no effect or may even increase the load on the spine.

In another study McGill, Norman, and Sharratt (1990) examined intra-abdominal pressure and myoelectric activity in the trunk musculature while six male subjects performed various types of lifts both wearing and not wearing an abdominal belt (a stretch belt with lumbar support stays, Velcro tabs for cinching, and suspenders for when subjects were not lifting). Wearing the belt increased intra-abdominal pressure by approximately 20%. Further, it was hypothesized that if belts were able to help support some of the low back extensor moment, one would expect to measure a reduction in extensor muscle activity. There was no change in activation levels of the low back extensors nor in any of the abdominal muscles (rectus abdominis or obliques).

In a recent study that examined the affect of belts on muscle function, Reyna and colleagues (1995) examined 22 subjects for isometric low back extensor strength and found belts provided no enhancement of function (although this study was only a four-day trial and did not examine the affects over a longer duration). Ciriello and Snook (1995) examined 13 men over a four-week period lifting 29 metric tonnes in four hours twice a week both with and without a belt. Median frequencies of the low back electromyographic signal (which is sensitive to local muscle fatigue) were not modified by the presence or absence of a back belt, strengthening the notion that belts do not significantly alleviate the loading of back extensor muscles. Once again this trial was not conducted over a very long period of time.

In 1986, Lantz and Schultz observed the kinematic range of gross body motions while subjects wore low back orthoses. While they studied corsets and braces rather than abdominal belts, they did report restrictions in the range of motion, although the restricted motion was minimal in the flexion plane. In a more recent study McGill et al. (1994) tested flexibility and stiffness of the lumbar torsos of 20 male and 15 female adult subjects, both while they wore and did not wear a 10 cm leather abdominal belt. The stiffness of the torso was significantly increased about the lateral bend and axial twist axes but not when subjects were rotated into full flexion. Thus, it would appear from these studies that abdominal belts assist to restrict the range of motion about the lateral bend and axial twist axes but do not have the same effect when the torso is forced in flexion, as in an industrial lifting situation. Posture of the lumbar spine is an important issue in injury prevention for several reasons, but in particular Adams and Hutton (1988) have shown that the compressive strength of the lumbar spine decreases when the end range of motion in flexion is approached. Therefore, if belts restrict the end range of motion one would expect the risk of injury to be correspondingly decreased. While, the splinting and stiffening action of belts occurs about the lateral bend and axial twist axes, stiffening about the flexion–extension axes appears to be less. A most recent data set presented by Granata et al. (1997) supports the notion that some belt styles are better in stiffening the torso in the manner described above, namely the taller elastic belts which span the pelvis to the rib cage. Furthermore, they also documented that a rigid orthopedic belt generally increased the lifting moment, while the elastic belt generally reduced spinal load, but a wide variety in subject response was noted (some subjects experienced increased spinal loading with the elastic belt). However, even in well-controlled studies, it appears that belts can modulate lifting mechanics in some positive ways in some people and in negative ways in others.

## 33.3  Physiological Studies

Blood pressure and heart rate were monitored by Hunter and colleagues (1989) while five males and one female subject performed dead lifts, bicycle riding, and one-armed bench presses, while wearing and not wearing a 10-cm weight belt. A load of 40% of each subject's maximum weight in the dead lift was held in a lifting posture for two minutes. The subjects were required to breathe throughout the duration so that no valsalva effect occurred. During the lifting exercise blood pressure was significantly higher (up to 15 mmHg), while heart rate also was higher when the belt was worn. Given the relationship between elevated systolic blood pressure and an increased risk of stroke, Hunter et al. (1989) concluded that individuals who may have cardiovascular system compromise are probably at greater risk when undertaking exercise while wearing back supports.

Recent work conducted in our own laboratory (Rafacz and McGill, 1996) investigated the blood pressure of 20 young men performing sedentary and very mild activities both with and without a belt (the belt was the elastic type with suspenders and Velcro tabs for cinching at the front). Wearing this type of industrial back belt significantly increased diastolic blood pressure for quiet sitting and standing both with and without a handheld weight, during a trunk rotation task, and during a squat lifting task. There is increasing evidence to suggest belts increase blood pressure!

Over the past three or four years I have been asked to deliver lectures and participate in academic debate on the back belt issue. On several occasions, occupational medicine personnel have approached me after hearing the effects of belts on blood pressure and intra-abdominal pressure, and have expressed suspicions that long-term belt wearing at their particular workplace may possibly be linked with higher incidents of varicose veins in the testicles, hemorrhoids, and hernias. At this point in time, there has been no scientific and systematic investigation of the validity of these claims and concerns. Rather than wait for strong scientific data to either lend support to these conditions or dismiss them, it may be prudent to simply state concern. This will motivate studies in the future to track the incidents and prevalence of these pressure-related concerns to assess whether they are indeed linked to belt wearing.

## 33.4    Psychophysical Studies

Studies based on the psychophysical paradigm allow workers to select weights that they can lift repeatedly using their own subjective perceptions of physical exertion. McCoy et al. (1988) examined 12 male college students while they repetitively lifted loads from floor to knuckle height at the rate of 3 lifts per minute for a duration of 45 minutes. They repeated this lifting bout three times, once without a belt and once each with two different types of abdominal belts (a belt with a pump and air bladder posteriorly, and the elastic stretch belt previously described in the McGill, Norman, and Sharratt [1990] study). After examining the various magnitudes of loads that subjects had self-selected to lift in the three conditions, it was noted that wearing belts increased the load that subjects were willing to lift by approximately 19%. There has been some concern that wearing belts fosters an increased sense of security, which may or may not be warranted. This evidence may lend some support to this criticism of wearing abdominal belts.

## 33.5    Back Belt Prescription

My earlier report (McGill, 1993) presented data and evidence that neither completely supported nor condemned the wearing of abdominal belts for industrial workers. Definitive laboratory studies that describe how belts affect tissue loading and physiological and biomechanical function have yet to be performed. In addition, clinical trials of sufficient scientific rigor to comprehensively evaluate the epidemiological risks and benefits from exposure to belts must be done. The challenge remains to arrive at the best strategy for wearing belts. Therefore, the available literature will be interpreted and given placement, and also combined with "common sense" to derive the most sensible position on prescription.

Given the available literature, it would appear the universal prescription of belts (i.e., providing belts to all workers in a given industrial operation) is not in the best interest of globally reducing both the risk of injury and compensation costs. Uninjured workers do not appear to enjoy any additional benefit from belt wearing and in fact may be exposing themselves to the risk of a more severe injury if they were to become injured and may have to confront the problem of weaning themselves from the belt. However, if some *individual* workers perceive a benefit from belt wearing, then they may be allowed to conditionally wear a belt, but only on trial. The mandatory conditions for prescription (*for which there should be no exception*) are as follows:

1. Given the concerns regarding increased blood pressure and heart rate, and issues of liability, all those who are candidates for belt wearing should be screened for cardiovascular risk by medical personnel.
2. Given the concern that belt wearing may provide a false sense of security, belt wearers must receive education on lifting mechanics (back school). All too often belts are being promoted to industry as a quick fix to the injury problem. Promotion of belts conducted in this way is detrimental to the goal of reducing injury as it redirects the focus from the cause of the injury. Education programs should include information on how tissues become injured, techniques to minimize musculoskeletal loading, and what to do about feelings of discomfort to avoid disabling injury.
3. No belts will be prescribed until a full ergonomic assessment has been conducted of the individual's job. The ergonomic approach will examine, and attempt to correct, the cause of the musculoskeletal overload and will provide solutions to reduce the excessive loads. In this way, belts should only be used as a supplement for a few individuals while a greater plant-wide emphasis is placed on the development of a comprehensive ergonomics program.
4. Belts should not be considered for long-term use. The objective of any small-scale belt program should be to wean workers from the belts by insisting on mandatory participation in comprehensive fitness programs and education on lifting mechanics, combined with ergonomic assessment. Furthermore, it would appear wise to continue vigilance in monitoring former belt wearers for a period of time following belt wearing, given that this period appears to be characterized by elevated risk of injury.

# References

Adams, M.A. and Hutton, W.C. Mechanics of the intervertebral disc, in *The Biology of the Intervertebral Disc*, Ed. P. Ghosh, Boca Raton, FL: CRC Press (1988).

Ciriello, V.M. and Snook, S.H. The effect of back belts on lumbar muscle fatigue. *Spine* 20(11):1271-1278 (1995).

Granata, K.P., Marras, W.S., and Davis, K.G. Biomechanical assessment of lifting dynamics, muscle activity and spinal loads while using three different style lifting belts, *Clin. Biomech.* 12(2): 107-115 (1997).

Harman, E.A., Rosenstein, R.M., Frykman, P.N., and Nigro, G.A. Effects of a belt on intra-abdominal pressure during weight lifting. *Med. Sci. Sports Exercise* 2(12):186- 190 (1989).

Hunter, G.R., McGuirk, J., Mitrano, N., Pearman, P., Thomas, B., and Arrington, R. The effects of a weight training belt on blood pressure during exercise. *J. Appl. Sport Sci. Res.* 3(1):13-18 (1989).

Kraus, J.F., Brown, K.A., McArthur, D.L., Peek-Asa, C., Samaniego, L., and Kraus, C. Reduction of acute low back injuries by use of back supports. *Int. J. Occup. Environ. Health* 2:264-273 (1996).

Lander, J.E., Hundley, J.R., and Simonton, R.L. The effectiveness of weight belts during multiple repetitions of the squat exercise. *Med. Sci. Sports Exercise* 24(5):603-609 (1992).

Lantz, S.A. and Schultz, A.B. Lumbar spine orthosis wearing. I. Restriction of gross body motion. *Spine* 11(8):834-837 (1986).

McCoy, M.A., Congleton, J.J., Johnston, W.L., and Jiang, B.C. The role of lifting belts in manual lifting. *Int. J. Ind. Ergonomics* 2:259-266(1988).

McGill, S.M. Abdominal belts in industry: A position paper on their assets, liabilities and use. *Am. Ind. Hyg. Assoc. J.* 54(12):752-754 (1993).

McGill, S.M. and Norman, R.W. Reassessment of the role of intra-abdominal pressure in spinal compression. *Ergonomics* 30(11):1565-1588 (1987).

McGill, S., Norman, R.W., and Sharratt, M.T. The effect of an abdominal belt on trunk muscle activity and intra-abdominal pressure during squat lifts. *Ergonomics* 33(2):147-160 (1990).

McGill, S.M., Seguin, J.P., and Bennett, G. Passive stiffness of the lumbar torso in flexion, extension, lateral bend and axial twist: The effect of belt wearing and breath holding. *Spine* 19(6):696-704 (1994).

Mitchell, L.V., Lawler, F.H., Bowen, D., Mote, W., Asundi, P. and Purswell, J. Effectiveness and cost-effectiveness of employer-issued back belts in areas of high risk for back injury. *J. Occup. Med.* 36(1):90-94 (1994).

Nachemson, A.L., Andersson, G.B.J., and Schultz, A.B. Valsalva maneuver biomechanics. Effects on lumbar trunk loads of elevated intra-abdominal pressures. *Spine* 11(5):476-479 (1986).

NIOSH, Workplace use of back belts, U.S. Department of Health and Human Services, Centres for Disease Control and Prevention. National Institute for Occupational Safety and Health. July, 1994.

Rafacz, W. and McGill, S.M. Abdominal belts increase diastolic blood pressure. *J. Occup. Environ. Med.* 38(9):925-927 (1996).

Reddell, C.R., Congleton, J.J., Huchinson R.D., and Mongomery J.F. An evaluation of a weightlifting belt and back injury prevention training class for airline baggage handlers. *Appl. Ergonomics* 23(5):319-329 (1992).

Reyna, J.R., Leggett, S.H., Kenney, K., Holmes, B., and Mooney, V. The effect of lumbar belts on isolated lumbar muscle. *Spine* 20(1):68-73 (1995).

Walsh, N.E. and Schwartz, R.K. The influence of prophylactic orthoses on abdominal strength and low back injury in the workplace. *Am. J. Phys. Med. Rehab.* 69(5):245-250 (1990).

# 34

# The Influence of Psychosocial Factors on Sickness Absence

Chris J. Main
*Hope Hospital, Salford, U.K.*

A. Kim Burton
*University of Huddersfield*

Michele C. Battié
*University of Alberta*

## 34.1   Introduction

Despite increasing advances in medical technology, the cost of musculoskeletal incapacity, particularly low back pain, in terms of sickness benefits, invalidity benefits, and associated allowances has led to a fundamental reconsideration of the nature of chronic incapacity. Recent reports from the United Kingdom (CSAG, 1994), the United States (Bigos et al., 1994), and New Zealand (ACC and the National Health Committee, 1997) in their recommendations for a comprehensive multidisciplinary assessment for patients still symptomatic at six weeks, are based on the clear assumption that a significant proportion of chronic incapacity is preventable. Such a proposition represents a fundamental challenge to much of current medical practice.

Incapacity, as defined clinically is seen as the result of a series of processes in which nature of injury, rate of recovery, and extent of recovery are significantly intermeshed with the nature of medical management and patients' beliefs about the nature of their injury, confusion about the nature of hurting and harming, and mistaken beliefs that physical activity will lead to further damage. Such beliefs may also be the product of the organizational culture in which they work and affected by workplace practices in which persistence of incapacity is viewed with suspicion and conditions of service dictate that "100%" recovery is required before the employer is willing to undertake the "risks" of permitting an injured employee to return to work.

There has been a tendency also to predicate beliefs about incapacity on an incorrect and outmoded view of the relationships among physical and psychological factors in the context of injury. In simple terms, the mind and the body have become divorced. Psychological influences are viewed with suspicion either as a sign of mental infirmity or as malingering. The adoption in particular of a blue-collar macho culture has made it difficult to admit that there are many factors affecting successful recovery, and these are not simply confined to the parameters of medical treatment and associated rehabilitative efforts.

The costs of incapacity have led to a considerable financial commitment to the investigation and amelioration of risk factors for chronicity. It has been argued in the previous chapter that the concept of risk has been confined almost entirely to biomechanical and ergonomic perspectives. Psychosocial factors, similarly, have been investigated principally as aspects of "stress," a multifaceted concept which has incorporated a wide range of perspectives including management style, cognitive demands of tasks, fatigue, the social climate of work (whether supportive or antagonistic), and the recognition of stress responses in individuals. A wide-ranging survey of the literature has revealed direct relevance to the *interaction* of occupational and clinical factors in recovery from injury and return to work.

It is the intention of this chapter to address such issues from the perspective specifically of *psychological obstacles to rehabilitation*. An attempt will be made to offer a new conceptual framework and suggest a set of specific procedures and guidelines which may be of assistance in considering afresh the problem of occupational incapacity, using pain-associated incapacity (and back pain in particular) as an example.

## 34.2 Workloss and Suboptimal Work Functioning: The Extent of the Problem

The extent of the problem has been addressed in our previous chapter, and the issue will be addressed here specifically only to the extent that a specific psychosocial perspective is relevant.

### Influence of Psychological Factors on Accidents and Reinjury

Psychological factors have been implicated in the occurrence of accidents, both in terms of their frequency and in the actual nature of the accidents. "Accident-proneness" has been appraised from both the statistical and the psychological or physiological points of view. A number of specific psychological characteristics have been associated with the occurrence of accidents. These can be conceptualized first in terms of central nervous system dysfunction in terms of the processing of information, and second in terms of performance or execution of tasks.

Apparent accident-proneness may be indicative of or may result from some sort of impaired function in the interpretation or execution of tasks (Porter, 1988). Fatigue and mood disturbance can affect concentration. Difficulties in concentration in turn can affect the accurate coding and decoding of information., which may lead to misinterpretation of information with associated performance compromise. She concludes that individuals do appear to differ in their propensity toward minor accidents, and this is most likely the result of an attention deficit, which is itself a consequence of stress. Such an effect may be particularly noticeable in complex tasks or tasks requiring sustained vigilance. The complexity of the task also appears to be important rather than the intelligence of the subject.

Execution of tasks can also be affected by impaired neuromuscular coordination as a result of fatigue or disturbances to muscle control. Possibly mediated by sympathetic over-activity such as heightened muscle tension or tremor. The specific importance of such factors, over and above central processing effects, on accident-proneness does not seem specifically to have been evaluated.

Whatever the scientific basis for the term "accident-proneness," it would appear to be the case, however, that some individuals *believe* themselves to be more prone to accidents, and as such engage in a variety of superstitious behaviors designed to reduce the likelihood of accidents. Such behaviors may range from relatively benign rituals to marked avoidance of a range of situations. To the extent that normal functioning is not compromised, such behavior can be viewed as harmless eccentricity, but where it develops a compulsive quality and interferes significantly with effective functioning or psychological well-being then it may constitute a clinical disorder necessitating treatment. Attempts to return to work may be hindered by such beliefs.

More common, however, are specific fears of hurting and harming when engaging in specific activities rather than worries about unforeseen accidents. Such fears can either be about the activity being painful or about the dangers of further damage or reinjury. Such fears may lead to excessive caution in the execution of tasks, disturbances to normal biomechanical rhythm, and thereby increased the actual risks of accident or re-injury. (The evaluation of the nature and significance of such beliefs and fears is discussed further below.)

## Costs and Effects of Long-Term Disability

The economic costs of pain-associated incapacity have been discussed in earlier chapters. In addition to the costs to industry of lost productivity and wage-replacement, there is of course a cost to the injured worker and his/her family. This "cost" has to be understood not only in financial terms but also in terms of its psychological effects. In addition to the pain and limitations on everyday living, absence from work can be demoralizing. Boredom, loss of identity, and diminished self worth can create stress and depression, which can have a major impact on the individual and the family. Major psychological effects may not be evident until the worker has been off work for some time and faces the prospect of losing his/her job. Persistence of pain, disturbance to normal activities (particularly sleep) and routines of daily living, and concerns about their recovery and satisfactory return to work can produce frustration and irritability even in the short term. In order to understand the complex interaction between medical, psychological, and occupational factors, it is necessary to reconsider the nature of incapacity and its relationship with sickness absence from work.

# 34.3   Models of Incapacity

The evidence for and against biomechanical and ergonomic models has been reviewed in our related chapter (Burton et al., 1997). Any workplace will have its own culture or *zeitgeist* concerning the nature of injury and the specific role of ergonomic and biomechanical factors. These maxims may be "enshrined" in specific documents concerning work practices or be the subject of health and safety training or procedures required by the organization to reduce the risk of injury. As mentioned, the actual *evidence* for the efficacy of primary prevention is in fact weak, and whether workers understand, adhere to, or even accept the assumptions behind such regulations is unknown. To the extent that workers do form a view about the nature of injury and the process of recovery, it seems likely that they would adopt a biomechanical or ergonomic view of causation rather than a more complex biopsychosocial formulation.

If a clearly identified injury has occurred, and this has resulted in a medical assessment, then the person may have been offered a diagnosis implying a structural basis to their complaint. It may be implied that some sort of permanent damage might have occurred and a series of impressive high-technology investigations may be recommended. Even if no permanent physical impairment in terms of structural damage to bone or nerves is identified, the notion of a significant weakness and therefore occupational vulnerability may be promulgated. Either of these views *may* of course have substance if a significant trauma has occurred, but usually have not. Even in the context of minor injury, it may be implied that the work accident (or incident) simply has been "the straw that broke the camel's back." It may be implied that there has been a cumulative impact of work leading finally to tissue breakdown. Osteoarthritic changes on X-ray may be adduced as evidence of this, despite the fact that only age-related changes may be apparent.

A worker may thus come to believe that his or her problem can only be understood in straightforward biomechanical, ergonomic, or medical terms. The worker may believe that the nature of the "injury" has specific implications for the nature of treatment and rehabilitation. His or her understanding of these issues concerning causality may make it difficult to accept the worth of and indeed necessity for, a speedy return to normal work. He may be resistant to rehabilitative efforts based on a self-help approach and may resist vigorously the suggestion that successful return to work may be hindered or indeed be prevented by psychosocial factors.

It is, in our view, necessary to consider disability from a wider perspective. Research into the nature of disability in clinical settings has suggested the need for a wider perspective on pain-associated incapacity than the traditional medical model. In formulating the biopsychosocial model of disability (Waddell, 1992), Waddell and his colleagues demonstrated that level of disability could by adequately understood only if consideration was given to psychological and social as well as physical factors. In an earlier publication (Waddell et al., 1984), they critically compared the influence of a large number of psychological factors on reported disability. They found that patients' current level of distress and behavioral responses to pain were far more important as predictors of disability than general personality traits, specific hypochondriachal fears, or generalized beliefs about patients' perceived degree of control over their lives. The biopsychosocial model, as it came to be known, was later further refined by the incorporation of beliefs about pain into a dynamic model addressing the transition from acute to chronic incapacity (Waddell et al., 1993). The relative importance of specific psychological factors is described in the next section. Recently it has been suggested (Main and Watson, 1996) that it may be helpful to consider incapacity as a developing process rather than as some sort of "end-state."

Most injuries seem to be over-reliant on concepts of tissue-healing time, where it is assumed that after about six weeks tissue should have "healed" and by implication the physical status should have returned to its preinjury status. This perception has the unfortunate concept that pain and pain-associated incapacity persisting beyond that time may be viewed with suspicion. Clearly a proportion of patients remain incapacitated after injury and theoretical models are needed which address the *process* of chronicity.

## Injury and Incapacity: A Dynamic Model

Such perspectives of chronic incapacity as a *dynamic* process have not to date been integrated adequately into the psychology of return-to-work after injury. Before addressing occupational factors specifically, it is necessary from the psychological point of view to consider further the nature and influence of psychological factors in relation to pain and incapacity.

## 34.4   Psychological Factors and Pain

### Nature of Psychological Factors

During the last 15 years, the role of psychological factors in the genesis and maintenance of pain problems has been increasingly recognized (Main and Spanswick, 1991). Early research into personality traits and identifiable psychiatric illness has been followed by investigation of more specific psychological characteristics such as psychological distress, pain behavior, beliefs about pain, and pain coping strategies. A range of psychometric tests, behavioral measures, and more general tools assessing function or impact of pain have been developed both for the assessment of pain and the investigation of specific psychological dimensions and mechanisms.

## Clinical Perspectives on Pain and Disability

There is a wide range of psychological tests available. For the purpose of this review, psychological aspects of pain will be considered under a number of headings. The focus of this review will be primarily on clinical and occupational applications. A comprehensive analysis of the relative merits of objective and subjective measures in chronic pain from a methodological point of view is presented elsewhere (Dworkin and Whitney, 1992).

### Evaluation of Pain

There is no simple relationship between the amount of tissue damage and the sensation of pain since the experience of pain is affected by a number of physiological and psychological factors. It cannot therefore be assessed as simply, for example, as temperature since it has both sensory and emotional components. Pain has been defined as: *An unpleasant sensory and emotional experience associated with actual or potential damage, or described in terms of such damage…pain is always subjective* (International Association for the Study of Pain, 1979). The emotional content can be appraised by investigation of the vocabulary patients use to describe their pain (Melzack, 1987; Melzack and Katz, 1992) or the way they complete pain drawings (Ransford et al., 1975; Parker et al., 1995). Thus, while the assessment of pain may appear to be fairly straightforward, an understanding of its significance requires a more comprehensive assessment of the pain *patient*.

### Influence of General Psychological Characteristics

In general clinical practice there is a range of goals for psychological assessment including influences on pain perception, adjustment to pain, selection for particular types of treatment, and the design of treatment interventions. Methods of assessment include administration of questionnaires, psychodiagnostic clinical interviews, videotape analysis, and behavioral observation. No single assessment tool or indeed method of assessment is suitable for all purposes. The use of interviews and self-report inventories in the evaluation of pain patients is reviewed comprehensively elsewhere (Bradley et al., 1992). The focus of many of these studies has been on the nature of *pain perception*. For the purpose of this review, specific attention will be directed toward influences on incapacity, recovery from injury, and *obstacles to rehabilitation*.

It is important to consider how patients react to the whole treatment process, and the establishment of a successful "therapeutic alliance" between doctor and patient is extremely important (Hazard et al., 1994). With regard to pain patients in particular they considered that this was best achieved by setting mutually agreed goals concerning pain relief, improved functional capacity, and occupational reengagement. Mayer and his colleagues (Mayer and Gatchel, 1988) have consistently advocated functional restoration as the *primary* goal for rehabilitation for pain-associated incapacity. While acknowledging the tremendous value and importance of functional restoration in treatment of the chronically incapacitated patient, a degree of preselection is involved in all treatment programs and the noncommitted might feel that it cannot be assumed that all psychological obstacles can necessarily be dissolved or rendered inert by a robust functional restoration approach. Nonetheless, the functional restoration approach has been hugely important in shifting emphasis from "psychopathology" to function.

Although the more psychiatrically derived and complex psychological dimensions have been as yet of only limited value in the prediction of outcome of treatment, it would appear that simpler measures of the individuals' response to pain and incapacity are helpful.

In back-associated disability, patients' level of distress, in the form of heightened somatic awareness and depressive symptomatology, has been shown to be more closely associated with reports of disability than general personality characteristics or hypochondriacal fears (Main and Waddell, 1987). The two distress questionnaires later were combined into the Distress Risk Assessment Method or DRAM (Main et al., 1992), which yields a fourfold classification of patients in terms of their level of distress. Although the test does not offer an evaluation specifically of the relationship between distress and incapacity, scores on this test have been shown to be predictive of the development and continuation of chronic incapacity in patients with back pain (Burton et al., 1995). The relationship between generalized distress and more specific fears of hurting and harming (see below) may merit further investigation.

### Pain Behavior

Although it is not possible to experience another's pain, people can communicate their pain either verbally or nonverbally. During the last 30 years, the importance of pain behavior has been increasingly recognized (Fordyce, 1976). It is important to recognize that the explanation for a pain behavior is not necessarily immediately apparent. Indeed a clear formulation is usually considered to require a detailed functional behavior analysis to identify the *precise* circumstances in which the behavior is occurring. The type of analysis will depend on the behavior in question. Typical pain behaviors might include verbalized pain responses, nonverbalized pain behaviors, patterns of activity, or health care usage (e.g., medication use). The behavioral view has been hugely influential as an alternative to the "disease-dominated" medical model and has led to significant changes in the manner and content of health care delivery (as evidenced by the development of pain management programs and functional restoration programs).

Pain behavior may be assessed in a number of ways, and a variety of methods have been devised for use in different settings. Clinical approaches have included assessment of pain behavior during clinical examination (Waddell et al., 1980; Waddell et al., 1984) and use of videotaped ratings of pain behavior during a standardized sequence of movements (Keefe and Hill, 1985), but simpler rating scales have also been devised (Richards et al., 1982).

Although the interpretation of pain behavior is not without controversy, it should form an important aspect of assessment following injury and may contain important features indicative of obstacles to successful rehabilitation. Only the most committed of behavioral theorists would probably now argue that pain behavior can be adequately understood without an evaluation of specific beliefs, fears, and coping strategies. Nonetheless, in many contexts, self-report may be insufficiently informative, and an adequate crystallization of the problem may require a systematic behavioral analysis.

### Cognitive Factors in Pain

The study of cognition has become increasingly complex, but it is possible in general to subdivide the field into three distinct fields of enquiry: beliefs or appraisal; cognitive processes; and coping strategies. Such perspectives have led to the investigation of cognitive factors in pain, adjustment to pain, and in the development of chronic incapacity. The complexity of these interrelationships is indicated in the following quotation:

"Patients who believe they can control their pain, who avoid catastrophizing about their condition; and who believe they are not severely disabled appear to function better than those who do not. Such beliefs may mediate some of the relationships between pain severity and adjustment" (Jensen et al., 1991, p. 249).

### Specific Beliefs About the Nature of Pain

There are a number of different types of belief or appraisal about the nature of pain, and the development of new tests assessing different aspects of beliefs has become something of a "growth industry." A comprehensive review of the field is presented elsewhere (DeGood and Schutty, 1992). While many of these tests seem to be of value, there is clearly much conceptual overlap among them and further research is needed to determine the precise value of each of these tests for particular purposes and contexts. For this review, a number of tests will be chosen to illustrate the types of belief which have been found in the literature to be of relevance in the assessment of the psychological impact of pain.

In the psychological literature there has long been interest in the extent to which individuals believe they can control or gain control over aspects of their life. Researchers have assessed specific beliefs about health (Wallston et al., 1978). The same scale was then adapted for the study of pain by simple substitution of "pain" for "health" throughout the questionnaire (Toomey et al., 1991; Crisson and Keefe, 1988). Other "pain locus of control" scales have been developed (Main and Waddell, 1991) and used specifically in occupational settings. In one study, negative pain locus of control beliefs were significantly associated with absence from work (Symonds et al., 1996), but in a related study, the introduction of educational leaflets produced increased optimism about the control of pain (Symonds et al., 1995).

Locus of control beliefs would seem to have some influence on response to treatment and return to work, but the relationship of such beliefs to other beliefs merits further examination, and a critical comparison of these questionnaires in a wider range of occupational contexts is needed. Their specific value in occupational rehabilitation has yet to be demonstrated.

## Specific Beliefs About Outcome of Treatment

Specific beliefs about treatment outcome can also be investigated. Such beliefs are often referred to as "self-efficacy beliefs." Self-efficacy has been defined in terms of the belief that one was capable of producing a behavior which was necessary to achieve a certain outcome (Bandura, 1977). This general idea has been investigated in terms of the relationship between such beliefs and resultant behavior in relation to treatment outcome.

According to DeGood and Shutty (1992), "These concepts suggest that pain patients who perceive themselves lacking the capacity to acquire self-management skills might be less persistent, more prone to frustration, and more apt to be noncompliant with treatment recommendations. Hence, some patients might demonstrate adequate understanding of particular treatment rationale, yet be noncompliant due to their perceived inability to produce the behavior necessary to follow treatment recommendations" (DeGood and Shutty, 1992, p. 221).

The Self-Efficacy Questionnaire or SEQ (Nicholas et al., 1992) is a 10-item scale yielding a single score giving a measure of the patient's confidence in the outcome of pain management (or rehabilitation in which they have to play an active part). This questionnaire seems promising but requires further scientific evaluation. Such beliefs would also seem to be important in the context of occupational rehabilitation. It has been shown, for example, that patients' beliefs about whether or not they would be able to return to work were the best predictor of return to work after rehabilitation (Sandstrom and Esbjornsson, 1986).

There are as yet few questionnaires specifically validated for the assessment of self-efficacy beliefs in occupational settings. Recently one such questionnaire has been developed specifically to investigate beliefs about back pain. Unlike the previous measures it has been used primarily in occupational settings. The Back Beliefs Questionnaire or BBQ (Symonds et al., 1996) includes scales specifically concerning the perceived inevitability of back pain and its future course. They found that workers with a previous history of back pain were more likely to believe their backs would be problematic in the future. More pessimistic beliefs about the controllability of pain and less willingness to take responsibility for it were associated with a higher number of spells of back pain and more back-associated work loss in the past.

## Specific Fears of Hurting, Harming, and Further Injury

It is crucial to recognize the role of fear and avoidance as obstacles to rehabilitation following injury. Behavioral theorists such as Fordyce (1976) explained the development of "avoidance learning" where successful avoidance of pain established a behavioral pattern which was successful in reducing pain but with the cost of maintaining the "disability." Letham et al. (1983) and Slade et al. (1983) incorporated these ideas into a "fear avoidance model of exaggerated pain perception" in chronic low back pain. In their view, such beliefs were central in the development of pain avoidance behavior. Letham et al. (1983) described patients as *confronters* or *avoiders*, and it has been observed (Waddell et al., 1993) that fear and avoidance of pain can become more disabling than pain itself, since although avoidance at early stages may reduce nociception, the avoidance behaviors may persist in anticipation of pain rather than simply as a response to it.

Fear, specifically of movement and reinjury, may also become established. If an injured worker mistakenly believes that resumption of a particular activity will actually lead to an increase in pain or produce further damage, this obstacle to rehabilitation will have to be removed if the worker is to be rehabilitated. The genesis of such fears has also been investigated. Vlaeyen et al. (1995a) found that fear of movement and reinjury was more related to depressive symptoms and catastrophizing than to pain itself. Finally, in a study of 300 patients attending their family doctor with acute back pain, fear–avoidance beliefs predicted outcome at two and twelve months (Klenerman et al. 1995). The importance of addressing such beliefs and associated psychological mechanisms within a reactivation framework would seem to be compelling, but development of accurate assessment instruments is still at an early stage.

The Fear–Avoidance Beliefs Questionnaire or F.A.B.Q (Waddell et al., 1993) has two scales. The first (physical activity scale) assesses beliefs about the influence of physical activity on pain, while the second (work scale) examines specifically beliefs about the influence of work on pain. The Tampa Scale of Kinesophobia or T.S.K. (Miller et al., 1991) has also been recently developed. A recent study (Vlaeyen et al., 1995b) has shown that fear of movement may have a major impact on behavioral performance. This topic would seem to merit further investigation.

### Coping Strategies

Coping can be considered in terms of what people do to try to diminish the occurrence or impact of unpleasant events of various sorts. Patients may employ a wide range of behavioral and coping strategies in order to limit the effects of pain. Choice of strategies will depend on patient's beliefs about pain, on their confidence in being able to influence events, and of course on their repertoire of coping behaviors. In addition to differentiating behavioral from cognitive coping strategies, researchers have also described coping behaviors in terms of their appropriateness and effectiveness, specifically in relation to pain and adjustment. Thus, Brown and Nicassio (1987) made the important distinction between active and passive coping strategies. Perhaps the most widely used questionnaire is the Coping Strategies Questionnaire or CSQ (Rosenstiel and Keefe, 1983). The 50 items give scores of six cognitive scales (diverting attention, reinterpreting pain sensation, coping self-statements, ignoring pain sensations, passive praying, or hoping and catastrophizing), two behavioral coping scales (increasing activity level and increasing pain behaviors), and two single items concerning perceived control over pain.

The questionnaire offers a general evaluation of the relative use of effective (or appropriate) coping strategies and ineffective (or inappropriate) coping strategies. Several studies have shown that negative or ineffective coping strategies such as catastrophizing ("fearing the worst") are associated with higher levels of self-reported disability or adjustment (Jensen et al., 1991; Keefe et al., 1989; Main and Waddell, 1991). The catastrophizing scale has also shown impressive predictive value in outcome of treatment, particularly in patients with acute pain (Burton et al., 1995). Brandtstadter (1992) distinguished two fundamentally different ways of coping with chronic pain: *assimilative coping,* involving active attempts to alter circumstances in line with personal preferences and *accommodative coping* or "downgrading" of goals or expectations when goals are seen to be unattainable through active coping efforts. Two scales: Tenacious Goal Pursuit (TGF) and Flexible Goal Adjustment (or FGA) have been developed to measure these aspects of coping (Brandstadter and Renner, 1990). In a recent study of coping style and pain-associated distress, it was concluded "accommodative coping functions as a protective resource by preventing global losses in the psychological functioning of chronic pain patients and maintaining a positive life perspective. Most important, the ability to flexibly adjust personal goals attenuated the negative impact of the pain experience (pain intensity, pain-related disability) on psychological well-being (depression). Furthermore, pain-related coping strategies led to a reduction of disability only when accompanied by a high degree of flexible goal adjustment" (Schmitz et al., 1996; p. 41).

These research findings are consistent with early work by demonstrating that depression is primarily a function of pain-associated incapacity (or interference with life) rather than a direct result of pain itself (Rudy et al., 1988)). These findings suggest that, in the design of a return-to-work rehabilitation program for the injured worker, a degree of flexibility should be seriously considered.

### Conclusion

There is now wide evidence that psychological factors influence pain perception, level of incapacity, adjustment to disability, and response to treatment in clinical settings. A number of key psychological dimensions have been identified. Such factors are important not only for understanding the impact of pain on the individual but also for understanding the nature of pain-associated incapacity and recovery from injury. A number of recent studies have suggested that a number of these psychological factors are also predictive of the *development* of chronic incapacity and as such can be considered as risk factors for chronicity.

## Determinants of Chronicity in Clinical Settings

Psychological factors, however, need to be understood not only in terms of their influence on pain perception and on response to treatment, but also on the process of chronicity (or conversely, recovery from incapacity).

Several studies have compared the *relative* importance of physical and psychological factors on the prediction of outcome of treatment. As has been stated elsewhere, most treatment and intervention is based on concepts of tissue damage and amelioration of biomechanical dysfunction following injury. Many studies have tried to identify the importance of all sorts of anthropomorphic, clinical, and biomechanical factors in terms of risk of injury, or risk of continued incapacity (Halpern, 1992). Multivariable analyses of such data suggest that the relative contribution of the different classes of variable to prediction of chronicity is particularly complex (Burton and Tillotson, 1991). In general, however, it may be the (clinical) historical and psychosocial information rather than the socio-occupational or biomechanical variables that are associated with lack of recovery (Burton et al., 1999).

In chronic back pain patients, levels of distress at time of initial assessment were found to be as important as known risk factors (such as previous lumbar surgery) in outcome from conservative treatment 2 to 4 years later (Main et al., 1992). The particular importance of psychological factors in *chronic* pain patients may not be particularly surprising, but recent studies on patients with acute pain have come up with similar findings. In a one-year follow-up cohort study of patients with back pain attending for osteopathic treatment, cognitive factors (specifically negative or inappropriate coping strategies) were found to be by far the most powerful predictors of outcome; particularly in the subgroup of patients with acute back pain in which it was possible to explain 47% of variance in outcome from negative cognitive coping strategies alone (Burton et al., 1995). In another study (Main et al., *submitted*) psychological factors had a powerful influence on outcome of treatment in patients with acute back pain, whether treated by McKenzie physiotherapy or by a nonsteroidal anti-inflammatory drug, both immediately following treatment and at one-year follow-up.

Such factors have not been investigated as yet with sufficient precision in the context of occupational rehabilitation, but seem to merit investigation. It would appear, nonetheless, that levels of distress or dysfunctional thoughts may be important factors in the *development* of chronicity and thus represent psychological obstacles to recovery which need to be addressed as part of the rehabilitative process.

## Mechanisms of Chronic Incapacity

The importance of psychological factors has been clearly shown, but most of the above studies have been either cross sectional in nature or involved analysis of outcome of treatment and have not addressed, therefore, the possible *mechanisms* involved. The original Gate Control Theory or GCT (Melzack and Wall, 1965) postulated an interaction between nociception and central psychological factors, but most research studies have treated physical and psychological factors as if they were quite independent. The relationship between emotion and physiological activity has been shown however in a series of investigations of peripheral and central physiological responses to experimental stimuli. (Flor et al., 1989). In an early experiment (Flor et al., 1985) they had shown heightened levels of muscle activity in the paraspinal muscles to personally relevant stressors in patients with low back pain.

Although there is some evidence for differences in resting levels of SEMG between back pain patients (Ahern et al., 1988), guarded movements of various sorts, while protective during the acute stages of recovery from injury, are clearly evident in many chronic pain patients. Guarded movements can be investigated using surface electromyography (SEMG) from the lumbar muscles, and recently a specific measure of the guarded movement during forward flexion the Flexion Relaxation Ratio (FRR) has been developed and validated for use with low back pain patients (Watson et al., 1997a). In a further study using the FRR, the relationship between muscle activity and response to treatment was investigated (Watson et al., 1997b).

A significant correlation was found initially between fear avoidance beliefs and abnormalities of muscle action prior to treatment (a three-week pain management program). Following the pain management program there were significant correlations between changes in the muscular abnormalities and changes in fear avoidance beliefs as well as between changes in self efficacy beliefs and changes in muscular abnormalities. The results raise the possibility that following injury, or a significantly painful episode, some people develop persistent abnormalities in muscle electrical activity suggestive of a persistent physiological abnormality around the painful site.

They conclude:

"The extent to which these abnormalities are mediated by fears of hurting and harming is not yet entirely clear, but changes in SEMG patterns towards the normal following pain management are associated with reductions in fear of hurting and harming and increase in self-confidence as measured by self-efficacy beliefs."

## Conclusions from Clinical Studies

It would seem, therefore, that several conclusions can be reached from the above research findings.

1. Psychological factors are implicated in the perception of pain.
2. Reactions to pain, in terms of distress, specific beliefs, coping strategies and the development of pain behavior, seem likely to be more influential than more widespread psychological constructs or perspectives.
3. It has been possible to identify specific psychological factors which place individuals at risk of the development of both the persistence of and the development of chronic incapacity.
4. It would appear that these psychological factors are more important than clinical history or physical examination factors in the prediction of outcome.
5. Psychologically oriented pain management program can reduce these risk factors.
6. Laboratory investigations have shown than reduction in specific fears of hurting and harming are associated with physiological changes in electrical activity in muscle.
7. It would appear that future research might usefully be directed at secondary prevention in patients in whom chronic incapacity has not yet become established.

# 34.5 Psychological Aspects of Work

## Difficulties in the Identification of Work-Related Stress

The complex sociopsychological characteristics of organizations has made the investigation of work-related stress extremely difficult. Most of the studies have tended to focus on specific features of the organization or particular job characteristics. Kasl (1992) identifies, for example: organizational size and structure, cumbersome or arbitrary procedures and practices, and role-related issues. Cox (1993) discusses the significance of role ambiguity, role conflict, and role insufficiency as aspects of the latter. Responsibility for people (Leiter, 1991), lack of job security, and failure to realize job expectations are among other factors considered to be of importance. A number of recent studies have examined the nature of work stress (Schaefer and Moos, 1993; Crum et al., 1995; Alliger and Williams, 1993). Other studies have examined more specific aspects, such as job satisfaction (Medcof and Hausdorf, 1995), job opportunity/frustration (Noack, 1994), and job predictability and turnover (Pearson, 1995).

Organizational stressors therefore can be identified, and certain characteristics of the working environment constitute risk factors in their own right. The extent to which an individual copes with such risk factors will influence job satisfaction, morale, and psychological well-being. Extended absence from work, however, can have a marked effect on the perception of the work environment. Interpersonal relationships at work for example may be an important determinant of successful return to work. Good relationships with colleagues, a sense of being valued, and worry about letting one's colleagues down may act as important incentives to return to work after injury. An unpleasant or difficult interpersonal environment may represent a significant obstacle to return to work. Appraisal of such factors, combined with a loss of self-confidence and anxiety about not being able to perform satisfactorily on return to work may also represent a specific hindrance. According to Bigos et al. (1990) "once an individual is off work, perceptions about symptoms, about the *safety* of return to work, and about the impact of return to work on one's personal life can affect recovery even in the most well-meaning worker," (p. 854; Bigos' italics).

## Approaches to the Definition and Study of Work Stress

For the purpose of this chapter an attempt will be made to examine the nature of work stress from a general perspective, derived from Cox (1993). Finally, it has to be recognized that although it is possible to identify work stress, there may be an interdependence with other life stresses. Long working hours, for example, can significantly interfere with family life.

There have been three different but overlapping approaches to the definition and study of stress (Cox, 1993). According to the "engineering model," stress is conceptualized as an aversive or noxious characteristic of the work environment and is construed essentially as an environmental *cause* of ill health. The model has been criticized on the grounds of oversimplicity, since specific stimuli such as noise may be stressful or beneficial depending on a number of other factors. Furthermore, according to Douglas (1992), perception of risk and risk-related behavior are not adequately explained by the natural science of objective risk and are strongly determined by group and cultural biases.

In the "physiological model," stress is defined in terms of the common physiological effects of a wide range of aversive and noxious stimuli; i.e., stress is the physiological *response* to a threatening or dangerous environment. The model has been criticized because differences in physiological responses to the same stressor, and difficulties in identifying unambiguously physiological responses which are specifically stress responses. According to Cox (1993), there are subtle but important differences so that noradrenaline activation may be more related to the physical activity inherent in various tasks, while adrenaline activation may be more related to feelings of effort and stress.

More fundamentally, both models have been criticized for their reliance on an over simplistic stimulus-response paradigm, while ignoring individual differences in the cognitive and perceptual processing of information; ignoring the interactions between the individual and his/her environment; and in their lack of a "systems-based" approach, ignore the psychosocial and organizational contexts of work stress.

The third model, the "psychological" model, conceptualizes stress in terms of the *dynamic interaction* between the person and the work environment. Interactional models focus on the structural features of the person's interaction with the work environment, while transactional models are concerned principally with psychological mechanisms such as cognitive appraisal and coping which underpin the interactions.

The Person–Environment Fit or P–E fit model (French et al., 1974) identifies two aspects of *fit*:

1. The degree to which an employer's attitudes and abilities meet the demands of the job, and
2. The extent to which the job environment meets the workers' needs.

Lack of fit, whether defined objectively or subjectively (in terms of the individual's perceptions), has been recognized as a potential stress. Job demands and job decision latitude have also been identified as powerful factors in psychological reaction to work (Karasek, 1979).

According to transactional models, stress involves elements of both cognition, such as appraisal, and emotion. According to Lazarus and Folkman (1984), *primary appraisal* involves monitoring the situation and may lead to identification of a problem. such problem recognition may be accompanied by unpleasant emotions or general discomfort. *Secondary appraisal*, following after recognition of the problem, involves attempts to generate a set of coping strategies. According to Cox (1993): "The experience of stress is therefore defined, first, by the realization that they are having difficulties in coping with demands and threats to their well being and, second, that coping is important and the difficulty in coping depresses or worries them" (p. 17).

Finally, the appraisal process may be thought of in five stages, beginning with no more than a hazy recognition of the existence of a problem, and leading via an *on going* process of interaction between coping strategies, reappraisal, and redefinition of the problem.

## Coping with Occupational Stress

A wide range of coping strategies has been identified. They have been differentiated, for example, into task-focused or emotion-focused (Dewe, 1987) and into strategies changing meaning or managing stress symptoms (Pearlin and Schooler, 1978). Individual characteristics such as informational styles (*blunters* and *monitors*) have been identified (Miller et al., 1988), and concepts of vulnerability or job hardiness (Kobasa, 1979; Kobasa et al., 1981) have also been considered. Finally, individual differences have been investigated both as *components* of the appraisal process, or as *moderators* of the stress–health relationship (Payne, 1988).

Behavioral characteristics such as "Type A behavior" (Friedman and Rosenman, 1974) were identified originally as risk factors for coronary heart disease. Since then concepts such as work commitment and time-urgency have become an integral part of the psychology or work. Gaining control is an important coping strategy in dealing with work difficulties. If as a coping strategy it is ineffective and the problem persists, frustration or anger can result in significant work stress, deterioration in health, and absence from work.

## Predictors of Chronic Incapacity in Occupational Settings

Cats-Baril and Frymoyer (1991), in a study of employees with between 2 and 6 weeks' incapacity, identified four main risk factors for long-term disability. These included job characteristics (work status at time of the survey, work history, and type of occupation); job satisfaction factors (including preretirement policies and benefits; perception about whether the injury was compensable; who was at fault and whether a lawyer had been contacted); past hospitalizations; and educational level. The utility of their predictive model, however, was criticized on methodological grounds specifically because of their failure to control for duration of disability, and because of the likely influence of psychological factors in terms of expectation.

"The population of patients who have already incurred two weeks of off-work time secondary to low back trouble probably are anticipating long-term disability, and, perhaps more importantly, so are their health care providers, employers, and insurers. These expectations may trigger illness behaviors that help establish long-term disability" (Lehmann et al., 1993, p. 1110).

In a recent review of risk factors in industrial low back pain, four major types of risk factors have been implicated in industrial low back pain (Bigos et al., 1990). These were individual factors (mainly demographic and anthropomorphic), physical findings (mainly radiographic), workplace factors (including various work characteristics), and psychological factors (determined from psychological tests or evidence of substance abuse). In their view, however, methodological problems significantly compromised accurate evaluation of these factors.

According to a recent review, "Monotonous work, high perceived work load, and time pressure are related to musculoskeletal symptoms. The data also suggest that low control on the job and lack of support by colleagues are positively associated with musculoskeletal disease. Perceived stress may be an intermediary in the process" (Bongers et al., 1993, p. 297). Specific psychological features have been found in a number of prospective studies. Carosella et al. (1994) found that patients with low return-to-work expectations, heightened perceived disability, pain, and somatic focus had problems complying with an intensive work rehabilitation program. Haazen et al. (1994) have shown that change in distorted pain cognitions, workers' compensation status, and use of medication were the most important predictors in behavioral rehabilitation of low back pain. (They were, however, pessimistic about the overall level of prediction achieved.) Bigos and colleagues (1991) found that psychological distress, low job satisfaction, and a history of back trouble were the strongest predictors of the report of back pain in the future. The above studies, although tantalizing, do not identify the putative psychological factors with sufficient degree of accuracy to evaluate their specific importance. There is a need for the development of psychometrically sound occupationally focused instruments to assess aspects of job satisfaction; perception of safety; and the perceived consequences of sickness absence for job security and financial stability. In addition to investigation of the employee, it may be necessary furthermore to consider characteristics of the organization.

In the clinical assessment of low back pain, the lack of a simple relationship between physical signs and resulting incapacity is well recognized and different sorts of factors have been adduced to explain the apparent mismatch. Recent clinical studies into the outcome of treatment for low back pain have however offered a more specific evaluation of the role of different sorts of psychological variables in the prediction of chronic incapacity (as determined by self-reported disability) than is currently available in most occupational studies. The findings of such studies for our understanding of the nature and development of chronic incapacity may have relevance also to secondary prevention in occupational settings and so some of these studies will now be reviewed.

In a study of attitudes, beliefs, and absenteeism among workers in a biscuit manufacturing factory, Symonds et al. (1996) showed that workers who had taken in excess of one week's absence due to low back trouble had significantly more negative attitudes and beliefs (when compared with workers who had taken shorter absences, or with those who reported no history of back trouble). Beliefs about the inevitability of back pain, fears of hurting or harming, and perceived disability were significantly associated with absenteeism. In an associated study (Symonds et al., 1995), introduction of a psychosocial pamphlet, designed to correct mistaken beliefs about back pain (e.g., confusing hurting with harming) and reduce avoidance behavior, successfully reduced extended sickness absence resulting from low back trouble.

## 34.6   Socioeconomic and Medicolegal Dimensions

Since the turn of the century, the nature of chronic incapacity has been considered in the medicolegal literature, but little systematic research into the role of socioeconomic factors has emerged. The dichotomous view of pain as *either* physical *or* psychological has prevailed. The persistence of pain without a clear and unambiguous physical lesion has been considered either as a manifestation of psychiatric disorder or as evidence of simulation (malingering). The conceptual confusion evident in the investigation of pain and incapacity in the legal system is reviewed elsewhere (Main and Spanswick, 1995). For the purpose of this chapter, it can be stated that in occupational rehabilitation, the specific influence of socioeconomic factors in general, and of medicolegal factors in particular awaits further research; both in terms of their specific influence and in terms of their possible interactions with physical and psychological factors.

## 34.7   Return to Work: Prevention of Chronic Incapacity

### The Role of Education

Results of back schools based mainly on medical and ergonomic principles have been somewhat disappointing. As far as *primary prevention* is concerned, according to Nordin et al. (1992) there is little evidence for sole efficacy of back schools, but they may be beneficial within the wider context of a company-wide effort (Wood et al., 1987) where management and unions participate and the program is geared to the needs of the employee (Versloot et al., 1992). Yet somewhat surprisingly, in a recent U.K. study carried out in a light industrial environment, mere distribution of a simple educational pamphlet designed specifically to reduce fear–avoidance beliefs, produced a substantial reduction in extended absence associated with LBT (Symonds et al., 1995). As far as *secondary prevention* is concerned, the picture is somewhat confused, perhaps because of the variety of approaches adopted. Bergquist-Ullman and Larsson (1977), in a randomized controlled trial, showed that back school was superior to physical therapy treatment or placebo; but recent studies (Berwick et al., 1989; Stankovic and Johnell, 1990) have failed to replicate the findings. Methodological difficulties in the evaluation of their efficacy are presented elsewhere (Schlapbach and Gerber, 1991). According to Scheer et al. (1995, p. 970): "Although an ergonomic education for recently injured workers makes inherent sense, there is *not* incontrovertible published evidence that back school is more efficacious than placebo for acute LBP."

## Physical Therapy

Physical therapists employ a wide range of techniques ranging from passive modalities such as heat treatment, ice packs, interferential stimulation, and a range of manipulative techniques. The approaches are detailed elsewhere (Tan et al., 1992; Ottenbacher and Fabio, 1985). As far as return to work is concerned, the evidence for efficacy is unproven, perhaps because there is no requirement for the individual to be *actively* involved in the treatment process.

## Functional Restoration (Sports Medicine Approach)

Functional restoration programs (FRP) place a particular emphasis on the restoration of strength and the development of fitness. As such, they are more appropriate for the chronically incapacitated worker in whom a "disuse syndrome" may have developed, than for a secondary prevention approach in the more recently incapacitated worker. They do have an advantage over the individualized physical therapy approach, however, in that they require a specific Functional Capacity Evaluation or FCE (Mayer and Gatchel, 1988) which can be more closely matched with specific occupational requirements than a more generalized and less focused clinical approach.

## Specific Psychological Approaches

A number of individual approaches to the management of pain are available (Gatchel and Turk, 1996) and may be offered as part of routine pain management following injury. As far as return to work is concerned, there is no more evidence for individualized psychological therapy such as psychotherapy or stress reduction as a *single* treatment approach than for individualized physical therapies. The recent development of graded activity programs with a behavioral therapy approach, however, has shown much more promise (Lindstrom et al., 1992). In the latter study, patients were randomized to either traditional physiotherapeutic care or to a graded activity program with a behavioral therapy approach. The graded activity program proved to be a successful method of restoring occupational function and facilitating return to work in subacute low back patients.

## Conclusion

It would seem that the traditional content of workplace education concerning injury, pain, and rehabilitation needs radical revision. Ergonomic interventions alone may be insufficient to prevent the development and/or persistence of musculoskeletal problems (Burton et al., 1997b). It is clear that the problem of return to work is multifaceted. It seems that to enable successful return to work, occupational rehabilitation requires a multidisciplinary perspective with an integrated model of work disability such as the "Rochester model" (Feuerstein, 1991). This "highlights the potential interaction of a patient's medical condition and ability to meet physical and psychological demands at work. It also emphasizes the importance of psychological factors (such as fear of reinjury, expectations of return to work, perception of disability, illness behavior, and pain and stress coping skills) in rehabilitation efforts directed at RTW (Feuerstein and Zastowny, 1996, p. 463).

## 34.8 Implications

### Reappraisal of the Nature of Work-Related Incapacity

In this chapter we have attempted specifically to consider the psychological influences of "injury" on absenteeism and return to work. They have been considered from a biopsychosocial framework, recognizing the interactions among physical, psychological, and socio-occupational factors. It is considered that models of incapacity constructed solely from biomechanical and ergonomic principles are inherently limited since they are based on a partial understanding of the nature of absenteeism (Burton et al., 1997b). It follows from the above that the concept of risk analysis merits fundamental reconsideration.

If an individual has no interest, for whatever reason, in successful and satisfactory return to work, then consideration of psychological obstacles to return to work is an irrelevance. It is important to distinguish generalized occupational stress from specific injury-associated distress. While both may present obstacles to return to work, they require different solutions. Furthermore, it is important to recognize that in the well-intentioned worker psychologically mediated work incapacity should not be confused with irredeemable work incapacity. It has been shown that lengths of absence from work are associated both with characteristics of the individual and features of the working situation. It is necessary therefore to consider absenteeism both from a system and an individual perspective (Versloot et al., 1992).

### Solutions for the Individual Worker

In general it is not in any worker's interests to become chronically incapacitated. Although most workers recover from injury, unnecessary delay in return to work is costly and potentially psychologically damaging. Once an individual has become injured, the assessment of the risk of chronic incapacity needs to considered. Such risks need to be understood not only from the biomechanical and ergonomic perspectives, but also from the psychological perspective. Clinical studies have identified early psychosocial risk factors for chronicity. Consideration should be given to a system of routine evaluation of such risks if the worker does not make a speedy recovery.

Assessment of the individual worker cannot be undertaken without consideration of the context of assessment. Whether or not an individual becomes chronically incapacitated depends not only on his/her actual and perceived clinical status, but also on the socioeconomic system. Differences in social policy, in remuneration for sickness, and in industrial agreements regarding the protection of the injured worker will all have a profound effect on sickness absence and chronicity (I.A.S.P., 1995). The importance of the socioeconomic implications of continued incapacity should not be underestimated.

Features of the occupational environment can be more or less conducive to return to work. The issue of whether it is right to expect someone to be *fully fit* merits some discussion. From an employer's perspective, in terms of job performance this is clearly desirable, since it obviates the need for special consideration, discussion of temporary restricted work, with the associated problem of preventing such arrangements being viewed adversely by workmates. The increase in all sorts of litigation in industrial society is a matter of fact, irrespective of one's political stance on it. The issue is increasingly seen as a risk management issue. The easiest way to eliminate all risk simply would be to refuse to employ or re-employ anyone who has had any sort of injury. Such a solution however, in a skilled workforce with a relatively high rate of minor job-associated injury, could be extremely expensive as well as socially unacceptable. On the other hand, identification of *specific role conflicts* likely to follow injury and subsequent absence from work could, at relatively low cost, be specifically addressed in a time-limited workplace rehabilitation program for patients with specific musculoskeletal rehabilitation needs. Success of such an endeavor would of necessity have to take into consideration reaction of managers, their management style, and the reaction of workmates as well as specific task functioning and general job performance.

Given the costs of extended absenteeism, specific attention might be given to a number of the other risk factors for occupational stress (discussed above). While accepting that a fully comprehensive psychological evaluation is not a practical proposition, specific attention might be directed at the assessment of *changes* in the perception of the work environment by the injured worker in terms of:

    a. Job satisfaction
    b. Job stress (including relationships with colleagues)
    c. Specific requirements regarding job performance
    d. Perceived role conflict

In fact a new instrument has been recently developed specifically for use in occupational settings. The Psychosocial Aspects of Work (PAW) questionnaire measures three different types of attitude about work. It contains a 7-item job satisfaction scale, a 4-item social support scale, and a 4-item mental stress scale (Symonds et al., 1996). Although at an early stage of development, this questionnaire would seem to have some potential as a general screening instrument.

The identification of specific psychological obstacles to work requires specific evaluation. Since such obstacles are probably evident soon after injury, they may be perhaps best tackled close to the work environment immediately after injury during the phase of active treatment. Since most musculoskeletal injury, however, if managed properly, should resolve successfully, provision of "front-line" treatment within occupational health settings has an obvious appeal, and many employers have employed physical therapists from various professional disciplines for this purpose. Such an approach, provided it is not seen as a thinly disguised management ploy to reduce costs or identify malingerers, has an intuitive appeal in that it is preventing the "distancing" between the injured worker and his/her worksite. Unfortunately, since such workplace-based treatment has been offered primarily with an ergonomic or biomechanical emphasis, specific psychological obstacles to return to work have not usually been incorporated into the format of the rehabilitation program.

Although a number of clinical measurement tools (as identified above) are now available, further research is needed into specific occupational applications. The following clinical dimensions would seem to be of particular importance and perhaps should be incorporated routinely when a worker has been off for more than a few days, or where occupational records indicate a pattern of increasing spells of work absence with nonspecific musculoskeletal complaints.

    1. General level of distress.
    2. Specific beliefs and fears about pain and rehabilitation
    3. Pain coping strategies
    4. Fear of physical activity

## Solutions for a "Sick Workplace"

There is an urgent need to develop sound measurement instruments for the perception of work. In development of new tests, however, it should be remembered that jobs differ significantly in their psychological characteristics and so occupationally sensitive instruments with adequate normative data will be required.

### Assessment of Psychological Risk Factors

Traditionally, assessment of risk in occupational settings has tended to focus on environmental characteristics, such as safety, or on ergonomic dangers, such as poor lifting and handling techniques. In this chapter it has been argued that the concept of psychological risk must be recognized. The reduction of unnecessary absenteeism needs to be considered both from an occupational and a clinical perspective.

### Reduction of Occupational Stress

It would seem likely on a simple intuitive level that perceived occupational stress will delay return to work after injury. While the widespread introduction of mental health promotion (Kasl, 1992) in industry as an answer to occupational stress in industry is perhaps more of a pious hope than a realistic possibility, specific attention to the company "ethos" regarding injury and pain-associated incapacity might merit consideration. It has recently been recommended that stress management programs which encourage individuals to change workplace factors through innovation would be a useful addition to interventions that emphasize individual adaptation to stressful work environments (Bunce and West, 1994). According to Cox and Griffiths (1995), "The challenge for the late 1990s is a practical one: to develop effective systems for assessing and managing the problem in organizations, and then to educate those organizations in using those systems" (p. 3). It is being suggested that workplace rehabilitation after injury *also* should incorporate such perspectives.

### Establishment of a Work-Based Rehabilitation Policy

Identification of possible changes to working practices necessary prior to full recovery and development of a clear work-based rehabilitation policy will reduce anxiety about return to work. Working practices clearly are job specific, but it is known that certain occupations carry higher risks for back injury, even though the physical demands may differ (Halpern, 1992). Functional Capacity Evaluation is an integral part of the "sports medicine approach" (Mayer and Gatchell, 1988), but the emphasis is very much on the capacity of the individual. Traditionally, graded work practices have taken place in rehabilitation settings, but it may be possible to establish closer integration between rehabilitation and occupational health in the design of *work-based* graded return to work, as has been evident in Scandinavia. While a clear understanding of the biomechanical demands and ergonomic features of the jobs concerned is essential, such considerations have to be located within an appropriate model of rehabilitation (Feuerstein, 1991; Feuerstein and Zastowny, 1996).

### Management Style and Conditions of Employment

Return to work is affected first and foremost by conditions of employment which may have a specific influence on rate of return to work irrespective of severity of injury. Management style is a key ingredient of both job satisfaction and job stress (Cox, 1993). Specific attention to the management of return to work may pay dividends. In one North American study (Wood, 1987), for example, a simple telephone call from a line manager offering good wishes and expressing concern for the absentee's welfare significantly increased rate of return-to-work.

### The Design of Context-Sensitive Interventions

A recent review of RCTs of early intervention in industrial low back pain was pessimistic (Scheer et al., 1995), but frequently the design of such interventions has been inappropriate in that they have been over-reliant on biomechanical and ergonomic principles (Burton et al., 1997). A number of recent studies have suggested that *specifically designed* and *appropriately targeted* interventions may be effective. Such interventions have ranged in complexity from the introduction of educational pamphlets designed to change beliefs (Symonds et al., 1995) and telephone calls (Wood, 1987); via graded exercise programs (Lindstrom et al., 1992) and occupationally specific secondary prevention programs (Linton et al.), to major tertiary rehabilitation programs (Mayer and Gatchell, 1988).

## 34.9 Conclusion

The rising costs of back-associated incapacity has led to a number of reports which have offered solutions to the problem of long-term incapacity (CSAG, 1994; Bigos et al., 1994; ACC and the National Health Committee, 1997). These reports essentially have addressed the problems of secondary incapacity rather than primary prevention, and made a number of clinical recommendations involving assessment and management. A recent report by the International Association for the Study of Pain (IASP, 1995) has offered a radical "solution" to work absence from nonspecific back pain. It sees long-term absenteeism primarily as a consequence of the "over-medicalization" of pain. It recommends that after six weeks' work absence, such patients be redefined as "activity-intolerant." The incapacity is then no longer defined as a medical problem. Whatever economic advantages such a "solution" may offer in answer to the increasing costs of back-pain associated incapacity (which has to be addressed), the report has not met with universal support. It has been criticized mainly on the grounds that it seems to rest on old-fashioned concepts of tissue-healing time and does not offer either an adequate clinical or an adequate occupational answer to the problem of persistent incapacity (Turk and Main, 1997).

Traditionally, clinical treatment has in general been considered in complete isolation of the occupational context. Conversely, the psychological aspects of return to work, if considered at all, have tended to omit consideration of such clinical factors. In terms of psychological mechanisms, however, it is clear that an effective rehabilitation program has to address both perspectives. In this chapter an attempt has been made to identify the sorts of psychological factors which influence return to work following injury. It would appear that the traditional separation into clinical and occupational factors needs to be reconsidered. The cost of unnecessary absenteeism is considerable. Adequate attention to the psychological risk factors highlighted in this chapter might be of considerable economic benefit in the reduction of unnecessary absence from work. The development of context-sensitive interventions, while recognizing the powerful influence of psychosocial risk factors within an overall disability management framework, may offer a better way forward than the narrow biomechanical and ergonomic-based approaches, or the fairly drastic "social policy" solutions offered by *Pain in the Workplace* (IASP, 1995).

## Acknowledgments

Sincere thanks to Dr. Ana-Paola Viera for assistance with the background literature.

## References

ACC and the National Health Committee. 1997. *New Zealand Acute Low Back Pain Guide.* Wellington, New Zealand.

Ahern D.K., Follick M.J., Council J.R., Laser-Wolston N., and Litchman H. 1988. Comparison of lumbar paravertebral EMG patterns in chronic low back pain patients and controls. *Pain* 34: 153-160.

Alliger G.M. and Williams K.J. 1993. Using signal-contingent experience sampling methodology to study work in the field: A discussion and illustration examining task perceptions and mood. *Personnel Psychol.* 46: 525-549.

Bandura A. 1977. Self-efficacy: Toward a unifying theory of behavioral change. *Psych. Rev.* 84: 191-215.

Bergquist-Ullman H. and Larsson U. 1987. Acute low back pain in industry. *Acta Orthop. Scand.* 170 (Suppl). 1-117.

Berwick D.M., Budman S., and Feeldstein M. 1989. No clinical effect of back schools in an HMO. A randomised prospective trial. *Spine* 14: 338-349.

Bigos S., Bowyer O., and Braen G. 1994. Acute low back problems in adults. *Clinical Practice Guidelines No. 14* AHCPR Publications No. 95-0645 Agency for Health Care Policy and Research. U.S Department of Human Health and Human Services.

Bigos S.J., Battié M.C., Nordin M., Spengler D.M., and Guy D.P. 1990. Industrial low back pain, in Weinstein J.N. and Weisel S.W. (Eds.) *The Lumbar Spine.* p 846-859. Philadelphia: WB Saunders, and Co.

Bigos S.J., Battié M.C., Spengler D.M., Fisher L.D., Fordyce W.E., Hansson T.H., Nachemson A.L., and Wortley M.D. 1991. A prospective study of work perceptions and psychological factors affecting the report of back injury *Spine* 11: 252-255.

Bongers P.M., de Winter C.R., Kompier M.A.J., and Hildebrandt V.H. 1993. Psychosocial factors at work and musculoskeletal disease. *Scand. Journ. Work Environ. Health* 19: 297-312.

Bradley L.A., Haile J.McD., and Jaworski T.M. 1992. Assessment of psychological status using interviews and self-report instruments, in Turk D.C. and Melzack R. (Eds.) *Handbook of Pain Assessment* pp 193-213. New York: The Guilford Press.

Brandstadter J. and Renner G. 1990. Tenacious goal pursuit and flexible goal adjustment: explication and age-related analysis of assimilative and accommodative strategies of coping. *Psychol. Aging.* 5: 58-67.

Brown G.K. and Nicassio P.M. 1987. The development of a questionnaire for the assessment of active and passive coping strategies in chronic pain patients. *Pain* 31: 53-65.

Bunce D. and West M. 1994. Changing work environments: innovative coping responses to occupational stress. *Work and Stress* 8: 319-331.

Burton A.K., Battié M.C., and Main C.J. 1999. The relative importance of biomechanical and psychosocial factors in low back injuries, in Karwowski W. and Marras W.S. (Eds.) *The Occupational Ergonomics Handbook.* Boca Raton, Florida: CRC Press Inc.

Burton A.K., Symonds T.L., Zinzen E., Tillotson K.M., Caboor D., Van Roy P., and Clarys J.P. 1997. Is ergonomic intervention alone sufficient to limit musculoskeletal problems in nurses? *Occup. Med.* 47: 25-32.

Burton A.K., Tillotson M.K., Main C.J., and Hollis S. 1995. Psychosocial predictors of outcome in acute and sub-chronic low back trouble. *Spine* 20: 722-728.

Burton A.K. and Tillotson K.M. 1991. Prediction of the clinical course of low back trouble using multivariable models. *Spine* 16: 7-14.

Carosella A.M., Lackner J.M., and Fuerstein M. 1994. Factors associated with early discharge from a multidisciplinary work rehabilitation program for chronic low back pain *Pain* 57: 69-76.

Cats-Baril W.L. and Frymoyer J.W. 1991. Identifying patients at risk of becoming disabled because of low-back pain: the Vermont Rehabilitation Engineering Center predictive model. *Spine* 16: 605-607.

Clinical Standards Advisory Group. 1994. *Back Pain: Report of a CSAG Committee on Back Pain,* London: HMSO.

Cox T. 1993. Stress research and stress management: putting theory to work. *HSE Contract Research Report no 61.* Sudbury, Suffolk: HSE Books.

Cox T. and Griffiths A. 1995. "Editorial: Guidance for U.K. employers on managing work-related stress. *Work and Stress.* 9: 1-3.

Crisson J.E. and Keefe F.J. 1988. The relationship of locus of control to pain coping strategies and psychological distress in chronic pain patients. *Pain* 35: 147-154.

Crum R.M., Muntaner C., Eaton W.W., and Anthony J.C. 1995. Occupational stress and the risk of alcohol abuse and dependence. *Alcoholism Clinical and Experimental Research* 19: 647-655.

DeGood D.E. and Shutty M.S. Jr. 1992. Assessment of pain beliefs, coping and self-efficacy, in Turk D.C. and Melzack R. (Eds.) *Handbook of Pain Assessment* pp 214-234. New York: The Guilford Press.

Dewe P. 1987. New Zealand ministers of religion: identifying sources of stress and coping strategies. *Work and Stress* 1: 351-363.

Douglas M. 1992. *Risk and Blame.* London: Routledge.

Dworkin S.F. and Whitney C.W. 1992. Relying on objective and subjective measures of chronic pain, in Turk D.C. and Melzack R. (Eds.) *Handbook of Pain Assessment,* pp 429-446. New York: The Guilford Press.

Feuerstein M. 1991. A multidisciplinary approach to the prevention, evaluation and management of work disability. *Journ. of Occup. Rehab.* 1: 5-12.

Feuerstein M. and Zastowny T.R. 1996. Occupational rehabilitation, in Gatchel R.J. and Turk D.C. (Eds) *Psychological Approaches to Pain Management: A Practitioner's Handbook.* New York: The Guilford Press.

Flor H. and Turk D.C.1989. Psychophysiology of chronic pain: do chronic pain patients exhibit symptom specific psychophysiological responses? *Psychol. Bull.* 105: 215-259.

Flor H., Turk D.C., and Birbaumer N. 1985. Assessment of stress-related psychophysiological stress reactions in chronic back pain patients. *Journ. of Consult. & Clin. Psychol.* 53: 354-364.

Fordyce W.E. 1976. *Behavioural Methods for Chronic Pain and Illness.* St. Louis: CV Mosby.

French J.P.R., Rogers W., and Cobb S. 1974. A model of person-environment fit, in Coehlo G.W., Hamburg D.A., and Adams J.E. (Eds.) *Coping and Adaptation.* New York: Basic Books.

Friedman M. and Rosenman R.H. 1974. *Type A: Your Behaviour and Your Heart.* New York: Knoft.

Gatchel R.J. and Turk D.C. (Eds). 1996. *Psychological Approaches to Pain Management: A Practitioner's Handbook.* New York: The Guilford Press.

Haazen I.W.C.J., Vlaeyen J.W.S., Kole-Snijders A.M.K., van Eek F.D., and van Es F.D. 1994. Behavioural rehabilitation of chronic low back pain: searching for predictors of treatment outcome *Journ. of Rehab. Sci.* 7: 34-43.

Halpern M. Prevention of low back pain: basic ergonomics in the workplace. 1992, in Nordin M. and Vischer T.L. *Clinical Rheumatology: International Practice and Research. Common Low Back Pain: Prevention of Chronicity* 6: 705-730. London: Bailliere Tindall.

Hazard R.G., Haugh L., Green P., and Jones P. 1994. Chronic low back pain: the relationship between patient satisfaction and pain, disability and impairment outcomes. *Spine*: 19: 881-887.

I.A.S.P., 1995. *Back Pain in the Workplace: Management of Disability in Nonspecific Conditions.* Seattle: IASP Press.

International Association for the Study of Pain. 1979. Pain terms: a list with definitions and notes on usage. *Pain* 6: 249

Jensen M.P., Turner J.A., Romano J.M., and Karoldy P. Coping with chronic pain: a critical review of the literature. *Pain* 1991: 249-283.

Kobasa S. 1979. Stressful life events, personality and health: an enquiry into hardiness. *Journ. of Pers. and Soc. Psychol.* 37: 1-13.

Karasek R.A. 1979. Job demands, job decision latitude and mental strain: implications for job redesign. *Admin. Sci. Quart.* 24: 205-308.

Kasl S.V. 1992. Surveillance of psychological disorders in the workplace, in Keita G.P. and Sauter S.L. (Eds.) *Work and Wellbeing. An Agenda for the 1990s.* Washington, D.C.: American Psychological Association.

Keefe F.J. and Hill R.W. 1985. An objective approach to quantifying pain behaviour and gait patterns in low back pain patients. *Pain* 21: 153-161.

Keefe F.J., Brown G.K., Wallaston K.A., and Caldwell D.S. 1989. Coping with rheumatoid arthritis pain: catastrophising as a maladaptive strategy *Pain* 37: 51-56.

Klenerman L., Slade P.D., Stanley I.M., Pennie B., Riley J.P., Atchison L.E., Troup J.D.G., and Rose M.J. 1995. The prediction of chronicity in patients with an acute attack of low back pain in a General Practice setting. *Spine* 20: 478-484.

Kobasa S., Maddi S., and Courington S.1981. Personality and constitution as mediators in the stress-illness relationship. *Journ. of Health and Soc. Behav.* 22:368-378.

Lazarus R.S. and Folkman S. 1984. *Stress, Appraisal and Coping.* New York, Springer Publications.

Lehmann T.R., Spratt K.F., and Lehmann K.K. 1993. Predicting long term disability in low back injured workers presenting to a spine consultant. *Spine* 18: 1103-1112.

Leiter M. 1991. The dream denied: professional burnout and the constraints of human service organisation. *Canad. Psychol.* 32: 547-558.

Letham J., Slade P.D., Troup J.D.G., and Bentley G. 1983. Outline of a fear-avoidance model of exaggerated pain perception. Part 1. *Behav. Res. and Ther.* 21: 401-408.

Lindstrom I., Ohlund C., and Eek C.E. 1992. Graded activity of subacute low back pain patients. A randomised prospective clinical study with an operant conditioning behavioural approach. *Physical Therapy* 72: 279-293.

Linton S.J., Bradley L.A., Jensen I., Spangfort E. and Sundell L. 1989. The secondary prevention of low back pain: a controlled study with follow-up. *Pain* 54: 353-359.

Main C.J., Hollis S. and Roberts. A. (submitted). Psychological predictors of outcome of acute low back pain. *Journal of Psychosomatic Research.*

Main C.J. and Waddell G. 1987. Personality assessment in the management of low back pain. *Clin. Rehab.* 1 139-142.

Main C.J., Wood P.L.R., Hollis S, Spanswick C.C., and Waddell G. 1992. The distress assessment method: A simple patient classification to identify distress and evaluate risk of poor outcome. *Spine* 17: 42-50.

Main C.J. and Spanswick C.C. 1991. Pain: psychological and psychiatric factors. *Br. Med. Bull.* 47:732-742.

Main C.J. and Spanswick C.C. 1995. Functional overlay and illness behaviour in chronic pain: distress or malingering? Conceptual difficulties in medico-legal assessment of personal injury claims. *Journ. of Psychosom. Res.* 39: 737-753.

Main C.J. and Waddell G. Cognitive measures in pain. 1991. *Pain* 46: 287-298.

Main C.J. and Watson P.J. 1996. Guarded movements and the development of chronicity. *Journ. of Musculoskeletal Pain* 4:163-170.

Mayer T. and Gatchel R.J. 1988. *Functional Restoration for Spinal Disorders: the Sports Medicine Approach.* Philadelphia: Lea and Febiger.

Medcof, J.W. and Hausdorf P.A. 1995. Instruments to measure opportunities to satisfy needs, and degree of satisfaction in the workplace. *Journ. Occup. and Organis. Psychol.* 68:193-208.

Melzack R. 1987. The short form McGill Pain Questionnaire. *Pain* 30: 191-197.

Melzack R and Katz J. 1992. The McGill Pain Questionnaire: Appraisal and current status, in Turk D.C. and Melzack R. (Eds.) *Handbook of Pain Assessment* pp. 152-168. New York: The Guilford Press.

Melzack R. and Wall P.D. 1965. Pain mechanisms: a new theory. *Science* 150: 971-979.

Miller R.P., Kori S.H., and Todd D.D. 1991. The Tampa Scale. Unpublished report. Tampa Florida (reported in Vlaeyen et al., 1995b).

Miller S., Brody D., and Summerton J. 1988. Styles of coping with threat: implications for health. *Journ. of Pers. and Soc. Psychol.* 54: 142-148.

Nicholas M.K., Wilson P.H., and Goyen J. 1992. Comparison of cognitive-behavioural group treatment and an alternative non-psychological treatment for chronic low back pain. *Pain* 48: 339-347.

Nordin M., Cedraschi C., Balangue F., and Roux E.B. 1992. Back schools in the prevention of chronicity, in Nordin M. and Vischer T.L. *Clinical Rheumatology: International Practice and Research. Common Low Back Pain: Prevention of Chronicity* 6: 685-703. London: Bailliere Tindall.

Ottenbacher K. and DiFabio R.P. 1985. Efficacy of spinal manipulation/mobilization therapy — a meta-analysis. *Spine* 10:833-837.

Parker H., Wood P.L.R., and Main C.J. 1995. The use of the pain drawing as a screening measure to predict psychological distress in chronic low back pain. *Spine* 20: 236-243.

Payne R. 1988. Individual differences in the study of occupational stress, in Cooper C.L. and Payne R. (Eds.) *Causes, Coping and Consequences of Stress at Work.* Chichester: Wiley and Sons.

Pearlin L. and Schooler C. 1978. The structure of coping. *Journ. of Health and Soc. Behav.* 19: 2-21.

Pearson C.A.L. 1995. The turnover process in organisations: An exploration of the role of met-unmet expectations. *Human Relations* 48: 405-420.

Porter C.S. 1988. Accident proneness: a review of the concept. *Int. Revs. of Ergonomics* 2:177-206.

Ransford A.O., Cairns D., and Mooney V. 1976. The pain drawing as an aid to the psychological evaluation of patients with low back pain. *Spine* 1: 127-134.

Richards J.S., Nepomuceno J.A., Riles M., and Suer Z. 1982. Assessing pain behaviour: the UAB pain behaviour scale *Pain* 14: 393-398.

Rosenstiel A.K. and Keefe F.J. 1883. The use of coping strategies in chronic low back pain patients: relationship to patient charactereistics and current adjustments. *Pain* 17: 33-34.

Rudy T.E., Kerns R.D., and Turk D.C. 1988. Chronic pain and depression: towards a cognitive mediation model. *Pain* 35: 129-140.

Sandstrom J. and Esbjornsson E. 1986. Return to work after rehabilitation: the significance of the patient's own prediction. *Scand. Journ. Rehab. Med.* 18:29-33.

Schaefer J.A. and Moos R.H. 1993. Relationship, task and system stressors in the health care workplace. *Journ. of Community and Applied Soc. Psychol.* 3: 285-298.

Scheer S.J., Radack K.L., and O'Brien D.R. 1995. Review article: Randomised controlled trials in industrial low back pain relating to return to work. Part 1. Acute interventions. *Arch. Phys. Med. Rehab.* 76:966-973.

Schlapbach P. and Gerber N.J. 1991. Back school. *Rheumatology* 14: 25-33.

Schmitz U., Saile H., and Nilges P. 1996. Coping with chronic pain: flexible goal adjustment as an interactive buffer against pain-related distress. *Pain* 67: 41-51.

Slade P.D., Troup J.D.G., Letham J., and Bentley G. 1983. The fear-avoidance model of exaggerated pain perception. Part 2. *Behav. Res. and Ther.* 21: 409-416.

Stankovic R. and Johnell O. 1990. Conservative treatment of acute low back pain: a prospective randomised trial. Mckenzie method of treatment vs. patient education in "Mini Backschool." *Spine* 15:120-123.

Symonds T.L., Burton A.K., Tillotson K.M., and Main C.J. 1995. Absence resulting from low back pain can be reduced by psychosocial intervention at the workplace. *Spine* 20: 2738-2745.

Symonds T.L., Burton A.K., Tillotson K.M. and Main C.J. 1996. Do attitudes and beliefs influence work loss due to low back trouble? *Occup. Med.* 46: 25-32.

Tan J.C., Roux E.B., Dunand J., and Vischer T.L. 1992. Role of physical therapy in the management of common low back pain, in Nordin M. and Vischer T.L. *Clinical Rheumatology: International Practice and Research. Common Low Back Pain: Prevention of Chronicity* 6: 629-655. London: Bailliere Tindall.

Toomey T.C., Mann J.D., Abashian S., and Thompson-Pope S. 1991. Relationship between perceived self-control of pain, pain description and functioning. *Pain* 45: 129-133.

Turk D.C. and Main C.J. 1997. Resource review: back pain in the workplace. *American Pain Society Bulletin.* 7(3): 11-13.

Versloot J.M., Rozeman A., vanSon A.M., and van Akkerveeken P.F. 1992. The cost-effectiveness of a back school program in industry. 1992. *Spine* 17: 22-27.

Vlaeyen J.W.S., Kole-Snijders A.M.K., Rotteviel A.M., Ruesink R., and Heuts P.H.T.G. 1995b. The role of fear of movement/(re)injury in pain disability. *Journ. Occup. Rehab.* 5: 235-252.

Vlaeyen J.W.S., Kole-Snijders A.M.J., Boeren R.G.B., and van Eek H. 1995a. Fear of movement/(re)injury in chronic low back pain and its relation to behavioural performance, *Pain* 62: 363-372.

Waddell G. 1992. Biopsychosocial analysis of back pain, in Nordin M. and Vischer T.L. (Eds.) *Common Low Back Pain: Prevention of Chronicity,* pp 523-557. London: Bailliere Tindall.

Waddell G., McCulloch J.A., Kummel E., and Main C.J. 1984. Symptoms and signs: physical disease or illness behaviour. *British Medical Journal* 289: 739-741.

Waddell G., Somerville D., Henderson I., Newton M., and Main C.J. 1993. A fear avoidance beliefs questionnaire (FABQ) and the role of fear avoidance beliefs in chronic low back pain and disability. *Pain* 52: 157-168.

Waddell G., McCulloch J.A., Kummell E., and Venner R.M. 1980. Nonorganic physical signs in low back pain. *Spine* 5:117-125.

Wallston K.A., Wallston B.S., and DeVellis R. 1978. Development of the multidimensional health locus of control scale (MHLC). *Health Educ. Monogs.* 6: 160-170.

Watson P.J., Booker C.K., Main C.J., and Chen A.C.N. 1997a. Surface electromyography in the identification of chronic low back pain patients: the development of the flexion relaxation ratio. *Clin. Biomechan.* 12: 165-171.

Watson P.J., Booker C.K., and Main C.J. 1997b. Evidence for the role of psychological factors in abnormal paraspinal activity in patients with chronic low back pain (CLBP). *Journ. of Musculoskeletal Pain.* 5: 41-56.

Wood D.J. (1987) Design and evaluation of a back injury prevention program within a geriatric hospital. *Spine* 12: 77-82.

# 35

# Back Pain in the Workplace: Implications of Injury and Biopsychosocial Models

Michele C. Battié
*University of Alberta*

Chris J. Main
*Hope Hospital, Salford, U.K.*

A. Kim Burton
*University of Huddersfield*

The growth of back pain complaints and associated long-term disability in the industrial insurance system defies simple explanations based on concomitant changes in physical environments, work demands, or increases in degenerative or pathological changes to the spine. As has been eloquently chronicled in an historical overview by Allen and Waddell (1989), back symptoms have been present throughout the ages. What appears to have changed dramatically in many societies is public perception of back pain as an "injury" or medical problem, and its effect on work and life in terms of disability.

Long-term disability from low back pain is a relatively new phenomenon in the history of Western civilization, and its dramatic growth in the last half century suggests that factors other than physical pathology may be influencing its development. This is not to say that back pain is not genuinely experienced by large numbers of people, or that severe symptoms do not occasionally cause physical limitations of some duration. On the contrary, numerous health surveys indicate that back symptoms are extremely common in both developed and third-world countries (Waddell, 1987; Anderson, 1984; Honeyman and Jacobs, 1996; Näyha et al., 1991; Bierine-Sørensen, 1984; Troup, 1987; Videman, 1989), and that they begin at an earlier age than once thought (Balague et al., 1988; Burton et al., 1996; Salminen et al., 1995). Yet, only a small percentage of persons experiencing back symptoms each year file industrial injury claims or seek medical care.

Back injury reporting and disability are clearly influenced by other factors than purely the presence of pain. For example, the severity of symptoms and the physical demands of the workplace are likely to interact to affect reporting and absenteeism. While an individual with light physical job demands may be able to continue work through an episode of back pain, the same condition may be sufficiently exacerbated by heavy loading or work in awkward postures, such that the individual is unable to continue with normal work tasks during recovery. In addition to the effects of symptom severity and job demands, other factors of likely importance to the onset and persistence of disability are cultural norms (Anderson,

1984; Waddell, 1987; Honeyman and Jacobs, 1996), socioeconomic conditions (Volinn et al., 1988; Volinn, 1991) and opportunities for compensation (Loeser et al., in press; Leavitt, 1992), attitudes and beliefs about back problems (Symonds et al., 1995, 1996), emotional distress (Bigos et al., 1992; Klitzman et al., 1990), and job satisfaction and the social work environment (Bigos et al., 1992; Linton et al., 1989; Linton and Kamwendo, 1989; Linton, 1990). Fears that continuing work activities will cause further harm also may influence the decision to file a complaint or miss work. Innumerable factors could affect the response to back symptoms, many relating to perceptions of the problem and the perceived benefits and costs of various options available in the social systems and cultural norms in which the person lives and works.

Considering the factors noted above, a biomedical model, focusing purely on physical pathology, has long been recognized as inadequate to explain back pain disability or formulate a sufficient management strategy to combat it. Focusing simply on physical pathology and biomechanical causes and solutions to the problem would appear to be a gross, and occasionally misleading, oversimplification of the back pain experience in the developed countries of the world. While such factors should not be ignored, a broader perspective is needed if growth in work-related back pain disability is to be curbed. Some suggest that biopsychosocial (Waddell, 1987) or biocultural models (Honeyman and Jacobs, 1996) may provide better frameworks for, and insights into, the development and management of this growing problem. In this chapter we will examine the industrial injury model and explore the shift in paradigms over the past decade toward biopsychosocial and sociocultural models of back pain disability and their implications.

## 35.1   Back Pain in the Workplace: An Injury Model

As more is learned about the epidemiology and natural history of back symptoms, that they are endemic and episodic in nature, the more difficult it becomes to apply a simple injury model. The challenges of applying such a model in the working environment were demonstrated by Robertson and Keeve (1983), who reported on the results of efforts to prevent industrial back injury claims by minimizing physical hazards at the workplace through establishing and enforcing safety and health regulations. Although these efforts led to a decrease in objectively verifiable injuries, such as fractures and lacerations, more subjective injury reports, including back strains, were unaffected.

Most back symptoms are of gradual onset, and are not associated with a traumatic accident or unusual activity (Videman et al., 1989; Rowe, 1969; Välfors, 1985). Therefore, it may be inappropriate to refer to all back symptoms occurring in the workplace as injuries, a common practice to meet the demands of the industrial insurance system. Among first-year nurses, who often have relatively high and unpredictable loading conditions in patient handling, symptoms were found to be of sudden onset with or without a specific inciting event in only 28% of cases cited (Videman et al., 1989). The great majority of back pain episodes were of gradual onset. Thus, it may be more appropriate to reserve the term "injury," and the images that the term elicits, for back symptoms of sudden onset following a specific inciting event, such as a fall, when direct trauma is more likely. As Allen and Waddell outline in their historical review of back pain, this condition has not always been viewed in an injury context (Allen and Waddell, 1989).

Hadler noted that once an inguinal herniation was termed a "rupture" and then considered an injury, and a ruptured disc was accepted as a source of back symptoms in the 1930s, back pain also received injury status in the workers' compensation system (Hadler, 1995). He goes on to note that since that time much of what is costly and contentious in workers' compensation systems relates to back injuries. Furthermore, Americans, in general, have come to perceive musculoskeletal discomfort in an injury context. As will be further discussed in later sections, the injury model has other consequences.

### The Influence of Physical Loading

If we followed the paradigm that back pain and resultant disability are consequences of injury caused by mechanical overload, to be prevented by lessening loads on the spine, then it would follow that populations with heavier loads would have a higher incidence of back pain and subsequent disability. This clearly is

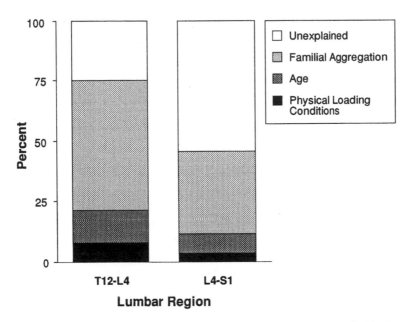

**FIGURE 35.1**   The graph displays the percentage of variability in disc degeneration in the $T_{12}$-$L_4$ and $L_4$-$S_1$ spinal regions explained by specific occupational and leisure physical loading variables relative to that explained by familial aggregation which represents a combination of genetic and shared childhood environmental influences (beyond that explained by age). (Adapted from Battié MC, Videman T, Gibbons LE et al.: Determinants of lumbar disc degeneration. *Spine* 1995; 20:2601-2612.)

not the case, however, as has been discussed previously in Chapter 20 (Burton et al.). It is puzzling that instead, back pain complaints and disability are highest in countries with generally less extreme occupational physical loading conditions and high utilization of medical services.

Volinn (1997) studied the point prevalence of low back pain in low and middle income countries, where a far greater portion of the population is engaged in heavy labor, and often into old age, as compared to relatively high income countries. He hypothesized that if back pain were caused by heavy physical loading, that low income countries with substantially more heavy labor would have a higher point prevalence. Instead, a considerably lower back pain prevalence was found among populations of low income farmers as compared to wealthy populations of Northern European countries that had two to three times higher prevalences. He concluded that "hard physical labor itself is not necessarily related to LBP." He did, however, find that workers from low income countries working in urban "enclosed workshops," such as sewing machine operators, had the highest prevalence of back pain.

As we discussed in Chapter 20 (Burton et al.), even when physical loading is investigated in the etiology of degenerative changes in the lumbar spine, its effects are seldom clear. In a recent study of determinants of disc degeneration (Battié et al., 1995), occupational and leisure time physical loading exposures in adulthood were found to have a minor influence on degree of degenerative findings, particularly when compared to familial aggregation, which represented a combination of genetic and childhood environmental influences (Figure 35.1). It was suggested that this relatively modest influence of physical loading may explain the mixed results of earlier studies of the effects of physical loading on degenerative findings. The investigators concluded, that disc degeneration may result primarily from a combination of genetic influences, childhood exposures, and unidentified factors, which may include complex, unpredictable interactions.

While the role of occupational physical loading may be modest in terms of accelerating disc degeneration, certainly some physical activities may aggravate back symptoms, and ergonomic interventions may make it easier for someone experiencing back problems to continue work or return sooner following an acute or recurrent episode. Appropriate interventions may include accommodations to the work environment or task that minimize extreme loads or awkward postures. It is important, however, to keep the role of ergonomic interventions in perspective when developing more comprehensive management strategies.

## Perceptions of Fault

The common view of back pain as an industrial injury may lead to the perception that fault lies with the work tasks or environment, particularly if symptoms occur during the course of normal work activities. Furthermore, affected workers may feel as though their "injuries" are the result of negligence on the part of their supervisors or employers for placing them in hazardous situations. It has been hypothesized that such perceptions of fault may influence the attitudes and expectations of workers experiencing back symptoms. Placing blame on external sources also may influence the affected worker's sense of responsibility in promoting the recovery process (Bigos and Battié, 1987).

DeGood and Kiernan (1996) were interested in investigating the influence of perceptions of fault on expectations of recovery, psychological distress, and behavioral disturbances. When they grouped patients under their care for chronic pain according to whether they believed that their employer, another (primarily doctors or another driver from a MVA), or that no one was really responsible for their pain, they found the groups were similar with respect to current pain intensity and level of activity limitations. However, those who blamed their employer for their current pain problems were significantly more likely to be unemployed, to report negative responses from past treatments and expect the same from future therapy, to perceive that they had been treated unfairly by their employer before and after the onset of their pain, and to exhibit greater psychological distress than those who felt that no one was to blame. The retrospective nature of this study limits the conclusions, but the findings do support the hypothesis that perceptions of fault may influence pain behavior, psychological disturbances, and poor expectations of recovery.

It also has been suggested that individuals who report back injuries may be more inclined to forget having had minor back symptoms related to performing common activities prior to their "injury," which may reinforce perceptions that the injury is responsible for any subsequent back symptoms (Fagan, 1995).

## Back Pain in the Workers' Compensation System

When a person experiences what is believed to be work-related back symptoms and chooses to file an injury claim he or she enters the workers' compensation paradigm. This paradigm defines work tasks as causal, infers structural damage, and requires medical testimony of injury. Unfortunately, conditions that are identified primarily through symptom complaints, where medicine has yet to clearly identify the underlying pathology, do not fit well into this paradigm. Affected individuals often find themselves in the position of having to demonstrate to health care providers and insurance adjusters that they are indeed injured and worthy of receiving the benefits awarded to such a designation. This situation can place the worker in an adversarial position where his or her integrity is questioned, such that an appropriate response is to demonstrate that the health problem is serious enough for acceptance by the system. Such patients are sometimes viewed with suspicion, and as Leavitt (1992) once stated, "The unfortunate problem is that stereotypes have consequences. Doubts raised by labels often shape evaluation and treatment of industrial workers in problematic ways to the extent that their integrity and status as patients are challenged." This is counter-productive to recovery.

Several studies have suggested poorer outcomes in patients receiving workers' compensation. Studies have found patients on compensation were less likely to benefit from lumbar interbody fusion for "degenerative disc disease" as compared to patients not receiving compensation (Flynn and Hoque, 1979). Compensation also has been associated with residual back pain following surgery (Hanley and Shapiro, 1989) and delayed recovery from conservative treatment (Greenough and Frazer, 1989). Leavitt (1992) pointed out, however, that researchers typically do not account for the influence of physical labor on the outcome of return to work, and that compensation and noncompensation groups usually differ on occupational physical job demands. In a study designed to determine the effects of compensation on work loss while controlling for differences in occupational physical demands, he determined that irrespective of the job performed work-related back pain was associated with prolonged disability duration (Leavitt, 1992).

A later study of 1,366 self-referred patients with back pain of less than 8 weeks' duration was conducted to investigate whether workers with acute low back pain were better served if they were insured through a workers' compensation insurance policy (Hadler et al., 1995). Those with and without workers' compensation insurance were deemed to be similar in clinical status and demographic variables at entry. Yet, while most in both groups returned to full function and work activities rapidly, those with workers' compensation insurance returned to work less convinced that they had recovered to the level of health experienced before the onset of their back pain episode. This perception of compromised health remained even after perceived job characteristics were controlled, diminishing the chances that this finding could be due to differences in work-related physical demands.

Loeser et al. (1995) note that discussions around the future of workers' compensation insurance programs center around understanding: (1) personal and social factors that lead an individual to injury or illness reporting and the adoption of a sick role, with its associated disability and health care utilization, and (2) factors that influence the behavior of workers in a compensation system. They proceeded to conduct a comprehensive review looking at the effects of wage replacement levels on claims behavior. As part of the review, they conducted some reanalyses of previously reported data in an effort to control for possible confounding factors that may have influenced the original interpretation of results. They included 24 studies that directly addressed the relationship between wage replacement level and workers' compensation claims incidence and duration.

In their synthesis of the literature Loeser at al. (1995) conclude that, when other factors are held constant, both the incidence and duration of industrial injury claims in the workers' compensation system are associated with the level of wage replacement benefits. The effect on duration appears to be somewhat larger than that for incidence. The methodologically stronger studies provided estimates of a 1.83 to 11.28% increase in claim duration as a result of a 10% increase in workers' compensation benefits. Evidence was mixed with respect to a relationship to underlying injury rates. The challenge would appear to be finding the appropriate balance between financially supporting an injured worker to prevent major financial hardship during recovery, while providing financial incentive for timely return to work and discouraging unnecessarily extended disability.

Hadler et al. (1995) urge an examination of whether or not regional backache belongs in the workers' compensation paradigm, under the "injury rubric." They suggest that the industrial injury distinction received by back symptoms has not led to a safer, healthier workplace, nor has it forwarded the understanding or management of this problem. Over past decades the frequency of workplace injuries has been relatively constant, but their consequences for workers have grown dramatically. As evidence, Hadler et al. note that disability from work-related injuries has steadily climbed, as measured through workdays lost. In addition, the cost of workers' compensation in the United States in the two decades following 1971 increased 12-fold, and went from 0.5% to 2.5% of the national payroll. Such increases are not exclusive to the United States; similar growth in back-related disability has been noted in other Western countries as well.

A controversial discussion document, *Back Pain in the Workplace: Management of Disability in Nonspecific Conditions* (1995), reached similar conclusions to those discussed above. Yet, reconstructing a more successful and humane system of dealing with disability or work incapacity associated with nonspecific symptom complaints faces many challenges. Hopefully, such documents as *Back Pain in the Workplace*, while stirring controversies, will also advance the discussion of constructive reforms in the medical management and social systems that have been dismally inadequate in dealing with this problem.

## 35.2   Back Pain in the Workplace: A Biopsychosocial Model

A decade ago, in a Volvo Award winning paper, Waddell (1987) presented a strong argument for the adoption of a biopsychosocial model of illness for the management of low back problems. At the core of this model is the acknowledgment and integration of physical, psychological, and social aspects of illness. Such a model demands a broad approach to considering factors influencing illness and recovery. Since the time Waddell encouraged the use of a biopsychosocial model, evidence supporting the importance of psychosocial factors in back pain reporting and disability has steadily grown. Much of this evidence has been presented in the previous chapters.

Most studies of back problems and suspected risk factors have been retrospective or cross-sectional in design and have been quite limited with respect to inferences regarding risk. Currently, a number of prospective cohort studies specifically investigating predictors of work-related back pain complaints have been reported (Battié et al., 1989a; Battié et al., 1989b; Battié et al., 1990; Bigos et al., 1992; Cady et al., 1979; Chaffin and Park, 1973; Dueker et al., 1994; Mostardi et al., 1992; Ready et al., 1993; Rossignol et al., 1993; Venning, 1987). Most of these studies have focused on the effects of individual physical factors and medical history. One of the most comprehensive prospective studies of predictors of work-related back pain complaints and disability was conducted among Boeing employees (Bigos, 1992). This study demonstrated that the role of psychosocial factors in risk of future injury complaints and disability was greater than that of individual physical factors, such as lifting strength, aerobic capacity, and range of motion which proved to be very poor predictors (Battié et al., 1989a, 1989b, 1990), findings that have been supported by subsequent prospective cohort studies (Mostardi et al., 1992; Ready et al., 1993; Dueker et al., 1994). Specifically, emotional distress, as determined from the Minnesota Multiphasic Personality Inventory, and poorer job task enjoyment and co-worker relationships were implicated in risk of reporting work-related back problems. A later analysis looking at perceptions of the work environment in a broader context of work-related injury and illness reporting revealed that lower job task enjoyment and co-worker relations had a similar influence on the reporting of other types of injuries, as well (Battié et al., 1993). Thus, there is no indication from the Boeing study that workers who file back injury claims differ significantly from workers claiming other types of injuries with respect to these work perceptions. It should be emphasized, however, that such workplace perceptions, while associated with a higher risk of reporting, are poor predictors of precisely who will and will not report a future industrial injury. While most studies that have included factors related to job satisfaction have found associations with back pain reporting and work loss (Bigos et al., 1992; Ready et al., 1993; Magnusson et al., 1996), all have not (Symonds et al., 1996). The inconsistency may be partly due to differences in question presentation and partly to differing philosophies among occupational groups; these issues require further study.

As was discussed in the previous chapter, other perceptions or beliefs have been linked to the development of long-term disability. The fear–avoidance model was first proposed by Lethem et al. in 1983 to explain exaggerated pain perception and behavior resulting from relatively minor nociception and underlying pathology. Fear of pain is the central concept of the model, with the two extreme responses being confrontation or avoidance. Confrontation results in resumed social and physical activities which regularly tests the pain relative to activities and allows the individual to calibrate a response, such that activity and pain level remain in synchrony. On the other hand, avoidance leads to withdrawal of normal activities and an increased likelihood of encountering and adapting to various positive and negative reinforcers of chronic pain. They hypothesized and provided data to support that confrontation or avoidance responses depend on the psychosocial context (stressful life events, personal pain history, personal coping strategies, and personal characteristics) in which the symptoms originate and management occurs (Lethem et al., 1983; Slade et al., 1983). Philips (1987) later hypothesized that "avoiding stimulation plays an active part in reducing the sufferer's sense of control over pain, and in increasing his or her expectation that exposure will increase pain. These cognitive changes encourage further withdrawal from normal activities and a growing intolerance of stimulation." Over the past decade the fear–avoidance model has been studied and supported as having an important role in the development of chronic pain (Klenerman et al., 1995).

Szpalski et al. (1995) conducted a population-based survey in Belgium that revealed that health beliefs related to whether back problems experienced would be lifelong or not was significantly associated with current health behaviors and history of health care utilization. Another study examining beliefs about back problems among British office and manufacturing workers found those who had missed more than one week of work due to low back problems, as compared to those with less time loss or no reported back problems, were significantly more likely to have negative attitudes and beliefs (Symonds et al., 1996). Specifically, they were more likely to believe that negative consequences of back pain were inevitable and to feel less control and responsibility over pain. It is not clear, however, from such a retrospective study design whether such beliefs influenced the time loss or whether the episodes that led to greater time loss led to more negative beliefs. However, a prospective study designed to change negative beliefs about back problems suggests that such a change may positively influence subsequent back-related work loss, strengthening the hypothesis that beliefs about back problems may directly influence illness behavior (Symonds et al., 1995).

## Sociocultural Influences

It is generally accepted that back pain is common in populations throughout the world, yet the interpretation of the experience and associated behaviors differ significantly depending upon cultural norms. When studying health problems in rural Nepal, Anderson (1984) reported that although back symptoms were very common, virtually no one sought care for the symptoms when medical services were made available. It appeared that back pain problems were not perceived as a medical problem, but rather a common consequence of normal living and aging. Later, Waddell (1987) reported that before the introduction of Western medicine in Oman, extended disability for back pain problems was virtually nonexistent. Similar to the experience in Nepal, while persons were crippled by such diseases as polio or tuberculosis and injuries involving spinal fractures, they did not stop daily life or become permanently disabled from idiopathic back pain problems. Back pain reportedly was, however, a very common condition there, as is the case in Western countries.

In a study of back pain in an Australian Aboriginal community (Honeyman and Jacobs, 1996), nearly one third of men and half of women reported problems with long-term low back pain when privately questioned. However, the Aboriginals did not perceive back pain as a health issue, and consequently did not report such symptoms openly, display pain-related behaviors, or seek medical care. The result of this study led Honeyman and Jacobs to suggest that "cultural beliefs and practices influence how people respond to back pain in themselves and in others, including how and whether they present to health professionals or seek involvement of others."

Sanders et al. (1992) investigated cross-cultural similarities and differences in Sickness Impact Profile scores of patients seeking help for back pain problems at pain clinics in the United States, Japan, Mexico, Columbia, Italy, and New Zealand. They found levels of psychosocial, vocational, and avocational impairment to be greatest in the Americans, followed by the New Zealanders and Italians. The authors note that many sociocultural factors could be responsible for these differences, such as work ethic, social support and acceptance, economic stability and entitlement issues, beliefs and coping strategies, to name a few. The numbers of representative patients included in the study were small, however, and further work will be needed to more clearly establish differences in chronic low back pain patients between these cultures and to explore the causes of such differences.

Similarly, an earlier study found more interference with social, sexual, recreational, and vocational activities in Americans attending a pain clinic for chronic low back pain than their counterparts in New Zealand (Carron et al., 1985). The study investigators hypothesized that substantial differences in the disability systems, rather than inherent cultural differences, may have accounted for the differences between patient groups. The study's limitations, however, make such inference speculative. Interestingly, while a similar percentage of patients from each group attributed the onset of their pain to a specific incident (74% vs. 71%), patients in the United States were three times as likely to cite a work-related accident, whereas New Zealanders, with a no fault insurance system were just as likely to cite a non-work-related incident.

## Work-Related Disability Management Interventions

There have been several interventions designed to minimize disability from back problems that appear promising and worthy of further examination. They are based on a biopsychosocial model and the belief that psychosocial factors, as well as physical factors, influence response to injury and illness, particularly return to work. The interventions have several commonalities; they are employer-based, involve changing attitudes and responses to back symptom complaints, and focus on decreasing the negative consequences of back problems once reported. They concentrate on improving the quality and quantity of communication with the "injured" worker and other parties involved, and support temporary work modifications, if necessary.

Two reports in the mid 1980s described a program designed to mitigate the negative consequences of back problems among employees of a shoe-parts manufacturer (Fitzler and Berger, 1982, 1983). The employer previously had been active in ongoing safety training and enforcing adherence to regulations and had redesigned work tasks to eliminate unnecessary or frequent heavy exertions, but continued to have significant problems with back injury complaints and subsequent disability. Therefore, the company began a new strategy that emphasized changing attitudes and responses of management personnel toward back pain reporting and encouraging prompt, appropriate medical attention. Management training emphasized that back pain symptoms are very common in the adult population and that malingering is rare. Recognizing employee needs also was emphasized, along with responding empathetically. To meet the goal of prompt medical attention, symptom relief agents were provided onsite through the employer's medical service. The affected employee also was provided with education in back care, including treatment and recovery expectations and a discussion of any activity restrictions. If in-house care was deemed insufficient, referral was made to an outside physician. Attempts were made to modify work activities, if needed, to allow the employee to continue working or return to work more quickly during recovery.

There was a dramatic and sustained decrease in workers' compensation costs for the three years following the implementation of the program, as compared to the three prior years. Workers' compensation costs decreased during the first year of the program to approximately 20% of what they had been in each of the prior three years, and further decreased to 10% during the second year. This decrease was sustained during the third year of the program. Because the efforts of the program were determined simply by examining the experiences with back problems within the company before and after the program was implemented, it is possible that other changes occurring at the same time may have been partly responsible for the apparent effects of the program. There is also the Hawthorne effect to consider, employee behavior may have been affected temporarily simply in response to a change in the system. Yet, there are several points in support of the effect being due to the intervention. First, there was a substantial decrease in related work loss and medical and indemnity costs, but not in the rate of back symptom reporting. This result supports the specific intent of the program to minimize the negative consequences of back problems once reported, and not to influence reporting. Second, the apparent effects of the program endured over the three years of observation following implementation.

Subsequently, Battié reported on an intervention with several similar traits conducted in a long-term health care facility (Battié, 1994). The goal was to investigate the effects of an intervention designed to create a more supportive and accommodating work environment following injury reporting. The intervention focused on improved communication and provisions for temporary accommodations in work activities, if needed, to allow an earlier return to the workplace. A system was put in place to ensure regular contact was maintained with employees throughout recovery. The contacts were made by either the employee's supervisor or the safety and health representative, and expressed concern for the employee's well-being, offered assistance, and let the employee know that he or she was missed. A special training session was held for all supervisors to improve their understanding of back problems and encourage empathetic attitudes and responses to back symptom complaints. There was a 77% reduction in the mean back injury claim cost during the intervention year, as compared to the claims filed in the prior three years. Furthermore, none of the back injury claims filed were over $5,000, as compared to 14% of claims filed in the prior three years at the intervention site or among other long-term health care facilities in the region during the year of intervention. This intervention was of a short duration (only one year) and results may be due, in part, to the normal fluctuations commonly experienced in injury rates and costs. It is interesting to note, however, that similar to the previously reported experience in the shoe-parts manufacturing company, the incidence of back injury claims appeared unaffected during the intervention. It was primarily work loss and associated costs that decreased, which were the outcomes specifically targeted by the intervention.

In contrast, Greenwood et al. (1990) reported on an early intervention program among coal miners, initiated by the West Virginia Workers' Compensation Fund that involved early case management by an independent rehabilitation firm that acted as an advocate for the patient. Unlike employer-based disability management approaches, the intervention failed to lower medical costs or prevent extended disability. It would appear that the support of the workplace may be of critical importance in the success of disability management approaches that rely on such elements as improved communication and the availability of modified work duties, when needed.

Wood (1987) also reported on the effects of an intervention in a group of long-term health care facilities designed to decrease the duration of wage loss claims, similar to the subsequent study by Battié (1994), which was described previously. The "personnel program," as it was called, focused on improving communications between all the parties involved in the recovery of the affected employee, including: the employee, his or her doctors, the Workers' Compensation Board, and the hospital personnel. A hospital representative immediately contacted the employee and the workers' compensation representative after the report of back symptoms. The employee then was contacted regularly. The tone of the communications was empathetic and helpful, and included the message, "You are a vital part of the hospital team. Your work is important and your job is waiting." The hospital representative served as a liaison between the various parties involved. After the adoption of the "personnel program," the proportion of high time-loss claims significantly decreased, as compared with the prior period (7.1 vs. 1.7%). There also was a reversal in the trend of increasing claims rates over the prior years.

Wood also examined the effects of a training program to improve work practices. Although the training program was popular among the staff, no added benefit beyond that received from the "personnel program" could be determined based on quantitative data. In general, despite its popularity, there has been little evidence to suggest that educational approaches presenting biomedical and ergonomic principles alone are effective in primary prevention of back symptom reports and disability (Scheer et al., 1995). There have been reports of other approaches to education, however, that appear more promising. They are based on a biopsychosocial model of the problem.

Symonds et al. (1995) implemented an education program among workers in a British biscuit manufacturing factory involving the distribution of written information aimed at changing common misconceptions about back problems. The information was based on the fear–avoidance model and aimed at enforcing active coping strategies, rather than passive, avoidance behavior. When comparing the employee group receiving the educational information to a control group, a decrease in incidence of back pain reports requiring brief absences was observed in both groups during the follow-up period, but the intervention group had significantly less extended sickness absence resulting from low back trouble.

Another educational program with encouraging results was reported by Versloot et al. (1992). The findings of the longitudinal field study in a Dutch bus company provided some evidence to support a decrease in overall length of absenteeism related to the program, although the incidence of time loss episodes was unaffected. The program emphasized psychosocial factors, including responsibility of one's own health, interactions between mind and body in relation to illness, managing stress and coping strategies, as well as ergonomic advice specifically tailored to work and leisure activities.

Recognizing that back problems are most often recurrent in nature, several worksite programs have aimed to decrease subsequent problems in employees with back problem histories by encouraging regular exercise. Although the benefits of exercise in the prevention and treatment of back problems can be debated, and appear to vary depending upon the stage of symptoms (Fass, 1996), several workplace exercise interventions targeting employees with previous back problems report positive effects (Donchin et al., 1990; Kellett et al., 1991; Gundewall et al., 1993). The various exercise programs were offered through sessions at the workplace during regularly scheduled work time. In two of the programs, exercise compliance and greater gains in "fitness" outcomes were found along with fewer symptoms and work loss (Donchin et al., 1990; Gundewall et al., 1993). However, another study reporting significantly less work loss attributable to program participation did not find a concomitant change in "fitness" measures (Kellett et al., 1991). This raises questions about whether physical or psychosocial aspects of the program were most responsible for the observed decrease in subsequent work loss. These interventions suggest that employer-supported activity programs specifically for employees with a history of back problems may decrease subsequent problems and be cost-effective. Whether apparent benefits are due to the physiological effects of exercise or the many possible psychosocial effects of the programs, including a more positive attitude toward the employer, is unclear.

Other interventions to control disability from work-related back problems have aimed at improving the quality of medical care through the use of quality-based, standardized diagnostic and treatment protocols. Among the motivations for developing such protocols was to minimize apparently haphazard variations in care. Wiesel and his coworkers (1984) were among the first to develop and investigate the effects of using such a system for the medical management of industrial back incident reports. They demonstrated a significant and sustained decrease in lost workdays and medical and compensation costs following the implementation of the program. Interestingly, fewer back injury incident reports also were reported, an unexpected finding given that the intervention was designed to improve care once back problems were reported. Since Wiesel et al. presented their approach, other similar models have been developed in an effort to improve worker/patient outcomes (Wiesel et al., 1985, 1994; Lonstein and Wiesel, 1988; Tufo et al., 1991), but not all have achieved similar findings (Battié et al., 1995). One aspect of the intervention reported by Wiesel et al. that may have significantly influenced the success of their program was their strong working relationship with the employers of the intervention sites. They worked closely with the employers and had support and cooperation when, for example, temporary work accommodations were needed to encourage earlier return to work.

## Practical Considerations for Workplace-Based Management of Low Back Problems

Judging the effectiveness of workplace interventions is challenging. There can be difficulties in selecting adequate control or comparison groups and good results often observed during short-term follow-up may be simply a brief manifestation of the "Hawthorne effect." The many unpredictable changes that can occur in the work environment related to labor–management relations, administrative processes in the industrial insurance system, as well as the imposition of other safety and health initiatives, to name just a few, further complicate the evaluation of specific interventions. Yet the examples mentioned above do warrant further attention and point to several practical considerations for safety and health professionals involved in workplace-based interventions to minimize back-related disability.

First they suggest the need to adopt the broader perspective of a biopsychosocial model in developing injury management strategies. Such a model recognizes that a variety of psychological and sociocultural factors, as well as physical factors, can influence back problems and related work loss. Several specific considerations follow.

1. Improve the quality and quantity of communication between the workplace and the employee following an injury report. Early contact should express genuine concern and offer assistance, if possible. Contact should be maintained with employees throughout recovery to let them know they are missed and will be welcomed back when well enough to return (Wood, 1987; Battié, 1994). Management training may be a necessary step toward changing negative attitudes and beliefs about back injury claimants to create a more responsive environment of trust and employee advocacy (Fitzler and Berger, 1982, 1983; Battié, 1994).
2. The importance of maintaining as normal a life style as possible during recovery should be stressed. Recent randomized clinical trials have demonstrated that even in the acute stages of back problems, advising the patient not to fear activity, but rather to maintain as much normal activity as tolerated

is more effective in promoting recovery than prescribing specific exercises or other common treatment approaches (Fass, 1996; Malmivaara, 1995). Along the same line, temporary work modifications should be made when possible to allow an employee to return to activities in the workplace as soon as medically advisable (Fitzler and Berger, 1982, 1983; Wood, 1987; Battié, 1994).

3. Educational approaches should include messages that promote realistic, positive attitudes and beliefs about back problems and their consequences. In particular, consider promoting positive coping strategies and an active approach to confronting and managing symptoms to discourage fear–avoidance reactions and maladaptive responses (Versloot et al., 1992; Symonds et al., 1995). Information on biomechanics and ergonomic advice should be practical and tailored specifically to the employees' work and leisure activities.

4. If health promotion programs involving exercise are being considered, it may be most cost-effective to target individuals with a history of recent or recurrent back problems (Keller et al., 1991; Gundewall et al., 1993). While the employee may be expected to do some exercise on his or her own time, having some sessions during work hours, or offering other types of employer support, may send a powerful message about the employer's commitment to employee well-being.

5. Consider ways of bolstering employees' perceptions of the workplace and their personal resources to discourage maladaptive approaches to handling problems. Management style and labor–management relations would appear to be important aspects of such perceptions, influencing both job satisfaction and stress (Cox, 1993).

6. Implement any new intervention in such a way that its effects can be evaluated. The interventions discussed earlier, while promising, are in need of further study to verify their effects in other environments and to better understand the specific mechanisms through which they influence back-related disability and work loss. There are many examples of approaches that have reported significant positive results in some environments, but have not been successful when duplicated in others.

## 35.3 Summary

The aim of this chapter was to highlight some of the apparent strengths and shortcomings of two competing models of back problems in the workplace, rather than offering an extensive literature review. We have had greater than a half-century of experience using the injury model, stressing the role of biomechanical factors in the etiology and management of back problems, and it has been an unequivocal failure. This is not to say that biomechanical considerations and ergonomic interventions have no value in the management of back problems in the workplace, but rather that they are insufficient in preventing the development and persistence of long-term work incapacity. It is time for a new paradigm. Many psychosocial and cultural factors influence the response to back symptoms, in terms of reporting and disability. Many of the factors appear to relate to attitudes and beliefs about back problems and the perceived costs and benefits of the various options available in the sociocultural systems in which we live and work. In this broader perspective of the problem lie opportunities to develop more effective approaches to curbing the impact of back problems in the workplace.

## References

Allen BA, Waddell G: An historical perspective on low back pain and disability. *Acta Orthop Scand Suppl* 1989; 60:1-23.

Anderson RT: An orthopaedic ethnography in rural Nepal. *Med Anthropol* 1984; 8:46-59.

Balague F, Dutoit G, Waldburger M: Low back pain in school children. *Scand J Rehab Med* 1988; 20:175-179.

Battié MC, Bigos SJ, Fisher LD, Hansson TH, Nachemson AL, Spengler DM, Wortley MD, and Zeh J: A prospective study of the role of cardiovascular risk factors and fitness in industrial back pain complaints. *Spine* 1989a; 14(2):141-147.

Battié MC, Bigos SJ, Fisher LD, Hansson TH, Jones ME, and Wortley M: Isometric lifting strength as a predictor of industrial back pain reports. *Spine* 1989b; 14(8):851-856.

Battié MC, Bigos SJ, Fisher LD, Spengler DM, Hansson TH, Nachemson AL, and Wortley MD: The role of spinal flexibility in back pain complaints within industry: A prospective study. *Spine* 1990; 15:(8):768-773.

Battié MC, Bigos SJ, Fisher LD, Fordyce WE, and Gibbons LE: *The Effect of Psychosocial and Workplace Factors on Back Related and Other Industrial Injury Claims.* The International Society for the Study of the Lumbar Spine, Marseilles, France, June 15-19, 1993.

Battié MC: *The Effect of Improved Communication and Work Accommodations on Back Injury Claims.* The International Society for the Study of the Lumbar Spine, Seattle, WA, June 21-25, 1994.

Battié MC, Videman T, Gibbons L, Fisher L, Manninen H, and Gill K: Volvo Award in Clinical Sciences 1995. Determinants of lumbar disc degeneration: A study relating lifetime exposures and MRI findings in identical twins. *Spine* 1995; 20:2601-2612.

Bergquist-Ullman M, Larsson U: Acute low back pain in industry. A controlled prospective study with special reference to therapy and confounding factors. *Acta Orthop Scand* 1977; 170:1-117.

Berwick DM, Budman S, Feldstein M: No clinical effect of back schools in an HMO. A randomized prospective trial. *Spine* 1989; 14(3):338-344.

Biering-Sørensen F: Physical measurements as indicators for low back trouble over a one-year period. *Spine* 1984; 9:106-119.

Bigos SJ, Battié MC: Acute care to prevent back pain disability: Ten years of progress. *Clin Orthop* 1987; 221:121-130.

Bigos SJ, Battié MC, Spengler DM, Fisher LD, Fordyce WF, Hansson T, Nachemson AL, Zeh J: A longitudinal, prospective study of industrial back pain reporting. *Clin Orthop Relat Res* 1992; 279:21-34.

Burton AK, Clarke RD, McClune TD, Tillotson KM: The natural history of low back pain in adolescents. *Spine* 1996; 21(20):2323-2328.

Cady LD, Bischoff DP, O'Connell ER, et al: Strength and fitness and subsequent back injuries in fire-fighters. *J Occup Med* 1979; 21:269-272.

Carron H, DeGood DE, Tait R: A comparison of low back pain patients in the United States and New Zealand: Psychosocial and economic factors affecting severity of disability. *Pain* 1985; 21:77-89.

Chaffin DB, Park KS: A longitudinal study of low back pain as associated with occupational weight lifting factors. *Am Ind Hyg Assoc J* 1973; 34:513-525.

Cox T: Stress research and stress management: putting theory to work. HSE Contract Research Report No. 61. Sudbury, Suffolk: HSE Books, 1993.

DeGood DE, Kiernan B: Perception of fault in patients with chronic pain. *Pain* 1996; 64:153-159.

Donchin M, Woolf O, Kaplan L, Floman Y: Secondary prevention of low back pain. A clinical trial. *Spine* 1990; 15(12):1317-1320.

Dueker JA, Ritchie SM, Knox TJ, et al: Isokinetic trunk testing and employment. *J Occup Med* 1994; 36:42-48.

Fagan JM, Rehm A, Ryan WG, Hodgkinson JP: The perceived relationship between back symptoms and preceding injury. *Injury* 1995; 26:335-336.

Fass A: Exercises: which ones are worth trying, for which patients, and when? *Spine* 1996; 21(24):2874-2879.

Fitzler SL, Berger RA: Attitudinal change: The Chelsea back program. *Occup Health Saf* 1982; 51:24-26.

Fitzler SL, Berger RA: Chelsea back program: One year later. *Occup Health Saf* 1983; 52:52-54.

Flynn JC, Hoque MA: Anterior fusion on the lumbar spine. End-result study with long-term follow-up. *J Bone Joint Surg* (Am) 1979; 61(8):1143-1150.

Fordyce WE (ed): *Back Pain in the Workplace: Management of Disability in Nonspecific Conditions.* Seattle, International Association for the Study of Pain Press, 1995.

Greenough CG, Fraser RD: The effects of compensation on recovery from low back injury. *Spine* 1989; 14:947-955.

Greenwood JG, Wolf HJ, Pearson JC, et al: Early intervention in low back disability among coal miners in West Virginia: Negative findings. *J Occup Med* 1990; 32:1047-1052.

Gundewall B, Liljeqvist M, Hansson T: Primary prevention of back symptoms and absence from work. A prospective randomized study among hospital employees. *Spine* 1993; 18(5):587-594.

Hadler NM, Carey TS, Garrett J: North Carolina Back Pain Project: The influence of indemnification by workers' compensation insurance on recovery from acute backache. *Spine* 1995; 20:2710-2715.

Hanley EN, Shapiro JA: The development of low back after excision of a lumbar disc. *J Bone Joint Surg* (Am) 1989; 71:719-721.

Honeyman PT, Jacobs EA: Effects of culture on back pain in Australian Aboriginals. *Spine* 1996; 21:841-843.

Kellett KM, Kellett DA, Nordholm LA: Effects of an exercise program on sick leave due to back pain. *Phys Ther* 1991; 71(4):283-291.

Klenerman L, Slade PD, Stanley IM, Pennie B, Reilly JP, Atchison LE, Troup JD, Rose MJ: The prediction of chronicity in patients with acute attack of low back pain in a general practice setting. *Spine* 1995; 20(4):478-484.

Klitzman S, House JS, Israel BA, et al: Work stress, non-work stress and health. *J Behav Med* 1990; 13:221-243.

Leavitt F: The physical exertion factor in compensable work injuries. A hidden flaw in previous research. *Spine* 1992; 17:307-310.

Lethem J, Slade PD, Troup JD, Bentley G: Outline of a fear-avoidance model of exaggerated pain perception. *Behav Res Ther* 1983; 21(4):401-408.

Linton AJ, Bradley LA, Jensen I, et al: The secondary prevention of low back pain: A controlled study with follow-up. *Pain* 1989; 36:197-207.

Linton SJ, Kamwendo K: Risk factors in the psychosocial work environment for neck and shoulder pain in secretaries. *J Occup Med* 1989; 26:23-28.

Linton SJ: Risk factors for neck and back pain in a working population in Sweden. *Work and Stress* 1990; 4:41-49.

Loeser J, Henderlite S, Conrad D: Incentive effects of workers' compensation benefits: A literature synthesis. *J Health Politics and Law* 1995; 52: 34-59.

Lonstein MB, Wiesel SW: Standardized approaches to the evaluation and treatment of industrial low back pain (review). *Occup Med* 1988; 3(1):147-156.

Magnusson ML, Pope MH, Wilder DG, Areskoug B: Are occupational drivers at an increased risk for development musculoskeletal disorders. *Spine* 1996; 21(6):710-717.

Main CJ, Spanswick CC: "Functional overlay" and illness behaviour in chronic pain: distress or malingering? Conceptual difficulties in medico-legal assessment of personal injury claims. *J Psychosomatic Res* 1995; 39(6):737-753.

Malmivaara A, Hakkinen U, Aro T, et al: The treatment of acute low back pain: Bedrest, exercises, or ordinary activity? *N Engl J Med* 1995; 332:351-355.

Mostardi RA, Noe DA, Kovacik MW, et al: Isokinetic lifting strength and occupational injury. A prospective study. *Spine* 1992; 17:189-193.

Näyha S, Videman T, Laasko M, Hassi J: Prevalence of low back pain and other musculoskeletal symptoms and their association with work in Finnish reindeer herders. *Scand J Rheumatol* 1991; 20(6):406-413.

Philips HC: Avoidance behaviour and its role in sustaining chronic pain. *Behav Res Ther* 1987; 25:273-279.

Ready AE, Boreskie SL, Law SA, et al: Fitness and life style parameters fail to predict back injuries in nurses. *Can J Appl Physiol* 1993; 18:80-90.

Robertson LS, Keeve JP: Worker injuries: the effects of Workers' Compensation and OSHA inspections. *J Health Politics, Policy and Law* 1983; 8(3):581-597.

Rossignol M, Lortie M, Ledoux E: Comparison of spinal health indicators in predicting spinal status in a 1-year longitudinal study. *Spine* 1993; 18:54-60.

Rowe ML: Low back pain in industry: Updated position. *J Occup Med* 1969; 11:161-169.

Salminen JJ, Erkintalo M, Laine M, Pentti J: Low back pain in the young. A prospective three-year follow-up study of subjects with and without low back pain. *Spine* 1995; 20(19):2101-2107.

Sanders SH, Brena SF, Spier CJ, Beltrutti D, McConnell H, Quintero O: Chronic low back pain patients around the world: Cross-cultural similarities and differences. *Clin J Pain* 1992; 8:317-323.

Scheer SJ, Radack KL, O'Brien DR Jr: Randomized controlled trials in industrial low back pain relating to return to work. Part 1. Acute interventions. *Arch Phys Med Rehabil* 1995; 76(10):966-973.

Schlapbach P, Gerber NJ: Back school. *Rheumatology* 1991; 14:25-33.

Slade PD, Troup JDG, Lethem J, Bentley G: The fear-avoidance model of exaggerated pain perception - II. Preliminary studies of coping strategies for pain. *Behav Res Ther* 1983; 21:409-416.

Stankovic R, Johnell O: Conservative treatment of acute low back pain. A prospective randomized trial: McKenzie method of treatment versus patient education in "mini back school." *Spine* 1990; 15(2):120-123.

Symonds TL, Burton AK, Tillotson KM, Main CJ: Absence resulting from low back trouble can be reduced by psychosocial intervention at the workplace. *Spine* 1995; 20(24):2738-2745.

Symonds TL, Burton AK, Tillotson KM, Main CJ: Do attitudes and beliefs influence work loss due to low back trouble? *Occup Med* 1996; 46:25-32.

Szpalski M, Nordin M, Skovron ML, Melot C, Cukier D: Health care utilization for low back pain in Belgium. *Spine* 1995; 20:431-442.

Troup JD, Foreman TK, Baxter CE, et al: 1987 Volvo award in clinical sciences: The perception of back pain and the role of psychophysical tests of lifting capacity. *Spine* 1987; 12:645-657.

Tufo HM, Rothwell MG, Frymoyer JW: Managing the quality of care for low back pain, in Frymoyer JW (Ed.): *The Adult Spine: Principles and Practice.* New York, Raven Press Ltd., 1991, pp 61-75.

Välfors B: Acute, subacute and chronic low back pain: Clinical symptoms, absenteeism and working environment. *Scand J Rehabil Med* 1985; Suppl 11:1-98.

Venning PJ, Walter SD, Stitt LW: Personal and job-related factors as determinants of incidence of back injuries among nursing personnel. *J Occup Med* 1987; 29:820-825.

Versloot JM, Rozeman A, van Son AM, van Akkerveeken PF: The cost-effectiveness of a back school program in industry. A longitudinal controlled field study. *Spine* 1992; 17(1):22-27.

Videman T, Rauhala H, Asp S: Patient-handling skill, back injuries and back pain. An intervention study in nursing. *Spine* 1989; 14:148-156.

Volinn E, Lai D, McKinney S, et al: When back pain becomes disabling: A regional analysis. *Pain* 1988; 33:33-39.

Volinn E: Back sprain in industry: The role of socioeconomic factors in chronicity. *Spine* 1991; 16:542-548.

Volinn E: The epidemiology of low back pain in the rest of the world: A review of surveys in low and middle income countries. *Spine* 1997; 22(15): 1798.

Waddell G: A new clinical model for the treatment of low back pain. *Spine* 1987; 12:632-644.

Wiesel SW, Feffer HL, Rothman RH: Industrial low back pain: a prospective evaluation of a standardized diagnostic and treatment protocol. *Spine* 1984; 9(2):199-203.

Wiesel SW, Feffer HL, Rothman RH: The development of a cervical spine algorithm and its prospective application to industrial patients. *J Occup Med* 1985; 27(4):272-276.

Wiesel SW, Boden SD, Feffer HL: A quality-based protocol for management of musculoskeletal injuries: a ten-year prospective outcome study. *Clin Orthop Relat Res* 1994; 301:164-176.

Wood DJ: Design and evaluation of a back injury prevention program within a geriatric hospital. *Spine* 1987; 12:77-82.

# 36

# Upper
# Extremity Support

Carolyn M. Sommerich
*North Carolina State University*

## 36.1 Introduction

A concise review of the literature on upper extremity supports is presented in this chapter. Broadly speaking, there are two categories of upper extremity supports. One category contains devices that are used as *medical treatment* to relieve symptoms or improve functionality. For example, wrist splints and elbow braces are commonly prescribed in the conservative treatment of carpal tunnel syndrome and lateral epicondylitis (tennis elbow), respectively. The other category contains items that are used as *workstation modification* in order to minimize joint and soft tissue loading or to reduce effects of physiologic tremor on precision tasks. Arm rests on chairs and palm rests on newer computer keyboards theoretically provide users with opportunities to reduce loads on postural muscles in the shoulder and forearm. Both aspects of upper extremity support, as medical treatment and as workstation modification, are discussed in this chapter. Situations wherein usage overlaps, such as wearing wrist splints during work, are discussed as well. Additionally, the use of arm supports for some specific work activities, including keyboarding and microsurgery, are discussed in some detail. Where sufficient evidence exists, recommendations are provided for circumstance-specific use of upper extremity supports.

## 36.2 Fundamentals

In this section, basic theories of upper extremity support are explored, including the ways in which various designs provide support, and under what circumstances they may be employed.

### Splinting and Bracing as Medical Treatment

*Wrist Splints.* In the treatment of fractures, splinting had long been known to provide rest and relief from pain and disability. By analogy, splinting has been applied to arthritis (Ehrlich, 1968), where the treatment is typically prescribed to relieve pain and inflammation, but may also be prescribed to prevent deformities (contracture), or increase function (Spoorenberg et al., 1994). Wrist splinting is also used in conservative treatment of carpal tunnel syndrome (CTS). When CTS is attributed to pregnancy or to work which is highly repetitive or requires sustained non-neutral wrist postures, splinting may reduce pain and para-esthesia (Ditmars, 1993). Symptoms would not be expected to respond to splinting when CTS is due to vibration exposure, lumbrical muscles entering the distal carpal tunnel during pinch grips, pathologic encroachment (carpal bone fracture or dislocation), or external pressure applied to the palm (from hand tools with sharp edges or short handles, or from repetitive activation of palm buttons).

Splinting the wrist in extension (using a cock-up splint) is a common practice. This practice is based on studies of wrist position during common activities, which have shown the centroid of motion for most common tasks to be an extended position (Palmer et al., 1985). However, minimum pressure in the carpal canal, and therefore on the median nerve, has been shown to occur when the wrist is in a neutral orientation (in both flexion–extension and radial–ulnar axes) (Weiss et al., 1995). Splinting the wrist in extension may be effective for treating early stage (fully reversible inflammatory) lateral epi-condylitis (Nirschl, 1985). Passive extension of the wrist relieves tensile stress at the common origin of the wrist and finger extensor muscles, the loci of pain in epicondylitis.

Splints are categorized as either rigid or flexible. Rigid splints do not permit any wrist motion, while flexible splints permit a limited range of motion, thereby preserving more hand function. Rigid splints may also be referred to as immobilizing, resting, or static splints. Flexible splints are also referred to as activity or working splints, or wrist supports. Splints usually have a plate or bar in or across the palmar region which may interfere with grasping activities. Both prefabricated and custom-made splints are used in treatment protocols, depending on the preference of the medical provider. Wearing recommendations may be for night use, usage during painful activity, or continuous usage, depending on the protocol recommended by the medical provider.

*Elbow Bracing.* A variety of occupations and tasks have been associated with tennis elbow, including meat processing, carpentry, plumbing, short cycle repetitive tasks, typing, and writing (Nirschl, 1985). Pain associated with tennis elbow (medial or lateral epicondylitis) may be relieved by wearing a support band just distal to the elbow, that covers the upper part of the forearm. Typically, about 5 cm in width, the band inhibits maximum contraction of the extrinsic wrist and finger muscles, which reduces tensile stress at the origins of those muscles (the tendons, aponeuroses, and epicondyles thought to be the sites of involvement). A band may be prescribed for use only during strenuous activity or throughout the day (Froimson, 1971; Valle-Jones and Hopkin-Richards, 1990).

## Arm Supports as Ergonomic Modifications to a Workstation

Work in constrained postures is an acknowledged risk factor for neck and shoulder disorders (Hagberg, 1984), particularly when joint angles deviate significantly from neutral (resting) orientations. In reviewing numerous studies of the role of posture in neck and upper limb disorders, Wallace and Buckle (1987) concluded that posture may not always be a sufficient cause, but may interact with or add to other physical or psychosocial risk factors. Constrained postures impose continuous loads (stress) on muscles, tendons, ligaments, and joints. When they are of sufficient duration, even low levels of stress can lead to adverse health outcomes. Aarås (1994) reported a significant reduction in lost time associated with musculoskeletal illness at an electronics assembly facility when static loads on workers' trapezius muscles were reduced from levels of 4 to 6% of maximum voluntary contraction (% MVC) to below levels of 2% MVC, through workstation modification. For reference, simply positioning the arm for typing (upper arm vertical, elbow bent 90 degrees) has been estimated to require trapezius muscle activation of 3% MVC (Hagberg and Sundelin, 1986).

Potential pathologic reactions from sustained work-related stress on various musculoskeletal elements in the neck and shoulder include degenerative joint disease; tendon disorders, including degeneration and tendinitis; and problems with muscles, including myofascial syndrome (Hagberg, 1984). Develop-ment of localized muscular fatigue may be an immediate consequence of sustained muscle loading. Symptoms of fatigue may include discomfort or pain, reduced strength capacity, increased time for hand-eye coordination, increased hand tremor, and, when severe, difficulties in hand positioning (Chaffin, 1973). Low level muscle contractions (5% MVC) sustained for one hour have been shown to result in a

12% reduction in MVC, and shifts in the spectral frequency of the electromyographic (EMG) signal that are associated with localized muscle fatigue (Jørgensen et al., 1988). Considering typical work break schedules (morning and afternoon work blocks, each four hr in length and each interrupted by a single 15 min break), maintaining a work posture for an hour would not be extraordinary, especially for individuals performing precise, hand-intensive work, or work requiring intense concentration.

Working with the upper arm unsupported and out of the vertical plane has been associated with symptoms of musculoskeletal disorders in the neck and shoulder region, with duration of exposure a key element (Melin, 1987; Jonsson et al., 1988). Ergonomic hazard surveillance tools assign penalties for arms positioned out of the vertical or not supported (McAtamney and Corlett, 1993). Working with arms unsupported, significant levels of static activity have been recorded from muscles in the shoulder and forearm of typists performing typical keyboard operations (Onishi et al., 1982; Sommerich et al., 1995). Positioning an unsupported arm or arm segment in any posture away from vertical requires activation of shoulder and arm muscles in order to oppose gravitational forces. Concomitant tensile loads are imposed on ligaments and tendons in the upper extremity, including the shoulder. These active and passive internal loads contribute, along with any external loads at the hand, to compressive and shear loading of the joints of the upper extremity. However, it is important to recognize that joint posture is not the sole factor affecting muscle activity requirements. Joint posture interacts with arm segment weight and the weight of any handheld objects. Specifically, it is the moments that result from these interactions that directly determine muscle activity requirements. These levels of muscle contraction, and associated soft tissue and joint loads, are necessary in order for an individual to accommodate the physical conditions of his or her work, including work space layout and tools. However, levels of muscle activity that exceed physical work requirements are not uncommon, and may occur due to operator training or technique, environmental conditions, task complexity, or personality (Lundervold, 1958; Westgaard and Bjørklund, 1987; Goldstein, 1964).

*Joint and Soft Tissue Stress Reduction.* The use of arm supports may be an effective method for reducing muscle activity requirements imposed by physical work conditions. Effectively, arm supports reduce arm segment weight. This reduces joint moments due to external forces, which, in turn, reduces moments required from muscles and may reduce stress on other soft tissues. Frequently supporting hands and arms on the work surface was found to be associated with lower incidence of pain in the neck, shoulder, and arms in groups of professional keyboard operators performing different types of keyboarding tasks (Hünting et al., 1981). In a laboratory-based study where subjects performed a simulated soldering task, Schüldt et al. (1987) demonstrated that supporting the upper extremity through suspending the elbow or resting the elbow on a support resulted in marked reductions in shoulder muscle activity. Muscle activity was reduced whether subjects sat upright or sat with trunks flexed over the work surface. Sitting in a slightly reclined position reduced muscle activity so significantly that arm support did not provide further reductions in activity in that condition.

Based on force plate measurements, Occhipinti et al. (1985) found that 4 to 7% of body weight was supported through resting wrists on the keyboard support surface, while 7 to 14% was supported when resting forearms on the same surface. Ranges reflect differences due to changes in unsupported trunk postures. Based on biomechanical modeling, they estimated reductions in spinal loading at the level of the third lumbar vertebra (L3) ranging from approximately 25 to 100 kg, depending upon type of arm support and trunk posture. In a related study, Colombini et al. (1986) reported 6.6% of body weight supported through a forearm support when the back was also supported.

*Methods of Supports.* There are many ways to provide direct support for the upper extremity, and many points at which a support can contact the upper extremity including the elbow, along the forearm, at the wrist, or at the base of the palm. Objective effectiveness and user acceptance are both important issues in determining which methods may be appropriate for a particular situation. For example, as mentioned above, researchers have demonstrated reductions in shoulder muscle activity when the elbow was supported via suspension, or through fixed support located directly under the elbow during a soldering task (Schüldt et al., 1987). Chaffin (1973) demonstrated that time to fatigue was extended by 2 to 4 times with the use of an elbow support. Alternative keyboards are now available with built-in wrist rests. In an early study of alternative keyboards, subjects tended to prefer a split keyboard with a large built-in forearm–wrist support over both a split keyboard with a small built-in support and a traditional keyboard with a large support (Nakaseko et al., 1985). Subjects were found to exert about twice as much force on the large support (about 35 N), compared to the small one (about 17 N), meaning that more upper body weight was supported by the larger support. Efficacy and preference for various types of supports will be examined in more detail in the Applications section of this chapter.

*Influence of Support on Posture.* Supports may be useful in helping muscles to maintain or improve the position of the supported limb. However, other body parts may be impacted by support, as well. For example, during a series of six 10-minute long typing periods, Weber et al. (1983) found arm abduction was greater when trained typists keyed with a forearm–wrist support, when compared to unsupported keying. Subjects also sat nearer to the keyboard support surface when arms were supported.

In theory, one effect of wrist pad usage should be a reduction of wrist extension (if the pad is positioned appropriately). However, in a study of 12 office employees, Paul and Menon (1994) reported wrist extension ranging from 29 to 41 degrees when subjects used five different wrist pads that varied in shape, width, and compressiveness. However, no unsupported condition was provided to show whether extension with any of the pads was different from extension when no pad was used.

## 36.3   Applications

The theoretical benefits of various upper extremity supports were discussed in the previous section. Results from numerous studies are summarized in this section in order to demonstrate how and under what circumstances various types of upper extremity supports have been shown to be effective.

### Splinting and Bracing

#### Effectiveness as Medical Treatment

*Wrist splints for disorders affecting the wrist.* There is a consensus that usage of a splint should be based on a medical opinion, and that the usage should be supervised by a medical provider (Falkenburg, 1987). However, usage protocols and treatment effectiveness seem to vary widely across providers. Results of several studies of various protocols are provided in the following paragraphs.

Use of a working splint for one week, at night and during stressful activities, was found to reduce, at the end of the week, subjective assessments of pain, numbness, and tingling in a group of CTS patients with abnormal nerve conduction study results, when compared with symptoms in a similar group of patients who did not receive splints (Dolhanty, 1986). Wrists were splinted at 0 to 5 degrees of flexion.

The efficacy of conservative treatment of CTS based on steroid injection followed by continuous use of a neutral position, rigid splint for four weeks was tested prospectively by Weiss et al. (1994). No patients with advanced CTS or associated medical conditions (such as diabetes, pregnancy, or arthritis) were admitted to the study. The authors found that only 13% of the treated hands were cured (defined as symptom free) at final follow-up, which occurred between 6 and 18 months after the injection.

A similar success rate (17%) was achieved by Banta (1994), with a regime of ibuprofen and three weeks of continuous, neutral positioned, wrist immobilization for 23 hands with early-mild CTS. A second stage treatment consisting of one week of iontophoresis and splinting followed by two more weeks of continuous splinting resulted in an improvement in 58% of the remaining hands. The author's overall success rate was 65%, based upon absence of symptoms at six months follow-up.

From a retrospective review of 105 CTS patients treated with a neutral angle rigid splint designed to permit hand function, Kruger et al. (1991) determined that 67% of the patients received symptom relief attributable to the splint. There was a significant decrease in the median sensory latency group average between pre- and post-treatment measurements. Splinting seemed to be most effective for patients who had experienced symptoms for a short period of time (1 to 3 months). Patients were told to wear the splints at night and during the day as much as possible, however, the authors were not able to confirm the wearing patterns of the patients. Though no lower limit was provided for post-treatment measurements, some occurred as long as 17 months after receiving the splint.

Based on a review of 363 hands with CTS, Kaplan et al. (1990) identified five factors which were related to the successful medical management of CTS. The five factors were age over 50 years, symptom duration exceeding 10 months, constant paresthesia, stenosing flexor tenosynovitis, and a positive Phalen's test in less than 30 s. Treatment success was defined as a patient remaining symptom free for six months following conservative treatment with a rigid wrist splint (neutral wrist position; worn at night and during the day when symptomatic) and anti-inflammatory medication. Follow-up averaged 15.4 months, but ranged from 6 to 48 months. Although their overall success rate was only 18.4%, there was a 66.7% success rate in the subgroup of patients who did not have any of the five factors. In the subgroup of patients who had one of the five factors, the success rate dropped to 40.4%; in the group with two, the rate was 16.7%, and was only 6.8% in the group with three factors. No patients with more than three factors were successfully treated.

Splinting may be specifically prescribed to relieve nocturnal CTS symptoms. This is based on the theory that patients sleep with wrists curled (flexed), which results in elevated pressure on the median nerve. Luchetti et al. (1994) found no statistically significant differences in carpal tunnel pressure (CTP) due to splinting, either during the day or throughout the night, between two groups of CTS patients (one splinted in 20 deg of extension, and the other not splinted). It is not known whether splinting would have had an effect if splints had maintained the wrists in neutral rather than extended postures.

The question of appropriate splint angle was addressed in a clinical trial in which patients' subjective responses indicated that a two-week regime of splinting in a neutral position was more effective in relieving CTS symptoms than was splinting in 20 degrees of extension (Burke et al., 1994). Both splint designs appeared to be more effective for relief of nocturnal symptoms than daytime symptoms. Also of interest was the finding that only 8% of patients who continued to use their splints for two months experienced further symptom relief.

Compliance is an important part of any medical treatment. In their study of splint usage in the treatment of rheumatoid arthritis, Spoorenberg et al. (1994) found that patients were less likely to adhere to wearing advice for an immobilization splint than for an activity splint, although none of the patients who were told to wear their splints at night did so. Only 17% of the patients with immobilization splints who rested often (and therefore had the opportunity to wear the splint often) did so. In contrast, 57% of the patients with activity splints wore them often. This is, undoubtedly, at least partly due to patients' perceptions of the splints. While 96% of the rheumatologist responding to the survey prescribed both immobilization and activity splints for reduction of pain, only 44% of patients with immobilization splints found they relieved pain, and only 26% found they improved hand function. Seventy-eight percent of those patients found the rigid splints both unwieldy and ugly. In contrast, 75% of the patients with activity splints found they relieved pain and improved hand function. Only 29% thought they were ugly, but 63% found even the activity splints to be unwieldy.

*Wrist splints for tennis elbow.* As in the case with most musculoskeletal disorders and their associated risk factors, unless exposure to the risk factors associated with tennis elbow are reduced or eliminated, after immobilization treatment pain returns once a patient resumes the activities that lead to the development of tennis elbow (Nirschl, 1985). Therefore, Nirschl (1985) advocated for alteration or elimination of abusive activity rather than for immobilization of either the elbow or wrist joint. He cautioned that rigid immobilization could result in muscle atrophy. He also recommended changes in training techniques and equipment, along with counterforce bracing once activities were resumed. Little (1984) did not find splints to be effective in the treatment of tennis elbow. In two groups of subjects who both received cortisone and ultrasound treatment for tennis elbow, he found essentially no difference in recovery time (time when subjects became symptom-free) though the patients in one group were given splints and told to rest, while members of the other group were told to continue their normal activities.

*Elbow bracing.* Chen (1977) found reduced pain in patients with tennis elbow who used an elbow brace. Patients were reported to be able to perform many normal activities without experiencing symptoms, and were even able to participate in racquet sports. Apparently, however, an elbow brace may not provide instant relief from tennis elbow pain. Wadsworth et al. (1989) were not able to demonstrate a statistically significant reduction in pain in a group of patients with tennis elbow who performed grip and extension strength tests, with and without an armband, although patients reported feeling more comfortable with the brace. For the arm affected by tennis elbow, patients demonstrated a significant increase in wrist extension strength, and a lesser increase in grip strength when wearing the armband. No such increases occurred in the unaffected (control) arm.

*Use on the Job.* The science of ergonomics seeks to match workplace requirements with human capabilities and limitations. The preferred method of matching is through engineering controls, by which the work or work environment is altered to fit the worker. Use of wrist splints on the job is, instead, a method of altering the worker. Falkenburg (1987) described wrist splinting as "an industrial treatment to keep the employee on a job." She specifically mentioned, however, that splinting interferes with hand grasp, thereby reducing hand strength and interfering with holding or grasping tasks. Such functionality problems are documented in some of the splint research studies that are summarized in the following paragraphs. Most of the research on splints has been conducted in laboratory settings using healthy subjects, so results are only suggestive of what may occur in actual employment situations.

In an early field study of splinting as an intervention for CTS, results from a battery of objective and performance-based tests demonstrated that CTS was actually aggravated by the use of a rigid splint during work (Armstrong, 1981). In contrast, time off and light duty were found to be effective intervention methods.

One reason splints might be thought to be useful on the job for patients with CTS would be to reduce pressure on the median nerve by restricting patients from deviating wrists too far from a neutral posture. However, in a laboratory-based study using only healthy subjects, Rempel et al. (1994) were not able to demonstrate any difference in carpal tunnel pressure during a hand-intensive task between conditions when subjects wore a flexible wrist splint and when they did not, even though wrist motion, in both flexion–extension and radial–ulnar directions, was restricted by the splint. Rempel, Manojlovic et al. (1994) did find, however, that simply putting on the splint significantly raised CTP in each subject.

Perez-Balke and Buchholz (1995) also observed reduced wrist motion in subjects performing a repetitive pick and place task while wearing a rigid splint (with a malleable metal bar across the forearm, wrist, and palm), compared to performing the same task without the splint. They found reductions in peak grip force when subjects were splinted. Female subjects were particularly handicapped by the splint. When wearing the splint, their grip strength was reduced by a greater percentage than the males in the study (Perez-Balke and Buchholz, 1994; 1995). Their results might be explained, at least in part, by the work of Fransson-Hall and Kilbom (1993), who recorded significantly lower hand pain threshold tolerances and shorter times to experience pain in female subjects who experienced pressure applied to various points over the palm, fingers, and thumb, compared with results from male subjects who experienced the same test protocol. Another important finding from Perez-Balke and Buchholz (1995) was the increase in activity in the deltoid muscle and forearm flexor muscle group during the pick and place task when the splint was worn. In other words, subjects were less efficient when wearing the splint (more input for the same output).

In another task-based study using healthy subjects, Carlson and Trombly (1983) found that wearing a rigid splint slowed the performance of several manual tasks. Increased performance times were recorded for each of seven different tasks in the Jebsen Hand Function Test (which includes writing, manipulating small and medium-sized objects, picking up heavy cans, and feeding) when subjects wore a rigid splint, compared to when no splint was worn. The authors suggested that increased task performance times would be particularly problematic for an employee wearing a splint on a job with speed requirements (such as a highly repetitive assembly task), or for an employee who fatigued easily. They also made note of remarks of fatigue in the shoulder and upper trunk from several subjects, following task performance with the splint. Obviously, whether or not a person is wearing a wrist splint, the hand still needs to be in the same spatial location to perform a given task. If the wrist is not able to position the hand, the positioning task falls to the more proximal joints (elbow, shoulder, spine), and the muscles that control those joints. This is another example of the worker adapting to the workplace.

Stern et al. (1994) also employed the Jebsen Hand Function Test to evaluate the impact on performance due to splint usage in a group of healthy subjects. In contrast to the previous study, these authors tested several styles of commercial, nonrigid splints, which they referred to as "static wrist extensor orthoses," devices designed to support rather than immobilize the wrist. Only in manipulating small objects was there a significant difference in performance between the splinted and free hand conditions. In the other six tests, performance when wearing at least one of the five test splints did not differ from performance in the free hand condition, though it was not the same commercial splint which matched barehanded performance in each test. Subjects also wore each splint for a day, and reported differences in subjective experiences among the different splints. Splints differed in terms of the types of daily tasks with which they interfered (such as hygiene, housekeeping, or driving). Splints also differed in terms of comfort, temperature, and pain experienced by the subjects.

*Possible Adverse Results.* No reports were found in the literature which associated wrist splint usage with accidental injury. However, one report described a fracture of the distal radius in a gymnast wearing a standard gymnast's wrist support, which caught while he worked on the high bar. This experience should cause workers, employers, and medical providers to consider whether a wrist support could pose a danger in some situations. Wearing a splint might increase the chances of an employee becoming caught in rotating machinery, either directly by catching the splint, or indirectly by restricting the employee's movements and thereby restricting his or her ability to avoid contact with the equipment.

## Arm Supports as Ergonomic Modifications to a Workstation

Arm support is typically considered for work that requires steady positioning of the hands for some period of time, such as electronics assembly, keyboarding, or surgery. One brief, but interesting mention was made by Jex and Magdaleno (1978) regarding their efforts to model the potential vibration damping effects of elbow rests for pilots flying high-performance aircraft. As in the case of wrist splints, few intervention studies have been performed to examine the effects of arm support on the job. Most of the research has been conducted in laboratories, on healthy subjects with limited exposure to each test condition, making it difficult to predict either short- or long-term effects from use of the various types of arm support. Results of many of those studies are summarized below.

### Support During Specific Work Tasks

*Assembly work.* Eighteen months after installing arm suspension devices at an electronics plant, Harms-Ringdahl and Arborelius (1987) found that workers continued to choose to use the devices, which may have been due to the reductions in shoulder and neck discomfort experienced by the workers. Compared to before the suspension was introduced, average discomfort intensity was reduced by one-half to two-thirds. Before the suspension was introduced, about 40% of the workers rated their afternoon shoulder and neck discomfort greater than 50 on a scale of 0 to 100, whereas after 18 months of suspension use, only about 14% rated their discomfort that high.

In making ergonomic modifications to several workstations in an electronics assembly plant, Aarås (1994) recognized the importance of reducing shoulder load moments in workers. They achieved this through a number of engineering controls, including the provision of arm rests, either positioned on the chair or work surface, for any tasks carried out above elbow height.

*Keyboarding.* When given an opportunity to work at an adjustable workstation, either with or without a forearm–wrist support, two-thirds of a sample of 67 keyboard operators preferred the support (Grandjean et al., 1983). Seventy-eight percent of that sample did not find the support hindered their work on the keyboard. Based on observations of the operators, when the forearm–wrist support was present 80% of the subjects used the support. When no formal support was present, 50% of the subjects still rested their forearms or wrists on the keyboard support surface. A recent cross-sectional epidemiological study of VDT work seemed to objectively confirm the importance of arm supports for keyboard operators. Bergqvist et al. (1995) found neck/shoulder discomfort in a group of VDT operators was associated with a combination of three factors: more than 20 hr per week of VDT work, limited rest break opportunity, and working with the lower arm unsupported.

In a study of the effects of palm rest height and profile, 40 typists were each allowed to work with nine different palm rests during a one-week period, and were asked to use the one they preferred for at least half a day (Parsons, 1991). Though it is unclear how the supports compared physically, unlike the outcome of the study by Grandjean et al. (1983), in this study only four of the 40 typists found the rests to be useful, while seven commented about increased discomfort. Several operators suggested that arm rests on their chairs might have been more comfortable because they would have provided support while enabling freedom of wrist and hand movement.

Relying on objective electromyographic data, two studies have demonstrated reduced activity in forearm extensor muscles with use of one particular alternative keyboard with a built-in wrist support (Smith and Cronin, 1993; Gerard et al., 1994), although a portion of that reduction was apparently due to reductions in typing speed which occurred with the alternative keyboard. However, palm rests were not shown to reduce activity in any of the forearm, shoulder, or back muscles studied by Fernström et al. (1994). The authors discussed a couple of important limitations of their study which may have impacted their results: none of their subjects had prior experience with palm rests; and both practice (10 to 20 min) and testing (5 min) times in the study were limited. The authors suggested that people unfamiliar with palm rests tend to avoid using them (which may be important to remember when evaluating other support studies, or when introducing supports into the workplace).

Bendix and Jessen (1986) also found that forearm muscle activity (extensor carpi radialis) was not affected by the use of a wrist support, in a situation wherein subjects typed on electric typewriters, although trapezius muscle activity actually increased somewhat with use of the support. Both performance and subjective assessments of conditions with and without the support were similar. All subjects in the study were professional secretaries who had experienced discomfort in the neck and shoulder or elbow region for a substantial portion of the twelve-month period preceding the study. Prior to data collection, subjects had been allotted two weeks to adapt to the wrist support. In spite of the lack of objective or subjective beneficial effects of the support, nine of the twelve subjects wanted to keep the support once the study was completed.

Unlike wrist pads which are fixed relative to the keyboard, full motion forearm supports are fixed relative to the user's forearms. They are designed specifically to facilitate unrestricted motion of the arm and hand in a fixed horizontal plane, while maintaining support of the forearm and hand. Nonetheless, Powers et al. (1992) found that subjects, described as office workers with substantial computer experience, using full motion forearm supports thought that the supports slowed their typing and increased their errors, although objective measurements showed this not to be the case. The authors did not find any postural benefits to using this type of support, based on a limited assessment of wrist and elbow postures.

Erdelyi et al. (1988) studied the effects of fixed arm supports and arm suspension on experienced keyboard operators, some of whom were experiencing shoulder and neck pain. Both support methods were effective in reducing trapezius activity in the group with pain. In contrast, the group of healthy subjects experienced either no reduction in muscle activity, or, in some conditions, an increase in activity with the supports. In each condition, pain sufferers tended to display higher levels of muscle activity than did healthy subjects. Neither group preferred working with the supports. Compared to these findings with experienced keyboard operators, Sihovonen et al. (1989) found a significant reduction in trapezius activity with the use of moveable arm supports in a group of healthy, novice word processors.

*Medical procedures.* Several different wrist and forearm supports have been designed for use during microsurgical, plastic, and ophthalmic surgical procedures (Halliday, 1988; Bustillo, 1968). The common objectives of each are to provide fatigue relief for shoulder and back muscles during procedures that require surgeons to remain almost immobile for long periods of time, as well as to provide a steady base of support for control of postural tremor. Unfortunately, there are other restrictions in an operating room which may diminish the impact of arm supports, including the adjustability and height requirements of the surgical microscopes and the height and thickness of the patient support surface.

An electromyographic study of dentists performing a variety of procedures on patients revealed fairly high levels of activity in the trapezius and extensor carpi radialis muscles (Milerad et al., 1991) A second study was performed in order to determine the effect of several factors on muscle activity during dental procedures, including different types of upper extremity support (Milerad and Ericson, 1994). Compared to a no-support condition, arm support provided at the elbow significantly decreased muscle activity in the deltoid and trapezius muscles. Support at the elbow appeared to be more effective than support applied at the hand in reducing shoulder muscle activity. Whether such a support would be acceptable in practice is not clear from this study, since dentists were not used as subjects.

*Possible Adverse Results.* Most concerns are for the effects of localized pressure at contact sites between the support and the user's body. Direct pressure applied to the base of the palm was shown to raise the pressure in the carpal tunnel in cadaver specimens (Cobb et al., 1995), with point of application appearing to be the crucial factor. A 1 kg force applied just proximal to the distal wrist crease resulted in a mean increase in carpal tunnel pressure of 9 mmHg (median value of 1 mmHg), whereas that same force applied just distal to the distal wrist crease resulted in a mean pressure of 77 mmHg (median value of 29 mmHg) (Cobb et al., 1995). Pressure in the range of 20 to 30 mmHg begins to impact nervous and circulatory function and performance. *In vivo*, CTP has also been shown to increase when typing while resting the wrist, either on a wrist rest or on the keyboard support surface, in comparison to unsupported typing (Horie et al., 1994). However, those authors did not report wrist posture in the three test conditions. If changes in wrist extension occurred between those three conditions, it would be difficult to determine whether the pressure change was due to external pressure on the wrist or change in wrist extension.

Pressure is also a concern at the elbow. Working with continuously flexed elbows, especially if supported at the elbow such that the ulnar nerve is compressed, may result in the development of cubital tunnel syndrome (compression of the ulnar nerve at the cubital tunnel) (McPherson and Meals, 1992).

## 36.4　Summary

In reviewing the literature there seems to be little consensus on the effectiveness or acceptance of upper extremity supports. However, some summary statements and recommendations can still be made.

Wrist splints and elbow braces are forms of medical treatment. As such, usage is an issue which should be decided between a medical provider and patient. Individuals should not self-prescribe these devices. If an individual's symptoms are the result of physical exposures, either in the workplace or outside of work, there is no reason to believe the symptoms will not return if the individual continues to be exposed to the same risk factors. Wrist splints may, in fact, cause a disorder to migrate to a more proximal joint that is forced to compensate for the reduced functionality imposed on the injured wrist by the splint. As such, the use of wrist splints at work should not be encouraged.

Wrist splints used at work would, at best, be considered an administrative control measure. However, nowhere are splints categorized as personal protective equipment (PPE), and they may actually make hand-intensive tasks more difficult to perform. The decision to wear a splint on the job should be made by an occupational health professional who has firsthand knowledge of the patient's workstation, tools, and tasks. Wrist splints may treat symptoms, but they do not address the root cause of an individual's discomfort. Thoughtful engineering controls are the most effective method for reducing an individual's exposure to physical risk factors associated with work-related upper extremity symptoms and disorders, such as carpal tunnel syndrome.

Wrist pads, forearm supports, and arm rests can all be categorized as engineering controls. Yet, based on the literature, their presence does not guarantee benefits or even usage. First, a worker's need for support must be balanced with the need for freedom of movement of the upper extremity. Continuous support methods would be appropriate if arms and hands are maintained in the same spatial location for extended periods of time. However, if movement is required, then the need is for a support that provides a resting point during mini-breaks (a few seconds in length) and longer pauses in activity. This resting support should not hinder the individual during work periods. Additionally, the worker's preference must be considered. For example, in modifying VDT workstations, once chair, monitor, and keyboard heights have all been established, operators should be provided with a variety of support options from which to choose. They should receive proper instructions in the ways the various supports should be used. Operators should also be encouraged to try out the supports at their own workstations.

There are a myriad of commercially available ergonomic support devices for keyboard workstations. The key to evaluating the efficacy of these products is to determine how and where the support is delivered, what postural changes might occur when using the device, and whether operators will find the support interferes with task performance. Based on the information in this chapter and some trial-and-error user-testing effective support devices may be found for those operators interested in utilizing them. For industrial settings, appropriate commercial supports may not exist. However, with some basic ergonomics training engineers, maintenance personnel, and operators may be able to devise unique support systems tailored to particular workstations or operators.

The goal of any upper extremity support is to aid an individual in achieving or maintaining a desired posture with less effort or less discomfort. Given that joints are designed to move, and most tasks require motion, there is often a conflict between the desire for support and the need for motion. Identifying a support which will provide support without interfering with motion requirements is important for the successful employment of upper extremity supports.

## References

Aarås, A. 1994. The impact of ergonomic intervention on individual health and corporate prosperity in a telecommunications environment. *Ergonomics*. 37(10):1679-1696.

Armstrong, T. 1981. *Investigation of Occupational Wrist Injuries in Women*. 2 R01 OH 00679, U.S. Department of Health and Human Services, Centers for Disease Control, National Institute for Occupational Safety and Health.

Banta, C.A. 1994. A prospective, nonrandomized study of iontophoresis, wrist splinting, and anti-inflammatory medication in the treatment of early-mild carpal tunnel syndrome. *JOM*. 36(2):166-168.

Bendix, T. and Jessen, F. 1986. Wrist support during typing — a controlled, electromyographic study. *Appl Ergo*. 17(3):162-168.

Bergqvist, U., Wolgast, E., Nilsson, B., and Voss, M. 1995. The influence of VDT work on musculoskeletal disorders. *Ergonomics*. 38(4):754-762.

Burke, D.T., Burke, M.M., Stewart, G.W., and Cambré, A. 1994. Splinting for carpal tunnel syndrome: in search of the optimal angle. *Arch Phys Med Rehabil*. 75(Nov):1241-1244.

Bustillo, J.L. 1968. Hand and arm support in ophthalmic surgery. *Am J Ophthal*. 66(2):345-346.

Carlson, J.D. and Trombly, C.A. 1983. The effect of wrist immobilization on performance of the Jebsen Hand Function Test. *Am J Occ Therapy*. 37(3):167-175.

Chaffin, D.B. 1973. Localized muscle fatigue — definition and measurement. *J Occup Med*. 15(4):346-354.

Chen, S. 1977. A tennis elbow support (Letter to the editor). *Br Med J*. (1 Oct):894.

Cobb, T.K., An, K.-N., and Cooney, W.P. 1995. Externally applied forces to the palm increase carpal tunnel pressure. *J Hand Surg*. 20A(2):181-185.

Colombini, D., Occhipinti, E., Frigo, C., Pedotti, A., and Grieco, A. 1986. Biomechanical, electromyographical, and radiological study of seated postures, in *The Ergonomics of Working Postures*, Ed. N. Corlett, J. Wilson, and I. Manenica, p. 331-344. Taylor & Francis, London.

Ditmars, D.M. 1993. Patterns of carpal tunnel syndrome. *Hand Clinics*. 9(2):241-252.

Dolhanty, D. 1986. Effectiveness of splinting for carpal tunnel syndrome. *CJOT*. 53(5):275-280.

Ehrlich, G., E. 1968. Splinting for arthritis. *Medical Times*. 96(5):485-489.

Erdelyi, A., Sihvonen, T., Helin, P., and Hänninen, O. 1988. Shoulder strain in keyboard workers and its alleviation by arm supports. *Int Arch Occup Environ Health*. 60:119-124.

Falkenburg, S.A. 1987. Choosing hand splints to aid carpal tunnel syndrome recovery. *Occupational Health & Safety*. (May):60, 63-64.

Fernström, E., Ericson, M.O., and Malker, H. 1994. Electromyographic activity during typewriter and keyboard use. *Ergonomics.* 37(3):477-484.

Fransson-Hall, C. and Kilbom, Å. 1993. Sensitivity of the hand to surface pressure. *Appl Ergo.* 24(3):181-189.

Froimson, A.I. 1971. Treatment of tennis elbow with forearm support band. *J Bone Joint Surg.* 53-A(1):183-184.

Gerard, M.J., Jones, S.K., Smith, L.A., Thomas, R.E., and Wang, T. 1994. An ergonomic evaluation of the Kinesis Ergonomic Computer Keyboard. *Ergonomics.* 37(10):1661-1668.

Goldstein, I.B. 1964. Role of muscle tension in personality theory. *Psych Bull.* 61(6):413-425.

Grandjean, E., Hünting, W., and Pidermann, M. 1983. VDT workstation design: preferred settings and their effects. *Human Factors.* 25(2):161-175.

Hagberg, M. 1984. Occupational musculoskeletal stress and disorders of the neck and shoulder: a review of possible pathophysiology. *Int Arch Occup Environ Health.* 53:269-278.

Hagberg, M. and Sundelin, G. 1986. Discomfort and load on the upper trapezius muscle when operating a wordprocessor. *Ergonomics.* 29(12):1637-1645.

Halliday, B.L. 1988. A new surgical head rest. *Br J Ophthal.* 72:284-285.

Harms-Ringdahl, K. and Arborelius, U.P. 1987. One-year follow-up after introduction of arm suspension at an electronics plant, in *Proceedings of the Tenth International Congress, World Confederation for Physical Therapy,* p. 69-73. Sydney.

Horie, S., Hargens, A., and Rempel, D. 1994. Effect of keyboard wrist rest in preventing carpal tunnel syndrome, in *Proceedings of the Marconi Keyboard Research Conference,* Marshall, CA.

Hünting, W., Läubli, T., and Grandjean, E. 1981. Postural and visual loads at VDT workplaces. 1. Constrained postures. *Ergonomics.* 24(12):917-931.

Jex, H.R. and Magdaleno, R.E. 1978. Biomechanical models for vibration feedthrough to hands and head for a semisupine pilot. *Aviat Space Environ Med.* (Jan):304-316.

Jonsson, B.G., Persson, J., and Kilbom, Å. 1988. Disorders of the cervicobrachial region among female workers in the electronics industry. A two-year follow up. *Intl J Indust Ergo.* 3:1-12.

Jørgensen, K., Fallentin, N., Krogh-Lund, C., and Jensen, B. 1988. Electromyography and fatigue during prolonged, low-level static contractions. *Eur J Appl Physiol.* 57:316-321.

Kaplan, S.J., Glickel, S.Z., and Eaton, R.G. 1990. Predictive factors in the non-surgical treatment of carpal tunnel syndrome. *J Hand Surg.* 15B(1):106-108.

Kruger, V.L., Kraft, G.H., Deitz, J.C., Ameis, A., and Polissar, L. 1991. Carpal tunnel syndrome: objective measures and splint use. *Arch Phys Med Rehabil.* 72(June):517-520.

Little, T. 1984. Tennis elbow — to rest or not to rest? (Letter to the editor). *The Practitioner.* 228:457.

Luchetti, R., Schoenhuber, R., Alfarano, M., DeLuca, S., De Cicco, G., and Landi, A. 1994. Serial overnight recordings of intracarpal canal pressure in carpal tunnel syndrome patients with and without wrist splinting. *J Hand Surg.* 19B(1):35-37.

Lundervold, A. 1958. Electromyographic investigations during typewriting. *Ergonomics.* 1(3):226-233.

McAtamney, L. and Corlett, E.N. 1993. RULA: a survey method for the investigation of work-related upper limb disorders. *Appl Ergo.* 24(2):91-99.

McPherson, S.A. and Meals, R.A. 1992. Cubital tunnel syndrome. *Orthop Clin of NA.* 23(1):111-123.

Melin, E. 1987. Neck-shoulder loading characteristics and work technique. *Ergonomics.* 30(2):281-285.

Milerad, E. and Ericson, M.O. 1994. Effects of precision and force demands, grip diameter, and arm support during manual work: an electromyographic study. *Ergonomics.* 37(2):255-264.

Milerad, E., Ericson, M.O., Nisell, R., and Kilbom, Å. 1991. An electromyographic study of dental work. *Ergonomics.* 34(7):953-962.

Nakaseko, M., Grandjean, E., Hünting, W., and Gierer, R. 1985. Studies on ergonomically designed alphanumeric keyboards. *Human Factors.* 27(2):175-187.

Nirschl, R.P. 1985. Muscle and tendon trauma: tennis elbow, in *The Elbow and Its Disorders,* Ed. B.F. Morrey, p. 537-552. W.B. Saunders, Philadelphia.

Occhipinti, E., Colombini, D., Frigo, C., Pedotti, A., and Grieco, A. 1985. Sitting posture: analysis of lumbar stresses with upper limbs supported. *Ergonomics.* 28(9):1333-1346.

Onishi, N., Sakai, K., and Kogi, K. 1982. Arm and shoulder muscle load in various keyboard operating jobs of women. *J Human Ergol.* 11:89-97.

Palmer, A.K., Werner, F.W., Murphy, D., and Glisson, R. 1985. Functional wrist motion: a biomechanical study. *J Hand Surg.* 10A(1):39-46.

Parsons, C.A. 1991. Use of wrist rests by data input VDU operators, in *Contemporary Ergonomics*, Ed. E.J. Lovesay, p. 319-321. Taylor & Francis, London.

Paul, R. and Menon, K.K. 1994. Ergonomic evaluation of keyboard wrist pads, in *Proceedings of the 12th Triennial Congress of the International Ergonomics Association*, p. 204-207. Toronto.

Perez-Balke, G. and Buchholz, B. 1995. A study of the effect of a wrist splint on extrinsic flexor and anterior deltoid electromyography during a pick and place task, in *Proceedings of the Human Factors and Ergonomics Society 39th Annual Meeting*, p. 958. San Diego.

Perez-Balke, G. and Buchholz, B.O. 1994. A study of the effect of a "resting splint" on peak grip strength, in *Proceedings of the Human Factors and Ergonomics Society 38th Annual Meeting*, p. 544-548. Nashville.

Powers, J.R., Hedge, A., and Martin, M.G. 1992. Effects of full motion forearm supports and a negative slope keyboard support system on hand-wrist posture while keyboarding, in *Proceedings of the Human Factors Society 36th Annual Meeting*, p. 796-800. Atlanta.

Rempel, D., Manojlovic, R., Levinsohn, D.G., Bloom, T., and Gordon, L. 1994. The effect of wearing a flexible wrist splint on carpal tunnel pressure during repetitive hand activity. *J Hand Surg.* 19A(1):106-110.

Schüldt, K., Ekholm, J., Harms-Ringdahl, K., Németh, G., and Arborelius, U.P. 1987. Effects of arm support or suspension on neck and shoulder muscle activity during sedentary work. *Scand J Rehab Med.* 19:77-84.

Sihovonen, T., Baskin, K., and Hänninen, O. 1989. Neck-shoulder loading in wordprocessor use: effect of learning, gymnastics and arm supports. *Arch Occup Environ Health.* 61:229-233.

Smith, W.J. and Cronin, D.T. 1993. Ergonomic test of the Kinesis keyboard, in *Proceedings of the Human Factors and Ergonomics Society 37th Annual Meeting*, p. 318-322. Seattle.

Sommerich, C.M., Marras, W.S., and Parnianpour, M. 1995. Activity of index finger muscles during typing, in *Proceedings of the Human Factors and Ergonomics Society 39th Annual Meeting*, p. 620-624. San Diego.

Spoorenberg, A., Boers, M., and van der Linden, S. 1994. Wrist splints in rheumatoid arthritis: a question of belief? *Clin Rheum.* 13(4):559-563.

Stern, E.B., Sines, B., and Teague, T.R. 1994. Commercial wrist extensor orthoses: hand function, comfort, and interference across five styles. *J Hand Ther.* 7(Oct-Dec):237-244.

Valle-Jones, J.C. and Hopkin-Richards, H. 1990. Controlled trial of an elbow support ('Epitrain') in patients with acute painful conditions of the elbow: a pilot study. *Current Med Res Op.* 12(4):224-233.

Wadsworth, C.T., Nielsen, D.H., Burns, L.T., Krull, J.D., and Thompson, C.G. 1989. Effect of the counterforce armband on wrist extension and grip strength and pain in subjects with tennis elbow. *JOSPT.* 11(5):192-197.

Wallace, M. and Buckle, P. 1987. Ergonomic aspects of neck and upper limb disorders, in *International Reviews of Ergonomics*, Ed. D.J. Oborne, p. 173-200. Taylor & Francis, London.

Weber, A., Sancin, E., and Grandjean, E. 1983. The effects of various keyboard heights on EMG and physical discomfort, in *Ergonomics and Health in Modern Offices*, Ed. E. Grandjean, p. 477-483. Taylor & Francis, London.

Weiss, A.-P.C., Sachar, K., and Gendreau, M. 1994. Conservative management of carpal tunnel syndrome: a reexamination of steroid injection and splinting. *J Hand Surg.* 19A(3):410-415.

Weiss, N.D., Gordon, L., Bloom, T., So, Y., and Rempel, D.M. 1995. Position of the wrist associated with the lowest carpal-tunnel pressure: implications for splint design. *J Bone Joint Surg.* 77-A(11):1695-1699.

Westgaard, R.H. and Bjørklund, R. 1987. Generation of muscle tension additional to postural muscle load. *Ergonomics.* 30(6):911-923.

# Index